Human Biology

Mike Boyle
Bill Indge
Kathryn Senior

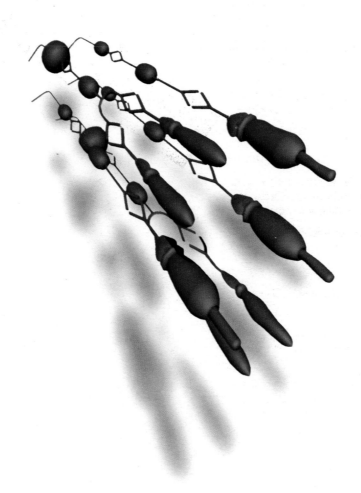

Collins Educational

An imprint of HarperCollins*Publishers*

Published by Collins Educational

An imprint of HarperCollins*Publishers* Ltd

77–85 Fulham Palace Road

London W6 8JB

First published 1999

ISBN 000 329 0956

British Library Cataloguing in Publication Data:
A catalogue record for this book is available from the British Library.

Editor: Kathryn Senior
Publishing coordinator: Pat Winter
Series designer: Glynis Edwards
Book designers: Glynis Edwards and Ken Vail Graphic Design
Cover design: Michael Faulkner
Artists: Barking Dog Art, Jerry Fowler, Peter Harper, Illustrated Arts, Pantek Arts, TTP International.
Picture research: Caroline Thompson
Indexer: Julie Rimington
Production controller: Anna Pauletti

Printed and bound by Rotolito Lombarda, Milan, Italy.

CONTENTS

About this book

Human biology: the science of people

DURING THE LAST couple of centuries, our scientific knowledge has expanded at a staggering rate. Two hundred years ago the average person in the UK had a life expectancy of about 45. Tuberculosis, smallpox, cholera, plague and typhoid all took their toll. People didn't know anything about bacteria or viruses; there was no effective sewage treatment and clean water was a rare commodity. Malnutrition was common: few people understood the need for a balanced diet that included all the main vitamins, fibre and protein. And as for surgery – with no anaesthetics or antiseptics – you can probably imagine the horrors involved.

This child has smallpox; the last recorded case of this viral disease was in 1977, in Somalia

At the turn of the millennium, we take for granted many of the greatest advances in biology and medicine. Relatively few people in the more developed countries now die from diseases caused by other organisms. The average life expectancy of someone born in the UK is currently about 80. Many killer diseases are either very rare or have been eradicated completely – as in the case of smallpox. Heart disease and cancer are now the biggest killers in the western world but progress in the treatment of both is advancing rapidly. In the next century, it might even become possible to control some aspects of the genetic causes of ageing, extending life expectancy beyond 100.

As the human population continues to grow and impact on the natural world, our understanding of

plant biology and our increasing awareness of the complexities of ecology is helping us to cope. Application of the latest biological techniques to agriculture, for example, is allowing us to develop new crop plants to increase the efficiency of food production and to grow crops with a higher resistance to drought and disease.

Genetic engineering of crop plants is a controversial development but it is one which could solve the problem of food shortage in many parts of the world

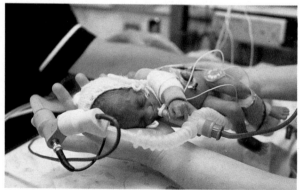

Fifty years ago, a baby born prematurely would have had little chance of survival. Today, advances in technology can save babies born as early as 23 weeks' gestation

Of course, any advance in knowledge brings its problems, and the biological revolution is no exception. As people live longer, an increasing number of older people now look forward to a retirement longer than their working life. As medical technology improves, so does the demand for its benefits. And then there are the moral and ethical dilemmas which surround, for example, genetic engineering of crop plants and infertility treatment.

So what will the next century bring? We are certain to see a continuation of the genetic revolution started by Crick and Watson when they worked out the structure of DNA. The Human Genome Project – a quest to map all human DNA sequences – is already underway. Its completion will be a major milestone: in the future we will be able to find out exactly what individual genes do, how they interact to switch each other on and off, and how they control growth and development.

To the student

We have written this book to give you a thorough introduction to Human Biology at advanced level – this may be A-level, Advanced GNVQ, Scottish Higher or International Baccalaureate. We also hope that other students such as nurses and first-year undergraduates will find it a valuable introduction to study at a higher level. The book contains the material which is part of any core syllabus, as well as many of the common options, such as physiology, ecology, exercise science and biotechnology.

But we have aimed to do more than just give you the facts that you need in order to cope well with your exam. We hope we have also been able to:

● support and reinforce your understanding of the subject rather than just bombard you with facts.

● convey our enthusiasm and enjoyment of this fascinating subject.

● de-mystify the jargon-filled areas of biology.

● emphasise the 'So what?' factor. We try to stress the applications of what you are learning so that you can relate the facts to the living world.

A recent survey showed that the average A-level human biology syllabus contains more new words than a GCSE language course. It often seems that scientists want to keep their subject to themselves by describing their work in Greek and Latin names. In this book, we do, of course, introduce you to new biological terms, but they are always explained carefully. We have kept the non-biological language very clear and straightforward and it has been written to give you a taste of the drama of the subject.

The 'guided tour'

Before you start using this book, it is a good idea to become familiar with its major features.

First, the sections. The book is divided into sections that cover eight themes:

● Section 1: Molecules and cells

● Section 2: Supply and demand

● Section 3: Homeostasis

● Section 4: Control systems

● Section 5: Human life span

● Section 6: The world we live in

● Section 7: Genes, genetics and evolution

● Section 8: Health and disease

Each section begins with a theme opener, which introduces the chapters in a section.

In our experience, many teachers start covering a subject at A level assuming that students are familiar with the 'basics'. In this book, we have tried to assume as little as possible and have written some introductory chapters to sections that are designed to be a gentle scene-setter giving an overview of major areas such as genetics, health and disease and ecology. The other chapters in each section have the following features in common.

The opener

Each chapter begins with a short piece of text that focuses on a practical application of the science covered later - allergies, cancer, the ageing process, the origins of life, genetic disease, the threat of over-population, sex and many more. Their aim is to satisfy the vital 'So what?' factor.

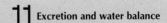

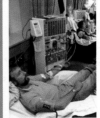

11 Excretion and water balance

IF ONE OF YOUR KIDNEYS were to stop working because of injury or disease, you could probably still lead a normal life. However, people whose kidneys both fail are faced with a crisis: water, urea and potassium build up rapidly in their body. They may continue to pass some urine, but they cannot get rid of all the waste produced by normal cell processes.

Most people suffering from kidney failure hope for a transplant, but there is a shortage of donors. Until a suitable organ becomes available, patients have to rely on dialysis to filter their blood and balance their fluid intake. As the photograph on the left shows, dialysis is uncomfortable and inconvenient. To reduce the time they need to spend in dialysis, kidney patients must stick to the strict Giovanetti diet. This comes of something of a shock. They must limit their fluid intake to only half a litre a day, about a quarter of the amount a human adult would normally drink. They must also control their protein intake to about 30 to 40 grams per day, the amount of protein in a small egg.

But perhaps the biggest problem is the need to regulate potassium. This ion is a normal constituent of the body, but large amounts cause serious problems, including heart failure. Potassium-rich foods include citrus fruits, bananas, instant coffee, peanuts, treacle and – a big blow for many people – chocolate. In this strange diet, carbohydrates are not a priority. Few patients feel like eating anyway, and this new restricted diet makes the task of finding appetising food even more difficult.

This patient is undergoing dialysis. For several hours, blood flows from the patient's forearm into the machine. Here, much of the waste, together with excess salt and water, is removed by filtration. Find out more about this process in the Assignment at the end of this chapter

1 WASTE AND WATER CONTROL

The text

The main text introduces ideas from scratch. Key words are highlighted in bold and explained. Throughout the chapter the text is supported by full colour diagrams and photos with explanatory labels, annotations and captions.

Marginal reminders

Marginal reminders are small boxes headed by a tick. They summarise an essential point in the main text or give you an instant reminder of information contained in other chapters or from your GCSE studies. Some are useful exam hints.

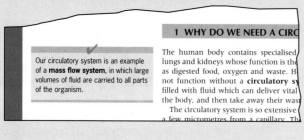

✓ Our circulatory system is an example of a **mass flow system**, in which large volumes of fluid are carried to all parts of the organism.

1 WHY DO WE NEED A CIRC

The human body contains specialised, lungs and kidneys whose function is the as digested food, oxygen and waste. H not function without a **circulatory sy** filled with fluid which can deliver vital the body, and then take away their wast
The circulatory system is so extensive a few micrometres from a capillary. Th

Self-test questions

Each chapter contains several self-test questions which students find valuable. Try to answer them as you work through the text: they test your progress and make you think. (The answers, thankfully, are at the back of the book – but no cheating – think first!)

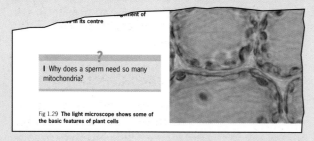

? I Why does a sperm need so many mitochondria?

Fig 1.29 **The light microscope shows some of the basic features of plant cells**

Feature boxes

The feature boxes contain topics of special interest – historical features, practical techniques or modern applications of science. Often, the information is not specifically related to the syllabus, but we have included it because we think it's interesting. Premature babies, brittle bones, hormone replacement therapy, cartilage and footballers' knees, hangovers – it's all good stuff.

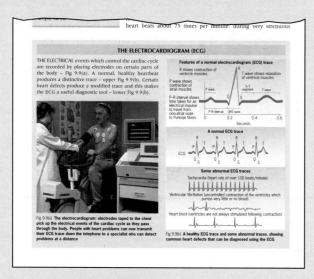

heart beats about 75 times per minute; during very strenuous

THE ELECTROCARDIOGRAM (ECG)

THE ELECTRICAL events which control the cardiac cycle are recorded by placing electrodes on certain parts of the body – Fig 9.9(a). A normal, healthy heartbeat produces a distinctive trace – upper Fig 9.9(b). Certain heart defects produce a modified trace and this makes the ECG a useful diagnostic tool – lower Fig 9.9(b).

Fig 9.9(a) **The electrocardiogram: electrodes taped to the chest pick up the electrical events of the cardiac cycle as they pass through the body. People with heart problems can now transmit their ECG trace down the telephone to a specialist who can detect problems at a distance**

Fig 9.9(b) **A healthy ECG trace and some abnormal traces, showing common heart defects that can be diagnosed using the ECG**

Extension boxes

These cover ideas which go beyond the normal A-level syllabus requirements. They are there to give you more in-depth knowledge to help you to better understand a concept that is central to the syllabus.

Cartilage

Cartilage is a tough, smooth and flexible connective tissue. There are three types of cartilage:
- **Hyaline cartilage**. This is a compressible and elastic tissue found in the nose and trachea, and at joints where it covers the ends of bones to provide smooth surfaces.
- **White fibrous cartilage**. This is found between the vertebrae and acts as a shock absorber.
- **Yellow elastic cartilage**. This is a very elastic tissue found in the ears, **pharynx** (throat) and **epiglottis** (the flap which closes over the airway during swallowing).

Fibrous cartilage is a good shock absorber because it is **compressible** (it can withstand pressure without being damaged) and it can absorb the impact of a rapid jolt which would damage a more brittle structure such as a bone. Studies into the structure of fibrous cartilage reveal how it is adapted to perform its function. The toughness, not surprisingly, is provided by **collagen fibres** (see Chapter 3). But a second type of chemical, **proteoglycan**, has proved to be especially interesting. Proteoglycan binds loosely to water molecules, but releases them when under pressure. So water seeps out of cartilage

when it is compressed, and goes back again when the pressure is released, allowing cartilage to act as a very tough sponge to reduce the shock on the rest of the body.

Despite this, fibrous cartilage cannot absorb the impact of strenuous exercise. When we run and jump, for example, much of the cushioning effect is due to the tendons, ligaments and muscles. To illustrate this, think about what happens when you do a standing jump. If you land on your toes, there is no problem. The shock is absorbed by the muscles and tendons in your feet and legs. If you land on your heels (even from a few centimetres) the jarring effect is very obvious. Don't try this!

Fig 2.5 **The knee joint has to take the weight of most of the body, and in strenuous sports such as football, the twisting and turning can easily damage the cartilage. This player is having keyhole surgery to remove a fragment of cartilage which has been torn**

Examples

Worked Examples take you through some of the more difficult exam questions. Examples often include calculations, and are designed to show you that using maths in human biology is easier than you thought.

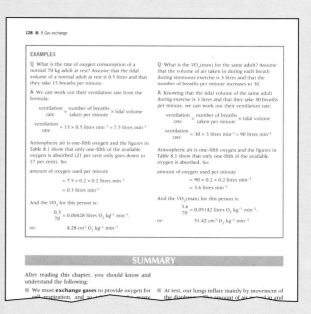

128 ■ 8 Gas exchange

EXAMPLES

Q What is the rate of oxygen consumption of a normal 70 kg adult at rest? Assume that the tidal volume of a normal adult at rest is 0.5 litres and that they take 15 breaths per minute.

A We can work out their ventilation rate from the formula:

$$\text{ventilation rate} = \frac{\text{number of breaths taken per minute}}{} \times \text{tidal volume}$$

$$\text{ventilation rate} = 15 \times 0.5 \text{ litres min}^{-1} = 7.5 \text{ litres min}^{-1}$$

Atmospheric air is one-fifth oxygen and the figures in Table 8.1 show that only one-fifth of the available oxygen is absorbed (21 per cent only goes down to 17 per cent). So:

amount of oxygen used per minute
= 7.5 × 0.2 × 0.2 litres min⁻¹
= 0.3 litres min⁻¹

And the VO_2 for this person is:

$$\frac{0.3}{70} = 0.00428 \text{ litres O}_2 \text{ kg}^{-1} \text{ min}^{-1},$$

or: 4.28 cm³ O₂ kg⁻¹ min⁻¹

Q What is the VO_2(max) for the same adult? Assume that the volume of air taken in during each breath during strenuous exercise is 3 litres and that the number of breaths per minute increases to 30.

A Knowing that the tidal volume of the same adult during exercise is 3 litres and that they take 30 breaths per minute, we can work out their ventilation rate:

$$\text{ventilation rate} = \frac{\text{number of breaths taken per minute}}{} \times \text{tidal volume}$$

$$\text{ventilation rate} = 30 \times 3 \text{ litres min}^{-1} = 90 \text{ litres min}^{-1}$$

Atmospheric air is one-fifth oxygen and the figures in Table 8.1 show that only one-fifth of the available oxygen is absorbed. So:

amount of oxygen used per minute
= 90 × 0.2 × 0.2 litres min⁻¹
= 3.6 litres min⁻¹

And the VO_2(max) for this person is:

$$\frac{3.6}{70} = 0.05142 \text{ litres O}_2 \text{ kg}^{-1} \text{ min}^{-1},$$

or: 51.42 cm³ O₂ kg⁻¹ min⁻¹

SUMMARY

After reading this chapter, you should know and understand the following:

■ We must **exchange gases** to provide oxygen for cell respiration, and to

■ At rest, our lungs inflate mainly by movement of the diaphragm. The amount of air

Summaries

The main text of each chapter ends with a Summary. This lists the main points of the text. Excellent for revision, or to gain an overview of the topics covered in the chapter.

Questions

Most chapters contain a selection of recent examination questions from the major examining boards. Research has shown that one of the biggest factors in examination success is practice of past paper questions.

References to specific questions are included at appropriate points in the text.

Do I need advanced chemistry and maths?

Generally – no. Recent A-level syllabuses have deliberately reduced their mathematical and chemical content to make the subjects more accessible.

In this book we reflect this. The chemical details are kept to a minimum. The mathematical aspects covered centre around the application of numeracy – a basic **key skill**. So, no calculus or complex algebra, but there are exercises in which you work out percentages or the scale of a diagram. The Examples take you through more complex maths where necessary. We look at the chemistry of water and carbon in a bit more detail in the Appendices at the end of the book.

Practical work

Practical work is a central feature of any biology course. Biological research is practically based, and students at A level should aim to be competent practical biologists. This book is not a practical manual and so does not contain the details for carrying out experiments, but we do feature some experimental biology as it is of such central importance in research.

People at the cutting edge of modern science make progress due to their expertise in one tiny area and their skills in the latest investigative techniques. As one famous scientist put it, 'Don't tell me what you've learned, tell me which techniques you've mastered'.

Assignments

Assignments appear at the end of most chapters. They focus on an application of the subject matter and contain questions which give the opportunity to practise key skills.

Reproduced sample page (Summary / Questions, p. 175)

11 Excretion and water balance ■ 175

SUMMARY

After reading this chapter, you should know and understand the following:

■ The kidneys remove metabolic waste and control water and solute levels in the body. As a result of these functions, the kidney also plays a vital role in the control of blood volume and pressure.

■ Each kidney is made from around one million **nephrons**, narrow tubules closely entwined with blood vessels.

■ At one end of the nephron, the **renal capsule**, the kidney filtrate is formed by ultrafiltration (pressure filtration) of the blood.

■ The first filtrate has the same composition as tissue fluid. As it passes along the nephron, the composition of filtrate is altered by various active transport mechanisms that reabsorb some substances whilst allowing others to pass into the urine.

■ A large amount of filtrate is formed, but over 99 per cent if it is reabsorbed in the **first convoluted tubule**, mainly by active transport mechanisms and osmosis. Usually, all glucose and amino acids are reabsorbed into the blood.

■ The movement of solutes from **loop of Henle** creates a region of high solute concentration in the medulla, through which the collecting ducts must pass. As filtrate (now urine) flows along the collecting ducts, water leaves by osmosis. The resulting urine is **hypertonic** (more concentrated than body fluids).

■ The **second convoluted tubule** alters the composition of urine according to the needs of the body.

QUESTIONS

1 The diagram in Fig 11.Q1 represents a nephron from a human kidney.

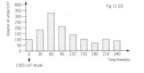

Fig 11.Q1

a) Name the part labelled **X**.

b) Sodium chloride is actively pumped out of **Z** into the medulla of the kidney. This sodium chloride moves back into **Y**.
Explain the effect of the sodium chloride concentration in the medulla of the kidney on the reabsorption of water from the collecting duct.

c) Most of the sodium chloride filtered into the glomerular filtrate is reabsorbed.
(i) From which parts of the nephron does this reabsorption take place?
(ii) How is the reabsorption of sodium chloride controlled?

[AEB 1996: Biology: Specimen paper, q.1]

2
a) If the glomerular filtration rate of the kidneys is 120 cm³ min⁻¹ and the tubular reabsorption rate is 114 cm³ min⁻¹, calculate the rate of urine formation per minute.

b) State **two** differences between tubular reabsorption of water in the first (proximal) convoluted tubule and the second (distal) convoluted tubule.

c) What might reduce the rate of tubular reabsorption of water?

d) The minimum rate of urine production is 300 cm³ per day. Explain why it is necessary for some urine to be produced each day.

[AEB 1995 Human Biology: Paper 1, q.6]

3 The graph of Fig 11.Q3 shows the volume of urine collected from a subject before and after drinking 1000 cm³ of distilled water. The subject's urine was collected immediately before the water was drunk and then at intervals of 30 minutes for several hours.

Fig 11.Q3

Reproduced sample page (Assignment, p. 56)

56 ■ 3 The chemicals of life

Assignment

DNA FINGERPRINTING

In 1987, Robert Melias made legal history in the UK when he was convicted of rape based on evidence obtained by **DNA fingerprinting**. This technique – more accurately called **DNA profiling** – was developed at Leicester University by Alec Jeffreys in 1984 and is proving to be an invaluable tool in forensic medicine.

The principles behind DNA profiling

Every human body cell contains 46 chromosomes. Each one consists of a single elaborately coiled piece of DNA which, if stretched out, can be as long as 5 cm. The structure of DNA can be used to identify individuals.

Between the many genes that occur along the DNA molecule are regions which code for nothing at all. Within this non-coding DNA are **hypervariable regions**, so called because they vary enormously in length from person to person. Hypervariable regions consist of particular base sequences called **core sequences**, which are repeated again and again. Different people have different numbers of repeats and so have differently sized hypervariable regions. When these are labelled and separated according to size, a pattern is produced. Each pattern is unique to each individual person and can be used as the basis of DNA profiling (Fig 3.A1).

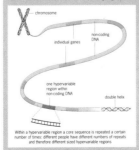

Fig 3.A1 **Although each chromosome has several thousand genes, they only account for about 10 per cent of the length; the rest is non-coding DNA. Within these segments are hypervariable regions. These are unique to each individual and, when they are separated out, the pattern they form provides the basis of DNA profiling**

1
a) What is a gene?
b) What is the name given to a fault which sometimes occurs when DNA is copied?
c) Faults accumulate in the hypervariable regions more frequently than they do in the genes. Why do you think this is?

Getting a sample of DNA

All body cells contain the same DNA, and so virtually any tissue sample can be used for DNA profiling. The amount of tissue required is very small: 0.05 cm³ of blood, 0.005 cm³ of semen or one hair root!

2
a) Which blood cells contain DNA?
b) Why do forensic scientists need a larger sample of blood than of semen?

Processing the DNA and separating the fragments

Once you have a DNA sample, the next stage is to cut the molecule using **restriction enzymes**, which act rather like molecular scissors (see Chapter 34). These enzymes cut DNA at specific base sequences, giving a complex mixture of DNA fragments, some of which contain the hypervariable regions.

Electrophoresis (see page 50) separates the fragments. The fragment mixture is placed in a trough in some gel. When a current is applied, the negatively charged DNA fragments move towards the positive terminal, or **anode**. The fragments move at different speeds according to their size: small ones move faster than larger ones. The fragments separate out into bands, as shown in Fig 3.A2 for blood.

The bands of DNA are transferred from the gel onto a nylon membrane by **Southern blotting**, a process which

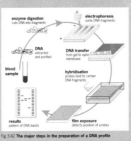

Fig 3.A2 **The major steps in the preparation of a DNA profile**

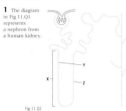

MOLECULES AND CELLS

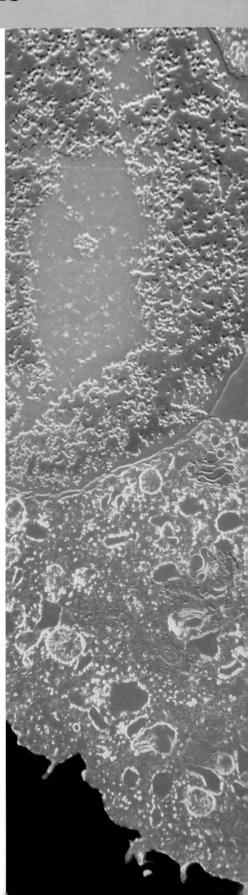

HUMAN BIOLOGY is about understanding how the human body works. It also includes some study of our place in the wider scheme of things. In history, as people studied the human body, they started with the overall structure and then moved in to discover more detail. In this century we have taken this further to discover the structure and function of individual cells and molecules.

To study A-level human biology today, you need to know a lot about the way cells and molecules interact in the major processes of life. You can't, for example, even start to understand how the lungs work without knowing how molecules diffuse across membranes. This itself is difficult until you know something about the detailed structure of a membrane.

So, while later in the book we look in detail at what many students consider is the really interesting stuff – the major processes that make us tick – we start by looking at the basic unit of life: the cell.

As you start your course, you probably have only a vague idea about what is in a cell, and so this is the subject of Chapter 1. In Chapter 2 we look at how cells are organised into whole organisms; in Chapters 3 and 4 we find out about the basic chemicals of life and their role in the structure of cells; in Chapter 5 we focus on membranes and how they control the materials which pass in and out of cells.

However, just because cell biology is a necessity which leads on to more advanced study of biological systems, don't get the idea that it's boring. A quick scan of the latest issue of New Scientist or Scientific American will reassure you that the study of molecules and cells is at the very forefront of modern biological and medical research, into AIDS, cancer, genetic disease, heart disease, the way the brain works... and a million more major projects world-wide.

1 What's in a cell?

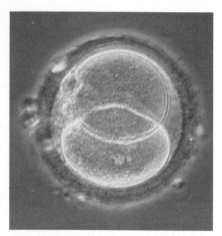

This human embryo, just 24 hours old, has already divided once. The genes contained in the sperm, together with those in the egg, have formed the genes of a new individual and have begun to control its development

THE CELLS IN YOUR BODY are tiny – a row of about 40 would fit into 1 mm. But, although they are too small to see with the naked eye, their complexity is remarkable. A single cell can be thought of as an organised chemical system, separated from its disorganised surroundings by a membrane.

You started life as a single fertilised cell that contained one set of genes from your mother and one set from your father. These genes controlled how you developed as an embryo, telling cells when to divide and what to develop into – nerve, muscle or bone, for example. By the time you were born you had grown into a unique individual made of billions of cells.

Even more remarkably, every one of your cells contains a complete set of your genes; enough to make a clone of you. In theory, we could take a cell from anywhere in your body, remove the nucleus and place it into a newly-fertilised cell whose own nucleus had been removed. Given the right conditions, your genes would then take over the embryo, and develop into what would be your identical twin.

Research on human cloning is – not surprisingly – presently illegal because it is fraught with moral and ethical problems. However, the cloning of plants and farm animals is becoming more common, allowing us to make copies of organisms with the most desirable features.

1 THE COMPLEXITY OF CELLS

A human body contains about 50 million million cells. Every day we make new skin cells to replace those that wear away, new red blood cells to replace those that die and new cells to replenish the lining of our digestive system. Not all cell types have the high turnover of these cells, but it is still easy to think of individual cells as small, simple and disposable. And, when you look at cheek cells under a light microscope, what you can see looks fairly uninspiring.

But these impressions are deceptive. As Fig 1.1 shows, individual **animal cells** are extremely complex in both their structure and function. So are **plant cells** (see page 20). Even something like a **bacterium**, a primitive single-celled organism, is far more complex than the fluid that surrounds it (see Figs 1.3 to 1.5).

2 THE CELL THEORY

The observations of early microscopists led to the development of the **cell theory**, a general acceptance that all living things are made of cells. Modern cell theory has three central ideas:

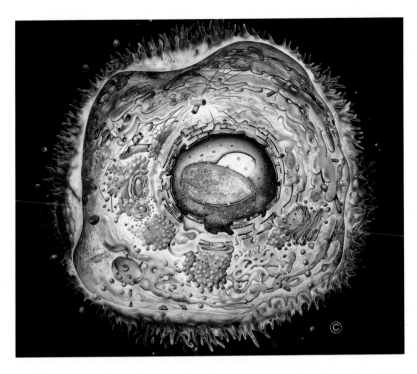

Fig 1.1 **This painting illustrates something of the complexity of an animal cell. It was commissioned by the Science Museum on behalf of the Biochemical Society to provide the centrepiece of a permanent exhibition opened in 1987**

- The cell is the smallest independent unit of life.
- The cell is the basic living unit of all organisms: all organisms are made up of one or more cells.
- Cells arise from other cells by cell division. They cannot arise spontaneously.

As you study biology in more detail, you will discover that there are exceptions to every rule. In the case of cell theory, that exception is the **virus**. Viruses do not have a cellular structure or organisation, and whether they are actually living organisms is a subject of debate. Find out more about them in Chapter 33.

A 'Spontaneous generation' was the belief that living things could be created from non-living ones. A recipe for the spontaneous production of mice was reported in the 1600s by the chemist J.B. van Helmont: 'If you press a piece of underwear, soiled with sweat, together with some wheat in an open-mouthed jar, after about 21 days the odour changes and the ferment, coming out of the underwear and penetrating through the husks of wheat, changes the wheat into mice.' Suggest what was really happening.

3 USING CELL STRUCTURE TO CLASSIFY ORGANISMS

Organisms can be classified on the basis of the internal organisation of their individual cells. With the exception of viruses, all organisms are either **prokaryotic** or **eukaryotic**. Prokaryotic cells are relatively simple, they have no separate nucleus and show little organisation. Bacteria are prokaryotes. In contrast, eukaryotic cells are larger and show much more internal organisation. Animals, plants, fungi and protoctists are all eukaryotes. The main differences between prokaryotes and eukaryotes are shown in Table 1.1.

Table 1.1 **Differences between prokaryotes and eukaryotes**

	Prokaryotes	Eukaryotes
Organisms	bacteria	plants, animals, protoctists, fungi
Diameter of cells	0.1–10 µm	10–100 µm
Site of genetic material	DNA in cytoplasm	DNA inside distinct nucleus
Organisation of genetic material	DNA is circular; no histone proteins; DNA does not condense at cell division	DNA is linear; attached to histone proteins; condenses into visible chromosomes before cell division
Internal structure	few organelles	many organelleswith complex membrane systems
Cell walls	always present	present in plants and fungi and some protoctists; never in animals
Flagella	have simple flagella	have modified cilia called undulipodia which consist of microtubules in a distinctive '9 + 2' arrangement (see Fig 1.29)

4 PROKARYOTES – THE FIRST ORGANISMS?

A few billion years ago, the first living organisms to evolve on Earth were probably prokaryotes. The term literally means 'before the nucleus' because the genetic material (DNA) of these organisms is not enclosed by a nuclear envelope.

It is tempting to think of prokaryotes as inferior to eukaryotes, but is some ways they have achieved greater success. They have been on Earth more than twice as long as eukaryotes, they are present in greater numbers (there are more bacteria living on your skin than there are people on Earth) and they occupy an enormous number of different habitats. Some bacteria, for example, are able to live in volcanic springs at temperatures as high as 90°C.

How small is small?

Cells and the molecules they contain are very small. When we study them, we must think about measurements that are minute beyond our imagination. It is easy, for example, to develop a mental block when confronted by the statement 'The nanometre is 10^{-9} m.'

The two units commonly used to describe microscopic objects are the **micrometre** (μm) and the **nanometre** (nm). Starting with a familiar unit, the **millimetre** (mm), one thousandth of one millimetre is known as a micrometre (μm). The micrometre is used to describe cells and organelles. An average animal cell is 30 to 50 μm across; the nucleus has a diameter of about 10 μm. Plant cells can reach 150 μm or more in length.

When describing small cellular components and molecules, the useful unit is the nanometre (nm). A nanometre is one thousandth of a micrometre. As a rough guide, the light microscope reveals structures that can be measured in micrometres, but you need an electron microscope to see objects measured in nanometres.

Fig 1.2 **Units of measurement; how they are used and how they relate to each other**

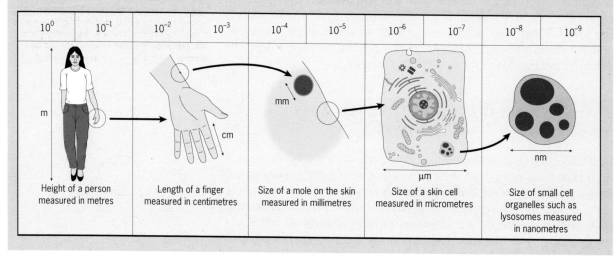

| 10^0 | 10^{-1} | 10^{-2} | 10^{-3} | 10^{-4} | 10^{-5} | 10^{-6} | 10^{-7} | 10^{-8} | 10^{-9} |

Height of a person measured in metres

Length of a finger measured in centimetres

Size of a mole on the skin measured in millimetres

Size of a skin cell measured in micrometres

Size of small cell organelles such as lysosomes measured in nanometres

Bacteria

Most bacteria (see Figs 1.3 – 1.5) are spherical or rod-shaped cells, several micrometres long. Their rigid protective **cell wall** is made of **peptidoglycan**, a substance unique to bacteria. Beneath the cell wall, the **cell surface membrane**, similar in structure to the membrane of eukaryotic cells, completely encloses the contents of the cell. Some types of bacteria, such as *Neisseria meningitidis*, which can cause meningitis, also have a **capsule**. This is a sticky coat outside the cell wall.that prevents them from drying out, or

B (a) Estimate the diameter (in micrometers) of a full stop on this page.

(b) How many nanometres are there in 1 millimetre?

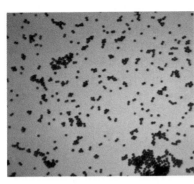
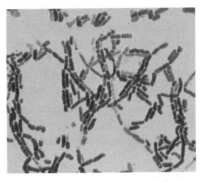

Fig 1.3 **Looking at bacteria with a light microscope reveals that there are two main forms: the spherical organisms are called cocci (singular coccus), and the rod-shaped forms are called bacilli (singular bacillus)**

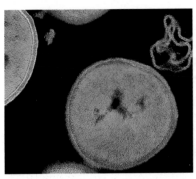
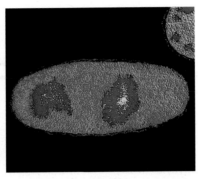

Fig 1.4 **Although bacteria look much simpler than animal or plant cells, their membrane – blue in cocci (left) and fine red outline in bacillus (right) – forms a compartment in which the complex chemical processes take place that enable them to feed, respire, grow, excrete, move, respond and reproduce**

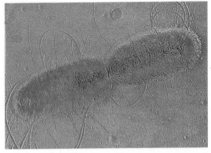

protects them from digestion by intestinal enzymes or from attack by a host's immune system.

Fig 1.5 shows the internal structure of a rod-shaped bacterium, *Escherichia coli* (often abbreviated to *E. coli.*)

Fig 1.5 ***E. coli* as seen in an electronmicrograph (above and below left), and in 3D (below right). *E. coli* is a bacillus found in the human colon. The presence of *E coli* in water or food can indicate contamination by human sewage**

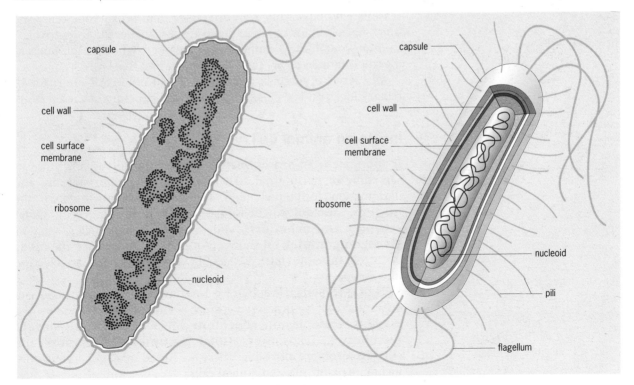

capsule

cell wall

cell surface membrane

ribosome

nucleoid

capsule

cell wall

cell surface membrane

ribosome

nucleoid

pili

flagellum

?

C Some bacteria can digest, or break down, oil and plastic. Suggest how this could be useful to people.

Inside the cell surface membrane, the bacterial cell is a single cytoplasmic compartment which contains DNA, RNA, proteins and small molecules. Bacteria have a circular piece of DNA in a region of the cytoplasm known as the **nucleoid**. There are smaller rings of DNA, known as **plasmids**, elsewhere in the cytoplasm. Plasmids are of great interest to biologists because they often contain genes which code for antibiotic resistance, and can be used to carry genes between cells in genetic engineering (see Chapter 31).

Bacteria feed by **extracellular digestion**. They release enzymes into the surrounding medium and absorb the soluble products. Bacteria can synthesise a wide variety of enzymes, and some species are able to digest unlikely substances such as oil and plastic. Protein are synthesised on **ribosomes** (page 14), and cell respiration occurs on **mesosomes**, inner extensions of the cell surface membrane.

Some bacteria are **motile**: they can swim. They have thin fibres called **flagella** (singular flagellum) that are corkscrew-shaped and that rotate, propelling the bacteria in different directions.

The roles of bacteria in disease and recycling are discussed in Chapters 33 and 23 respectively.

5 EUKARYOTES

The term **eukaryote** means 'true nucleus', because the DNA of eukaryotic cells is confined to a definite area inside the cell enclosed by a **nuclear envelope**.

Eukaryotic cells also have other **organelles** which form compartments. By being in a compartment the chemicals involved in a particular process, such as respiration or photosynthesis, are kept separate from the rest of the cytoplasm. This allows the chemical reactions of the process to take place quickly and efficiently. This high degree of internal organisation is one of the reasons why eukaryotic cells are larger than prokaryotic cells. The fluid that occupies the space between organelles is the **cytosol**, a solution containing a complex mixture of enzymes, the products of digestion (amino acids, sugars etc) and waste materials.

All species of animals, plants, fungi and protoctists are made of eukaryotic cells. Let's look in detail at the structure of these cells.

Inside an animal cell

If you look at an **animal cell** under a light microscope you can see some of its internal structure. With the right staining and illumination techniques (and a good quality microscope), the **nucleus**, **nucleolus**, **chromosomes** (in a dividing cell), **Golgi complex**, **mitochondria** and food storage particles such as **glycogen granules**, all show up quite well. But, to see the inside of a cell in more intricate detail, you really need an electron microscope.

The electron micrograph in Fig 1.6 shows the internal structure of an animal cell. You can see far more of the cell's components – the **nucleus**, **endoplasmic reticulum**, **mitochondria**, **cell surface membrane**, **lysosomes**, **Golgi complex**, **ribosomes** and **cytoskeleton** are all visible. These are found in all eukaryotic cells, including plant cells and animal cells.

✓

1 mm = 1000 µm
1 µm = 1000 nm

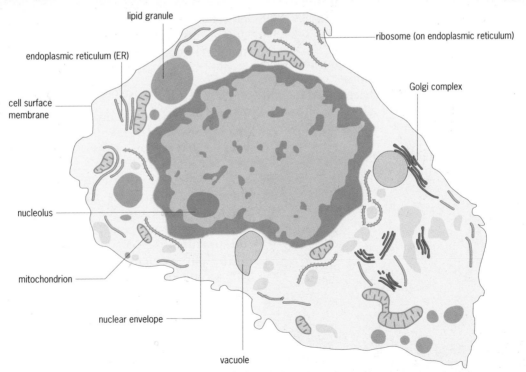

We shall look at the detailed structure of each individual organelle and find out how each type contributes to the function of the cell as a whole.

Fig 1.6 **A micrograph of a human cell with an interpretive diagram. It shows that the cytoplasm of animals cells contains a complex system of membrane-bound organelles**

Using microscopes

The light microscope

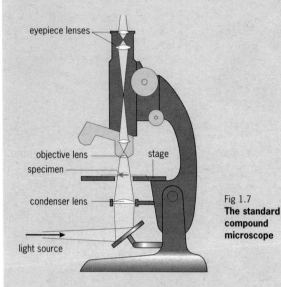

Fig 1.7
The standard compound microscope

Fig 1.7 shows the structure of a **light** or **optical microscope**. This instrument is also known as a **compound microscope** because two lenses, the **eyepiece lens** and the **objective lens**, are combined to produce a much greater magnification than is possible with a single lens. The total magnification is calculated by multiplying the magnification of the two lenses together. For example, if the eyepiece lens has a magnification of ×10 and the objective lens is ×50, the total magnification is ×500.

The light microscope has powers of magnification of up to ×1500, good enough to see cells, larger organelles and individual bacteria, but not powerful enough to reveal smaller structures such as cell surface membranes, viruses or individual

Table 1.2 **Structures normally visible in animal and plant cells with the light microscope**

Organelle	Function	Animal cell	Plant cells
Nucleus	control of cell activities	✓	✓
Vacuole	storage and support	✗	✓
Chloroplasts	photosynthesis	✗	✓
Cell wall	support	✗	✓

molecules. Table 1.2 shows organelles in animal and plant cells which are visible with a light microscope. Fig 1.8 shows a light micrograph of an animal cell.

An important feature of a microscope is its **resolving power**, which should not be confused with the magnification. Two objects close together may appear as one single image when viewed under the light microscope. Increasing the magnification does not allow you to **resolve** the two objects into separate images; the objects just appear to be a larger single image. Resolving power is as important as magnification when investigating structural details.

The limitation of the light microscope is due to the nature of light itself. The wavelength of light determines the maximum effective magnification and the resolving power. The wavelength of visible light is around 500–650 nm and the resolving power – the resolution – of the light microscope is 200 nm (0.2 µm), so two objects separated by less than 200 nm appear as one object.

The electron microscope (EM)

Fig 1.9 shows the essential features of an electron microscope (EM). Invented in the 1930s, the present-day version of the EM can magnify up to 500 000 times and has a resolution of 1 nm. In other words, the EM can resolve two objects that are only 1 nm apart. As many biological molecules are larger than 1 nm, the EM can be used to study the arrangement of individual molecules that make up structures in a cell.

The development of the EM has had a huge impact on biology. A magnification of half a million means that an object the size of a full stop becomes over 200 m in diameter – the length of two football pitches. Organelles that are only blurred images when viewed with a light microscope can now be studied in great detail. Many new cell structures have been discovered using electron microscopy.

While the light microscope uses lenses to focus a beam of light, the electron microscope uses

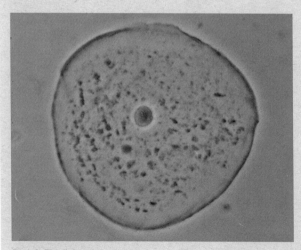

Fig 1.8 **The light microscope can reveal the basic features of animal cells**

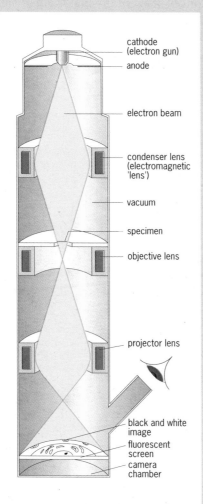

Fig 1.9 **The basic features of the transmission electron microscope. The beam of electrons is focused by electromagnets before passing through the specimen, and the image appears on a fluorescent screen**

cathode (electron gun)
anode
electron beam
condenser lens (electromagnetic 'lens')
vacuum
specimen
objective lens
projector lens
black and white image
fluorescent screen
camera chamber

Table 1.3 **Comparing light and electron microscopes**

	Light microscope	Electron microscope
Illumination	light	electrons
Focused by	lenses	magnets
Maximum magnification	×1500	×500 000
Resolving power	200 nm (0.2 μm)	1 nm
Specimens	living or dead	dead
Preparation of specimens	often simple	more complex
Cost of equipment	relatively cheap	very expensive

Scanning and transmission electron microscopes

There are two main types of electron microscope – the **transmission electron microscope** and the **scanning electron microscope**.

In a transmission electron microscope (TEM), a beam of electrons is transmitted through the specimen. The specimen must be thin and it is stained using electron-dense substances such as heavy metal salts. These substances deflect electrons in the beam and the pattern that the remaining electrons produce as they pass through the specimen is converted into an image. The electron micrograph in Fig 1.6 was produced using the TEM.

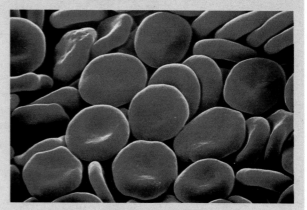

Fig 1.10 **Human red blood cells photographed with the help of a scanning electron microscope**

electromagnets to focus a beam of electrons. The wavelength of the electrons is much smaller than the wavelength of light, the resolving power is so much greater.

The main disadvantage of the EM is that the electron beam must travel in a vacuum because, being so small, electrons are scattered when they hit air molecules. Specimens for the EM must therefore be prepared (killed, dehydrated and fixed) so that they retain their structure inside a vacuum. Such harsh preparation methods can damage cells, and cause **artefacts** – features that do not exist in the living cell – to appear. For example, microsomes, tiny vesicles surrounded by ribosomes, were seen when animal cells were examined using the electron microscope. At first, cell biologists thought that these were organelles they had not noticed before, but they later realised that microsomes were fragments of endoplasmic reticulum (see page 13) produced by the fixing process.

The main differences between electron microscopy and light microscopy are shown in Table 1.3.

We can use scanning electron microscopes (SEMs) to study relatively large three-dimensional objects. Thin sections are not necessary because the SEM records the electrons that bounce off the surface of the object rather than passing through it. Fig 1.10 shows an example of the detailed three-dimensional image that an SEM produces. Although the SEM does not have the resolving power of the TEM, it is more versatile and can be used to observe many kinds of intact structures. The person operating the SEM can move the specimen about, to look at the surface of the structure from a variety of angles – a bit like taking aerial views of the countryside.

?

D If a light microscope had an eyepiece lens of ×25 and an objective lens of ×40, what would the total magnification be?

Table 1.4 summarises the functions of some of the major organelles and structures in the eukaryotic cell.

Table 1.4 **Summary of the functions of major eukaryotic cell organelles and structures**

Organelle	Occurrence	Size	Function
Nucleus	usually one per cell	10 µm	site of the nuclear material – the DNA
Nucleolus	inside nucleus	1–2 µm	manufacture of ribosomes
Mitochondrion	numerous in cytoplasm; up to 1000 per cell	1–10 µm	aerobic respiration
Rough endoplasmic reticulum	continuous throughout cytoplasm	extensive membrane network	isolation and transport of newly synthesised proteins
Smooth endoplasmic reticulum	usually small patches in cytoplasm	variable	synthesis of some lipids and steroids
Ribosome	free in cytoplasm or attached to rough ER	20 nm	site of protein synthesis
Golgi body	free in cytoplasm	variable	modification and synthesis of chemicals
Lysosome	free in cytoplasm	100 nm	digestion of unwanted material
Chloroplast	cytoplasm of some plant cells, eg mesophyll	4–10 µm	site of photosynthesis
Vacuole	usually large, single fluid-filled space in plant; smaller and more numerous in animals	up to 90 per cent of volume of whole plant cell	storage of salts, sugars and pigments; creates turgor pressure by interaction with cell wall
Cell surface membrane	encloses the cytoplasm of all cells	7–10 µm	exchange and transport of materials into and out of the cell
Cell wall	surrounds all plant cells	thickness varies	provides rigidity and strength

See questions 1 and 2. ■

The cell surface membrane

The **cell surface membrane** is the boundary between the cell and its environment. It has little mechanical strength but plays a vital role in controlling which materials pass in and out of the cell.

Although basically a double layer of phospholipid molecules, arranged tail to tail, the cell surface membrane is a complex structure, studded with proteins. These can be embedded in the membrane or they can penetrate the bilayer forming **pores** (holes or channels – see Fig 1.11). Its structure and function are covered in detail in Chapter 5.

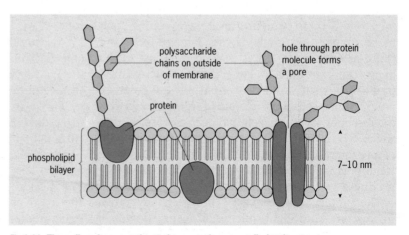

Fig 1.11 **The cell surface membrane is a complex organelle but its structure can be represented by this simplified diagram**

The nucleus

The **nucleus** is the largest and most prominent organelle in the cell. Almost all eukaryote cells have a nucleus – red blood cells in mammals are an exception. Every nucleus is surrounded by a **nuclear envelope**. As Fig 1.12 shows, this consists of two membranes that are separated by a gap of 20 to 40 nm.

The nucleus is usually spherical and about 10 μm in diameter. It contains the cell's DNA, which carries information that allows the cell to divide and carry out all its cellular processes. From the micrographs in Fig 1.13 you can see that, in a dividing cell, the DNA is highly condensed into thread-like structures called **chromosomes** but, at other times, it is spread throughout the nucleus as **chromatin**.

Fig 1.12 **As the electron micrograph and the diagram show, the nucleus is bounded by the nuclear envelope, a double membrane which contains many pores, each one about 100 nm in diameter. The nuclear pores represent about 15 per cent of the total surface area of the nuclear envelope, indicating the heavy traffic of materials in and out of the nucleus. The nucleus contains the cell's genetic material which exists as chromatin, loosely packed DNA attached to proteins called histones. The nucleolus is clearly visible**

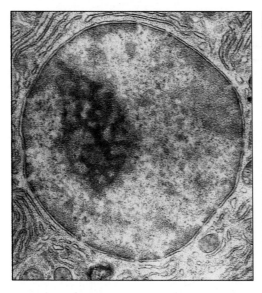

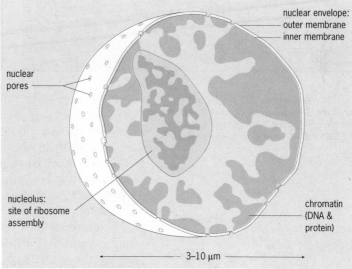

Fig 1.13 **Before a cell divides, its DNA condenses into visible chromosomes. This figure shows cells at various stages of cell division. For further information about cell division, see Chapter 27**

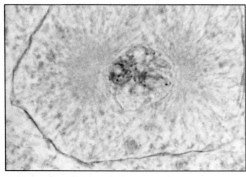

1 DNA condenses into chromosomes

2 Chromosomes move to the centre of the cell

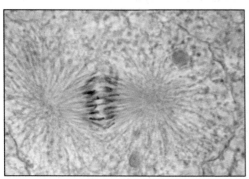

3 Chromosomes are pulled apart

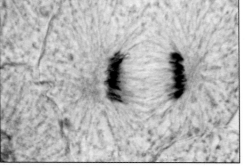

4 Each new cell receives an identical set of chromosomes

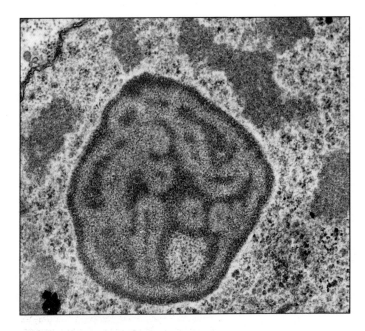

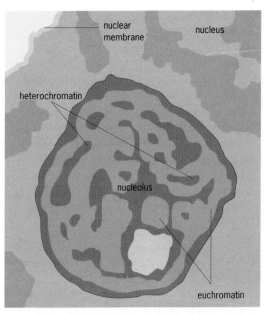

Fig 1.14 **The nucleolus has a highly organised structure. The pale-staining part at lower right contains DNA which is being 'read' to produce the RNA that will be used to make new ribosomes**

Nuclei also have one or more **nucleoli** (Fig 1.14). These dark-staining, spherical structures are ribosome-producing centres: they synthesise ribosomal RNA and package it with ribosomal proteins to make ribosomes.

A dividing cell condenses and packages its DNA into chromosomes, to be more easily transported inside the cell. The chromosomes must be moved to opposite ends of the cell so that, when the cell splits, each daughter cell receives the correct amount of DNA.

Most human body cells contain two copies of each chromosome – they are said to be **diploid**. **Mitosis**, the type of division that occurs during the growth of an organism, produces two daughter cells genetically identical to the parent cell. The parent cell duplicates its DNA, condenses it into chromosomes and then splits to form two new cells. Each one receives a copy of each chromosome pair. **Meiosis**, the other type of cell division, produces **gametes** (sex cells). A cell dividing by meiosis duplicates its DNA, condenses it into chromosomes, as in mitosis, but then usually splits to form four new cells. Each of these contains only one copy of each chromosome, half the number in the parent cell, and is said to be **haploid**.

Cell division is covered more fully in Chapter 27.

In a cell that is not dividing, close examination of the chromatin reveals two different levels of density. Dark-staining chromatin, consisting of tightly packed DNA, is known as **heterochromatin**; the lighter, more loosely packed material is called **euchromatin**: Fig 1.4. In this loosely packed state, DNA is accessible to proteins that 'read' genes to produce molecules of **messenger RNA**. Messenger RNA passes into the cytoplasm where it is used by ribosomes as a **template** to build **proteins** from **amino acids**.

Individual segments of DNA called **genes** contain the information necessary to make individual proteins, including the enzymes that control most of the cell's activities. In fact, a central concept in biology that is true for all cells, prokaryotes and eukaryotes, is that:

Many genes code for making enzymes which, in turn, control the activities of the cell.

You can study enzymes in Chapter 4 and we look at protein synthesis in Chapter 28.

Endoplasmic reticulum (ER)

The nuclear envelope joins with the membrane of the **endoplasmic reticulum** (**ER**), a system of complex tunnels that are spread throughout the cell.

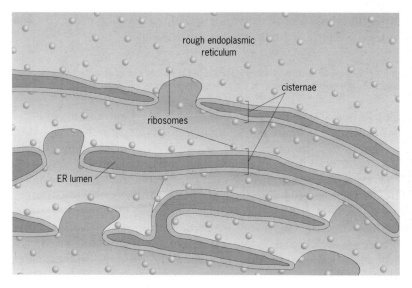

Fig 1.15 **Electron micrograph (above) and interpretive diagram (left) showing rough ER. The ER is a large sheet of membrane that is folded over on itself many times, forming stacked layers called** cisternae. **The space inside the cisternae, the ER lumen, forms an extensive transport system throughout the cytoplasm**

On much of the outside surface of the ER in a eukaryotic cell are the sites of attachment for ribosomes. This gives it a grainy appearance (Fig 1.15) and its name, **rough ER**.

The main function of rough ER is to keep together and transport the proteins made on the **ribosomes**. Instead of simply diffusing away into the cytoplasm, newly made proteins are threaded through pores in the membrane and accumulate in the space which is called the **ER lumen** (Fig 1.16). Here, they are free to fold into their normal three-dimensional shape. Not surprisingly, a mature cell that makes and secretes large amounts of protein – such as one that makes digestive enzymes – has rough ER that occupies as much as 90 per cent of the total volume of the cytoplasm. The rough ER is also a storage unit for enzymes and other proteins.

Small vesicles containing newly synthesised proteins pinch off from the ends of the rough ER and either fuse with the Golgi complex or pass directly to the cell surface membrane.

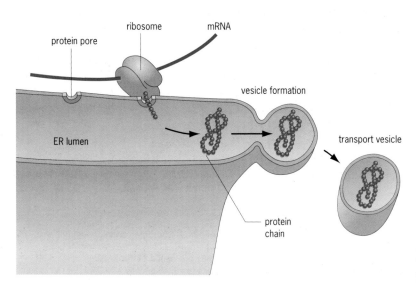

Fig 1.16 **Accumulation of proteins in the ER lumen. Many proteins made on the ribosomes are threaded through pores in the ER membrane**

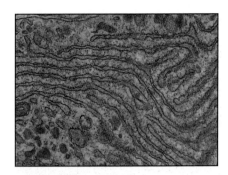

ER with no ribosomes attached is known as **smooth ER** (Fig 1.17). Smooth ER tends to occur in small areas that are not continuous with the nuclear membrane. Smooth ER is not involved in protein synthesis but is the site of steroid production (many hormones are steroids). It also contains enzymes that **detoxify**, or make harmless, a wide variety of organic molecules, and it acts as a storage site for calcium in skeletal muscle cells.

Fig 1.17 **Smooth ER has no attached ribosomes. It is usually not as abundant as rough ER, although it is common in the cells of the liver, gut and some glands**

Ribosomes

Ribosomes are small, dense organelles, about 20 nm in diameter, present in great numbers in the cell. Most are attached to the surface of rough ER but they can occur free in the cytoplasm, as in Fig 1.18. This artist's impression of protein synthesis shows the ribosome's distinctive shape. Ribosomes are made from a combination of ribosomal RNA and protein (65 per cent RNA: 35 per cent protein).

Ribosomes are involved in protein synthesis. We say that ribosomes are the site in the cell where amino acids are assembled in the right order to produce new proteins. The code on messenger RNA (mRNA) is used by the ribosome to put amino acids together

E What function might the cells of the liver, gut and glands have in common? How could you recognise such cells from electron micrographs?

F Why was ER not discovered until after the invention of the electron microscope?

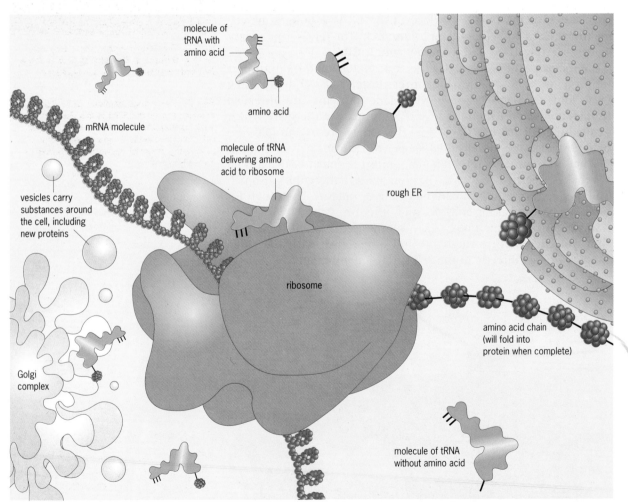

See question 2.

Fig 1.18 **An artist's impression of protein synthesis. A ribosome can be thought of as a giant enzyme on which a protein is assembled. Transfer RNA molecules (tRNA) bring specific amino acids to the ribosome where they are added to the growing amino acid chain according to the code on the mRNA which is a mobile copy of a gene (not to scale)**

in chains to form a specific proteins. This is another central concept in biology that you will learn more about later on:

> **A gene is a piece of DNA that codes for a particular protein. A copy of the gene in the form of messenger RNA passes out of the nucleus and travels to the ribosome where it controls protein synthesis.**

Protein synthesis is covered in detail in Chapter 28.

Generally, proteins that are to be used inside the cell are made on free ribosomes (Fig 1.18), while those that are to be secreted out of the cell are made on ribosomes that are bound to ER membranes (Fig 1.16).

The Golgi complex

The **Golgi complex**, also called the **Golgi apparatus**, is a tightly-packed group of flattened cavities or vesicles (Fig 1.19). The whole organelle is a shifting, flexible structure; vesicles are constantly being added at one side and lost from the other. Generally, vesicles fuse with the **forming face** (the one nearest to the nucleus) and leave from the **maturing face** (the one nearest to the cell surface membrane).

Fig 1.20 summarises the relationship between the rough ER and the Golgi complex.

The Golgi complex appears to be involved with the synthesis and modification of proteins, lipids and carbohydrates. Studies have shown that proteins made on the ribosomes are packaged into vesicles by the ER. Some of the vesicles join with the Golgi complex and the proteins they contain are modified before they are secreted out of the cell.

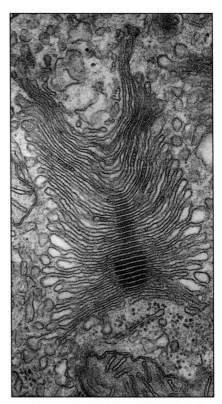

Fig 1.19 **The Golgi complex with a very large number of vesicles. The forming face is at the foot and secretory vesicles are at the top**

Fig 1.20 **The various functions of the Golgi complex are summarised in this diagram. Vesicles from the ER or from outside the cell bring material to the Golgi complex. After processing, material passes out of the cell or enters lysosomes**

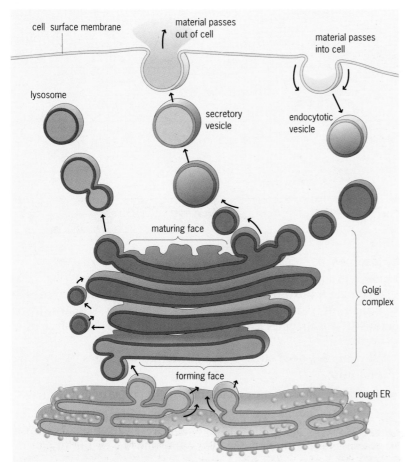

A vesicle is a small spherical organelle, bounded by a single membrane, which is used to store and transport material around the cell.

■ See questions 4 and 5.

Cell fractionation

In order to study the function of a particular organelle it is often helpful to isolate it from the rest of the cell. This can be done by cell fractionation, a process which is outlined in Fig 1.21.

Fig 1.21 **Cell fractionation: how to isolate particular types of organelle from liver tissue**

1 Liver tissue is placed in an ice-cold liquid. When the cell is broken up, membranes are ruptured and many chemicals which do not normally mix are brought together. The low temperature minimises unwanted reactions, including self-digestion by digestive enzymes.

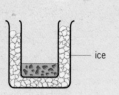   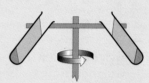

2 Tissue cut into small pieces   **3** Tissue put into blender/homogeniser to break up whole cell   **4** Mixture filtered to remove debris   **5** Filtrate spun in centrifuge

Principle: The organelles will separate out in a particular order, according to their density and shape.
After a time, the sediment containing a particular type of organelle can be separated from the supernatant (the liquid which contains the remaining organelles). The exact times vary from tissue to tissue.

	Organelle		Centrifuge setting (g)	Time (minutes)
First to separate out	Nuclei 		800–1000	5–10
	Mitochondria		10 000–20 000	15–20
	Lysosomes			
	Rough ER 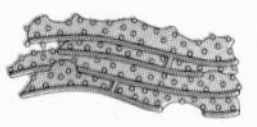		50 000–80 000	30–50
	Plasma membranes		80 000–100 000	60
	Smooth ER			
Last to separate out	Free ribosomes 		150 000–300 000	> 60

Lysosomes

Lysosomes are small vesicles 0.2 – 0.5 μm in diameter (Fig 1.22) that contain a mixture of digestive enzymes called **lytic enzymes**. It is important that the membrane of lysosomes remains intact because if the enzymes leak out they could digest vital molecules in the cell.

Why do cells need such potentially lethal structures? They have several uses:

● To supply the enzymes which destroy old or surplus organelles.

● To digest material taken into the cell. After a white cell has engulfed a bacterium, for example, lysosomes discharge enzymes into the vacuole (Fig 1.23) and digest the organism. This process is called **phagocytosis**.

1 mm = 1000 μm
1 μm = 1000 nm

- To destroy whole cells and tissues. Parts of tissues and organs often need to be removed after they have performed their function. The muscle of the uterus is reduced after giving birth and milk-producing tissue is destroyed after weaning. Bone is also constantly made and reabsorbed throughout life.

Lysosomes are similar in structure to many other vesicles in the cell and are thought to be made in the same way: inactive digestive enzymes from the rough ER pass through the Golgi apparatus where they are activated and packaged into vesicles.

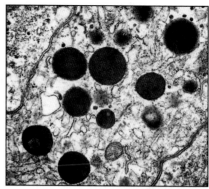

Fig 1.22 **Lysosomes are simply bags of digestive enzymes. They can be distinguished from other vesicles in the cell only by using a stain specific for the chemicals inside the lysosomes**

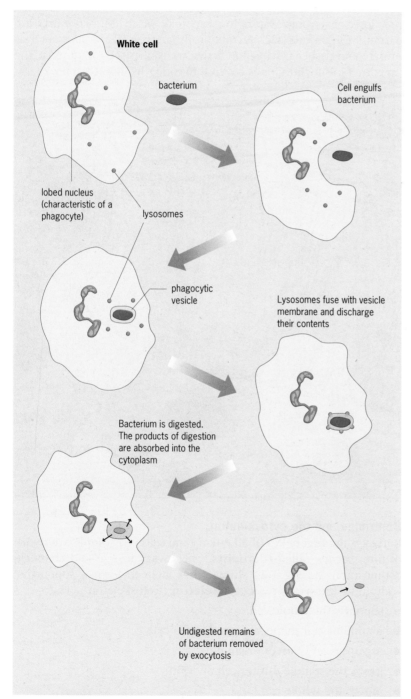

White cell

bacterium

Cell engulfs bacterium

lobed nucleus (characteristic of a phagocyte)

lysosomes

phagocytic vesicle

Lysosomes fuse with vesicle membrane and discharge their contents

Bacterium is digested. The products of digestion are absorbed into the cytoplasm

Undigested remains of bacterium removed by exocytosis

Fig 1.23 **One of the functions of lysosomes is to digest material taken into the cell by the process of phagocytosis (see Chapter 26). Here, a white cell, known as a phagocyte, is ingesting a bacterium. Lysosomes discharge their enzymes into the temporary vacuole and the bacterium is digested**

G What would happen to an organism if it lost control of its lysosomes?

H If we could control lysosome activity in specific tissues, how could this be used in the treatment of cancer?

I Why are digestive enzymes synthesised in an inactive form?

Mitochondria

Mitochondria are relatively large, individual organelles which occur in large numbers in most cells. They are usually spherical or elongated (sausage-shaped) and are 0.5–1.5 µm wide and 3–10 µm long. Their function is to make **ATP** via the process of **aerobic respiration**. ATP is a molecule that diffuses around the cell and provides instant chemical energy to the processes that require it (see Chapter 22).

As Fig 1.24 shows, mitochondria have a double membrane; the outer membrane is smooth while the inner one is folded. This arrangement gives a large internal surface area on which the complex reactions of aerobic respiration can take place. Mitochondria are particularly abundant in metabolically active cells, tissues such as muscle and tissues involved in active transport (see Chapter 5). The function of mitochondria is covered in more detail in Chapter 22.

Fig 1.24 **Mitochondria are the sites of aerobic respiration within cells. The inner membrane is folded into cristae, which give a large surface area for attachment of some of the enzymes involved in cell respiration**

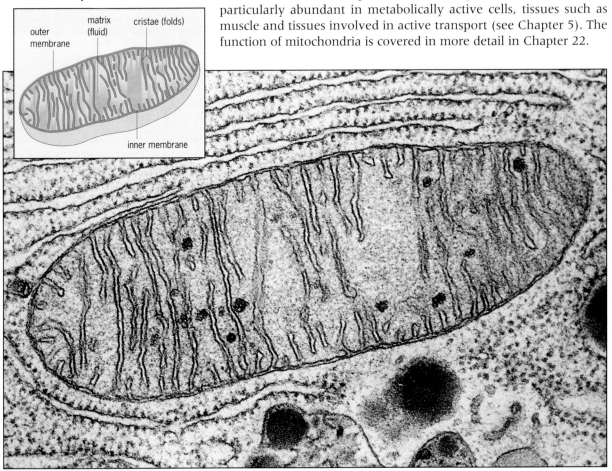

Centrioles and the cytoskeleton

Between the organelles of all eukaryotic cells is a complex network of fine threads called **filaments.** These vary in size and chemical composition. As Fig 1.26 shows, they help to form a supportive scaffolding known as the **cytoskeleton** ('cell skeleton'). This:

● supports the whole cell,

● maintains cell shape, as in red blood cells,

● organises and moves organelles,

● forms the spindle during cell division,

● moves the whole cell.

This last function indicates that the cytoskeleton is not a rigid structure; it can be assembled or dismantled in seconds – as happens in a moving white blood cell.

AUTORADIOGRAPHY

ELECTRON MICROSCOPES are useful tools with which to investigate cell structure, but they do not reveal much about the processes that go on inside a living cell. After World War II, work on the atomic bomb produced chemicals called **radioactive isotopes** as a by-product. These are very useful for studying metabolism because they behave in the same way as non-radioactive atoms and, in the quantities in which they are used, they are unlikely to harm the cell. However, their movements through the cell can be traced (Fig 1.25).

After cells are exposed to a **radioactive isotope** such as **carbon-14**, they are washed to remove excess label and then fixed, embedded, sectioned and mounted on microscope slides. The slides are then taken to a dark room and coated with photographic emulsion that has been heated to make it melt. After coating, the slides are cooled and kept in the dark for various time periods from a few days to several weeks. During this time, radioactive decay causes a chemical change to occur in the film of photographic emulsion. When this is developed using standard photographic techniques and the slide is examined under the light microscope, silver grains can be seen in the areas exposed by the radioactivity. The pattern of grains shows the structures of the cell that have incorporated the radioactive label.

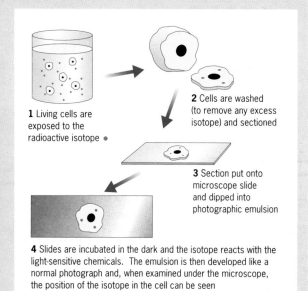

1 Living cells are exposed to the radioactive isotope •

2 Cells are washed (to remove any excess isotope) and sectioned

3 Section put onto microscope slide and dipped into photographic emulsion

4 Slides are incubated in the dark and the isotope reacts with the light-sensitive chemicals. The emulsion is then developed like a normal photograph and, when examined under the microscope, the position of the isotope in the cell can be seen

Fig 1.25 **The basic method of autoradiography**

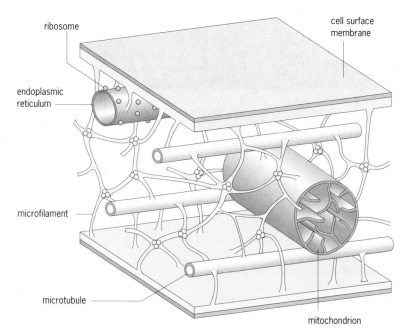

Fig 1.26 **Inside the cell is a network of tubules which forms the cytoskeleton**

labels: ribosome, cell surface membrane, endoplasmic reticulum, microfilament, microtubule, mitochondrion

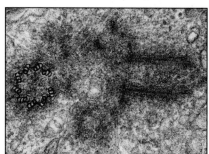

Centrioles are short bundles of filaments, set at right angles to each other (Fig 1.27). They are found in a clear area of cytoplasm known as the **centrosome**. Centrioles occur in all animal cells but are absent from the cells of plants. For a long time their function was thought to be just the formation of the **spindle** – the cradle of threads that guide chromosomes during cell division. In addition to spindle formation, the centrioles act as the centre of formation for the whole cytoskeleton and they have therefore become known as **microtubule organising centres**.

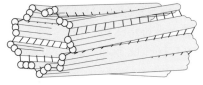

Fig 1.27 **The centrioles are two bundles of fibres that form the cytoskeleton of the cell and also give rise to the spindle during cell division**

See question 6.

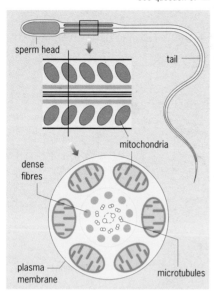

Fig 1.28 **A cross-section of the tail of a human sperm showing the '9 + 2' arrangement of microtubules in its centre**

?

I Why does a sperm need so many mitochondria?

Fig 1.29 **The light microscope shows some of the basic features of plant cells**

Microtubules make up more complex external cell structures called **cilia**. These tiny hair-like projections are used for cell movement. The tail of a sperm is, for example, a modified cilium. In human cells, cilia contain a bundle of microtubules in the '9 + 2' arrangement shown in Fig 1.28.

Plant cells

Typical leaf cells from a plant, seen with the light microscope, are shown in Fig 1.29.

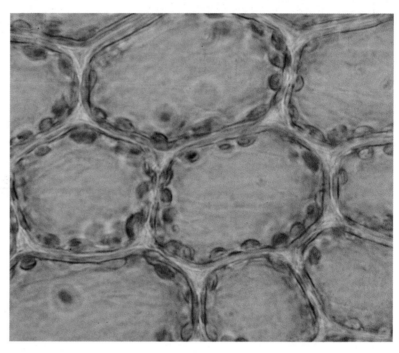

Feature	Cell: Animal	Plant
Nucleus	✓	✓
Plasma membrane	✓	✓
Mitochondria	✓	✓
Rough ER	✓	✓
Smooth ER	✓	✓
Golgi bodies	✓	✓
Lysosomes	✓	✓
Cell wall	✗	✓
Plastids, eg chloroplasts	✗	✓
Vacuoles	✗	✓
Centrioles*	✓	✓*

Table 1.5 **Comparing plant cells and animal cells (*centrioles not present in plants)**

?

J How many of the following cells would fit into a line 1 cm long?

(a) Animal cells of 20 μm

(b) Plant cells of 50 μm

(c) Bacterial cells of 2 μm

Plant cells have several features that are not found in animal cells (see Table 1.5). A plant cell is surrounded by a cell wall made of cellulose. A large proportion of the inside of the cell is taken up with a fluid-filled compartment known as the vacuole. Together, the wall and vacuole maintain the shape of the whole cell.

As you can see from Fig 1.30, plant cells have specialised organelles, the chloroplasts, which enable them to make their own food by photosynthesis. Chloroplasts bring together all the chemicals involved in photosynthesis, and provide sites for the chlorophyll molecules so that they can absorb the maximum amount of light.

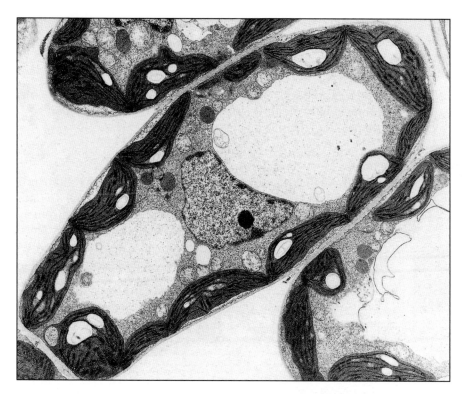

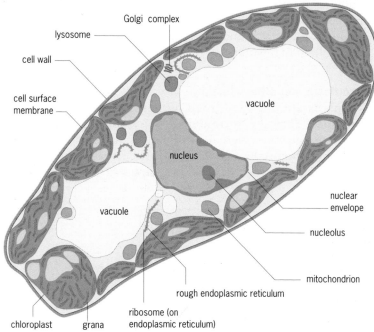

Fig 1.30 **Transmission electron micrograph (above left) and explanatory diagram (below left) showing a cell from a soya bean leaf**

K Which parts of a plant would be unlikely to contain chloroplasts, and why?

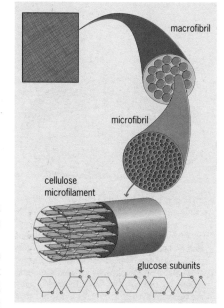

Fig 1.31 **The cellulose cell wall is a composite material. Fibrils containing thousands of cellulose molecules are cemented together by a 'glue', a complex mixture of compounds**

The cellulose cell wall

Unlike animal cells, plant cells are enclosed by a **cell wall** which consists of many **cellulose fibres** cemented together by a mixture of other organic substances.

Cellulose is a **polysaccharide** (a sugar polymer) which consists of long straight chains of glucose molecules bonded to adjacent molecules (see Chapter 3). In the cell wall, around 2000 parallel cellulose molecules are packed together to form **microfibrils**. These in turn are bundled together to form **fibrils**, as you can see in Fig 1.31. Like fibreglass, the cell wall has great strength because of the many strong fibres and the 'glue' that holds them together.

The main functions of the cell wall are:

- to provides rigidity and strength to the cell. Cell walls need to be strong enough to resist expansion to allow the cell to become **turgid**; see Chapter 5.
- to force the cell to grow in a certain way. A particular arrangement of fibrils causes the cell to assume a particular shape – for example a long, thin tube.

Vacuoles

Plant cells also have a **vacuole**, a large, fluid-filled cavity. In mature cells the vacuole can occupy over 80 per cent of the cell volume, and is filled with a fluid called **cell sap** (Fig 1.32). The sap consists of a complex mixture of sugars, salts, pigments and waste products in water.

Vacuoles have several functions:

- They absorb water by **osmosis** and therefore swell, pushing the cytoplasm against the cell wall. In this state, a plant cell is said to be turgid.
- They store food substances such as sugars and mineral salts.
- They store pigments that give colour to plant structures such as petals.
- They can accumulate waste products and by-products of metabolism. In some cases, these chemicals may be toxic or have an unpleasant taste and are used by the plant to make it less palatable to herbivores.

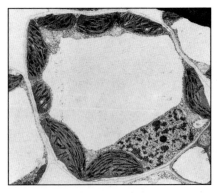

Fig 1.32 **In a mature plant cell, the vacuole becomes so large that the cytoplasm is reduced to a thin layer around the cell**

Chloroplasts and other plastids

Chloroplasts are one of a group of plant cell organelles known as **plastids**. They are surrounded by a double membrane and contain an elaborate internal membrane system that houses the chemicals of photosynthesis (Fig 1.33). The structure and function of chloroplasts is covered in detail in Chapter 22.

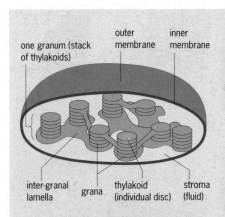

one granum (stack of thylakoids)
outer membrane
inner membrane
inter-granal lamella
grana
thylakoid (individual disc)
stroma (fluid)

Fig 1.33 **Chloroplasts are organelles in which all the chemicals associated with photosynthesis are brought together. Chlorophyll molecules are housed on membranes to maximise light absorption. In a flowering plant, most chloroplasts are found in the palisade cells of the leaves**

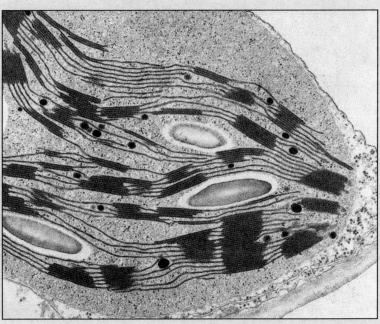

SUMMARY

When you have finished this chapter you should know and understand the following:

■ All living organisms (except viruses) are composed of either **prokaryotic** or **eukaryotic** cells.

■ Bacteria are prokaryotes; plants, animals, protoctists and fungi are eukaryotes.

■ Prokaryotic cells are relatively small, with very little internal membrane organisation.

■ Eukaryotic cells are relatively large, with a high degree of internal organisation.

■ Eukaryotic cells have a true, membrane-bound **nucleus** and many **organelles** – membrane-bound compartments in which particular chemical processes take place.

■ The nucleus, **ribosomes**, **rough ER**, **vesicles** and **Golgi complex** in eukaryotic cells all take part in different stages in the making and packaging

of chemical products. Some chemicals are used within the cell, others – such as hormones or digestive enzymes – are released.

■ **Mitochondria** are the site of production of most of the eukaryotic cell's **ATP**, which provides the energy for other cellular processes.

■ Eukaryotic cells possess a **cytoskeleton** – a network of fine **tubules** that supports and shapes the cell.

■ Plant cells contain structures not found in animal cells: **chloroplasts**, a **cell wall** and a **vacuole**.

■ The cell can be studied using a variety of experimental techniques: the **electron microscope** reveals the structure of the cell, **cell fractionation** isolates individual organelles and, with the use of **radioactive isotopes** in **autoradiography**, the passage of materials through the cell can be followed.

QUESTIONS

1 The photograph is an electron micrograph of part of a cell from the pancreas of a mammal.

a) (i) Name features labelled **A**, **B** and **C**.

(ii) Briefly describe the part played by features **A** and **B** in the production of protein granules in the cytoplasm.

(iii) Some of the features labelled **C** are elongated; others are more or less circular. Explain this difference in shape.

b) (i) The magnification of this photograph is 2350 times. Calculate the actual maximum diameter of the labelled protein granule **D**. (Show your working.)

(ii) Give **one** way, apart from size and shape, by which the protein granules might be distinguished from mitochondria.

(iii) Name **one** protein that is secreted by pancreatic cells.

[AEB 1996 Biology/Human Biology AS: Specimen Paper 1, q.12]

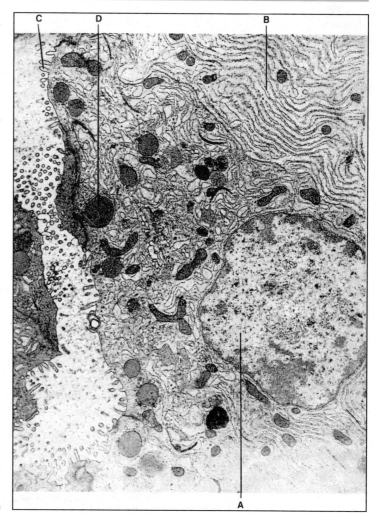

(Animal Physiology Unit, Babraham, Cambridge)

2 The arrows in the diagram show the path followed by a protein produced in a secretory cell.

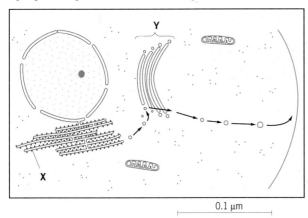

0.1 μm

a) Calculate the magnification. Show your working.

b) Identify organelle **X**.

c) With reference to the protein being produced:
   (i) identify **one** function of organelle **Y**;
   (ii) explain how the protein reaches the outside of the cell from organelle **Y**.

[AEB: 1995 Human Biology: Paper 1, q.2]

3 Liver cells were ground to produce an homogenate. The flow chart shows how centrifugation was used to separate organelles from liver cells.

A spun at low speed   **B** spun at medium speed   **C** spun at high speed

supernatant

supernatant   supernatant

sediment

sediment examined using electron microscope   sediment examined using electron microscope   sediment examined using electron microscope

Drawings of electron micrographs of three organelles separated by the centrifugation are shown below. The drawings are **not** to the same scale.

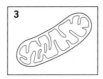

a) Complete the table below.

Electron micrograph	Name of organelle	Centrifuge tube in which the organelle would be the main constituent of the sediment
1		
2		
3		

b) Explain why it is possible to separate the organelles in this way.

[NEAB Feb 1995 Biology: Modular Life Processes Test, q.2]

4 A sample of animal tissue was treated in order to separate the cell components. A chemical analysis of the pure fractions shown in the table was carried out.

Cell component	DNA	RNA	Protein	Phospholipid
Cell surface membrane				
Rough endoplasmic reticulum				
Mitochondria				
Nuclei				

a) Complete the table to show the results of the analysis. Mark the box with a tick if you think that the chemical was present or with a cross if the chemical was absent.

b) Describe concisely how the cell components were separated.

[AEB 1994 Human Biology: Paper 1, q.2]

5 Actively secreting cells growing in tissue culture were provided with a dose of radioactive amino acids for a very short period. At various times, samples of the cells were taken and homogenised and fractionated so that different parts of the cells could be measured for their radioactivity. The following table shows the results of this experiment.

	Radioactivity present in the following cell organelles		
Time after radioactive aminoacids given (minutes)	Rough endoplasmic reticulum	Golgi apparatus	Secretory vesicles
1	123	21	7
20	84	42	7
40	39	84	7
60	28	77	7
90	27	49	28
120	24	38	56
180	28	21	63
240	18	11	20

a) Using the data provided, draw a graph to show the changes in radioactivity in the different organelles in the cells. Put all three lines on one graph.

b) Describe the role of the three organelles mentioned in the production of material to be secreted by the cells

[National Extension College Flexible Learning Pack: Biology. Assignment 1, Part 2, q.1]

Assignment

RED BLOOD CELLS

Human blood contains two main types of cell: red blood cells (RBCs) and white cells (see Chapter 34). Red blood cells are relatively simple cells that transport oxygen from the lungs to the tissues of the body. They contain neither nuclei nor mitochondria and they are very well adapted to their function in the body.

In this chapter you looked at different ways to study cell structure. This Assignment shows you how different types of microscope can be used to look at red blood cells. Fig 1.A1 shows how to make a blood smear using a drop of blood.

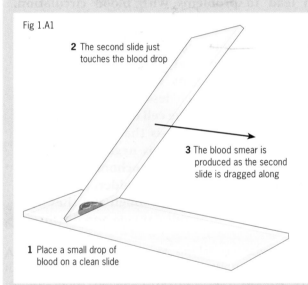

Fig 1.A1

2 The second slide just touches the blood drop

3 The blood smear is produced as the second slide is dragged along

1 Place a small drop of blood on a clean slide

1 Suggest one reason why it might not be safe to use human blood when making a smear in a school or college laboratory.

Fig 1.A2 shows how red blood cells would look if the blood smear were examined with a light microscope.

2 Suggest two reasons why the centre part of each cell appears lighter in colour than the outer part of the cell.

3 The magnification of this photograph is ×1500. Calculate the diameter of the red blood cell labelled A. Give your answer in micrometres.

Before making a smear, the drop of blood can be mixed with various different solutions. Table 1.A1 shows the appearance of a smear made from blood mixed with a strong solution of sodium chloride and blood mixed with a detergent solution. (Detergent dissolves lipids.)

Table A1.1

Solution added to blood smear	Appearance of blood
Strong salt solution	Red blood cells appear small in size and crinkly in outline.
Detergent solution	No cells visible. Smear appears a yellowish red colour.

4 What additional information does the table give about red blood cells? Explain your answer.

Now look back at Fig 1.10. This is an electronmicrograph of red blood cells taken with a scanning electron microscope.

5
a) What evidence can you find that this image has been taken with the aid of a scanning electron microscope?
b) To what extent does the appearance of the cells in this photograph support your answer to Question 2?

When you look at any cell with a microscope, you have to take care in interpreting what you see. This is particularly true when an electron microscope is used. The processes involved in preparing the material for examination can alter the appearance of the cell. The photograph in Fig A1.3 was taken with a transmission electron microscope.

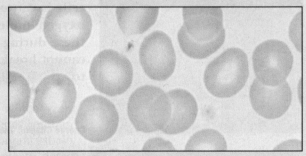

Fig 1.A2 **A light microscope image of red blood cells**

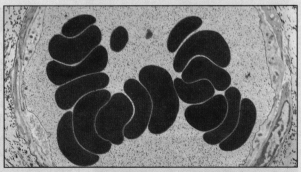

Fig 1.A3 **Transmission electron micrograph of red blood cells in blood**

6 Suggest an explanation for each of the following observations:
a) The red blood cells appear a uniform dark colour.
b) The red blood cells appear to be different sizes and shapes.

Human tissues

In humans, as in all mammals, each organ of the body is made up from a combination of these four basic tissue types:

- **Epithelial tissue** forms thin sheets which line and cover body structures. Epithelial tissue lines the intestines, for example.

- **Connective tissue** is tough and fibrous and forms structures which hold the body together. Ligaments and tendons are mainly connective tissue.

- **Muscular tissue** can contract to produce movement.

- **Nervous tissue** has the ability to conduct impulses, allowing communication between different parts of the body.

Epithelial tissue

Epithelial tissues form continuous sheets which line or cover most structures and cavities in the body. Different types are classified according to the size and shape of their cells, and the number of cell layers they contain (Fig 2.4). Epithelial tissues have very little mechanical strength; they are supported and attached to underlying connective tissue by a **basement membrane,** a continuous sheet consisting of collagen and other proteins.

Epithelial tissues form barriers which keep different body systems separate. They also have many other functions. For instance, the epithelial cells which line the mammalian respiratory tract are **ciliated**: they form a 'carpet' of tiny hair-like processes which trap inhaled dust and sweep it away from the lungs and out towards the throat where it is swallowed. Other epithelial tissues, such as the lining of the mammalian intestine, secrete mucus.

?

A Smoking has been found to affect the lining of the respiratory system. What type of tissue is found here?

Fig 2.4 **The different types of epithelial tissues in the mammalian body. There are basically two groups;** simple (these have one layer of cells) and stratified (these have more than one layer of cells). **Note that the cells of the top layer of the stratified forms of the three epithelia are very similar to the simple forms. You don't need to learn all the different forms: it is more important to see how structure relates to function. For example, simple squamous epithelium is thin and is ideally suited to be an interface where substances are exchanged, such as at the surface of alveoli. Stratified squamous epithelium is thicker and can be replaced continuously, making it perfect as a protective covering to the skin**

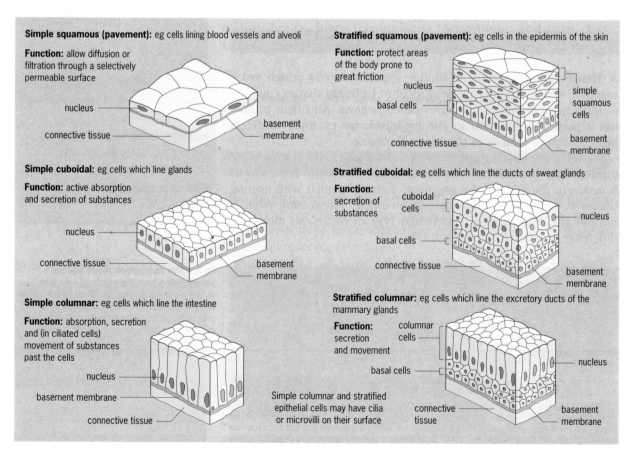

Simple squamous (pavement): eg cells lining blood vessels and alveoli

Function: allow diffusion or filtration through a selectively permeable surface

nucleus

connective tissue

basement membrane

Simple cuboidal: eg cells which line glands

Function: active absorption and secretion of substances

nucleus

connective tissue

basement membrane

Simple columnar: eg cells which line the intestine

Function: absorption, secretion and (in ciliated cells) movement of substances past the cells

nucleus

basement membrane

connective tissue

Stratified squamous (pavement): eg cells in the epidermis of the skin

Function: protect areas of the body prone to great friction

nucleus

basal cells

simple squamous cells

connective tissue

basement membrane

Stratified cuboidal: eg cells which line the ducts of sweat glands

Function: secretion of substances

cuboidal cells

nucleus

basal cells

connective tissue

basement membrane

Stratified columnar: eg cells which line the excretory ducts of the mammary glands

Function: secretion and movement

columnar cells

nucleus

basal cells

Simple columnar and stratified epithelial cells may have cilia or microvilli on their surface

connective tissue

basement membrane

Connective tissue

As their name suggests, connective tissues usually *connect* structures together. Connective tissue cells are not packed closely together but are separated by a **non-cellular matrix** (a mesh) which they secrete. (Notice the contrast to epithelial tissue where the cells *are* tightly packed.) The nature of the matrix accounts for the property of the tissue. In bone, for example, the matrix is a hard mineral strengthened by fibres of collagen, a tough protein (see Chapter 3).

The main types of connective tissue found in the human body are listed in Table 2.1. **Cartilage**, **tendons**, **ligaments**, **blood** and **bone** are all connective tissues. Each one is adapted to perform a particular function. All except blood have a structural role in the body. Blood is described as a fluid connective tissue because the cells are separated by a liquid, the **plasma**. Because connective tissues are mainly structural, their **metabolic demands** (their need for food and oxygen) are relatively small.

> ✔ Connective tissue consists of cells in a matrix which usually contains strengthening fibres.

Table 2.1 **The main types of mammalian connective tissue**

Connective tissue	Function
Bone	Forms the skeleton. Protects, supports the main organs of the body. Anchors the muscles
Tendon	Attaches muscles to bones
Ligament	Attaches bones to bones and provides support at joints
Cartilage	Smoothes surfaces at joints. Prevents collapse of trachea and bronchi
Adipose tissue	Stores fat and provides insulation
Blood	Transports substances around the body (see Chapter 9)
Areolar tissue	Protects organs, blood vessels and nerves. Gives strength to epithelial tissue. General 'packing tissue'

> ✔ Tendons are made from fibrous connective tissue which contains many closely packed collagen fibres. In some situations, such as in the legs during vigorous exercise, the tendons withstand enormous forces.

Cartilage

Cartilage is a tough, smooth and flexible connective tissue. There are three types of cartilage:

- **Hyaline cartilage**. This is a compressible and elastic tissue found in the nose and trachea, and at joints where it covers the ends of bones to provide smooth surfaces.
- **White fibrous cartilage**. This is found between the vertebrae and acts as a shock absorber.
- **Yellow elastic cartilage**. This is a very elastic tissue found in the ears, **pharynx** (throat) and **epiglottis** (the flap which closes over the airway during swallowing).

Fibrous cartilage is a good shock absorber because it is **compressible** (it can withstand pressure without being damaged) and it can absorb the impact of a rapid jolt which would damage a more brittle structure such as a bone.

Studies into the structure of fibrous cartilage reveal how it is adapted to perform its function. The toughness, not surprisingly, is provided by **collagen fibres** (see Chapter 3). But a second type of chemical, **proteoglycan**, has proved to be especially interesting. Proteoglycan binds loosely to water molecules, but releases them when under pressure. So water seeps out of cartilage when it is compressed, and goes back again when the pressure is released, allowing cartilage to act as a very tough sponge to reduce the shock on the rest of the body.

Despite this, fibrous cartilage cannot absorb the impact of strenuous exercise. When we run and jump, for example, much of the cushioning effect is due to the tendons, ligaments and muscles. To illustrate this, think about what happens when you do a standing jump. If you land on your toes, there is no problem. The shock is absorbed by the muscles and tendons in your feet and legs. If you land on your heels (even from a few centimetres) the jarring effect is very obvious. Don't try this!

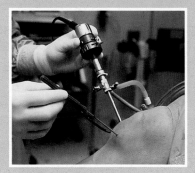

Fig 2.5 **The knee joint has to take the weight of most of the body, and in strenuous sports such as football, the twisting and turning can easily damage the cartilage. This player is having keyhole surgery to remove a fragment of cartilage which has been torn**

Fig 2.6 **Micrographs of muscle. Top: skeletal, middle: smooth, and bottom: cardiac muscle. See Table 2.2 for descriptions**

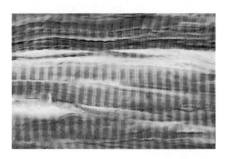

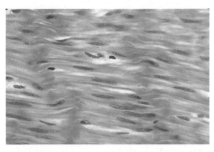

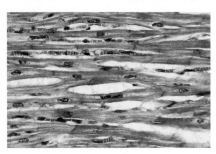

?

B Which type of muscle would you expect to have the fewest mitochondria? Explain.

C What features of skeletal muscle make it unsuitable to form the muscles of the heart?

✔

The process of photosynthesis is covered in Chapter 22.

Muscular tissue

Muscular tissues are **contractile**: they contain protein fibres, **actin** and **myosin**, which produce movement when they slide over each other (see Chapter 18).

As Fig 2.6 shows, there are three types of muscle: **skeletal muscle**, **smooth muscle** and **cardiac muscle**. The properties of the three types are summarised in Table 2.2.

Table 2.2 **Comparing the three types of mammalian muscle**

Muscle:	Skeletal	Smooth	Cardiac
Other names	striped, striated, voluntary	unstriped, unstriated, involuntary, visceral	heart
Site in the body	attached to skeleton	tubular organs; gut, reproductive system, glands, bronchioles	heart
Function	movement and maintenance of posture	usually controls movement of substances along tubes	heartbeat
Control	voluntary	involuntary (not under conscious control)	myogenic (self-generating)
Speed of contraction	rapid	slow, usually known as peristalsis (rhythmic squeezing movements (see page **XXX**)	rapid
Speed of fatigue	rapid	slow	slow

A large proportion of the body mass of most vertebrates is due to their muscles. As you might expect, active muscle tissue has a very high metabolic rate, and many mitochondria are packed between the muscle fibres.

Nervous tissue

Nervous tissue contains specialised cells called **neurones** which transmit electrical impulses. Nervous tissue controls and coordinates the activities of the body. Changes in its internal and external environment act as **stimuli** which are detected by **receptor cells** in **sense organs**. From here, impulses travel along **sensory neurones** to the **central nervous system** (CNS). The CNS consists of the brain and spinal cord, both made up mainly of neurones. The brain processes the information it receives and decides which response to make, often taking into account previous experience (memory). Impulses are sent out along motor neurones to **effector organs**: the muscles and glands.

Control and coordination in the nervous system is covered in Chapters 14 and 15.

Plant tissues

Like animals, plants are made up of different tissues which together form a whole functioning organism. There are, of course, basic differences between plants and animals, and plant tissues reflect these differences. Tissues of a multicellular plant are adapted for the following functions:

- photosynthesis
- transport
- support
- storage
- protection
- reproduction

3 ORGANS AND BODY SYSTEMS IN ANIMALS

An **organ** is a collection of tissues which work together to perform a particular function. For example, think about that most delicate and sophisticated of organs, the eye.

- The **iris** contains muscular tissue, as do the **suspensory ligaments** that control the **lens.**

- The **retina** is made from nervous tissue which detects light and sends information to the brain.

- Epithelial tissues make up blood vessels and other structures such as the **capsule** around the lens, and the **conjunctiva** which covers the front of the eye.

- The whole organ is held together and protected by connective tissue. The **sclera,** in particular, forms a tough, protective outer coating.

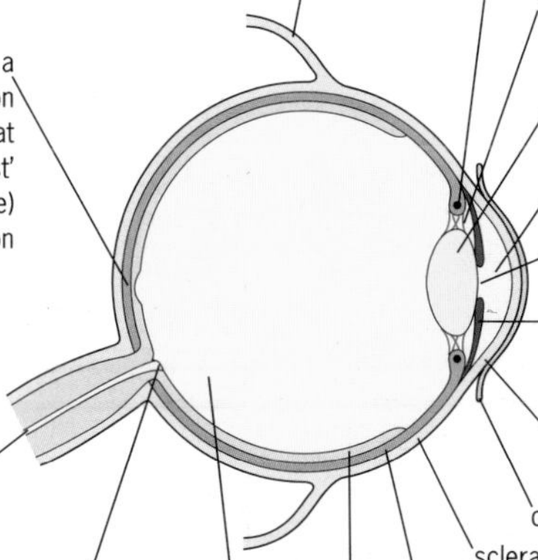

Fig 2.7 **The human eye. The eye is a sphere, about 2.5 cm across. Its outer wall has three layers: an inner layer, the retina, a middle layer, the choroid, and a tough outer layer, the sclera**

In a whole organism, organs form **systems** which carry out life processes such as digestion, excretion, reproduction. The combined systems of the body also need to create a stable internal environment to keep cells bathed in fluid which has the optimum temperature and composition.

Here is a short summary of the main systems in the human body together with their primary functions:

- The **digestive system** extracts from food the simple food molecules which cells need.

- The **excretory system** removes the cell's waste products from the body.

- The **respiratory system** provides oxygen and removes carbon dioxide.

- The **circulatory system** transports all necessary substances to the cells, and removes waste products.

- The **muscles**, **skeleton** and **nervous system** combine to ensure that the organism obtains food and avoids danger.

?

D Which body system – not mentioned in the list on this page – does not contribute directly to the maintenance of internal conditions in the body?

SUMMARY

When you have studied this chapter, you should know and understand the following:

- The cells which make up the human body are **specialised** for a variety of different functions. This is called **division of labour.**

- Specialised cells form **tissues**, tissues form **organs** and organs form **organ systems**.

- A tissue is a collection of similar specialised cells.

- The tissues of the human body can be classified into four broad groups: **epithelial**, **connective**, **muscular** and **nervous**.

- The whole organism consists of all the systems working together to produce an individual which is able to control its internal conditions.

Assignment

BURNS AND TISSUE GRAFTS

The severity of a burn can be measured on a three-point scale:

First degree burns are minor or 'partial thickness' burns (part-way through the skin), often caused by a mild scald or over-exposure to sunlight. The main symptom is redness of the skin. Though painful, they usually heal quickly and need minimal medical treatment.

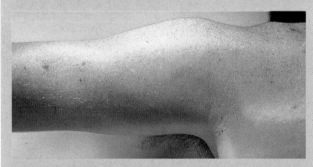

Second degree burns are partial thickness burns caused by severe scalding, contact with flames or very severe sunburn. The dermis of the skin is damaged as well as the epidermis, and redness is accompanied by blistering. This type of burn is extremely painful and usually needs medical attention.

Third degree burns are severe, full thickness burns. Though they may not be painful, third degree burns completely destroy the skin and often some of the underlying tissue such as fat, muscle and bone.

Fig 2.A1

1 Why might a patient with third degree burns often be in less pain than one with less severe burns?

Estimating the damage

The extent of a burn is expressed as a percentage of the total body area the burn covers. As a rough guide, medical staff use the rule of nines, shown in Fig 2.A2.

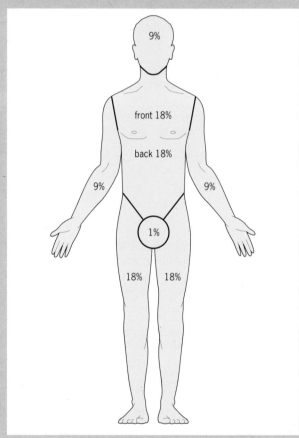

Fig 2.A2 **The extent of a burn can be estimated by using the 'rule of nines'**

2

a) A patient has burns to the whole of his back and all of both legs. What percentage of his body surface is this?

b) From your knowledge of the functions of skin, list some of the problems which will result from the loss of a large area. Refer to Chapter 12 if you need a reminder about skin function.

Treatment

For many years, the main treatment available for burns, where the damage was not too extensive, was a **skin graft**. Skin from another region of the body was transferred onto the damaged area where it would eventually grow and provide an acceptable, if often unsightly, working replacement (Fig 2.A3). Burns covering up to 40 per cent of the skin surface can usually be treated by grafting techniques.

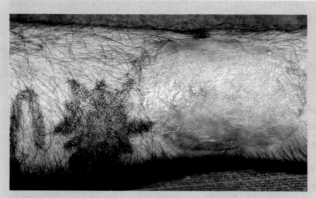

Fig 2.A3 **A skin graft on a man's forearm (to remove tattoos)**

3 Why is it desirable to use a patient's own skin rather than a graft from another person?

New techniques now allow a sample of the patient's own skin to be removed and grown outside the body. If a 2 cm^2 sample is kept in the right conditions, it grows into a sheet of skin which may measure as much as 10 000 times the original area – more than enough to cover a whole body. Burns of 70 per cent and more, previously thought to mean certain death, can now be treated in this way.

Tissue culture

Once removed from the body, the healthy skin sample must be kept in the correct conditions for growth. Cells must be surrounded by a culture medium which contains all the amino acids necessary for protein synthesis, and salts, glucose and several vitamins. Human cells also need an extract of human blood serum. It appears that the serum contains one or more proteins which act as growth factors, stimulating cell division and so rapid growth.

4

a) Predict the optimum temperature essential for the growth of skin cells.

b) Can you think of any problems which may arise if skin cells are provided with the ideal conditions for growth? How do you think these problems may be overcome?

Although skin grows much more rapidly in culture than it could grow in the patient, the process is not immediate. The patient therefore requires a temporary skin covering and this is usually tissue taken either from a pig or from a dead person.

5 Think of any problems that could be associated with using these two materials.

When the patient is stable and their wound is free from infection, the cultured skin is put in place. The patient must be kept sedated to give the new skin a chance to 'take.' Reports from medical centres who have pioneered this work report a success rate of as high as 80 per cent.

There are problems associated with this technique: the new skin may fail to adhere (stick) to the underlying tissue, and it may be too thin and fragile. In addition, the cost is very high, anything from £650 to £5 500 for every 1 per cent of body area covered!

The future

Tissue culture techniques are advancing rapidly. In the near future it is quite likely that skin grafts will adhere reliably and function like the original tissue, even after the most extensive burns.

An interesting advance on the use of skin from dead people (cadavers) has been developed in the USA. Skin consists of fibroblast cells which secrete a protein (collagen) matrix or 'scaffolding'. It is the cells themselves which stimulate the immune reaction which causes rejection. Researchers have been experimenting with cadaver skin in which the cells have been removed, leaving only the protein matrix. Once in place, the matrix releases growth factors which stimulate growth of the victim's own fibroblasts. Thus the healing process is greatly speeded up and the patient leaves hospital earlier.

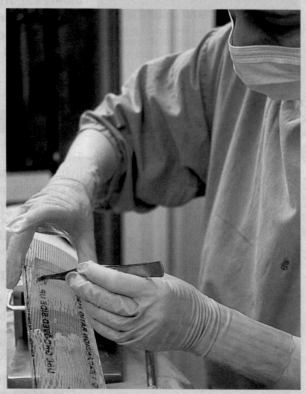

Fig 2.A4 **New tissue culture techniques allow large areas of skin to grow from a few healthy skin cells. Here, the cultured skin is sliced to stretch over three times its area**

Skin is not the only tissue type to be cultured; cartilage, bone and liver have all been successfully grow outside the body.

6 List some of the medical problems which could be treated if we could reliably replace cartilage, bone and liver.

Fig 3.3 **The term carbohydrate covers a range of chemicals which include sugars, starches and cellulose**

(a) **Athletes such as this racing cyclist use glucose drinks to restore their energy. Glucose is absorbed straight into the bloodstream; other carbohydrates such as starch have to be digested first**

(b) **Sweets like these are very popular but they cause tooth decay and they contain what we sometimes call 'empty sugar'. It requires very little digestion, it provides a lot of calories and it does not supply other essential nutrients such as protein, vitamins or minerals. Fruit contains plenty of sugar and vitamins and minerals too. Because the sugars are locked inside cells, they are released into the bloodstream much more gradually**

(c) **Carbohydrate-rich foods, particularly those which contain mainly complex carbohydrates such as potatoes, bread, pasta and rice, are a major component of the human diet**

(d) **In contrast to most other carbohydrates, cellulose is not easily digestible and is therefore not usually a source of energy. Instead, the strength of cellulose is used in products such as paper, cotton, lycra and nail varnish**

Carbohydrates are the first molecules made in photosynthesis. Lipids, proteins and nucleic acids are formed from carbohydrates.

Table 3.1 **Some common carbohydrates**

CHO type	Compound	Sub-units	Occurrence in living things
Monosaccharide	glucose		widespread
	fructose		sweet fruits
	galactose		milk
Disaccharide	maltose	2 × glucose	germinating seeds
	sucrose	glucose + fructose	fruit
	lactose	glucose + galactose	milk
Polysaccharide	starch	glucose	plants (storage)
	glycogen	glucose	animals (storage)
	cellulose	glucose	plant cell walls

Monosaccharides and disaccharides

Monosaccharides and disaccharides are classed as sugars and usually have names ending in -ose, such as sucrose and lactose. In monosaccharides, the three elements carbon, hydrogen and oxygen are always present in the same ratio and they have the basic formula $CH_2O)_n$. Monosaccharides are classified according to the number of carbon atoms they have: 3, 5 and 6 are the most usual (Table 3.2). In glucose, for example, n is 6, so its formula is $(CH_2O)_6$ or $C_6H_{12}O_6$.

Table 3.2 **Common monosaccharides**

No. of C atoms	Name	Common examples
3	triose	glyceraldehyde (see Chapter 22, Part III)
5	pentose	ribose, deoxyribose (in RNA and DNA)
6	hexose	glucose, fructose, galactose

Glucose

It is useful to begin a study of carbohydrates with **glucose**: it is the main source of energy for many organisms, and most of the common polysaccharides are glucose polymers.

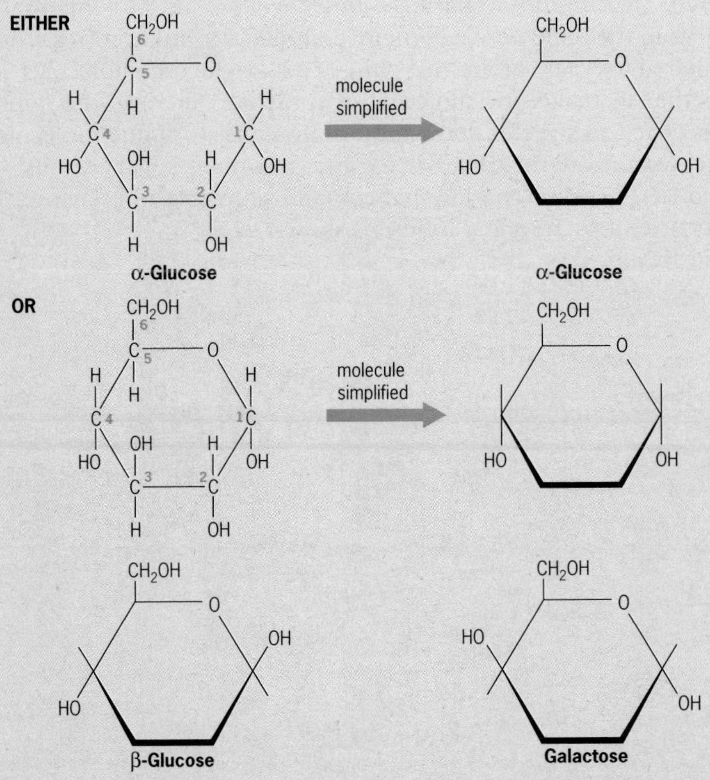

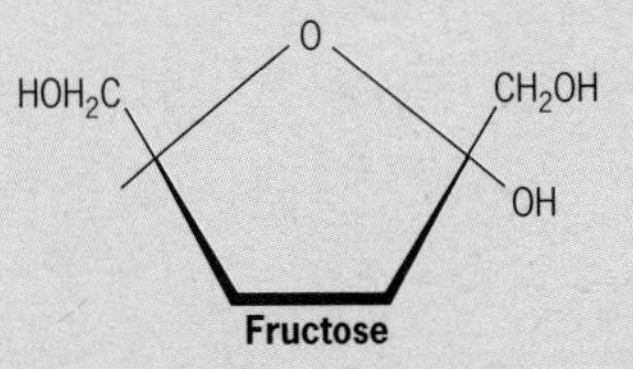

EITHER

CH₂OH

α-Glucose

molecule
simplified

α-Glucose

OR

CH₂OH

molecule
simplified

β-Glucose

Galactose

Fructose

Fig 3.4 **Variations on a common theme: common hexose sugars.** A molecule of a glucose is shown (top left) with all the atoms in place. This can be simplified to allow us to focus on the functional parts of the molecule — the OH groups on carbons 1 and 4. It is these groups that react most commonly with other sugars. Where H on its own is joined to a C, the H can be omitted in the simpler form.

The other three monosaccharides shown below share the formula $C_6H_{12}O_6$ but, as you can see, they all have a slightly different structure. They are said to be isomers of each other. You might think these differences are too small to affect the molecule, but they do greatly affect properties such as taste (sweetness), digestibility and the nature of the polymers formed when the monomers join together

Fig 3.5 **The Benedict's test can be used to estimate the amount of reducing sugars present in foods such as fruit juice.** Samples with no reducing sugars remain blue, as left; those with a low concentration produce a green suspension, those with more produce yellow and orange suspensions, and juices very rich in reducing sugars produce an orange precipitate, as shown right

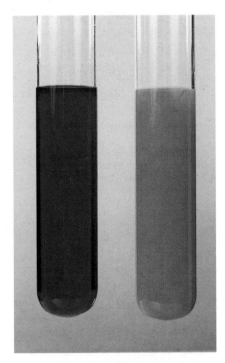

Glucose is a **hexose** (6-carbon) sugar which has the formula $C_6H_{12}O_6$. All other hexose sugars, such as **fructose** and **galactose**, have the same formula (Fig 3.4).

What is a 'reducing sugar'?

The term 'reducing sugar' reflects the fact that some sugars can **reduce** other chemicals. This basically means that the sugar can donate electrons to other substances. For a fuller explanation of reduction–oxidation reactions (redox reactions), see Chapter 22.

A standard test for a reducing sugar involves boiling the sample with **Benedict's solution**, a blue solution that contains copper sulphate. If a reducing sugar is present, the Cu(II) ions in copper sulphate are reduced to Cu(I) ions, resulting in an orange-red precipitate (Fig 3.5). Glucose, fructose, galactose, maltose and lactose are all reducing sugars, but sucrose is not. However, after sucrose is boiled with dilute acid to hydrolyse (split) it into its monosaccharides, it does produce a positive result.

?

A In the Benedict's test, why does sucrose give a negative result before hydrolysis but a positive result after hydrolysis?

Monosaccharides link by means of glycosidic bonds

When two monosaccharides join together in a condensation reaction, the bond between them, a **glycosidic** link, centres around a shared oxygen atom (Fig 3.6). Two α-glucose molecules join together to make one molecule of **maltose**. Sucrose, the familiar sugar used to sweeten food and drinks, consists of one molecule of glucose and one of fructose. Lactose, the main sugar in milk (see page 39), is a disaccharide that contains glucose and galactose. The structure of sucrose and lactose is shown in Fig 3.7.

Fig 3.6 **Two glucose molecules join to form maltose. Like many anabolic (building-up) reactions, this is a** condensation **reaction which involves the production of water**

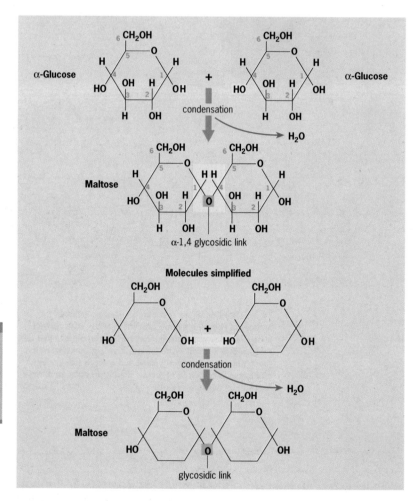

?

B Fig 3.6 shows an α-1,4-glycosidic link. What do the numbers refer to?

C What type of glycosidic linkages are found in sucrose and lactose?"

Fig 3.7 **The structure of sucrose and lactose. Sucrose is made by the condensation of glucose and fructose. Lactose is made by the condensation of glucose and galactose**

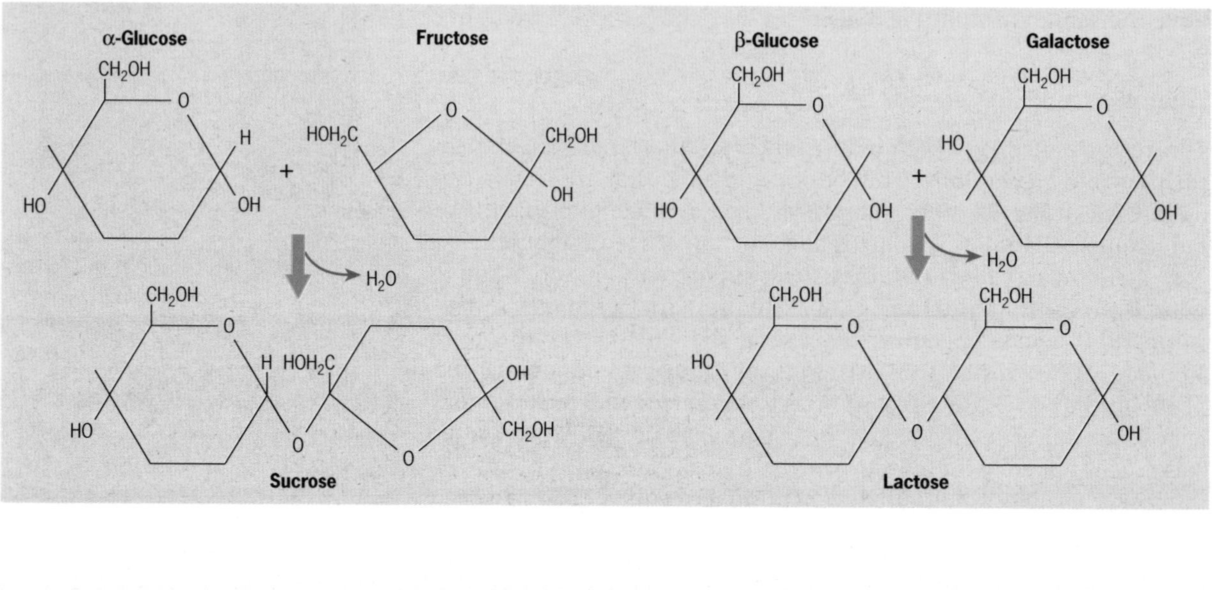

LACTOSE

LIKE ALL MAMMALS, humans feed their young on milk (Fig 3.8). The main carbohydrate in breast milk is the disaccharide lactose. But why should a mother go to the trouble of combining two monosaccharides into lactose when the baby will simply have to break it down again to use it as an energy source?

The answer lies in the size of the lactose molecule. In the mother, milk is made in the mammary gland and is stored there between feeds. Since it is a disaccharide, lactose stays in the milk, rather than diffusing into the surrounding tissue as smaller monosaccharides would tend to do. In the baby, lactose remains in the gut for the same reason and is broken down only gradually to form glucose. So there is a steady absorption of glucose into the infant's blood, rather than a sudden surge at feeding time.

Fig 3.8 **Milk contains all of the food chemicals that a new baby needs. The main carbohydrate in milk is lactose**

Several medical conditions can arise because of a failure to digest lactose. In adult life, some people stop making the enzyme **lactase** and so cannot break down lactose. Undigested lactose cannot be absorbed and it accumulates in the gut, encouraging the growth of bacteria which produce large amounts of carbon dioxide and lactic acid. The result is diarrhoea and wind. Affected people are said to be **lactose intolerant**.

A type of lactose intolerance that results in far more serious problems is found in people with an inherited condition called **galactosaemia**. Affected people can break down lactose into glucose and galactose but their liver cannot convert galactose to glucose. Galactose builds up to dangerous levels and the condition can be fatal.

Polysaccharides

Starch

Starch is the most abundant storage chemical in plants (Fig 3.9) and it is the single largest provider of energy for most of the world's population.

Starch has the three properties that are necessary for a storage compound. It is:

- compact,
- insoluble,
- readily accessible when needed.

Starch is a mixture of two compounds, **amylose** and **amlyopectin**. Amylose is an unbranched polymer in which glucose monomers are joined by α-1,4-glycosidic linkages. These bonds bring the monomers together at a slight angle and, when they are repeated many times, a spiral molecule is produced. In amylose there are six glucose residues in a turn of the spiral.

The glucose chains of amylopectin have α-1,4-glycosidic linkages *and* α-1,6-glycosidic linkages. This allows branching (Fig 3.10).

?

D What would be a suitable dietary treatment for galactosaemia?

E If a storage compound was soluble, what would it do to the cytoplasm of the cells in which it was stored?

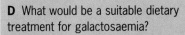 See question 1.

Fig 3.9 **In many plants, starch is the main storage carbohydrate. Foods such as rice, pasta and potatoes all have a high starch content. When treated with an iodine/potassium iodide solution, a blue-black starch-iodide complex is formed. This reaction has been used here to demonstrate the presence of starch in a potato**

?

F Why do cellulose-based materials such as cotton take a long time to rot?

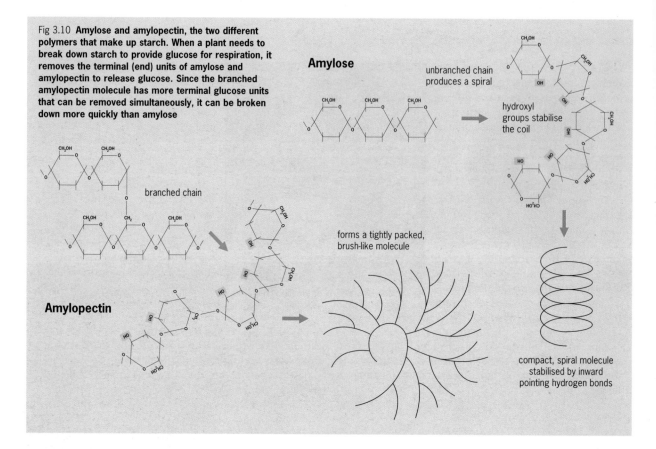

Fig 3.10 **Amylose and amylopectin, the two different polymers that make up starch. When a plant needs to break down starch to provide glucose for respiration, it removes the terminal (end) units of amylose and amylopectin to release glucose. Since the branched amylopectin molecule has more terminal glucose units that can be removed simultaneously, it can be broken down more quickly than amylose**

Amylose

unbranched chain produces a spiral

hydroxyl groups stabilise the coil

forms a tightly packed, brush-like molecule

Amylopectin

branched chain

compact, spiral molecule stabilised by inward pointing hydrogen bonds

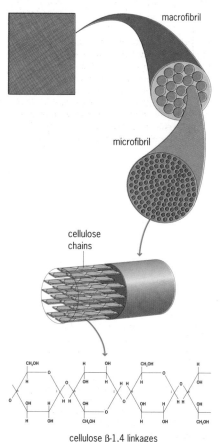

macrofibril

microfibril

cellulose chains

cellulose β-1,4 linkages

Fig 3.11 **Cellulose and the structure of the plant cell wall**

Glycogen

In humans, glycogen is the main storage carbohydrate. Its structure is similar to amylopectin, but it is even more frequently branched. In humans, glycogen is stored in large amounts in the liver and the muscles. During prolonged exercise, when the immediate supply of glucose is used up, the body restores its supplies by breaking down glycogen. If an average person goes without food, his or her glycogen stores last for about a day, but prolonged exercise such as marathon running can use all of the body's glycogen in a few hours. When glycogen runs out, the body turns to using its lipid stores. This is why eating less whilst taking more exercise is the quickest way to lose weight.

Cellulose

Cellulose is a structural polysaccharide: it gives strength and rigidity to plant cell walls. Individual cellulose molecules are long unbranched chains containing many β-1,4-glycosidic linkages (Fig 3.11). The molecules are straight and lie side by side, forming hydrogen bonds along their entire length. This results in strong bundles of chains called **microfibrils**.

Cellulose is probably the most abundant structural chemical on Earth, but few animals can digest it because they do not make the necessary enzyme, **cellulase**. Herbivorous animals, whose diet contains large amounts of cellulose, can deal with it because they have cellulase-producing microorganisms in their digestive system. Humans cannot digest cellulose, but we make good use of it in other ways (Fig 3.3(d)).

■ See question 2.

3 LIPIDS

Lipids are a varied group of compounds that include the familiar **fats** and **oils**. As they are non-polar molecules, most lipids are insoluble in water but soluble in non-polar solvents such as alcohol and ether. Important exceptions are phospholipids, which have polar heads. The **emulsion test** for lipids, shown in Fig 3.12, is based on the solubility of lipids in ethanol.

Lipids contain the elements carbon, hydrogen, oxygen and sometimes phosphorus and nitrogen. They are intermediate-sized molecules that do not achieve the giant sizes of the polysaccharides, proteins and nucleic acids.

Polar and ionic chemicals have particular areas of positive and/or negative charge and so will dissolve in polar solvents such as water. Non-polar molecules are not charged. Lipids are non-polar (although some have polar groups) and so generally do not dissolve in water.

Lipid structure and function

The **triglycerides**, which act mainly as energy stores in animals and plants, are a large important group of lipids.

Fig 3.12 **The emulsion test for lipids. The food to be tested is broken up into very small pieces, mixed with pure ethanol and shaken vigorously. Any lipid present dissolves in the alcohol. This top layer is poured off, mixed with water and the mixture is shaken vigorously again. If lipid is present, the mixture turns white as an** emulsion, **a suspension of fine lipid droplets forms. If no lipid is present, the mixture remains clear**

(a) **The body shape of the model on the right might be considered 'ideal' by many but it is a shape which few girls can acheive. It is also debatable whether such a low level of body fat is actually healthy**

(b) **This painting by Reubens shows the fuller female figure that was fashionable in previous centuries**

Fig 3.13 **Lipids are vital chemicals in all living organisms. In humans, triglycerides occur mainly in** adipose (fat storage) **tissue which forms under the skin and around internal organs. As well as storing energy, adipose tissue insulates against the cold, protects organs against physical damage, and contributes to a person's overall shape. The female body shape is, to a large extent, determined by adipose tissue distribution. There has always been considerable debate about the ideal body; it often appears to be dictated by fashion and culture, rather than commonsense**

Assignment

DNA FINGERPRINTING

In 1987, Robert Melias made legal history in the UK when he was convicted of rape based on evidence obtained by **DNA fingerprinting**. This technique – more accurately called **DNA profiling** – was developed at Leicester University by Alec Jeffreys in 1984 and is proving to be an invaluable tool in forensic medicine.

The principles behind DNA profiling

Every human body cell contains 46 chromosomes. Each one consists of a single elaborately coiled piece of DNA which, if stretched out, can be as long as 5 cm. The structure of DNA can be used to identify individuals.

Between the many genes that occur along the DNA molecule are regions which code for nothing at all. Within this non-coding DNA are **hypervariable regions**, so called because they vary enormously in length from person to person. Hypervariable regions consist of particular base sequences called **core sequences**, which are repeated again and again. Different people have different numbers of repeats and so have differently sized hypervariable regions. When these are labelled and separated according to size, a pattern is produced. Each pattern is unique to each individual person and can be used as the basis of DNA profiling (Fig 3.A1).

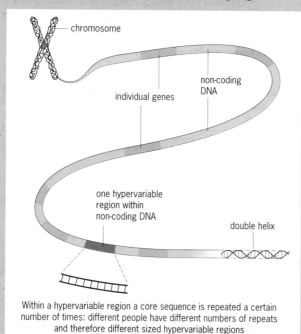

Within a hypervariable region a core sequence is repeated a certain number of times: different people have different numbers of repeats and therefore different sized hypervariable regions

Fig 3.A1 **Although each chromosome contains several thousand genes, they only account for about 10 per cent of the length: the rest is non-coding DNA. Within these segments are** hypervariable regions. **These are unique to each individual and, when they are separated out, the pattern they form provides the basis of DNA profiling**

1

a) What is a gene?
b) What is the name given to a fault which sometimes occurs when DNA is copied?
c) Faults accumulate in the hypervariable regions more frequently than they do in the genes. Why do you think this is?

Getting a sample of DNA

All body cells contain the same DNA, and so virtually any tissue sample can be used for DNA profiling. The amount of tissue required is very small: 0.05 cm^3 of blood, 0.005 cm^3 of semen or one hair root!

2

a) Which blood cells contain DNA?
b) Why do forensic scientists need a larger sample of blood than of semen?

Processing the DNA and separating the fragments

Once you have a DNA sample, the next stage is to cut the molecule using **restriction enzymes**, which act rather like molecular scissors (see Chapter 34). These enzymes cut DNA at specific base sequences, giving a complex mixture of DNA fragments, some of which contain the hypervariable regions.

Electrophoresis (see page 50) separates the fragments. The fragment mixture is placed in a trough in some gel. When a current is applied, the negatively charged DNA fragments move towards the positive terminal, or **anode**. The fragments move at different speeds according to their size: small ones move faster than larger ones. The fragments separate out into bands, as shown in Fig 3.A2 for blood.

The bands of DNA are transferred from the gel onto a nylon membrane by **Southern blotting**, a process which

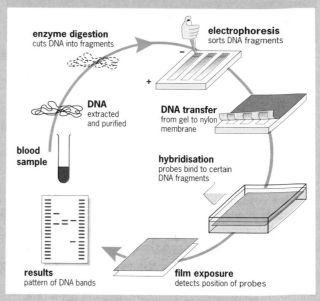

Fig 3.A2 **The major steps in the preparation of a DNA profile**

works by capillary action. At this stage the bands are still invisible, and must be stained so that the hypervariable regions can be seen.

Labelling the fragments

Within hypervariable regions are core sequences that are common to all humans. It is the number of times the sequences are repeated which varies from person to person.

Pieces of DNA complementary to these core sequences have been isolated and are produced in bulk for use as **genetic probes**. They are labelled with a marker chemical, commonly the enzyme **alkaline phosphatase** which fluoresces (produces light) when a particular substrate is added.

When the probes, complete with enzyme, are added to the DNA sample, they attach to the core sequences, thus marking the hypervariable regions. Excess probe is washed off, substrate for the enzyme is added, and bands which contain hypervariable regions fluoresce. If the blot is exposed to an X-ray film, dark lines appear wherever bands in the blot have emitted light, forming the familiar DNA profile.

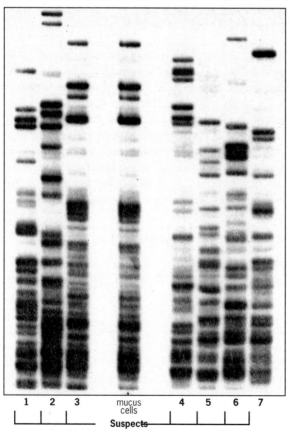

1   2   3   mucus   4   5   6   7
                cells
            |_____| Suspects |_____|

Fig 3.A3 **shows the DNA profile of six suspects compared with the sample taken from the tissues**

3
a) Why do the DNA bands fluoresce?
b) Do all the DNA bands fluoresce? Explain your answer.

Case studies

You have covered the basic theory of DNA profiling: now you can put it into practice. Here are two case studies for you to interpret.

Case 1: Whodunnit?

In 1989, a man robbed a bank at gunpoint whilst suffering from a particularly heavy cold. The mucus-filled tissues found on the scene provided enough DNA for a DNA profile.

Look at Fig 3.A3.

4 On the basis of the evidence, which suspect robbed the bank?

Case 2: Whose baby?

Bands in an individuals DNA profile must have come from either the mother or the father. When a child's DNA profile is compared to the mother's, it is easy to see which bands have been inherited. Any bands in the child's profile which did not come from the mother must have come from the father. Use this information to solve the next case study.

In this case study, the mother M is claiming that Mr Z, a famous rock musician, is the father of her 14-year-old daughter C. She is seeking a large maintenance settlement. Predictably, Mr Z denies all knowledge of this and his lawyer demands that a DNA profile be carried out. Fig 3.A4 shows the results.

M   C   Mr Z

Fig 3.A4 **The DNA profiles from Case Study 2, for the mother (M), child (C) and alleged father (Mr Z)**

5
a) Which of the daughter's DNA bands were inherited from her mother?
b) Do the remaining bands match those from Mr Z, showing that he is the father?

Questions for discussion

6 What would be the advantages and disadvantages of compulsory DNA testing for the whole UK population, so that the police have comprehensive files?

7 Would you like details of your DNA to be available to businesses, in the way that financial records are today? What problems could this cause?

4 Enzymes and metabolism

IT IS EASY to see how confectioners make chocolates with hard centres: they simply pour molten chocolate over the centre and wait for it to set. But what about the soft centres? Surely it isn't possible to pour liquid chocolate over a liquid centre and still keep the shape. So, how do they make the runny, yolk-like inside to chocolate eggs?

Chocolate lovers everywhere may be surprised to learn that the answer is: use an enzyme. To start with, the centre is solid and contains an enzyme and a polysaccharide. After the chocolate coating has set, the enzyme breaks down the long polysaccharide chains, turning the hard centre into the familiar runny filling. Increasingly, enzymes are being used in medicine, industry and biotechnology.

> **Metabolism** is a general term that describes the complex and interrelated chemical reactions that take place in an organism. Metabolic reactions are controlled by **enzymes**. Enzymes, like all proteins, are built up according to the code contained in genes.

1 WHAT ARE ENZYMES?

Enzymes are complex chemicals that control reactions in living cells. They are biochemical **catalysts**, speeding up reactions that would otherwise happen too slowly to be of any use to the organism. This definition is a bit of an over-simplification. An active enzyme may speed up a particular reaction, but living organisms do not need all reactions to be going at the maximum rate all of the time. The key word is *control*. It is more accurate to say that enzymes interact with other molecules to produce an ordered, stable reaction system in which the products of any reaction are made *when* they are needed, in the *amount* needed.

The role of enzymes in an organism

The metabolic pathway chart, part of which is shown in Fig 4.1, shows some of the large number of different but interconnected chemical reactions in a living cell. Charts like this are too complex to learn but they illustrate several important points:

● Many of the complex chemicals that living organisms need cannot be made in a single reaction. Instead, a series of simpler reactions occurs, one reaction after another, forming a **metabolic pathway**. A single pathway may have many steps in which each chemical is converted to the next. A specific enzyme controls each reaction.

● Each individual step in the chart represents one of the simple chemical reactions. At first glance, there seem to be a great variety of reactions, but look more closely and you will see that the same *types* of reaction occur again and again.

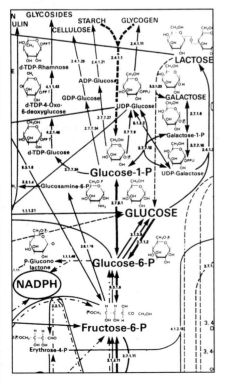

Fig 4.1 **Part of a much larger metabolic chart showing some of the biochemical reactions that occur in living cells. Some reactions carry a number: this identifies the enzyme that catalyses a particular reaction**

- Enzymes control cell metabolism by regulating how and when reactions occur. Using this very simple pathway as an example,

$$A \rightarrow B \rightarrow C \rightarrow D$$

the final product is substance D, the chemical needed by the living organism. The pathway needs three different enzymes and, when D is no longer needed, or if too much has been produced, one of the three enzymes is 'switched off'.

One enzyme, one reaction, but what a difference!

When carbon dioxide dissolves in water, a small proportion of the carbon dioxide molecules combine with water to form carbonic acid.

$$CO_2 + H_2O \rightarrow H_2CO_3 \rightarrow H^+ + HCO_3^-$$

It is a rather slow reaction, but, in the presence of the enzyme **carbonic anhydrase**, it is speeded up about 10^7 (ten million) times. One molecule of carbonic anhydrase can convert 600 000 molecules of carbon dioxide into carbonic acid every second. In living cells, this reaction allows other processes to occur much faster:

- In red blood cells, the enzyme speeds up the production of acid, which in turn causes oxyhaemoglobin to give up its oxygen. Without the enzyme, delivery of oxygen to the tissues would be much slower.

- In certain cells in the stomach lining, the activity of carbonic anhydrase in a pathway allows hydrochloric acid to be secreted rapidly. This creates the acidic conditions necessary for digestive enzymes in the stomach to work properly.

- In the cells of the kidney tubule, carbonic anhydrase speeds up excretion of excess acid, and so helps to maintain the pH of the body at the correct level.

A By what process are enzymes made in the cell?

2 THE CHEMICAL NATURE OF ENZYMES

Enzymes are globular proteins. They have a complex tertiary structure (see page 48) in which polypeptides are folded around each other to form a roughly spherical, or **globular** shape (Fig 4.2). This overall three-dimensional shape of an enzyme molecule is very

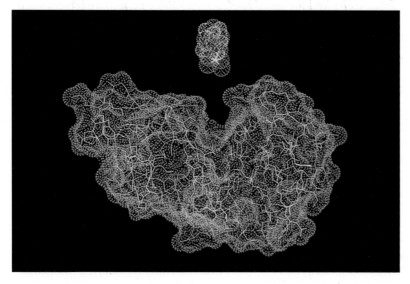

Fig 4.2 **The three-dimensional shape of ribonuclease A, an enzyme that helps break up mRNA in the cytoplasm of bacteria. The** active site, **the 'pocket' into which the substrate fits, is clearly visible**

Fig 4.3 **This schematic diagram shows the three-dimensional structure of an enzyme molecule. Hydrogen bonds are shown in red. The shape and the electrical charges on the substrate closely match those of the active site**

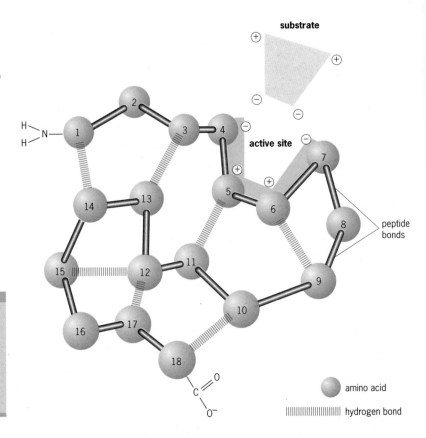

An enzyme acts on a chemical known as its **substrate**. The name of an enzyme often comes from substituting or adding *-ase* in the name of the substrate, so, for example, lactose is the substrate of the enzyme lactase.

B Predict substrates for the following digestive enzymes: protease, lipase, nucleotidase.

important: if it is altered, the enzyme cannot bind to its substrate and so cannot function. Enzyme shape is maintained by hydrogen bonds and ionic forces (Fig 4.3) and their function can be affected by changes in temperature and pH.

Enzymes have several important properties:

● Enzymes are **specific**: each one catalyses only one reaction.

● Enzymes combine with their substrates to form temporary enzyme–substrate **complexes**.

● Enzymes are not altered or used up by the reactions they catalyse, so can be used again and again.

● Enzymes work very rapidly and each has its own **turnover number** (see page 63).

● Enzymes are sensitive to temperature and pH.

● Enzyme function can be slowed down or stopped by **inhibitors.**

The specificity of enzymes

Taking digestive enzymes, you might find it difficult to believe that each enzyme catalyses only one reaction. Trypsin, for example, can begin the digestion of a wide variety of foods rich in protein: eggs, pork, chicken and soya, for example. But when you look at how trypsin works at the molecular level, you can see that this enzyme *is* **specific.** Trypsin cuts an amino acid chain at a point between two particular amino acids, arginine and lysine, and nowhere else. Most proteins have these two amino acids next to each other at some points in their polypeptide chain, and so can be partly digested by trypsin.

Several scientists in biochemistry have devised mechanisms – often called **models** – to explain how enzymes work. In coming up with ideas, they have had to take account of enzyme specificity.

Two models that are used to explain how enzymes work are the **lock and key hypothesis** and the **induced fit hypothesis**.

The lock and key hypothesis

This idea assumes that enzyme function depends on an area on the molecule known as the **active site**, highlighted in Figs 4.2 and 4.3. The active site is a groove or pocket in the surface of the enzyme into which the substrate molecule fits. Typically, the active site is formed by 3 to 12 amino acids. The size, shape and chemical nature of the active site corresponds closely with that of the substrate molecule, so they fit together like a key fits into a lock (Fig 4.4) or, perhaps more realistically, like two pieces in a three-dimensional jigsaw.

Although this model helps us to understand some of the properties of enzymes, it is now generally accepted that a modified version, known as the induced fit hypothesis, better represents what happens when an enzyme catalyses a reaction.

?

C If the enzyme amylase is specific, how can it catalyse the digestion of bread, potatoes and rice?

✓

The active site of an enzyme is part of the enzyme molecule and not part of the substrate.

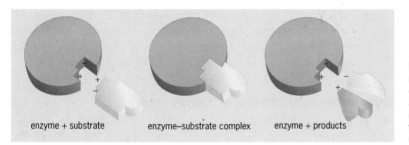

enzyme + substrate   enzyme–substrate complex   enzyme + products

Fig 4.4 **The lock and key hypothesis. The active site is a particular shape (the lock) into which only one substrate (the key) will fit. The enzyme and substrate combine for an instant to form an enzyme–substrate complex. The formation of this complex brings about the desired chemical reaction, converting substrate into product(s)**

The induced fit hypothesis

Experimental evidence suggests that the active site in many enzymes is not exactly the same shape as the substrate, but moulds itself around the substrate as the enzyme–substrate complex is formed (Fig 4.5). Only when the substrate binds to the enzyme is the active site the correct shape to catalyse the reaction. As the products of the reaction form, they fit the active site less well and fall away from it. Without the substrate, the enzyme reverts to its 'relaxed' state, until the next substrate moecule comes along.

Both models show why enzymes are not altered by the reactions that they catalyse: they bind to a substrate momentarily, allowing a reaction to happen, but do not themselves undergo any chemical change.

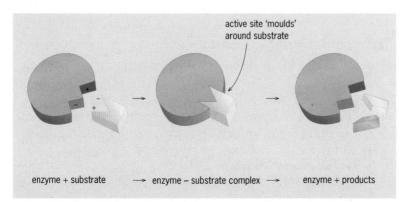

active site 'moulds' around substrate

enzyme + substrate ⟶ enzyme – substrate complex ⟶ enzyme + products

Fig 4.5 **The induced fit hypothesis. Before substrate binding, the enzyme's active site is 'relaxed'. When the substrate binds, the active site is pulled into the correct shape by molecular interactions between the two molecules, and an enzyme–substrate complex forms. As the products fall away from the active site, the molecule becomes 'relaxed' again**

The chemistry of enzyme action

In biology almost all reactions are reversible. If they were not, there would be no recycling of molecules because once large molecules had been built up they could not be broken down again (see Chapter 23).

A simple reversible reaction can be expressed as:

$$A + B \leftrightarrow C + D$$
$$\text{reactants} \qquad \text{products}$$

In this reaction, the **reactants** (A and B) combine to give the **products** (C and D). If this reaction were to take place in a test-tube, a proportion of the products would react to form A and B again. Eventually, an equilibrium would be reached in which the relative proportion of reactants and products would remain the same.

In theory, an enzyme allows a reversible reaction to reach equilibrium more quickly. An enzyme can speed the reaction in either direction; the way it does depends on whether there are more reactants or products present at the start. In living organisms, however, enzymes usually speed up reactions in one particular direction. This is because, in any organism, an enzyme acts only if the product of the reaction that it catalyses is needed. And, as the product is used as soon as it is made, equilibrium is rarely reached.

Enzymes work by lowering activation energy

For any chemical reaction to take place, bonds must be broken before new ones can form. The energy needed to break these bonds, and so set the reaction in motion, is the **activation energy**.

Many reactants need a large amount of energy to push them to a state where they can take part in a reaction, so many reactions take place only at high temperature. Many substances burn, for instance, but only after the initial activation energy has been supplied, perhaps by lighting a match. In the presence of enzymes, the activation energy is greatly lowered and this allows reactions

Fig 4.6 **In order for a reaction to happen, activation energy must be supplied. Then the rest of the reaction proceeds, just as the boulder rolls down the hill once the energy has been supplied to push it to the top. Catalysts such as enzymes work by lowering the activation energy**

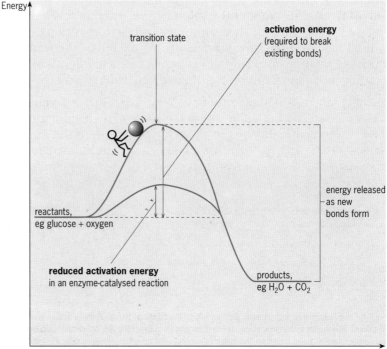

to take place at the relatively low temperatures normally found in living organisms. The 'half-way' point in a reaction is called the **transition state**, and this is represented by the top of the curve in Fig 4.6. The transition state represents the stage when the old bonds have been broken in order to allow new ones to form. Enzymes lower the activation energy by making it easier to achieve the transition state.

Enzymes speed up reactions in a number of ways:

● They can hold the substrates close together at the correct angle – this would otherwise have to occur by chance collision.

● They can position any charged groups on the active site to help the reaction to occur.

● By acid or base catalysis: the active site of the enzyme can behave as an acid or a base, donating or accepting protons (hydrogen ions, H^+) from the substrate.

● Through structural flexibility: the flexible shape of the enzyme can change during catalysis. This ensures that the substrates are brought together in the correct sequence for the reaction.

Many metabolic reactions require energy in addition to the presence of the relevant enzyme. This energy is supplied by ATP made by cell respiration (see Chapter 22, Part III). Reactions that need energy are made to happen by coupling them with ATP breakdown, and so are called **coupled reactions**.

How fast do enzymes work?

The speed at which an enzyme works is expressed as its **turnover number**. This is usually defined as the number of substrate molecules turned into product in one minute by one molecule of enzyme. Values range from less than a hundred to many millions. Some examples are given in Table 4.1.

Naming and classifying enzymes

Older enzyme names such as **pepsin**, **catalase** and **trypsin** give no clues about the nature of the reaction they catalyse. To cope with the rapidly expanding number of new enzymes, the International Union of Biochemistry has developed a scheme for naming and classifying enzymes. Very generally, enzymes are named be adding the suffix -ase to the name of their substrate. The rest of the name attempts to indicate the nature of the reaction taking place. Alcohol dehydrogenase, for example, catalyses the removal of hydrogen from alcohol (ethanol). Further examples are given in Table 4.2.

?

D The enzyme lysozyme helps to split bacterial cell walls. Where would you expect this enzyme to be present in the human body?

Table 4.1 **Some turnover numbers**

Enzyme	Turnover number
carbonic anhydrase	36 000 000
catalase	5 600 000
β-galactosidase	12 000
chymotrypsin	6 000
lysozyme	60

(Source: Biochemical society guidance notes 3, *Enzymes and their role in biotechnology*.)

Table 4.2 **Some common enzymes and their substrates**

Enzyme	Substrate	Reaction catalysed
maltase	maltose	hydrolysis of maltose to glucose
amylase	starch	hydrolysis of starch to maltose
alcohol dehydrogenase	alcohol (ethanol)	removal of hydrogen from alcohol
DNA ligase	DNA	joining together two DNA strands
RNA polymerase	nucleotides that make RNA	synthesis of mRNA on a DNA molecule
glycogen synthetase	glucose	polymerisation of glucose into glycogen
ATPase	ATP	synthesis or splitting of ATP

Although there are many different enzymes, they can be put into one of six main categories according to the type of reaction they catalyse:

- **Oxidoreductases** These catalyse oxidation and reduction (redox) reactions. In aerobic respiration, most of the cell's ATP is generated by redox reactions – see Chapter 22, Part III.

- **Transferases** These catalyse the transfer of a chemical group from one compound to another, such as the transfer of an **amino group** from an amino acid to another organic acid in the process of **transamination**.

- **Hydrolases** These catalyse **hydrolysis** (splitting by use of water) reactions. Most digestive enzymes are hydrolases.

- **Lyases** These catalyse the breakdown of molecules by reactions that do not involved hydrolysis.

- **Isomerases** These catalyse the transformation of one isomer into another, for instance the conversion of glucose 1,6-diphosphate into fructose 1,6-diphosphate. This is one of the first reactions in **glycolysis**, the first stage of respiration (Chapter 22, Part III).

- **Ligases** These catalyse the formation of bonds between compounds, often using the free energy made available from ATP hydrolysis. DNA ligase, for example, is involved in the synthesis of DNA.

?

E People with a great deal of physical stamina have been found to have a lot of **glycogen synthetase** in their muscles. Suggest what this enzyme does.

Factors affecting enzyme activity

The factors that affect enzyme activity also affect the functions of the cell and, ultimately, the organism. Enzymes are proteins and their function is therefore affected by:

- Temperature
- pH
- Substrate concentration
- Enzyme concentration
- Inhibitors

?

F Suggest which type of bonds are broken when an enzyme is denatured. (Look at Fig 4.3 for help.)

✔

Heating *denatures* enzymes; it does not kill them.

Temperature

For a non-enzymic chemical reaction, the general rule is: the *higher* the temperature, the *faster* the reaction: Fig 4.7(a). This same rule holds true for a reaction catalysed by an enzyme, but often only up to about 40 to 45 °C. Above this temperature, enzyme molecules begin to vibrate so violently that the delicate bonds that maintain tertiary and quaternary structure are broken, irreversibly changing the shape of the molecule. When this happens the enzyme can no longer function and we say it is **denatured**.

The effect of temperature on a reaction can be expressed by the **temperature coefficient**, commonly known as the Q_{10}.

Where t is the chosen temperature, the formula for the Q_{10} is:

$$\frac{\text{rate of reaction at } t + 10 \text{ °C}}{\text{rate of reaction at } t \text{ °C}}$$

Fig 4.7(b) shows how to calculate the Q_{10}. To avoid denaturing the enzyme, the values for living organisms need to fall in the range 4 to 40 °C, so we have chosen t as 20 °C.

$$Q_{10} = \frac{\text{rate at 30°C}}{\text{rate at 20°C}} = \frac{6}{3} = 2$$

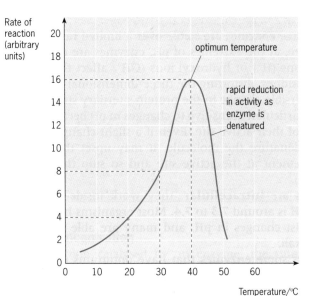

Fig 4.7(a) **Up to about 40 °C, the rate of enzyme-controlled reactions increases with temperature. The optimum temperature for many enzymes is about 40 °C, although the activity of an individual enzyme may increase up to about 50 °C or beyond. However, as the temperature passes 43–44 °C, most enzymes lose their activity**

Fig 4.7(b) **Not all enzymes have the same Q_{10} values. This variation is important because enzymes, even those in the same pathway, can vary significantly in their sensitivity to temperature. Outside an organism's normal temperature range, the enzymes may begin to work at different rates. This causes a metabolic imbalance that may be lethal, and may explain why some organisms die at temperatures that would seem to be relatively mild. Some Antarctic fish, for example, live in water that remains very constant at around −2 °C, and die if placed in water above 6 °C**

In practice, most enzymes have a Q_{10} between 2 and 3. A value of 2 means that the rate of reaction doubles with a 10 °C temperature rise, 3 means that it triples.

Some organisms have enzymes that are less sensitive to heat than those found in mammals. For example, certain bacteria can survive in hot volcanic springs and deep-sea hydrothermal vents (Fig 4.8) at temperatures of over 90 °C, so their enzymes must be active at these extreme temperatures.

?

G Egg albumen, when heated, goes solid and white. What do you think has happened to the proteins in the egg white, and why does the white not become runny again when the egg is cooled?

STABLE ENZYMES FOR WASHING POWDERS

ENZYMES ARE UNSTABLE, particularly at high temperature, so their commercial usefulness is limited. Many industrial processes need to take place in 'unnatural' environments, at high temperature and extremes of pH. But there are organisms that can thrive at high temperatures. Heat-loving bacteria have thermostable enzymes that are also more resistant to extremes of pH and other unfavourable conditions, such as organic solvents.

Genes that code for some thermostable enzymes have been transferred into bacteria that are easy to grow in large quantities, using recombinant DNA technology (see Chapter 31). As the bacteria that have received a particular gene multiply, the gene is translated into protein and large amounts of the enzyme are produced for commercial use.

An example of a thermostable enzyme is the alkaline protease **subtilisin**, the famous stain digester in biological washing powders. This enzyme is produced on a grand scale by the bacterium *Bacillus subtilis*, and is active in alkaline environments, and so is compatible with the other ingredients in washing powder. It is active at temperatures up to 60 °C, allowing it to be used in a wide variety of wash programmes.

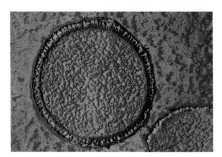

Fig 4.8 **This bacterium *Staphylothermus marinus* lives near deep sea hot-water vents. It thrives at temperatures of up to 98 °C. Such heat-loving bacteria have thermostable enzymes (enzymes that are very resistant to heat damage). These could have great potential in industry (see Feature box, left)**

?

H Why is it an advantage for humans to have a constant body temperature of around 37 °C?

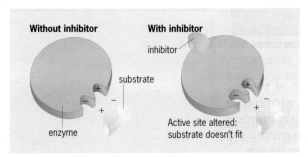

J Write down the main differences between competitive inhibitors and non-competitive inhibitors.

Non-competitive inhibitors

Non-competitive inhibitors bind to the enzyme away from the active site but change the overall shape of the molecule, modifying the active site so that it can no longer turn substrate molecules into product (Fig 4.14). Non-competitive inhibition has this name because there is no competition for the *active site*. The presence of a non-competitive inhibitor has the same effect as lowering enzyme concentration: all inhibited molecules are taken out of action completely. Fig 4.14(b) shows the effect of a non-competitive inhibitor.

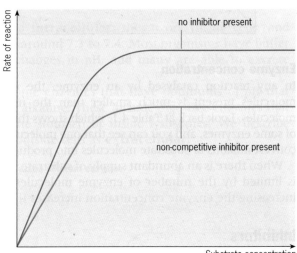

Fig 4.14(a) **Non-competitive inhibitors attach to enzyme molecules and alter the overall shape, so that the active site cannot function. Although the substrate can still bind, forming an *enzyme–inhibitor substrate* complex, the substrate cannot be turned into product. When the inhibitor molecule is removed, normal function is restored**

Fig 4.14(b) **Non-competitively inhibited enzymes show no activity at all, but unaffected ones work normally. Thus, *the maximum rate of reaction* is lowered, as it would be if we used less enzyme**

K What would the graph in Fig 4.14(b) look like if a *large* amount of a non-competitive inhibitor was added?

Cyanide is an irreversible inhibitor of **cytochrome oxidase**, one of the enzymes involved in respiration. Organisms poisoned with cyanide die because they are deprived of ATP, their immediate energy source.

Irreversible inhibitors

Irreversible inhibitors bind permanently to the enzyme, rendering it useless. For obvious reasons organisms rarely produce this type of inhibitor for their own enzymes, but they are splendid weapons to use against other organisms. A wide variety of natural toxins are irreversible inhibitors, as are many pesticides.

How inhibitors help to control metabolism

Many metabolic pathways are self-controlling: when a substance is needed a particular pathway is activated to produce it. When enough has been produced, the pathway is deactivated.

This happens because some enzymes in a metabolic pathway are inhibited by the end-product. As Fig 4.15 shows, if too much product begins to accumulate, this inhibits one of the enzymes in the pathway. When the product is once more in short supply, the inhibition is lifted and the pathway becomes active again. This self-regulation is an example of **negative feedback**. This is a fundamental principle important in homeostasis (see Chapter 10).

Fig 4.15 **A metabolic pathway can be self-regulating by having the end-product act as a non-competitive inhibitor on one of the enzymes in its production pathway**

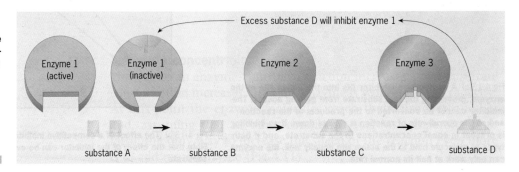

See questions 2 and 3 ■

3 ENZYMES AND BIOTECHNOLOGY

The ability of enzymes to catalyse specific chemical reactions at body temperature makes them very useful tools in the commercial world. Table 4.3 outlines some of industrial applications of enzymes. This is a rapidly changing field, and new applications of enzyme technology appear all the time.

Table 4.3 **Some applications of enzymes**
(Source: Biochemical Society Guidance Notes 3, *Enzymes and their role in biotechnology*)

Enzyme	Reaction	Source of enzyme	Application
Industrial applications			
α-amylase	breaks down starch	bacteria	converts starch to glucose in the food industry
glucose isomerase	converts glucose to fructose	fungi	production of high fructose syrups
proteases	digest protein	bacteria	washing powder
rennin	clots milk protein	animal stomach linings; bacteria	cheese making
catalase	splits hydrogen peroxide into $H_2O + O_2$	bacteria; animal livers	turns latex into foam rubber by producing gas
β-galactosidase	hydrolyses lactose	fungi	in dairy industry, hydrolyses lactose in milk or whey
Medical applications			
L-asparginase	removes L-asparagine from tissues – this nutrient is needed for tumour growth	bacteria (*E. coli*)	cancer chemotherapy – particularly leukaemia
urokinase	breaks down blood clots	human urine	removes blood clots, eg in heart disease patients
Analytical applications			
glucose oxidase	oxidises glucose	fungi	used to test for blood glucose, eg in Clinistix™ diabetics
luciferase	produces light	marine bacteria; fireflies	binds to particular chemicals indicating their presence, eg used to detect bacterial contamination of food
Manipulative applications			
lysozyme	breaks 1–4 glycosidic bonds	hen egg white	disrupts bacterial cell walls
endonucleases	break DNA into fragments	bacteria	used in genetic manipulation techniques, eg gene transfer, DNA fingerprinting

Industrial applications of enzymes

Enzymes are both specific and sensitive: this makes them ideal for use in analysis, which often involves very small samples. One such application is in the analysis of glucose – an important technique in both medicine and industry. Some readers may have heard of Clinistix™ (Fig 4.16) – the sticks used to test for the presence of glucose in urine. Clinistix contain two enzymes: glucose oxidase and peroxidase.

Glucose oxidase catalyses the following reaction:

glucose + oxygen → gluconic acid + H_2O_2 (hydrogen peroxide)

In a simple, visible test, the production of peroxide is coupled to the production of a coloured dye or **chromagen**.

The second enzyme, peroxidase, catalyses the following reaction:

$$DH_2 + H_2O_2 \rightarrow 2H_2O + D$$

D stands for the chromagen, a colourless hydrogen donor. When it loses its hydrogen it becomes coloured. The intensity of colour indicates the amount of glucose present.

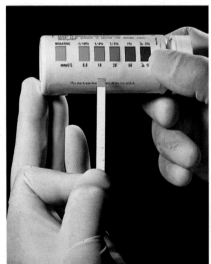

Fig 4.16 **The urine glucose test for diabetes. Glucose oxidase, peroxidase and DH_2 are fixed on a cellulose fibre pad. When this is added to a sample of urine, the colour reaction gives a quantitative measure of glucose**

SUMMARY

After reading this chapter, you should know and understand the following:

■ Enzymes are **globular** proteins with a precise, but delicate, 3D shape maintained by ionic and **hydrogen bonds**.

■ During a reaction, the substrate fits into a region on the enzyme surface called the **active site**.

■ Enzymes are **specific**: each enzyme catalyses one particular reaction.

■ Enzymes speed up reactions by lowering the **activation energy** needed to get the reaction started.

■ Between 4 °C and 40 °C, the rate of an enzyme-controlled reaction increases between two- and three-fold for every 10 °C rise in temperature. Increase in temperature beyond 40 °C usually **denatures** the enzyme. Activity is lost when its three-dimensional structure is destroyed.

■ Most enzymes have an optimum pH. For **intracellular** enzymes this is usually about 7.35. **Extracellular** enzymes such as those found in digestive juices may have optimums of extreme pH values.

■ The rate of an enzyme controlled reaction is limited by the supply of substrate, enzyme or the enzyme cofactor.

■ **Inhibitors** are substances that slow down or stop enzyme activity. They may be **reversible** or **non-reversible**. Reversible inhibitors may be **competitive** or **non-competitive**.

■ Competitive inhibitors tend to be similar in structure to the substrate and compete for the active site. Non-competitive inhibitors do not bind to the active site itself but alter the shape of the enzyme so that the active site is no longer functional.

QUESTIONS

1

a) Explain how the structure of an enzyme is responsible for its mode of action.

b) Discuss the effect of **three** factors which affect the rate of an enzyme-catalysed reaction.

[UCLES June 1997 Sciences: Biology Foundation Section B, q.1]

2

a) Explain how enzyme–substrate complexes are formed. Explain how this allows enzymes to act.

b) Explain in terms of molecular shapes how the following factors affect the rate of enzyme action: **(i)** temperature; **(ii)** competitive inhibition; **(iii)** non-competitive inhibition.

[NEAB February 1995 Processes of Life (BY1) module test, q.9

3
The graph in Fig 4.Q3(a) shows the results of an investigation into the effect of a competitive inhibitor on an enzyme-controlled reaction over a range of substrate concentrations.

a) Give **one** factor which would need to be kept constant in this investigation.

b) **(i)** Explain the difference in the rates of reaction at the substrate concentration of 10 μmol cm^{-3}.
(ii) Explain why the rates of reaction are similar at the substrate concentration of 30 μmol cm^{-3}.

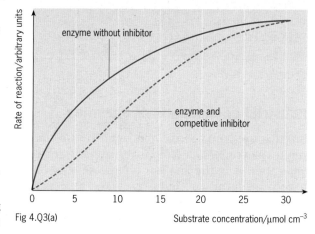

Fig 4.Q3(a)

c) Fig 4.Q3(b) represents a metabolic pathway controlled by enzymes.

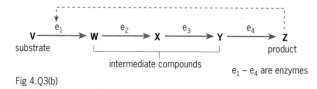

Fig 4.Q3(b)

(i) Name the type of control mechanism which regulates production of compound **Z**.
(ii) Explain precisely how an excess of compound **Z** will inhibit its further production.

[NEAB February 1997 Processes of Life (BY1) module test, q.3]

Assignment

ENZYMES AT WORK

Enzymes have enormous commercial potential. Since they can catalyse particular reactions at relatively low temperatures, they are more versatile and much cheaper than inorganic catalysts. Once a suitable enzyme is found, it is made on a large scale and then purified.

There is a vast world-wide demand for sweeteners, mainly for confectionery and soft drinks. Traditionally, manufacturers have used sucrose extracted from sugar beet or cane but, in recent years, **high fructose syrup**, a cheaper sugar beet product has become more common. Fructose is sweeter than sucrose, and can be made from the starch in sugar beet. About 40 million tonnes are now produced each year.

1 What sort of carbohydrate is starch?

Fructose is produced by an enzyme-catalysed process: the starch is first broken down into glucose and then glucose is converted to fructose.

2 What *type* of enzyme catalyses the conversion of:
a) starch to glucose?
b) glucose to fructose?

Three enzymes are involved in the production of high fructose syrups: **bacterial β-amlyase, fungal amyloglucosidase** and **bacterial glucose isomerase**. In the first step, the starch paste is heated to 105 °C. β-amylase is then added to reduce the **viscosity** (thickness) of the paste. The process results in a mixture of **maltodextrins** – branched sugars.

3
a) How do we describe an enzyme that works at high temperatures?
b) What is the advantage of higher temperatures in industrial processes?
c) Why does the action of the enzyme reduce the viscosity of the paste?

The next step is the hydrolysis of the maltodextrins to glucose. The substrate is cooled to 55–60 °C and acidified to pH 4.5. The enzyme amyloglucosidase, obtained from the fungus *Aspergillis niger*, is added. This is an **exoenzyme**: it removes the terminal glucose units from the maltodextrins.

4
a) What is the difference between an endoenzyme and an exoenzyme?
b) In the hydrolysis of polymers like starch, why is it an advantage to add an endoenzyme before using an exoenzyme?
c) Sketch graphs to show the likely effect of variations in temperature and pH on the activity of amyloglucosidase.

Amyloglucosidase acts on maltodextrins to produce glucose syrup. Although not as sweet as sucrose, it still has its uses in the food industry. The production of the extra-sweet fructose needs the enzyme glucose isomerase, which is obtained from bacteria (*Bacillus* or *Streptomyces* sp.). This enzyme is fixed on rigid granules and packed into a column. The glucose syrup then flows between the granules and the enzyme converts the glucose into fructose.

5 What is the advantage of using a fixed (immobilised) enzyme?

The end product is high fructose syrup – a mixture of about 42 per cent fructose and 55 per cent glucose. In these proportions it has the same sweetening power as sucrose.

An alternative sweetener is **aspartame**, a dipeptide that is 180 times sweeter than sucrose. The commercial production of aspartame again involves enzymes – particularly **aspartase**. The ease and cost of aspartame production affects the demand for high fructose syrups.

6 What is the main dietary advantage of using aspartame instead of sucrose?

7 All of the enzymes mentioned in this assignment come from organisms such as yeast and bacteria. Why are these organisms particularly suitable for large-scale enzyme production?

8 Imagine that a scientist isolates a human gene which codes for an enzyme that catalyses a reaction giving a very valuable product. List, and perhaps discuss, the steps that would have to be taken in order to get this product into commercial production.

Fig 4.A1 **Many soft drinks contain a high concentration of fructose syrup**

5 Movement in and out of cells

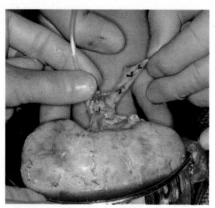

A kidney waiting to be transplanted. The pale pink colour changes to a much darker red when the new blood supply is connected

THE LIST OF PEOPLE awaiting an organ transplant is growing. Improving surgery and and reducing organ rejection rates has made transplants increasingly successful, leading to a greater demand for kidneys, hearts, livers and other body parts.

Surgeons face problems: an organ can become available in Dundee, but the most suitable recipient might be in Norwich. So, the organ usually has to be transported. But cells in tissues that are separated from their blood supply cannot carry out their normal metabolic functions if they are unable to exchange materials with blood. They run out of oxygen, accumulate waste and can die within minutes.

The solution is to cool the cells so that their metabolic demands are lowered. Organs packed in ice can survive a journey of several hours. During the transplant, the kidney is warmed to normal body temperature and connected without delay to the recipient's blood system.

1 THE CELL AND ITS ENVIRONMENT

Each living cell in the human body is a dynamic system which can exchange a volume of fluid several times bigger than its own volume every second! This happens only because cells are very small: if they were any larger the nutrients could not be taken up quickly enough to satisfy demand and waste could not be expelled efficiently enough to prevent poisoning of the cell. All organisms larger than about 1 mm have developed strategies to increase the exchange of materials, to meet the needs of *all* their cells.

Material passes into or out of cells by these basic processes:
- diffusion (and facilitated diffusion),
- osmosis,
- active transport,
- endocytosis and ectocytosis.

In this chapter, we look at each of these in detail but, since they all involve the cell surface membrane, we first take a detailed look at this important structure.

2 THE CELL SURFACE MEMBRANE

The cell surface membrane, often called the **cell membrane** and sometimes known as the **plasma membrane** or **plasmalemma**, is the boundary between a cell and its surroundings. It has little physical strength, but it plays a vital role in regulating the materials that pass in and out of the cell. Many of the organelles of the eukaryotic cell are also made up of membranes (Fig 5.1) with the same basic structure as the cell surface membrane.

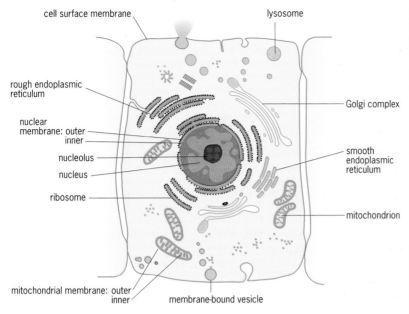

Fig 5.1 **Whenever you draw a eukaryotic cell, nearly every line you draw represents a membrane. In this diagram of an animal cell, all the membranes are drawn as red lines**

Fig 5.2 **The phospholipids in a membrane are arranged tail-to-tail, forming a bilayer**

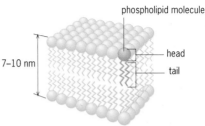

The structure of the cell surface membrane

As we saw in Chapter 1, the cell surface membrane is basically a double layer of **phospholipid** molecules about 7 to 10 nm thick (Fig 5.2). It cannot be seen with a light microscope and so its structure could not be studied directly until the electron microscope was developed. Instead, early cell biologists deduced its structure by investigating its properties.

The fluid mosaic theory

In 1972, aided by electron microscope studies and evidence from other techniques, Singer and Nicholson put forward the **fluid mosaic theory** (Figs 5.3 and 5.4). Today, most scientists accept it as the model that best represents the structure of living cell membranes.

A The cell surface membrane of a human cell is about 10 nm thick. Assume that a typical human cell is about 50 μm in diameter. How thick would you make a model cell surface membrane if it had to fit around a model cell that was 5 metres across?

Remember: 1 mm = 1000 μm
1000 μm = 1 nm

Fig 5.3 **The fluid mosaic model of the structure of a cell surface membrane. The membrane has been described as a collection of 'protein icebergs in a lipid sea'. Alternatively, you might find it easier to think of the phospholipids as a double layer of ping-pong balls with tails floating on the surface of a swimming pool, with the proteins as larger balls punctuating the phospholipid layer**

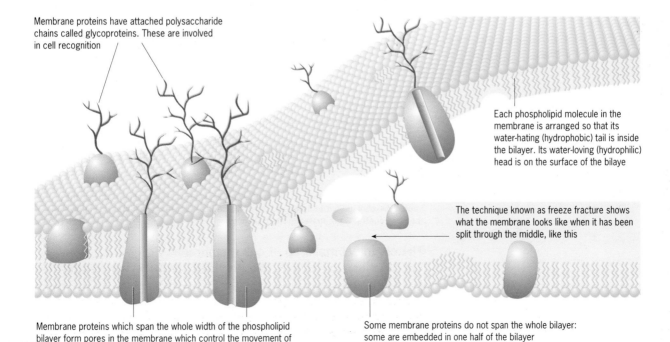

Membrane proteins have attached polysaccharide chains called glycoproteins. These are involved in cell recognition

Each phospholipid molecule in the membrane is arranged so that its water-hating (hydrophobic) tail is inside the bilayer. Its water-loving (hydrophilic) head is on the surface of the bilaye

The technique known as freeze fracture shows what the membrane looks like when it has been split through the middle, like this

Membrane proteins which span the whole width of the phospholipid bilayer form pores in the membrane which control the movement of substances into and out of the cell

Some membrane proteins do not span the whole bilayer: some are embedded in one half of the bilayer

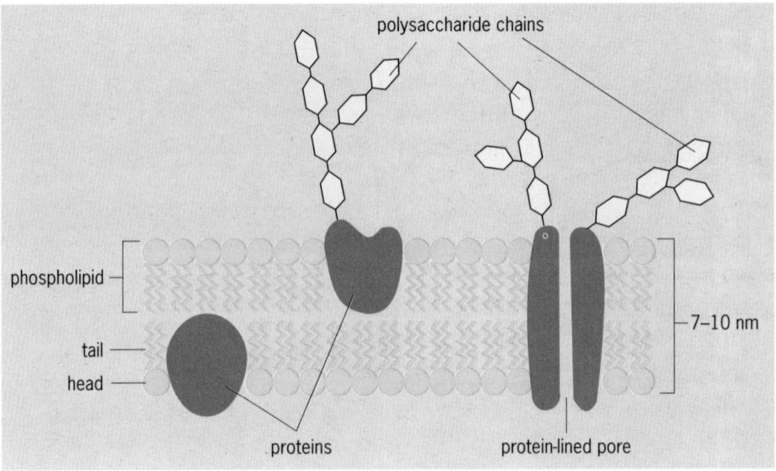

Fig 5.4 **A simplified diagram of a cell surface membrane, showing the features described in the fluid mosaic theory. This level of detail should be enough to enable you to explain the essential properties and functions of the membrane**

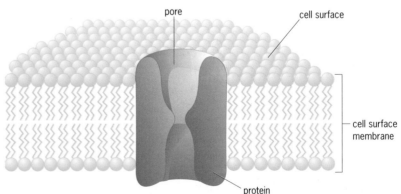

According to the fluid mosaic model, the cell membrane consists of a double layer of phospholipid molecules (known as a **lipid bilayer**), studded with proteins and other molecules. The name **fluid mosaic** is used because the bilayer is a very fluid structure (the phospholipid molecules are in constant sideways motion) and it contains a 'mosaic' of protein molecules.

The protein molecules may be found only in the top or only in the bottom layer of **lipids**, or they can span the entire membrane. There might also be **cholesterol** molecules that fit in between the phospholipids (see Fig 5.14), and **polysaccharides** that are attached to the **membrane proteins** or the lipids (Figs 5.3 and 5.4).

The lipid bilayer is a barrier to water and anything that is water soluble. But the majority of the chemicals that need to pass in or out of the cell *are* water soluble. The protein molecules in the membrane act as **hydrophilic pores**, water-filled channels that allow water-soluble chemicals to pass through (Fig 5.5). Pores are usually small and highly selective: they allow only specific molecules or ions through.

?

B Explain the terms hydrophilic and hydrophobic. How do these terms relate to a phospholipid molecule?

✔

Water is a **polar molecule** (it has regions of positive and negative charge) and is a solvent for **polar substances** such as sugars, charged ions (Na^+, Cl^-, Ca^{2+}, K^+), B and C vitamins and amino acids. Polar substances do not dissolve in lipid and so can cross a cell membrane only by going through pores.

Most fats, oils and lipids are **non-polar molecules** (they do not have charged regions) and do not dissolve in water. Other non-polar substances (such as vitamins A, D, E and K) can dissolve in lipids and so can cross cell membranes without going through pores.

Fig 5.5 **Some of the proteins in the membrane form pores. These allow particles that cannot dissolve in lipid to enter and leave the cell. Membrane pores are not simply 'holes'; they can control what passes through them**

The chemical make-up of cell membranes

Cell membranes contain phospholipids, proteins, cholesterol and polysaccharides.

Phospholipids are a major constituent of cell membranes. They form membranes naturally because in water they automatically arrange themselves into a bilayer that is impermeable to water and water-soluble substances (see Fig 5.2).

Membrane proteins have several functions. They:
- form pores through which water and water-soluble chemicals can pass,
- act as carriers in active transport,
- form receptor sites for hormones,
- are important in cell recognition.

Cholesterol, a chemical often mentioned in relation to heart disease (Fig 5.6), is actually a vital constituent of animal cell membranes.

Polysaccharides, branched polymers of simple sugars, stick out from the outer surface of some membranes like antennae (see Figs 5.3 and 5.4). They attach to lipids forming **glycolipids,** or to proteins forming **glycoproteins**. Glycolipids and glycoproteins help cells to recognise each other. This is important, for instance, in allowing the immune system to tell the difference between body cells and invading bacteria.

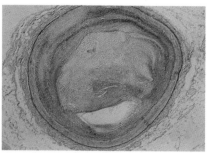

Fig 5.6 **The process in which fatty deposits form in the wall of an artery can lead to a heart attack or a stroke. Some research suggests that people with high blood cholesterol levels (cholesterol molecules that float freely in the blood) run a greater risk of developing these deposits. Although heart disease develops because of many different factors, it is probably unwise to eat a diet high in animal fat and cholesterol – see Chapter 6**

3 DIFFUSION

Diffusion is basically a 'mixing of molecules'. In a gas or liquid the molecules or ions move continuously, randomly bumping into each other and changing direction. In this way, particles tend to **diffuse**, or spread, so that they are spaced evenly in a gas or liquid, rather than being concentrated in one place (Fig 5.7).

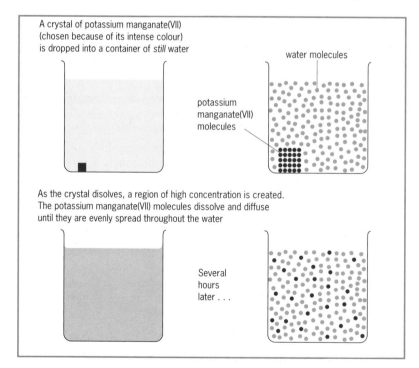

A crystal of potassium manganate(VII) (chosen because of its intense colour) is dropped into a container of *still* water

water molecules

potassium manganate(VII) molecules

As the crystal disolves, a region of high concentration is created. The potassium manganate(VII) molecules dissolve and diffuse until they are evenly spread throughout the water

Several hours later . . .

Fig 5.7 **If you drop a crystal of potassium manganate(VII) (chosen for its visibility) into still water, it dissolves and the ions diffuse until evenly spread. This takes several hours (depending on temperature) but can be speeded up by stirring. Diffusion is a passive process which requires no** *input* **of energy: it does however depend on the** *kinetic energy* **(energy of movement) of the molecules in a gas or a liquid**

This is a working definition:

Diffusion is the movement of particles within a gas or a liquid from a region of high concentration to a region of lower concentration.

Diffusion is the main process by which substances move over short distances (Fig 5.8). It is too slow to move substances efficiently over distances much greater than a fraction of a millimetre. In practice, substances usually diffuse over very small distances. In the

?

C Why can't solids diffuse?

See the Assignment on page 86 which deals with diffusion in organs of the human body.

mammalian lung, for example, oxygen diffuses through the thin epithelium of the alveolus and into the blood, a journey of usually less than a hundredth of a millimetre (10 μm). It is then carried away to other parts of the body by the circulatory system. The rapid movement of materials around an organism in a stream of fluid is called **mass flow**.

Diffusion and energy

The difference in concentration of a substance between two regions is called a **concentration gradient**. Particles that are free to move have **kinetic energy**. The region of a gas or liquid with the highest concentration of particles of a particular substance has the highest kinetic energy for that substance. Particles move down a concentration gradient by diffusion, until they are spread evenly.

The most important point to remember is that diffusion is a **passive process**: it requires no *input* of energy (see Fig 5.7). It follows that movement of a substance *against* a concentration gradient – active transport (see page 82) – requires energy.

Factors that affect the rate of diffusion

The rate at which molecules of a substance diffuse from one region to another has a great influence on the design of cells, organs and whole organisms (Fig 5.8). Several factors affect the rate of diffusion:

● The surface area between the two regions. The greater the surface area, the greater the rate of diffusion.

● The distance over which diffusion occurs. This is known as the *length of the diffusion pathway*.

● The concentration gradient – the relative concentrations of the substance in the two areas. Diffusion is more efficient if the concentration gradient can be maintained: this is achieved by transporting the substance away from the immediate area (eg by blood) once it has diffused, or by combining it with another chemical to prevent it from diffusing back.

● The size and nature of the particles. Fat-soluble substances can diffuse through the lipid bilayer of the membrane. Water-soluble substances must pass through the protein pores which tend to be small and selective. Generally, very large molecules cannot diffuse into or out of cells at all.

● The temperature at which the process takes place. At higher temperatures, molecules in a liquid or a gas have more kinetic energy and so diffuse more quickly.

Fig 5.8 **A schematic diagram to show some of the features which increase the rate of diffusion. An efficient exchange of materials requires a large surface area, maintenance of a diffusion gradient and a thin cell surface membrane. Increasing the temperature would also speed up diffusion, but this is not an option**

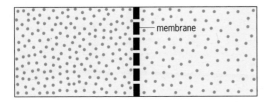

(a) Small concentration gradient and thick, unfolded membrane, so diffusion is slow.

(b) • Diffusing molecules are taken away (by blood, for example), so there are always more molecules on the left than on the right. Therefore the diffusion gradient is maintained.

• Thin membrane, so diffusion across is faster.

• Membrane folded into microvilli, creating a greater surface area for diffusion.

Calculating the rate of diffusion

The rate at which substances diffuse can be estimated from a simple formula which takes into account the factors that affect diffusion. Rate of diffusion is proportional to:

$$\frac{\textbf{surface area} \times \textbf{concentration difference}}{\textbf{length of diffusion pathway}}$$

This relationship is known as **Fick's law**. For efficient diffusion, the values on the top line of this equation should be as large as possible and the value on the bottom line should be as small as possible.

Diffusion and the human body

Diffusion is not fast enough to be effective over distances greater than a few millimetres. The body has many organs which are adapted for the exchange of materials – the lungs, intestines and kidneys are obvious examples. However, these organs are of little use without a circulatory system which transports the exchanged materials to all other areas of the body. The **blood system** is an example of what is known as a **mass flow** system.

The main function of the circulatory system is to bring materials to within diffusing distance of living cells, and to take away cell products. In the human body, no cell is more than a few micrometres away from a capillary (see Chapter 9).

4 FACILITATED DIFFUSION

Some substances enter and leave cells much faster than you would expect if only diffusion occurred. We now know that some membrane proteins assist, or **facilitate**, the diffusion of some substances across the cell membrane. Two types of protein are responsible for **facilitated diffusion**:

● Specific **carrier proteins** take particular substances from one side of the membrane to the other.

● **Ion channels** are proteins which open and close to control the passage of selected charged particles.

Carrier proteins

Until the 1970s, cell biologists thought carrier proteins worked by rotating within the membrane, like turnstiles. Newer research points to a different explanation (Fig 5.9). As soon as the diffusing molecule binds to the carrier protein, the protein undergoes a change in shape so that the diffusing molecule ends up at the other side of the membrane, where it is released.

Like diffusion, facilitated diffusion involves movement *down* a concentration gradient and requires no *input* of metabolic energy.

Diabetes and carrier proteins

The concentration of our blood sugar is kept relatively constant by controlling how much glucose passes from the blood into cells. After a meal, when blood glucose levels are high, the hormone insulin is released. This hormone activates a mechanism in the cell membrane which facilitates the diffusion of glucose into the cell, so lowering the blood sugar concentration.

?

D Phosphorylated chemicals (those with a phosphate group added) cannot pass easily through cell membranes. Suggest why glucose molecules are phosphorylated as soon as they enter a cell.

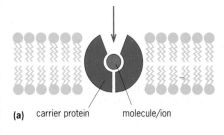

(a) carrier protein   molecule/ion

(b)

Fig 5.9 **Facilitated diffusion using a carrier protein. The diffusing molecule interacts with the carrier protein, causing a change in shape which 'squeezes' the molecule through the channel**

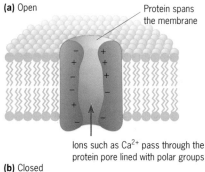

(a) Open

Protein spans the membrane

Ions such as Ca^{2+} pass through the protein pore lined with polar groups

(b) Closed

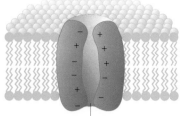

Protein changes shape, so pore becomes too narrow to allow ions through

Fig 5.10 **An ion channel. Because they are charged particles, ions cannot easily pass through the non-polar lipid bilayer. Specific membrane proteins form polar pores through which ions can pass. These channels are usually specific for one type of ion and can open and close according to the needs of the cell. When fully open, over 1 million ions per second can flow through a single channel**

One form of diabetes in humans is caused by a gene mutation which alters the structure of glucose carrier proteins in cell surface membranes. An affected person cannot get enough glucose into their cells. Unlike diabetes that results from the inability to make insulin (see Chapter 10), this form of the disease is difficult to treat, because injecting insulin has little effect.

Ion channels

Ion channels are proteins with a central 'hole' lined with polar groups (Fig 5.10). Ion channels facilitate the diffusion of charged particles such as Ca^{2+}, Na^+, K^+ and Cl^- ions. Many are **gated**, so can open or close. Cells use ion channels to control the movement of ionic substances between themselves and other cells, and to regulate the ionic composition of their cytoplasm.

A difference in concentration of ions may lead to a net positive or negative charge in a particular region. This type of concentration gradient is called an **electrochemical gradient**. Charged particles move towards regions of opposite charge. This is important in the process of transmission of a nerve impulse (see Chapter 14) and in the process by which red blood cells exchange materials with the tissues (see Chapter 9).

5 OSMOSIS

Sea water, because of the salt it contains, is about three times more concentrated than human blood. If someone who is already dehydrating is desperate enough to drink sea water, the salt causes water to move out of the blood and into the stomach by **osmosis**, causing further dehydration. So victims of shipwrecks can die of dehydration, even though they are surrounded by water.

A working definition of the process is as follows:

> **Osmosis is the diffusion of water only. It is the net movement of water molecules from a region of their higher concentration to a region of their lower concentration, through a partially permeable membrane.**

In biology, we usually talk about the diffusion of substances which are dissolved in water. But what about the water molecules themselves? Does water diffuse down a diffusion gradient? The answer is yes, and the diffusion of water is known as osmosis.

To explain osmosis, look at Fig 5.11. The solute molecules cannot diffuse in either direction because they are too big to pass through the membrane. But water molecules *can* get through and water diffuses down its concentration gradient.

Water molecules move from the left, where there is a higher concentration of water molecules, to the right, where there is a lower concentration of water molecules. (The solution on the left is what we normally think of as the less concentrated solution, determined by the solute concentration.)

When trying to understand osmosis, it helps to be clear about what happens when a substance dissolves in water. Because they carry a charge, the solute molecules become surrounded by a shell of water molecules (chemists say they are **hydrated**). When this

movement of water ⟹

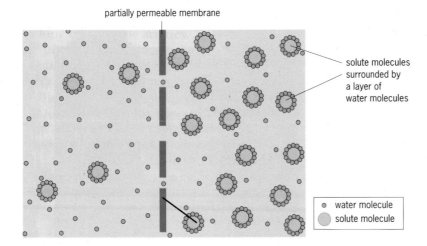

partially permeable membrane

solute molecules
surrounded by
a layer of
water molecules

○ water molecule
◉ solute molecule

Fig 5.11(a) **A 'weak' solution is one with a low concentration of solute molecules but a high concentration of water molecules. It has more 'free' water molecules than the 'concentrated' solution on the right and so has a high water potential. By** *high* **we mean a negative value close to zero**

Fig 5.11(b) **A more 'concentrated' solution has a higher concentration of solute molecules and a lower concentration of water molecules. It therefore has a low water potential. By** *low* **we mean a more negative value.**
   Overall, the water moves from a region of high water potential to an area of lower water potential

happens, the water molecules that form the 'shell' are no longer free to move around as they were before: they have been 'tied up' by the solute molecules. The region with the higher solute concentration 'ties up' more water molecules which are then not free to diffuse across the membrane. Thus, osmosis results from a net mvement of water molecules into the region of high solute potential.

Osmosis in living cells

All cells contain cytoplasm, a complex solution separated from its surroundings by a partially permeable cell surface membrane. So, all cells have the potential to gain or lose water by osmosis. The overall tendency for water to enter or leave a cell is determined by water potential.

Water potential

If no other factors are involved, a cell that is placed in a solution of the same concentration will neither gain nor lose water. Water molecules pass equally in either direction and there is no *net* change in the cell volume: Fig 5.12(a).

A cell placed in a solution that contains less solute (is more dilute) than the cytoplasm will gain water by osmosis: Fig 5.12(b). If the surrounding solution contains more solutes (is more concentrated) than the cytoplasm, the cell will lose water by osmosis: Fig 5.12(c).

The water potential is the potential of any system to absorb water molecules by osmosis.

Water potential is measured in units of pressure, **kilopascals**, **kPa**, and is a negative scale. For example, a relatively weak solution of

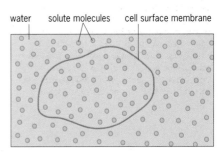

water   solute molecules   cell surface membrane

(a) **When the solution surrounding the cell contains the same amount of solute as the cytoplasm, the cell neither gains nor loses water**

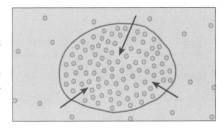

(b) **When the solution surrounding the cell has more solutes in it, the cell loses water by osmosis**

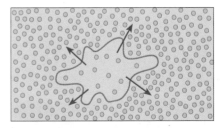

(c) **When the cell's cytoplasm contains more solutes than the surroundings, the cell gains water by osmosis**

Fig 5.12 **The factors that determine whether water enters or leaves a cell by osmosis**

How water molecules cross cell membranes

We have already said that the cell surface membrane is impermeable to water. In fact, it is not 100 per cent impermeable. Some water (a tiny amount) should be able to pass through a lipid bilayer. But recent studies have shown that water passes through purified phospholipid bilayers, with all the proteins removed, at rates approximately 100 to 1000 times faster than you might expect.

The exact reason for this is not known, but scientists think the constant sideways motion of the phospholipids, coupled with the flexing of the fatty acid tails, creates temporary holes (Fig 5.13)

through which the water molecules can slip. These holes appear in only one half of the membrane at a time, so the water molecules 'wait' at the half-way point until a hole appears in the other side, rather like crossing a busy road, one lane at a time.

Membranes with a high proportion of cholesterol are less permeable to water, and to simple ions such as sodium and chloride. It could be that cholesterol has a stabilising effect on the phospholipids. By slotting in between them, the cholesterol molecules minimise the creation of temporary holes (Fig 5.14).

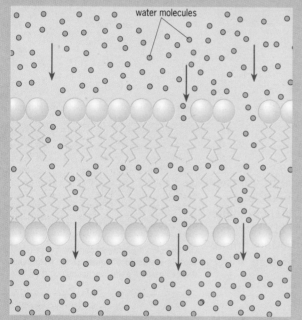

Fig 5.13 **A possible explanation for the fact that lipid bilayers are permeable to water. Temporary holes, created by the random sideways movement of the lipid molecules, act as channels through which water can pass**

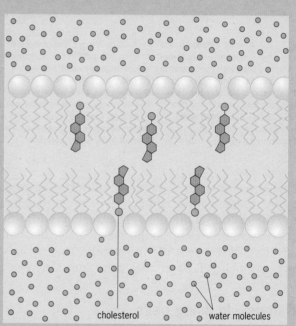

Fig 5.14 **Cholesterol is an important constituent of the membranes of many types of cell. The effect of cholesterol is to reduce leakage of water and ions through the lipid bilayer, probably by reducing the sideways motion of the lipid molecules**

See question 1. ■

?

E Preserves such as jam and marmalade have a high sugar content. Suggest why bacteria do not thrive when they land on these foods?

sucrose can have a water potential of −200 kPa while a more concentrated solution can have water potential of −500 kPa. Pure water has the highest possible water potential: zero.

If the two solutions described above were separated by a partially permeable membrane, water would move from the more dilute solution into the more concentrated one. So, if there are no other factors involved (such as physical pressure), water will move from a region of high (less negative) water potential to a region of lower (more negative) water potential.

If you have a problem with negative scales (and many people do), compare it to temperature. Of two solutions, one at −40 °C and the other at −60 °C, it is obvious which is the colder. If the two solutions were placed in contact with each other, heat would pass from the warmer to the colder until they were both about equal. It's the same with the movement of water in osmosis.

Osmosis in animal cells

If you place a human cell, or any animal cell in distilled water, it absorbs water by osmosis and swells up. As cell surface membranes have virtually no physical strength, the cell often bursts.

Fig 5.15 shows what happens to red blood cells placed in hypertonic, isotonic and hypotonic solutions. When blood plasma becomes hypertonic to red blood cell cytoplasm, as it does during severe dehydration, water is lost from the red blood cells, which shrink and become crinkled, or **crenated**. At the other extreme, hypotonic plasma causes red blood cells to swell and burst, leaving 'ghosts' of cell surface membrane. This process in known as **haemolysis**: literally 'blood splitting'.

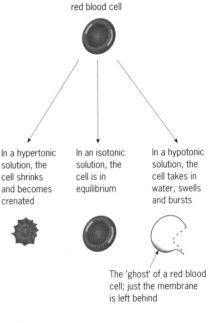

Fig 5.15(a) **All animal cells are prone to the same effects as these red blood cells when placed in solutions of different concentrations**

red blood cell

In a hypertonic solution, the cell shrinks and becomes crenated

In an isotonic solution, the cell is in equilibrium

In a hypotonic solution, the cell takes in water, swells and bursts

The 'ghost' of a red blood cell; just the membrane is left behind

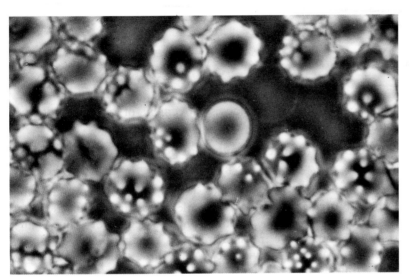

(b) **Red blood cells become crenated in a hypertonic solution**

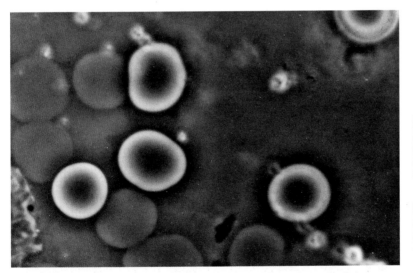

(c) **Red blood cells take in water and burst in a hypotonic solution (five red cells have not yet burst)**

In order to protect our cells from the effects of osmosis, we must maintain our body fluids at the same solute concentration as our cell cytoplasm. We do this by **osmoregulation**. In the human body, the kidneys are responsible for osmoregulation (see Chapter 11).

?

F Why is it incorrect to say that 'sea water is a hypertonic solution'?

✔

The following terms are useful when describing osmosis in human cells.

Hypertonic: a greater solute concentration.
Hypotonic: a lower solute concentration.
Isotonic: an equal solute concentration.

These are relative terms and so can only be used to compare solutions. For example, sea water is *hypertonic* to human blood plasma, distilled water is *hypotonic* to human blood plasma.

Assignment

EXCHANGE PROCESSES IN THE HUMAN BODY

All organisms exchange materials with their surrounding. The amount of material an organisms needs is proportional to its volume – the mass of living tissue it possesses. However, the amount of material is able to exchange is proportional to its surface area – the area of tissue in contact with the environment. Humans are, biologically speaking, huge demanding organisms. The aim of this Assignment is to develop an appreciation of the ways in which the human body is designed to maximise the exchange processes.

The surface area and volume problem
Imagine a cube-shaped multicellular organism that exchanges materials over its surface. (Obviously, no cubic organisms exist, but this keeps the calculations simple.)

1 Copy and complete the following table:

Size of each side of the organism (units)	2	5	10	?
Surface area (for a total of 6 sides (units^2)	24	?	600	60 000
Volume (units^3)	8	125	?	?
Surface area: volume ratio	?	1.2:1	?	?

The important principle shown by the table is that as organisms get larger, their volume increases faster than their surface area.

The consequence of this is that organisms must develop ways of increasing their surface area so that they can exchange more materials. So larger organisms like ourselves can only survive if they evolve ways to optimise the exchange of materials in spite of this. Organisms can:

- increase their surface area.
- make membranes/barriers as thin as possible.
- maintain a diffusion gradient.

As soon as materials are absorbed they are moved away from the site of absorption, often by a circulatory system to ensure that an equilibrium is never reached.

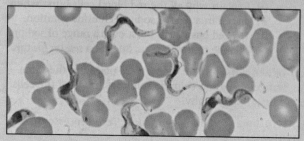

Fig 5.A1 **Trypanosomes in blood**

2 Fig 5.A1 shows a simple unicellular organism called a **trypanosome**, the parasite responsible for sleeping sickness. Does this organism need special adaptations to increase its surface area to volume ratio? State a reason for your answer.

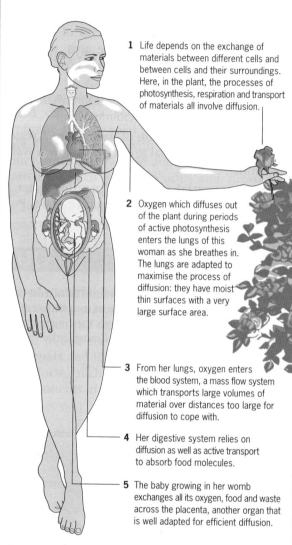

1 Life depends on the exchange of materials between different cells and between cells and their surroundings. Here, in the plant, the processes of photosynthesis, respiration and transport of materials all involve diffusion.

2 Oxygen which diffuses out of the plant during periods of active photosynthesis enters the lungs of this woman as she breathes in. The lungs are adapted to maximise the process of diffusion: they have moist thin surfaces with a very large surface area.

3 From her lungs, oxygen enters the blood system, a mass flow system which transports large volumes of material over distances too large for diffusion to cope with.

4 Her digestive system relies on diffusion as well as active transport to absorb food molecules.

5 The baby growing in her womb exchanges all its oxygen, food and waste across the placenta, another organ that is well adapted for efficient diffusion.

Fig 5.A2 **Many of the vital life processes depend directly on diffusion**

Fig 5.A2 shows some of the organs designed to maximise the exchange of materials including the lungs, digestive system and placenta.

3

a) Briefly state the function of the lungs, digestive system and placenta in Fig 5.A2.
b) List the features common to these three organs, and explain how each helps the exchange process.
c) The human circulatory system is described as a **mass flow** system. What does this mean and why are such systems necessary?

SUPPLY AND DEMAND

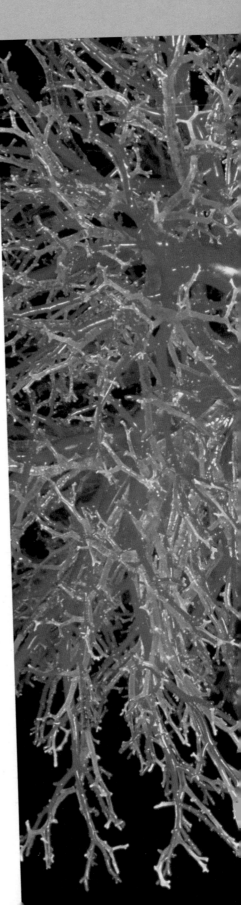

IN 1995 AMERICAN NEWSPAPERS reported that the world's fattest man had died, weighing 465 kilograms (73 stone). The reports stated that it took a fork-lift truck to move his body from his bedroom, but only after they had demolished the wall to get him out.

You might think that people as obese as this must eat all the time, but in daily terms, the excess dietary intake that is required to put on weight is minimal. If he started out as a 70 kg 16-year-old, he would only have had to put on 37 grams per day – about the weight of a chocolate bar or 1.5 bags of crisps – in order to reach his final weight at age 45.

Although a great deal of fuss is made about gaining weight and getting fat, most of us manage to remain at a relatively constant size. This is possible because we match our energy intake to our energy output with remarkable precision. For instance, if an average person takes in over 400 kg of food (and over 800 litres of fluid) in the course of a year, and only gains 0.5 kg (around 1 lb), they are regulating their metabolism within very narrow limits.

In this section we look at the way in which we nourish our bodies, by providing all the cells that make it up with the materials they need – matching supply to demand. We also look at the way our body deals with waste. We begin in Chapter 6 with diet and its implications for a healthy lifestyle (providing a good foundation for the study of lifestyle diseases in Chapter 35). The subsequent chapters in this section look at the major body systems that carry out digestion (Chapter 7), breathing (Chapter 8) and blood circulation (Chapter 9).

6 Nutrition

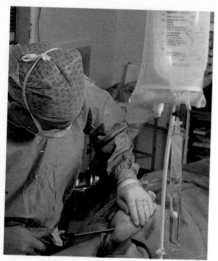

In the increasingly popular cosmetic treatment of liposuction, the fatty adipose tissue is literally hoovered out of 'problem' areas

IN THE AFFLUENT COUNTRIES of the western world a lot of people eat too much. The number of overweight people is on the increase, and at the same time, others have made dieting an obsession.

We know that to lose weight we have to eat less food and exercise more, but the trick is to find the self discipline to do it. One glance at the newsagent's shelves shows that there are many magazines devoted to slimming. Many of the articles are basically the same – diets, exercise regimes and inspirational stories such as: I lost ten stones and so can you.

If will-power fails, there are more drastic methods. Liposuction removes excess fatty tissue, usually from the buttocks and thighs, while 'beer bellies' can be removed in the same way. To cut down on intake, desperate people have even had their jaws wired up or part of their stomach surgically stapled so that it cannot hold much food.

1 THE IMPORTANCE OF FOOD

Humans, like all organisms, need food:

- as a source of energy
- to supply raw materials for repair, growth and development of body tissues
- to supply vital vitamins and minerals

The old cliché 'we are what we eat' is perfectly true. As developing babies, we were built when molecules from the food our mothers ate were assembled according to the genes we inherited, and ever since we have been taking in the food we need to build and power our bodies.

In Chapter 3 we looked at the chemicals of life: carbohydrates, proteins, lipids and nucleic acids. This chapter looks at how we provide our bodies with these and other essential substances.

2 HUMAN NUTRITION: FOOD AND DIET

A healthy diet provides us with a balanced selection of nutrients which our bodies need in order to carry out vital life processes. People who are **malnourished** are short of some of the types of nutrients which they need to stay healthy. Someone who is **starving** cannot obtain enough energy to meet the demands of the body.

Fig 6.1 **Eating for two. A developing baby is fed via the placenta, an organ which takes the molecules from food directly out of the mother's blood. When the infant is born, the mother will need even more extra food in order to produce milk**

abou
suga
N
sho
unt
hur
con
the
rap
bas
an

for
rel
diş
lei

Pr
Tl
th
in
ai
b
s

2
ε
ı
ı
ٍ

A Can a person be malnourished without starving? Explain your answer.

Fig 6.2 **There is a massive range of foods available all around the world, yet they all contain the same basic chemicals: carbohydrates, proteins, lipids, vitamins, minerals and water**

The energy in food

All of our food comes directly or indirectly from plants. The energy contained in food is derived originally from the plants in the process of **photosynthesis** (see Chapter 22). Food consists mainly of large organic molecules whose energy can be released by the process of **respiration** (see page Chapter 22).

As you sit here reading this book, you might think that you are not using much energy, but even at rest, your body has a steady energy demand. Think about some of the processes going on:

- Your heart is beating.
- You are breathing.
- Food is being pushed along your intestines.
- Food is being absorbed from your intestines into your blood.
- Urine is moving from your kidneys to your bladder.
- Nerves are taking information from your sense organs to your brain, keeping you informed of changes in your environment.
- You are thinking: your brain is processing information coming in from the sense organs, sorting out what is important and what can be ignored.
- Unless you are somewhere very hot, you are probably losing heat to your surroundings. You need to replace the heat you lose in order to maintain a constant internal body temperature.
- Some of the molecules from digested food, such as amino acids, are being built up into molecules that will help to make new cells and tissues.

All these processes (and many others) require energy which is released from food molecules by respiration, a process which goes on constantly in all of our cells.

The amount of energy a person needs depends on:
- their age,
- their level of activity,
- their size,
- their genetic background (some people inherit a very high or a very low metabolic rate).

Assignment

OBESITY AND BODY MASS INDEX (BMI)

Obesity is a term which means 'grossly overweight' and, despite more people understanding the problems of the condition, the proportion of overweight and obese people is on the increase.

Fig 6.A1 **The United States has the highest proportion of obese people in the world**

Table 6.A1 **The increasing incidence of obesity**

Country	weight level	Age range	Year	Incidence/%	
				men	women
England:	obese (BMI >30)	16–64	1980	6	8
			1994	13	16
England:	overweight (BMI 25–30)	16–64	1980	35	24
			1991	40	26
Germany:	obese (BMI >30)	25–69	1985	15	17
			1990	17	19
USA:	obese* (white people)	20–74	1978	24	24
			1988–91	34	34
USA:	obese* (black people)	20–74	1978	26	45
			1988–91	32	49

*In the USA survey the threshold for obesity was BMI >27.8 men, >27.5 for women

Sources: Obesity Research Information Centre, London, and Dept of Health Report 46 (1994) Nutritional Aspects of Cardiovascular disease, HMSO

1 Given the present rate of increase, how many men and women in England will be obese by the year 2008?

Body mass index (BMI)

2 Obesity could be defined in terms of weight alone. What would be wrong with this?

In 1869 a Belgian astronomer Quetelet worked out that people's body mass varied not in proportion to their height, but in proportion to the square of their height.

So the formula for body mass index is:

$$\frac{mass}{height^2}$$

For example, a man who is 1.83 m tall (6 feet) and weighing 82 kg would have a BMI of

$$\frac{82}{(1.83)^2} = \frac{82}{3.35} = 24.47$$

3 Work out your own BMI.

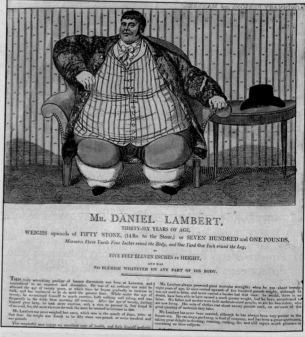

MR. DANIEL LAMBERT,
THIRTY-SIX YEARS OF AGE,
WEIGHS upwards of FIFTY STONE, (14lbs to the Stone,) or SEVEN HUNDRED and ONE POUNDS,
Measures Three Yards Four Inches round the Body, and One Yard One Inch round the Leg,
IS
FIVE FEET ELEVEN INCHES IN HEIGHT,
AND HAS
NO BLEMISH WHATEVER ON ANY PART OF HIS BODY.

Fig 6.A2 **This is Daniel Lambert (1770–1809), once thought to be the heaviest man in England. He weighed 336 kg (52 stone 11 lb) and was 180 cm (6 ft) tall**

4 Work out the BMI for Daniel Lambert.

5 List some of the problems that a person as heavy as Daniel Lambert would face in their everyday life. Assume that they are alive today, not in the seventeenth century

a) What sorts of things would it be impossible for them to do?

b) Can you think of any advantages of being this weight?

The BMI is now universally used to define underweight, overweight and obesity, as shown in Table 6.A1.

BMI/kg m^{-2}	Description
less than 20	underweight
20–24.9	normal
25–29.9	overweight
30–40	obese
over 40	severely obese

Table 6.A1

The problems of obesity

There was a time when fat people were looked upon just as happy, jolly folk. Nowadays, we know the problems associated with obesity, and obese people have to deal with the intolerance of others towards their size and body image.

One look at the population of an old people's home will support the blunt statement that 'old people aren't fat and fat people aren't old'. Obesity puts a great strain on the body that more often than not leads to severe medical problems and a premature death. Consider some of the problems:

Coronary heart disease. Obesity increases the levels of triglyceride and cholesterol in the blood (see Chapter 35).

6 Suggest why obesity puts a strain on the heart. Why does it have to work harder?

Diabetes – the inability to control blood sugar levels. In an obese person, insulin becomes less effective than normal, a problem known as **insulin resistance**.

High blood pressure.

Joint problems. Excess weight puts a lot of strain on some joints.

7 Suggest which joints will be affected most.

Accidents Obese people tend to be clumsier than less heavy people, and unable to move quickly.

Depression and suicide – responses to a poor body image.

The causes of obesity

Many obese people claim that eating too much is not the problem; it's in their genes. Recent research suggests that there may be some truth in this, but we can only inherit a *tendency* towards obesity or 'leanness'. We must still eat the food which allows us to fulfil our potential, and studies suggest that social and environmental factors play a huge part. Children born to overweight parents tend to be given larger portions and generally follow the influence of their parents.

Lifestyle

People are generally eating more and exercising less, using the car for even the shortest journeys and taking part in leisure activities with minimal energy needs.

8
a) Suggest how the leisure activities of children have changed over the last few decades.
b) Imagine you are a family doctor. How would you persuade an obese couple to change their lifestyle?

Drugs

Research has shown that adipose (fat storage) tissue gives off a hormone, **leptin** (Greek, leptos = slender/thin) which suppresses the appetite centre of the brain. There are artificial appetite suppressants available, but their effect is only temporary. Clearly, there would be a huge market for an effective drug.

9 What properties would the perfect weight control drug have?

Surgery

Adipose tissue can be surgically removed; the process of liposuction is one of the cosmetic surgeons' biggest money-spinners. However, all surgery carries some risk and, without a long-term change in eating and exercise habits, the problem will return.

7 The human digestive system

IN JUNE 1822, Alexis St Martin, an American army porter, was accidentally shot in the stomach from a musket fired at close range. This did him no good at all. He sustained several wounds, including a large hole out of which poured his recently eaten breakfast. This horrific injury was a turning point in our understanding of digestion.

Alexis was strong and his wounds healed, except for the large hole. William Beaumont, an army surgeon, tried to close the wound for nearly 10 months. He used a tight bandage to plug the wound and eventually a skin flap grew over the hole, forming a sort of valve which kept the stomach contents in. The hole, however, could be reopened at any time.

Beaumont had a walking, talking experiment! He used the opportunity to the full, conducting many investigations of which he kept meticulous records. These tell how he could press on the abdomen above the liver to obtain bright yellowy-green bile that had been released into the duodenum. And how he placed pieces of different kinds of food in the stomach and checked them at various times afterwards to see the digestive processes in action. The strange partnership between doctor and patient lasted for nine years.

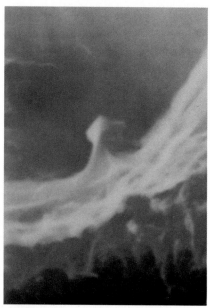

Nowadays, more sophisticated methods can be used to study problems of the digestive tract. Here, the patient has swallowed a barium meal so that an X-ray of the stomach shows an ulcer which appears as a white mushroom shape on the stomach's folded lining

1 THE ALIMENTARY CANAL

Our diets consist of a wide range of simple and complex chemicals, along with a significant amount of bacteria. The overall function of the digestive system is to break down the larger molecules such as protein and starch, turning them into a form which is simple, soluble and therefore easily absorbed.

The intestine can be thought of as a long coiled tube which runs through the middle of the body. The food we eat is subjected to 'conveyor-belt' food processing as it passes through the different regions of the gut. A common misconception is that the food we eat is 'inside' us, when really it travels through a cavity in the middle of the body. If we cannot digest the food, it cannot be absorbed into the body and passes straight through.

The alimentary canal, or 'gut', is a muscular tube which leads from the mouth to the anus. The overall process of nutrition can be divided into several stages:

- **Ingestion**, taking in food. Food is propelled through the alimentary canal by **peristalsis** (rhythmic contractions of the gut wall).

- The **mechanical breakdown** of food material into smaller pieces, mainly by the action of chewing or the churning action of the stomach.

- **Digestion:** the chemical breakdown of complex food molecules such as carbohydrates, proteins and fats into simpler ones.
- **Absorption:** passage into the bloodstream of simple food molecules such as amino acids, sugars and fatty acids, vitamins, minerals and water.
- **Egestion:** the elimination of undigested food material from the body.

Humans are **heterotrophs**; we cannot make our own food and so must obtain it in the form of large organic molecules from other organisms. All of these molecules were originally made by green plants.

2 THE HUMAN DIGESTIVE SYSTEM

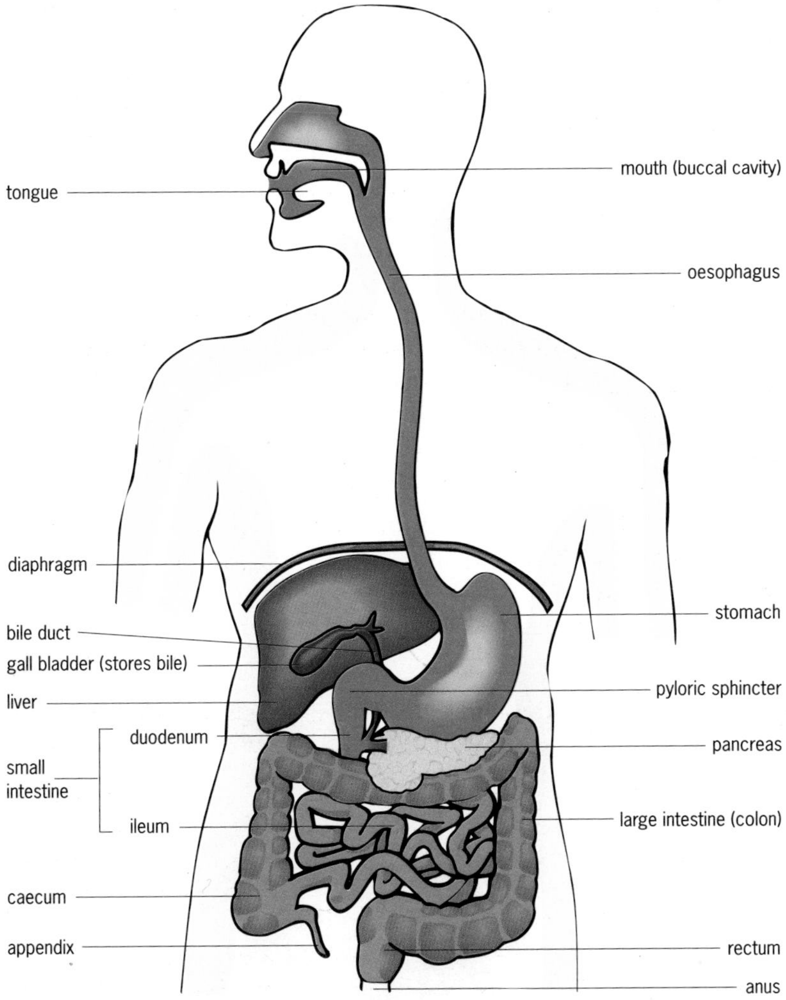

Fig 7.1 **The human digestive system**

The human **digestive system** (Fig 7.1) consists of the alimentary canal and its associated glands, the **salivary glands**, the **liver** and the **pancreas**.

The alimentary canal begins at the **mouth** and ends at the **anus**. Between the two openings is a long convoluted tube, organised into several distinct regions. The **oesophagus** carries food from the mouth to the **stomach**, a muscular bag or sac which stores food and is the first significant site of digestion.

Beyond the stomach is the **small intestine** which has two main parts, the **duodenum** and the **ileum**. These are the main sites of digestion and absorption. The **large intestine** includes the **appendix,** the **colon** whose main function is to absorb water, and the **rectum,** and ends at the anus.

3 FROM THE MOUTH TO THE STOMACH

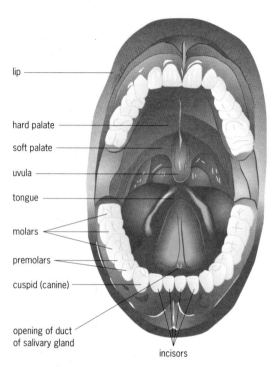

Fig 7.2 **There are three pairs of salivary glands in the mouth and one pair can easily be seen. If you look under the tongue and dry the area, you will be able to see saliva oozing out**

The mouth or **buccal cavity** is where food processing starts. The lips and tongue combine to mix food with saliva and push it between the teeth which begin mechanical breakdown of food. The mouth is lined by **stratified epithelium** (epithelial cells stacked on top of each other – see page 28), which protects the mouth from friction damage as food is chewed. There is a very high turnover of the epithelial cells: we replace the lining of our mouth – and the rest of the gut – every 24 hours or so.

The action of chewing, or **mastication**, breaks down food mechanically, making it easy to swallow. The powerful jaw muscles generate enormous force. Saliva lubricates and softens the food. Saliva is secreted continuously, but we produce more if we see, smell, taste, or even just think about food. (A summary of the mechanisms which control the release of all digestive secretions is shown later in Fig 7.15.) Overall, we produce 1 to 1.5 litres of saliva each day.

Saliva is mainly water (99.5 per cent) with some dissolved substances (0.5 per cent) including:

- **mineral salts** (phosphates and hydrogencarbonates)
- **salivary amylase** (a starch-digesting enzyme) that breaks molecules of starch into maltose.
- **mucin** (a slimy glycoprotein lubricant)
- **lysozyme** (an enzyme which kills bacteria)

Salivary amylase, the starch-splitting enzyme in saliva, acts on food pieces that are produced by chewing, beginning

The senses of taste and smell

Gustation (taste) and **olfaction** (smell) are the body's chemical senses. There are four basic sensations of taste: *sweet, salt, sour* and *bitter*. On its own the tongue is poor at distinguishing between different foods; a slice of apple and a piece of raw onion both taste sweet, and without the extra help of the nose you would have difficulty telling which was which. You may have had this sort of problem when your nose is blocked. It is the nose which plays the major role in our appreciation of food.

Receptor cells for taste are located mainly in taste buds on the upper surface of the tongue. Certain regions of the tongue appear to react more strongly than others to particular taste sensations (Fig 7.3).

We can only smell a substance that is **volatile** – one which evaporates easily. **Olfactory receptor cells** (smell detectors) are located in the roof of the nose cavity. Molecules of the odour stimulate these olfactory cells, causing them to send impulses to the brain for processing. Although the tongue can detect only the four basic chemical types, the nose can detect thousands of different chemicals. It is this subtlety which allows us, for instance, to

tell the difference between wines from different vineyards, or different types of coffee.

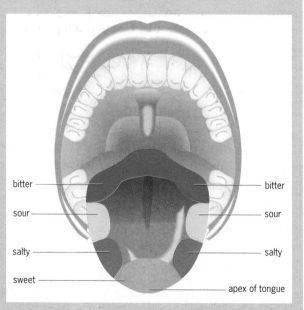

Fig 7.3 **The taste receptors for sweet, salt, bitter and sour are located in well-defined regions along the outer edge of the tongue**

the process of chemical breakdown. However, the speed at which most people chew and swallow their food means that salivary amylase has little chance to act. Research has shown that a significant minority of people produce no salivary amylase (a genetic curiosity) but this condition is likely to go undetected.

Swallowing

Swallowing is not a voluntary action, although it can be initiated voluntarily. It is a reflex response to food or liquid touching the back of your throat (the act of moving food to the back of your throat is the voluntary action). The act of swallowing is illustrated in Fig 7.4.

Theoretically, when the tongue forces food or liquid to the back of the throat, the food is able to travel in one of four directions:

- back out of the mouth
- into the nasal cavity
- into the trachea (windpipe)
- into the oesophagus (where it should go)

When we swallow normally, the pathways which would allow the first three options are closed off, and food or liquid is forced into the oesophagus. But sometimes it doesn't work out like that. If you have ever been laughing, eating crisps and drinking fizzy liquids simultaneously you may have experienced fizzy liquid going 'up your nose' or crisp particles going 'down the wrong way' (into your trachea).

Looking at teeth

Humans have two sets of teeth during their life. **Milk dentition** appears from the age of about 6 months and lasts until the age of 5 or 6 (Fig 7.5). Humans have 20 milk teeth which are gradually replaced (they are actually forced out) by a second set, the **permanent dentition**. Our 32 adult teeth should last for the rest of our lives, although **tooth decay** often prevents this.

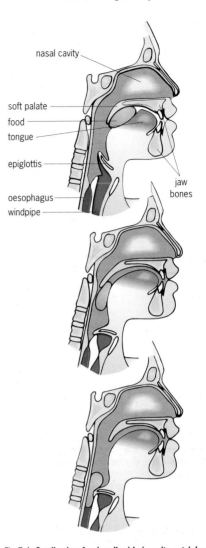

Fig 7.4 **Swallowing food or liquids is quite a tricky procedure. Since you can drink a glass of water standing on your head, it's not just a question of allowing material to drop into the stomach. The liquid is forced out of the mouth and into the oesophagus by the action of muscles (peristalsis)**

?

A Look at Fig 7.4. Describe how food is prevented:
(a) from entering the nasal cavity,
(b) from entering the trachea,
(c) from being forced back out of the mouth.

?

B The notation for human adult dentition is:
$$\frac{2\ 1\ 2\ 3}{2\ 1\ 2\ 3}$$
Suggest what this means. Why is the total number of teeth only 16?

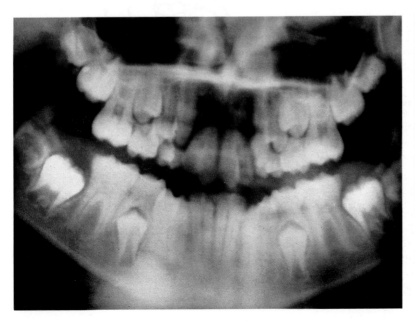

Fig 7.5 **An X-ray image of a six-year-old girl showing her current (milk) teeth and the new (adult) teeth waiting to push through**

Fig 7.6 **A human adult incisor tooth** (a) **and molar tooth** (b) **are shown in vertical section**

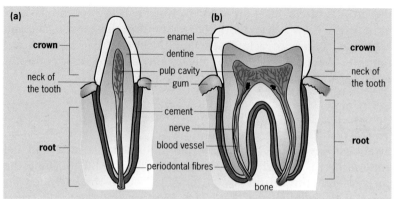

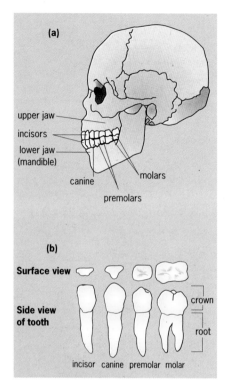

Fig 7.7 **Humans have heterodont dentition (different types of teeth).** (a) **shows the position of the four types of human teeth in the skull.** (b) **shows what each type of tooth looks like (surface and side views)**

Tooth structure

A tooth typically consists of three parts: the **crown**, the part above the **gum**; the **root**, the part hidden below the gum (embedded in a **socket**) and the **neck** of the tooth (Fig 7.6).

Teeth are made of a bone-like material, **dentine**. This surrounds a **pulp cavity** into which blood vessels and nerves run. The blood supply nourishes the living bone, supplies oxygen and removes waste products. The nerves allow us to sense pressure and touch, and intense pain if the nerve is exposed by damage. **Tooth enamel**, the hardest material in the body, overlies the crown. **Periodontal fibres** help secure the tooth, and a layer of **cement** around the root fixes the tooth in its bony socket.

Types of teeth

The number, size and shape of teeth that a mammal has depends on the diet. As Fig 7.7 shows, humans, who are omnivores, have four types of teeth:

- **Incisors**: chisel-shaped front cutting teeth.
- **Canines**: pointed teeth used for grasping and tearing.
- **Premolars**: small chewing teeth with two **cusps** (projections) on the crown.
- **Molars**: large back teeth used for chewing and grinding. These have four cusps.

TEETH AND DISEASE

THERE ARE TWO main types of dental disease, **dental caries** and **periodontal disease**.

Dental caries (tooth decay) – probably the most widespread disease in the western world – results when the dentine and enamel on a tooth is gradually broken down. The process begins when bacteria such as *Streptococcus mutans* release acids onto teeth as they feed on the sugars that remain there after we have eaten. The bacteria become fixed to the teeth in a capsule formed by the sticky polysaccharide **dextran**. The bacteria, dextran and other debris are collectively known as **dental plaque**.

The acids produced by bacterial digestion start to break down tooth enamel, and plaque may react with chemicals in saliva, hardening to form **calculus**. If this condition remains untreated, microbes can reach the pulp cavity of the tooth, causing inflammation, infection and pain.

Periodontal disease is a gum disease which affects the **cement** and **periodontal fibres**. Teeth eventually become loose as the gums recede.

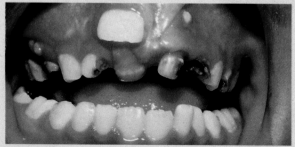

Fig 7.8 **Lack of proper care and attention to dental hygiene can lead to dental caries (cavities which need filling) and periodontal disease (loosening of the teeth)**

LOOKING INSIDE THE ALIMENTARY CANAL

IT IS POSSIBLE to look inside the alimentary canal of a living person (Fig 7.9) using an **endoscope**, a flexible tube with a light and a miniature camera at the end.

The mouth (left) clearly shows the complex apparatus needed to break down food and then propel it into the oesophagus (above right).

The stomach (below right) has deep ridges called **rugae**. Cells lining the **gastric pits** produce a strong acid. The stomach lining also secretes large quantities of mucus, which prevent its own tissues being damaged.

The walls of the intestine are more gently folded, and are also awash with digestive secretions. The velvety appearance (below) of the lining is due to the enormous number (4 to 5 million) of tiny projections – the **villi** – which line the wall. Villi are only 0.5 to 1 mm high, but they can move from side to side to improve the contact they have with the food in the intestine. They increase the surface area of the lining; this is necessary to ensure efficient absorption of nutrients.

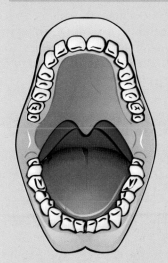

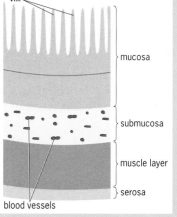

The lining of the intestine, like the lining of the entire canal, is made up of four layers:

An inner **mucosa**. This secretes digestive juices or absorbs food products. In areas prone to damage (by bone fragments in the oesophagus, for example) the **epithelial cells** (cells of the lining) are **stratified**, or layered, for protection.

A **submucosa**. This contains lots of blood vessels to carry away absorbed food. It is also rich in nerves which coordinate the actions of muscles (that produce the squeezing movements of peristalsis) with the release of digestive secretions from the canal wall.

A thick muscle layer. Muscle fibres are positioned in rings around the intestine and along the length of the wall. This muscle is **involuntary** (see Chapters 2 and 18).

An outer **serosa**. This is a tough layer of connective tissue, and extensions of it form sheets that attach the canal to the wall of the abdomen. The sheets which support the intestine are called the **mesenteries** (see Fig 7.13), the layers which line the abdomen are called the **peritoneum**.

Our journey ends at the colon (left). There are fewer secretions, but lots of mucus here. The colon does not have villi or permanent folds. Instead, it has a 'pinched' appearance due to thin bands of muscle. These produce a strong churning movement. Bacteria on the walls of the colon cause some of the non-digested food to ferment, forming gases whose composition depends on the nature of the bacteria and of the food eaten.

Fig 7.9 **A remarkable journey through the alimentary canal. Note the rich blood supply to the gut wall and the glistening digestive secretions. The diagrams adjacent to the photographs show the structure of the alimentary canal wall at each of the locations chosen**

The oesophagus, route to the stomach

The oesophagus is about 25 cm long and 2 cm in diameter. After swallowing, smooth muscles in the oesophageal wall contract rhythmically to propel food, by peristalsis, from the **pharynx** (throat) to the stomach. Elastic tissue in the walls enables the oesophagus to expand as food passes along it. As in the mouth, stratified epithelium protects against friction damage.

4 THE STOMACH

The stomach is a muscular sac under the **diaphragm**. When empty, it is the same size as a large sausage, but it can stretch to the size of a melon. The stomach is the first area of any significant digestion.

Fig 7.10 **Structure of the human stomach**

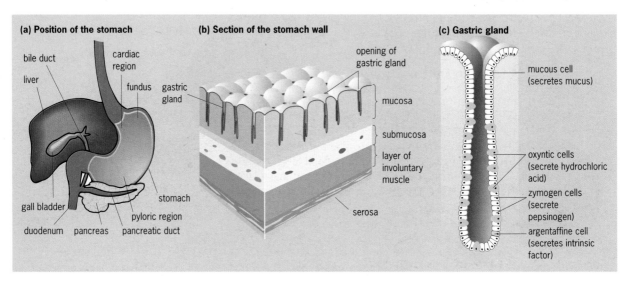

(a) Position of the stomach

bile duct
liver
cardiac region
fundus
gastric gland
gall bladder
stomach
duodenum pancreas pyloric region
pancreatic duct

(b) Section of the stomach wall

opening of gastric gland
mucosa
submucosa
layer of involuntary muscle
serosa

(c) Gastric gland

mucous cell (secretes mucus)
oxyntic cells (secrete hydrochloric acid)
zymogen cells (secrete pepsinogen)
argentaffine cell (secretes intrinsic factor)

Food remains in the stomach (Fig 7.10) for between 30 minutes and 4 hours depending on the type of meal (fatty meals stay there the longest). The food is digested mechanically and chemically. The resulting semi-liquid material, **chyme**, passes into the duodenum, the first part of the small intestine. Overall, the stomach:

- mixes food with gastric juice by muscular action,
- retains food, giving enzymes time to act,
- digests proteins through the action of the enzyme **pepsin**,
- curdles milk with the enzyme **rennin**,
- absorbs some simple chemicals such as water, salts as ions, and alcohol.

Gastric juice is secreted by specialised groups of cells inside **gastric pits** in the mucosa (see Fig 7.10). Each type of cell produces a specific secretion:

- **Oxyntic cells** secrete a solution of hydrochloric acid which brings the pH of gastric fluid down to between pH 2.0 and 3.0.
- **Zymogen cells** (peptic cells) secrete the enzyme **pepsinogen** which is later converted into the protein-splitting enzyme, **pepsin**.
- **Mucous cells** secrete the **mucus** that protects the stomach lining from the digestive action of its own secretions.

Pepsin and pepsinogen

One of the main functions of the stomach is to begin to digest proteins (a process completed in the small intestine). But the stomach must avoid digesting its own tissue. So its zymogen cells secrete pepsin – the enzyme which breaks proteins and large polypeptides into smaller polypeptides – in an *inactive* form, pepsinogen, which is only converted to pepsin in the lumen of the stomach after contact with hydrochloric acid. The hydrogen ions in the acid cause the pepsinogen to unfold and become pepsin, the active form of the enzyme.

Pepsin is a powerful **endopeptidase enzyme**: it breaks specific peptide bonds in the middle of the protein chain, turning protein molecules into polypeptides (see Chapter 3 for detail on the structure of proteins). The process of protein digestion is completed in the small intestine where **exopeptidase** enzymes remove amino acids by hydrolysis from the ends of the short chains produced by the endopeptidase. The soluble amino acids can then be absorbed, along with some di- and tripeptides.

■ See question 2.

Milk digestion

Caseinogens, the proteins in milk, are *water soluble*. They are valuable nutrients (milk is the sole source of food for young mammals) but, if they remained in their soluble form, they would leave the stomach before protein digestion had finished. To avoid this, the stomach produces **rennin**. This **curdles** milk, converting soluble caseinogen into insoluble **casein**. Like pepsin, rennin is also secreted in an inactive form, **prorennin**. Like pepsinogen, prorennin is converted to its active form by contact with stomach acid.

> **?**
> **D (a)** Would you expect rennin production to increase or decrease with age in mammals? Give a reason for your answer.
>
> **(b)** Would you expect rennin to be found in animals other than mammals?

The acid conditions of the stomach

The hydrochloric acid in gastric fluid:

- provides the optimum pH for pepsin and rennin,
- denatures proteins and helps to soften tough connective tissue in meat,
- is a strong **bactericide** (it kills bacteria) and so protects the body from some of the harmful microbes which might enter the body in food.

Control of gastric secretion

Since we do not eat food continuously, it is important that we produce digestive enzymes only when food is in the gut. If large quantities of the acids and protein-attacking enzymes were released into an empty stomach, there would be a danger of **autolysis** (self-digestion).

Nerves and hormones control the secretion of gastric juice in a process which has three distinct phases (Fig 7.11):

- **Nervous phase** The sight, smell or taste of food initiates a **nerve reflex** in which impulses from the brain trigger gastric glands to release their secretions.
- **Gastric phase** (hormonal) Food in the stomach stimulates the lining to secrete the hormone **gastrin**. Gastrin increases gastric juice secretion through direct action on the gastric glands.

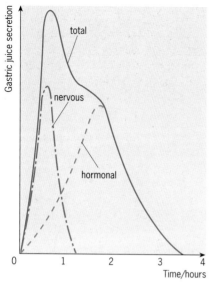

Fig 7.11 **Secretion of gastric juice is partly controlled by a nerve (the** vagus nerve**) and partly by a hormone (gastrin). Gastrin stimulates gastric glands, causing them to release their secretions. It increases movement in the tract and relaxes the sphincters which control the movement of food into and out of the stomach**

See question 3. ■

● **Intestinal phase** (hormonal) The duodenal lining, stimulated by partially digested food, produces a second hormone, **enteric gastrin**. This hormone also acts on the stomach's gastric glands, producing further small amounts of gastric juice.

A summary of the mechanisms which control the release of all digestive secretions is shown later in Fig 7.15.

Pressure sensors and chemical sensors in the stomach detect stomach stretching and the presence of chyme. The resulting nerve impulses, together with the action of gastrin, direct the stomach to start emptying its contents into the duodenum.

5 THE SMALL INTESTINE

The small intestine is about 5 metres long and is made up of two main parts, the **duodenum** and the **ileum**.

The duodenum takes up the first 25 cm or so of the small intestine, and food entering from the stomach receives secretions from two sources: **bile** (page 114) from the liver and **pancreatic juice** from the pancreas. These juices contain the many enzymes necessary to complete digestion, see Table 7.1. While the duodenum is the main site of digestion, most absorption takes place in the ileum, which makes up the remainder of the small intestine. The ileum joins the large intestine at the caecum, near the appendix (refer back to Fig 7.1).

The traditional view of events in the small intestine was that chyme arriving from the stomach receives *three* digestive juices: bile, pancreatic juice and intestinal juice (succus entericus) from the wall of the small intestine itself. Thereafter, digestion occurs in the lumen of the small intestine and the soluble products are absorbed. Recently, it has become clear that the situation is more complex.

The fluid secreted by the intestine contains no enzymes, but consists of water, hydrogencarbonate ions and mucoprotein. This juice serves to neutralise the stomach acid, protects the duodenal wall from digestive enzymes and generally provides enough water for optimum digestive conditions.

Secondly, much of the digestion of food take place not in the gut lumen but *in the epithelial cells themselves*. While the pancreatic enzymes bring about digestion in the gut lumen, enzymes embedded in the epithelial cell surface membrane and in the cytoplasm complete the process. Products of digestion such as disaccharides and di- and tripeptides are digested by these cell surface enzymes and are released back into the lumen before being absorbed through pores elsewhere in the membrane.

Table 7.1 **A summary of the main human digestive enzymes**

Secretion	Enzymes produced	Site of production	Site of activity	pH	Substrate	Products
Saliva	salivary amylase	salivary glands	mouth	6.5–7.5	starch	maltose
Gastric juice	pepsin rennin	stomach	stomach	2.0	proteins, polypeptides caseinogen (milk protein)	small polypeptides
Pancreatic juice	trypsin chymotrypsin carboxypeptidase pancreatic amylase lipase	secretory cells of the pancreas (acini)	duodenum	7.0	proteins, polypeptides proteins polypeptides starch fats and oils	short polypeptides polypeptides dipeptides and amino acids maltose fatty acids, glycerol

The small intestine:

- moves food from the stomach to the large intestine by peristalsis. (Food moves through the intestine at about 1 cm per minute.)

- secretes large amounts of water necessary for most of the chemical reactions involved in digestion.

- completes the digestion of carbohydrates, proteins and fats.

- absorbs the vast majority of the small soluble food molecules produced by digestion.

- secretes the hormones **CCK-PZ** (page 113) and **secretin.**

- protects against infection: it is closely associated with intestinal lymph nodes.

The small intestine has walls which are structured in much the same way as the walls of other parts of the alimentary canal (see the Feature box on page 107), but it has some obvious modifications to enable it to perform its function as the main site of digestion and absorption.

?

E The average length of the small intestine is about 6.35 metres. How long does it take for food to travel from one end to the other?

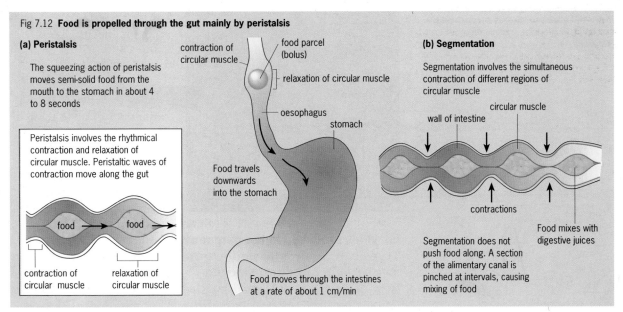

Fig 7.12 **Food is propelled through the gut mainly by peristalsis**

(a) Peristalsis

The squeezing action of peristalsis moves semi-solid food from the mouth to the stomach in about 4 to 8 seconds

Peristalsis involves the rhythmical contraction and relaxation of circular muscle. Peristaltic waves of contraction move along the gut

food → food →

contraction of circular muscle

relaxation of circular muscle

contraction of circular muscle

food parcel (bolus)

relaxation of circular muscle

oesophagus

stomach

Food travels downwards into the stomach

Food moves through the intestines at a rate of about 1 cm/min

(b) Segmentation

Segmentation involves the simultaneous contraction of different regions of circular muscle

wall of intestine

circular muscle

contractions

Food mixes with digestive juices

Segmentation does not push food along. A section of the alimentary canal is pinched at intervals, causing mixing of food

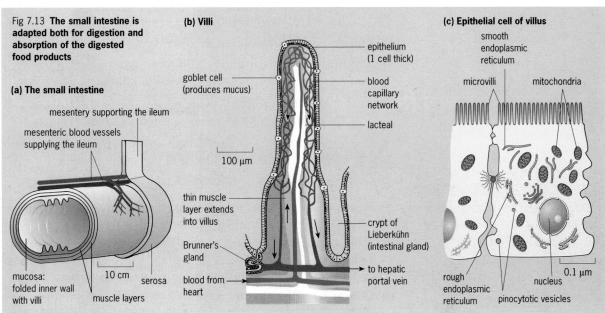

Fig 7.13 **The small intestine is adapted both for digestion and absorption of the digested food products**

(a) The small intestine

mesentery supporting the ileum

mesenteric blood vessels supplying the ileum

mucosa: folded inner wall with villi

10 cm

serosa

muscle layers

(b) Villi

goblet cell (produces mucus)

epithelium (1 cell thick)

blood capillary network

lacteal

100 μm

thin muscle layer extends into villus

Brunner's gland

blood from heart

crypt of Lieberkühn (intestinal gland)

to hepatic portal vein

(c) Epithelial cell of villus

smooth endoplasmic reticulum

microvilli

mitochondria

0.1 μm

rough endoplasmic reticulum

nucleus

pinocytotic vesicles

The pancreas and liver both have other functions that are not connected directly with digestion. The pancreas produces the hormones insulin and glucagon (Chapter 10) and the liver has many regulatory – homeostatic – functions (also Chapter 10).

Fig 7.14 **The pancreas is a compound gland: it is both an endocrine gland (releasing products – hormones – to the inside of the body) and an exocrine gland (releasing products to the outside of the body). The endocrine function of the pancreas is performed by small patches of cells, the islets of Langerhans, which secrete insulin and glucagon. Most of the pancreas is take up by acini which produce digestive secretions**

Pancreatic juice

Pancreatic juice is a secretion of the pancreas which enters the duodenum via the **pancreatic duct**. Our pancreas secretes over a litre of alkaline pancreatic juice every day. The fluid contains several enzymes which are listed in Table 7.1 and are discussed in detail below.

Secretory cells, **acini** (Fig 7.14) make up most of the glandular tissue of the pancreas. They secrete digestive juice, discharging their contents into a pancreatic duct. Hydrogencarbonate ions make pancreatic juice slightly alkaline (pH = 7.1 to 8.2), allowing it to neutralise the stomach acid and helping to create optimum conditions for the intestinal digestive enzymes.

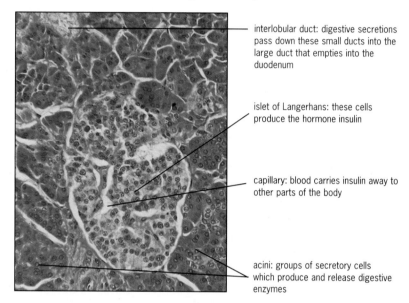

interlobular duct: digestive secretions pass down these small ducts into the large duct that empties into the duodenum

islet of Langerhans: these cells produce the hormone insulin

capillary: blood carries insulin away to other parts of the body

acini: groups of secretory cells which produce and release digestive enzymes

Enzymes produced in the pancreas

Trypsin is a powerful **endopeptidase enzyme** secreted in an inactive form called **trypsinogen**.

● **Enterokinase**, an enzyme released by the intestinal wall, converts the inactive trypsinogen into its active form:

$$\text{trypsinogen} \xrightarrow{\text{enterokinase}} \text{trypsin}$$

Trypsin breaks down proteins into polypeptides:

● **Chymotrypsin** continues the digestion of proteins into polypeptides. This enzyme is secreted in an inactive form and is converted by trypsin:

$$\text{chymotrypsinogen} \xrightarrow{\text{trypsin}} \text{chymotrypsin}$$

● The polypeptide-digesting enzyme **carboxypeptidase** is also activated by trypsin in the lumen of the small intestine:

$$\text{procarboxypeptidase} \xrightarrow{\text{trypsin}} \text{carboxypeptidase}$$

An exopeptidase enzyme, carboxypeptidase converts polypeptides into smaller peptides and amino acids.

● **Pancreatic amylase** completes the breakdown of starch to maltose started by salivary amylase in the mouth. A second amylase enzyme is needed as salivary amylase is inactivated by stomach acid.

● **Pancreatic lipase** continues the breakdown of fats into fatty acids and glycerol.

F Why do the protein-splitting enzymes trypsin and chymotrypsin need to be produced in an inactive form?

G What is the difference between an endopeptidase and an exopeptidase enzyme?

Control of pancreatic secretions

Pancreatic secretion, like gastric secretion, is controlled by both nervous and hormonal mechanisms. Impulses travel from the brain down the vagus nerve to trigger the secretion of pancreatic juice. A summary of the mechanisms which control the release of all digestive secretions is shown in Fig 7.15.

The lining of the duodenum secretes two hormones in response to acidic chyme arriving in the lumen of the duodenum. **Secretin** stimulates the pancreas and liver to secrete pancreatic juice rich in hydrogencarbonate ions. **CCK-PZ (cholecystokinin-pancreozymin** – once considered to be two separate enzymes) stimulates the gall bladder to release bile and the pancreas to release digestive enzymes (see Fig 7.15).

■ See question 4.

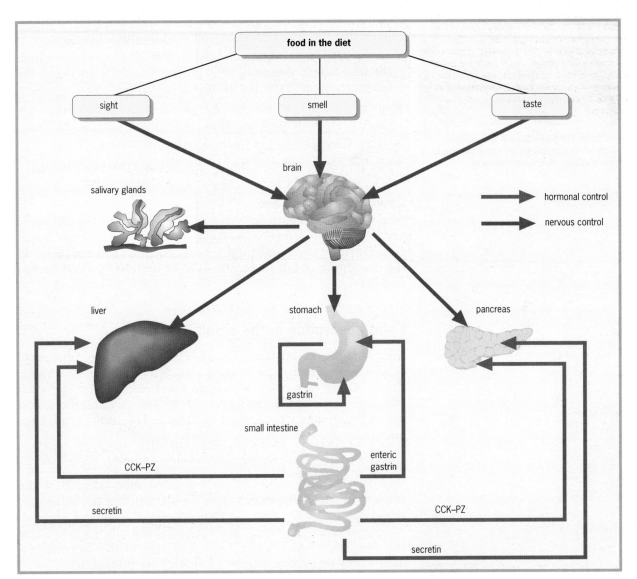

Fig 7.15 **The digestive tract is supplied by nerves from the autonomic nervous system (see Chapter 16). Particularly important is the vagus nerve which stimulates the secretion of digestive juices. Hormones travel to the digestive tract in the blood.**

Information coming into the body from food (about its appearance and smell) travels to the hypothalamus and cerebral cortex in the brain. These send signals back to the body to stimulate the salivary glands, stomach, liver and pancreas to start releasing their secretions. As food enters the stomach, gastrin is released and this encourages more gastric juice secretion. Similarly, the presence of food in the duodenum causes the release of gastrin, secretin and CCK-PZ.

We do not carry on eating until we burst because in humans, as in other animals, food intake is controlled by hunger and satiety centres in the hypothalamus. A full stomach and high blood glucose levels stimulate the satiety centre, so that we do not feel hungry any more

Bile

Liver cells called **hepatocytes** produce thick yellow-brown or olive green bile which contains:

- water,
- bile salts (sodium glycocholate and sodium taurocholate),
- bile pigments (breakdown products of red blood cells, for example **bilirubin**) and some mucus,
- cholesterol.

Liver cells produce around 0.8 to 1.0 litre of bile daily. Secretions from individual cells pass into tiny canals called **bile cannaliculi**. These lead to the gall bladder, a small sac-like organ which stores the bile until it is needed. Bile is released into the duodenum when CCK-PZ stimulates the muscle wall of the gall bladder to contract. Bile reaches the duodenum through the **bile duct**. Bile:

- **emulsifies** fats (breaks large fat or oil droplets into an emulsion of microscopic droplets). This process massively increases the surface area available for fat digesting enzymes.
- neutralises the (acidic) chyme from the stomach and creates the ideal pH for intestinal enzymes.
- stimulates peristalsis in the duodenum and ileum.
- allows the excretion of cholesterol, fats and bile pigments.

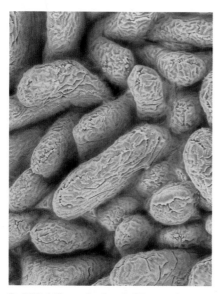

Fig 7.16(a) **The projections in this scanning electron micrograph are intestinal villi which have a velvety covering of microvilli**

Control of bile secretion

Bile secretion is controlled in a number of ways. The hormone **secretin** acts together with nervous stimulation (by the vagus nerve) to increase the rate of bile secretion. The acidity of chyme in the duodenum and the hormone CCK-PZ stimulate the gall bladder to contract.

Food absorption in the small intestine

Digestion breaks up large insoluble food molecules into small soluble molecules that can be easily absorbed through the gut lining. About 90 per cent of all absorption takes place within the small intestine. (A small amount occurs in the stomach and the rest in the large intestine.) Undigested material travels through the large intestine and is expelled from the body.

The internal lining of the small intestine is well adapted to its function. The epithelial lining has a much greater surface area than that of an equivalent smooth-sided tube because of the three structural features shown in Fig 7.16(b). The epithelial cells absorb amino acids, monosaccharides, fatty acids and glycerol, water and other substances including vitamins, nucleic acids, ions and trace elements (see Fig 7.17). The type of transport process used varies depending on the substance, but diffusion, facilitated diffusion and active transport are all involved (see Chapter 5).

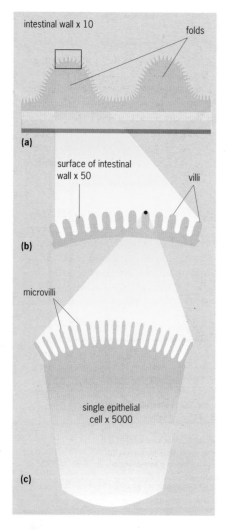

Fig 7.16(b) **An average 70 kg man has approximately 100 square metres of absorbing surface in his small intestine. This increased surface area of the intestinal lining is due to (a) folds in the inner surface of the intestinal wall, (b) moveable projections (0.5–1.0 mm in length) called villi which are present on the folded surface of the wall, and (c) microscopic projections called microvilli on the cell surface membranes of epithelial cells which line the villi**

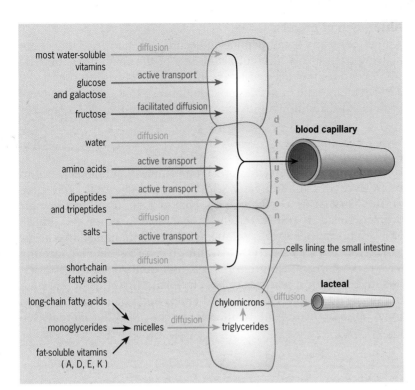

Fig 7.17 **The absorption of the products of digestion through the lining of the intestine**

EXAMPLE

Q Explain how the structure of the ileum relates to its function of absorbing the products of digestion.

A Relating structure to function is one of the basic skills a biologist must master. The structure of the small intestine is a classic example. In this example, refer to Fig 7.16 as you read.

When faced with this question, many students will simply state a well-learned GCSE list of 'large surface area, good blood supply, thin walls etc', and expect full marks. At A-level a little more depth is required. Why is a large surface area vital? What does a good blood supply do?

'Large surface area'

The basic idea behind a large surface area is that a lot of intestine is in contact with a lot of digested food molecules. These are absorbed by diffusion or active transport and so the more membrane there is, the more pores and carrier proteins there are available. The fact that the intestine is long is the most obvious way to increase the surface area, but there are also the villi and the microvilli.

'Good blood supply'

Like most organs which are adapted for the exchange of materials, there is also a way of maintaining the diffusion gradient so that there is always more digested food on one side of the absorbing membrane than the other, thus ensuring that absorption is continuous and as rapid as possible. As soon as the digested food molecules are absorbed into the blood, they are taken away and replaced with blood containing fewer of these molecules.

'Thin walls'

Fick's law (Chapter 5) states that the speed of diffusion is inversely proportional to the diffusing distance. Put simply, the thinner the membrane, the faster the diffusion, and so the barrier between food and blood is as thin as possible, being as little as a few micrometres – the thickness of the epithelial cell and the cell lining the capillary.

The small intestine also has another feature of note: the epithelial cells contain many mitochondria which make the ATP needed to power the active transport mechanisms.

Absorption of carbohydrates

Glucose is actively transported into the cells of the small intestine and across the walls of blood capillaries. Active transport is necessary because diffusion alone would be too slow to supply the body's needs. Also, there would be diffusion out of the epithelial cells if the concentration of food in the intestines was very low. Energy for the transport is obtained from the hydrolysis of ATP.

Absorption of proteins

Amino acids and small peptide molecules (di- and tripeptides) are actively transported into epithelial cells of the small intestine. The small peptides are digested by enzymes either on the epithelial cell surface membrane or inside the cytoplasm so that eventually every protein is broken down into its constituent amino acids.

Absorption of fats

Bile salts and lipase enzymes break up complex lipid molecules into monoglycerides and free fatty acids. The monoglycerides and some of the fatty acids combine with bile salts to form microscopic droplets called **micelles**. Micelles are soluble in water and diffuse easily into epithelial cells, along with the remaining fatty acids. Inside the cells, the fatty acids and glycerol recombine forming triglycerides which acquire a protein coat that stops them sticking together, and form particles called **chylomicrons**. It is in this state that lipids leave epithelial cells and enter, not a blood capillary this time, but a branch of the lymphatic system called the **lacteal** (see Fig 7.17).

Transport of absorbed food products

Amino acids, dipeptides and simple sugars diffuse out of epithelial cells in the small intestine, directly into blood capillaries within the villi. From here they pass to the liver via the **hepatic portal vein**. The liver processes absorbed food, converting some into storage products. Glycogen, copper, iron, vitamins A, D, E and K can all be stored. The liver breaks down other food products, including excess amino acids which are **deaminated** – their **amine groups** are removed. These breakdown products are then passed to the kidneys for excretion. See Chapter 10 for more about liver function and Chapter 11 for details on excretion.

The lymphatic system plays a major role in the transport of absorbed lipids. Chylomicrons which have entered the lacteals remain in suspension, giving the **lymph fluid** a milky white appearance. From here, the chylomicrons move into larger lymphatic vessels which eventually drain into a large duct that empties into the blood. In the blood, these complexes are broken down into fatty acids and glycerol and enter cells to be used in the synthesis of complex lipids.

> **?**
>
> **H** Why are the nutrients produced by colon bacteria (in the large intestine) of less use to us than nutrients provided by bacteria in the small intestine?

BACTERIA IN THE GUT

AROUND 10^{14} BACTERIA live in a healthy human body, mostly in the digestive tract. The bulk of these – they include at least 400 species – inhabit the large intestine. These microorganisms are **mutualistic**, which means that they share our food. In return, they may supply useful products such as vitamins, or help the digestive process, as they do in ruminants. Some bacteria can also contribute to an animal's resistance to disease by competing with potentially harmful bacteria for sites in the gut.

Successful digestive tract bacteria tolerate a wide range of pH conditions, resist the effects of antibodies secreted by intestinal tissue and survive the constant motion of the intestinal contents.

Although gut bacteria are **non-pathogenic** (they do not cause disease in a healthy body), infection can arise after, for example, damage to the large intestine. Bacterial contamination of the peritoneum (the lining of the abdominal cavity) can lead to **peritonitis**, a potentially fatal infection of the abdomen.

6 THE LARGE INTESTINE: DEALING WITH UNDIGESTED FOOD

The large intestine is about 1.5 m long and extends from the ileum to the anus. It has four sections, the **caecum**, the **colon**, the **rectum** and the **anal canal**. The large intestine:

- absorbs water,
- is the site of manufacture of certain vitamins (microorganisms within the colon of some animals produce vitamin K and folic acid),
- forms and expels undigested food residue as **faeces**, in the process of **egestion**.

The caecum receives material from the ileum. In humans, the bottom of the caecum is attached to a small blind tube, the **appendix**. This tube is twisted and coiled and is about 8 cm long. The human appendix plays no part in digestion, but can become inflamed, causing **appendicitis**.

Much of the water which is poured onto food during its passage through the digestive tract (as saliva, gastric juice, pancreatic juice, for example) is reabsorbed. This reabsorption is vital: the digestive system pours up to 8 litres of fluid into the gut each day. If we lost all this fluid we could dehydrate. Most water absorption occurs in the small intestine, but of the litre or so that enters the large intestine, all but 100 cm^3 is reabsorbed by the colon. Minerals (as ions) can also diffuse or be actively transported into the bloodstream from this segment of the large intestine.

Diseases such as cholera and dysentery which cause severe diarrhoea can result in potentially lethal dehydration within a very short time. To combat this, **oral rehydration therapy** can be given, a mixture of glucose and salts in water – see the Feature box on page 168.

The final sections of the large intestine, the rectum and anus, are concerned with the compaction of faeces and the act of **defaecation**. Faeces contain some water, mineral salts, undigested food, bile pigments, products of bacterial decomposition and epithelial cells that have become detached from the digestive tract.

SUMMARY

By the end of this chapter, you should know and understand the following:

- The human **alimentary canal** is a tube which leads from the **mouth** to the **anus**. It is associated with glands such as the **salivary glands**, **pancreas** and **liver**.

- The alimentary canal, or digestive system, is responsible for the **ingestion**, **digestion**, **absorption** and **egestion** of food.

- Food is broken down *mechanically* by the chewing action of teeth and by the muscular churning of the alimentary canal.

- **Digestive juices** are poured onto food as it travels down the alimentary canal. These soften and lubricate the food and also contain digestive enzymes which break the food down *chemically*.

- The muscles of the alimentary canal 'squeeze' the food – a process called **peristalsis** – and push it along from one end to the other.

- Large food molecules are broken down to their building blocks: carbohydrates are digested to form simple sugars, proteins are broken into amino acids, and fats are converted to fatty acids and glycerol.

- Digestion of food in the small intestine is due to the actions of bile and pancreatic juice, and the process is completed inside or on the surface of the epithelial cells themselves.

- Digested food products are absorbed into the body across the wall of the alimentary canal. Most enter the bloodstream, but the products of fat breakdown enter the **lymphatic system**.

- A sophisticated control system ensures that digestive enzymes are released in the right place at the right time.

QUESTIONS

1 Fig 7.Q1 shows the human digestive system.

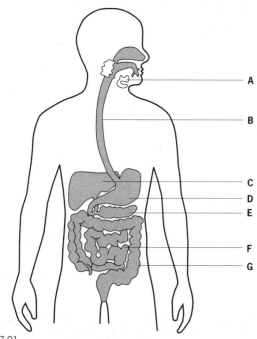

Fig 7.Q1

a) Give the letter of an organ where each of the following is produced: **(i)** endopeptidase, **(ii)** maltase.

b) Name the compounds which are produced by the digestion of triglyceride.

c) Describe **one** role of bile in digestion of triglycerides.

d) **(i)** Name the hormone which causes the gall bladder to contract.
(ii) What is the stimulus that initiates the release of this hormone?

[NEAB June 1998 Physiology Module Test, q.1]

2

a) Fig 7.Q2(a) shows two amino acids joined together by condensation.

Fig 7.Q2(a)

Draw a diagram to show how the molecule in Fig 7.Q2(a) could be hydrolysed to produce two amino acid molecules.

b) Fig 7.Q2(b) shows part of a polypeptide molecule.

Name the type of protein-digesting enzyme which would hydrolyse the polypeptide in Fig 7.Q2(b) at the points shown by the arrows.

c) Describe **two** ways in which the epithelial cells lining the ileum are adapted for their function of absorbing the molecules produced by digestion.

[NEAB June 1997 Human Systems Module Test, q.3]

3 Below is a section through the human stomach wall.

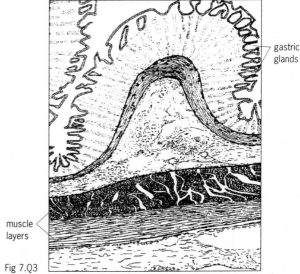

Fig 7.Q3

(Reproduced from *Atlas of Histology*, Freeman and Bracegirdle, by permission of Heinemann Educational Publishers.)

a) Describe **one** function of the muscle layers in the stomach wall.

b) **(i)** Name the type of protein-digesting enzyme secreted by the gastric glands.
(ii) Describe the function of this type of enzyme.
(iii) Name the hormone which stimulates the release of this enzyme.

c) Explain briefly why a different enzyme is required for the digestion of each type of food substance.

[NEAB March 1998 Physiology Module Test, q.2]

4 Fig 7.Q4(a) and Fig 7.Q4(b) are electron micrographs showing different parts of a secretory cell from the pancreas.

a) Give **one** piece of evidence, visible in Fig 7.Q4(a) which supports the view that:
(i) the image was taken with the aid of an electron microscope;
(ii) this is a eukaryotic cell;
(iii) the main function of this cell is to produce proteins.

b) **(i)** Describe **one** characteristic feature of structure **A** which would identify it as a mitochondrion.
(ii) Suggest an explanation for the fact that there are relatively large numbers of mitochondria in this cell.

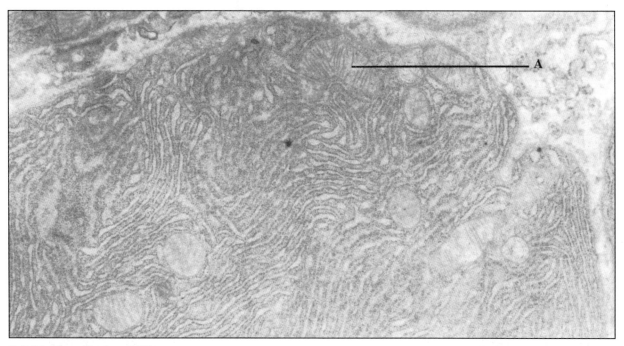

Fig 7.Q4(a) **(magnification ×1600)**

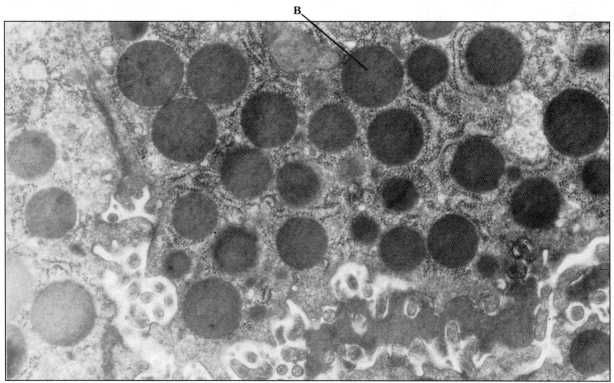

Fig 7.Q4(b) **(magnification ×26 700)**

c) Feature **B** in Fig 7.Q4(b) is a vesicle containing digestive enzymes.

    (i) The image of Fig 7.Q4(b) has been magnified 26 700 times. Calculate the diameter of the vesicle labelled **B** in micrometres. Show your working.

    (ii) Explain why the vesicles appear a uniform dark colour when seen with an electron microscope.

    (iii) Explain how the presence of partly digested food in the duodenum normally leads to the release from the pancreas of the enzymes in these vesicles.

[AEB Summer 1997 Human Biology Paper 2, q.3]

8 Gas exchange

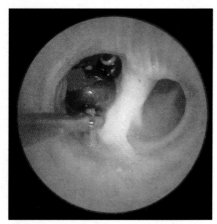

This is an endoscope picture: a child has swallowed the top of a cocktail stirrer shaped like a frog's head, it found its way down the trachea and is stuck at the entrance of one of the bronchi. A tool at the tip of the endoscope is being used to pull it out

DISEASES OF THE LUNG such as asthma, bronchitis and emphysema can be investigated at first hand thanks to the endoscope, a camera on the end of a flexible tube. This technology allows doctors to investigate parts of the body – such as the lungs, the stomach and the intestines – without the need for surgery. Not only does it relay enlarged pictures to a TV screen, but tools on the end of the device can take samples of tissue (such as tumours), remove blockages and even perform small operations.

Having inserted an endoscope into a person's airway, the doctor's first decision is whether to turn into the right or left lung. Here, he or she could look for obstructions, tissue damage or tumours. In practice, an endoscope can be used only for the larger branches, the trachea and the bronchi. If the camera was small enough to go right into a terminal alveolus, the endoscope operator would face another 23 decisions of left or right, such is the extent of the branching of the airways.

In total, there are about 2^{23} terminal bronchioles that lead from the trachea to the alveoli of the human lung. The number of alveoli is even larger, about 300 million. The combined surface area for gas exchange is huge, about the size of a badminton court.

1 WHY DO WE EXCHANGE GASES?

Humans, like all living organisms, need energy to carry out the processes of life. In particular, we need energy for movement, growth and repair and to keep our body temperature stable. We obtain this energy from the oxidation of organic molecules such as glucose in the process of **respiration** (see Chapter 22). Respiration uses glucose and oxygen and produces carbon dioxide, water and energy in the form of adenosine triphosphate (ATP). This goes on constantly in every living cell of our bodies and, to keep the process going, we must obtain a constant supply of oxygen and expel the carbon dioxide as waste. We humans are are also large warm-blooded organisms, and so we need a massive amount of oxygen to supply our metabolic demands.

Fortunately, air contains a lot of oxygen, and it is relatively easy for us to move a large amount of air into and out of our bodies in a short time: our hugely developed lungs are adapted to bring enough air sufficiently close to the blood to allow diffusion to take place.

In this chapter we look at the structure of the lungs, the mechanism of ventilation, the process of gas exchange and, finally, the way in which our breathing rate changes according to the varying demands of the body.

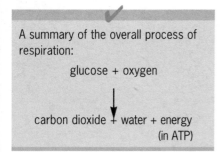

A summary of the overall process of respiration:

glucose + oxygen

↓

carbon dioxide + water + energy
(in ATP)

Fig 8.1 **The structure of the human gas exchange system**

(a) **The overall structure**

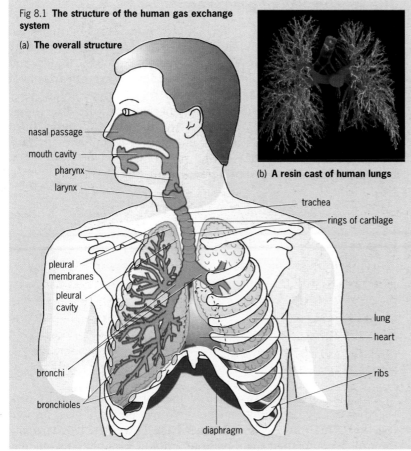

(b) **A resin cast of human lungs**

(c), (d) and (e) **The fine structure of the alveolus and the mechanism by which gas is exchanged**

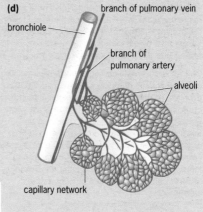

2 GAS EXCHANGE ORGANS

The intestines, kidneys and lungs, organs that are adapted for exhange of materials, all show several features that maximise the exchange process. The **lungs** are responsible for exchanging gases between the blood and the air outside the body. Oxygen enters the body by diffusion, and carbon dioxide leaves in the same way. The lungs increase the efficiency of diffusion by having:

- **A large surface area.** The huge surface area for gas exchange is is the result of many little air pockets called alveoli, which are surrounded by many blood capillaries.

- **Thin, permeable walls.** The gases have only to pass through two thin cell layers, thus ensuring that the diffusing distance is as short as possible.

- **A good blood supply.** The lungs receive as much blood per minute as the rest of the body put together. Blood whisks incoming oxygen away quickly to a different part of the body. This ensures that a diffusion gradient is always present: there is always less oxygen in the blood that flows through the lungs than there is in the air.

- **Efficient ventilation.** This is a physical pumping mechanism, usually called breathing, that continually brings fresh air to the gas exchange surfaces in the alveoli and keeps the diffusion gradient as large as possible.

Alveoli don't collapse when we breathe out because their surface is covered by an anti-sticking chemical called **surfactant** (see the Feature box on page 124).

The mechanism of breathing

The lungs themselves are spongy – they cannot move on their own – so how do we breathe? Lungs **inflate** and **deflate** because they are *expanded* and *compressed* by movements of the ribcage and diaphragm. This is possible because two **pleural membranes** hold the lungs to the ribcage and diaphragm. The outer membrane lines the **thoracic cavity**: the inner membrane encloses the lungs. Between the two is a very narrow space, the **pleural cavity**, filled with **pleural fluid**. This allows the pleural membranes to slide smoothly over each other during breathing, and prevents them from separating. Fig 8.2 shows how we breathe in and out.

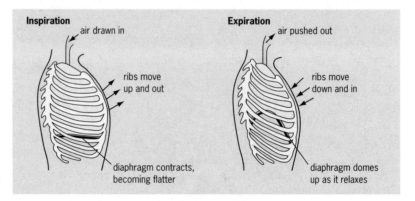

Fig 8.2 **The mechanism of breathing**

(a) Inspiration **(breathing in)** happens when the external intercostal muscles contract and pull the ribcage upwards and outwards, away from the spinal column. At the same time, the diaphragm contracts and flattens, pushing down on the abdominal organs such as the liver. These movements increase the volume and therefore lower the pressure in the thorax. As the pressure in the thorax falls below that of the atmosphere, air is forced into the lungs to equalise the pressure

(b) Expiration **(breathing out)** is normally a passive process: it uses no energy. When inspiration is over, the muscles in the thorax relax and breathing out follows due to a combination of gravity, the elastic recoil of the connective tissues of the lungs and the pressure exerted by the abdominal organs. We can consciously speed up expiration by forcing air out of our lungs using our internal intercostal muscles. This happens, for instance, when we blow up a balloon or play a wind instrument

To understand why two pleural membranes continue to stick together during breathing, imagine two wet pieces of glass pressed together. They can easily slide over each other, but it is virtually impossible to pull them apart without introducing air into the middle. If air is introduced into the pleural cavity, after a stab wound, for example, the lung collapses, a situation known as a **pneumothorax**.

Using a spirometer to determine lung volume

Fig 8.3 shows a **spirometer**, a device for measuring the volumes of air that a person breathes in and out. The trace that this apparatus produces tells us a lot about lung volume.

First of all, your lungs have a **total capacity**, the maximum amount of air that the lungs can hold during the deepest possible breath. This is not the total volume of the lungs, however, because the lungs

Fig 8.3 **A spirometer and the trace it produces**

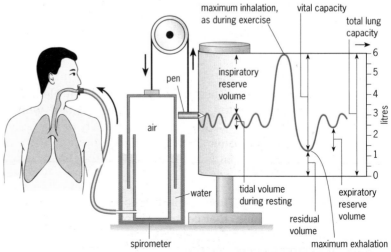

Vital   = total lung – residual volume
capacity   capacity

Ventilation = number of × tidal
rate   breaths taken   volume
               per minute

are never totally emptied. Even when you have exhaled as far as possible, there is still some air in the alveoli, and in the bronchi and trachea, which are held open permanently by rings of cartilage. The volume of air that remains in the lungs after breathing out is called the **residual volume**. The maximum usable lung volume (the total lung capacity minus the residual volume) is called the **vital capacity**. The average vital capacity is 4.5 to 5 litres for men and 3.5 to 4 litres for women.

Secondly, during normal breathing, the volume of air that moves in and out of the lungs in each breath is called the **tidal volume**. In a normal adult at rest, this is about 0.5 litres.

Thirdly, the trace shows that after breathing in at rest, the person could inhale an extra 3 litres, a value known as the **inspiratory reserve volume** (**IRV**). He or she could also breathe out another litre or so: the **expiratory reserve volume** (**ERV**). These are the extra volumes of air that we breathe in and out during exercise.

Finally, we can use the trace to work out the **ventilation rate**: the volume of air taken into the lungs in one minute. To do this, multiply the number of breaths taken in a minute by the tidal volume.

Using a spirometer to determine oxygen consumption

We can use a spirometer to estimate a person's rate of oxygen consumption. If the person re-breathes the air in the closed system of the spirometer, the composition of that air changes: the oxygen level decreases and the carbon dioxide level increases. Soda lime placed in the apparatus absorbs the carbon dioxide and so the volume of air in the spirometer decreases in volume as the oxygen is used up. Fig 8.4 a shows a trace produced in this way.

To calculate the amount of oxygen used, we measure the volume decrease in a given time. In our example, the trace shows that the air volume fell from the 1600 cm³ mark to the 1300 cm³ mark in one minute. So, this person used up 300 cm³ of oxygen in one minute.

WARNING!! Re-breathing your own air can be dangerous. You should never do investigations like this without close supervision.

Gas exchange at the alveoli

Gas exchange between air and blood occurs at the alveoli (Fig 8.5). These tiny air sacs create a huge surface area: 1 cm³ of frog lung tissue has a surface area of about 20 cm², but the corresponding figure for a mouse lung is over 800 cm². Other mammals have a similar value. The total surface area of one human lung is about 70 to 100 m².

A If the average breathing rate is 15 breaths per minute, and the tidal volume is 0.5 litre, calculate the ventilation rate.

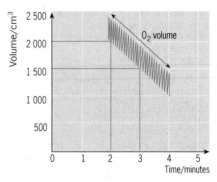

Fig 8.4 **In addition to lung volume, the spirometer can also be used to measure the volume of oxygen used. If the carbon dioxide is absorbed as soon as it is exhaled, the total volume of air falls, showing how much oxygen is being used. We can use the measurements to determine how much oxygen is used at rest, and to see how this changes with exercise**

■ See questions 1 and 2.

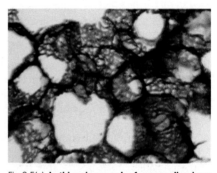

Fig 8.5(a) **In this micrograph of mammalian lung tissue, the alveoli and capillary networks that surround them can clearly be seen**

Fig 8.5(b) **Gas exchange at the alveolar surface. In the human lung, the alveolar endothelium, the membrane which lines the air sacs, is only 0.2 µm thick. This minimises the distance oxygen has to diffuse to enter the blood. Breathing maintains a fresh supply of air and the circulatory system takes away oxygenated blood so that constant gas exchange is possible**

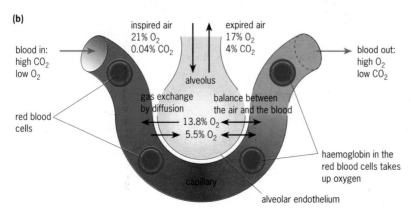

(b)

inspired air
21% O₂
0.04% CO₂

expired air
17% O₂
4% CO₂

blood in:
high CO₂
low O₂

blood out:
high O₂
low CO₂

alveolus

red blood cells

gas exchange by diffusion

balance between the air and the blood

13.8% O₂
5.5% O₂

haemoglobin in the red blood cells takes up oxygen

capillary

alveolar endothelium

B The alveolar membrane is 0.2 µm thick. What fraction of a mm is that?

Fig 8.6 **The air we exhale is saturated with water vapour. When you wake up on a cold day to see condensation on the windows, it is likely that a significant amount of that water came from your lungs**

?

C Why would it be impractical to breathe under water using a twenty-foot snorkel?

As we breathe in, fresh air enters the lungs and passes into individual alveoli. Oxygen diffuses rapidly through the walls of the alveoli and into the blood. Here, most of it combines with haemoglobin in red blood cells (see page 145). At the same time, carbon dioxide diffuses out of the blood and into the alveoli. It is breathed out during the next few expirations.

Table 8.1 **The composition of inhaled, alveolar and exhaled air**

Type of air	Percentage of total volume			Water vapour	Temperature /°C
	Oxygen	Carbon dioxide	Nitrogen + inert gases		
Atmospheric	21	0.04	79	variable	variable
Alveolar	13.8	5.5	80.7	saturated	37
Exhaled	17	4	79.6	saturated	37

Table 8.1 shows the composition of atmospheric, alveolar and exhaled air. Each value for exhaled air is an average of the values for inhaled and alveolar air because exhaled air is a mixture of the two.

The composition of exhaled air varies during the course of a single expiration. The first air to emerge has a very similar composition to atmospheric air because it has been nowhere near the alveoli – it has simply filled the **dead space** in the trachea and bronchi. As the exhalation continues, air that has been deep inside the alveoli is breathed out. The composition of this air has been altered by gas exchange: it contains more carbon dioxide and less oxygen than atmospheric air.

THE LUNGS OF PREMATURE BABIES

ALVEOLI ARE MINUTE bubble-like air sacs lined with moisture, and are liable to collapse because of surface tension. To prevent this, the alveolar epithelium secretes a **surfactant**, a mixture of phospholipids, which greatly reduces the surface tension and keeps the alveoli open.

Without surfactant the lungs cannot function effectively and severe breathing problems can develop. An unborn baby does not start to secrete surfactant until about the 22nd week of pregnancy and the lungs have not accumulated enough surfactant to cope with breathing until about the 34th week.

Any babies born before this have immature lungs and suffer from a condition called **respiratory distress syndrome**. The effort needed to inhale and inflate the collapsed alveoli becomes too great and, without medical help, the baby can die from exhaustion and suffocation. Surfactant can now be made artificially. It is introduced into the lungs of premature babies to help them to begin breathing. This major breakthrough

means that babies as young as 23 weeks (17 weeks premature) now have more of a chance of survival.

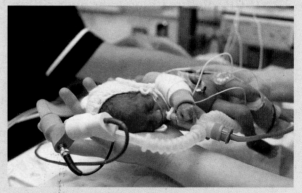

Fig 8.7 **This baby was born at 24 weeks, but survived due to the treatment of her lungs with surfactant. Only a tiny amount (0.5 cm³) of surfactant is needed, enough to line the alveoli with a layer one or two molecules thick**

The effect of smoking on gas exchange

Smoking is the largest cause of preventable death in the western world, and is responsible for about 100 000 deaths per year in the UK. In addition to lung cancer and coronary heart disease (see Chapter 35), smoking causes bronchitis and generally reduces the gas exchange capacity of the lungs.

Research has shown that there are two distinct types of smoke.

CLIMBING MOUNT EVEREST – AN EXPLANATION OF PARTIAL PRESSURES

AS YOU STUDY the workings of the lungs and blood in more detail, you will come across the term **partial pressure**. To explain this, it helps to think about someone climbing Mount Everest (height 8848 metres).

At sea level, there is a lot of air pushing down on us, and this **atmospheric pressure** has a value of about 100 000 pascals (Pa) or 100 kPa. We can therefore say that the **barometric pressure** is 100 kPa. Dry air is 20.9 per cent oxygen at sea level so the partial pressure of oxygen (pO_2) is 20.9 per cent of 100, which is about 20.9 kPa.

As our mountaineer progresses up the mountain, the atmospheric pressure becomes less because there is less air pushing down on him. The partial pressure of oxygen decreases accordingly (Fig 8.8).

Oxygen passes into the lung tissues because of the different concentration in the alveolar air and the blood, and at higher altitudes the difference is smaller, making it difficult to take in enough oxygen to meet demands. At 5300 metres the pO_2 is only half of that at

sea level, and by the time the summit is reached it is only about 8 kPa – just over one third that at sea level. This is why most mountaineers who attempt Everest do so with pressurised oxygen containers, and to conquer the mountain without the help of additional oxygen – as has been done – is a remarkable feat.

Fig 8.8 **The effect of altitude on the partial pressure of oxygen. Aircraft cabins must be pressurised and moun-taineers usually need the help of pressurised oxygen canisters. The human body is able to acclimatise to lower partial pressures of oxygen by increasing the amount of haemoglobin in the blood – see the Chapter 9 Assignment for more details**

■ See question 3.

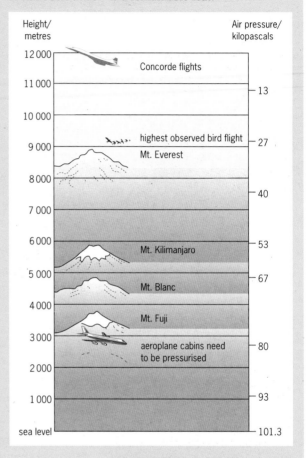

Mainstream smoke passes into the smoker through the filter, **sidestream** smoke comes out of the burning end of the cigarette. In a room of smokers, about 85 per cent of the smoke is sidestream, which means that passive smokers don't even get the benefit of the filter. Cigarette smoke contains three substances that have a direct effect on the lungs: nicotine, carbon monoxide and tar.

Nicotine is not the most harmful chemical in smoke, but it is highly addictive, and has a powerful effect on the nervous system (see page 218). It is thought to increase the stickiness of blood platelets, increasing the risk of blood clots forming inside arteries and veins. If these clots break away they can cause a heart attack or a stroke (see Chapter 35).

Carbon monoxide binds to haemoglobin molecules, preventing then from carrying oxygen. Smokers therefore have a lower oxygen-carrying capacity than non-smokers.

Tar is a complex mixture of chemicals that paralyses the cilia but stimulates the goblet cells (Fig 8.9), resulting in a build-up of mucus that traps dust and pathogens. This results in 'smoker's cough' and increases the smoker's chance of developing chronic bronchitis.

Fig 8.9 **The lungs have an efficient cleaning system: goblet cells (orange) secrete mucus that traps dust, while a carpet of cilia (tiny hairs, green) waft the mucus to the throat where it is swallowed (or to the nose, from where it is blown). Smoking disrupts this cleaning mechanism, and often leads to chronic bronchitis**

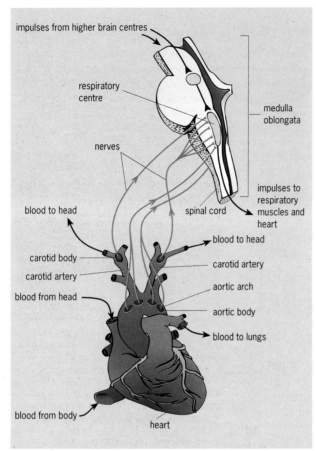

impulses from higher brain centres

respiratory centre

medulla oblongata

nerves

impulses to respiratory muscles and heart

spinal cord

blood to head

carotid body

carotid artery

blood from head

blood to head

carotid artery

aortic arch

aortic body

blood to lungs

blood from body

heart

Fig 8.10 **Regulation of the levels of oxygen and carbon dioxide in the blood. Chemoreceptors (carbon dioxide/pH sensitive cells) occur in both the carotid and aortic bodies, as well as in the respiratory centre itself**

See questions 5, 6 and 7. ■

The terms 'lactic acid' and 'lactate' are often used interchangeably. In water, lactic acid forms lactate ions and hydrogen ions (H^+), so in this chapter we refer to lactate. The same goes for other organic acids, such as pyruvic acid/pyruvate.

Together, the carotid and aortic bodies are known as the **peripheral chemoreceptors**. These cells sense changes in carbon dioxide and pH levels and, to a lesser extent, they are also sensitive to changes in oxygen levels.

In the long term, the epithelia become damaged and replaced by scar tissue, cilia are destroyed, and some of the smaller airways become blocked by mucus. Chronic bronchitis develops with its persistent cough that produces large quantities of phlegm (mucus, bacteria and white cells). Tar also contains carcinogens and heavy smokers have a greatly increased chance of developing lung cancer.

The control of breathing

Control of breathing is **involuntary**: we don't have to think continually about breathing in and out, and our breathing rate is automatically matched to our needs. Breathing rate changes as our bodies detect the physical and chemical variations that occur when we carry out different activities.

Setting a regular pattern

Breathing is controlled by the **respiratory centre**. This is located in the brain in an area called the **medulla oblongata** (Fig 8.10).

Regular nerve impulses travel down **efferent** (outgoing) nerves that pass from the respiratory centre out to both the external intercostal muscles and the diaphragm. These muscles then contract, initiating inhalation. As air enters the lungs, stretch receptors in the airways start firing and feed information to the brain about how the inflation of the lungs is progressing. The more the lungs inflate, the faster the stretch receptors feed back impulses. When the lungs are sufficiently inflated, signals from the respiratory centre stop for a short time and exhalation follows automatically.

Changing breathing rate to meet demand

Ensuring that the body has a constant supply of oxygen is obviously an important aspect of homeostasis (see Chapter 10). But, surprisingly, the body is relatively insensitive to falling oxygen levels. It is much more sensitive to an increase in carbon dioxide and so this is a much more reliable indicator of the need for oxygen. The levels of oxygen in the arterial blood vary very little, even during exercise, but the carbon dioxide levels vary in direct proportion to the level of exertion. The heavier the exercise, the greater the carbon dioxide concentration of the blood. Lactate levels also increase during exercise. Any increase in carbon dioxide or lactate concentration in the blood lowers its pH.

Chemoreceptors (see Fig 8.10) are extremely sensitive to the composition of the blood which flows past them, and can detect very small changes in pH. They occur at three sites:

- **Central receptors** in the medulla oblongata. These are sensitive to the carbon dioxide concentration in the blood that flows through this region of the brain.

- The **carotid bodies** in the wall of the **carotid artery**.

- The **aortic bodies** situated on the **aortic arch**, just above the heart.

When chemoreceptors register a change in carbon dioxide level or pH, they send nerve impulses to the respiratory centre in the brain. This responds by sending more frequent impulses to the external intercostal muscles and diaphragm. When this happens, our ventilation rate increases: you breathe harder and faster. Heart rate also increases (see Chapter 9) and so the body automatically increases oxygen delivery at the same time as removing the extra carbon dioxide.

The control of breathing rate is very similar to the control of heart rate (see page 138) with one important difference: we can control our breathing rate by thinking about it. This suggests that the higher, 'conscious' centres of the brain are more closely linked to the respiratory centre than they are to the cardiovascular centre, the part of the brain that controls heart rate. Also, research shows that pulse and ventilation rate change dramatically during exercise, even *before* the concentration of blood gases has a chance to change. It is as if the body predicts what is about to happen. How this works is not understood and is an active area of research.

The effects of oxygen deprivation

In some situations, at high altitudes for example, oxygen levels can fall without carbon dioxide levels increasing. In a rarefied or artificial atmosphere, normal breathing flushes carbon dioxide out of the lungs but there may not be enough oxygen to replace it. When this happens, the chemoreceptors often fail to register that anything is wrong, and the brain can become starved of oxygen.

The first symptoms of oxygen starvation are feeling ridiculously happy, having impaired senses and lacking judgement. When mountaineers, fighter pilots or deep sea divers start giggling and making stupid mistakes, it is a sure sign that they are not getting enough oxygen. It is also a signal for their colleagues to act fast, if they can, and provide them with emergency oxygen. If they don't get help quickly, they soon lapse into unconsciousness, and brain damage and death follow (Fig 8.11).

Fig 8.11 **In 1875, three French physiologists decided to investigate the effects of low oxygen levels on the human body. The easiest way to get oxygen-poor air was to go up in a hot-air balloon to observe the effects of the rarefied atmosphere on each other. At first there were no obvious effects and they happily continued to throw out ballast, going up to 8000 metres. At this point they all fainted. The balloon eventually came down on its own, and one of the scientists woke up to find the other two dead**

Athletes and VO_2(max)

Many physical activities such as jogging, swimming and team sports, rely on energy released by the aerobic pathway of cell respiration (see Chapter 13). The level of performance an athlete can achieve is largely governed by how fast oxygen gets to the muscles.

The rate at which a person uses oxygen is called the **VO_2** and is measured in terms of the volume of oxygen consumed (cm^3), per kilogram of body weight, per minute.

The **VO_2(max)** is the maximum rate at which oxygen is consumed and is the amount of oxygen that can be delivered to the tissues when the lungs and heart are working as hard as possible. Athletes use a knowledge of VO_2(max) in their training, as a measure of how hard they are working. A training schedule, for example, might require an athlete to work at 65 per cent of their VO_2(max) for a set length of time.

The Example shows how to calculate VO_2 and VO_2(max) for an average person. (You can work out your own, if you are able to measure your tidal volume, the number of breaths you take per minute and your weight in kg.) You can see that the amount of oxygen consumed during exercise increases ten-fold or 1000 per cent from 4.28 to 51.42 cm^3 O_2 kg^{-1} min^{-1}.

EXAMPLES

Q What is the rate of oxygen consumption of a normal 70 kg adult at rest? Assume that the tidal volume of a normal adult at rest is 0.5 litres and that they take 15 breaths per minute.

A We can work out their ventilation rate from the formula:

$$\frac{\text{ventilation}}{\text{rate}} = \frac{\text{number of breaths}}{\text{taken per minute}} \times \text{tidal volume}$$

$$\frac{\text{ventilation}}{\text{rate}} = 15 \times 0.5 \text{ litres min}^{-1} = 7.5 \text{ litres min}^{-1}$$

Atmospheric air is one-fifth oxygen and the figures in Table 8.1 show that only one-fifth of the available oxygen is absorbed (21 per cent only goes down to 17 per cent). So:

amount of oxygen used per minute

$$= 7.5 \times 0.2 \times 0.2 \text{ litres min}^{-1}$$

$$= 0.3 \text{ litres min}^{-1}$$

And the VO_2 for this person is:

$$\frac{0.3}{70} = 0.00428 \text{ litres O}_2 \text{ kg}^{-1} \text{ min}^{-1},$$

or: $\qquad$ $4.28 \text{ cm}^3 \text{ O}_2 \text{ kg}^{-1} \text{ min}^{-1}$

Q What is the $VO_2(\text{max})$ for the same adult? Assume that the volume of air taken in during each breath during strenuous exercise is 3 litres and that the number of breaths per minute increases to 30.

A Knowing that the tidal volume of the same adult during exercise is 3 litres and that they take 30 breaths per minute, we can work out their ventilation rate:

$$\frac{\text{ventilation}}{\text{rate}} = \frac{\text{number of breaths}}{\text{taken per minute}} \times \text{tidal volume}$$

$$\frac{\text{ventilation}}{\text{rate}} = 30 \times 3 \text{ litres min}^{-1} = 90 \text{ litres min}^{-1}$$

Atmospheric air is one-fifth oxygen and the figures in Table 8.1 show that only one-fifth of the available oxygen is absorbed. So:

amount of oxygen used per minute

$$= 90 \times 0.2 \times 0.2 \text{ litres min}^{-1}$$

$$= 3.6 \text{ litres min}^{-1}$$

And the $VO_2(\text{max})$ for this person is:

$$\frac{3.6}{70} = 0.05142 \text{ litres O}_2 \text{ kg}^{-1} \text{ min}^{-1},$$

or: $\qquad$ $51.42 \text{ cm}^3 \text{ O}_2 \text{ kg}^{-1} \text{ min}^{-1}$

SUMMARY

After reading this chapter, you should know and understand the following:

■ We must **exchange gases** to provide oxygen for cell respiration, and to take away the waste carbon dioxide that this process produces.

■ Lungs are **respiratory organs** that increase the surface area for gas exchange. These organs have a large surface area and thin membranes. Ventilation and a good blood supply help to maintain a high diffusion gradient.

■ Our breathing movements serve to ventilate air sacs called **alveoli**, bringing in a continuous supply of fresh oxygen and flushing out waste carbon dioxide.

■ The lungs cannot move on their own, they inflate when pulled outwards by the **external intercostal muscles** and the **diaphragm**. The lungs are attached to the inside of the ribcage by attraction between the two pleural membranes and the pleural fluid between them.

■ At rest, our lungs inflate mainly by movement of the diaphragm. The amount of air moved in and out at rest is called the **tidal volume** and the **ventilation rate** can be worked out by multiplying the tidal volume by the number of breaths taken per minute. The total amount of air which we can inhale, after exhaling as much as possible, is called the **vital capacity**.

■ The rate of breathing is controlled by the **respiratory centre** in the **medulla oblongata** of the brain. This sets a regular breathing pattern that is modified according to information received from **chemoreceptors**. These detect the rise in carbon dioxide in the blood and the fall in pH that indicate the body's need for oxygen.

QUESTIONS

1 The graph of Fig 8.Q1 shows the changes in pressure inside the lungs of a person breathing normally.

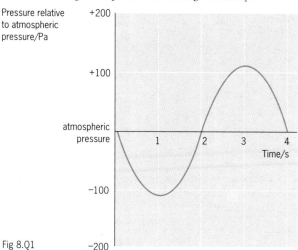

Fig 8.Q1

a) Use the graph to calculate the rate of breathing in breaths per minute.

b) (i) Between what times is this person breathing out?
(ii) What is the evidence from the graph to support your answer to **b)(i)**?

c) Explain how the pressure inside the lung may be decreased.

[NEAB June 1997 Biology: Human Systems Module Test, q.2]

2

a) Fig 8.Q2 shows part of the human airway system.
(i) Name the parts labelled **A**, **B** and **C**.
(ii) Give a distinctive feature of each of the parts labelled **A**, **B** and **C**.

b) The table below shows the effect of different types of breathing on ventilation.

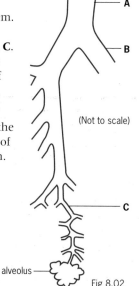

(Not to scale)

alveolus

Fig 8.Q2

(i) Suggest what is meant by the term *dead space volume*.
(ii) **Pulmonary ventilation = Tidal volume × Respiratory rate.**
Using the data above, derive a similar word equation to show how the rate of alveolar ventilation has been calculated.
(iii) Explain why alveolar ventilation decreases with shallow rapid breathing.
(iv) What will happen to a person who continues to ventilate by shallow, rapid breathing?

[OCSEB January 1997 Structured Science Scheme Unit B4 Physiology of Animals and Plants, Section A, q.1]

3 At high altitudes, atmospheric pressure is lower than at sea level. This means that people living at high altitude have less oxygen available to them compared with those living at low altitude.

The table below shows the results of three measurements made on blood of normal individuals resident at each of three different altitudes.

Altitude /m	Percentage saturation of arterial blood with oxygen	Oxygen content of arterial blood/cm^3 per 100 cm^3	Haemoglobin in blood /g per 100 cm^3
150	94.2	18.3	15.5
3700	88.2	19.9	17.4
4375	83.5	20.9	18.6

a) (i) How does the percentage saturation of arterial blood with oxygen change with increasing altitude?
(ii) Using the information in the table, explain how people living at high altitude have adapted to the low oxygen availability.
(iii) State *one* feature, other than those shown in the table, which enables high altitude residents to overcome the problem of low availability of oxygen.

b) State *three* environmental factors, other than availability of oxygen, which vary between sea level and high altitude sites at the same latitude.

c) A visitor to high altitude from sea level may suffer from mountain sickness. State *three* symptoms of mountain sickness.

[Edexcel June 1998 Human Biology Module Test HB2, q.6]

Breathing type (all at rest)	Tidal volume /cm^3 breath^{-1}	Resp. rate /breaths min^{-1}	Dead space volume/cm^3	Pulmonary ventilation /cm^3 min^{-1}	Alveolar ventilation /cm^3 min^{-1}
quiet	500	12	150	6000	4200
deep, slow	1200	5	150	6000	5250
shallow, rapid	150	40	150	6000	0

4

a) Describe how the human respiratory surface is ventilated.

b) Describe the role of the nervous system in:
- **(i)** maintenance of regular breathing whilst the body is at rest;
- **(ii)** increasing the rate of ventilation of lungs during exercise.

[NEAB June 1998 Physiology Module test, q.9]

5

The graph of Fig 8.Q5 shows the effects of changes in the concentration of respiratory gases in the air on the volume of air inhaled in one minute by a person.

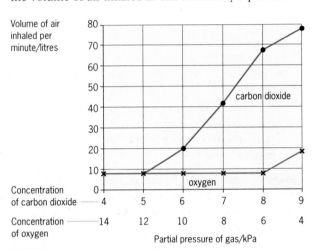

Fig 8.Q5

a) Calculate the percentage increase in the volume of air inhaled per minute when the partial pressure of carbon dioxide rises from 5 kPa to 8 kPa.

b) Describe the effect of changing the oxygen concentration on the volume of air inhaled per minute.

c) The volume of air inhaled per minute is regulated by the respiratory centre. In which part of the brain is the respiratory centre located?

[NEAB March 1998 Physiology Module Test, q.3]

6

a) Both muscle and bones are made up of cells. Explain why muscle cells may use oxygen at a faster rate than bone cells do.

b) Describe what causes air to enter the lungs.

c) Describe the role of the nervous system in:
- **(i)** maintenance of regular breathing whilst the body is at rest;
- **(ii)** increasing the rate of ventilation of lungs during exercise.

[NEAB June 1998 Biology: Human Systems Module Test, q.9]

7 [See Chapter 9 for help with oxygen dissociation.]

a) Fig 8.Q7 shows a typical oxygen dissociation curve for human haemoglobin.

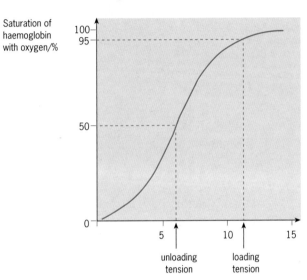

Fig 8.Q7   Oxygen tension/kPa

- **(i)** Where in the body would you find an oxygen tension of 15 kPa?
- **(ii)** What effect would an increase in blood CO_2 concentration have on the unloading tension?
- **(iii)** Explain how an increase in blood CO_2 may affect the supply of oxygen to tissues.

b) Distinguish between: bronchioles and alveoli; dead space air and residual volume; normal breathing and forced breathing.

c) Inspiration of air is under the control of the ventilation centre.
- **(i)** Where is the ventilation centre located?
- **(ii)** What is the most important factor stimulating the ventilation centre to increase inspiration?
- **(iii)** A person voluntarily breathed deeply and rapidly over a period of two minutes. Explain what would happen as a result of this overbreathing.

[OCSEB June 1998 Structured Science Scheme Unit B4 Physiology of Animals and Plants, q.4]

Assignment

ASTHMA

Asthma is a common childhood ailment, affecting at least one in ten children, and a large number of adults. The number of asthma cases has risen dramatically in recent years, fuelling speculation that an increase in air pollution is to blame.

To fully understand why asthma happens, you need to be familiar with the fine structure of the lungs – see page 121.

1
a) What is the difference in structure between the bronchi and terminal bronchioles?
b) What type of muscle lines the bronchi and terminal bronchioles?

What is asthma and what causes it?
Put simply, asthma is a difficulty in breathing caused when the smooth muscles of the bronchioles contract, narrowing the airways that lead to the alveoli. Asthma sufferers then find it hard to breathe and have to make far more effort to deliver a normal amount of air to the lungs.

Many factors can cause asthma, including an allergic reaction (an 'over-reaction' of the immune system to a substance which, in non-sufferers, has no effect). Common allergens are house dust, fur, feathers and pollen.

Fig 8.A1 **The house dust mite, *Dermatophagoides farinae*, is a normal inhabitant of our mattresses. Feeding on the large deposits of skin which accumulate there, the faeces they produce can cause an allergic reaction which leads to asthma in some people**

2 Suggest what measures could be taken to reduce the problems caused by the house dust mite shown in Fig 8.A1.

Treating asthma
Several approaches are used together to help control asthma. A doctor's first priority with all but very young children is to explain what is happening inside the lungs. Sufferers will be less frightened and can to come to terms with their problem. The worst physical effects of the condition are then controlled by a combination of drugs and prevention.

3 What advice would you give to a patient suffering from pollen-induced asthma?

Generally, two types of drugs are prescribed to asthma sufferers: **bronchodilators** and **steroids**. Bronchodilators give instant relief from chest tightness and wheezing because they contain chemicals which relax the bronchiole walls. Many asthma sufferers carry an inhaler around with them. It contains the hormone adrenaline which relieves asthma. This is very useful in the treatment of severe attacks, but is unsuitable for regular use.

4
a) What are the effects of the hormone adrenaline on the body? (You might need to look at page 248.)
b) Suggest what happens when an asthma sufferer takes an adrenaline-based drug.
c) Suggest why adrenaline is not suitable for use in long-term asthma treatment.

Steroids such as **Becotide™** act by reducing the degree of inflammation of the bronchioles. Steroids are preventative: asthma sufferers take regular doses morning and night to reduce the problem of over-reaction. The amount of steroids taken is minimal but they are effective if taken according to instructions.

Fig 8.A2 **Using an inhaler is easy, once you have mastered the right technique. Many people have had the wrong technique for years, with the result that their inhalers are not as effective as they might be**

5
a) Suggest why it is an advantage to inhale steroids rather than inject them.
b) The girl in Fig 8.A2 is using an inhaler. To get the full benefit from each dose – and not just coat the inside of her mouth – she must learn the correct technique. Find out what this is.

6 Describe how you would you investigate the hypothesis that air pollution is linked with the measured increase in asthma cases.

9 The circulatory system

Despite a high cholesterol diet, the Masai people have low rates of coronary heart disease

IN CORONARY HEART DISEASE, now one of the commonest causes of death in the western world, the arteries supplying the heart muscle become narrowed by a fatty deposit. This reduces the blood supply to the heart muscle, leading to angina. If the coronary arteries become completely blocked, the result is a heart attack.

You might expect someone who ate meat, blood and milk almost exclusively to have a high cholesterol level and a high risk of developing coronary heart disease. However, despite eating exactly this diet, the Masai people of Kenya and Tanzania have *low* blood cholesterol. The Masai can eat a huge amount of cholesterol per day – up to 2000 milligrams – and still maintain a healthy blood cholesterol level of around 3.5 millimoles per litre. How do they do it?

Investigations into the Masai diet suggest that the answer could lie in some of the tree barks that the Masai add to their food. These have been found to contain cholesterol-lowering chemicals called saponins. These chemicals have been found in more than 100 different plant species including legumes such as chick peas and soya. It is thought that the saponins work by binding to cholesterol in the gut, preventing its absorption into the bloodstream.

1 WHY DO WE NEED A CIRCULATORY SYSTEM?

Our circulatory system is an example of a **mass flow system**, in which large volumes of fluid are carried to all parts of the organism.

The human body contains specialised organs such as intestines, lungs and kidneys whose function is the exchange of materials such as digested food, oxygen and waste. However, these organs could not function without a **circulatory system**: a network of tubes filled with fluid which can deliver vital materials to all the cells of the body, and then take away their waste.

The circulatory system is so extensive that every living cell is only a few micrometres from a capillary. The blood that passes nearby provides the cell with the materials it needs, takes away its waste and generally bathes it in favourable conditions.

Our circulatory system has three components:

- A fluid, *blood*, which flows in the system, carrying materials around the body.

- A system of tubes or vessels – *arteries, veins, capillaries* – which carry the fluid.

- A pump, *the heart*, which keeps the fluid moving through the vessels.

In this chapter, we look at these three components in detail.

Our double circulation

Fig 9.1(a) shows that humans, like other mammals have a double circulation: a **pulmonary circulation** that carries blood between the heart to the lungs and a **systemic circulation** that carries blood between the heart and the rest of the body.

To understand the advantage of a double circulation, you might find it helpful to consider the situation in fish (Fig 9.1(b)). Fish have a single circulation and a two-chambered heart. Blood first travels to the gills where it passes through thin systemic capillaries, picking up oxygen *but losing pressure*. Blood continues to travel around the body of the fish, back to the heart, but more slowly because it is at low pressure. This situation works for fish but is no use in warm blooded animals such as ourselves; the slow single circulation would not be able to supply all our cells with sufficient amounts of the freshly oxygenated blood they need.

With a double circulation, the right side of the heart receives deoxygenated blood from the body and pumps it to the lungs. Here, the blood gains oxygen and again loses pressure. However, the blood returns to the heart which gives it a boost so that it can reach all the body parts quickly. The volume of blood going around each circulation in any given time is the same, but the systemic circulation is pumped under greater pressure by the heart because the blood has to reach all the extremities of the body.

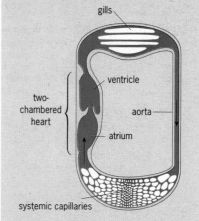

Fig 9.1 **Double and single circulations compared**

■ deoxygenated blood

■ oxygenated blood

(a) **Humans have a double circulation and a four-chambered heart. Deoxygenated blood, coloured blue in the diagram, passes into the right side of the heart and is pumped to the lungs where it picks up oxygen and releases carbon dioxide. Oxygenated blood, coloured red, returns to the heart to be pumped to all parts of the body except the lungs**

(b) **The single circulation of the fish is a simple circuit in which blood loses pressure as it passes through the gills and so moves slowly around the rest of the body.**

2 THE HUMAN CIRCULATORY SYSTEM

The human heart beats over 100 000 times a day, creating the pressure to force blood through more than 80 000 kilometres of arteries, veins and capillaries. Fig 9.2 illustrates the main vessels that make up the human circulatory system.

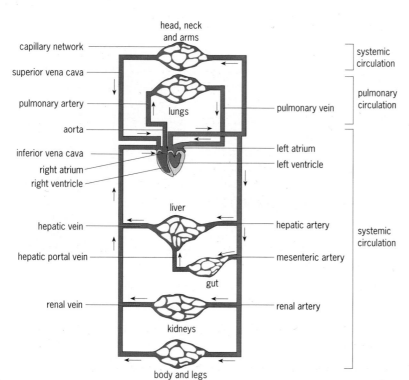

Fig 9.2 **An overview of human circulation. Deoxygenated blood (coloured blue) passes into the right side of the heart through the superior vena cava and the inferior vena cava. The right ventricle pumps it into the pulmonary circulation and it picks up oxygen and releases carbon dioxide as it passes through the lungs.**

Now oxygenated (coloured red), the blood returns to the heart. The left ventricle then pumps it around the systemic circulation and it travels to all parts of the body except the lungs

HEART DISEASE

HEART DISEASE accounts for 26 per cent of all deaths in the western world. The underlying causes of heart disease are structural problems, **ischaemia** (an inadequate coronary blood supply) or faulty electrical conduction.

Structural problems

These are defects in the anatomy of the heart, often present from birth. A fetus has a modified circulation because the placenta is doing the job of the fetal lungs, intestines and kidneys (Fig 9.3). Fetal circulation differs from that of an adult by having a **foramen ovale** and a **ductus arteriosus**.

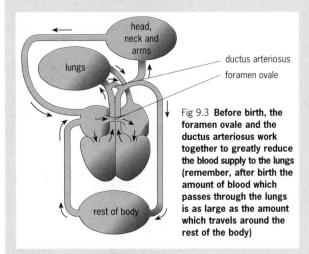

ductus arteriosus
foramen ovale

Fig 9.3 **Before birth, the foramen ovale and the ductus arteriosus work together to greatly reduce the blood supply to the lungs (remember, after birth the amount of blood which passes through the lungs is as large as the amount which travels around the rest of the body)**

The foramen ovale is a hole between the right and left atria which closes immediately after birth. If it fails to close, the baby has a 'hole in the heart'. This allows oxygenated and deoxygenated blood to mix, reducing the efficiency of the circulation. A person can lead a normal, healthy life with a small hole in the heart, but larger ones need surgery.

Faulty valves allow blood to flow backwards and so reduce the efficiency of the heart. The result is a **heart murmur** that is heard with a stethoscope as a 'lub-swish' instead of lub-dup (see page 137). Artificial valves are used to correct very bad heart murmurs, but many valve faults do not need treatment.

Inadequate coronary blood supply

By far the most widespread heart condition is **coronary heart disease (CHD)**. This is also known as **ischaemic heart disease**. The coronary arteries narrow when a fatty deposit called an **atheroma** builds up in them (see Chapter 35). Atheromas can reduce, or even block, the blood supply to heart muscle, causing symptoms ranging from mild chest pain to a full heart attack, or **myocardial infarct**.

Research shows that people who develop heart disease usually have one or more of the following: high blood cholesterol, high blood pressure, a history of cigarette smoking, diabetes mellitus, or a genetic predisposition (they have inherited genes that make them more likely to develop heart disease. They might also be obese (fat) and might not have exercised regularly.

CHD leads to two common conditions, angina and heart attacks. **Angina pectoris** literally means 'chest pain'. Blood flow to the heart muscle is sufficient at rest, but the coronary arteries are so blocked up that they cannot deliver the extra oxygen needed to cope with exercise. Any exertion leads to chest pain, so the lifestyle of the sufferer is significantly restricted.

In a heart attack, the blood flow to a particular part of the heart is blocked and that region of **myocardium** (heart muscle) dies. The body can recover if only a small area of muscle dies, but extensive infarctions are often fatal.

Faults in the conducting system

Contrary to popular belief, the heart doesn't have to stop beating to cause death; it just has to stop pumping effectively. Many deaths are due to **arrhythmias**, disruptions in the normal rhythm of the heart. For example, **ventricular fibrillation** (rapid and uncontrolled contraction of the ventricles) is always fatal if untreated. Ventricular fibrillation is caused by disease or by traumas such as electrocution. It can be corrected with a **defibrillator**, a machine which delivers a very strong current across the heart for a short period of time. This depolarises all the heart cells at the same time. All activity stops for 3 to 5 seconds, allowing time for the heart to re-establish normal electrical activity.

If a heart attack damages the conducting system of the heart, the coordination of heartbeat can be disrupted. This condition is often called a **heart block**. A patient with a permanent heart block can be fitted with an artificial pacemaker (Fig 9.4).

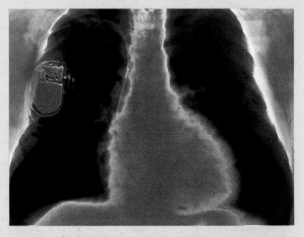

Fig 9.4 **An artificial pacemaker is basically a small box which houses small batteries and complex electronics. The device is fitted under a muscle in the upper thorax, and a wire leads down a vein into the heart. The wire ends in an electrode, seen at the base of the heart. The electrode touches the heart muscle, bringing about contraction. New, sophisticated pacemakers can sense changes in breathing, movement and body temperature, and can adjust heart rate accordingly**

Structure of the heart

The heart is composed mainly of **cardiac muscle**, a specialised tissue that contracts automatically, powerfully and without fatigue, throughout our lives (Fig 9.5).

Fig 9.5(a) **A longitudinal section of the heart**

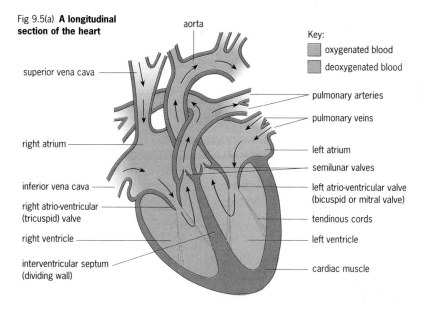

Key:
- oxygenated blood
- deoxygenated blood

aorta
superior vena cava
right atrium
inferior vena cava
right atrio-ventricular (tricuspid) valve
right ventricle
interventricular septum (dividing wall)

pulmonary arteries
pulmonary veins
left atrium
semilunar valves
left atrio-ventricular valve (bicuspid or mitral valve)
tendinous cords
left ventricle
cardiac muscle

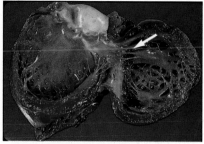

Fig 9.5(c) **The heart is a slightly twisted, asymmetrical organ: the best way to appreciate its three-dimensional structure is to dissect a pig's or sheep's heart or to study a model**

Fig 9.5(b) **The fine structure of cardiac muscle. The rapid spread of impulses through cardiac muscle is possible because individual cells are connected by specialised junctions called intercalated discs. This system of communication ensures that a wave of contraction passes through the cardiac muscle, so that the chambers contract at the right time**

(d)

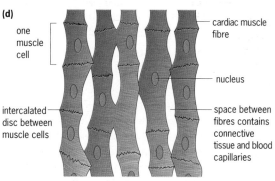

one muscle cell
intercalated disc between muscle cells
cardiac muscle fibre
nucleus
space between fibres contains connective tissue and blood capillaries

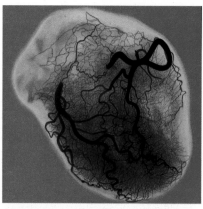

Fig 9.5(d) **The coronary circulation contains all the blood vessels which supply the heart muscle. In the photograph, these blood vessels are easy to see because they have been filled with a dye that is opaque to X-rays**

The thickness of the walls in the different heart chambers reflects their function. The atria are thinly muscled: they pump blood the short distance to the ventricles directly below them. The right ventricle is more heavily muscled than either of the atria: it has to force blood a much further distance to the lungs. The left ventricle has the thickest wall: it has to push blood all around the body.

It is important that blood flows through the heart in one direction only. Two sets of valves close to prevent backflow:

- The **atrio-ventricular valves** (**AV valves**) lie between the atria and the ventricles to prevent blood from returning to the atria when the ventricles contract. The **tricuspid valve** on the right side has three **cusps**, or flaps. The **bicuspid valve**, or **mitral valve**, on the left has two cusps. Both atrio-ventricular valves are subjected to great pressure and ultra-tough tendinous cords, arrowed in Fig 9.5(c), prevent them turning inside-out.

- The **semi-lunar valves** guard the openings to the pulmonary artery and the aorta and prevent backflow of blood into the ventricles. Both sets of valves have three semi-lunar (half moon-shaped) cusps.

You can estimate the weight of your own heart. Generally, it is 0.59 per cent of your total body weight.

Valves are simply strong flaps of tissue: they cannot move on their own. A common mistake is to credit valves with the active control of blood flow. As blood begins to flow back, the valve is forced shut and this prevents any further backflow.

How the heart beats

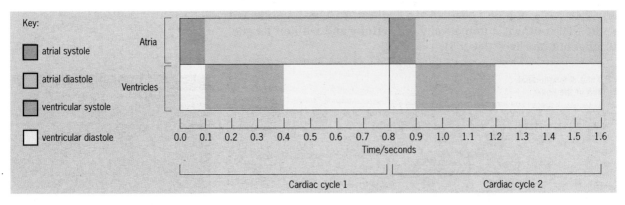

Key:

| □ atrial systole |
| □ atrial diastole |
| □ ventricular systole |
| □ ventricular diastole |

Fig 9.6 **The time scale of the cardiac cycle at rest. During exercise the heart rate can increase to over 150 beats per minute, so the cycle only takes half the time**

The heart continues to beat when removed from the body. Individual heart muscle cells grown in culture beat on their own! Because of this, we say that the heart is **myogenic**: the stimulus which drives it to beat originates in the muscle itself. (Some animals have hearts that are **neurogenic**: they beat only when stimulated by external nerves.)

The sequence of events in a single heartbeat is known as the **cardiac cycle** (Fig 9.6). The cycle involves **systole**, or contraction, and also **diastole**, or relaxation, of the atria and ventricles. The cycle has four overlapping stages:

- **Atrial systole.** Both atria contract, forcing blood into both of the ventricles. This stage lasts 0.1 seconds.
- **Ventricular systole.** Both ventricles contract, forcing blood through the pulmonary artery to the lungs and through the aorta to the rest of the body. This takes 0.3 seconds.
- **Atrial diastole.** The atria relax, although the ventricles are still contracted. Blood enters the atria from the large veins coming from the body. This takes about 0.7 seconds.
- **Ventricular diastole.** The ventricles relax, and become ready to fill with blood from the atria as the next cycle begins. This takes about 0.5 seconds.

Given an average heart rate of 75 beats per minute, each cycle takes 0.8 seconds.

Control of the cardiac cycle

Heart beat must be carefully controlled so that each chamber contracts only when full of blood. To achieve this, the events of the cardiac cycle are carefully coordinated (Fig 9.7).

A single heartbeat starts with an electrical signal from a region of specialised tissue called the **sino-atrial node** (**SAN**), on the wall of the right atrium. This is the 'pacemaker', or heartbeat regulator. The electrical signal spreads out over the walls of the atria, causing them to contract.

From there, the signal does not pass directly to the ventricles. If it did, the ventricles would begin to contract before they had filled with blood. Instead, the impulse is delayed slightly. A second node, the **atrio-ventricular node** (**AVN**) picks up the signal and channels it down the middle of the **ventricular septum** through a collection of specialised cardiac muscle fibres called the **bundle of His**. From here the signal spreads throughout the wall of the ventricles, through the **Purkyne** (or **Purkinje**) **fibres**, and the ventricles contract *after* they have filled with blood.

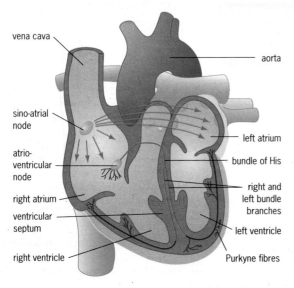

Fig 9.7 **The conduction system of the human heart initiates and controls a heartbeat. Cells in the sino-atrial node act as a pacemaker, initiating impulses which spread through the walls of the atria and, after a delay, through the walls of the ventricles via the atrio-ventricular node**

Electrical events and volume/pressure changes in the cardiac cycle

We have looked at what happens to the blood which passes through the heart during the cardiac cycle, and at the electrical events that coordinate the contraction and relaxation of the atria and ventricles. Fig 9.8 shows how these events correspond to pressure and volume changes in the heart and blood vessels, and also to the sounds that you hear when you listen to someone's heart beating. You will need to refer to this figure as you read on.

During **atrial systole**, the atria fill with blood from the vena cava and the pulmonary vein. Some of the blood that enters the atria flows straight into the ventricles, without any need for contraction. Atrial systole is initiated when the SAN sends out an electrical signal that spreads out over both the atria, causing contraction and forcing the remainder of the blood into the ventricles. There are no heart sounds. Atrial systole is very short, little more than a 'twitch'. Atrial relaxation, or **atrial diastole,** lasts for the remainder of the cycle. All the rest of the 'action' is in the ventricles.

Ventricular systole begins when the ventricles have filled with blood. The AVN picks up the signal from the SAN and then conducts impulses down through the bundle of His and on through the Purkyne fibres in the walls of the ventricles. This stimulates the ventricles to contract. Blood is forced upwards, forcing the semi-lunar valves open and the AV valves shut. The closing of the AV valves causes the 'lub' of the 'lub-dup' heart sound. Pressure in the arteries rises sharply as blood is forced into them.

Fig 9.8 The cardiac cycle. Many exam questions use this diagram, without labels, to test understanding of the events of one heartbeat. It is important to remember what is cause and what is effect, ie:

- **Pacemaker cells initiate systole (contraction).**

- **The squeezing of the muscle walls reduces the volume and so increases the pressure in the chambers, forcing blood in a particular direction.**

- **The direction of blood flow causes the valves to open or close. The valves ensure that blood flow through the heart is in one direction only**

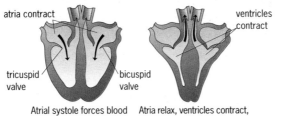

atria contract

tricuspid valve

bicuspid valve

Atrial systole forces blood into ventricles. Tricuspid and bicuspid valves open

ventricles contract

Atria relax, ventricles contract, valve close. Blood goes into aorta and pulmonary artery

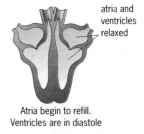

atria and ventricles relaxed

Atria begin to refill. Ventricles are in diastole

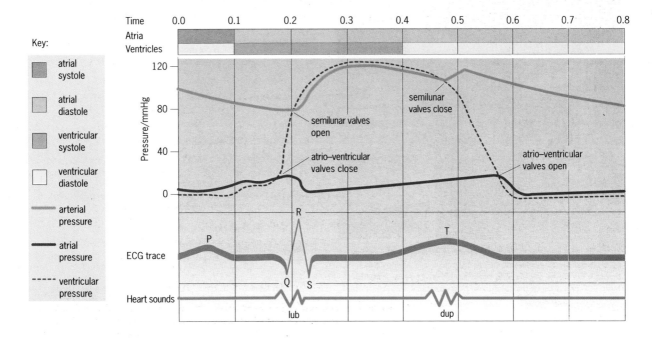

> Cardiac output = stroke volume ×
> heart rate
> (measured in litres: 1000 cm³ = 1 litre)

> **A** Calculate the cardiac output:
>
> **(a)** at rest when the stroke volume is
> 80 cm³ and the heart rate is 75 beats
> per minute;
>
> **(b)** during vigorous exercise when the
> stroke volume is 100 cm³ and the
> heart rate is 150 beats per minute.

See question 1. ■

> The QRS complex is associated with
> the electrical impulse passing through
> the AVN, the bundle of His and the
> Purkyne fibres.

During **ventricular diastole**, the ventricle walls relax, arterial pressure falls and blood begins to flow back into the ventricles. This reversal of the flow causes the semi-lunar valves to shut, causing the second, 'dup', heart sound. Meanwhile, the atria have been filling with blood and, as the ventricles relax, blood flows from the atria into the ventricles, forcing the AV valves open again.

Stroke volume and cardiac output

The volume of blood pumped by the heart during one cardiac cycle is the **stroke volume**. A typical stroke volume in an adult is about 80 cm³: every time the heart beats, 80 cm³ of blood is forced through the pulmonary artery to the lungs and 80 cm³ is forced into the aorta to the body. So, the heart pumps over 8 500 litres of blood per day. Stroke volume increases during exercise, and regular exercise results in a permanent resting increase to 110 cm³, or more.

The volume of blood pumped in one minute is called the **cardiac output**. It is calculated by multiplying the stroke volume by the heart rate and is expressed in litres of blood per minute.

Control of heart rate

Heart rate is modified according to the needs of the body. It increases during physical exercise to deliver extra oxygen to the tissues and to take away excess carbon dioxide. At rest, a normal adult human heart beats about 75 times per minute: during very strenuous

THE ELECTROCARDIOGRAM (ECG)

THE ELECTRICAL events which control the cardiac cycle are recorded by placing electrodes on certain parts of the body – Fig 9.9(a). A normal, healthy heartbeat produces a distinctive trace – upper Fig 9.9(b). Certain heart defects produce a modified trace and this makes the ECG a useful diagnostic tool – lower Fig 9.9(b).

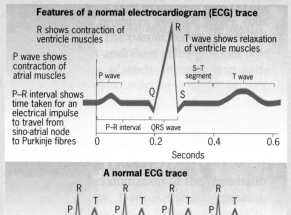

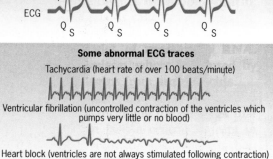

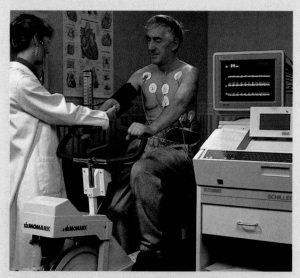

Fig 9.9(a) **The electrocardiogram: electrodes taped to the chest pick up the electrical events of the cardiac cycle as they pass through the body. People with heart problems can now transmit their ECG trace down the telephone to a specialist who can detect problems at a distance**

Fig 9.9(b) **A healthy ECG trace and some abnormal traces, showing common heart defects that can be diagnosed using the ECG**

exercise it might beat 200 times per minute. Heart rate is controlled by the sino-atrial node. The rate goes up or down when the SAN receives information via two autonomic nerves which link the sino-atrial node with the **cardiovascular centre** in the brain (Fig 9.10):

- A **sympathetic** or **accelerator nerve** speeds up the heart. The synapses at the end of this nerve secrete **noradrenaline**.

- A **parasympathetic** or **decelerator nerve**, a branch of the **vagus nerve**, slows down the heart. The synapses at the end of this nerve secrete **acetylcholine**.

A **negative feedback system** controls the level of carbon dioxide and, indirectly, the level of oxygen in the blood. During exercise, the blood level of carbon dioxide starts to rise. This is detected by **chemoreceptors** (cells sensitive to chemical change) situated in three places: the carotid artery, the aorta and the medulla. Nerve impulses travel from these receptors to the cardiovascular centre. In response, the cardiovascular centre sends impulses down the sympathetic nerve to increase the heart rate.

When the carbon dioxide level drops, the cardiovascular centre responds by sending impulses down the parasympathetic nerve, and the heart rate returns to normal. The control of heart rate is closely related to the control of breathing (see Chapter 8).

Several factors affect heart rate:

- Secretion of **adrenaline** in response to stress, excitement or other emotions. Adrenaline is the hormone that prepares the body for action, and one of its effects is to increase heart rate.

- Movement of the limbs, as in exercise. It is thought that stretch receptors in the muscles and tendons relay information to the brain, telling the cardiovascular centre that oxygen levels will soon fall and that carbon dioxide will soon build up. This initiates signals which speed up heart rate and breathing rate.

- The level of respiratory gases in the blood, as described above.

- Blood pressure. When this gets too high, a fail-safe mechanism prevents any further increase in heartbeat.

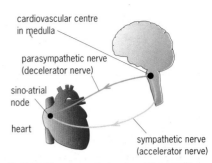

Fig 9.10 **The nerves that connect the cardiovascular centre in the brain to the heart. The accelerator and decelerator nerves are part of the autonomic nervous system**

3 BLOOD VESSELS

There are three types of blood vessels: **arteries**, **veins** and **capillaries**. The structure of each is closely related to its function (see Fig 9.11 and Table 9.1).

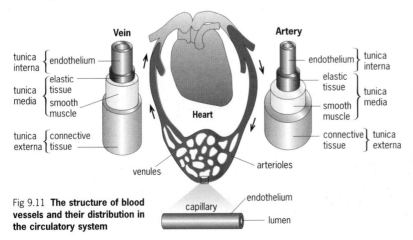

Fig 9.11 **The structure of blood vessels and their distribution in the circulatory system**

Arterial blood is under high pressure and surges occur every time the ventricles contract. The walls of the arteries are able to absorb and smooth out these pulse waves so that by the time blood reaches the arterioles, it is flowing steadily.

Table 9.1 **A comparison of arteries, veins and capillaries**

■ tunica externa of collagen fibres
■ lumen
▨ tunica media of smooth muscle and elastic fibres
— tunica intima of endothelium

Arteries carry blood away from the heart towards other organs of the body. Arteries branch into smaller **arterioles** which branch into tiny capillaries. Capillaries are **permeable** ('leaky') vessels whose walls are one cell thick, allowing exchange of materials between blood and nearby cells. Blood flows from capillaries into **venules**, which drain into larger veins.

Vessel	Cross section	Direction of flow	Pressure	Oxygen content	Size of lumen	Presence of valves	Properties of wall
Arteries		away from heart	high	high*	relatively small	no	tough, powerful elastic recoil
Veins		back to heart†	low	low*	relatively large	yes	thin, distend easily
Capillaries	endothelial cell	through organs	medium	oxygen lost through wall	small, about the diameter of one red blood cell	no	one cell thick, permeable

* Except in pulmonary circulation, where oxygenated blood is carried in the veins and deoxygenated blood is carried in the arteries.
† Except portal veins, eg the hepatic portal vein, which carries blood **between** organs and not to or from the heart.

As Fig 9.11 and Table 9.1 show, the walls of arteries and veins have the same three layers: the **tunica interna**, the **tunica media** and the **tunica externa**. The relative thickness and composition of these layers varies according to the function of the vessel. Arteries, which have to withstand pulses of high pressure, have a thick tunica media containing smooth muscle and elastic fibres. Veins have a thinner tunica media and a larger lumen and carry slower flowing blood at low pressure. Valves prevent blood flowing backwards.

Capillary circulation

The circulatory system keeps all cells bathed in **tissue fluid**. The composition of tissue fluid stays reasonably constant because permeable capillaries allow exchange of materials between tissue fluid and the blood.

> ✔ All living cells are surrounded by tissue fluid. This is also known as **interstitial** or **intercellular fluid**.

How tissue fluid is formed

Fig 9.12 shows what happens in living tissue. Blood from the arterioles is under high **hydrostatic pressure**. When blood enters the capillary, substances begin to leak out through the permeable capillary wall. The capillary walls act as filters and a proportion of all chemicals below a particular size are squeezed out, forming tissue fluid. The composition of tissue fluid is very similar to that of plasma, but tissue fluid lacks most of the large proteins which cannot pass through the capillary wall. Plasma proteins remain in the blood where they play an important part in the drainage of tissue fluid.

> ✔ Hydrostatic pressure is physical pressure. The contraction of the ventricles of the heart creates hydrostatic pressure in the blood vessels.

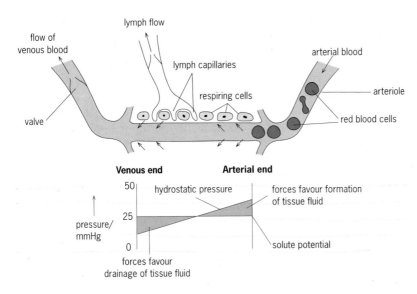

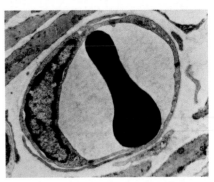

Fig 9.12 **The drainage and formation of tissue fluid. At the arterial end of the capillary, the high hydrostatic pressure forces water and small molecules out of the blood, providing the tissues with a fluid that contains nutrients and oxygen. The hydrostatic pressure decreases as blood flows along the capillary. When it falls, water drains back into the blood by osmosis, taking with it wastes such as urea and carbon dioxide**

Fig 9.13 **Capillary walls are about 1 mm thick. Most capillaries have a lumen which is about the same diameter as the red cells which have to squeeze through them in single file as they unload their oxygen**

Venous end		Arterial end	
Hydrostatic pressure	= 16	Hydrostatic pressure	= 40
Water potential (osmotic force)	= 25	Water potential (osmotic force)	= 25
Net difference	= –9	Net difference	= +15
Tissue fluid drains into blood		Tissue fluid leaves blood	

How tissue fluid returns to the blood

Two main forces act on the blood as is passes along the capillary:

- The hydrostatic pressure of the blood: this tends to force water and solutes *out* of the capillary.

- Osmosis: this tends to draw water *into* the capillary.

As blood flows along a capillary, fluid passes to the tissues, and the volume and hydrostatic pressure of the blood in the capillary decreases (Fig 9.13). As the water potential created by the large plasma proteins is relatively constant, water begins to drain back into the blood when hydrostatic pressure falls below the water potential. Cell waste, such as urea, and substances that have been secreted into the tissue fluid, diffuse into the capillary, contributing to the composition of venous blood.

Blood flow in veins

Blood which drains into the venules is deoxygenated, under low pressure and contains many wastes products and cell secretions. Several features allow blood to return to the heart:

- Valves in veins close and prevent backflow.

- Working muscles surrounding the veins squeeze blood along as they contract. The action of muscles, especially those in the legs, is so important that it has been called the 'secondary heart' or the 'venous pump' (Fig 9.14).

- Gravity helps blood to flow from organs 'above' the heart.

- The negative pressure created in the thorax during **inspiration** (breathing in) draws blood from surrounding veins.

- The action of the heart. After systole (contraction), the elastic walls of the chambers recoil, drawing blood in from the veins.

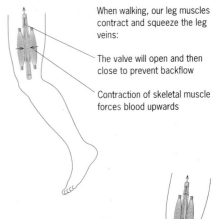

Walking

When walking, our leg muscles contract and squeeze the leg veins:

The valve will open and then close to prevent backflow

Contraction of skeletal muscle forces blood upwards

Passive upright position

When a person is standing upright, the blood pressure at the base of the large veins in the leg may rises

The blood is almost static due to the pressure caused by the height of the column of blood above

Fig 9.14 **The action of the leg muscles squeezes blood along veins, and the valves ensure that flow can only be in one direction. This venous pump is very important to the circulation, and regularly standing still for any length of time can lead to problems. Millions of people suffer from varicose veins or from haemorrhoids (piles), damaged veins whose walls have been stretched by pools of accumulated blood**

Control of blood flow

The body often needs to alter blood flow to different areas, according to circumstances. For example, when we are hot, we lose excess heat because more blood flows to the surface of the skin. Blood flow can be modified by **sphincters** and **shunt vessels**, or by altering the diameter of the vessel itself.

A sphincter is a ring of muscle around a blood vessel which can contract, reducing the lumen size of the vessel and so reducing or preventing blood flow to a particular area. Sphincters can redirect blood into shunt vessels and these by-pass a particular area. For instance, when we are cold, sphincters reduce blood flow to the skin surface, redirecting it along shunt vessels that keep the blood deeper in the body (see Chapter 12).

Many blood vessels, particularly arterioles, contain smooth muscle fibres which can contract to reduce blood flow through the vessel; this is **vasoconstriction**. Blood flow increases when the muscle fibres relax; this is **vasodilation**. Blood pressure can be regulated by altering the degree of vasoconstriction or vasodilation.

4 THE LYMPHATIC SYSTEM

The lymphatic system is part of the immune system (Chapter 34). It is also part of the circulatory system: it returns to the heart the small amount of tissue fluid that cannot be returned by the veins.

We can think of the lymph system as an extra set of veins (Fig 9.15). Flow through lymphatic vessels is slow, but important. Just how important is obvious when you see someone with the disease elephantiasis (Fig 9.16).

Lymphatic flow begins in the capillary beds, where small amounts of tissue fluid drain into tiny lymphatic capillaries (Fig 9.17). The walls of these vessels are more permeable than blood capillaries to lipids and large molecules such as proteins, and so lymph contains a high proportion of these substances. Many cells secrete substances which are too large to enter the blood directly, and so can only pass into the general circulation via the lymphatics. The lymph capillaries drain into larger lymph vessels which look like thin, transparent veins. These vessels have valves to prevent backflow. Lymph contains no red blood cells, and so is pale and clear. All lymph vessels flow towards the upper chest, passing through numerous lymph nodes which filter out bacteria and cell debris.

Peripheral vasoconstriction refers to constriction of the blood vessels in the skin (periphery = outer region).

?

B What effect do the following have on blood pressure?

(a) vasoconstriction, **(b)** vasodilation.

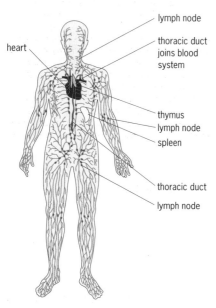

Fig 9.15 **The major lymphatic vessels in the human body. The lymph system drains into the blood system in the upper thorax. A large lymph vessel, the thoracic duct, connects with the subclavian vein (sub-clavicle = under collar bone). Here lymph mixes with blood before entering the vena cava on its way to the heart**

Fig 9.16 **Lymphatic drainage takes about 120 cm³ of fluid out of the tissues every hour. When a lymphatic vessel is blocked, the limb quickly swells until the pressure interferes with normal blood flow. This photograph shows someone suffering from elephantiasis, a condition caused by a parasitic worm, *Wuchereria bancrofti*, which blocks lymphatic vessels**

Fig 9.17 **Far right: A longitudinal section through a lymphatic capillary. The cells that form these capillary walls overlap, forming tiny valves that allow flow of lymph in one direction only (here, upwards). Lymph capillaries in the villi of the intestines are called lacteals, and are important in the drainage of lipids and large molecules**

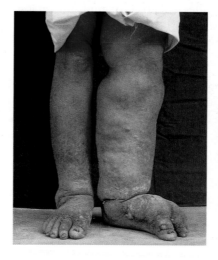

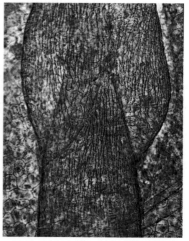

5 BLOOD PRESSURE

The term 'blood pressure' refers to the physical pressure of blood flowing through the arteries. This depends on the force created by the pumping of the heart, the volume of circulating blood and the size of the blood vessels. The body must keep blood pressure inside fairly strict limits. It must be high enough to force blood through all the capillaries so that all cells of the body are well nourished, but not too high because this would make the heart work unnecessarily hard and risk damage to blood vessels. High blood pressure is a risk factor for atherosclerosis (hardening of the arteries; see Chapter 35), and it causes a variety of other health problems. Fig 9.18 shows how we measure blood pressure.

Have you ever stood up very quickly, and then felt faint or dizzy? This happens because the rapid change in posture has altered the body's blood distribution, causing a temporary lack of blood to the brain. Fortunately, body mechanisms quickly bring the blood flow back to normal.

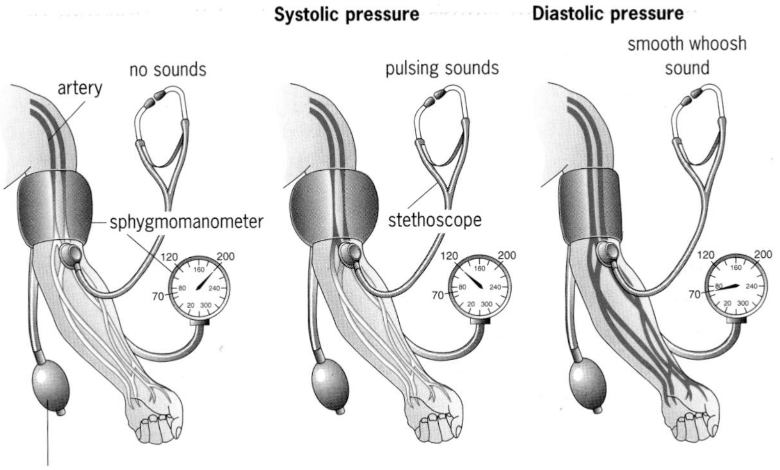

Systolic pressure

no sounds

artery

sphygmomanometer

Diastolic pressure

pulsing sounds

smooth whoosh sound

stethoscope

hand pump

Fig 9.18 **Blood pressure is usually measured using a sphygmomanometer. An inflatable cuff is placed around the upper arm and inflated until all blood flow, in or out of the arm, stops. The blood flow in the brachial artery (at the elbow) is monitored using a stethoscope.**

After inflation, there is no sound, but, as air escapes from the cuff, the pressure decreases until it falls just below that created by the heart as it contracts. At this point, blood is heard spurting through the constriction in the artery. This pressure is the systolic value.

Pressure in the cuff continues to drop until blood can be heard flowing constantly. This is the diastolic value and represents the pressure to which arterial blood falls between beats

■ See questions 2, 3, 4, and 5.

The diastolic pressure is used as an indicator for medical problems: a normal reading is between 60 and 80 mmHg, with anything over 90 regarded as high. Elsewhere in the book we talk about pressure in terms of kPa, but blood pressure is still measured in the old mmHg units.

Control of blood pressure

A negative feedback system keeps blood pressure inside safe limits. The pressure of the blood is detected by the **carotid sinus** (Fig 9.19), a small swelling the carotid artery. When blood pressure gets too high, the walls of the sinus expand and **stretch receptors** in the artery wall inform the cardiovascular centre in the medulla. Signals from here then lower heart rate and vasodilation, so lowering blood pressure.

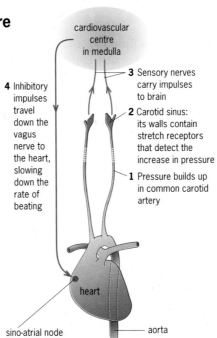

cardiovascular centre in medulla

4 Inhibitory impulses travel down the vagus nerve to the heart, slowing down the rate of beating

3 Sensory nerves carry impulses to brain

2 Carotid sinus: its walls contain stretch receptors that detect the increase in pressure

1 Pressure builds up in common carotid artery

heart

sino-atrial node

aorta

Fig 9.19 **Regulation of the heart rate is influenced by the carotid sinus which detects blood pressure**

Many mechanisms work together to keep blood pressure constant. The size of blood vessels, the heart rate and the volume of circulating blood can all change to increase or decrease blood pressure, according to the needs of the body. The kidneys are particularly important because they control how much fluid we lose. A drop in blood volume and pressure causes more fluid to be retained in the blood and less to be lost in the urine (see Chapter 11).

C 'Blood pressure one twenty over seventy', shouts the doctor. What does this mean?

6 THE BLOOD

Blood is the fluid which flows through blood vessels, and the sight of blood is a sure sign that a vessel has ruptured. Losing large amounts of blood has serious consequences: death can result if the ruptured vessels are not sealed and blood volume is not rapidly returned to normal. In an emergency, when there is no time to find out the patient's blood group, accident victims are given fluids called **plasma expanders**. These are isotonic fluids that contain no cells but they still increase blood volume.

Blood is a complex mixture containing cells, cell fragments and a range of dissolved molecules. It does the following:

● Transports materials. Blood transports digested food and oxygen to respiring tissues. It also takes carbon dioxide and waste products away from respiring cells to the various organs that remove them. It carries hormones from endocrine glands to target organs.

● Distributes heat. The blood helps to keep body temperature stable by distributing heat from metabolically active organs, such as the working muscles, to the rest of the body (see Chapter 12).

● Provides pressure. Many organs of the body depend on the physical pressure of the blood to carry out their function. For example, filtration in the kidney, formation of tissue fluid and erection of the penis all depend on blood pressure.

● Acts as a buffer. The blood contains many proteins and ions which act as buffers, keeping the pH constant by 'mopping up' any excess acid or alkali. Haemoglobin in red blood cells is also an important buffer.

● Defends the body against infection (see Chapter 34).

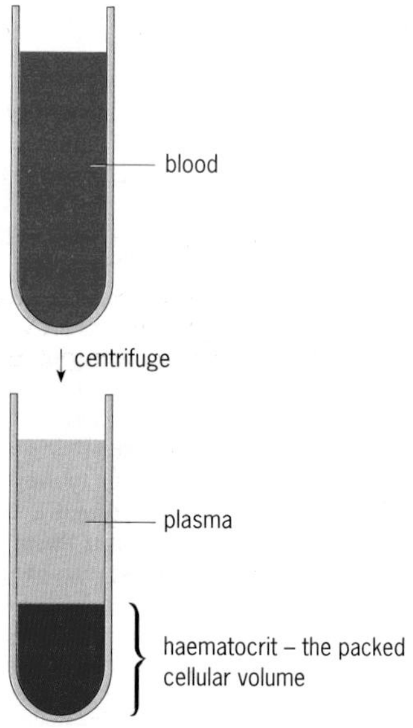

What's in blood?

Fig 9.20 shows that centrifuged blood separates out into two distinct layers: a **cellular portion** called the **haematocrit**, and the **plasma**. The cellular portion contains red cells, white cells and platelets.

Plasma

Plasma is the fluid part of blood. Its main constituents are shown in Table 9.2. The exact composition of the plasma varies greatly. For instance, in the hepatic portal vein, which leads from the intestines to the liver, the plasma is much richer in dissolved foods such as sugars, vitamins and amino acids than the plasma in any other vein.

Fig 9.20 **The percentage of blood volume taken up by cells is known as the haematocrit. A haematocrit of 39 would indicate that 39 per cent of the blood is composed of cells, mostly red cells. The average value for men is about 42 while women average 38. These values are affected by factors such as anaemia or high altitude**

D Suggest how **(a)** high altitude and **(b)** anaemia will affect the haematocrit.

	Components	Function
Proteins:	albumin antibodies fibrinogen	osmotic balance immunity blood clotting
Salts:	sodium, potassium, chloride hydrogencarbonate, calcium	osmotic balance, conduction of nerve impulses carriage of CO_2, buffering, blood clotting
Products of digestion:	glucose, amino acids, fatty acids, glycerol, vitamins	nourishment of cells
Hormones:	protein, eg insulin; lipid (steroids),eg testosterone	communication
Heat	distributed around body to maintain constant body temperature	
Oxygen	vital in aerobic cell respiration	
Waste products:	urea, CO_2	none: they must be removed by excretion (see Chapters 10 and 11)

Table 9.2 **The main constituents of plasma**

Red blood cells

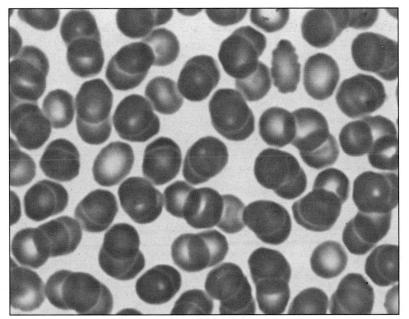

Fig 9.21(a) **If you look at a drop of blood under the microscope, the most obvious feature is the mass of red blood cells**

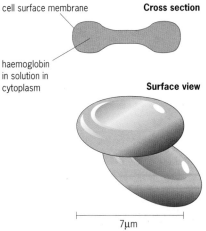

cell surface membrane   **Cross section**

haemoglobin in solution in cytoplasm

Surface view

7μm

Fig 9.21(b) **Red blood cells are biconcave discs filled with haemoglobin and other chemicals which help the loading and unloading of oxygen**

In this chapter we study red blood cells only (Fig 9.21): white cells and platelets are covered in Chapter 34.

Also known as **erythrocytes**, red blood cells are by far the most numerous in the blood. A single cubic millimetre of blood contains around 5 million, sometimes more. Red blood cells have one function: they carry the respiratory gases. The **haemoglobin** (Hb) they contain picks up oxygen in the lungs and swaps it for carbon dioxide in the respiring tissues.

Red cells have no nucleus: they are packed with Hb and the enzymes and chemicals that allows the Hb to carry oxygen effectively. Having Hb inside red blood cells rather than in solution in the plasma gives several advantages:

- A much greater volume of Hb can be carried in cells than could be dissolved in plasma.

- Hb can be kept in a favourable chemical environment to allow faster loading and unloading of respiratory gases.

- Hb molecules of a particular age are kept together and can be easily replaced when old.

- Hb in cells does not affect the osmotic properties of the blood (free Hb would).

- Hb in cells cannot be lost by excretion.

Red blood cells have a regular shape, described as a biconcave disc. This shape is maintained by the **cytoskeleton,** a complex but flexible internal scaffolding made from protein fibres. When you think about what red blood cells do it is easy to see why they have this shape. They must have a large enough volume to carry useful amounts of oxygen but they also need a large enough surface area to load and unload it quickly. The biconcave shape is a good compromise between the maximum volume of a sphere and the maximum surface area of a flat disc.

Red blood cells are distorted when they have to squeeze through capillaries, which often have lumens slightly smaller than the diameter of the red cell. Red cells rarely cause blockages because they have a flexible shape and a smooth membrane.

Red blood cells are made in the bone marrow of the vertebrae, ribs and pelvis by specialised **stem cells**. The turnover of red cells is very rapid: we make about 20 million red cells per second! Red blood cells circulate for about 100 to 120 days before they are destroyed in the liver by the phagocytes called **Kupffer cells** (Fig 10.5(e)) and also in the spleen.

Blood as a transport medium

Oxygen is carried in the blood in two ways: 98 per cent travels as **oxyhaemoglobin**, the oxygen–haemoglobin complex. The remaining two per cent is dissolved in the plasma. Carbon dioxide is carried in three ways: as HCO^{3-} ions in the plasma (70 per cent), combined with haemoglobin as a **carbamino** compound (23 per cent) and in simple solution in the plasma (7 per cent).

The structure of haemoglobin

The Hb molecule is a large conjugated protein with a molecular mass of about 64 500 kilodaltons. Each Hb molecule consists of four **globin** sub-units consisting of a polypeptide chain and a prosthetic group called **haem**. At the centre of each haem is an iron ion (Fe^{2+}) which combines with oxygen. There are four haem groups, so the overall equation for the reaction is:

$$Hb + 4O_2 \rightarrow HbO_8$$
haemoglobin + oxygen → oxyhaemoglobin

The polypeptide chains hold the haem groups in place (see Fig 3.28) and help to load oxygen. When the first haem group combines with an oxygen molecule, the Hb molecule alters its shape so that the next haem group is exposed, making the loading of the second oxygen easier.

Haemoglobin in action

There are many substances which react readily with oxygen, but Hb is one of the few which can combine with oxygen where it is abundant and then release it when the concentration falls. This property is illustrated in a graph called an **oxygen dissociation curve** (Fig 9.22). This is plotted by analysing the percentage of Hb saturated with oxygen at different concentrations of oxygen. The graph shows that Hb becomes fully saturated with oxygen at the concentrations found in the lungs, and gives up a relatively large proportion of its oxygen in the lower oxygen concentrations that occur in the tissues.

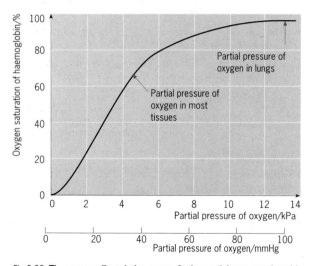

Fig 9.22 **The oxygen dissociation curve. At the partial pressure found in the lungs, haemoglobin becomes 97–99 per cent saturated with oxygen. Surprisingly, haemoglobin releases only about 23 per cent of its oxygen in the respiring tissues, so the blood returning in veins is still about 75 per cent saturated. This suggests that three out of the four haem groups are still bound to oxygen. This allows great flexibility: if a tissue such as a working muscle becomes particularly oxygen starved, the blood can release large amounts of extra oxygen**

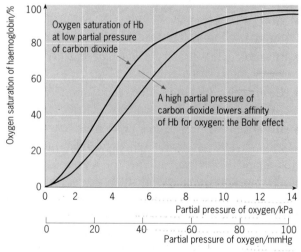

Fig 9.23 **If the oxygen dissociation curve is plotted at higher carbon dioxide concentrations, it moves to the right, showing that haemoglobin has a reduced affinity for oxygen: this is called the Bohr effect**

The dissociation curve in Fig 9.22 was plotted using mixtures of gases in which the concentration of carbon dioxide was constant. Fig 9.23 shows that, at higher levels of carbon dioxide, the curve moves to the right.

This is a vital point: it shows that carbon dioxide lowers the **affinity** of Hb for oxygen. This means that when carbon dioxide concentration is higher, haemoglobin does not hold on to its oxygen quite as well: Hb therefore tends to give up oxygen in areas of high carbon dioxide – such as in the respiring tissues which need it most. This lowering of affinity by carbon dioxide is called the **Bohr effect**, or the **Bohr shift**.

Fetal haemoglobin

Before birth, a fetus must obtain oxygen from its mother via the placenta. If fetal Hb had the same affinity for oxygen as the mother's Hb, no transfer of oxygen would be possible. But, as Fig 9.24 shows, fetal Hb has a higher affinity for oxygen than adult Hb It can therefore pick up oxygen in the same conditions which cause the maternal blood to release it. After birth, the baby's body makes adult Hb which gradually replaces the fetal version.

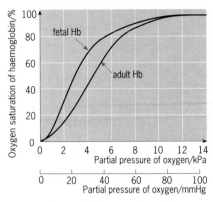

A human baby is known as an embryo up to about eight weeks of development. After this, it is known as a fetus for the rest of the pregnancy.

Fig 9.24 **The dissociation curve for fetal haemoglobin is to the left of the adult version. So, at the oxygen concentrations found at the placenta there is an efficient transfer of oxygen from mother to fetus**

Unloading oxygen at the tissues

The series of events that result in haemoglobin unloading its oxygen to supply respiring tissues is shown in Fig 9.25.

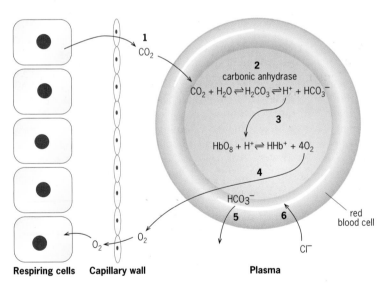

Fig 9.25 **Unloading oxygen at the respiring tissues**

Step 1 **Carbon dioxide diffuses from the respiring tissue, through the capillary wall and the plasma, into the red blood cell**

Step 2 **Inside the red cell the enzyme carbonic anhydrase catalyses the conversion of carbon dioxide into carbonic acid (H_2CO_3)**

Step 3 **The hydrogen ions released by the carbonic acid makes the haemoglobin molecules less stable, causing them to release the four oxygen molecules**

Step 4 **The oxygen is free to diffuse into the respiring tissues**

Step 5 **The accumulating HCO_3^- ions diffuse out into the plasma. leaving the inside of the red cell with a net positive charge**

Step 6 **To maintain a neutral charge, Cl^- ions (the commonest negative ions in plasma) diffuse into the cell; this is called the chloride shift**

Myoglobin and muscles

Mammalian muscle contains the respiratory pigment, myoglobin (myo = muscle). The myoglobin molecule can be thought of as one quarter of a haemoglobin molecule: it consists of one polypeptide chain and one haem, and combines with one oxygen molecule. Myoglobin has a higher affinity for oxygen than haemoglobin and so it can pick up oxygen from haemoglobin and store it until the muscle becomes short of oxygen. When this happens (during exercise, for example), the oxygen tension in the muscle drops. Haemoglobin has no more oxygen to deliver and myoglobin then releases its oxygen.

■ See question 6.

SUMMARY

When you have read this chapter, you should know and understand the following:

■ Humans have a double circulation: the **pulmonary circulation** to the lungs and the **systemic circulation** to the rest of the body.

■ The human **heart** is a four-chambered muscular pump made of **cardiac muscle**. This can contract powerfully and without fatigue.

■ A single heartbeat, the **cardiac cycle**, consists of **atrial systole** (contraction) followed by **ventricular systole**, then **atrial** and **ventricular diastole** (relaxation).

■ The heart is **myogenic**: the electrical impulses which control the cardiac cycle arise from the muscle itself.

■ From the heart, blood passes into vessels in the following order: **arteries**, then **arterioles**, then **capillaries**, then **venules**, then **veins**.

■ The **lymphatic system** acts as an extra set of veins, helping to drain fluid from the tissues back into the bloodstream.

■ Blood consists of **red cells**, **white cells**, **platelets** and **plasma**.

■ Red blood cells, or **erythrocytes**, contain **haemoglobin**, a conjugated protein that can pick up oxygen where it is abundant (lungs) and release it where it is needed (respiring tissues).

■ Blood carries carbon dioxide in three ways; as hydrogencarbonate (HCO_3^-), combined with haemoglobin, and as simple solution in the plasma.

QUESTIONS

1 A mammalian heart continues to beat after it is deprived of nerve connections. The sino-atrial node (SAN) in a mammalian heart acts as the pacemaker. Cardiac muscle fibres serve as the conducting channels for a wave of excitation which spreads from the SAN to both atria. The wave eventually arrives at the base of the right atrium where there is a second node, the atrio-ventricular node (AVN). The wave reaches the AVN 0.045 seconds after leaving the SAN. A bundle of large muscle fibres (purkyne or purkinje fibres) arises from the AVN. This bundle of fibres (the bundle of His) forks into right and left branches which pass into the walls of the respective ventricles where fibres can be traced to all parts. A time delay of 0.12 seconds occurs at the AVN before the wave of excitation passes into the purkyne fibres. The right and left branches conduct the wave of excitation either side of the septum to the base of the ventricles, before spreading up the lateral walls to reach the top of the ventricles. The wave reaches the base of the ventricles 0.04 seconds after leaving the AVN and the top of the ventricles 0.08 seconds after leaving the AVN.

a) Copy and complete the table below to summarise the timing of the passage of a single wave of excitation from origination to its arrival at the top of the ventricles.

Position	Time from origination of wave/s
SAN 0.00	
AVN	
beginning of bundle of His	
base of ventricles	
top of ventricles	

b) State **one** piece of evidence that suggests that the origin of heart rhythm is an inherent property of the heart muscle itself.

c) Suggest, giving a reason, what the effect would be on heart rate of specifically cooling the SAN.

d) Suggest why it is important that there is a 0.12 second delay before the wave passes from the AVN into the purkyne fibres.

Experimental clamping of purkyne fibres squashes them and prevents them from conducting excitation waves.

e) Predict the effect of clamping the right branch of the purkyne fibres on ventricular systole.

f) Ventricular contraction starts at the base of the ventricles and not at the top. Suggest how this benefits the flow of blood through the heart.

[UCLES March 1995 Modular Sciences: Biology, Paper 2, q.3]

2 A student cycled strenuously on an exercise bicycle for ten minutes. During this time, blood pressure measurements were taken. These measurements continued for a further five minutes after the exercise had finished. After the ten minutes' cycling, the student was near exhaustion and felt very faint. The changes in the student's blood pressure are shown in Fig 9.Q2.

a) Distinguish between *systolic* and *diastolic* blood pressure.

b) With reference to Fig 9.Q2, describe the changes in systolic blood pressure between: **(i)** 2 and 12 minutes; **(ii)** 12 and 14 minutes.

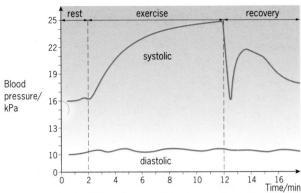

© David R Lamb, *Physiology of Exercise, Responses and Adaptations*, adapted by permission of MacMillan.

Fig 9.Q2

c) Explain the change in systolic blood pressure during exercise.

d) With reference to blood pressure, suggest why the student felt faint immediately after exercise.
Suggest how the student might avoid feeling faint at the end of strenuous exercise.

[UCLES Spring 1995 Modular Sciences: Biology, Paper 2, q.4]

3 Fig 9.Q3 shows cross-sections of three different types of blood vessel. They are not drawn to the same scale.

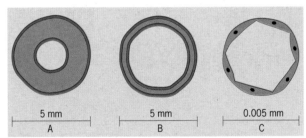

Fig 9.Q3

a) Identify blood vessels **A**, **B** and **C**.

b) State *two* ways in which vessel **A** is adapted for its functions.

[ULEAC 1996 Biology Specimen paper: Module Test B3, q. 1]

4 Fig 9.Q4 shows the pathways for the conduction of electrical impulses during the cardiac cycle.

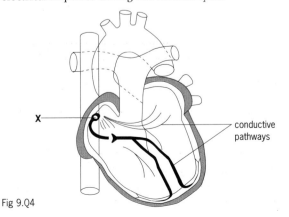

Fig 9.Q4

a) **(i)** Name the structure labelled **X**.
 (ii) In the wall of which chamber of the heart is structure **X** located?
 (iii) Describe the role of structure **X** in the control of the cardiac cycle.

b) The table shows the pressures in the left atrium, left ventricle and aorta during a single cardiac cycle.

| Stage | Pressure/kPa | | |
	Left atrium	Left ventricle	Aorta
1	0.5	0.4	10.6
2	1.2	0.7	10.6
3	0.3	6.7	10.6
4	0.4	17.3	16.0
5	0.8	8.0	12.0

Give the number of **one** stage when
(i) blood flows into the aorta,
(ii) the valve between the atrium and the ventricle (bicuspid valve) is open,
(iii) the bicuspid and aortic valves are closed.

[NEAB March 1998 Biology: Physiology Module Test, Section A, q.5]

5 The graphs of Fig 9.Q5 show how the pressure of blood in two arteries in a healthy person varies with time. The brachial artery supplies blood to the muscles of the arm.

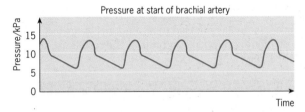

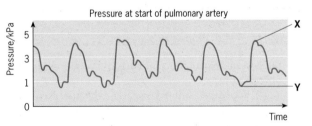

Fig 9.Q5

a) Which chamber of the heart pumps blood into the pulmonary artery?

b) Describe what is happening in the heart **(i)** at the time labelled **X** on the graph; **(ii)** at the time labelled **Y** on the graph.

c) **(i)** Explain what causes the difference in the maximum pressure recorded in the two arteries.
 (ii) Suggest **one** advantage of this difference in maximum pressure.

[NEAB June 1998 Biology: Physiology Module Test, q.3]

6 Haemoglobin is an oxygen carrying pigment found in red blood cells. A molecule of haemoglobin consists of an iron-containing haem group and a protein called globin. Fig 9.Q6(a) shows how haemoglobin is produced in the human body.

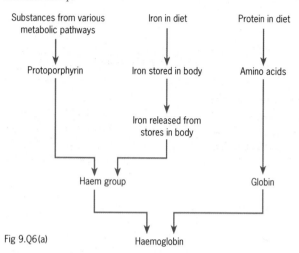

Fig 9.Q6(a)

a) Describe the part played in protein digestion by:
(i) exopeptidases; **(ii)** the epithelial cells of the small intestine.

b) Suggest why the level of protoporphyrin in the red blood cells of people suffering from anaemia may be higher than normal.

In a study of three groups of women, the mean blood loss during menstruation was measured. Each of the women in group A was fitted with a standard intra-uterine device (IUD). Those in group B where fitted with an IUD which continually released small amounts of progesterone. Those in group C acted as a control. The results are shown in the table.

Group	Treatment	Mean blood loss during menstruation/cm³
A	Standard IUD	91
B	IUD which releases progesterone	19
C	Control group	41

c) Suggest an explanation for each of the following observations:
(i) anaemia is often associated with the use of intra-uterine devices;
(ii) the mean blood loss during menstruation of the women in group **B** was lower than that of the women in the control group.

d) There are 12.5 g of haemoglobin in 100 cm³ of blood. The iron content of haemoglobin is approximately 0.3 per cent. Calculate the mean amount of iron lost by the women in group C. Show your working.

e) The graph of Fig 9.Q7(b) shows the changes which took place in the haemoglobin concentration of the blood and the total blood volume of a woman during a normal pregnancy.

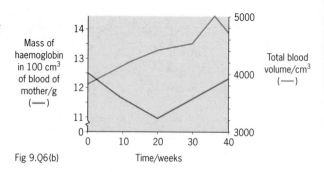

Fig 9.Q6(b)

(i) Suggest an explanation for the change in haemoglobin concentration which took place between weeks 0 and 20.
(ii) Explain the advantage of the change in haemoglobin concentration which took place between weeks 20 and 40.

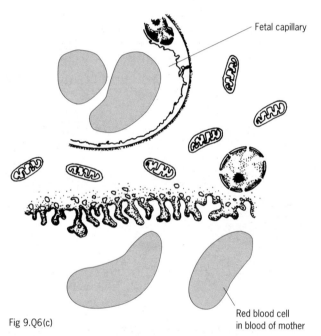

Fig 9.Q6(c)

f) The drawing of Fig 9.Q6(c) shows an electron micrograph of part of a human placenta.

The placenta absorbs iron from the blood of the mother mainly by active transport. Explain how **two** features, visible in the drawing, are adaptations for this function.

[AEB June 1998 Human Biology Paper 2, q.1]

Assignment

ALTITUDE TRAINING

This Assignment links aspects of breathing (Chapter 8) and aspects of the circulatory system that we have covered in this chapter. So to answer the questions, you will need to refer to both chapters to find the relevant text.

Working out

Which sport is the most demanding? Is it football? Or marathon running? It might surprise you, but its neither. These two sports are just a stroll compared with the most gruelling events that you can do. Top of the list are cross-country skiing and long-distance bike racing (such as the Tour de France).

But how do we compare the demands of different sports and events? One way is to measure oxygen consumption. In Chapter 8 we saw that a person's oxygen consumption can be expressed in terms of VO_2 and $VO_2(max)$.

Fig 9.A1 **Cross-country skiers and long distance cyclists top the table for maximum oxygen consumption**

a) Suggest what does the term VO_2 means.
b) What is the difference between VO_2 and $VO_2(max)$?
c) What are the average values for VO_2 and $VO_2(max)$, as shown in Chapter 8? (Remember to include the units.)

Studies show that competitors in cross-country skiing and long-distance cycling can have oxygen consumptions as high as 90 cm³ per minute per kilogram.

In events covering many kilometres over many hours, just a few seconds' advantage can mean the difference between success and failure. Not surprisingly, there is a search for ever more effective training methods which can increase athletes' oxidative capacity – their ability to get oxygen to their muscles. For several decades, one of the favoured methods of increasing the oxygen-carrying capacity of the blood has been to train at altitude.

Altitude training

In Chapter 8, we saw that mountaineers who ascend without oxygen apparatus experience problems as the air gets thinner and the partial pressure of oxygen decreases.

a) Explain what is meant by partial pressure of oxygen.
b) Explain what happens to the partial pressure of oxygen as you go up a mountain.

The human body is able (get acclimatised) to adapt to low oxygen levels if given time – usually several weeks. The count of red cells in the blood increases from about 5 000 000 per cubic millimetre to around 7 000 000.

a) What percentage increase in this?
b) What is the advantage of this change?

Acclimatisation

The mechanism of this acclimatisation not completely understood, but this is what we think happens. The kidney detects the lowered oxygen levels in the blood and responds by releasing a hormone, **erythropoetin**, which targets cells in the bone marrow. In response, the bone marrow increases red cell production.

There are several ways of measuring the oxygen-carrying capacity of the blood. In addition to the number of red cells per cubic millimetre, there is the haematocrit and the amount of haemoglobin per litre. The haematocrit is the percentage of the blood taken up by cells (average values 41–42 per cent for males, about 38 per cent for females). The average amount of haemoglobin per litre is about 16 g for males and 14 g for females.

4

a) Write a flow diagram to summarise the mechanism of acclimatisation.
b) A person has trained at altitude and has returned to sea level. What effect will this have on their resting pulse rate? Explain your answer.
c) What do you think happens to the red blood cell count and resting pulse rate in the long term, if the athlete remains at sea level?

For a long time, athletes have realised the great benefit of training at altitude: if you go up into the mountains to train, the increased oxygen-carrying capacity of your blood gives you a distinct advantage over those competitors who have remained at sea level.

However, a problem comes from the fact that exercise done at altitude it not as beneficial at that done at sea level – you can't work the muscles as hard. So the problem is to 'live high and train low'.

There are two ways to achieve this. One is to have 'the Alps in a caravan' and live at low altitude in a mobile home with a controlled atmosphere, so that the partial pressure of oxygen is lowered to simulate high-altitude conditions.

The other solution is the highly illegal, unfair and dangerous practice of taking genetically engineered erythropoietin (RhEPO – recombinant human erythropoietin), which will mimic the effects of altitude training without the time and expense.

5 Outline the steps involved in making genetically engineered human erythropoietin. (You will need to look at Chapter 31 to answer this).

In the last decade, the deaths of several top cyclists have been linked to RhEPO abuse and the resulting ultra-high haematocrit. The main danger comes from the increased risk of an embolism, a blood clot which circulates before becoming lodged in a vital blood vessel, such as in the heart, lungs or brain.

6 Suggest how an increased haematocrit can lead to an embolism.

7 Read the following extract and then answer the questions which follow.

Human proteins are boosting performance: the evidence

Genetic engineers have found a way to mass-produce erythropoietin (EPO), which is a hormone and is a front-line drug for treating anaemia. Use of EPO boosts red blood cell production and consequently blood haemoglobin levels.

The increase in haemoglobin is a powerful lure for athletes in endurance events. Evidence for the use of EPO has come from comparing two cross-country skiing events at Lahti and Thunder Bay. EPO is detectable in the body for only six to eight hours but its effect lasts as long as a red blood cell – around 120 days.

(Reproduced with permission from an article in *The Daily Telegraph*)

a) Explain how the graph of Fig 9.A2(a) provides evidence that EPO has been used by some cross-country skiers.

b) Explain how increased haemoglobin concentration might lead to increased performance in endurance events.

The graph of Fig 9.A2(b) shows the oxygen dissociation curves for haemoglobin as blood passes through capillaries in the lungs and in the skeletal muscles of an athlete.

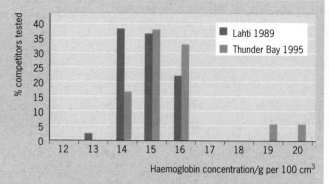

Fig 9.A2(a) **Comparing the haemoglobin concentration in the blood of cross-country skiers from two eevents, one in 1989 and one six years later, in 1995**

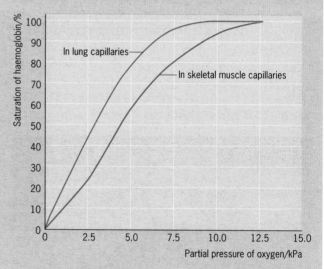

Fig 9.A2(b) **The oxygen dissociation curves for haemoglobin in blood passing through capillaries in the lungs and in the skeletal muscles of an athlete**

c) Explain how features of the oxygen dissociation curves for haemoglobin in the lungs and in the skeletal muscles benefit an athlete.

[Question 7 is an original exam question: NEAB March 1998 Biology: Physiology Module Test, Section B, q.8]

Catching the cheats

So how can you tell when an athlete has cheated and used RhEPO? It is a natural hormone and so you would expect to find it in the blood, especially in those who have trained at altitude. The authorities have decided that for the present the only practical way to detect abuse is to measure the *effect* of the RhEPO, that is, the amount of haemoglobin in the blood. The assumption is that unnaturally high levels are indicators of abuse.

Recently the cycling authorities set the level at a haematocrit of 50 per cent, and the International Skiing Federation set haemoglobin levels at 18.5 grams per litre for males and 16.5 for women. Predictably, athletes have complained that it is perfectly possible to naturally have levels this high, and the debate looks set to continue.

HOMEOSTASIS

ONE OF THE most remarkable characteristics of the human body is its ability to control its internal environment, no matter what the conditions are like outside. You could be sitting in a sauna or swimming in a glacial lake, dieting or eating everything in sight, resting or running a marathon. Whatever the conditions, the organs and systems of your body usually make the adjustments necessary to ensure that all individual cells inside your body are kept under optimum conditions.

The downside of this teamwork is that if a particular organ becomes diseased and no longer functions properly, homeostasis of the whole body can fail, and cells can start to die. In order to remain healthy, the gut must keep on absorbing food, the lungs must constantly exchange gas, the heart must pump and the kidneys must remove waste whilst balancing the volume and concentration of body fluids. There is no single 'most important' organ in your body – each has a role to play.

In this section we focus on the basic concepts and mechanisms of homeostasis, and then look at some examples including blood sugar, temperature and the workings of the liver and kidney. Finally, we investigate exercise physiology – the short term and long term changes that happen to our bodies when we exercise. The fact that our muscles can work twenty times as hard and yet our body can still maintain acceptable internal conditions is one of the most impressive feats of homeostasis.

10 Homeostasis

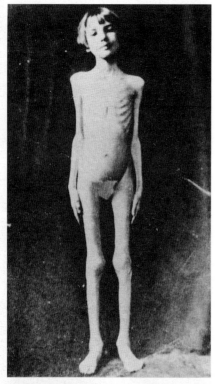

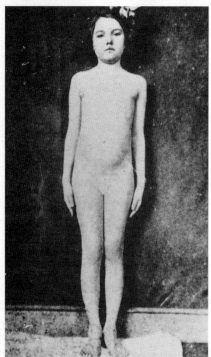

For Catherine Hayes, the discovery of insulin came just in time. Daily injections of insulin allowed the cells of her body to absorb glucose effectively, so nourishing her tissues and lowering her blood sugar level

MILLIONS OF PEOPLE world-wide are unable to control their blood glucose level. They have a condition known as diabetes mellitus. Early symptoms include a raging thirst, extreme fatigue and excessive urination. If sufferers are not treated, they lose weight and eventually die.

Until 1922, doctors regarded the symptoms of diabetes mellitus as a death sentence, especially when they occurred in children. Physicians confirmed a diagnosis by tasting the patient's urine to find out if it contained sugar (today, there are better methods). One noted, 'In children the disease is rapidly progressive, and may prove fatal within a few days... As a general rule, the older the patient at the onset, the slower the course.'

In 1923 a 16-year-old Canadian girl, Catherine Hayes, wrote, 'Only a year ago I was a human skeleton. Can you imagine being 5 feet 4 inches and weighing only 55 pounds [less than 4 stone]? My doctor had placed me on a starvation diet, the only available treatment at the time. I became too weak to engage in any physical activity, even walking, and I eventually lost most of my muscle tissue. My skin became so dry that it flaked and peeled. It was not only painful, but embarrassing, and I wondered how my school friends could bear to look at me. Many of them could not, and in the last year, I spent much of my time alone.

'However, what was even harder to bear was the knowledge that my life could end at any time. It was very difficult to live with the thought that if I took a turn for the worse I could lapse into a coma and die.

'Then came the news about the discovery of insulin...'

1 THE CONCEPT OF HOMEOSTASIS

The word **homeostasis** means 'steady state'. Homeostasis describes how the body regulates its processes to keep its internal conditions as stable as possible. Homeostasis is necessary because cells, especially those of humans and other higher animals, are efficient but very demanding. To function properly they need to be bathed in tissue fluid that can provide the optimum conditions. Nutrients and oxygen must be delivered and waste needs to be removed. The concentration, temperature and pH of the fluid between cells must also be kept at levels that guarantee efficient cell functioning.

However, the phrase 'steady state' is a bit misleading. The conditions inside our bodies are not constant, but are kept within a narrow range. Some factors, such as core temperature and blood pH, fluctuate only slightly, while others, such as blood glucose, vary considerably throughout a normal day without producing any harmful effects.

In this chapter we look at the control of blood glucose and at the role of the liver. Other aspects of homeostasis are covered elsewhere in the book. Table 10.1 shows you where to find them.

The mechanism of homeostasis

When you start to study how the body controls a physiological factor such as temperature, blood glucose or blood pressure, it is important to organise your thoughts by asking the following questions:

● What conditions bring about a change in the factor being considered?

● What detects the change?

● How is the change reversed?

You will soon notice a pattern. Whenever a physiological factor changes, the body detects the change and then, by using nervous or hormonal signals, or both, it reverses the change. The extent of the correction is monitored by a system called **negative feedback**. This makes sure that, as levels return to normal, corrective mechanisms are scaled down.

Control of body temperature illustrates this mechanism well (see Chapter 12). When we are in a very hot environment or when we have been doing strenuous exercise, our body temperature rises. The brain detects this and sends signals to the body to bring the temperature down using various corrective mechanisms such as sweating and increased blood flow to the skin. As the body cools, the drop in temperature is monitored by the brain, which begins to send out fewer signals. Sweating then decreases.

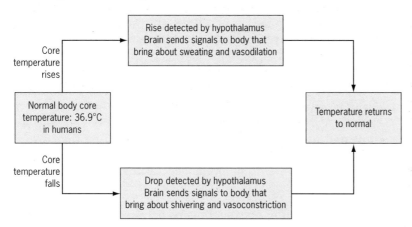

The opposite of negative feedback is **positive feedback**. In this situation, a change is amplified rather than returned to normal. Positive feedback in living systems is rare, but there are a few examples. Damage to a blood vessel causes a cascade reaction that brings about blood clotting: a few molecules of a substance become activated and each one then activates many more (see Chapter 34). The result is a tangled mesh of protein fibres that plugs the hole in the blood vessel.

Positive feedbacks also occur in abnormal situations and when normal homeostatic mechanisms get out of control. Elderly people whose sensory systems have deteriorated can suffer from **hypothermia** (see the Assignment at the end of Chapter 12).

Table 10.1 **Homeostatic mechanisms covered in this book**

Homeostatic mechanism	Covered in
Temperature	Chapter 12, pp 180–191
Blood glucose	Chapter 10, pp 154–165
Blood pressure	Chapter 9, pp 132–152
Solute concentration of blood	Chapter 11, pp 166–179
Blood pH	Chapter 11, pp 166–179
Blood volume	Chapter 11, pp 166–179
Blood hormone levels	Chapter 16, pp 240–251 Chapter 19, pp 278–295

?

A Why are homeostatic mechanisms often described as detection–correction systems?

Fig 10.1 **The mechanism of homeostasis as illustrated by temperature control. All other examples follow the same general pattern**

■ See question 1.

When they start to become cold, their body systems fail to respond, and the drop in body temperature goes uncorrected. As they grow colder, their metabolic rate decreases still further, they produce less heat and so in cold conditions they continue to cool down at an ever-increasing rate. Death occurs when their core temperature falls to about 25 °C.

Homeostatic mechanisms at the molecular level

We can also look at homeostasis at the level of molecules. In Chapter 4 we saw that enzymes control metabolic pathways, and that particular products are made, step by step, in a series of carefully controlled reactions. In a metabolic pathway, any product is made only as fast as it is needed. Excess product often prevents further quantities being made by inhibiting one of the enzymes in the pathway. When the excess has been used up, the inhibition is lifted and production continues. The process is self regulating and so we describe it as a **homeostatic mechanism**.

Glucose is one of the most abundant substances in our diet. Plant material contains **starch** and **cellulose,** and meat contains **glycogen**. All three are glucose polymers. During digestion, both starch and glycogen are broken down into glucose, which then passes into the blood in large amounts. Humans cannot digest cellulose, and this constituent of plant material forms much of our dietary **fibre**.

The terms used to describe glucose metabolism – glycogenesis, glycogenolysis and gluconeogenesis – can be confusing. Try remembering the origins of the words that make them up:

Glyco, gluco = sugar
lysis = splitting
neo = new
genesis = generation, or formation

2 THE CONTROL OF BLOOD GLUCOSE

In normal circumstances, we obtain most of our energy by respiring glucose. Cells therefore need a regular supply of this simple sugar. Some vital organs, notably the brain, cannot do without it even for a short time: lack of glucose can cause brain damage. Other cells and tissues, such as muscles, can respire lipids or even proteins for a short time if glucose is unavailable.

Blood glucose comes from:

● Digestion of carbohydrates in the diet.
● Breakdown of **glycogen** (see Chapter 3). This storage polysaccharide is made from excess glucose in a process called **glycogenesis**. Glycogen is particularly abundant in liver and muscle cells. When needed, glycogen can be broken down quickly to release glucose in the process of **glycogenolysis**.
● Conversion of non-carbohydrate compounds. Following deamination, the acid part of the amino acid is converted to glucose. Pyruvate and lactate (see Chapter 3) can also be converted to glucose. The conversion process in either case is called **gluconeogenesis**, which means literally 'the generation of new glucose'. During prolonged fasting, blood glucose is maintained by conversion of the body's protein and lipid stores.

The blood of a healthy person contains between 80 and 90 mg of glucose per 100 cm³. This normal value is maintained even during prolonged fasting. But the value rises to around 120 to 140 mg per 100 cm³ shortly after a meal, when carbohydrate digestion is in full swing. Feedback mechanisms bring the levels back to normal in about two hours. Under normal conditions, the kidney is able to reabsorb all of the blood glucose passing through it, preventing any from being lost in the urine.

The mechanism of blood glucose control

The control of blood glucose level is a good example of homeostasis. A negative feedback mechanism operates to detect and correct the level of blood glucose, maintaining it within 'safe' limits.

THE DISCOVERY OF INSULIN

IN THE EARLY DAYS of science, one of the most direct ways to find out an organ's function was to remove it surgically and then observe the effects of its loss on the organism. If the symptoms could be relieved by injecting a ground-up extract of the same gland, then the organ in question was an endocrine gland – a gland that releases hormones into the bloodstream.

By the early years of this century, this effective but less-than-subtle approach had been used to clarify the role of several organs, including the thyroid gland. But, although researchers suspected that the pancreas was an endocrine gland, they ran into problems when they tried to use this method to demonstrate that the pancreas made a hormone responsible for controlling glucose metabolism. When the pancreas of a dog was removed, the animal developed the symptoms of diabetes. But injecting an extract of ground-up pancreas failed to relieve the symptoms, a result that recurred in several experiments.

make pancreatic juice, degenerated, leaving just the islets of Langerhans functional. These dogs did not develop diabetes. The pancreases of the group 2 dogs were removed, and these animals did become diabetic.

Banting then made a pancreatic extract from the group 1 dogs and injected it into the diabetic dogs of group 2. The result was dramatic – an instant reduction of blood sugar that could be achieved repeatedly in several different animals. There was great excitement, and then

Fig 10. 3 **Banting and Best. In 1923 the Nobel prize was awarded to Banting and Macleod, but Banting shared his prize with Best, and Macleod did likewise with Collip**

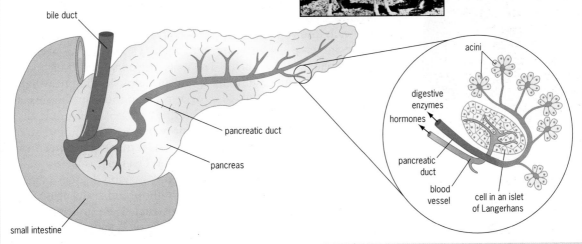

Fig 10.2 **The major part of the pancreas makes a juice containing digestive enzymes, but small patches of cells, called islets of Langerhans, produce the hormones insulin and glucagon**

Frederick Banting, a Canadian doctor, first had the idea that the digestive enzymes also made by the pancreas could be destroying any other active substances that were being made there. He persuaded the head of the physiology department at the University of Toronto, John JR Macleod, to let him have a laboratory, ten dogs and an assistant – Charles Best.

The dogs were subjected to one of two treatments. The pancreatic ducts of the dogs in group 1 were tied so that the animals could not produce pancreatic juice. Over the course of several weeks the acini, the cells that

disappointment as the extracts were tried in humans and found to be too impure (they produced fever). Banting and Best enlisted the help of the biochemist James Collip. He went on to make a purer extract than had previously been possible, that could be used for people.

Some people feel that animal experiments are a bad idea, but the few dogs used in these early experiments enabled Banting's team to change a progressive and fatal illness into a chronic, manageable condition. As a direct result of these experiments, insulin became available to the world's diabetics. Today there are over 300 000 insulin-dependent diabetics in the UK, and as many as 30 million world-wide. Without insulin, they would not be alive.

The pancreas plays a central role in the control of blood glucose. The digestive functions of the pancreas are covered in Chapter 7 but in this section we are concerned with its **endocrine role**: how it produces the hormones **insulin** and **glucagon** to control blood glucose (Fig 10.2).

B During their experiments, Banting and Best noticed that flies congregated around the urine of the diabetic dogs. Suggest an explanation for this.

See questions 2 and 3. ■

The pancreas itself detects any change in the level of blood glucose. If blood glucose becomes too high, β **cells** in the **islets of Langerhans** respond by releasing insulin. This hormone travels to all parts of the body in the blood, but exerts an effect mainly cells in muscles, liver and adipose (fat storage) tissue. Insulin lowers blood glucose by making cell surface membranes more permeable to glucose. It activates transport proteins in the membranes, allowing glucose to pass into cells. Insulin also activates enzymes inside the cells. Some of these enzymes convert the glucose to glycogen, others increase protein and fat synthesis.

If the levels of blood glucose get too low, α **cells** in the islets of Langerhans secrete glucagon. This hormone fits into receptor sites on cell surface membranes, and activates the enzymes inside the cells that convert glycogen to glucose. The glucose then passes out of the cells and into the blood, raising blood glucose levels.

When control of blood glucose fails

People with the disease **diabetes mellitus** are unable to control the level of glucose in their blood. This produces a range of symptoms. Blood glucose levels can get too high because the affected person produces little or no insulin. Without insulin, glucose cannot pass into cells and remains in the blood. The solute concentration of the blood increases, interfering with effective circulation and making the individual very thirsty. The cells are starved of their main fuel and are forced to respire lipids and proteins, leading to weight loss and eventual starvation. Glucose appears in the urine because blood sugar levels are so high that the kidney cannot reabsorb it all.

Today, the Clinistix™ system provides a rapid, quantitative test for glucose in the urine (see page 69), removing the need to taste the urine to diagnose diabetes (see the Opener).

Diabetes has a variety of causes and varying degrees of severity. In the UK, about 25 people in every thousand suffer from diabetes in one form or another. That means that there are over a million sufferers. About a third of these have Type I diabetes, roughly two-thirds have Type II (see below).

Types of diabetes mellitus

There are two main types of diabetes: **Type I** and **Type II**.

Type I diabetes, also known as 'early onset diabetes', occurs when the body cannot make insulin. This is often caused by an auto-immune reaction which attacks and destroys cells in the islets of Langerhans. This form of the disease usually appears before the age of 20, and the onset is sudden. Sufferers have this condition for the rest of their lives, but they can be treated by regular injections of insulin (Fig 10.4) matched to their glucose intake (in diet) and expenditure (in exercise). Before insulin was available, untreated Type I diabetes was usually fatal within a year of diagnosis. Most diabetics today need insulin injections, but they lead normal lives.

Type II diabetes, also called 'late onset diabetes' is more common than type I, accounting for about 70 per cent of the cases in the UK. It tends to begin during middle age and is more common in those who are overweight. This form of diabetes is due to a decline in the efficiency of islet cells, or to a failure of the cell surface membranes to respond to insulin. Fortunately, in many cases it can be controlled by regulating the diet.

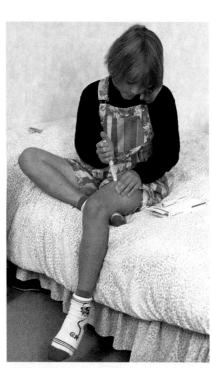

Fig 10.4 **This girl has diabetes and has to inject insulin every day. For many years, insulin for human treatment was derived from cows or pigs, but both are slightly different from human insulin and can cause an immune reaction, reducing their effectiveness. Today, diabetics use human insulin produced by genetically engineered bacteria (Chapter 31)**

Diabetes mellitus is describes an inability to control blood glucose levels. It is not the same as **diabetes insipidus**. This form of diabetes is caused by a lack of anti-diuretic hormone in the body (see page 247).

3 THE LIVER PLAYS A CENTRAL ROLE IN HOMEOSTASIS

The liver receives blood from two sources: from the intestines via the hepatic portal vein, and from the hepatic artery (see Fig 10.5(b)). The composition of blood that flows into the liver can fluctuate greatly, depending on factors such as the timing and nature of the last meal, but the content of blood leaving the liver is remarkably constant.

The structure of the liver

Before we go on to look at the functions of the liver in more detail, it is important to understand its structure (Fig 10.5).

Although control of blood sugar is a 'whole-body' process, the liver plays a central role. It is the first organ to receive the blood from the intestines that contains high levels of food molecules such as sugars following a meal. Liver cells are acutely sensitive to insulin and are also particularly rich in the enzymes involved in glucose metabolism.

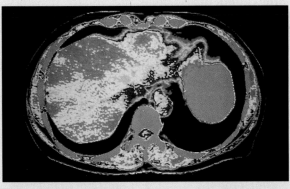

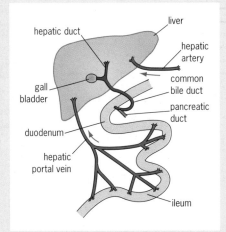

(b) **The blood supply to the liver and associated organs. The liver weighs about 1.5 kg in a normal adult**

Fig 10.5 **Structure of the liver**

(a) **This CAT (computerised axial tomography) scan through the abdomen of someone lying on their back shows just how much space the liver takes up: the liver is yellow and orange and the stomach is pink**

(e) **The fine structure of the liver. Blood is delivered to the liver cells in branches of the hepatic artery and the hepatic portal vein. As blood flows along the sinusoids, some chemicals are removed while others are added by secretion. The liver secretes bile into the canaliculi. These drain into small branches of the bile duct**

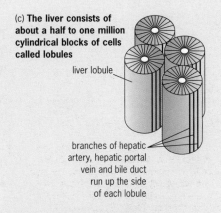

(c) **The liver consists of about a half to one million cylindrical blocks of cells called lobules**

liver lobule

branches of hepatic artery, hepatic portal vein and bile duct run up the side of each lobule

(d) **A micrograph of liver lobules. Radiating sinusoids drain into the central veins**

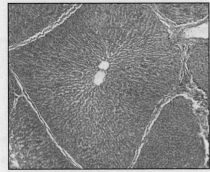

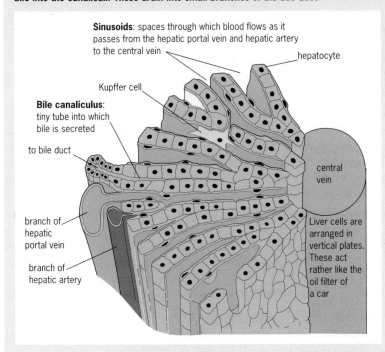

Sinusoids: spaces through which blood flows as it passes from the hepatic portal vein and hepatic artery to the central vein

hepatocyte

Kupffer cell

Bile canaliculus: tiny tube into which bile is secreted

to bile duct

branch of hepatic portal vein

branch of hepatic artery

central vein

Liver cells are arranged in vertical plates. These act rather like the oil filter of a car

Blood is brought into the liver by two blood vessels. The 30 per cent that arrives in the **hepatic artery** is oxygenated blood, while the 70 per cent delivered by the **hepatic portal vein** contains relatively little oxygen but is rich in nutrients that have been absorbed from the intestines. The **hepatic vein** removes blood from the liver.

The liver consists of hundreds of thousands of **lobules**, each about 1 mm in diameter, which surround branches of the hepatic vein. Channels called **sinusoids** radiate out from each central vein. These are surrounded by rows of liver cells called **hepatocytes**. These apparently unspecialised cells perform the majority of the liver's functions, and the composition of the blood changes as it flows along each sinusoid towards the central vein.

Dotted along the sinusoids are numerous white cells called **Kupffer cells** (see Fig 10.5(e)). These are phagocytes (see Chapter 34) that engulf bacteria and debris. Parallel to the sinusoids are fine channels called **bile canaliculi** (singular: *canaliculus*). Hepatocytes secrete the constituents of bile into the canaliculi, and this drains into the gall bladder. Bile is covered in detail in Chapter 7.

The main functions of the liver

The thousands of different chemical functions performed by the liver contribute greatly to the overall composition of the blood. They can be grouped under a few basic headings:

- Control of blood glucose levels
- Control of amino acid levels
- Synthesis of plasma proteins
- Synthesis of fetal red blood cells
- Destruction of red blood cells
- Detoxification
- Production of bile
- Control of lipid levels
- Storage of vitamins
- Cholesterol formation

We will now go on to look at some of them in detail.

The control of amino acid levels

The human body cannot store proteins. Every day an adult needs a minimum amount (40 to 60 grams, about the weight of one egg) to provide the amino acids the body needs to repair and grow new cells. Most people take in more than this in their diet, and the liver breaks down any amino acids that are not used. Obviously, growing children and pregnant or breastfeeding mothers need more protein.

Amino acids, like many other digested foods, reach the liver via the hepatic portal vein. They can be:

- **Deaminated**. This process removes the amino group (NH_2) of an amino acid and forms ammonia (NH_3): Fig 10.6(a). The organic acid residue is usually respired, while the toxic ammonia is quickly converted into a more harmless substance, **urea**, via the **ornithine cycle**: Fig 10.6(b). The kidneys remove urea from the body (see Chapter 11).

- **Transaminated**. There are eight essential amino acids (10 in children) that must be present in the diet. The remaining 12 are termed non-essential amino acids because they can be made in the liver by transamination. This process involves the transfer of an amino group from an amino acid to an acid (derived from carbohydrate metabolism), thereby making a new amino acid.

- Used to synthesise plasma proteins such as fibrinogen.

- Released unchanged into the general circulation (most cells need a supply of amino acids for protein synthesis).

C Hepatocytes have many mitochondria and microvilli. What does this suggest about their function?

D Why can't we take in a week's supply of protein in a single meal by eating one large steak or omelette?

Fig 10.6 **The ornithine cycle is a series of reactions in which toxic ammonia is combined with carbon dioxide to form urea. Effectively, a molecule of urea is assembled on the ornithine. When complete, the urea will become detached, leaving ornithine to continue the cycle. The basic steps are as follows:**

1 **One molecule of CO_2 and one of NH_3 are added to the molecule of ornithine, forming citrulline**

2 **A second molecule of NH_3 is added to citrulline, forming arginine**

3 **Arginine is split to produce one molecule of urea and one molecule of ornithine to continue the cycle**

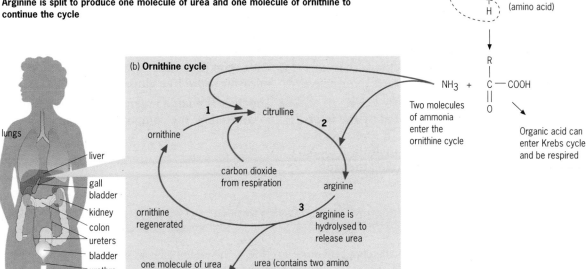

(a) **Deamination**
Amino group and hydrogen removed from amino acid as ammonia

$\frac{1}{2}O_2 + NH_2 \overset{R}{\underset{H}{\mid}} C - COOH$ (amino acid)

$NH_3 + \overset{R}{\underset{O}{\mid\mid}} C - COOH$

Two molecules of ammonia enter the ornithine cycle

Organic acid can enter Krebs cycle and be respired

(b) **Ornithine cycle**

lungs

liver
gall bladder
kidney
colon
ureters
bladder
urethra

ornithine

1 citrulline

2

carbon dioxide from respiration

ornithine regenerated

arginine

3 arginine is hydrolysed to release urea

one molecule of urea produced

urea (contains two amino groups)

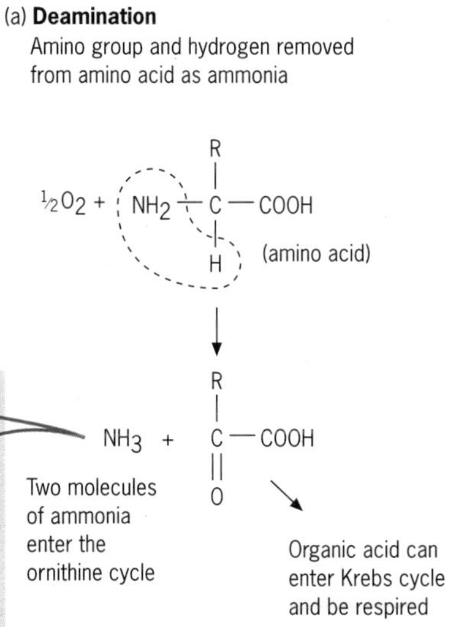

Detoxification and hormone breakdown

The liver concentrates and **detoxifies** (breaks down) many harmful chemicals. Some, such as hydrogen peroxide, are produced by the body itself. Others, alcohol and food additives for example, come from outside. Drinking large amounts of alcohol over a long period of time can cause the death of liver cells, followed by replacement with connective tissue ('scar' tissue). This is known as **cirrhosis** of the liver (Fig 10.7).

The liver also breaks down many circulating hormones. Removal of hormones from the blood is an important aspect of the control process as it ensures that hormones act only for as long as they are needed. Insulin, for instance, is rapidly metabolised by the liver and has a half-life of about 10 to 15 minutes.

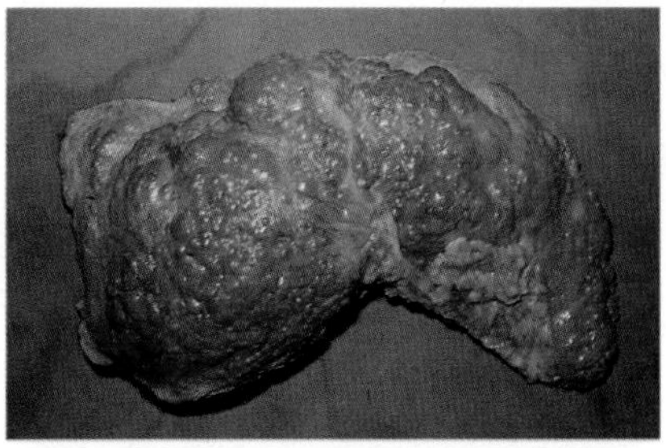

Storage

The liver stores relatively large amounts of vitamins A, D and B_{12}, enough to supply the body for several months. Other vitamins, such as most of the B complex, and the minerals copper and iron, are also stored but in smaller amounts. This high vitamin and mineral storage capacity explains why eating liver is good for you, even though some people don't find it a pleasant experience.

Manufacture of blood proteins and blood cells

The liver makes many important blood proteins including the most abundant plasma protein, **albumin**, fibrinogen and other substances involved in blood clotting (see Chapter 34). Given a supply of vitamin K, the liver can make prothrombin and other blood clotting factors.

The liver of a fetus manufactures red blood cells, but this complex process is taken over by the bone marrow after birth.

Fig 10.7 **The liver of an alcoholic who developed cirrhosis of the liver. In the liver, alcohol (ethanol) is converted into ethanal by the enzyme alcohol dehydrogenase. Ethanal is toxic and seems to be responsible for much of the long term alcoholic liver damage, although the exact mechanisms involved are unclear**

See questions 4 to 7.

SUMMARY

When you have studied this chapter, you should know the following:

■ The word **homeostasis** means 'steady state'. Homeostatic processes keep conditions in the body within narrow limits.

■ Homeostasis is usually maintained by **negative feedback mechanisms**. The body detects a change in a particular internal factor, such as core body temperature, and then activates a corrective mechanism to reinstate the normal level.

■ Blood glucose levels are monitored by the **islets of Langerhans** in the pancreas. If levels get too high, β cells in the islets secrete **insulin**. If blood glucose gets too low, α cells secrete **glucagon**.

■ **Insulin**, a peptide hormone, increases the permeability of cell surface membranes to glucose. It also appears to activate intracellular enzyme systems that convert glucose to glycogen, fat and protein.

■ **Glucagon** promotes the breakdown of **glycogen**, releasing more free glucose into the blood.

■ The liver is a large organ that plays a major role in controlling the composition of the blood.

QUESTIONS

1 The diagrams in Fig 10.Q1 show the difference between negative feedback and positive feedback.

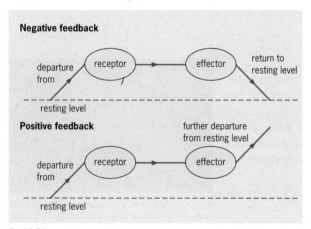

Fig 10.Q1

a) Suggest why negative feedback is frequently involved in homeostasis.

b) Explain how negative feedback enables the carotid and aortic bodies and the medulla to maintain a constant blood carbon dioxide level.

c) Explain why the mechanism involved in the initiation of an action potential is an example of positive feedback.

[NEAB June 1997 Biology: Human Systems Module Test, q.7]

2 Fig 10.Q2 shows part of a muscle cell membrane, including a membrane receptor for insulin.

a) Name the molecules labelled A and B.

b) Explain how glucose enters a muscle cell.

c) How is the rate of glucose uptake by a muscle cell affected by insulin?

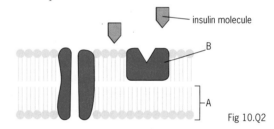

Fig 10.Q2

[NEAB February 1995 Modular Biology: Physiology Test q. 5]

3 An investigation was carried out to find the effect of diet on the rate at which muscle glycogen was replenished after exercise. Three athletes each exercised for two hours. During the subsequent recovery period each athlete received a different diet. The results are shown in Fig 10.Q3.

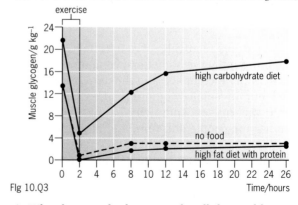

Fig 10.Q3

a) What features do the curves for all three athletes have in common?

b) Suggest **two** improvements that could be made to the design of this investigation to give more reliable results.

c) Explain how insulin is involved in the recovery of muscle glycogen levels in the athlete who was given the high carbohydrate diet.

d) In the athlete given the high-fat plus protein diet, much of the protein and fat was not converted into glycogen. Describe briefly what happens in the body to excess amino acids from the protein.

[NEAB June 1995 Modular Biology: Physiology Module Test q. 4]

4 Fig 10.Q4 shows part of a liver lobule.

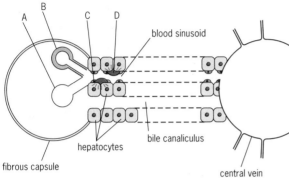

Fig 10.Q4

a) (i) Name vessels **A** and **B**.
   (ii) Describe the part played by vessels **A** and **B** in liver function.

b) (i) Draw an arrow in the sinusoid to show the direction of blood flow in the lobule.
   (ii) Explain the function of cells **C** and **D**.

[AEB 1994 Biology: Specimen paper 1, q. 4]

5 Write an essay on homeostasis (This should take you 40 mins max.)

[AEB 1991 Human Biology Paper 2, q. 4]

6 Fig 10.Q6 shows the main blood vessels going to and coming from the liver.

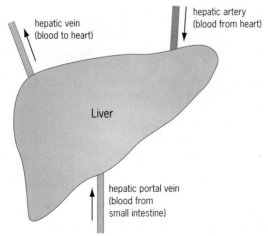

Fig 10.Q6

a) In a healthy person the blood glucose level in the hepatic vein fluctuates much less than that in the hepatic portal vein. Explain why this is so.

b) Blood sugar level is more or less constant, even if a person has not eaten for several days. How does gluconeogenesis help maintain this constant blood sugar level?

c) Suggest why people suffering from diabetes are advised to eat their carbohydrate in the form of starch rather than as sugars.

[NEAB June 1997 Biology: Physiology Module Test, q.2]

7 Below is an entry from a dictionary of biological terms.

blood glucose pool: the total amount of glucose in the blood at any one time. Accurate control of blood glucose is very important. If the concentration falls too low, the central nervous system ceases to function correctly. If it rises too high, then there will be a loss of glucose from the body in the urine. Hormones including *insulin* and *glucagon* play an important part in maintaining a constant level of glucose in the blood. Although the overall level stays within narrow limits, it is important to realise that glucose is always being added to and removed from the blood glucose pool. Digestion and absorption of carbohydrates and conversion from the body's stores of glycogen and fats tend to increase the blood glucose level while such processes as respiration decrease it. This is summarised in Fig 10.Q7.

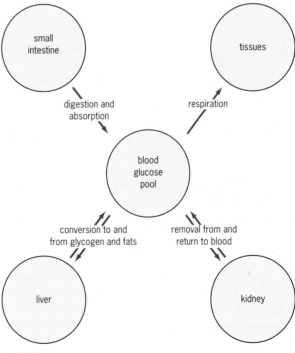

Fig 10.Q7

a) Describe the part played by insulin and glucagon in maintaining a constant level of glucose in the blood.

b) In the kidney, glucose is removed from and returned to the blood glucose pool. Explain how this occurs.

[NEAB June 1997 Biology: Human Systems Module Test, q.8]

Assignment

A DAY IN THE LIFE OF A DIABETIC

Being a diabetic is a way of life. Unlike other chronic conditions such as asthma, a diabetic cannot just take their medication and forget there was ever a problem. Instead, they must become experts in the management of their own condition, matching up dietary intake with insulin doses and exercise regimes in order to keep their blood glucose within definite limits.

This is an account of the way in which one person, Anna, deals with her diabetes. It should be stressed from the start that the condition affects different people in different ways. A blood sugar level that would have one person fainting on the floor may produce no symptoms in another. Bear this in mind as you read on.

1 Explain how a non-diabetic person responds to a rise in blood sugar level. Your answer should include the effect of insulin on cells.

2 What is the difference between a chronic and an acute condition?

Types of diabetes

There are two basic types of diabetes; **Type 1** and **Type 2**. Type 1 is also known as 'early onset' or 'insulin dependent' as it appears in early life and controlling it usually requires insulin. At the root of the problem is thought to be an auto-immune reaction in which the body's antibodies destroy its own insulin-producing cells. As these cells cease to produce insulin, less insulin is released into the blood after a meal. Without sufficient insulin, most of the glucose that enters the blood from the intestines after digestion cannot get into cells. Glucose accumulates in the blood and the cells, which are starved of their main fuel, turn to alternatives – lipid and protein.

Five of the major symptoms of Type 1 diabetes are:

Excessive thirst.

Excessive urinating.

Weight loss.

Glucose in the urine.

The fruity smell of ketones on the breath (a by-product of lipid metabolism).

3 Using the information given above, suggest explanations for each of the five symptoms.

The story of Anna

Anna, who is now in her forties, was diagnosed as a diabetic at the age of 18 months. This is early, but her brother had also developed diabetes (when he was 4) and so her mother was acutely aware of the symptoms.

4 Which of the symptoms of diabetes would be apparent in Anna when she was only a toddler?

5 It is thought that people can inherit a susceptibility to diabetes, but it needs an environmental trigger of some sort, such as exposure to a virus. Does Anna's story support this idea?

Anna has come to terms with her diabetes and has found that management is easiest within a routine. Overall, the aim is to keep her blood glucose within the range of 4 to 11 millimoles per litre. (A non-diabetic maintains their blood glucose to within about 4 to 9 millimoles per litre.)

When she wakes up in the morning, Anna tests her blood sugar. This is done by pricking her finger and putting a drop of blood on a test strip, similar to that on page 69. The level of blood glucose is then obtained by either matching up the colour to the chart or by placing the strip into a machine which takes the reading automatically. Anna would expect her glucose levels to be about 3 at this time in the day. Anything less than 4 is regarded as hypoglycaemic, although Anna gets no symptoms until the value approaches 1.

6

(a) What fraction of a mole is a millimole?
(b) Suppose Anna has a blood glucose reading of 10 milli-moles. If the relative molecular mass of glucose is 180, how many grams of glucose in one litre of her blood?

After taking the reading, Anna injects herself with insulin. There are now two types of insulin available; fast acting and slow acting. Anna injects 2 units of fast acting and 14 of slow acting. Fast acting is normal, soluble insulin and works straight away. Slow acting insulin is attached to a retarding agent that releases the insulin slowly over the next few hours.

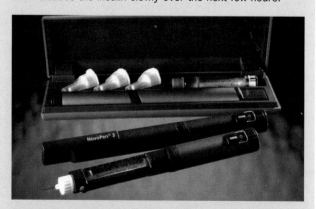

Fig 10.A1 **For diabetics, there is now an alternative to syringes. Special pen-like injection devices such as this Novopen are widely available. They are very accurate and discreet to use**

7
(a) Suggest the advantage to diabetics of having slow acting insulin.
(b) Suggest why you must have some fast acting insulin.

Until relatively recently the insulin was obtained from animal pancreas. This was less than perfect because the non-human insulin slowly brought about an immune reaction. Eventually, the 'foreign' insulin had little effect. With modern genetic engineering, however, it is now a relatively simple task to produce human insulin on a large scale – see Fig 10.A2.

8 Why can't insulin be taken orally?

After breakfast, Anna goes to work. In her handbag are glucose tablets and some biscuits as well as her insulin and testing equipment. Anna has a sweet drink and a couple of biscuits at break.

9 It is frequently said that 'diabetics can't eat sugar' but this is obviously not true. Suggest why Anna needs the mid-morning biscuits.

Going 'hypo' and 'hyper'
A Type 1 diabetic who takes no insulin experiences the five symptoms listed on the opposite page – they have gone 'hyper'. In Anna's case, the early warning signs are headaches and feeling tired or lethargic.

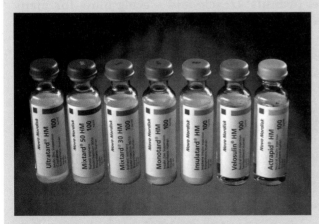

Fig 10.A2 **Human insulin is made on a large scale by genetic engineering and is supplied in different doses**

On the other hand, once a diabetic has taken insulin they must take care to keep their blood glucose levels up or they might go 'hypo'. For Anna, symptoms of too little blood sugar are similar to going hyper: tiredness, irritability and an inability to concentrate. Often, other people notice the symptoms first.

10 The terms hypo and hyper are short for hypoglycaemia and hyperglycaemia. What do these words mean?

11 If Anna finds that she is going hypo, suggest what she can do.

Anna has a normal lunch and another snack in mid-afternoon. At tea time she takes her blood sugar again – it is normally about 8 millimoles. After this she takes more insulin, 2 units of short acting and 10 slow acting. If she is going out for a meal, and will eat more than normal, she will take an extra two units of short acting insulin. She must also be careful what she drinks, because alcohol has a powerful effect on glucose metabolism.

Before bed time, Anna has a sweet drink, usually chocolate, and checks her blood sugar again. It needs to be around 8 millimoles because she will not eat for another 8 hours or so.

12
(a) Why is it important that Anna does not go 'hypo' in the night?
(b) Suggest what Anna does if her reading is, say, 6 units?

13 From what you have read of Anna's daily routine, sketch a graph of her blood glucose levels over a 24 hour period.

Type 2 diabetes
Many diabetics have Type 2 diabetes, This is also known as 'late onset' and the treatment depends on the underlying cause. In most Type 2 cases, the condition is due to a combination of one or more of the following:

Muscle and fat cells do not respond to insulin.

The body can't produce enough insulin to meet demand.

Liver cells release too much glucose from their stores.

For many Type 2 diabetics, a diet and exercise plan may be enough to keep blood sugar levels within normal limits.

11 Excretion and water balance

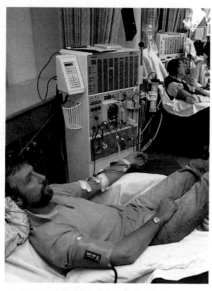

This patient is undergoing dialysis. For several hours, blood flows from the patient's forearm into the machine. Here, much of the waste, together with excess salt and water, is removed by filtration. Find out more about this process in the Assignment at the end of this chapter

IF ONE OF YOUR KIDNEYS were to stop working because of injury or disease, you could probably still lead a normal life. However, people whose kidneys both fail are faced with a crisis: water, urea and potassium build up rapidly in their body. They may continue to pass some urine, but they cannot get rid of all the waste produced by normal cell processes.

Most people suffering from kidney failure hope for a transplant, but there is a shortage of donors. Until a suitable organ becomes available, patients have to rely on dialysis to filter their blood and balance their fluid intake. As the photograph on the left shows, dialysis is uncomfortable and inconvenient. To reduce the time they need to spend in dialysis, kidney patients must stick to the strict Giovanetti diet. This comes of something of a shock. They must limit their fluid intake to only half a litre a day, about a quarter of the amount a human adult would normally drink. They must also control their protein intake to about 30 to 40 grams per day, the amount of protein in a small egg.

But perhaps the biggest problem is the need to regulate potassium. This ion is a normal constituent of the body, but large amounts cause serious problems, including heart failure. Potassium-rich foods include citrus fruits, bananas, instant coffee, peanuts, treacle and – a big blow for many people – chocolate. In this strange diet, carbohydrates are not a priority. Few patients feel like eating anyway, and this new restricted diet makes the task of finding appetising food even more difficult.

1 WASTE AND WATER CONTROL

The human body consists mainly of water. A person weighing 65 kg contains about 40 litres of water, of which 28 litres is **intracellular** (inside cells). The rest, the **extracellular** fluid is made up from about 9 to 10 litres of tissue fluid and 2 to 3 litres of blood plasma.

The kidneys play a major role in regulating the volume and composition of these body fluids. They excrete or conserve water and salt so that the volume and composition of the blood and body fluids remains more or less constant. This is a vital aspect of homeostasis.

The kidneys also ensure that waste products do not build up by filtering the blood. Waste is allowed to pass through and then out of the body, whilst important substances such as glucose are reabsorbed and conserved. The removal of metabolic waste from the blood is called **excretion**.

Fig 11.1 **When we drink a large amount of fluid, our kidneys have the job of getting rid of the excess water, to prevent body fluids from becoming too dilute. This man is drinking beer and the alcohol it contains also affects his kidney function – more of that later**

Fig 11.2 **Adult humans need about 40 to 50 grams of protein per day, and so there is enough here for over a week! However, the body cannot store protein and so the daily excess is broken down in the liver. The resulting urea is excreted in the urine**

Urine is the end product of all processes that occur in the kidney.

Most excreted waste leaves the body in the urine, but some waste is also lost in **sweat** and in air that we breathe out. Sweat contains mainly salt and water, while **exhaled air** contains carbon dioxide and water vapour.

In this chapter we concentrate on **nitrogenous excretion**, the removal of waste compounds that contain nitrogen. The main nitrogenous compound excreted by humans is **urea**. This is made in the liver following the breakdown of excess amino acids. Enzymes in the liver remove the amine (NH_2) group from amino acids in the process of **deamination**. The ammonia formed is a highly toxic compound that must not be allowed to build up. It immediately enters a series of reactions called the **ornithine cycle**, which produce the relatively harmless compound urea:

$$2NH_3 \quad + \quad CO_2 \quad \rightarrow CO(NH_2)_2$$
$$\text{ammonia + carbon dioxide} \rightarrow \quad \text{urea}$$

See page 161 for more details on the ornithine cycle. Urea is carried in the blood from the liver to the kidneys to be excreted.

In this chapter we look at the role of the kidneys in homeostasis. The kidneys selectively eliminate water and solutes, such as sodium, potassium and chloride ions, so that the water and solute balance of the body is kept at the correct level. The need to balance the solute concentration of body fluids is called **osmotic regulation**, or **osmoregulation**.

Osmoregulation is the maintenance of a constant solute concentration within the body. In Chapter 5 we saw why this is important: if we place animal cells in a *hypertonic* solution, they lose water by osmosis and shrivel. If we put them in a *hypotonic* solution, they gain water and may burst. Both situations are harmful to cells, and so it is important that we maintain the solute concentration of our body fluids within narrow limits.

?

A If a person has 3 litres of blood plasma, and their haematocrit (the volume occupied by cells in the blood) is 40 per cent, what is their total blood volume?

✔

Excretion is the removal of chemical waste from the body, waste that is produced by the metabolic processes within cells and which would be toxic if allowed to accumulate. You should not confuse excretion with **egestion** or **defaecation**, the removal of undigested food and other debris from the intestine.

✔

Different fluids are often compared with respect to their solute concentrations. A solution that has a higher solute concentration than another is said to be **hypertonic**. A solution that has a lower solute concentration than another is **hypotonic**. A solution that has exactly the same solute concentration is **isotonic**. So, for example, we can say that seawater is hypertonic to human blood plasma, because seawater has a much higher concentration of dissolved salts.

ORAL REHYDRATION THERAPY (ORT)

MOST PEOPLE will have experienced the unpleasant symptoms that accompany a bout of mild food poisoning. It could be due to a dodgy late night kebab, an undercooked chicken leg at a barbecue, or those prawns that lurked in the fridge just a little too long.

Vomiting is caused by bacterial toxins that irritate the gut lining. In the intestine, the frequency of peristalsis increases and the contents move along the gut rather more rapidly than usual. This doesn't give the large intestine enough time to absorb water from the waste and the result is **diarrhoea**. Fortunately, for most of us, the symptoms are short lived.

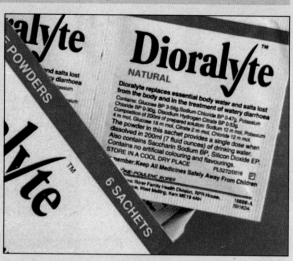

Fig 11.4 **Sachets of oral rehydration salts. Those we can buy in the UK contain more than the essential ingredients**

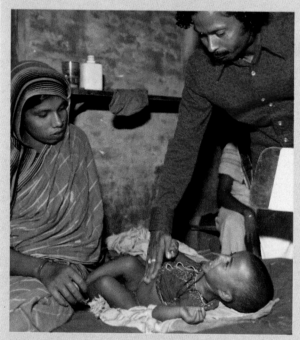

Fig 11.3 **The baby in this picture has bacterial dysentery that he developed after drinking water contaminated with faeces. He risks death from dehydration if he does not receive rehydration therapy.**
An oral rehydration mixture is easy to make up: one level tablespoonful each of glucose and salt dissolved in a pint of boiled and cooled water

However, both vomiting and diarrhoea can be deadly if severe and prolonged. Dysentery and cholera can rapidly lead to death by dehydration as they cause the body to lose fluid faster than it can be replaced.

Every day the average person consumes about 2 to 3 litres of fluid in one form or another, and we also pour a huge volume of fluid (over 8 litres) into our intestines in the form of digestive juices. Diarrhoea does not allow for the efficient reabsorption of these fluids. Also, vital ions such a sodium, potassium and chloride, collectively known as **electrolytes**, are lost. If untreated, this loss can lead muscle spasms, cramps, coma and heart failure.

Oral rehydration therapy can be used to treat dehydration. This does not involve expensive drugs, simply a mixture of glucose and salt in water. In cases where the patient cannot keep anything down, the rehydration solution can be given directly into the bloodstream via a drip. This simple treatment has saved millions of lives in places where dysentery and cholera are very common.

2 AN OVERVIEW OF THE HUMAN URINARY SYSTEM

The human urinary system is shown in Fig 11.5. The kidneys lie at the back of the abdominal cavity, just below waist level, where they are protected to some extent by the spine and the lower part of the ribcage. Usually, the left kidney is slightly above the right.

The kidneys receive blood from the two **renal arteries** that branch off the aorta (see Chapter 9). The kidneys receive the largest blood supply of any organ, per gram of tissue. About 1200 cm³ of blood flows to each of them every minute. Incoming blood must be at high pressure to ensure proper kidney function: they can't filter blood effectively if the pressure drops. Blood leaves the kidneys in the **renal veins**.

The urinary system is also known as the **renal system**.

If you place your hands on your hips, your thumbs show the position of your kidneys.

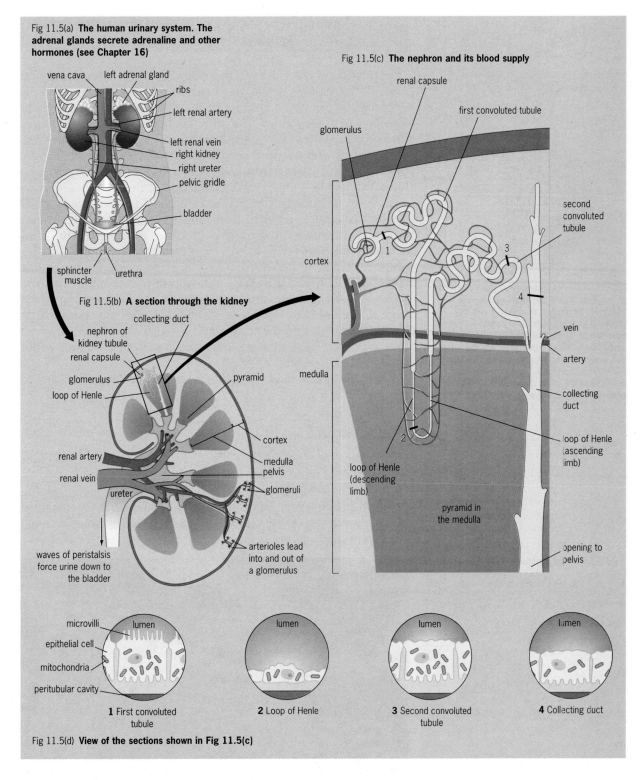

Fig 11.5(a) **The human urinary system. The adrenal glands secrete adrenaline and other hormones (see Chapter 16)**

vena cava
left adrenal gland
ribs
left renal artery
left renal vein
right kidney
right ureter
pelvic gridle
bladder
sphincter muscle
urethra

Fig 11.5(c) **The nephron and its blood supply**

renal capsule
first convoluted tubule
glomerulus
second convoluted tubule
cortex
vein
artery
medulla
collecting duct
loop of Henle (ascending limb)
loop of Henle (descending limb)
pyramid in the medulla
opening to pelvis

Fig 11.5(b) **A section through the kidney**

collecting duct
nephron of kidney tubule
renal capsule
glomerulus
loop of Henle
pyramid
cortex
medulla
pelvis
glomeruli
renal artery
renal vein
ureter
arterioles lead into and out of a glomerulus
waves of peristalsis force urine down to the bladder

microvilli
lumen
epithelial cell
mitochondria
peritubular cavity

1 First convoluted tubule

lumen

2 Loop of Henle

lumen

3 Second convoluted tubule

lumen

4 Collecting duct

Fig 11.5(d) **View of the sections shown in Fig 11.5(c)**

Urine made by the kidneys is pushed down muscular tubes, the **ureters**, by peristalsis (rhythmic muscular contractions). The ureters empty into the **bladder**, a muscular bag that stores urine until it is convenient to release it. The capacity of the human bladder varies from 400 to 700 cm^3 or more. When it begins to get full, stretch receptors in the walls inform the brain of the urgency of the situation. **Urination**, or **micturition**, happens when the **sphincter muscles** relax, allowing urine to pass out of the body through the **urethra**.

?

B A person has 5 litres of blood, and the kidneys filter 1.2 litres per minute. How many times on average does the total volume of blood in the body pass though the kidneys every hour?

3 STRUCTURE AND FUNCTION OF THE MAMMALIAN KIDNEY

The structure of the kidney is shown in Fig 11.5(b). The functional unit of the kidney is the **nephron** (or **kidney tubule**). It is important to know the position of the nephrons in relation to the overall plan of the kidney. The outer **cortex** of the kidney contains the **renal capsules** (also called **Bowman's capsules**) and the **first convoluted** and **second convoluted tubules** (also called the **proximal** and **distal tubules**), while the **medulla** houses the **loops of Henle** and the **collecting ducts**. Bundles of collecting ducts form **pyramids** that deliver urine into an open space called the **pelvis**. From here, urine flows down the ureters to the bladder.

Kidney function involves two processes: **ultrafiltration** and **active transport** (Fig 11.6). Ultrafiltration is filtration under pressure: blood is 'squeezed' to form a fluid called **glomerular filtrate** (usually just **'filtrate'**). Active transport then modifies the filtrate, secreting some substances into it and reabsorbing others from it, according to the needs of the body. The end result is that blood flows back into the body without much of its harmful waste. This waste, a solution containing urea, salts and various other chemicals, is the urine.

Let's now look at how individual parts of the kidney contribute to this overall process.

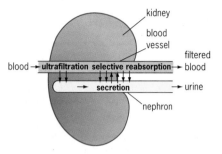

Fig 11.6 **A simple summary of kidney function. At the near end of the nephron, blood is filtered to produce a fluid that is virtually identical to tissue fluid. The filtrate is then modified by active transport. This involves active secretion of substances into the filtrate and active reabsorption of substances into the blood. These processes are possible only because of the close association between the nephron and blood system (see Fig 11.5(c))**

The nephron

Each human kidney contains about a million nephrons, together with a maze of blood vessels and some connective tissue (Fig 11.7). In this section we deal with the function of each region of the nephron in sequence, but you must remember that the nephron functions as a whole: the activities of one region are essential to the effectiveness of others.

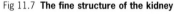

Fig 11.7 **The fine structure of the kidney**

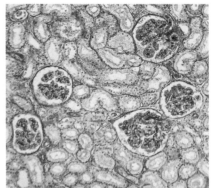

(a) Under the microscope, five renal capsules (containing glomeruli) can be seen, surrounded by sections of tubules. An underwater dissection allows some of the fine nephrons to be teased out and viewed individually without the aid of a microscope. Each nephron, although only 60 μm in diameter, can be over 14 cm long when uncoiled

(b) A scanning electron micrograph of a glomerulus with part of the torn renal capsule (whitish) round it

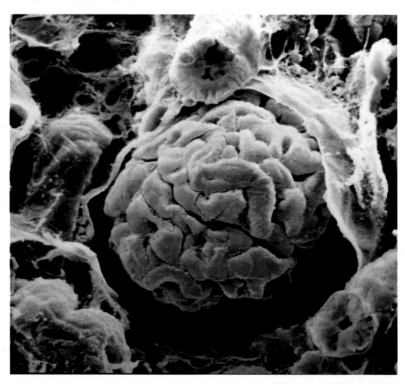

Ultrafiltration in the renal capsule

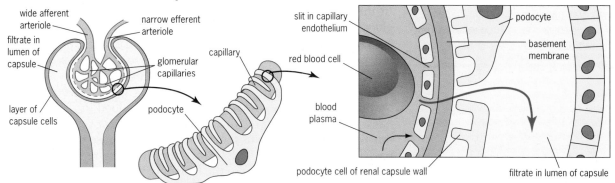

The walls of capillaries in the renal capsule are much more permeable than those of normal capillaries: the cells do not fit tightly together, but have thin slits in between, through which all the constituents of the plasma can pass

The renal capsule is lined with unique cells called **podocytes** ('foot-cells'). These cells, like those of the capillary, do not fit tightly together, but form a network of slits that fit over the capillary

Between these two relatively coarse filters is the continuous basement membrane. This finer filter prevents the passage of all molecules with a relative molecular mass greater than about 68 000 kilodaltons, so the larger molecules (mainly proteins) remain in the blood

Fig 11.8 shows the **renal capsule** in detail. It is shaped rather like a wine glass, with a central knot of blood vessels called the **glomerulus**. This area of the nephron filters blood by ultrafiltration (filtration under pressure). Obviously, this requires two things: a means of creating pressure and a filter.

The kidneys receive blood from the first branch off the aorta, so the blood is already under pressure when it reaches the nephron. This pressure is maintained and enhanced because the **afferent arteriole**, the blood vessel that takes blood into the glomerulus, is short and has a larger diameter than the longer **efferent arteriole** that takes blood away. This physical or **hydrostatic pressure** forces blood against a filter that consists of three layers:

- The lining or **endothelium** of the glomerular blood vessels.
- The **basement membrane**.
- The cells of the renal capsule itself.

Fig 11.8 shows how these membranes are arranged. The middle basement membrane acts as a fine filter and is therefore mostly responsible for the chemical composition of the filtrate. At this stage the filtrate is identical to tissue fluid (see Chapter 9).

The rate of filtrate production is high: about 125 cm³ per minute. Obviously, we don't produce anything like this volume of urine or we would be constantly in the loo and would dehydrate rapidly On average we produce about 1 cm³ of urine per minute. The rest, over 99 per cent of the filtrate, is *reabsorbed*. In fact, after the renal capsule, the rest of the nephron is concerned with adjusting the volume and composition of the filtrate. Necessary substances are reabsorbed; toxic compounds and excess solutes and water are removed.

Several forces act on the fluids in the renal capsule, opposing or encouraging the filtration process. The high hydrostatic pressure of blood is the dominant force but it is opposed by the hydrostatic pressure and solute concentration of the filtrate (Table 11.1).

Fig 11.8 **The fine structure of the renal capsule: a region of the kidney adapted for ultrafiltration of the blood. Note the difference in size between the afferent and efferent arteriole. The blood is filtered through three layers of cells: the endothelium of the capillaries, the basement membrane and the capsule wall**

?

C If an individual lost a large amount of blood in an accident, what effect would this have on kidney function?

Force acting	Opposes or encourages filtrateformation	Approximate value/kPa
Hydrostatic pressure of blood	encourages	8.0
Hydrostatic pressure of filtrate in capsule	opposes	−2.4
Solute concentration of blood	opposes	−4.3
Overall filtration pressure	encourages	1.3

Table 11.1 **Summary of the forces acting in the renal capsule**

?

D The loop of Henle is sometimes describes as a **hairpin counter-current multiplier**. Explain this description.

The first convoluted tubule

In the **first convoluted tubule** (Fig 11.9), many solutes such as glucose and amino acids are totally reabsorbed into the blood by active transport. Normal urine should not contain glucose. Only when blood glucose becomes very high, due to diabetes for example, does the reabsorption mechanism fail (Fig 11.10).

In addition, this part of the tubule is very close to blood vessels that carry blood away from the glomerulus. Blood in these vessels has a *low* hydrostatic pressure but a relatively *high* solute potential due to the plasma proteins that remain there because they could not pass through the filter. This allows the blood to reabsorb a large percentage of the water from the first convoluted tubule by osmosis.

As you can see from Fig 11.9, the cells that line the first convoluted tubule show all the classic adaptations to active transport: a large surface area, provided by microvilli, and many mitochondria to provide ATP to power the process.

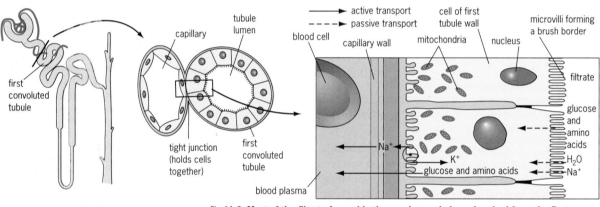

Fig 11.9 **Most of the filtrate formed in the renal capsule is reabsorbed from the first convoluted tubule. Amino acids, glucose and sodium are removed from the tubule by active transport, and water follows passively by osmosis**

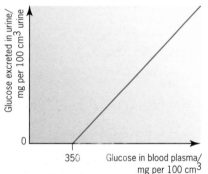

Fig 11.10 **Normally the kidneys are able to reabsorb all of the glucose into the blood. Only when blood sugar exceeds a threshold of about 350 mg per 100 cm³ does glucose begin to appear in the urine**

The loop of Henle

The **loop of Henle** is a long U-shaped region of the nephron that descends deep into the medulla and then returns to the cortex. The loop creates a region of high solute concentration in the medulla. The collecting ducts pass through this region, and the osmotic gradient between the inside of the collecting duct and the outside draws water out of the duct by osmosis. Consequently, the urine becomes more and more concentrated (compared to body fluids) as it passes down the duct. So the loop is a vital adaptation that benefits humans and other land-living organisms: it allows us to get rid of waste without losing too much water.

Fig 11.11 outlines how the loop of Henle works. Fluid in the two limbs of the loop flows in opposite directions. We describe this sort of arrangement as a **counter-current system**. As fluid travels up the ascending limb, sodium chloride (NaCl) is transported actively out of the limb into the surrounding area. This causes water to pass out of the descending limb by osmosis. The net result is that the solute concentration at any one level of the loop is slightly lower in the ascending limb than in the descending limb. The longer the loop, the more chance there is for this mechanism to build up a high sodium chloride concentration. If the loop in Fig 11.11 were only half the length shown, sodium chloride would accumulate to only about 600 units.

?

E People on survival courses are taught to assess their level of dehydration by looking at the colour of their urine. Explain how they would do this.

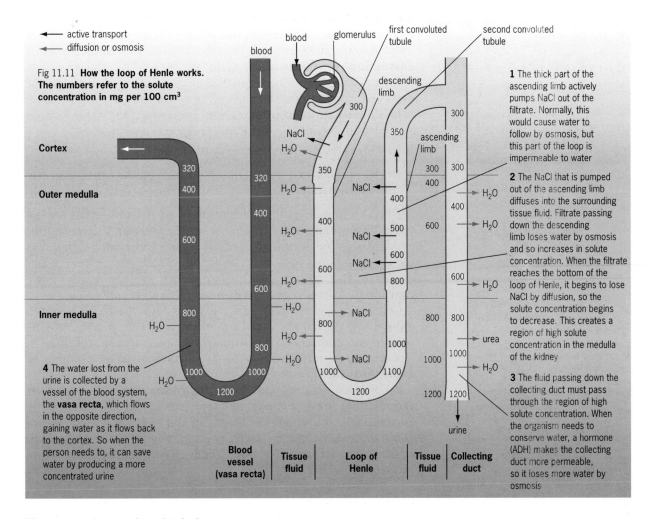

Fig 11.11 **How the loop of Henle works. The numbers refer to the solute concentration in mg per 100 cm³**

→ active transport
→ diffusion or osmosis

1 The thick part of the ascending limb actively pumps NaCl out of the filtrate. Normally, this would cause water to follow by osmosis, but this part of the loop is impermeable to water

2 The NaCl that is pumped out of the ascending limb diffuses into the surrounding tissue fluid. Filtrate passing down the descending limb loses water by osmosis and so increases in solute concentration. When the filtrate reaches the bottom of the loop of Henle, it begins to lose NaCl by diffusion, so the solute concentration begins to decrease. This creates a region of high solute concentration in the medulla of the kidney

3 The fluid passing down the collecting duct must pass through the region of high solute concentration. When the organism needs to conserve water, a hormone (ADH) makes the collecting duct more permeable, so it loses more water by osmosis

4 The water lost from the urine is collected by a vessel of the blood system, the **vasa recta**, which flows in the opposite direction, gaining water as it flows back to the cortex. So when the person needs to, it can save water by producing a more concentrated urine

The second convoluted tubule

Whilst the first convoluted tubule is reabsorbing most of the filtrate, the second convoluted tubule 'fine tunes' the remaining fluid, according to the immediate needs of the body. This tubule plays an important role in the regulation of pH, salt and water balance.

■ See questions 1–3 and 5–7.

Table 11.2 **The water balance sheet for a 24-hour period**

Water gain	Volume/ cm³	Water loss	Volume/ cm³
Food and drink	2100	through skin	350
Metabolic water	200	sweat	100
		in breath	350
		urine	1400
		faeces	100
Total	2300	Total	2300

4 THE ROLE OF THE KIDNEY IN HOMEOSTASIS

The kidney contributes to several vital homeostatic mechanisms. One of the the most important is regulation of water content and blood volume.

Water balance in humans

Table 11.2 shows a typical water balance sheet for an average person, assuming normal activity and a comfortable external temperature.

On average, we get almost two-thirds of our water from drinks and a third from food. We obtain a small but important proportion of water from metabolic reactions, notably cell respiration.

Some of the water loss shown in Table 11.2 is unavoidable. Metabolic waste must be removed in solution, and so some water loss in urine is inevitable. Similarly, water is always lost from the lungs as we breathe out. A significant amount of water is also lost by diffusion through our skin (this is not the same as sweating).

✔ When we sweat, we produce a salty liquid to keep our body temperature constant. On a hot day, or during exercise, we lose more water as sweat. Unless we drink more, our water balance is maintained by producing a smaller quantity of more concentrated urine.

?

F How would the water balance sheet change if the individual was suffering from diarrhoea?

G Which areas of the nephron carry out active transport?

H In desperation, a castaway drinks seawater. Why is this not a good idea?

See question 4. ■

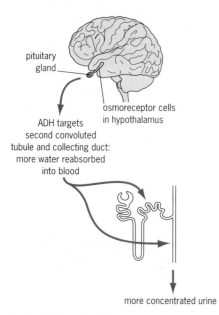

pituitary gland

osmoreceptor cells in hypothalamus

ADH targets second convoluted tubule and collecting duct: more water reabsorbed into blood

more concentrated urine

Fig 11.12 **When the solute concentration of the blood rises, the osmoreceptor cells in the hypothalamus stimulate secretion of ADH from specialised nerve cells in the posterior lobe of the pituitary gland. ADH makes the second convoluted tubule and the collecting duct more permeable to water, and allows more water to pass from the filtrate and into the blood. In this way, the blood becomes more dilute and blood volume increases**

The mechanism of water balance

Like most homeostatic mechanisms, maintenance of water balance involves a negative feedback loop that consists of a **detector** and a **correction mechanism** (see Chapter 10).

A part of the brain, the **hypothalamus,** contains **osmoreceptor cells** that are sensitive to the solute concentration of the blood. When the solute concentration rises, indicating that water loss has exceeded intake, the hypothalamus responds in two ways:

● It stimulates the thirst centre in the brain.

● It stimulates the pituitary gland to release **anti-diuretic hormone (ADH)**.

Fig 11.12 summarises the mechanism of ADH action. ADH acts on the kidney to reduce the volume of urine produced. It achieves this by increasing the permeability of the second convoluted tubule and the collecting duct to water. The action of ADH causes more water to leave the tubule and re-enter the blood. Much more concentrated urine is produced and vital water is conserved.

Conversely, when fluid intake exceeds loss, the blood becomes more dilute. When the hypothalamus detects this, it reduces ADH production. The action of ADH on the kidneys lessens, resulting in less water reabsorption and the production of larger volumes of dilute, or **insipid** urine.

People with the disease **diabetes insipidus** cannot produce ADH because they have a faulty pituitary gland. Once known as the 'pissing evil', this condition results in the constant production of dilute urine, leaving the sufferer permanently thirsty and unable to venture very far from the toilet. Today, it can be treated by giving extracted or synthesised ADH.

Control of blood volume and pressure

Since it regulates water reabsorption, ADH also regulates blood volume. A drop in blood volume leads to a drop in blood *pressure* that is detected by stretch receptors in the walls of the aorta and carotid arteries. Impulses from these detectors pass to the hypothalamus, which then triggers the secretion of more ADH. This acts on the kidneys and causes them to retain more water, so increasing blood pressure.

HANGOVERS!

SOONER OR LATER, many people experience the unpleasant 'morning after' feeling which tends to follow a bout of drinking too much alcohol. Many hangover symptoms are due to dehydration, rather than to the toxic effects of the alcohol or other ingredients. Research has shown that alcohol inhibits the production of ADH, causing water that the body needs to be lost in the urine. Many of the symptoms disappear when the body is rehydrated.

Fig 11.13 **Had he known about the dehydrating effects of alcohol, this man could have minimised his headache by having a long drink of water before he went to bed (assuming, of course, that he could find the tap)**

SUMMARY

After reading this chapter, you should know and understand the following:

■ The kidneys remove metabolic waste and control water and solute levels in the body. As a result of these functions, the kidney also plays a vital role in the control of blood volume and pressure.

■ Each kidney is made from around one million **nephrons**, narrow tubules closely entwined with blood vessels.

■ At one end of the nephron, the **renal capsule**, the kidney filtrate is formed by ultrafiltration (pressure filtration) of the blood.

■ The first filtrate has the same composition as tissue fluid. As it passes along the nephron, the composition of filtrate is altered by various active transport mechanisms that reabsorb some substances whilst allowing others to pass into the urine.

■ A large amount of filtrate is formed, but over 99 per cent if it is reabsorbed in the **first convoluted tubule**, mainly by active transport mechanisms and osmosis. Usually, all glucose and amino acids are reabsorbed into the blood.

■ The movement of solutes from **loop of Henle** creates a region of high solute concentration in the medulla, through which the collecting ducts must pass. As filtrate (now urine) flows along the collecting ducts, water leaves it by osmosis. The resulting urine is **hypertonic** (more concentrated than body fluids).

■ The **second convoluted tubule** alters the composition of urine according to the needs of the body.

QUESTIONS

1 The diagram in Fig 11.Q1 represents a nephron from a human kidney.

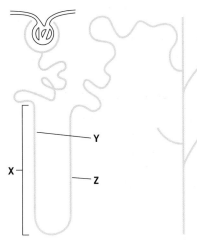

Fig 11.Q1

a) Name the part labelled **X**.

b) Sodium chloride is actively pumped out of **Z** into the medulla of the kidney. This sodium chloride moves back into **Y**.
   Explain the effect of the sodium chloride concentration in the medulla of the kidney on the reabsorption of water from the collecting duct.

c) Most of the sodium chloride filtered into the glomerular filtrate is reabsorbed.
   (i) From which parts of the nephron does this reabsorption take place?
   (ii) How is the reabsorption of sodium chloride controlled?

[AEB 1996: Biology: Specimen paper, q.1]

2

a) If the glomerular filtration rate of the kidneys is 120 cm³ min⁻¹ and the tubular reabsorption rate is 114 cm³ min⁻¹, calculate the rate of urine formation per minute.

b) State **two** differences between tubular reabsorption of water in the first (proximal) convoluted tubule and the second (distal) convoluted tubule.

c) What might reduce the rate of tubular reabsorption of water?

d) The minimum rate of urine production is 300 cm³ per day. Explain why it is necessary for some urine to be produced each day.

[AEB 1995 Human Biology: Paper 1, q.6]

3 The graph of Fig 11.Q3 shows the volume of urine collected from a subject before and after drinking 1000 cm³ of distilled water. The subject's urine was collected immediately before the water was drunk and then at intervals of 30 minutes for several hours.

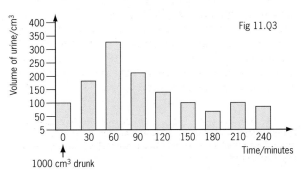

a) **(i)** Describe the changes in urine output during the period of the experiment.
(ii) Name the process responsible for these events.

b) Explain the difference in the volume of urine collected at 60 minutes and at 90 minutes.

c) The experiment was repeated, but this time the subject exercised vigorously for 10 minutes before drinking the water. How would you expect the results of the second experiment to differ from those in the graph? Explain your answer.

d) If, in the original experiment, the subject had drunk 0.9 per cent NaCl solution instead of 1000 cm³ of distilled water, would the same volumes of urine have been collected? Explain your answer.(Note 0.9 per cent NaCl solution is isotonic with blood plasma, i.e. it has the same osmotic potential as blood plasma.)

[AEB 1995 Human Biology: Paper 1, q.8]

4 Nephrosis is a kidney condition in which damage to the glomeruli results in large quantities of protein passing into the glomerular filtrate. This protein finally appears in the urine.

a) Suggest why this protein is not reabsorbed into the blood in the proximal convoluted tubule of the nephron.

b) As a result of nephrosis, large amounts of tissue fluid accumulate in the body, especially in the ankles and feet.
(i) Explain why the loss of protein from the blood results in the accumulation of tissue fluid.
(ii) Suggest why this fluid accumulates especially in the ankles and feet.

c) **(i)** Explain how the action in the kidney of the hormone aldosterone controls the sodium content of the blood.
(ii) Suggest why little aldosterone is produced by a person suffering from nephrosis.

[NEAB 1995 Modular Biology: Physiology Test, q.6]

5 Fig 11.Q5(a) shows a vertical section through a mammalian kidney.

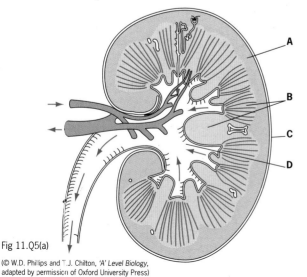

Fig 11.Q5(a)

(© W.D. Phillips and T.J. Chilton, 'A' Level Biology, adapted by permission of Oxford University Press)

a) Label **A** to **D**.

The effective filtration pressure (EFP) in the glomerulus of the mammalian nephron depends on three factors. These are the three forces shown labelled on the left of Fig 11.Q5(b).

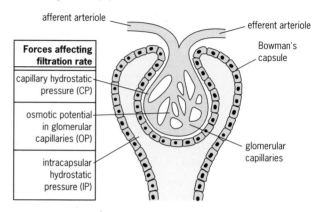

Fig 11.Q5(b)

(© M. Stewart (ed.) *Animal Physiology*, adapted by permission of Hodder & Stoughton/The Open University)

c) **(i)** Suggest which blood component is likely to provide the major contribution to the osmotic potential in the glomerular capillaries.
(ii) Complete the table below, which summarises the effect of the three forces on glomerular filtration, by placing a tick in each appropriate space.

Force	Glomerular filtration	
	encouraged	opposed
CP		
OP		
IP		

(iii) Construct a word equation that shows how the three forces interact to generate the effective filtration pressure.
(iv) Suggest **two** situations, or conditions, where the blood pressure in the glomerulus will fall, leading to impaired renal function.

c) Explain **briefly** the significance of microvilli and mitochondria in the functioning of the proximal convoluted tubule.

d) Outline **briefly** the function of the loop of Henle.

[UCLES 1995 Modular Biology: Transport, regulation and control Paper, q.3]

6 The diagram of Fig 11.Q6 shows a kidney tubule and its blood supply.
Sketch the diagram and use the guide lines and the letters provided to indicate an area where:

A ADH affects the permeability of cell surface membranes.

B active transport of chloride ions occurs.

C reabsorption of glucose occurs.

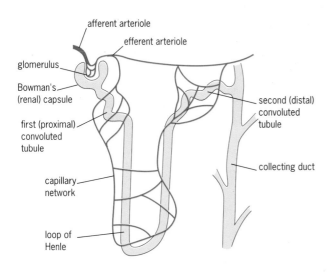

Fig 11.Q6

D there are podocyte cells and ultrafiltration takes place.

E the concentration of blood plasma proteins is highest.

[NEAB June 1998 Biology: Human Systems Module Test, q.4]

7 Fig 11.Q7 is an extract from an information sheet about the drug Frusemide.

a) Name the process described in the box labelled **Y** on the diagram.

b) Explain what causes water to pass out of the descending limb.

c) Sodium and chloride ions normally pass out of the ascending limb by active transport.

Suggest **two** ways in which Frusemide might prevent this process occurring.

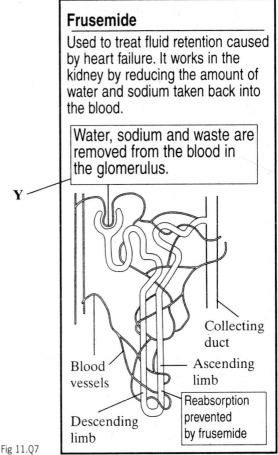

Fig 11.Q7

(Reproduced from an information sheet in the *Education* supplement by permission of *The Guardian*)

d) Explain how the rate of reabsorption of mineral salts from the collecting duct is controlled in a healthy person.

[NEAB March 1998 Biology: Physiology Module Test, q.6]

Assignment

TREATING KIDNEY FAILURE

In many cases of **chronic kidney failure** there is a gradual decline in kidney function, giving the patient plenty of warning. **Acute renal failure** is a crisis in which all kidney function effectively stops. There are many causes of acute renal failure, but generally they can be placed into the following categories:

Sudden loss of large amounts of fluid (blood or tissue fluid)

Inadequate blood flow to the kidneys

Bacterial infection in the kidneys

Effect of toxins

A blockage in the urinary tract caused, for example by damage to the ureter

 Why would a reduced blood flow to the kidneys interfere with kidney function, even though enough blood could get through to provide the kidney cells with the nutrients and oxygen they need to stay alive?

As we saw at the start of this chapter, the immediate problems of kidney failure are a build-up of fluid, urea and potassium. To minimise these problems, patients must follow the Giovanetti diet.

2

a) What are the essential features of this diet? (see Opener)

b) Suggest why protein intake needs to be limited.

Dialysis and kidney machines
Dialysis is a method of separating small molecules from larger ones using a partially permeable membrane. Blood dialysis, or **haemodialysis**, separates the smaller constituents of plasma such as urea and solutes from the larger ones such as proteins.

Blood is taken from the patient, usually from a vein in the forearm, and passed into the machine, where it runs through minute artificial capillaries. These are made from a partially permeable plastic that filters the blood. While blood flows inside the artificial capillaries, a special fluid, the **dialysate**, flows round the outside in the opposite direction.

In dialysis, molecules are exchanged between the blood and the dialysate. The composition of the dialysate is carefully controlled so that there is a net movement of urea, water and salts *out* of the blood.

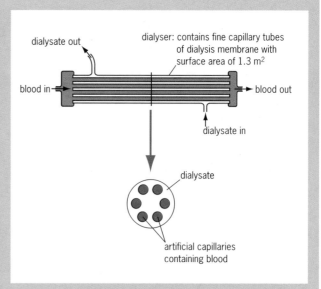

Fig 11.A1 **The fine tubes seen in this dialysis filter are artificial capillaries. Blood flows through the middle of these tubes, while the dialysing fluid flows along the outside in the opposite direction. Each filter is an expensive piece of precision engineering, but it can be used for only a few dialysis sessions and must then be discarded**

3

a) By what physical process do solutes enter or leave the blood during dialysis?

b) Why do the blood and dialysate flow in opposite directions?

c) Suggest two problems which might occur if the dialysate was pure water.

d) Why must the dialysate contain glucose, amino acids and salt?

e) Why is there no urea in the dialysate?

Fig 11.A2 shows the circuit taken by the blood as it passes through the dialysis machine.

4

a) Calculate the volume of blood which is processed by the dialyser in four hours.

b) Why is heparin, an anti-coagulant, added to the blood?

c) Why is heparin not given in the last hour of dialysis?

d) Why is a filter included in the blood circuit?

e) Suggest why the omission of the bubble trap could prove dangerous to the patient.

f) What would a positive reading on the haemoglobin sensor indicate?

g) The dialysis fluid is maintained at approximately 40 °C. Suggest two reasons for this.

h) Excess water can be removed from the blood in a number of ways. One way is to increase the amount of glucose in the dialysing fluid. Explain how this method would work.

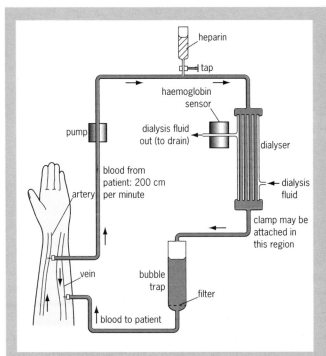

Fig 11.A2 **The essential features of a haemodialysis machine**

i) Another method involves partially clamping the blood tube at the region shown in Fig 11.A2. Explain the principle involved in this method of removing water.

CAPD

CAPD stands for **continuous ambulatory peritoneal dialysis**. In this fairly new treatment, individuals with kidney failure can use one of their own membranes, the **peritoneum**, as a dialysing membrane (Fig 11.A3). Over 3000 people in the UK are presently using this technique because of the advantages it offers over conventional dialysis.

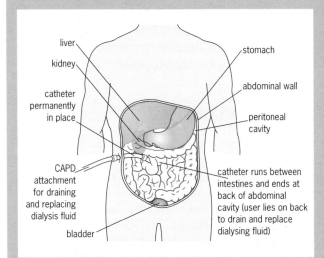

Fig 11.A3 **The peritoneum is a semipermeable membrane that lines the abdominal cavity and organs such as intestines. Dialysate introduced into this cavity draws waste and excess water out of the blood**

The basic principle is simple: patients have a hole, or **stoma**, made in their abdomen wall near the navel, through which a large volume of dialysing fluid is introduced through a tube or **catheter**. The patient is then free to walk around (hence the ambulatory part of the name) while dialysis occurs across the peritoneum between the blood and the dialysate. Every four to six hours the dialysate is replaced. This is a relatively simple exchange procedure that patients can carry out them-selves after some basic training.

5

a) Suggest the biggest problem with the exchange procedure.
b) Explain why it would be no use leaving the dialysate inside the body for longer than the recommended time.

Kidney transplants

When a donor kidney becomes available, it is a relatively simple operation to transplant it into another body. Surprisingly, the old kidneys are left in place: they are rather inaccessible and so are difficult to remove, but they do no harm. The new kidney is placed in the lower abdomen. Surgeons choose this site because the new kidney can be attached easily to a large artery (the **femoral artery**, supplying the leg) and is usefully right next to the bladder.

6 What is the main problem with a transplant once it has been carried out?

Finding a suitable donor for an organ transplant is difficult. Although several hundred thousand people die each year in the UK, only a tiny fraction can provide organs for transplant. For example, accident victims can become organ donors only if their injuries have not affected the organ itself. Due to road safety improvements, the number of serious accidents is decreasing. This is good news, but it means that the number of organs available for transplant is getting ever smaller.

A further complication is that only a minority of people carry donor cards, and permission to used body parts from recently deceased people has to be given by distressed relatives, who often say no. The problems of transplants are covered in more detail in Chapter 34, but you might want to find out more.

7

a) Discuss or jot down some ideas about what could be done to encourage more people to become organ donors.
b) It has been suggested that everyone should automatically be an organ donor, unless they take the decision to opt out of the scheme. Discuss whether this idea would work or not.

12 The skin and temperature control

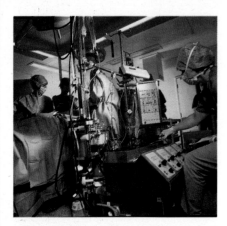

In a heart bypass operation, the heart is stopped for the time it takes to sew on new blood vessels (often veins from the leg). The heart–lung machine at the front takes blood from the vena cava and oxygenates it before returning it to the aorta, always at the right temperature and pressure

THERE ARE MANY recorded cases of 'near drowning', incidents in which people have been plucked from water and resuscitated, and subsequently restored to full health. Rescuing the victim quickly – within minutes – lessens the chance of brain damage as a result of oxygen starvation.

Things are different, however, when the water is very cold. People have been pulled from freezing water – having fallen through thin ice, for example – and have later recovered completely, despite being under water for as long as *one hour*. How is this possible?

It seems that two factors contribute. Firstly, a remarkable response known as the diving reflex occurs in people (especially infants) who are suddenly immersed in cold water. Heartbeat and breathing decrease and blood is redirected so that the organs at the core of the body – the heart, lungs and brain – receive much more blood than the limbs. In this way the vital organs get what little oxygen is available.

Secondly, the rapid cooling effect of the water, known as immersion hypothermia, reduces the oxygen demand of the brain and other organs. This means that people can survive for several minutes, perhaps even as long as an hour after the heart and breathing have stopped.

The same principle is used in surgery. Heart operations often involve stopping the heart and then connecting the patient to a heart–lung machine to supply the body with oxygenated blood. The heart itself, however, receives no blood and so to prolong the time available to operate, the core temperature of the patient's body is lowered to 27 °C. This reduces the oxygen demand sufficiently to double the time the surgeon has before the heart needs to be reconnected.

1 HUMANS AND HEAT

Fig 12.1 **Humans are unique in the animal kingdom in standing upright, having very little hair and being covered in sweat glands. This suggests that humans evolved in a hot climate**

Some animals, such as polar bears, are obviously adapted to withstand the cold. In contrast, humans almost certainly evolved in Africa, and, as a species, we are adapted to a warm climate. We have a thin covering of insulating fat and our body hair is sparse (Fig 12.1). Most hairs are tiny and no use for insulation. We are one of the few animals to be covered in sweat glands, and our skin can make melanin, a dark pigment that blocks out harmful ultraviolet light.

It seems that we were able to spread to the colder areas of the planet only because of our ability to control our environment. We could build shelters, make clothes and control fire. These skill more than made up for the fact that we had few physical features to allow us to cope with the cold.

2 THERMOREGULATION: THE BASICS

Humans, like most mammals, can maintain a constant core body temperature, despite changes in the surroundings. This is known as being warm-blooded or **homiothermic**.

The core temperature of the human body remains reasonably constant at around 37 °C. It can fluctuate by a degree or so (more during fever) but it is generally very stable. Fig 12.2 is a thermal image of an adult male, showing the definition of the body core; this includes the trunk, the head and the upper part of the arms and legs. Body temperature is taken by placing a clinical thermometer in sites that detect the temperature of the core (Fig 12.3).

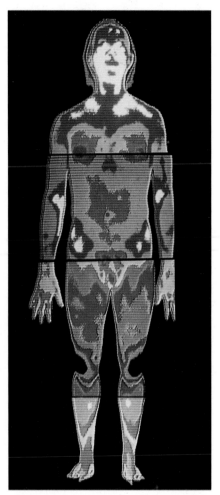

Fig 12.2 **This is a thermal image of a man taken with a camera sensitive to infrared radiation. Areas of higher skin temperature look red or orange, cooler areas look blue or purple. The temperature of the skin can vary greatly, but the core temperature remains more or less stable**

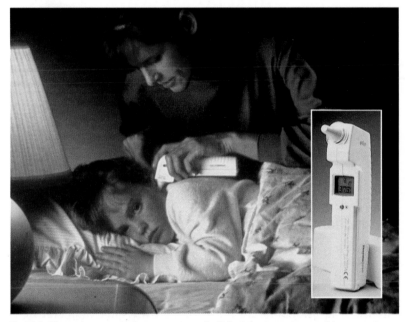

Fig 12.3 **Measuring core body temperature. A clinical thermometer provides a good estimate of core temperature when it is placed in the mouth, armpit or rectum. This more recent thermal probe allows an instant reading of the temperature of the blood near to the brain, without damaging the ear**

Heat is produced inside the body as a by-product of metabolic reactions. Heat production occurs throughout the body, and is especially high in working muscles. It is often stated that the liver is a particular source of heat, but tests have shown that the temperature of blood in the hepatic vein is no higher than in other vessels, suggesting that heat production occurs more generally in the organs of the body. Since body heat is a by-product of metabolism, the amount of heat produced depends on the metabolic rate. This can be increased by doing more exercise and by the secretion of hormones such as thyroxin or adrenaline (Chapter 16).

However, the principal way we control our body temperature is by increasing or decreasing heat loss to the environment, according to our needs. In humans, the basic mechanism that underlies temperature control, or **thermoregulation**, involves part of the brain called the **hypothalamus**. This acts as a thermostat. It can detect the temperature of the blood that passes through it and, if the temperature of the blood increases or decreases even slightly, the hypothalamus initiates corrective responses such as sweating or shivering.

Fig 12.4 **In a cold room, we may automatically assume a fetal position in bed to keep warm: when the arms and legs are drawn into the body this reduces the surface area that can lose heat. In a hot room we do the opposite, spreading out as much as possible**

When we encounter a particularly warm or cold environment, temperature receptors in the skin inform the hypothalamus. They also stimulate the higher, **voluntary**, centres of the brain. This means that we 'feel' hot or cold and decide to do something about it like changing position (Fig 12.4), changing our clothing or turning the heating up or down. Often, this *behavioural* response corrects the situation without the need for any *physiological* response.

Before we can look at temperature control in detail, we need to understand a little about the processes of heat transfer, and to study the structure of the human skin.

3 HEAT: WHERE DOES IT COME FROM? WHERE DOES IT GO?

For any organism to maintain a stable body temperature, its heat loss must equal its heat gain. **Homiotherms** like ourselves, who live in a temperate climate, produce our own internal body heat to keep warm but we also control the amount of heat we lose to the environment.

Heat can be gained or lost in four different ways:

- conduction
- convection
- evaporation
- radiation

Conduction

Conduction involves the transfer of heat between two objects that are in contact with each other. Heat is always conducted from a region of higher temperature to a region of lower temperature. When you sit on a cold seat, for example, heat is conducted from your body into the seat, until both are approximately the same temperature.

The efficiency with which a material conducts heat is called its thermal conductivity. Different materials have different conductivities. Air has a low thermal conductivity, water has a much higher one (Figs 12.5 and 12.6).

A clothed human walking in air at 15 °C could maintain body temperature comfortably. If immersed in water at that temperature, it would not be long before their core temperature dropped and hypothermia began to set in. The heat loss into the water would be

Fig 12.5 **When the Titanic sank in the North Atlantic, the water temperature was similar to that of the air – a little above freezing. However, as we now know, those who remained dry in the lifeboats had a far greater chance of survival. The high thermal conductivity of the water drew the heat out of the unfortunate victims in the sea, most of whom died from hypothermia**

much greater – water can 'draw out' heat approximately 25 to 30 times faster than air at the same temperature.

Materials with a low thermal conductivity are very good insulators. Animals with fur keep warm largely by trapping air between the hairs. Humans rely more on fatty (adipose) tissue for insulation. This has a lower thermal conductivity than other body tissues and so conducts heat more slowly to the surroundings.

Convection

Convection is heat transfer due to currents of air or water. A person immersed in cold, absolutely *still* water would be able to heat up the water immediately next to the skin, and could reduce their heat loss to some extent by not moving.

However, this situation does not happen in real life. For a person who has fallen into the sea or a river, there are usually strong currents that continually move the water over the skin, causing water to be lost quickly by convection. Similarly, fast moving air causes greater heat loss by convection. You have probably heard of the **wind chill factor**. A cold windy day feels a lot worse than a cold still day, even when the actual air temperature is the same.

In air, our clothes reduce heat lost by convection by trapping a layer of air next to the skin. In water, divers reduce the risk of heat loss by convection by wearing wet suits and dry suits (Fig 12.6).

Fig 12.6 **Most divers wear wet suits, even in tropical waters, because the body loses heat rapidly to any water that is below 29 °C. The wet suit traps a layer of water between the skin and the rubber. This warms up quickly as a result of conduction from the skin. The warm water layer cannot escape and so further heat loss is slow.**

For very cold water, *dry* suits are available. These trap a layer of air next to the skin. This provides even better thermal insulation

Evaporation

Evaporation is the change in state from a liquid to a gas. The evaporation of water uses up a large amount of energy, and this is known as the **latent heat of evaporation.** What this means for everyday life is that evaporation of water from a surface has a great cooling effect. Anyone who has ever stepped out of the bath and stood in a draught will have felt the power of this cooling. It is also why sweating is usually an effective way of losing heat when we get too hot (Fig 12.7). Even hot air – such as from a hair dryer – can have a cooling effect provided that the skin is wet so that evaporation can take place.

Radiation

Radiation is the loss of infra-red heat into the surroundings. A human sitting in a room at 20 °C radiates heat into the surrounding air, particularly from the exposed skin on the head, neck and hands. Under normal situations, at rest at comfortable room temperature, most of our heat loss is due to radiation. The heat that we radiate is in the infrared range. This is why infrared cameras can be used to search out humans and other warm-blooded animals from their colder surroundings.

In conduction, convection and evaporation, heat transfer occurs as a result of the movement of molecules. Radiation is fundamentally different: it does not depend on molecular movement. This explains why radiated heat can pass through a vacuum.

Fig 12.7 **When the air temperature is at or above body temperature, the *only* way we can lose heat is by evaporation. This is easy when the air is dry, but when humid, the air is already saturated with water vapour and so none can evaporate from the skin. This makes humid heat very uncomfortable**

In Chapter 4 we saw that enzymes, the chemicals that control metabolism, are temperature sensitive. Enzyme activity increases with temperature and this increases our metabolic rate. The Q10 is a value that describes the effect of a 10 °C rise in temperature on the rate of reaction. Most enzymes have Q10 values of between 2 and 3. This means that they double or triple their activity with every 10 °C rise in temperature, up to the temperature when the enzymes are denatured by heat.

4 THE PROBLEMS OF TEMPERATURE EXTREMES

Humans can maintain a stable body temperature over a wide range of external temperatures. But what happens when we are no longer able to thermoregulate effectively, and our internal temperature changes? What happens to people stranded in the desert under a baking sun, or to those trapped in freezing water?

Experiments have been carried out in which a naked man is asked to sit in a room while the temperature around him is gradually raised or lowered. Fig 12.8 shows a generalised graph that plots what happens.

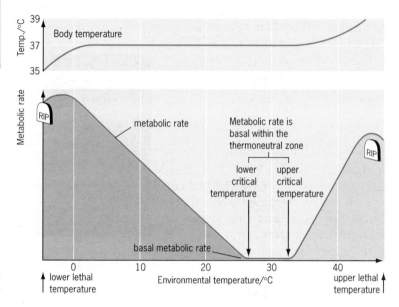

Fig 12.8 **The effect of temperature on metabolic rate**

Throughout the range 15 to 30 °C he is able to maintain a constant core temperature. A point is reached, however, when his sweating and vasodilation are no longer able to keep him cool and he is said to have reached the **upper critical temperature**. After this, his metabolic rate starts to rise, but the experiment is always stopped at this point. If it were not, a positive feedback would begin. The man's enzymes would work faster and consequently make even more heat, raising his temperature further. At about 42 to 44 °C, the **upper lethal temperature** would be reached and he would die.

Why do humans die when the upper lethal temperature is reached? A common theory says that vital enzymes become denatured, but this cannot be the whole story. It is more likely that death from excess heat is due to a metabolic imbalance, caused by enzymes working at different rates. As the temperature increases, some enzymes work faster than others: some intermediate chemicals build up while some vital end-products become scarce. The end result is that cells, notably those in the brain, become irreversibly damaged.

If the same type of experiment is done, but temperature is lowered, the room eventually becomes so cold that the body is no longer able to maintain a constant core temperature: the **lower critical temperature** has been reached. At this point the metabolic rate begins to rise, and this helps the situation by generating more heat. Again, the experiment is always halted at this stage but, if heat loss were allowed to continue, a **lower lethal temperature** would be reached and the subject would die. Find out more about this in the Assignment at the end of the chapter.

?

A Why does the upper critical temperature vary with humidity?

FREEZING HUMANS

CAN WE KEEP PEOPLE ALIVE by freezing them? Can people suffering from incurable diseases be put into deep-freeze until such time as a cure is found, and then revived? The simple answer is no.

Interestingly, human life can survive sub-zero temperatures in a metabolically inactive **dormant state**. Sperm, eggs and even embryos kept in liquid nitrogen at −196 °C, can be thawed out and used successfully in infertility treatment (see Chapter 19). However, it is not possible to freeze people and bring them back to life. One of the main problems is that the formation of ice crystals destroys the cells, but a critical factor in maintaining life seems to be the timescale involved: ice crystals take time to develop. When very small samples are immersed in liquid nitrogen, the freezing process happens rapidly: all the molecules simply 'lock' in position and ice crystals do not form. The cells and organelles are not damaged, and, when thawed, function normally. This technique, however, only works with tiny samples of tissue, that freeze instantly and entirely.

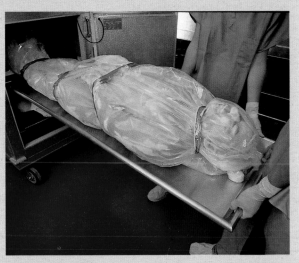

Fig 12.9 **If we developed a way of instantaneously freezing all the cells of a fully developed human, could they be thawed out and live to tell the tale? Only time will tell**

5 THE STRUCTURE OF HUMAN SKIN

The skin is in direct contact with the external environment and so plays a central role in maximising or minimising heat loss. Fig 12.10 shows the structure of mammalian skin.

Structurally, the skin is divided into two layers, the outer **epidermis** and the **dermis** underneath. Forming the boundary between the two is the **Malpighian layer**, whose function is to produce new epidermal cells by mitosis. These cells make the protein keratin, and once they are pushed up into the epidermis they flatten, die and dry up because they have no blood supply. We are constantly losing epidermal cells in a process called **desquamation**. House dust is mainly desquamated skin cells, and a significant proportion of the mass of our mattresses and pillows (about 10 per cent) is due to accumulated skin cells.

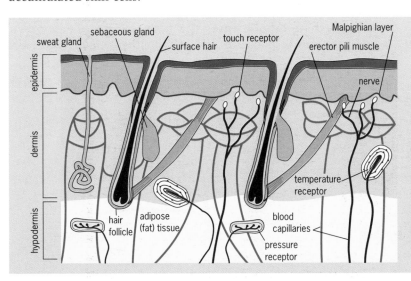

Fig 12.10 **The structure of human skin. Human skin differs from that of most other mammals by being quite bare. We do not necessarily have fewer hairs, but most are tiny and no use for protection or insulation. We are also covered in sweat glands, and can lose heat by evaporation over most of our body surface. In contrast, most mammals have far fewer sweat glands, because evaporation from wet fur is not effective as a cooling mechanism**

?

B Are tattoos places above or below the Malpighian layer? Explain.

The dermis is much thicker than the epidermis, and contains many different structures such as nerve endings, hair follicles and blood vessels, held together with elastic connective tissue. Beneath the dermis is the **hypodermis** that usually contains at least some **subcutaneous fat**. This fat storage tissue, called **adipose tissue**, determines human body shape to a large extent.

Human skin has several functions:

- It detects stimuli using cells that are sensitive to heat, cold, touch, pressure, and pain.
- It prevents excessive water loss or gain.
- It plays a role in thermoregulation by adjusting heat loss according to the circumstances.
- It prevents entry of microorganisms.
- It secretes hair, fingernails and toenails.
- It allows subtle forms of communication. We secrete natural scents, **pheromones**, from modified sweat glands (see Chapter 16).

See question 1. ■

6 THERMOREGULATION

Physiological responses to cold

There are four main physiological responses when the hypothalamus detects a drop in blood temperature:

- We **shiver** when muscles contract and relax rapidly. Shivering muscles give out four or five times as much heat as resting muscles.
- **Vasoconstriction** occurs when the arterioles that lead to the capillaries in the surface layers of the skin **constrict** (narrow), so reducing blood flow to the skin (Fig 12.11). This cuts down the amount of heat lost through the skin. Vasoconstriction is controlled by sympathetic nerves that pass from the **vasomotor centre** in the brain (Chapter 15). This centre, in turn, receives information from the hypothalamus.
- **Piloerection** means literally 'erection of hairs' and involves a reflex. In most mammals, piloerection makes the fur 'thicker', so that it traps more air to provide extra insulation. In humans, the **erector pili** muscles in the skin (see Fig 12.10) pull our tiny hairs upright, but only succeed in creating goose pimples.
- **Increased metabolic rate**. The body secretes the hormone adrenaline in response to cold. This raises metabolic rate and

Fig 12.11 **Blood flow to the skin can be altered by controlling the flow of blood in the arterioles leading to the surface capillaries.**
   When tiny sphincter muscles in the arteriole walls contract to prevent blood flow to the surface, blood is forced through shunt vessels and away from the skin, so less heat is lost

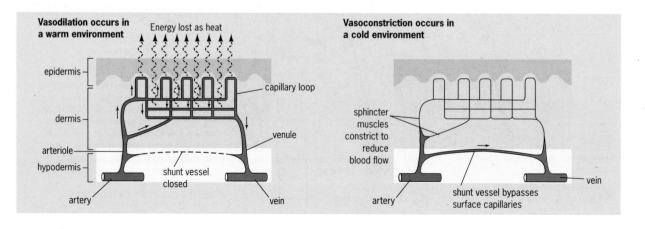

therefore increases heat production. People who live in cold conditions for a period of several weeks or months show a more permanent increase in metabolic rate due to the secretion of thyroxin (see Chapter 16).

Physiological responses to heat

There are two main physiological responses to heat:

- **Vasoconstriction.** This occurs when arterioles that lead to skin capillaries dilate and shunt vessels are closed off, resulting in a greatly increased blood flow to the skin (see Fig 12.11). As a result, more heat can be lost to the environment. Coupled with sweating, vasoconstriction is a very effective cooling mechanism.

- **Sweating.** Sweat is a salty solution made by sweat glands. Evaporation of sweat from the skin's surface leads to cooling. The efficiency of sweating depends on the humidity. In dry air, humans can tolerate temperatures of 65 °C for several hours. In humid air, however, when the sweat cannot evaporate, temperatures of only 35 °C (lower than the core body temperature) cause overheating.

C Which of the responses to cold will be the least effective in humans? Explain your answer.

See question 2.

A common misconception is that blood vessels move to the surface of the skin. Blood vessels are fixed. It is the blood flow through them which can be changed.

D Why do you think that people in the tropics are recommended to take salt tablets regularly?

See question 3.

7 SKIN COLOUR AND SUNLIGHT

The colour of human skin is determined genetically. This is an example of continuous variation (Chapter 26), which suggests that there are several genes operating which give rise to a whole range of skin colour from the palest to the darkest. The darkness of skin is due to the amount of **melanin** present (Fig 12.12). This pigment is able to absorb ultraviolet light, thus protecting the DNA in skin cells from damage (Fig 12.13).

Humans originated in Africa, and needed deeply pigmented skin to protect them from the fierce tropical sun. Under such circumstances, those with darker skin would have had an advantage over paler skinned individuals. It is more difficult, however, to see why white skinned races evolved. Scientists think that the reason for this may be to do with the ability to synthesise vitamin D.

Fig 12.12 **Skin contains cells called melanocytes that produce structures called melanosomes, which contain the melanin. Dark skin contains no more melanocytes than pale skin, but the melanosomes are larger and more evenly spread. Each melanocyte can protect the nuclei of several neighbouring cells. All but the darkest skins will respond to sunlight by producing more melanin**

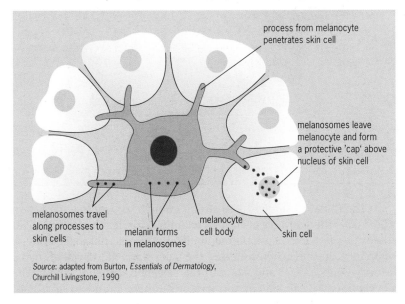

process from melanocyte penetrates skin cell

melanosomes leave melanocyte and form a protective 'cap' above nucleus of skin cell

melanosomes travel along processes to skin cells

melanin forms in melanosomes

melanocyte cell body

skin cell

Source: adapted from Burton, *Essentials of Dermatology*, Churchill Livingstone, 1990

Fig 12.13(a) **Black skin colour is due to the pigment melanin, a substance which protects skin from the damage by ultraviolet light. Skin cancer is rare in dark skinned races. Fair skinned people are not adapted to cope with strong sunlight**

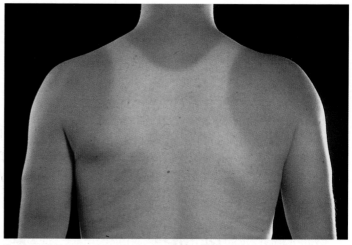

Fig 12.13(b) **This is not a tan, but a first degree burn! When the redness dies down the skin begins to make melanin, giving the desired tan. However, the tan wears away in a couple of weeks. People with pale skin lack the quantities of melanin that provide protection against the genetic damage that ultraviolet light can cause and so run a high risk of developing skin cancer if they sunbathe excessively**

Sunlight and the Earth

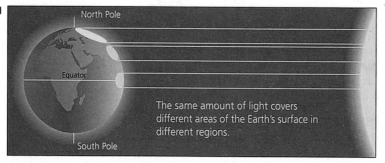

The same amount of light covers different areas of the Earth's surface in different regions.

Skin pigmentation in humans

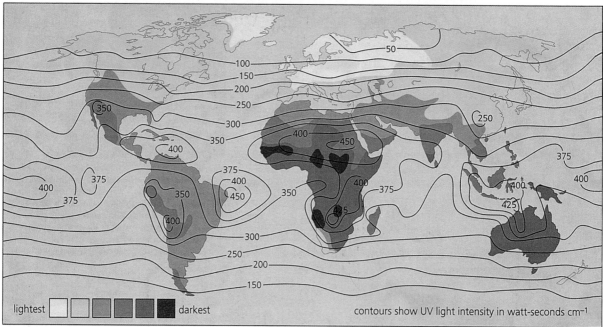

lightest ☐☐☐☐☐ darkest

contours show UV light intensity in watt-seconds cm⁻¹

Fig 12.14 **As a general rule, the further you go from the equator, the paler the skin colour. The map shows the relationship between skin colour and ultraviolet light intensity. (Adapted from Jolly and Plug, *Physical Anthropology and Archaeology*, 2nd edn, Knopf 1979, and Goldsby, *Race and Races*, Macmillan 1971)**

At the Equator, sunlight was strong enough to penetrate dark skin and enable the synthesis of enough vitamin D for health. At lower light intensities, however, dark skinned individuals could not make enough vitamin D. In these circumstances, the paler skinned individuals had a selective advantage and so natural selection brought about the development of pale skin in northern or southern latitudes (Fig 12.14).

THE EFFECT OF SUNLIGHT ON SKIN

THE SUN PRODUCES electromagnetic radiation of many different wavelengths. Some ultraviolet radiation is filtered out by the ozone layer, but some reaches the Earth's surface. Ultraviolet light is a potential problem because it is mutagenic – it damages DNA.

Ultraviolet radiation (Fig 12.15) is split into three categories:

– UVA (long wavelength, 315–400 nm) is the least likely to cause sunburn, and it cannot stimulate melanin production . However, it can damage DNA and lead to gene mutations that can give rise to skin cancer (Fig 12.16).
– UVB (medium wavelength, 295–314 nm) causes sunburn and stimulates melanin production.
– UVC (short wavelength, 200–294 nm) is very damaging to skin but fortunately much of it is filtered out by the ozone layer.

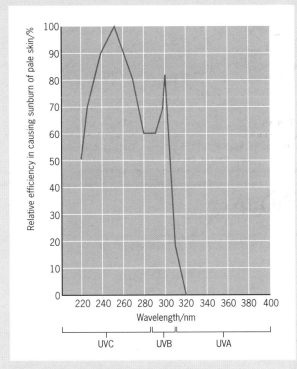

Fig 12.15 **This graph shows the effectiveness of different wavelengths of ultraviolet light in causing sunburn**

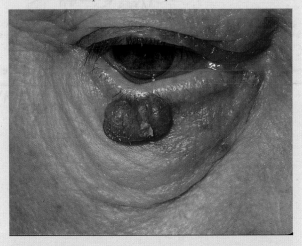

Fig 12.16 **Skin cancer: this is a malignant melanoma, a common cancer especially where light skinned people are exposed to strong sunlight. If the cancer can be removed before it spreads to other parts of the body, the chances of survival are over 90 per cent**

SUMMARY

After studying this chapter, you should know and understand the following:

■ Humans are warm blooded, or **homiothermic**. This means that we can maintain a stable core body temperature despite changes in the external temperature.

■ To maintain a stable body temperature, heat loss must equal heat gain.

■ Unless our climate is very hot, we regulate body temperature by producing heat as a result of metabolic reactions, and by controlling how much of it we release to the environment. When we are too hot, we maximise heat loss; when we are cold, we try to lose as little as possible.

■ Heat is transferred in four ways: **conduction**, **convection**, **evaporation** and **radiation**.

■ Temperature receptors in skin provide an early warning system for temperature change, allowing us to thermoregulate by our behaviour (altering position, clothing, etc). This often removes the need for a physiological response.

■ The **hypothalamus** can detect the temperature of the blood in the core of the body. If the blood temperature drops, the hypothalamus initiates corrective responses such as **vasoconstriction** and **shivering**. It the blood temperature increases, we respond by **vasodilation** and **sweating**. These are physiological responses.

■ Extremes of both heat and cold can cause death. Death as a result of cold occurs as a result of **hypothermia**. This is common in people who have fallen into freezing water.

■ The pigment **melanin** protects the skin from ultraviolet radiation. This type of radiation is **mutagenic** and exposure to some types of UV light can cause skin cancer.

QUESTIONS

1 [See also Chapters 10 and 14 to answer this question.]

a) Explain what is meant by *homeostasis*.

Fig 12.Q1(a) shows part of a generalised negative feedback system.

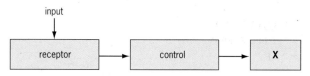

Fig 12.Q1(a)

b) with reference to Fig 12.Q1(a),
  (i) state what X represents
  (ii) on the diagram, draw an arrow to show where negative feedback takes place.

Negative feedback systems often form part of other control systems in the body. Fig 12.Q1(b) shows part of the control system for temperature regulation in mammals.

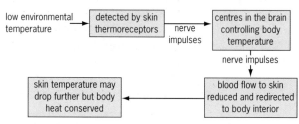

Fig 12.Q1(b)

c) With reference to Fig 12.Q1(b), explain why this part of the system does not show negative feedback.

Homeostatic mechanisms are responsible for the control of blood carbon dioxide concentration. At high altitudes, an increase in breathing rate occurs which decreases the carbon dioxide concentration of the blood, and leads to an increased urine production.

d) Explain the effect of high altitude on urine production.

[UCLES June 1997 Sciences: Transport, Regulation and Control, q.7]

2 [See also Chapter 10 to answer this question.]

a) Explain:
  (i) what is meant by homeostasis;
  (ii) why homeostasis is important in living systems.

b) Hill walkers can encounter extreme changes in environmental conditions. Describe the processes involved in thermoregulation when a walker responds to a rapid fall in external temperature.

[NEAB June 1995 Biology: Processes of Life Module Test, q.9]

3

a) Describe how thermoregulation in a mammal prevents an excessive rise in body temperature during exercise.

[NEAB March 1998 Biology: Processes of Life Module Test, q.10]

4 [See also Chapter 15 to answer this question]

The brain plays an important part in controlling the temperature and solute concentration of the blood. This is summarised in Fig 12.Q.4.

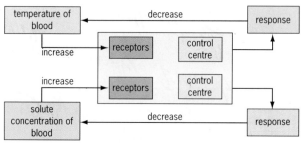

Fig 12.Q4

a) Name the part of the brain which is involved in controlling the temperature and solute concentration of the blood.

b) Use one of the examples in the diagram to explain the meaning of negative feedback.

c) The sympathetic nervous system is involved in the regulation of blood temperature.
  (i) What is the main role of the sympathetic nervous system in the body?
  (ii) Suggest one advantage of regulating blood temperature by the nervous system rather than by hormones.

[NEAB June 1996 Biology: Biological Basis of Behaviour Module Test, q.4]

Assignment

HYPOTHERMIA: NO ONE IS DEAD UNTIL WARM AND DEAD

The definition of hypothermia is a drop in core temperature to 35 °C or below. This condition is very common, and is regularly seen in elderly people or in people who have been immersed in cold water – such as those caught in shipwrecks or people who fall through ice.

Fig 12.A1 **A mountain rescue team uses an insulated bag to restore the core temperature of an accident victim**

Fig 12.A2 **Elderly people whose sensory systems have deteriorated can suffer from hypothermia. When they start to become cold, their body systems fail to respond and the drop in temperature goes uncorrected. As the temperature drops, their metabolic rate decreases and they produce less heat, and so they cool down even more quickly. Death occurs when their core temperature reaches about 25 °C**

People can be revived from the most extreme states of hypothermia, even when all the vital signs point to death. People who are cold and blue, showing no heartbeat or breathing and with fixed and dilated pupils, have been revived and have later recovered, hence the above title.

1

a) What do we mean by core temperature?
b) What temperature is normal core temperature?
c) Why are people more likely to suffer hypothermia when wet or immersed in water than when they are just out in the cold air?

The signs of hypothermia

There are several stages of hypothermia:

Impending hypothermia

The core temperature drops to 36 °C. The person begins to feel uncomfortable and the skin may become pale, numb and waxy. Muscles become tense and can shiver if the patient is not moving. Fatigue and signs of weakness begin to show.

Mild hypothermia

Core temperature drops below 35 °C. Uncontrolled, intense shivering begins. Victims are alert and aware of the situation and so may still be able to help themselves. Movements become clumsy and the cold causes pain and discomfort.

Moderate hypothermia

Core temperature drops below 33 °C. Shivering slows or stops, muscles begin to stiffen and mental confusion and apathy begin. Speech becomes slow and difficult to understand. Drowsiness and eccentric behaviour can follow.

Severe hypothermia

Core temperature falls below 31 °C. The skin is cold and often bluish-grey. The eyes are dilated and victim may appear drunk and deny that there is a problem. There is a gradual loss of consciousness. In extreme cases the victim can appear dead, being rigid and not breathing.

a) Describe the relationship between temperature and enzyme activity (refer to page 70 in Chapter 4).
b) What happens to the oxygen demand of the body as the temperature gets lower?

Treating hypothermia

The basic rationale behind hypothermia treatment is stop the victim's heat loss, allowing the core temperature to return to normal gradually. The basic goals of early care are to stabilise the core temperature and to prevent cardiac arrest. One of the body's first responses to cold is peripheral vasoconstriction and one of the first measures with a hypothermia victim is to wrap the victim in a blanket and allow them to warm up gently while preventing excessive heat loss. Surprisingly, heating the skin – by placing them near a fire, for example – is not a good idea. This will cause vasodilatation that will cause cold blood to return to the body core, resulting in an **afterdrop**. This can be fatal as it can cause the heart to stop.

a) What is peripheral vasoconstriction, and how does it help to conserve heat?
b) Chemicals such a caffeine and alcohol often give a person a warm 'glow' and this is because they cause vasodilation. Explain why coffee or alcohol should not be given to someone suffering from hypothermia.

13 Exercise physiology

To many people, the ultimate sporting achievement is to run a marathon, and it is certainly true that to run for over 26 miles is a remarkable physical feat. Yet there are people for whom a marathon would be just a warm-up. The real endurance record breakers are cross-country skiers, long distance cyclists and tri-athletes

WHAT DO WE MEAN by fitness? Overall, it refers to a person's ability to perform well in their chosen activity. But there are several different aspects to the concept of 'fitness'.

There is strength and speed. Both aspects depend largely on the power in the muscles. Some men can lift 600 kilograms – about the weight of a small car – above their heads, and there are top sprinters who can cover 100 metres in less than 10 seconds – by running at a speed of about 36 kilometres per hour. A person's strength depends to some extent on the cross-sectional size of their muscles, but other factors such as the shape of their skeleton, good technique and motivation, are all important. So is skill. Some people are naturally gifted; their brain is able to control their muscles in just the right way to perform a particular activity. Of course, skill levels are further improved through coaching, training and practice.

Yet another factor is suppleness, the flexibility of the body. Often an underestimated aspect of fitness, most sports require a degree of suppleness, and performance can be improved by incorporating suppleness work into a training programme.

In this chapter we mainly focus on a fourth element of fitness: stamina – the body's 'staying power'. We look at the way in which the body gets its energy, and the different types of 'energy currency' that are important in different sports.

1 GEARING UP FOR EXERCISE

As you sit and read this book, your body is ticking over nicely. Oxygen demand is low, and can be met comfortably by relatively shallow breathing and low pulse rate. Blood is delivering oxygen and glucose to your cells and waste products are being taken away. The levels of these chemicals remain relatively constant, and homeostasis, that all-important steady state (see Chapter 10), is being maintained with relatively little fuss. There is, literally, 'no sweat'.

All of this is rudely interrupted when you begin to exercise (Fig 13.1). The metabolic rate of the muscles increases by up to 20 times, or by 2000 per cent. To fuel this frantic activity, and to maintain some sort of stability, your body must adapt. The overall response of the body to exercise is an excellent example of how the different systems work together to carry on maintaining homeostasis.

Fig 13.1 This is Catherina McKiernan who can run a marathon in a time of 2 h 26 min 26 s. Throughout the 26 miles of this race, her body is able to keep her muscles supplied with oxygen and fuel (glucose and lipid), while ensuring that the waste products are eliminated before they build up. This is a magnificent feat of homeostasis

Homeostasis and exercise physiology

Exercise physiology is the study of the responses of the body to exercise. The effects are familiar: in the short term we sweat, pant and go red, while in the long term we 'get fit', improving our muscle tone, strength, stamina and general well-being. This chapter builds on your knowledge of the human body, pulling together many different areas to show you that homeostasis is a whole-body process involving many different systems. Table 13.1 summarises the relevant topics that are covered elsewhere in the book.

Before we look at the ways in which the body responds to exercise, we need to set the scene by looking at some basic principles:

● The process of **cell respiration** (see Chapter 22) releases the energy in organic molecules such as glucose and lipids, and transfers the energy to a chemical called **adenosine triphosphate**, **ATP** (Fig 13.2).

● ATP provides the energy for muscular contraction; it allows the fibres to slide over each other (see Chapter 18).

● When ATP splits by **hydrolysis** into ADP and P_i (phosphate), energy is released (Fig 13.3).

● The purpose of cell respiration is therefore to re-synthesise ATP from ADP and phosphate.

● ATP splitting is a **coupled** reaction. When we exercise, ATP hydrolysis is coupled with muscular contraction.

● When ATP is split, the some of the energy is used to power the muscle but, as no energy transfer is 100 per cent efficient, some is always lost as heat. This is why vigorous exercise produces large amounts of heat that must escape from the body.

The word **homeostasis**, meaning 'steady state', refers to the fact that conditions within the body need to be kept within certain limits. Homeostatic mechanisms include the control of blood gases (oxygen and carbon dioxide), blood pH, body temperature and blood glucose levels.

Table13.1 **Aspects of exercise physiology in this book**

Topic	Found in
Homeostasis and negative feedback systems	Chapter 10
The cardiac cycle and its control	Chapter 9
Breathing an its control	Chapter 8
Respiration – the production of ATP	Chapter 22
Temperature control	Chapter 12
Movement of muscle/muscle types	Chapter 18
Electrocardiograms	Chapter 9
Artificial pacemakers	Chapter 9
Heart defects and their repair	Chapter 9
Blood transfusions	Chapter 34
Dehydration and water balance	Chapters 7 and 11
Diet and energy content of food	Chapter 6

Energy and exercise

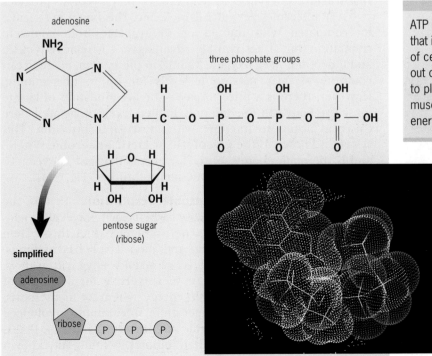

ATP is a relatively small, simple molecule that is generated inside the mitochondria of cells. It can diffuse rapidly in and out of the cells and tissues, and moves to places where it is needed, such as muscle fibres. ATP delivers instant energy in small, usable amounts.

Fig 13.2 **The structure of ATP. In the computer graphics image, adenosine is blue, pentose is white and the three phosphate groups are red. This molecule is present in all cells and is used to provide energy. The purpose of respiration is to re-synthesise ATP as fast as it is made, although during strenuous exercise this is not possible**

Fig 13.3 **When the ATP molecule is split (hydrolysed), the terminal phosphate is removed and combined with water. This reaction releases energy, which is used to drive energy-requiring reactions such as muscular contraction**

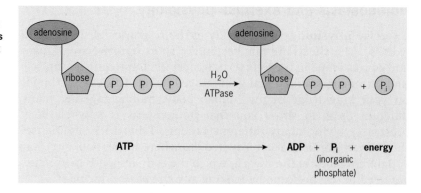

If the movement of muscles requires ATP, it follows that the ability of an athlete to move his or her muscles for any length of time requires a continued supply of this essential chemical. Muscles have *three* sources of ATP:

● **The ATP already present**. This provides instant energy and allows us move on demand. When we contract our muscles as hard and as fast as possible, we use ATP far faster than it can possibly be made. So we rely on the ATP that has accumulated during periods of relative rest. During maximum effort there is only enough ATP for about three seconds, but there is a back-up chemical, **creatine phosphate**, CP. The energy in CP can be used to instantly re-synthesise more ATP, allowing maximal exercise to continue for up to 10 seconds. This is called the **ATP/CP** system, or the **alactic anaerobic** system. *Alactic* means there is no build-up of **lactic acid**, or more accurately, **lactate ions**, and *anaerobic* means that this system does not require oxygen. All events that require explosive bursts of energy, such as weightlifting or short sprints, use the alactic anaerobic system.

● **The ATP provided by glycolysis, the first phase of respiration**. In Chapter 22 you can see that glycolysis provides two ATP molecules per glucose molecule. This might not seem a lot compared with the 36 or so available from complete respiration, but it has two big advantages: it is relatively quick and it does not need oxygen. Thanks to this system, exercise can continue at near maximum levels for up to one minute. However, there is a price to pay – the accumulation of lactate ions. The body can tolerate only limited levels of lactate before this acid begins to interfere with muscular contraction. This system is known as the **glycolytic** or **lactic anaerobic** system, and is the main energy source for events that last between 10 and 60 seconds, such as the 400 metres hurdles (Fig 13.4).

● **The ATP provided by aerobic respiration**. This is the complete breakdown of glucose, when each molecule yields about 36 molecules of ATP. The problem is that this system takes time to provide energy, and then it still has its limits. However, provided that the level of activity stays within those limits, the aerobic system can fuel exercise for a couple of hours of more. This is the main energy system for many sports: all those which last for longer than one minute. Prolonged exercise that increases heart rate and breathing rate, but that is sustainable for hours rather than minutes, is commonly known as aerobics (Fig 13.5).

> ✓ The stores of ATP and CP are known collectively as the **phosphate battery**, so called because they need to be re-charged after a bout of maximum exertion.

Fig 13.4 **Sprinters and hurdlers rely on strength and are usually heavily muscled. To power their muscles during short races they rely on the ATP/PC system for up to about 10 seconds, and for the rest of the race the glycolytic system provides the energy**

Fig 13.5 **A familiar sight all over the western world, sustained aerobic exercise can strengthen the heart, lungs and circulation. As an added bonus, it also uses lipid (fat) as its main fuel, so aerobics are a vital part of many weight-loss programmes**

The energy continuum

From what we have learned so far, it might appear that there are three separate energy systems. But, as we all know, we don't have to stop exercising after ten seconds and then again after one minute to wait for the next energy system to cut in and give us some ATP. All three energy systems blend smoothly together, one taking over from the other. This phenomenon is known as the **energy continuum** (Fig 13.6).

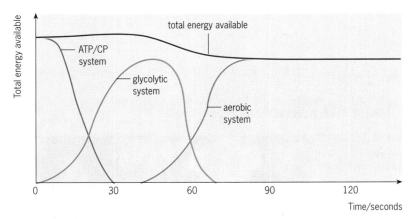

Fig 13.6 **The energy continuum. The graph shows that one system gradually takes over from another, allowing us to move our muscles continuously**

Fuels for exercise

Respiration is the release of the energy contained in organic molecules such as glucose. All foods – carbohydrates, lipids or proteins – can be respired when the need arises. As a rough guide, the body uses the following fuels:

● Glucose. This supplies the normal, everyday energy needs when we are eating regularly (ie not dieting or starving). Anaerobic exercise always depends on glucose, because lipids cannot be used to fuel glycolysis.

● Lipids. Most lipid is stored in fat-storage, or **adipose** tissue. It takes a while to mobilise the lipid stores, but once the process has begun, this reservoir can fuel the body for as long as it lasts (for some of us almost indefinitely). Lipids cannot be used for short term exercise, but they can be used to fuel aerobic exercise. This is why doing an aerobics class regularly can help to get rid of stored fat.

● Protein. The body usually gets around 5 per cent of its energy from protein. The proportion increases significantly only during starvation, when there is no other fuel available. Recent research suggests that protein can be used to fuel endurance events, but to what extent remains uncertain.

Overall, the main fuel used during exercise depends on the intensity of the activity. If you take running as an example, the main fuels for long, gentle jog is lipid. But, as the intensity increases, so does the proportion of energy that comes from glucose.

Table 13.2 **A summary of the energy systems used in exercise**

Energy system	Anaerobic alactic	Anaerobic lactic	Aerobic alactic
Other names	ATP/CP; phosphate battery	glycolytic; lactic system	aerobic system
Energy comes from	ATP/CP already in muscles	glycolysis	complete respiration; glycolysis, Krebs cycle, ET chain
Nature of energy supply	allows maximum strength but is short lived	allows exercise to continue at near-maximum for up to 1 minute	allows long term exercise at lower intensity
Timescale	up to 10 seconds	10 seconds to 1 or 2 minutes	anything longer than 1 to 2 minutes
By-product	no lactate	lactate	no lactate; carbon dioxide
Activity	explosive 'full strength'	longer sprint	endurance events such as cycling and marathons
Examples	sprints up to 100 metres; weight-lifting and throwing events	200 to 400 metre sprints	800 metres, marathons
Training needed	to improve speed or strength	to improve lactate tolerance	to increase endurance capacity and to strengthen cardiovascular system

DIET FOR ATHLETES

IT HAS LONG been known that the correct diet can improve an athlete's performance, and in recent years the diets of sports stars such as footballers have been increasingly in the spotlight (Fig 13.7).

The best approach is a **holistic**, or whole-body one. The most dedicated athletes pay close attention to their diet, taking in the right amount and types of food at the right time. In addition, they ensure that they have the right amount of sleep and minimise their alcohol intake, especially before important events.

How do you plan a diet to maximise performance? A widely used technique is known as **carbo-loading** or **glycogen loading**. About six days before an event, the athlete goes on to a low carbohydrate, high protein diet. This depletes the body's glycogen reserves. Then, for the three days preceding the event the athlete goes on a high carbohydrate diet, eating 8 to 18 grams of carbohydrate per kilogram of body weight per day. This significantly increases the amount of glycogen stored in the muscles, and this obviously helps during endurance events. It is of little help, however, if the event lasts for less than 90 minutes, and the effect wears off with repetition. Generally, athletes should aim for a diet in which 60 to 70 per cent of their energy comes from carbohydrate, increasing this figure in the day or two before an important event.

Fig 13.7 **The pre-match meal of footballers has traditionally been steak, but in recent years much more attention has been paid to their diet. Nutritionists recommend an easily digested high carbohydrate diet, such as pasta, along with a little low-fat protein such as chicken or fish. This will leave the athlete feeling neither hungry nor bloated, and will top up their glycogen reserves**

DEHYDRATION AND ISOTONIC DRINKS

LONG TERM EXERCISE leads to prolonged sweating, and this can easily dehydrate an athlete. By 'dehydrate' we mean that the blood and body fluids become too concentrated. In the correct biological jargon we say that the *water potential is lowered*. The practical consequence of this is that the blood becomes more viscous and cannot flow as easily. In addition, sweating loses vital ions such as sodium, potassium and chloride, collectively known as electrolytes. A loss of electrolytes can lead to muscular cramps and a greatly reduced performance.

Fig 13.8 **Isotonic drinks contain a solution of electrolytes and glucose or short chain polysaccharides, which can be easily digested and absorbed into the blood**

But, it is not too serious; all the athlete needs to do is to replace the lost water, electrolytes and glucose. This is easy enough if the athlete happens to be at home, but dehydration usually strikes when you are three-quarters of the way through a sporting event. Getting a drink that can do the trick takes a bit of planning.

The fastest way to combat dehydration is to drink pure water. This way, water is absorbed as fast as possible by osmosis. However, if you add glucose and salts to the drink you lower the water potential and therefore slow down the water uptake. In fact, if the drink becomes too concentrated it actually makes matters worse by drawing water *out* of the blood, in the same way as drinking seawater would.

The best compromise is to take an isotonic drink, a mixture that has the same water potential as body fluids. This way, all three components are absorbed as rapidly as possible. So do you have to spend money on expensive isotonic drinks? The simple answer is no. It is easy to make up an isotonic mixture on the same principle as **the oral rehydration therapy** given to people suffering from diarrhoeal diseases (see page 168). For events lasting less than 90 minutes it is debatable whether isotonic drinks are any benefit at all. The fluid can be replaced by that cheapest of drinks, water, and the electrolytes and glucose are replaced later by any sensible meal.

2 SHORT TERM RESPONSES TO EXERCISE

Exercise places great demands on the body. Think about the changes which occur when we exercise:

- Oxygen levels fall.
- Carbon dioxide and lactate levels increase.
- Body temperature increases.
- Blood glucose and glycogen levels fall.
- Fluid and electrolytes (salts) are lost as we sweat.

All this presents a great challenge to the homeostatic mechanisms of the body. In an attempt to maintain some sort of stability, the body responds by:

- **Increasing heartbeat and breathing rate**. The increased respiration of the muscles raises the carbon dioxide level in the blood, and consequently lowers the pH (carbon dioxide is an acidic gas). These changes are detected by sensitive cells – **chemoreceptors** – situated in the **carotid** and **aortic bodies** in major blood vessels (see page 126). In turn, these receptors inform the cardiovascular and respiratory centre in the brain, which responds by increasing the heart rate and the ventilation rate. The result is that gas exchange increases, as does the delivery of oxygen to the tissues and the removal of carbon dioxide.

- **Increased blood flow to the skin**. Muscle movement creates heat that must be lost. By increasing the diameter of the blood vessels that carry blood to the skin surface – a process called

Ventilation rate is the amount of air taken in by our lungs in one minute. It can be calculated as depth of each breath (the tidal volume) × no of breaths per minute. At rest, this could be about 0.5 litres × 12 breaths = 6 litres per minute.

peripheral vasodilation – the blood can push heat to the body's surface where it can be more easily lost to the environment. This is why we 'go red' during and after exercise.

- **Increased sweating**. Sweat glands secrete a salty solution that evaporates from the skin, taking heat with it. Coupled with vasodilation, sweating is a very efficient means of removing excess heat (see Chapter 12 for the science of heat transfer). The problem with excess sweating, of course, is that we also lose water and salts. If these are not replaced, the body becomes dehydrated and an athlete's performance can be impaired.

- **Increased mobilisation of glycogen**. In order to keep the muscles supplied with glucose, the glycogen stores in the muscles and the liver start to be mobilised. Glycogen is a highly branched polymer of glucose, which breaks down rapidly to release glucose (see Chapter 3). After prolonged exercise, glycogen stores are replenished from carbohydrate in the diet.

The recovery process

After exercise, following the short term changes listed above, the body does not immediately return to normal. Each system gradually returns to resting levels. Generally, the fitter the individual, more quickly the resting state is achieved.

Fig 13.9 shows the pulse rate of an athlete before, during and after a session on an exercise bike. Following exercise, pulse rate follows a classic pattern, a rapid initial fall followed by a slower return to normal. The shape of the curve results from two processes: the re-charging of the ATP/CP system – this is known as the **alactacid component** of recovery – and the removal of lactate, called the **lactacid component** of recovery. Both occur because the body was not able to deliver enough oxygen to keep up with demand. This shortfall is known as the **oxygen debt** and is repaid after exercise stops. On a longer term basis, the body must also restore its glycogen levels. After a particularly gruelling exercise session, restoring the glycogen can take up to 48 hours and is not associated with raised heart or breathing rate.

We shall look at each of these components in turn.

Fig 13.9 **The effect of pulse rate on exercise. Note that the resting pulse is just under 60, normal for a trained athlete. The increase in heart rate just before exercise starts is due to anticipation, caused by the release of adrenaline: the more important the event, the more noticeable is the effect of adrenaline. When exercise starts, oxygen demand is greater than supply, and an oxygen debt builds up. This is paid off after the activity has finished**

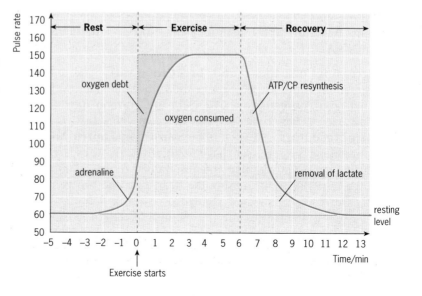

The alactacid oxygen debt component

This is the recharging of the ATP/CP system and takes 1 to 2 minutes to complete. Even fit people can't sprint continually, but after a short 'burst' it takes only a matter of seconds before there is enough ATP/CP to allow them to sprint again. In sports such as football, rugby, tennis or hockey, there are bursts of intense activity followed by periods of relative rest, during which the ATP/CP stores are restored by normal aerobic respiration.

The lactacid oxygen debt component

This is the removal of the lactate ions that accumulated during the exercise. This acidic chemical lowers the pH of the blood, interfering with the enzymes in aerobic respiration and so reducing ATP supply to muscles. A significant build-up is painful, and there is only so much muscle fatigue we can take before we simply have to stop (Fig 13.10). Interestingly, one of the main effects of training is to increase our tolerance to lactate, so that we can continue to exercise despite higher lactate levels.

Lactate builds up in the muscles and diffuses into the blood. It then has five possible fates:

● 65 per cent is oxidised to carbon dioxide and water

● 20 per cent is converted to glycogen

● 10 per cent is converted to protein

● 5 per cent is converted to glucose

● a trace is excreted in urine and sweat

The list clearly shows that the anaerobic system is rather wasteful in terms of energy. Only two ATP molecules are made per glucose, instead of a potential 36 (as in aerobic respiration). The production of 2 lactate molecules therefore represents a great waste of potentially useful energy: when each pair of lactate molecules are broken down in the liver, 34 molecules of ATP fail to be made available to the muscles.

The removal of lactate is speeded up by gentle exercise following the main activity. The **warm down** has the effect of reducing muscle soreness by keeping the capillaries dilated and therefore flushing oxygenated blood through the muscles.

Table 13.3 shows the recovery times that are recommended after exercise. This information is used by trainers to work out the optimum intervals between training sessions, and between training and events. The aim is to avoid chronic (long-term) fatigue that can have a drastic effect on an athlete's performance.

Fig 13.10 **This person has decided to improve his fitness by a little gentle jogging. He is comfortable at walking pace, but as soon as he starts to jog, he passes his** *anaerobic threshold*, **meaning that his muscles do not get oxygen quickly enough and lactate begins to accumulate, making his muscles ache. He will have to slow down until the lactate levels fall, then he will be able to run again. However, if he maintains regular exercise, he can raise his anaerobic threshold, allowing him to run for longer and/or faster without discomfort**

Table 13.3 **Recommended recovery times after exhaustive exercise**

Aspect of recovery	Recommended recovery time/min	
	minimum	maximum
Restoration of muscle ATP and CP	2 minutes	3 minutes
Repayment of alactacid oxygen debt	3 minutes	5 minutes
Restoration of oxygen myoglobin	1 minutes	2 minutes
Restoration of muscle glycogen (after prolonged exercise)	10 hours	48 hours
Removal of lactic acid from muscle & blood	1 hour *	2 hours*
Repayment of lactacid oxygen debt	30 minutes	1 hour

* = speeded up by a warm-down, ie if the muscles are kept working gently, the lactate is removed more quickly

?

A Why does lactate accumulate when oxygen is not supplied quickly enough?

3 THE LONG TERM RESPONSES TO EXERCISE

If we exercise regularly, our body adapts and we 'get fit'. We feel better, look better and are able to cope easily with exercise which, a few months earlier, would have had us gasping in a heap on the floor. A remarkable feature of the human body is its ability to respond to exercise, making us more able to cope with our chosen activity.

So what is actually happening?

Observable long term responses to exercise include changes to the heart, lungs and muscles, although the extent of the effects depend of the type of exercise done. For training to have an observable benefit, it must be above a certain intensity. Muscles must be **overloaded** before they begin to **adapt**.

For instance, if you were going to train for a rugby team, a brisk walk would be totally useless, because it would not overload your muscles. Rugby, like many sports, uses a combination of all the energy systems, and the training should reflect this balance. You would need to do endurance (cardiovascular) work as well as short and long sprints, along with training (eg with weights) that would exercise all of the relevant muscle groups. As the muscles adapt, the intensity of training must be increased continually, to ensure muscles are still overloaded. This is the concept of **progressive resistance** and ensures that the body continues to adapt and improve.

Changes to the heart

The heart responds to exercise like any other muscle; it enlarges, although the nature of the enlargement depends on the type of exercise done. Generally, there is an increase in the size of the **myocardium** (all the heart muscle) and therefore an increase in the size of the chambers. Consequently the **stroke volume** – the volume of blood pumped with each beat – increases. As a general guide, an untrained person has a stroke volume of about 90 cm^3, increasing to over 120 cm^3 after training.

When the heart can pump more blood per beat, it does not have to beat as often when the body is at rest. This is why getting fitter causes a decrease in resting pulse. This can go down from about 70 in the average untrained person to less than 50. Some of the world's top endurance athletes have a resting pulse of about 35 beats per minute.

Research also shows that exercise actually increases the strength of the blood vessels, allowing them to withstand higher pressures and reducing the risk of atherosclerosis (hardening of the arteries) later in life.

Changes to the lungs

The strength of the respiratory muscles (internal and external intercostals and the diaphragm) is increased, allowing a greater volume of air to be forced in and out. This means that **ventilation rate** improves. The total lung volume (the vital capacity) and surface area of the alveoli also increases, resulting in a greatly improved rate of gas exchange. Overall, the improvements in the circulation and gas exchange systems results in an increased **VO_2(max)** (see Chapter 8 for a revision of these important terms).

Muscular contraction and the different muscle types are covered in Chapter 18.

A lot of damage can be done by **over-training**, and the effects of this can be worse than undertraining. Over-training, especially in young athletes who are still growing, can damage muscles, joints, ligaments and tendons. In the long term, the immune system can be damaged, leaving athletes susceptible to infection. For this reason training is usually cyclical. Intense periods of training are used to reach peak fitness for an important event, while at other times, the athlete simply maintains fitness.

Changes to the muscles

The way in which the muscles adapt to exercise depends on the type of muscle fibres involved and the nature of the exercise. Here are some of the improvements that can occur:

- As a general rule, repetitive exercise with moderate resistance (such as aerobics, or light weights) improves muscle tone and stamina. Training with greater resistance (such as heavy weights) brings about an increase in muscle size, a process known as **hypertrophy**. Muscular hypertrophy involves an increase in the cross-sectional size of existing muscles. This is due to an increase in the number of myofibrils, the sarcoplasmic volume, and in the amount of connective tissue (tendons and ligaments).

- A general improvement in performance occurs because of an improved coordination of the motor units. A significant aspect of improving skill involves the antagonistic muscles. Studies have shown that as we practise particular movements, such as a tennis serve, we get smoother and more efficient. Not only do we train the active muscle to work better, we also learn to completely suppress the antagonistic muscle, preventing it from interfering with the action.

- With training, muscles increase their **oxidative capacity** – their capacity for respiration. This is achieved by an increase in the number of mitochondria in the muscle cells, an increased supply of ATP and CP and a rise in the quantity of the enzymes involved in respiration. Also increased is the ability of the muscles to store glycogen, the amount of myoglobin and the ability to use lipid as an energy store.

- The density of the capillaries running through muscles increases in response to exercise, a process called **capillarisation**. This allows a more efficient exchange between working muscles and the blood.

■ See questions 1 and 2.

SUMMARY

By the end of this chapter you should know and understand the following:

■ Exercise places great metabolic demands on the body. Homeostatic mechanisms must cope with a much greater oxygen demand as well as a build-up of carbon dioxide and heat. In the long term, there may be a significant loss of water and salts (electrolytes) through sweating.

■ Movement of muscles requires **adenosine triphosphate**, **ATP**, the chemical that supplies the muscles with the energy released in respiration.

■ There are three sources of ATP: these comprise three energy systems. The most immediate is the **ATP/CP**, or **alactic anaerobic** system: ATP already present in the muscles is backed up by creatine phosphate, which is used to make more ATP instantly. The second system that kicks into action is the **glycolytic** or **lactic anaerobic** system: ATP is supplied by glycolysis and this system results in a build-up of lactate ions. The third is the **aerobic** system in which energy comes from the complete oxidation of glucose or lipid.

■ The duration of an event determines the type of energy system used: Up to 10 seconds, the ATP/CP system; from 10 to 60 seconds, the glycolytic system; and more than 60 seconds, the aerobic system.

■ In the short term, the body responds to exercise by maintaining homeostasis. There is an increase in heartbeat and ventilation rate, coupled with vasodilation and sweating.

■ During exercise we can build up an **oxygen debt**. This is paid off after exercise has finished, or during rest periods within the activity. This is why pulse and ventilation rate remain higher than normal even when activity has stopped. Extra oxygen is needed to replace the ATP/CP stores and to oxidise accumulated lactate.

■ The long term responses of the body to exercise are generally those that improve the body's fitness. These include strengthening of the cardiovascular system (heart and blood vessels) and making the lungs and muscles more efficient. The overall result is a body that is better able to cope with the chosen activity.

QUESTIONS

Parts of questions marked * require a knowledge of training not covered in this chapter.

1

a) Define the term *anaerobic metabolism*.

b) Fig 13.Q1 shows the relationship between oxygen consumption and time, before, during and after maximal exercise.

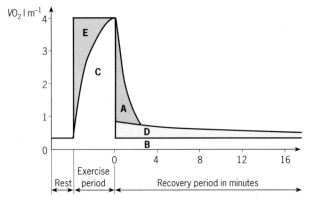

Fig 13.Q1

Using the letters **A**, **B**, **C**, **D** and **E** which label each section, identify:

    (i) the lactacid oxygen debt recovery;
   (ii) the oxygen debt;
  (iii) the alactacid oxygen debt recovery.

c) Describe the process of *ATP production* which restores the oxygen debt.

d) A student performs an interval training session during which the rate of *muscle phosphagen levels* during the recovery period was recorded. The results from this training session are given in the table below.

Recovery time/seconds	Muscle phosphagen restored
10	10%
30	50%
60	75%
90	87%
120	93%
150	97%
180	99%
210	101%
240	102%

    (i) Using the results shown in the table, plot a graph of recovery time against the percentage of muscle phosphagen restored.
   (ii) What resting interval would you recommend for a full recovery?
  (iii) What would be the effect of restarting the exercise after 30 seconds?

e)   **(i)** Part of the recovery mechanism after anaerobic exercise involves *myoglobin*. Explain the function of myoglobin during the recovery process.

   (ii) Explain the importance of the *cool-down* in the assistance of lactacid oxygen debt recovery and in the avoidance of muscle soreness.
  (iii)* How could information on oxygen debt recovery be of use to an athlete and coach in the design of training sessions?

[AEB June 1995 Physical Education Paper 2 Section A Physiology of Exercise, q.1]

2

a) (i) Describe the lactic acid system of ATP production and give an example *within a team game* when this energy system is predominantly used.
   (ii) A build-up of lactic acid will eventually cause muscle fatigue. Explain the effect that lactic acid has on skeletal muscle and give reasons why an endurance athlete would want to avoid an early build-up of lactic acid.

b)* Circuit training is a popular method of strength training for many activities. Identify the main characteristics of circuit training and the general guidelines you would follow when planning circuit.

c) Define the term 'carbo-loading', explain how it is achieved and give an example of the type of athlete who would benefit most from this practice.

d) The amount of oxygen available not only has a direct effect on whether energy is released aerobically or anaerobically, but also affects the type of food fuel used. Under what circumstances is carbohydrate used as the predominant food fuel and why?

[UCLES June 1998 Physical Education, Paper 2 Optional Topics, Section A: Scientific topics, Topic 4: Exercise Physiology, q.1]

3

During exercise, physiological changes take place within the body.

a)   **(i)** What is the difference between a short term physiological response and a long term physiological adaptation? Give examples to support your answer.
   (ii) If an athlete completed a twelve week programme of intense aerobic training, what physiological adaptations would you expect to take place within the cardiovascular system?

b)* The Multi-stage fitness test is commonly used by athletes to assess their level of endurance. State what this test measures and discuss the advantages and disadvantages of this method of assessment.

c) Myoglobin is found in the sarcoplasm of the muscle cell. Explain why an increase in myoglobin is beneficial to an endurance athlete.

[UCLES June 1998 Physical Education, Paper 2 Optional Topics, Section A: Scientific topics, Topic 4: Exercise Physiology, q.1]

Assignment

ENERGY SYSTEMS AND SPORT

In this chapter we have seen that the stamina of an athlete depends, to a great extent, on their ability to ability to supply their muscles with sufficient oxygen to make ATP by aerobic respiration. (You might need to revise respiration by looking at Chapter 22 before attempting this Assignment.)

1

a) What is the essential difference between aerobic and anaerobic respiration?

b) How many molecules of ATP are produced, per molecule of glucose:
   (i) from anaerobic respiration (glycolysis);
   (ii) from complete aerobic respiration (glycolysis, link reaction, Krebs cycle and electron transport chain)?

For athletes, it would seem that the aerobic pathway is the most important, but we must take into account the time taken to complete the process. Complete aerobic respiration takes between 50 and 60 seconds, but glycolysis takes only about 10 seconds to produce ATP. Glycolysis is therefore very useful to an athlete because it is quick and requires no oxygen, so can still provide ATP even when the circulation can't supply oxygen quickly enough for aerobic respiration.

If it takes 10 seconds for glycolysis to produce ATP, how can we move instantly? The answer is that there is already ATP in the muscles: it was stored up during periods of inactivity. During strenuous exertion, such as when we sprint flat out or lift a very heavy weight, we use our ATP reserves in about 3 seconds. This leaves 7 seconds, which we can get through, thanks to a back-up chemical called creatine phosphate (CP). CP is similar to ATP but yields even more energy when broken down, and can be used to instantly re-synthesise ATP, allowing maximum exertion to continue for another 6 or 7 seconds.

2

a) Suggest why we have enough ATP and CP to last for about 10 seconds.

b) Name the chemical produces as a consequence of anaerobic respiration. Explain its effect on the muscles.

The duration of a sport determines the main source of ATP used. The ATP/CP system is used for events lasting up to about 10 seconds, while the anaerobic lactic system (glycolysis) is used for events lasting between 10 and 60 seconds. Beyond about 60 seconds, the aerobic system provides most of the ATP.

3 Copy and complete the table below, identifying the energy system which provides *most* of the ATP for that particular activity.

Sport/Event	ATP/CP	Anaerobic	Aerobic
Javelin			
200 m sprint			
800 m run			
Shot put			
Marathon			
100 m sprint			
400 m sprint			
5 000 m run			
Soccer			

Fig 13.A1 **Sprinting is an anaerobic activity which relies on strength. Sprinters therefore tend to be heavily muscled**

Fig 13.A2 **Prolonged exercise relies on the aerobic system to provide a continual supply of ATP. Generally, fatigue sets in only when fuels such as glycogen begin to run out. Endurance specialists such as this marathon runner tend to be of slight build, with small muscles**

Sports such as soccer, hockey and rugby use a mixture of all the systems because the activity is prolonged but not constant. There are periods of intense activity followed by periods of relative rest.

4

a) Suggest why the 400 metres (world record under 40 seconds) is thought by athletes to be a particularly painful event?

b) What use would the information in the completed table (above) be to someone who wanted to improve his or her fitness for a particular event?

c) Suggest why sprinters usually have large muscles while endurance athletes are generally quite thin?

d) An amateur footballer wishes to get fit for the new season, so he jogs 5 miles each day. Explain why this would be an insufficient preparation and suggest the additional training he would need to do.

e) You can occasionally see joggers our for a run with their dogs, although it is rare for a cat to do the same. Cats in the wild hunt their prey by pouncing or using a short sprint, but hunting dogs rely on stamina in a long chase. Which is the predominant energy system in dogs and cats? Explain why dogs need more regular exercise than cats.

Fig 13.A3 **Wild dogs using stamina to run down wildebeest**

Aerobics and weight loss

Many people want to improve their overall fitness. But can exercising help them to lose weight as well? To understand more about aerobics and weight loss, we need to think about the type of fuel being used in respiration. The way we use respiratory fuels has been likened to the way in which we use money.

We all need spending power, and a common form of payment is cash. When we have no cash, we have to go to the cashpoint or bank and get some more out instantly. When our bank account is low, we can turn to our savings account. This may take a while to get at, but it provides a useful cushion. When we are completely broke, we will have to sell our belongings and property just to survive. Eventually, with no injection of cash, we will be declared bankrupt.

5 Rewrite the last paragraph, applying the analogy to human energy sources. Include the terms lipid, protein, glucose, dead, energy, food and glycogen. When you have done this, write down or discuss the limitations of this analogy.

It takes time to run down our glycogen store and start to use fats. This is a shame, because otherwise you could simply go for a run and return slimmer, if a little baggy. In practice, it takes several sessions – and often a reduced calorific intake – to bring about an improvement. Generally, exercise is beneficial because it strengthens the cardiovascular system and increases metabolic rate, so you use up more calories all the time, not just while exercising.

6

a) What is meant by **(i)** metabolic rate and **(ii)** cardiovascular system?

b) Imagine you are a GP faced with a rather fat couple in their forties. Write down or discuss how you would try to persuade them to take exercise. What exercise would you recommend and what precautions would you include?

CONTROL SYSTEMS

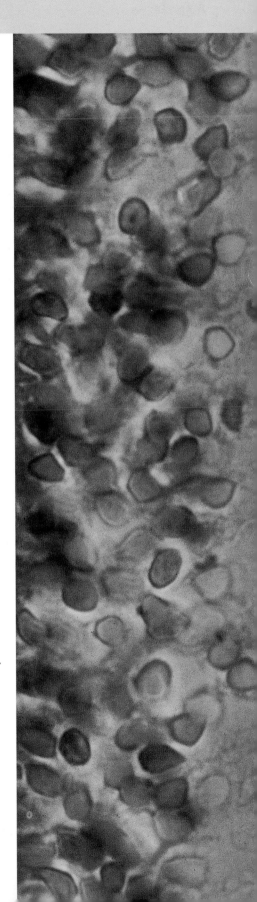

THERE ARE MORE CELLS in your body than there are people on Earth, so how do they all communicate effectively? How, for instance, do the cells of the kidney know when to save water and when to lose it? How do the sweat glands know when to sweat? These functions, and many more, rely on communication between different parts of the body.

This section looks at coordination – the way in which the body detects external and internal changes, and how this information is combined with memory to enable us to decide how to respond. Firstly, in Chapter 14 we look at the organisation of the nervous system, and at how the remarkable thread-like nerve cells can pass information from one end to the other.

Then, in Chapter 15, we look at the organ that makes humans human – the brain. Other mammals have similar body plans to humans, their organs and bio-chemical processes work in much the same way, but they are not human. It is our brain that sets us apart, and that is largely responsible for our success.

In the next chapter, we also look at hormones, the chemical communicators within the body. These work in a variety of ways to control our moods, feelings, stress levels and sexual activity as well as more basic physiological factors such as salt balance, water retention and blood glucose levels.

Chapter 17 covers the sense organs, which allow us to detect changes in our environment and to be aware of our surroundings. We look in detail at that most delicate and subtle of all sense organs – the eye.

Finally, we end this section with a chapter on movement that describes how our skeleton, joints, ligaments, tendons and muscles enable us to put thoughts into action.

14 The nervous system

The Amazonian Indians, not being analytical chemists, assess the strength of their preparation by how long it takes a monkey to fall out of the tree. 'One-tree curare' is the most powerful; the monkey only escapes to the next tree before the paralysing toxin brings it tumbling to the forest floor. 'Two-tree curare' is obviously weaker and 'three-tree curare' is the weakest preparation that is useful. After that, the monkey travels too far and, although it still dies from the action of the poison, it's usually impossible to find

SOME AMAZONIAN INDIANS have a very effective way of hunting monkeys and other animals. They tip their blowpipe arrows in a preparation made from the bark of one of the local trees. The substance, known as *curare*, is a powerful neurotoxin.

For many centuries, the exact nature of curare was a mystery. In 1814, in an attempt to confirm his suspicions, the explorer Charles Waterton injected a donkey with curare. Within a few minutes, the donkey appeared dead. Waterton cut a small hole in her throat and inflated her lungs with a pair of bellows. According to his account, the donkey regained consciousness and looked around. This crude method of artificial respiration was continued for a couple of hours until the effects of the curare had worn off.

Analysis of curare has since shown that it prevents nerve impulses getting through to the muscles, causing paralysis. In addition to affecting the movement of skeletal muscles, curare interferes with heartbeat and breathing, although the exact effects depend on the dosage and whether the curare is taken orally or by injection. If heartbeat and breathing are severely affected, the result is usually fatal.

In 1939, the active ingredient in curare was isolated, and in 1943 doctors started to use it as an anaesthetic. This gift from the rainforest has proved to be a vital tool in surgery, relaxing muscles and generally making the surgeon's job a lot easier. It is also useful for treating conditions in which the muscles go into spasm – polio, tetanus, epilepsy and cholera. Today, synthetic analogues of the active ingredient in curare, such as d-tubocurarine, are used widely in medicine.

1 AN OVERVIEW OF THE NERVOUS SYSTEM

The nervous system is an incredibly complex system, made up of specialised cells that allow the different parts of the body to communicate. Ultimately, the nervous system ensures that the body responds appropriately to the external conditions at any given time. It allows us to:

- gather information. Sense organs called **receptors** detect **stimuli** from the internal and the external environment (see Chapter 17).
- transmit sensory information to the **central nervous system** by means of the **sensory nerves** (see Chapter 15).
- coordinate information. Incoming information travels to the brain via the spinal cord. The brain then decides what to do. Decisions are often based on **memory**, the result of our past experience.

- transmit the information to the **effectors**: the muscles and glands. Impulses pass from the central nervous system to the effectors via **motor nerves** (Fig 14.1).

The basic unit of the nervous system is the specialised, elongated cell called the **neurone**. Neurones are said to be **excitable**. This means that they can transmit impulses from one part of the body to another. We concentrate on the structure and function of these remarkable cells in this chapter. In Chapter 15 we go on to look at how they work together in the nerves (Figs 14.2 and 14.3), the central nervous system and in the brain.

How information travels around the body

Information passes along neurones in the form of electrical signals called **nerve impulses.** A nerve impulse, known as an **action potential**, is not a message, nor is it an electrical current (a flow of electrons). It is more a change in ion balance in the nerve cell, which spreads rapidly from one end to the other, like the fire travelling along a burning fuse. It is little more than an electrical 'blip' – but the brain can make sense of the blips because they vary in frequency and arrive down specific nerves (see Chapter 15).

So what happens when the nerve impulse reaches the end of the neurone? It connects with other neurones at junctions called **synapses**. A nerve impulse crosses a synapse usually by means of a **chemical transmitter**. The whole of nervous system therefore communicates by a mixture of electrical and chemical signals. This allows information to travel around an organism with far greater speed and precision than if only chemical signals – such as hormones – were used.

It is important to remember that the human body also communicates by hormonal means. Overall, coordination in humans is achieved by a mixture of nervous and hormonal communication, and the two systems are closely linked (see Chapter 16).

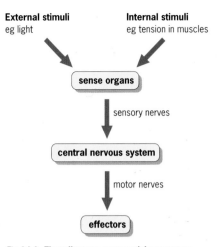

Fig 14.1 **Flow diagram summarising nervous system function**

The entire human body is made up from just four basic types of tissue (see Chapter 2). Epithelial tissue forms coverings and linings, connective tissue holds other tissues together, muscle tissue has the ability to contract and nervous tissue is excitable – it can transmit impulses. All of our organs are formed from these four basic tissue types.

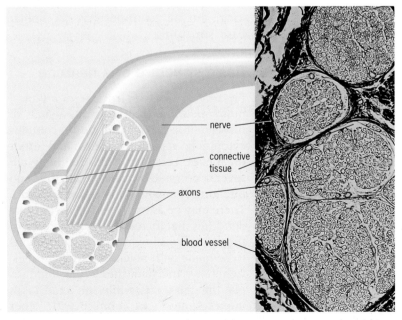

Fig 14.2 **A nerve is a bundle of axons, together with connective tissue and blood vessels. The micrograph shows a cross section of a nerve**

nerve

connective tissue

axons

blood vessel

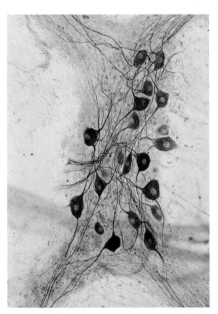

Fig 14.3 **The brown threads (axons) and their cell bodies (see Fig 14.4) form bundles that make up nerves**

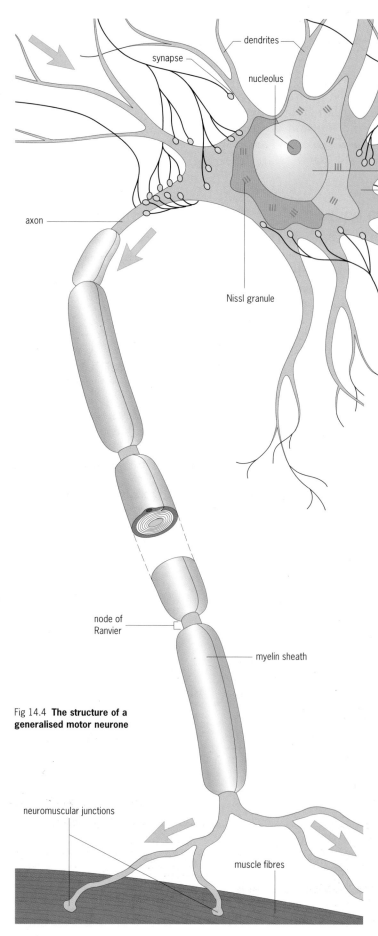

direction of
information transfer

dendrites

synapse

nucleolus

nucleus

cell body

synaptic bulbs at
axon terminals

axon

Nissl granule

node of
Ranvier

myelin sheath

Fig 14.4 **The structure of a
generalised motor neurone**

neuromuscular junctions

muscle fibres

2 THE NEURONE: BASIC UNIT OF THE NERVOUS SYSTEM

The neurone is the functional unit of the nervous system. It connects with many other neurones, receiving and transmitting information. Neurones vary in shape and size depending on their position and function (those extending down your leg can be over one metre long) but all neurones have a similar basic structure.

Basic structure of a neurone

Fig 14.4 shows the structure of a typical motor neurone that transmits signals to muscle fibres. The **cell body** contains cytoplasm, a large nucleus and other organelles. Leading off the cell body are several processes, called **dendrites** and one long elongated limb called an **axon**. There may be as many as 200 thread-like dendrites that increase the surface area of the cell body, allowing many connections to be made with neighbouring neurones. Generally, the dendrites bring impulses *into* the cell body, while the axon takes impulses *away*, to connect with other neurones, or with effectors such as muscles and glands.

A distinctive feature of neurones is that their cell bodies contain **Nissl granules**. These are the site of protein synthesis in the neurone and are thought to be part of a maintenance system that monitors the state of the cell. The cytoplasm, or **axoplasm**, extends throughout the neurone, into the dendrites, cell body, axon and synaptic bulbs. Materials reach different parts of the cell by **axoplasmic transport**.

Types of neurone

Neurones can be classified according to how many processes they have, as shown in Fig 14.5. They can also be classified according to their role in the body.

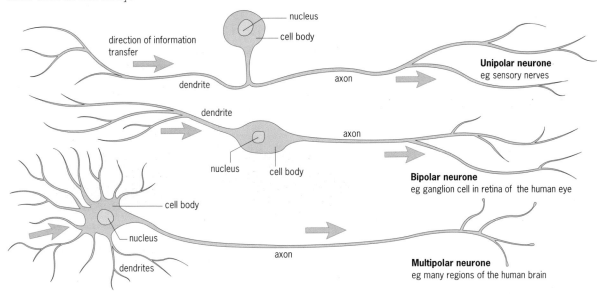

Fig 14.5 **Different types of neurones**

Unipolar neurones have a *single* process that comes out of the cell body and then divides into two branches, a dendrite and an axon. Unipolar neurones are very common in the **peripheral nervous system** (see Chapter 15). The sensory neurone, for example, has its incoming dendrite in the outer parts of the body, while the other branch, the axon, extends into the spinal cord.

Bipolar neurones have *two* processes that come out of the cell body: the dendrite and the axon. All neurones in the nervous system of a human embryo are bipolar, but most then develop into the other two types, so there are few bipolar neurones in the nervous system of an adult. They occur in the retina of the eye, the cochlea in the inner ear and in the nerves serving smell receptor cells in the nose.

Multipolar neurones have *many* processes coming out of the cell body – several dendrites and one axon. Multipolar neurones are the most common type of neurone in the brain and spinal cord.

Cells associated with neurones

Glial cells are packed between neurones, making up tissue called the neuroglia. Glial cells:

- give mechanical support to the neurone network,
- provide electrical insulation by coming between the neurones,
- form myelin sheaths around myelinated neurones,
- have a metabolic role: they control the nutrient and ionic balance round neurones, and break down neurotransmitter substances after signals are transmitted (see page 217).

Glial cells make up about half the weight of the human brain, outnumbering neurones 50 to 1. In other parts of the nervous system, the proportion is much lower, at about 10 to 1.

✔
The pale, creamy colour of myelinated nerves is due to the fatty (lipid) nature of the myelin that surrounds them.

Myelin – insulation for neurones

As Fig 14.6 shows, specialised glial cells called **Schwann cells** wrap themselves round the axons of some neurones. The Schwann cells form a thick, lipid-rich insulating layer called the **myelin sheath**. This insulates the axon electrically, rather like the plastic layer round a copper wire in an electrical flex. Neurones with myelin sheaths are said to be **myelinated**.

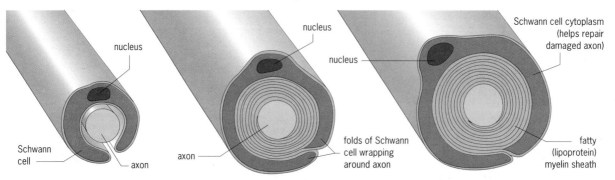

Fig 14.6 **The development of the myelin sheath around an axon. The Schwann cells wrap around many times, and when the cytoplasm is squeezed out the result is many layers of cell surface membrane. Thus the myelin sheath is mainly composed of phospholipid**

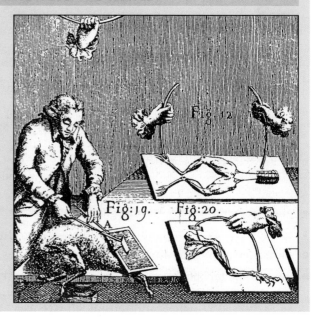

Fig 14.7 **The mouse on the right is a genetic mutation called a 'shiverer'. None of its nerves are myelinated, so they lack insulation. Consequently, impulses in axons can set up impulses in adjacent neurones. Muscles are constantly activated, with the result that the mouse shivers. People with multiple sclerosis (a** demyelinating **disease) have a similar problem. Their myelin sheaths become damaged and they have difficulty in controlling muscle movements**

Schwann cells cover most of the axon, but leave bare sections between the cells called **nodes of Ranvier** (see Fig 14.4). Nerve impulses that travel along a myelinated axon 'jump' between these gaps by **saltatory conduction** (see page 215). Myelinated neurones can conduct impulses twice as fast as unmyelinated neurones.

EARLY THEORIES ABOUT HOW SIGNALS TRAVEL

ABOUT FOUR HUNDRED years ago, people began to study how information travels around the body. Early scientists confidently believed that a fluid, the **succus nerveus**, was the message carrier. However, when dissection showed the absence of cavities in nerves and the lack of a free-flowing nerve 'juice' from the cut ends, this theory was abandoned.

Luigi Galvani (1737–1798) discovered the link between electricity and the action of nerve and muscle in living organisms. Galvani found that he could use static electricity to stimulate nerves in dissected frogs and produce muscle contractions, giving rise to the expression 'galvanise into action'. By 1849, Emil du Bois Reymond (1818–1896) had developed a machine that could detect a difference in electrical potential between the inside and outside of muscle fibres. He also showed the existence of electrical changes in nerves.

Fig 14.8 **Luigi Galvani used a small electrical current to stimulate the muscles of frogs and other animals, and so produced muscle contraction**

DISCOVERING HOW NEURONES CARRY INFORMATION

IN THE 1950s, two British physiologists, Alan Hodgkin (born 1914) and Andrew Huxley (born 1917) showed that there is a negative electrical potential between the inside of a neurone and the outside. With the equipment available at the time, they were unable to work with the very tiny axons found in mammals. Instead they used giant axons found in squid. These have a diameter of up to 1 mm (several hundred times wider than a mammalian axon), making it a relatively easy job to insert a tiny microelectrode.

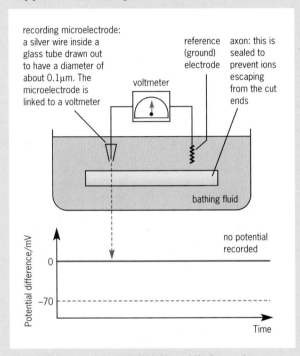

Fig 14.9 **The apparatus which Hodgkin and Huxley used to investigate the nature of the nerve impulse**

Fig 14.9 shows one of their key experiments. Part of an axon is bathed in an isotonic saline solution similar to the fluid that surrounded it in the living squid. A reference electrode – the ground electrode – is suspended in the bathing fluid. The voltmeter measures the electrical potential difference as a voltage between the two electrodes. When the microelectrode is inserted into the axoplasm, as shown in Fig 14.10, the voltage falls to –70 mV. This indicates that the inside of the squid axon is electrically negative with respect to the fluid surrounding it. The trace is produced by an **oscilloscope**, a device that allows us to view electrical changes on a screen.

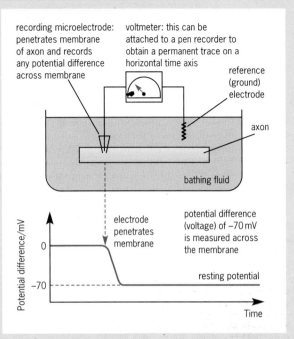

Fig 14.10 **Inserting a microelectrode into a squid axon to measure its resting potential. The oscilloscope trace shows the magnitude of the resting potential**

This experiment was later modified to find out what happened to the electrical potential when a signal was transmitted along the squid axon. Two electrodes were placed inside the axon, one at each end (this is shown in Fig 14.11), and a brief pulse of electrical current was applied through the stimulating electrode to mimic a nerve impulse. The recording electrode's voltmeter registered a change of about 90 mV, from –70 mV to +20 mV. Then almost immediately, the reading reverted back to –70 mV.

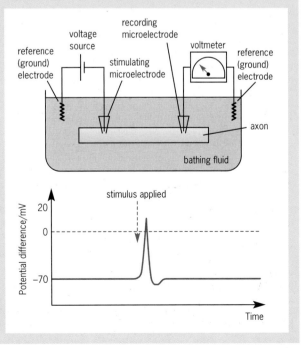

Fig 14.11 **Recording the action potential that results when an electrical stimulus is applied to the axon. The action potential appears as a spike on the oscilloscope trace**

3 HOW DO NEURONES CARRY INFORMATION?

Neurones have two properties that enable them to carry information. They are **excitable** (they can detect and respond to stimuli) and they are **conductive** (they can transmit a signal from one end to the other). Before we look at how a neurone transmits information, let's find out what is going on in the neurone before information arrives.

The neurone at rest

At any given moment a neurone needs to be ready to conduct impulses. This state of readiness is called the **resting potential**, and, at this point, the axon membrane is **polarised**. This means that fluid on the inside is negatively charged with respect to the outside. This difference in charge, about −70 mV, results from an unequal distribution of ions known as an **electrochemical gradient**.

The resting potential

Fig 14.12 **How the resting potential is established. In each square micrometre of axon membrane surface, there are between 100 and 200 sodium–potassium pumps. Each pump actively transports about 200 sodium ions out and about 130 potassium ions in, every second, causing positive ions to accumulate on the outside of the axon membrane. Potassium ions diffuse out through passive ion channels more readily than sodium ions can diffuse in. So overall, an electrochemical gradient develops in which the outside of the axon is positive relative to the inside**

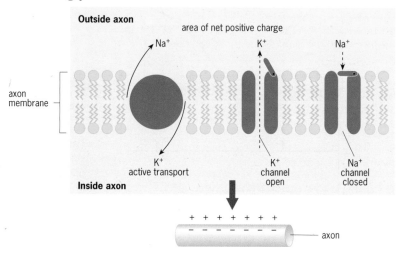

What causes the resting potential, and how is it maintained? Like most cells, neurones have selectively permeable membranes that contain specialist proteins. Some of these proteins act as passive ion channels; others are active transport mechanisms that pump ions. The steps that establish a resting potential are shown in Fig 14.12 and are summarised as follows.

1 An active transport mechanism in the neurone membrane swaps sodium ions for potassium; Na$^+$ ions are moved out, while K$^+$ goes inwards. On its own, this mechanism simply swaps one positively charged ion for another and so does not create the resting potential.

2 A neurone also has passive Na$^+$ channels and K$^+$ channels. In a resting neurone the K$^+$ channels are generally open while the Na$^+$ channels stay shut. *Thus potassium is able to diffuse out faster than sodium can diffuse in*, and diffusion via the passive ion channels is unable to balance out the active transport mechanism. Positive ions accumulate on the outside of the membrane and this then becomes positively charged compared to the inside.

Overall, the resting potential results from an active transport mechanism, which uses the energy in ATP to create an ionic imbalance. If the active transport mechanism failed, the potential would quickly be lost, and the neurone would no longer be excitable.

A (a) What are the essential features of an active transport mechanism.

(b) How is it different from diffusion?

The neurone in action

So what exactly is a nerve impulse? In one sentence, it is a temporary reversal of the resting potential; a change which spreads rapidly along the axon. A schematic diagram explaining how this happens is shown in Fig 14.13, while Fig 14.14 is an oscilloscope trace of an action potential, sometimes referred to as a 'spike' because of its shape.

From the point at which the action potential starts, the wave of electrical activity is propagated (travels) along the axon at great speed. Although the process is similar in myelinated and unmyelinated nerves, there are some important differences. Let's look first at what happens in an unmyelinated neurone.

Depolarisation

When a stimulus reaches the resting neurone, the Na⁺ ion channels open at the site of stimulation. The ions move in relatively slowly at first, making the membrane potential less negative. Then, at a threshold of about $-50\,mV$, a much more rapid inrush of Na^+ ions is triggered, and the inside suddenly becomes positive ($+40\,mV$) with respect to the outside. As the electrical charge across the membrane is reversed, we say that the membrane becomes **depolarised**.

Depolarisation initiates a nerve impulse. This is an **all or nothing** signal: if the threshold is not reached, depolarisation does not occur and no signal can be propagated.

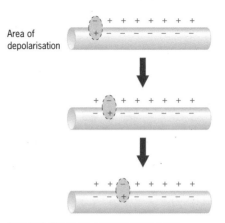

Area of depolarisation

Fig 14.13 **The basic concept of the nerve impulse – a wave of depolarisation which spreads along the axon. The active transport mechanism immediately re-establishes the resting potential as soon as the action potential has passed**

B (a) Suggest why it is important for neurones to have a threshold.

(b) What would be the effect of having no threshold?

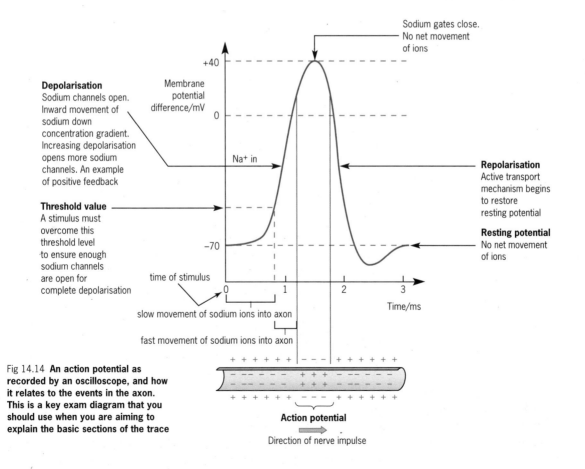

Fig 14.14 **An action potential as recorded by an oscilloscope, and how it relates to the events in the axon. This is a key exam diagram that you should use when you are aiming to explain the basic sections of the trace**

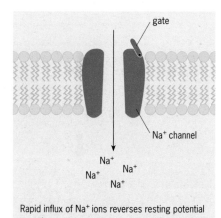

Rapid influx of Na⁺ ions reverses resting potential

Fig 14.15 **A diagrammatic model to show the ion channels in the axon membrane. When the Na⁺ channels open the rapid influx of ions causes the action potential by reversing the resting potential**

The action potential

So why does the area of depolarisation flow along the axon? When a patch of the axon membrane depolarises, a flow of current is produced locally as electrically charged ions move. It is thought that this current stimulates the next patch of membrane, whose Na^+ channels also open, as shown in Fig 14.15. In this way the action potential moves along the axon (in the direction away from the cell body).

The change in voltage (the amplitude) of the action potential is the same at all points along the axon. It does not increase or fall off as the signal travels. After a very brief moment – about 1 millisecond – the Na^+ channels close and the resting potential is established once again. The Na^+ channels cannot re-open until after the resting potential has been re-established.

The refractory period

Nerves conduct messages by 'firing' repeated action potentials along the nerve fibres. The time delay between one action potential and the next is called the **refractory period**. This has two phases.

- The **absolute refractory period**. During this time, immediately after the sodium channels close, *no further impulse can be conducted.*

- The **relative refractory period**. During this time, the membrane begins to recover and becomes increasingly responsive. *It is possible to initiate another action potential provided that the stimulus is greater than normal.*

The refractory period imposes a limit on the frequency of nerve firing. Large nerve fibres recover in 1 millisecond and could theoretically propagate 1000 impulses per second. Small fibres take longer to recover – about 4 milliseconds – and so could propagate about 250 impulses per second.

The refractory period ensures that each action potential is separated from the next, with no overlapping of signals. We can think of the information the signal conveys as coded information. The refractory period also ensures that a nerve impulse flows in one direction only: the wave of depolarisation can only move away from the refractory region, towards the axon terminal, and therefore onwards to the next neurone in the pathway.

The speed of an action potential

The speed at which a nerve impulse or action potential travels is known as its **conduction velocity**. In human nerve fibres, values range from 1 to 3 metres per second in unmyelinated fibres, and between 3 and 120 metres per second in myelinated fibres. In general, conduction velocity depends on the following factors.

- **Axon diameter**. The larger the axon, the faster it conducts.

- **Myelination of the neurone**. A nerve impulse travels faster in a myelinated nerve than an unmyelinated nerve. Myelin, a fatty material, acts as a barrier to the movement of ions across the membrane. Depolarisation occurs at the small gaps (nodes of Ranvier) between the Schwann cells. The action potential therefore 'jumps' from one gap to the next, a process known as saltatory conduction (see the next page).

- **Number of synapses involved**. Communication between neurones across the tiny gaps at the synapses involves chemical release and a brief time delay. The greater the number of synapses in a series of neurones, the slower the conduction velocity.

In myelinated nerves, impulses can travel at up to 120 metres per second – the equivalent of a person running the length of a football pitch in one second, ten times quicker than the best sprinters can run.

C What effect does a myelin sheath have on conduction velocity?

Saltatory conduction in myelinated neurones

Saltatory nerve conduction takes place in myelinated nerve fibres. The insulating Schwann cells round the axon allow ions to cross the membrane only at the nodes – the gaps between the Schwann cells. As a result, action potentials arise only at the nodes, and conduction occurs in a series of saltatory 'jumps' from node to node, as in Fig 14.16. (Saltare is Latin, meaning 'to leap'.) Saltatory conduction has two advantages:

- The conduction of a nerve impulse is fast. In human unmyelinated fibres, nerve impulses travel at 1 to 3 metres per second, while myelinated fibres conduct at speeds of up to 120 metres per second.

- Metabolically, saltatory conduction is quite economical, because fewer ions move across the membrane, so the ion pumps need less energy to restore the ionic balance.

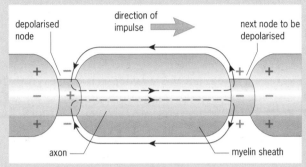

Fig 14.16 **Saltatory conduction in a myelinated nerve: the depolarisation 'jumps' from node to node. The arrows show that local currents are set up between the nodes**

4 COMMUNICATION BETWEEN NEURONES: THE SYNAPSE

When an action potential reaches the end of an axon it is passed on to the next neurone, or on to an effector cell such as a muscle or gland. The axon of one neurone does not usually make direct contact with the cell body of the next; the two cells are separated by a gap called a **synapse**. The structure of a synapse is shown below in Fig 14.17. Fig 14.18 is an electron micrograph of a synapse.

Fig 14.17 **The basic structure of a synapse**

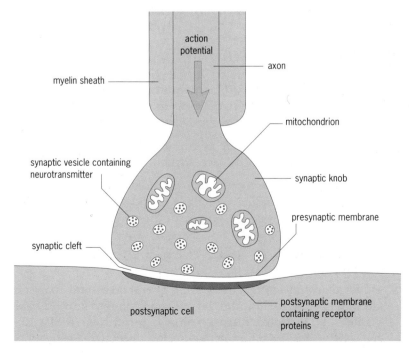

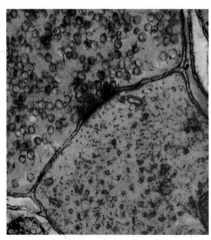

Fig 14.18 **Electron micrograph of a synapse. The presynaptic cell (purple) contains neurotransmitter vesicles (blue). The postsynaptic cell is pink**

The main features of a chemical synapse

The axon terminal of a presynaptic neurone is swollen, and is often called the **synaptic bulb** or **synaptic knob** (See Fig 14.17). It meets the cell body or a dendrite of the next axon, leaving a gap or **synaptic cleft** of about 20 nm between. This is less than one five-hundredth the width of a human hair. Synapses have a high electrical resistance, so that the action potential cannot simply jump from one neurone to the next.

The synaptic bulb contains many mitochondria, which provide energy for the manufacture of chemical transmitters. **Synaptic vesicles** are temporary vacuoles (membrane-bound spheres) that store **neurotransmitter** chemicals, the most common being **acetylcholine**. Neurotransmitters are small molecules that can diffuse easily across the synaptic cleft.

The cell that carries a signal *towards* a synapse is a **presynaptic cell**; the cell carrying the signal *away* from the synapse is a **postsynaptic cell**. Presynaptic cells are always neurones but postsynaptic cells can be either neurones or effector cells.

Neurones usually transmit information across synapses using transmitter chemicals also called **neurotransmitters**. Molecules of these chemicals cross the synaptic cleft and cause electrical changes in the membrane of the postsynaptic cell. If an action potential is produced, this then passes along the axon to the next neurone in the series. Direct communication between neurones is rare, but **electrical synapses** do exist; see page 219.

What happens at the synapse?

As you read the next section, follow the stages of chemical transmission at a synapse numbered in Fig 14.19.

1 An action potential arrives at the synaptic bulb.

2 This opens calcium channels in the presynaptic membrane. As the Ca^{2+} ion concentration inside the membrane is lower than outside, Ca^{2+} ions rush in.

3 As the Ca^{2+} concentration increases, synaptic vesicles move towards the membrane.

4 The vesicles fuse with the membrane, releasing neurotransmitter into the synaptic cleft.

5 The short journey across the synapse takes about a millisecond, longer than an electrical signal takes to travel the same distance. This time is therefore called the synaptic delay.

6 At the postsynaptic cell, the neurotransmitter binds to receptors on the postsynaptic cell surface membrane.

7 Some neurotransmitters open sodium channels in the postsynaptic membrane, causing an inflow of Na^+ ions. This creates an **excitatory postsynaptic potential** (**EPSP**) in the membrane. This potential lasts for only a few milliseconds and can travel only a short distance, but it makes the membrane more receptive to other incoming signals.

8 Other neurotransmitters open chloride and potassium channels, causing Cl^- ions to flow into the cell and K^+ ions to flow out. This creates the **inhibitory postsynaptic potential** (**IPSP**) that makes the postsynaptic membrane less receptive to incoming

D The synaptic cleft has a high electrical resistance but is very narrow. Suggest reasons for these two observations.

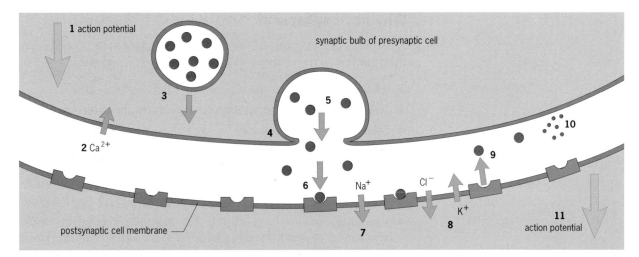

Fig 14.19 **The sequence of events in chemical transmission at a synapse**

signals. If more EPSPs than IPSPs are produced in the postsynaptic membrane, the change in potential can exceed the threshold potential needed to create a new action potential.

Receptor binding can also lead to the formation a second messenger (a transmitter substance) such as **cyclic AMP** (cAMP). This also changes the ionic permeability of the membrane, but it has a *longer lasting* metabolic effect on the ion channels. Such long-term changes to brain neurones are thought to underlie memory.

9 Once the neurotransmitter has acted on the postsynaptic membrane, it is immediately broken down by an enzyme on the postsynaptic membrane. If the transmitter remained, it could continue to stimulate the neurone, even without new impulses coming from the presynaptic cell.

10 The enzyme **acetylcholinesterase** splits acetylcholine, into choline and ethanoic acid. These components then diffuse back into the presynaptic membrane, when they are re-synthesised to acetylcholine using the ATP from the mitochondria.

11 An action potential is set up in the postsynaptic cell.

Types of neurotransmitter

More than 50 neurotransmitters have been identified and there are certainly more to find. There are four main groups:

- **acetylcholine** (neurones which release acetylcholine are described as **cholinergic**).
- **amino acids** such as gamma-aminobutyric acid (GABA), glycine and glutamate.
- **monoamines** such as noradrenaline, dopamine and serotonin (neurones which release noradrenaline are described as **adrenergic**).
- **neuropeptides** (chains of amino acids) such as endorphins (see Chapter 16 for more about endorphins).

Acetylcholine also acts throughout the brain, modifying the activity of other neurotransmitters. Nerve pathways in which acetylcholine is a neurotransmitter seem to be involved in motivation and memory. Acetylcholinesterase, the enzyme that breaks down acetylcholine at the postsynaptic membrane, does so in one five-hundredth of a second. The chemicals in some 'nerve gases' work by inhibiting acetylcholinesterase.

?

E Why can transmission only occur one way across synapses?

F If the EPSPs generated in an neurone are cancelled out by the IPSPs, what will be the response of the neurone?

■ See questions 1 to 5.

Why have synapses?

Synapses are important because they allow the transfer of information in nerve networks to be controlled. Synapses:

- allow information to pass from one neurone to another.
- help ensure that a nerve impulse travels in one direction only.
- allow the next neurone to be excited or inhibited.
- can amplify a signal (make it stronger).
- protect nerve networks by not firing when over-stimulated. When this happens the synapse is said to be fatigued. Over-stimulation might damage muscle or gland tissue.
- can filter out low-level stimuli. For example, you fail to notice the sound of a clock ticking because synapses are 'filtering out' the signal of sound.
- aid information processing by the action of summation (adding together the effect of all impulses received, see page 220).
- are modifiable and can form a physical basis for memory.

See questions 6 and 7. ■

Many drugs and poisons exert their effect because they interfere with the functioning of synapses – see the Assignment at the end of this chapter.

Overall, the significance of synapses cannot be over-emphasised. They allow us to select particular neural pathways. The process of learning is largely one of educating the synapses. People can play the violin or piano, or play tennis because their synapses allow their brains to coordinate their senses and muscles in the right way. Your memories, too, have a basis in synapses choosing specific pathways. If you are asked, 'What's the capital of France?' your synapses will (we hope) select a pathway of neurones in your brain which will lead you to the answer 'Paris'.

NICOTINE – THE DRUG THAT MIMICS ACETYLCHOLINE

DO YOU WONDER why people continue to smoke, even though they know the increased risk of lung cancer and heart disease associated with their habit? The answer lies partly with one of the components of tobacco: nicotine.

Nicotine is very addictive. It affects the brains of smokers, making them feel less stressed, better able to concentrate and less likely to eat sweet foods. Smokers become tolerant to nicotine over time, needing to smoke more to achieve the same effects. But how does nicotine cause addiction?

Studies carried out in the early 1980s using nicotine labelled with a radioactive tracer showed that it is taken into the brain very rapidly. Once there, it binds tightly to acetylcholine receptors, fooling postsynaptic cells into 'thinking' they are being stimulated. It also binds to other receptors which normally accept another neurotransmitter, dopamine.

The action of nicotine on both types of receptor in specific areas of the brain causes long-lasting changes to cell connections and may explain why it is addictive. We know that dopamine receptors, in particular, are involved in addictions to other substances such as amphetamines and cocaine.

Because nicotine is addictive but not carcinogenic (its the chemicals in tobacco tar that have been shown to cause cancer), smokers keen to kick the habit can get help. They can buy skin patches and gum which deliver nicotine to the brain, but not tar to the lungs.

Although patches and gum do help smokers to cut down or stop smoking altogether, they are only a partial solution. Nicotine affects the acetylcholine receptors in the parasympathetic nervous system that are involved with the constriction of blood vessels. Over time, circulatory problems and heart disease can result, and it is best to avoid these effects altogether.

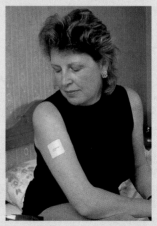

Fig 14.20 **Transdermal patches can be used almost anywhere on the skin. They slowly release substances such as nicotine into the bloodstream**

Electrical synapses

A few nerve cells communicate directly, by electrical synapses. Such cells are connected by hollow protein channels, which allow an action potential to travel directly from one cell to the other. Electrical synapses were first found in the crayfish, and later in vertebrates. They occur in human retinal ganglion cells. Because there is no synaptic delay at electrical synapses, signals pass almost instantaneously from one cell to the next.

5 COMMUNICATION BETWEEN NEURONES: PROCESSING INFORMATION

Neurones do not exist in isolation; they form complex networks that behave like integrated circuits in electronics. (An integrated circuit is defined as 'an assembly of electronic components that cannot be subdivided without destroying its function'.)

Synapses in action; facilitation and summation

Does the arrival of an impulse at a synapse mean that an action potential is always generated on the postsynaptic neurone? The answer is no, because it would lead to chaos; all neurones would automatically connect to others. Synapses therefore have a vital role in information processing. Transmission of information across synapses is *graded*. They can amplify or damp down the information they receive. In many cases, they will not transmit it at all.

A neurone can be fed information by both **excitatory** synapses, **producing excitatory postsynaptic potentials** (**EPSPs**) and **inhibitory** synapses producing **inhibitory postsynaptic potentials** (**IPSPs**) (Fig 14.21). Whether the cell develops an action potential is determined by the sum of all the excitatory and inhibitory synapses at any particular moment. Put simply, impulses arriving at some synapses will 'excite' the cell, while others will 'calm it down'. Whether or not a neurone generates an action potential depends on the balance of the two types.

Imagine a synapse discharging its transmitter onto a postsynaptic neurone. This will set up an EPSP, but if it is not big enough to reach the threshold, no action potential is generated. However, if other

Fig 14.21 (a) **Facilitation and** (b) **summation**

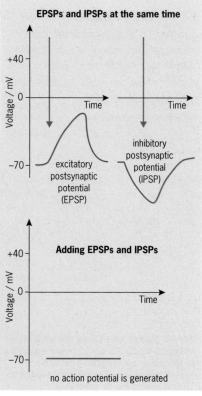

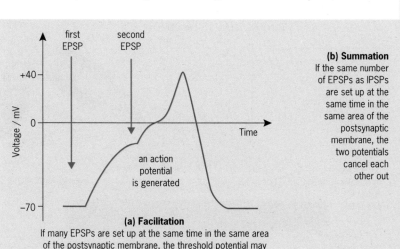

(b) Summation
If the same number of EPSPs as IPSPs are set up at the same time in the same area of the postsynaptic membrane, the two potentials cancel each other out

(a) Facilitation
If many EPSPs are set up at the same time in the same area of the postsynaptic membrane, the threshold potential may be reached and an action potential may result

synapses discharge their transmitter at the same time, or shortly after, the EPSPs will add up, or **summate**, until an action potential is generated. Generally, there are two types of summation:

Temporal summation – summation of two or more impulses that arrive rapidly one after the other down the *same* neurone (temporal = related in time).

Spatial summation – summation of two or more impulses arriving down *different* neurones at the same time (spatial = related in space). One neurone can receive information from many others – this is **synaptic convergence**. It follows that the arrival of one excitatory impulse will leave the neurone more responsive to another one. This is known as **facilitation**, and results from the summation of two or more synapses discharging their transmitter substance at the same time.

As a simple example of this idea, imagine the touch receptors from one area of skin feeding into one sensory neurone. An action impulse down just one receptor is almost certainly an insignificant stimulus, and can be ignored. It will not create an EPSP large enough to generate an action potential in the sensory nerve. However, if several touch receptors are stimulated at the same time, they will summate and produce a sensory impulse.

Note that facilitation is not a result of temporal summation, which is simply the accumulation of EPSPs because impulses arrive before the preceding EPSPs have died down.

Neuromuscular junctions

When a motor neurone terminates on a muscle, it branches into many specialised synapses called **neuromuscular junctions**. Fig 14.22 shows a typical neuromuscular junction. These structures are wider than ordinary synapses and come into close contact with the surface membrane of the muscle, the **sarcolemma**. The area of sarcolemma in contact with the synapse is called the **motor end-plate**, and contains acetylcholine receptor sites. When an action potential arrives at a neuromuscular junction, vesicles of acetylcholine are released in the usual way. The transmitter changes the permeability of the motor end-plate to Na$^+$ and K$^+$, creating an **end-plate potential** (**EPP**) which results in an action potential passing along the sarcolemma. This impulse brings about the contraction of muscles fibres in that area.

Chapter 18 gives more detail about muscular contraction, and the Opener and Assignment of this chapter illustrate how chemicals can affect the neuromuscular junction.

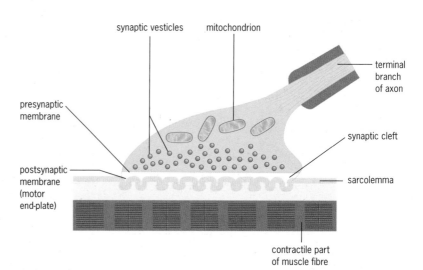

Fig 14.22 **Studies have shown that each vesicle contains about 10 000 acetylcholine molecules, and that about 100 vesicles need to be released before an action potential (leading to a muscle twitch) can be generated**

SUMMARY

When you have read this chapter you should know and understand the following:

■ The basic unit of the nervous system is the nerve cell, or neurone. This is a specialised cell with **dendrites** that take impulses into the cell body, and greatly elongated axons that take impulses away.

■ The axon membrane is able to use an active transport mechanism to establish a **resting potential**. This is an electrical charge across the membrane caused by an unequal distribution of ions.

■ The nerve impulse itself, called the **action potential**, is a momentary reversal in the resting potential, caused by a sudden rush of sodium ions into the axon. The action potential spreads rapidly along the axon.

■ The action potential lasts for only a millisecond or so, after which the resting potential is re-established. When an action potential has passed, there is a brief period of time – the refractory period – during which it is impossible to generate another action potential.

■ The nerve impulse passes from one nerve to another (or from a neurone to a muscle) by means of **synapses**.

■ Synapses are vital in selecting some neural pathways and not others. As such, synapses play a vital role in memory, skill and coordination of the body's activities.

■ Transmission across synapses occurs when a chemical, the neurotransmitter, is released by the **presynaptic membrane**. This diffuses across the gap and changes the permeability of the **postsynaptic membrane**, generating an **excitatory postsynaptic potential (EPSP)**. If the EPSPs are sufficiently large, an action potential is generated in the next neurone.

■ **Summation** means that the effect of several action potentials can add up to produce transmission at a particular synapse.

■ Many drugs and poisons work by affecting synaptic transmission.

QUESTIONS

1

a) Describe how an axon maintains a resting potential across the cell surface membrane.

b) The diagram of Fig 14.Q1 shows part of the cell surface membrane of an axon.

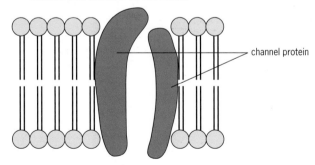

channel protein

Fig 14.Q1

    (i) During an action potential, sodium ions pass through this channel protein. Explain the link between the passage of sodium ions and the potential difference across the membrane.
    (ii) There are different types of channel protein in the cell surface membrane of an axon. Explain how the structure of protein molecules means that different types of channel protein can exist.

[NEAB February 1996 Biology: Biological Basis of Behaviour Module Test, q.5]

2

a) Describe how differences in permeability of the cell surface membrane to particular ions gives rise to a resting potential in an axon.

b) In the first stage of an action potential, the potential difference across the axon membrane changes from around −60 mV to approximately +40 mV. Explain how this happens.

c) Suggest:
    (i) why, although nerve impulses can travel in both directions in an isolated nerve cell, in a whole animal they travel only in one direction.
    (ii) how a single nerve cell can convey information about the strength of a stimulus.

[NEAB June 1995 Biology: Biological Basis of Behaviour Module Test, q.7]

3
The graph of Fig 14.Q3 shows the change in potential difference at a point in a neurone during the propagation of a nerve impulse.
a) From this graph, give the value of:
    (i) the resting potential of the neurone;
    (ii) the maximum change which occurs in the potential difference across the membrane.
b) Explain, in terms of ion movements, the change in potential difference which takes place between **(i)** points **A** and **B**; **(ii)** points **C** and **D**.

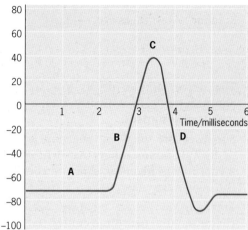

Fig 14.Q3

c) An action potential is sometimes referred to as an 'all or nothing' process.

   (i) Explain what is meant by describing an action potential as an 'all or nothing' process.

  (ii) Explain how this property relates to the way in which a nerve impulse conveys information about the strength of a stimulus.

[NEAB June 1996 Biology: Biological Basis of Behaviour Module Test, q.6]

4

a) Draw a labelled diagram to show the structure of a cell surface membrane.

Fig 14.Q4 shows the changes in membrane potential which occur during the transmission of a nerve impulse.

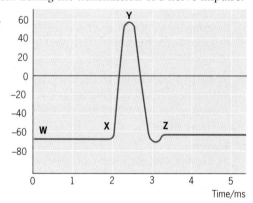

Fig 14.Q4

b)   **(i)** State which letters on the diagram correspond with the process of depolarisation of the axon membrane.

  (ii) State the direction in which sodium ions will move across the membrane during depolarisation.

 (iii) Explain how the impermeability of the axon membrane to sodium ions helps to maintain the resting potential at **W**.

Mammals have myelinated axons, whereas invertebrates, such as squids, have non-myelinated axons.

c) Explain the advantage of having myelinated axons. The table shows the relationship between axon diameter and speed of conduction in a squid axon and that of a cat.

Axon	Diameter/µm	Conduction velocity/m s⁻¹
squid	650	24
cat	4	26

Suggest why it is possible for both animals to conduct impulses with similar velocity.

[UCLES June 1997 Sciences: Biology Foundation, q.2]

5

a) Most synapses and neuromuscular junctions involve the use of substances called neurotransmitters. State what is meant by the term *neurotransmitter*.

Fig 14.Q5(a) shows a synapse. The diagram is not to scale.

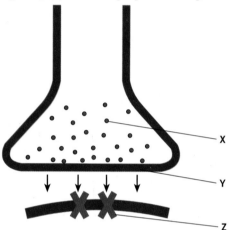

Fig 14.Q5(a)

b) Identify the parts labelled **X**, **Y** and **Z**.

c) Outline the mechanism by which information passes from one cell to another across a synapse.

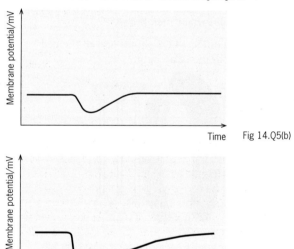

Fig 14.Q5(b)

Fig 14.Q5(c) **Effect of valium**

In the central nervous system (CNS) nerve cells have many synaptic connections with neighbouring cells. Also, there are many different types of synapse in the CNS and these have different physiological roles. Fig 14.Q5(b) shows the changes in the membrane potential at the postsynaptic membrane when a certain type of synapse in the CNS is stimulated.

Anti-anxiety drugs such as valium alter the events in this type of synapse when the neurotransmitter is released. This is shown in Fig 14.Q5(c).

d) **(i)** Suggest how **this** synapse affects the activity of the postsynaptic nerve cell.
(ii) Explain why the events shown in Fig 14.Q5(b) are unlikely to involve **sodium** ions.
(iii) State **two** significant differences between the events shown in Figs 14.Q5(b) and 14.Q5(c).
(iv) What physiological effect will valium have on the postsynaptic nerve cell?

[OCSEB January 1998 Structured Science Scheme: Unit B4 Physiology of Animals and Plants, q.2]

6

a) What happens in a neurone membrane when an action potential passes?

b) The neurotransmitter serotonin is involved in controlling appetite. The increase in blood sugar concentration after a meal causes certain cells in the brain to release serotonin. This binds to neurones which then inform the appetite centre in the brain that enough food has been eaten.

Fig 14.Q6(a) shows the normal events involving serotonin.

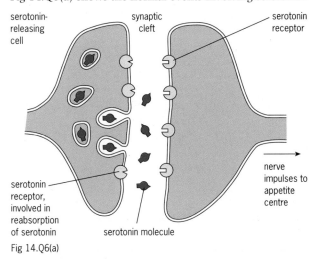

Fig 14.Q6(a)

(i) Serotonin is a neurotransmitter. Use the diagram and your knowledge of synapses to explain how information is transmitted across this synapse.
(ii) Adifax is a drug which can be used to affect people's eating habits. Fig 14.Q6(b) represents a serotonin molecule and an Adifax molecule. Explain how Adifax might affect the action of serotonin and the appetite of the person.

Fig 14.Q6(b)

[NEAB March 1998 Biology: Biological Basis of Behaviour Module Test, Section B, q.8]

7 Read the following passage.

Food was the most basic requirement of primitive humans, and arrow poisons were widely used for hunting. Many of the most powerful poisons such as curare originated from South America. There was total ignorance concerning their mode of action until Charles Waterton carried out his experiments in the 1820s. He gave curare, the drug in arrow poison, to a donkey which appeared to die ten minutes later. The animal was then revived by artificial ventilation of its lungs with a pair of bellows and went on to make a full recovery. The experiment showed that injection of arrow poison into the bloodstream causes death by respiratory failure.

It is known that curare competes with acetylcholine molecules for receptor sites at a neuromuscular junctions.

Adapted from: Mann J. *Murder, Magic and Medicine*

a) Describe the sequence of events that normally take place when a nerve impulse reaches a neuromuscular junction.

b) In the light of our present knowledge, explain how 'injection of arrow poison into the bloodstream causes death by respiratory failure' (end of paragraph 1).

c) A number of poisonous snakes produce venom which contains chemicals very similar in action to curare.

(i) Explain the advantage of this to the snake.
(ii) Explain how natural selection might have resulted in an increase in the toxicity of snake venom.

[NEAB June 1995 Biology: Biological Basis of Behaviour Module Test, q.8]

Assignment

TOXINS AND DRUGS

A huge variety of chemicals have an effect on the human body, and many act by interfering with the ways nerves or synapses work. In particular, they affect the transmission of impulses at synapses. In this Assignment we take a brief look at toxins and drugs.

1 Write a simple flow diagram outlining the sequence of events at a normal synapse.

Toxins

A toxin, by definition, is a poisonous substance of biological origin. The curare featured in the Opener is a toxin. In nature, animals and plants produce a huge variety of chemicals that organisms have developed to use against each other. Various species of spiders, frogs, jellyfish, snakes, molluscs, fish and plants – to name but a few – all make toxins that they use either to protect themselves or to kill prey.

Snake venom is of particular interest to neuroscientists. Venomous snakes have evolved poison glands as modifications of their salivary glands. Not only does the toxin subdue or kill the victim, preventing the snake from the serious injuries that might result from a violent struggle, some of them also start breaking down the prey's tissues. So when the snake gets round to eating it, the dead animal doesn't take so much digesting.

Fig 14.A1 **The krait (*Bungarus multicinctus*). The krait's venom contains a powerful neurotoxin, bungarotoxin. Cobras produce similar substances in their venom**

One of the first neurotoxic snake venoms was isolated from the krait (Fig 14.A1). In 1963, a small polypeptide consisting of about 70 amino acids was purified from the whole venom and was given the name **bungarotoxin**. An antibody to bungarotoxin was made and then attached to a fluorescent 'tag' so that its destination could be followed in the body of an animal injected with the krait's venom. The bungarotoxin was found almost exclusively at the neuromuscular junctions where it blocked the acetylcholine receptor sites on the postsynaptic membrane.

2 Predict the symptoms of krait bite. Explain your answer.

Bacterial toxins

Why is it that some bacteria cause disease while others are harmless? One of the key features is that **pathogenic** (disease-causing) bacteria produce an **exotoxin** that interferes with the host's metabolism in some way. A few of these toxins affect the nervous system. **Botulism**, a particularly nasty if rare form of food poisoning, is caused by the bacterium *Clostridium botulinum*. Botulinum toxin prevents the release of acetylcholine from the presynaptic membrane.

3 The symptoms of botulism include a weakness of the facial muscles, dilation of the pupils and difficulty in swallowing and breathing. Suggest why the botulinum toxin produces these effects.

Most people in this country are vaccinated against **tetanus** (Fig 14.A2). Also known as lockjaw because it causes muscle paralysis in the face and neck, tetanus results from infection from the bacterium *Clostridium tetani*. The tetanus toxin that this bacterium produces prevents the release of the substance at the neuromuscular junction that breaks down neurotransmitter. As a result, the effects of acetylcholine persist much longer than required.

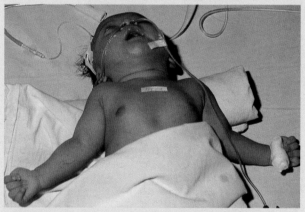

Fig 14. A2 **A baby with muscle stiffness caused by tetanus. The toxins produced by the bacterium are causing the baby's muscles to go into spasm**

4 What is the essential difference between the actions of tetanus toxin and botulinum toxin?

Drugs

Something that we touched on in this chapter and will go on and study in more detail in the next, is the brain. This organ is incredibly complex. It contains billions of nerve cells, each with thousands of connections to other cells, allowing for an infinite variety of pathways. There are many different transmitter substances in the brain that mediate synaptic transmission through these pathways. It should be no surprise that chemicals that interfere with neurotransmitters in the brain often produce sensory effects such as hallucinations.

These effects can occur as a side effect of a legal,

prescription drug that could be used to reduce pain, to induce sleep, or to reduce tension or depression. There are also substances that, since they have been recognised to alter the way the brain processes information, have been abused. Alcohol and nicotine both act on the brain and can be abused, but they are legal. The so-called 'recreational' drugs are illegal and often highly addictive. They interact with the brain and central nervous system and can have extremely dangerous side-effects.

5 Table 14.A1 includes the main classes of abused drugs, their mode of action and the common names of some individual drugs. However, they have been mixed up. Copy the table, matching up the correct examples and descriptions.

Table 14.A1 **The main classes of abused drugs, the effect they have and some examples**

Group	Effect	Examples
Depressants	Interfere with brain function. Distort perception and induce a dream-like state.	A variety of preparations of the plants *Cannabis sativa* and *Cannabis indica*. AKA hasish, marihuana.
Stimulants	Similar to alcohol, they sedate the central nervous system.	LSD. Mescaline. Magic mushrooms.
Hallucinogens	Pain killers.	Cocaine, ecstasy, amphetamine.
Cannabis	Keep the user awake, alert, excited.	A variety of anti-anxiety drugs, sleeping tablets including benzodiazepines (eg Valium)
Analgesics	Produce a relaxed, mild euphoria.	Morphine, heroin, methadone

Becoming dependent on drugs

If illegal drugs had a well understood mode of action, were very short lived in the body, had no side effects and did not induce dependency, or addiction, few people would worry. There are people who argue that drugs, if used 'sensibly', are not dangerous. However, the evidence is overwhelmingly against this.

With repeated use, many drugs can lead to addiction. This may be psychological, or physical. Psychological dependence is the most widespread. The person develops a strong desire for the state induced by the drug – elation, euphoria, stimulation, sedation, hallucination – and the state becomes much more desirable than normality. In many cases, the person experiences tolerance to the drug. Any attempt to kick the habit can result in severe symptoms of depression.

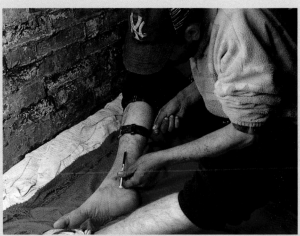

Fig 14.A3 **Abuse of drugs brings many dangers. Drugs bought on the street might be mixed with a variety of other substances that can be lethal, and the sharing of needles can spread infections such as AIDS and hepatitis**

6 How is a drug user most likely to respond when they start to become tolerant to the drug they are abusing?

Some drugs can, if used repeatedly, produce physiological changes and, after a time, the body cannot function 'normally' if the drug is not taken. Often a drug mimics one of the body's own products and this is no longer made as long as the drug is being taken. Therefore, when the drug is stopped, the body cannot respond immediately and start making the required product again. This situation occurs with heroin, which stops the body making its own endorphins. This is physical dependence and an attempt to stop will lead to dangerous and painful withdrawal symptoms.

7 Find out what some of the common withdrawal symptoms are.

8 Controversially, some drugs – notably alcohol and tobacco – are legal, while those listed in Table 14.A1 are not. What are the arguments for and against the legalisation of these drugs?

THE HUMAN BRAIN is an organ of great mystery. In addition to being unbelievably complex, it is very difficult to investigate experimentally. We now know that different areas of the brain have particular functions, but how did we find out? Some of the earliest clues came from accidents in which people suffered damage to a particular part of the brain.

Consider the tale of Phineas Gage, a US railroad worker. He was a popular and reliable man, polite and responsible, and he had been made a foreman. In September 1848, he was jamming a stick of dynamite into a hole using a tamping iron. A spark from the iron ignited the dynamite, and the metal rod came out of the hole a lot faster than it went in.

The rod entered Gage's face below his left cheekbone, passing through the eyeball and through the top of his skull before landing several yards away. Gage fell back in a heap, as you would, but remarkably, he did not die. He did not even pass out. He was driven by oxcart to a local physician, John Harlow. As the doctor stuck his fingers into the holes in Gage's head until the tips met, Gage asked when he would be ready to return to work. Within a couple of months he had recovered physically, but was no longer himself. Instead of being a gentle, honest, conscientious worker, Gage became 'a foul-mouthed and ill-mannered liar'.

He lived on as something of a celebrity for another 13 years before dying from an epileptic fit. On hearing of the death, Harlow managed to get Gage's body donated to medical research. He believed that the changes in Gage's behaviour had been caused by the damage to the frontal lobes of the brain. 'The equilibrium... between his intellectual faculties and animal propensities seems to have been destroyed,' Harlow wrote.

130 years on, scientists were able to use computer modelling to trace the damage done to Gage's brain. The metal rod missed the areas associated with language and motor function, but destroyed the ventromedial region on the left side of his frontal lobe. This is what made Gage so antisocial. People who have had tumours in this area of the brain have undergone the same sort of transformation.

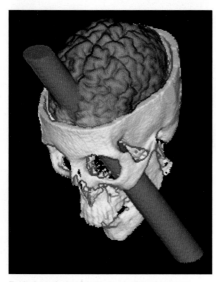

Early knowledge about brain function was pieced together with the help of accidents such as those suffered by Phineas Gage. The metal rod missed the vital blood vessels and the crucial areas of the brain that keep us alive. It did, however, damage the frontal lobes, and the subsequent changes in Gage gave us important clues about the function of this part of the brain

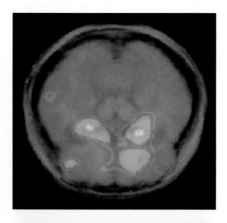

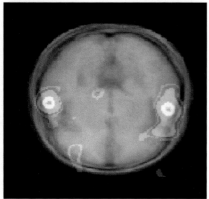

Today, there are more sophisticated ways of studying the brain. The upper PET scan shows which neurones are firing in the brain when a person is smelling something; the lower scan is of a person listening. With PET scans, therefore, we can pinpoint specific parts of the brain that are associated with body activities and behaviour

1 AN OVERVIEW OF THE CENTRAL NERVOUS SYSTEM

The human brain weighs about 1.5 kilograms, is mostly water and has the consistency of thick blancmange. It is, however, the most complex material known to man. It copes with a huge amount of information from the various senses, deciding what is important and what can be ignored. It stores thousands and thousands of memories for decades, and can sort them into chronological order. It allows us to control the complex functions of the body, while at the same time allowing us to maintain posture, read, write and talk. Can we make a computer to do all of this? Not a chance.

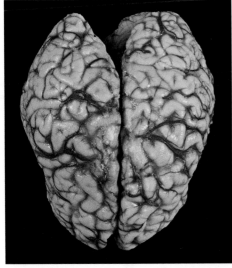

Fig 15.1 **The most obvious feature of the human brain (seen here from above) is the two huge cerebral hemispheres. These are the site of higher conscious functions such as memory, language and emotion**

The brain is the organ that makes humans human. We owe our success to this remarkable mass of nerves which takes over 20 years to mature, and which allows us to make tools and use language to an extent that far surpasses our nearest relative, the chimpanzee.

Fig 15.2 on the next page shows the overall structure of the nervous system. The human nervous system can be divided into two parts: the **central nervous system** (**CNS**) and the **peripheral nervous system**. The CNS consists of the brain and spinal cord, and is enclosed within the protective bone of the **cranium** and **vertebral column**. The peripheral nervous system brings information from the sense organs into the CNS, and then relays information out to the structures that bring about responses: the muscles and glands. The simplest type of behaviour, which illustrates the principles of the nervous system as a whole, is the **reflex arc**.

The reflex arc

A reflex response is a rapid, automatic response to a stimulus. Pulling your hand away from a hot pan-handle is a reflex action, as is blinking when an insect comes near your eye. A simple reflex involves communication between neurones in the peripheral nervous system and the spinal cord. The brain may be informed, but does not take part in the actual response. In fact, the response has usually taken place before the conscious brain is aware of the stimulus. Although many reflexes have a protective function, they can also help the body to coordinate complex muscular events, such as swallowing and maintaining posture. The nerve pathway involved in a reflex action is called a **reflex arc**.

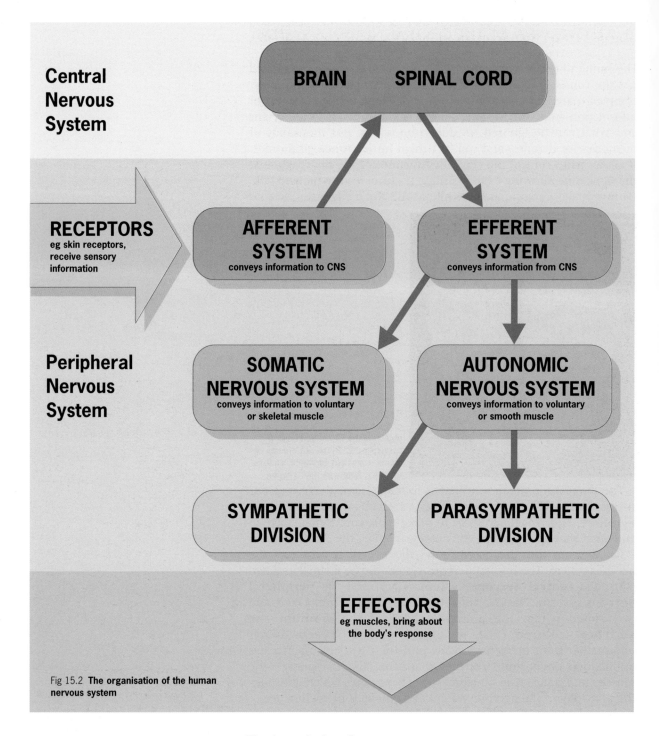

Fig 15.2 **The organisation of the human nervous system**

The knee-jerk reflex

The basic components of the knee-jerk reflex arc are shown in Fig 15.3. They include a receptor, a sensory neurone, a motor neurone and an effector. In everyday life, the knee-jerk reflex helps the body to control the tension in the thigh muscle, an important aspect of walking and running. As it starts with stretching of the muscle, it is also known as a **stretch reflex**.

Stimulus
Receptor
sensory neurone
spinal cord of central nervous system
Effector
Response
motor neurone

Fig 15.3 **The basic components of the knee-jerk reflex**

A tap on the patellar (knee-cap) tendon with a small percussion hammer makes the front thigh muscles stretch very slightly. Specialised stretch receptors in the muscles, **spindles**, respond to the change in length by generating nerve impulses, which pass along sensory neurones to the spinal cord. Here, the sensory neurones make synapses with motor neurones, and a message is sent straight back down the leg to the thigh muscles. The muscles contract, causing the lower leg to swing forward and giving the familiar knee jerk.

The knee jerk involves only a sensory and a motor neurone and there is no direct communication with the brain. Because it has only a single synapse, this type of reflex is called **monosynaptic**. In medicine, testing reflexes can help to detect disorders of the nervous system, allowing injured tissue along a nerve pathway to be located.

Other spinal reflexes

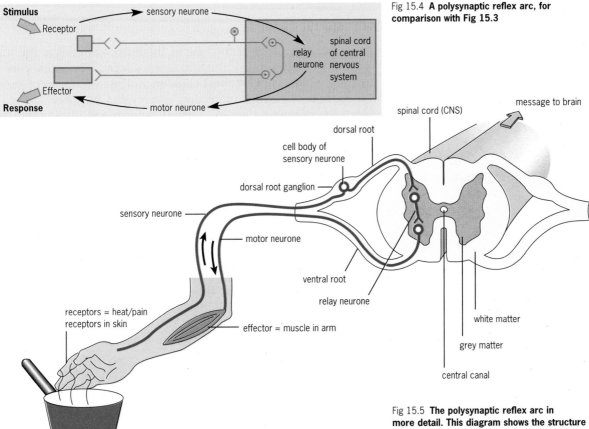

Fig 15.4 **A polysynaptic reflex arc, for comparison with Fig 15.3**

Fig 15.5 **The polysynaptic reflex arc in more detail. This diagram shows the structure of the spinal cord. Sensory information arrives and passes into the spine via the dorsal root. The ganglion is a swelling caused by the many sensory neurone cell bodies that all occur in the same place. Inside the spinal cord, the sensory neurone connects with an intermediate, or** relay **neurone, before impulses pass out to effectors via the ventral root. The reflex arc represents the most direct way of connecting up a sense organ to an effector, thus producing the fastest possible response**

Polysynaptic reflexes (Fig 15.4) have more than one synapse and are more common than monosynaptic reflexes. Examples include blinking when a foreign object enters the eye and the withdrawal reflex, shown in Fig 15.5. Pulling your hand away from a hot pan involves a circuit containing sensory receptors, sensory neurones, **spinal relay neurones** (interneurones), motor neurones and effector muscles. The principle is the same: the heat of the pan stimulates receptors, and impulses pass along sensory neurones towards the spinal cord. Here, instead of making synapses with motor neurones, they pass their signals on to relay neurones. These connect with motor neurones that cause muscles to contract, moving your hand away from the pan.

You know that you have touched the hot pan because some impulses do travel to the brain, but the movement that is part of the reflex is **involuntary**, not under your conscious control. It is more difficult to persuade people that the swearing which accompanies such an event is also involuntary.

2 THE HUMAN CENTRAL NERVOUS SYSTEM IN DETAIL

In this section, we look at the structure and function of the human spinal cord and the brain, and we see how both of them function in relation to the peripheral nerves and to the body's effectors. Knowing how neurones transmit impulses and understanding synaptic transmission (see Chapter 14) is an important preparation for taking this more 'holistic' view of the body's nervous system.

The spinal cord

The spinal cord starts at the base of the brain and ends at the first **lumbar vertebra** (this is roughly at waist level). The spinal cord is enclosed within the **vertebral column** and has a diameter of about 5 millimetres. Further protection is provided by three layers of tough membranes called **spinal meninges**, and by the **cerebrospinal fluid** which cushions the cord, acting as a shock-absorber. Each vertebra has an opening on its right and left sides to let spinal nerves pass through. These nerves extend into the body, forming the **peripheral nervous system**.

The brain

The brain receives stimuli from inside and outside the body (see Table 17.1, page 253). It maintains basic involuntary body 'housekeeping' functions such as heart rate, breathing rate and temperature control. It also coordinates the semi-automatic muscular actions of the body such as swallowing, and it initiates and controls voluntary activities such as walking and running. The human brain is the site of higher mental functions such as reasoning, emotion and personality.

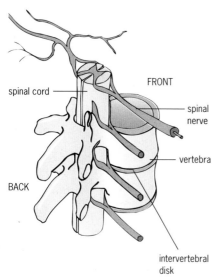

Fig 15.6 **The peripheral nerves emerging from the spine. etc etc - depends on photo**

See question 4. ■

Fig 15.7 **A vertical section through the human brain, showing the main areas**

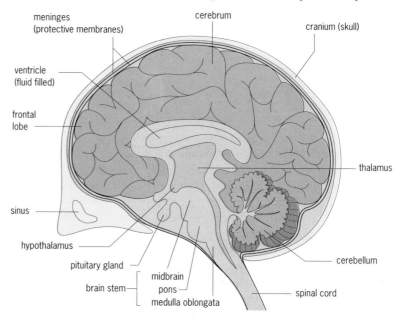

In primitive vertebrates such as fish, the brain is just a swelling of the front end of the spinal cord. It consists of three parts, a **hindbrain**, **midbrain** and **forebrain**, which correspond to basic functional regions. This three-part pattern can still be seen in humans, but it is only noticeable in the early stages of development of the embryo.

The hindbrain has three distinct structures. The **medulla oblongata** is a swollen portion at the bottom of the brain stem that houses vital centres controlling heart rate, breathing and blood supply. The **cerebellum** controls body movement and maintains balance, and the **reticular activating system** (a collection of neurones in the centre of the brain stem) filters incoming stimuli and controls wakefulness and sleep.

As in other animals, the midbrain in humans links the forebrain to the hindbrain (see Fig 15.7). Our emotions, which are located in the forebrain, can affect basic functions of the hindbrain such as control of blood vessel diameter, heart rate and sweating. When we are worried about something – exam results for example – the forebrain interprets this as stress and brings about the release of **adrenaline**, a hormone that prepares the body for action – see Chapter 17.

The forebrain has two main parts, the **cerebrum** (Fig 15.8) and a region containing the **thalamus** and the **hypothalamus**. The cerebrum is made up of two large **cerebral hemispheres**. They have a thin outer layer, the cortex, which is thrown into many folds with **fissures** (grooves) between. The cortex is the surface layer of the hemispheres, and most of our conscious thought takes place here. The cerebral cortex has three main roles:

● It controls and initiates voluntary muscle contraction.
● It receives and processes information from the senses (see Chapter 17) and all the body's receptors.
● As Fig 15.8 shows, it carries out the 'higher' mental activities of reasoning, and is regarded as the site of personality and emotion. Simply, it is what sets humans apart from other species.

> ✔ The brain stem is the medulla oblongata plus the pons plus the midbrain.

■ See questions 1, 2, 3 and 6.

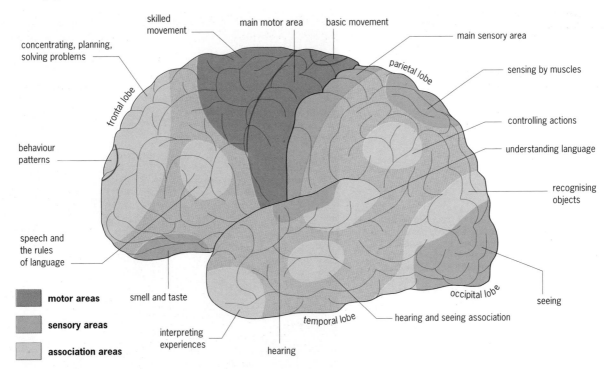

Fig 15.8 **The sensory, motor and association areas in the human cerebrum**

Different areas of the cerebral hemispheres are associated with different sensory and motor functions. Association areas link sensory and motor information and store memory. The areas are mapped out in Fig 15.8 on the previous page. The cerebrum of mammals is very large compared to the forebrains of other vertebrates. It is the dominant feature of the human brain and it encases other brain structures. Folds in the surface of the cerebrum increase the surface area for centres of control, where incoming nerve impulses are interpreted, or integrated, in the light of information already stored in the brain.

The **hypothalamus** is a key area. It receives a huge amount of internal and external sensory information and acts as a coordinating centre between the nervous system and the endocrine (hormonal) system. The hypothalamus is responsible for sensations such as hunger and thirst. It also helps control the autonomic nervous system since it regulates body temperature (see Chapter 10) and the balance of water and salts in the blood (see Chapter 11). It is linked to the pituitary gland and is involved in the release of hormones, including the antidiuretic hormone (which controls water reabsorption in the kidneys). We deal with the endocrine system in more detail in Chapter 16.

The **thalamus** (see Fig 15.7) directs sensory information from the sense organs to the correct part of the cerebral cortex.

The brain of a human weighs about one-fiftieth of body weight. Its delicate tissues are protected by the skull or **cranium** and by the **cranial meninges**, membranes which are continuous with the **spinal meninges**. **Cerebrospinal fluid** bathes the outside of the brain and fills the chambers – the **ventricles**. Twelve pairs of cranial nerves innervate (supply nerves to) various regions of the head. The human brain is thought to contain ten thousand million (10^{10}) neurones. Each neurone may be in contact with a thousand other cells, providing an immense number of different communication routes.

> ✔
> Meningitis is a potentially fatal disease characterised by inflammation of the meninges – the protective membranes that surround the brain and spine.

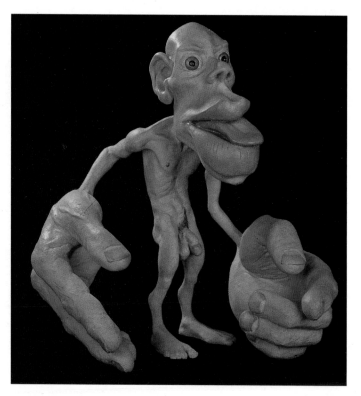

Fig 15.9 **No, you won't meet him on a dark night. Some parts of the body contain more receptors than others, and consequently a larger area of brain is devoted to them. In this distorted model the man's anatomy is proportional to the area in the brain which deals with the sensory information. Thus the hands, mouth and tongue are huge – they have many receptors per square centimetre – while other areas such as the torso and legs are relatively small. The eyes, however, are not in proportion in this model. These take up more of the human brain than the rest of the body put together**

EPILEPSY AND THE TWO SIDES OF THE BRAIN

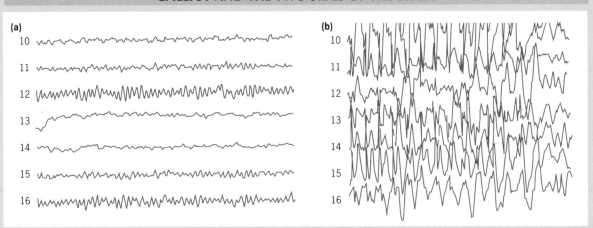

Fig 15.10 **EEG traces taken** (a) **during normal brain activity and** (b) **taken during an epileptic seizure. The electrical activity is picked up by electrodes taped to the head**

EPILEPSY is a common disorder the brain. Symptoms range in severity from a mild loss of concentration, known as an absence or petit mal, to full-blown convulsive fits (grand mal) in which the subject blacks out and falls to the floor. These can be dangerous if the sufferer lashes out – injuring themselves and others – or bites their own tongue.

The underlying cause of epilepsy is random, uncontrolled activity of some cells in the brain. This chaotic activity in both sensory and motor nerves causes patients to see and hear a variety of strange things such as flashing lights and bells, while muscles jerk uncontrollably.

A diagnosis of epilepsy can be confirmed and studied using a machine called an electroencephalograph (EEG). This measures electrical activity in the brain. Fig 15.10 shows two traces from a person with epilepsy: a normal reading and one taken during a seizure.

Epilepsy can often be successfully controlled by drugs. However, in extreme cases, the condition is treated by brain surgery, and one such operation has given us a fascinating insight into the workings of the brain. The cerebral hemispheres have been described as two separate brains, and in order to work effectively as a whole, the two halves must communicate. The bridge between the two halves is known as the corpus callosum (Fig 15.11). Neuroscientists discovered that the corpus callosum was involved in the spread of epileptic seizures.

In a seemingly drastic operation, surgeons sever most of the corpus callosum. This often causes the seizures to be less intense and dangerous. However, there are other amazing consequences. Initially, subjects appear to be perfectly normal: they can talk and read, and have no problems in recognising the world around them. However, if they close their right eye, and are given a familiar object such as a comb, they cannot put a name to it. Open the other eye, however, and, 'Ah, its a comb!'

The same happens with words. If a word such as 'TIGER' is looked at it with the left eye only, the patient can't read it. If they open the right eye, they can read the word immediately. This is because the left eye supplies information to the right side of the brain, and that is not where the language centre is situated. The right eye supplies information to the left side of the brain, to the language-processing neurones.

From studies of split-brain patients, and other studies, it appears that different sides of the brain have different functions. The left hemisphere contains the language centre, and the three R's – reading, writing and arithmetic. The right side, in contrast, is responsible for our imagination and sense of humour. It can also appreciate form, geometry and music. It cannot, however, put words to things. If the right hemisphere needs a word, it has to put in a call to the left side, via the corpus callosum.

Split-brain patients do not experience the symptoms forever. Within a few months the right hemisphere develops more language skills and can function on its own. It has even been suggested that split-brain patients could read two books simultaneously with different eyes!

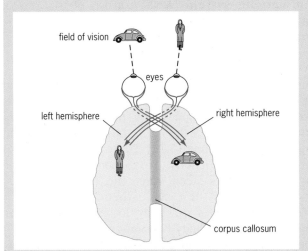

Fig 15.11 **The corpus callosum is a broad, thick mass of nerve fibres connecting the right and left cerebral hemispheres. The left eye supplies information to the right hemisphere, and vice versa. The corpus callosum allows the two sides of the brain to communicate**

White matter and grey matter

Different areas of nerve tissue in the central nervous system are either greyish or creamy-white. **Grey matter** contains nerve cell bodies; their nuclei are responsible for the grey colour. Grey matter also contains glial cells and some nerve fibres (these are mainly unmyelinated). **White matter** consists largely of myelinated fibres (their fatty sheaths are creamy-white), some unmyelinated fibres and glial cells.

White matter and grey matter are distributed in specific areas. Grey matter forms the surface layer of the brain, regions deep in the brain known as the basal ganglia, and the central column of the spinal cord. White matter forms a layer between the two areas of grey matter in the brain and encloses the column of grey matter in the spinal cord: look back at the spinal cord shown in Fig 15.4.

The peripheral nervous system

The nerves of the peripheral nervous system behave like major road systems, carrying traffic in and out of the central nervous system. **Afferent nerves**, also called **sensory nerves**, carry information from sensory receptors into the CNS. **Efferent nerves**, also called **motor nerves**, carry information from the CNS out to effector organs. The efferent system can be further subdivided into the **somatic** and **autonomic** systems. These differ in their function, rather than their structure or position in the body.

The somatic nervous system

The somatic nervous system contains both afferent and efferent nerves. It receives and processes information from receptors in the skin, voluntary muscles, tendons, joints, eyes, tongue, nose and ears, giving an organism the sensations of touch, pain, heat, cold, balance, sight, taste, smell and sound. It also controls voluntary actions such as the movement of arms and legs.

The autonomic nervous system

The **autonomic nervous system** (**ANS**) consists of two sets of *involuntary* nerves which act antagonistically – they have opposing effects. The system is entirely motor, made up of efferent nerves only. It does not carry sensory information: feedback from muscles and glands travels via the somatic system.

The ANS controls basic 'housekeeping' functions such as heart rate, breathing, digestion and blood flow. Heart rate, for example, can increase or decrease. This and many functions like it are controlled by the two branches of the ANS, the **sympathetic** and the **parasympathetic** systems (Fig 15.12). These two sets of nerves have **antagonistic** (opposite) actions.. So, for example, while the sympathetic system increases heart rate, the parasympathetic system lowers it. Generally, the sympathetic system has a stimulatory effect, and prepares the body for action, while the parasympathetic system returns body functions to normal.

Normally, the activity of both systems is balanced. But if the body is stressed, then the 'fear, flight and fight' reactions of the sympathetic nervous system take over, causing an increase in heart rate, faster breathing, an increase in blood pressure and an increase in blood sugar level. This makes the body ready for sudden

Afferent means 'incoming' while *efferent* means 'outgoing'. You can refer to 'afferent nerves' or 'efferent blood vessels', for example.

There is a parallel between the nervous and muscular systems of the human body. In the same way that we have voluntary and involuntary components of our nervous system, we also have skeletal (striped) muscle to deal with voluntary muscle contraction, and visceral (smooth) muscle to deal with involuntary muscle contraction. The different muscle types are described in Chapter 18).

strenuous activity. When the emergency is over, the parasympathetic system takes over. It decreases the heart and breathing rates and diverts blood supply back to 'housekeeping' activities such as digestion and food absorption.

The autonomic system was originally thought to be independent of the rest of the nervous system, hence the term autonomic, meaning 'on its own'. Now we appreciate that it is not autonomous, not 'self-governing', but is regulated by areas within the central nervous system, including the hypothalamus, cerebral cortex and the medulla oblongata.

■ See question 5.

Fig 15.12 **The autonomic nervous system, showing the opposing effects of the sympathetic and parasympathetic nervous systems. Note that the sympathetic system generally prepares the body for action, and as such has similar effects to the hormone adrenaline. However, the hormone produces all of the effects at once, while the sympathetic system can alter just one. It can, for instance, increase the heart rate without changing breathing rate.**
If you look at the structure of the two systems, you should see that the parasympathetic nerves lead straight out of the brain while those ofthe sympathetic system come out from the spine, passing out of the vertebral column at the appropriate level

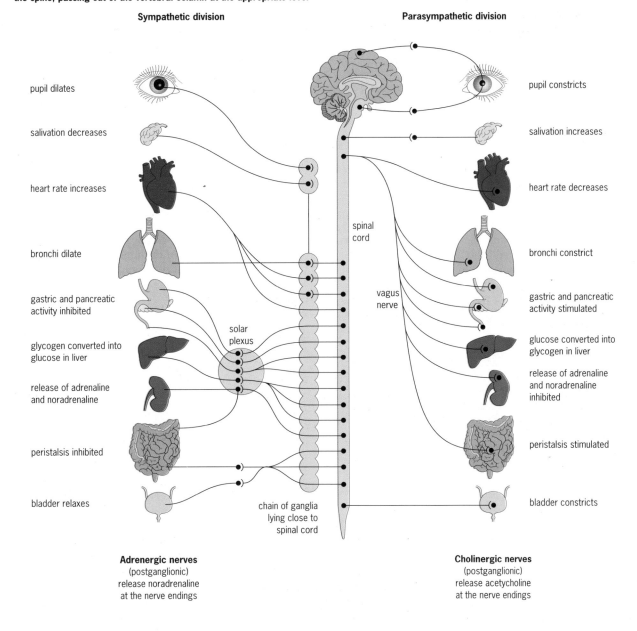

Sympathetic division

Parasympathetic division

pupil dilates — pupil constricts

salivation decreases — salivation increases

heart rate increases — heart rate decreases

spinal cord

bronchi dilate — bronchi constrict

gastric and pancreatic activity inhibited — vagus nerve — gastric and pancreatic activity stimulated

solar plexus

glycogen converted into glucose in liver — glucose converted into glycogen in liver

release of adrenaline and noradrenaline — release of adrenaline and noradrenaline inhibited

peristalsis inhibited — peristalsis stimulated

bladder relaxes — chain of ganglia lying close to spinal cord — bladder constricts

Adrenergic nerves
(postganglionic)
release noradrenaline
at the nerve endings

Cholinergic nerves
(postganglionic)
release acetycholine
at the nerve endings

SUMMARY

When you have finished this chapter you should know and understand the following:

■ The **central nervous system** (CNS) consists of the brain and spinal cord.

■ The simplest coordinated response is the **reflex arc**. In the case of a **monosynaptic** reflex arc such as the knee jerk, the pathway consists of the receptor in the tendon, the sensory nerve that connects directly to the motor nerve in the spine, and the effector, the thigh muscle.

■ More complex reflex arcs such as blinking are **polysynaptic**. These have more intermediate connections inside the CNS.

■ The human brain is divided into **hindbrain**, **midbrain** and **forebrain**. Generally, the midbrain and hindbrain are concerned with 'housekeeping' functions, while the **cerebral hemispheres**, which make up much of the forebrain, are responsible for the conscious functions such as language and memory.

■ The **peripheral** nervous system consists of the **somatic** and **autonomic** nervous systems.

■ The autonomic nervous system consists of the **sympathetic** and **parasympathetic** nervous systems. The two sets of nerves act antagonistically to control a variety of functions that are not under conscious control. Generally, sympathetic stimulation prepares the body for action while the parasympathetic system returns it to normal.

QUESTIONS

1

a) What is the function of each of the following parts of the cerebral hemispheres?
  (i) sensory areas
  (ii) association areas
  (iii) motor areas

b) In 1861 a man called Leborgne suffered a stroke. As a result of this he was unable to speak and was paralysed on his right side. Fig 15.Q1 shows the location of the damage found in Leborgne's brain.

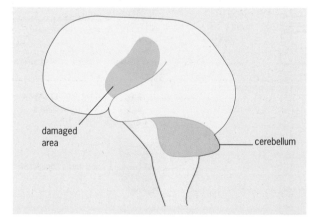

Fig 15.Q1

Suggest an explanation for:
  (i) Leborgne being unable to speak;
  (ii) Leborgne being paralysed on his right side.

[NEAB June 1998 Biology: Biological Basis of Behaviour Module Test, q.7]

2

a) Describe one function of each of the following regions of the brain.
  (i) cerebellum
  (ii) hypothalamus
  (iii) medulla

b) An external stimulus can result in a response involving movement. Describe the part played by the cerebral hemispheres in this process, starting with the arrival of the sensory information in the brain.

[NEAB March 1998 Biology: Biological Basis of Behaviour Module Test, q.3]

3 Fig 15.Q3 shows a vertical section through a human brain.

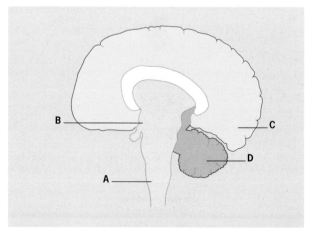

Fig 15.Q3

a) Give the letter of the region of the brain which:
   (i) regulates coordination of the solute concentration of the plasma;
   (ii) coordinates the control of the heart rate;
   (iii) receives sensory input from the eyes.

b) What is the function of the visual association area?

c) A stroke results from the bursting of a blood vessel on the surface of the brain. As a result of the damage that this causes to the brain, stroke victims may be paralysed in part of the body. Describe how observations made on stroke victims could confirm the function of particular areas of the cerebral hemispheres.

[NEAB June 1995 Biology: Biological Basis of Behaviour Module Test, q.6]

4 The diagrams **A** to **F** represent reflexes and control mechanisms found in the human body.
Select the diagram that represents each of the following.
You may use each letter once, more than once or not at all.

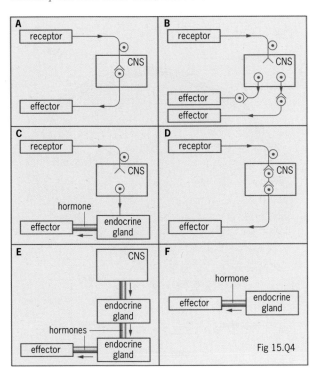

Fig 15.Q4

a) A withdrawal reflex;

b) a knee-jerk reflex;

c) the control of the iris diaphragm of the eye;

d) the control of plasma thyroxine levels.

[AEB 1990 Human Biology Paper 1, q.7]

5 The table compares some effects of the sympathetic and parasympathetic systems.

Feature	Sympathetic	Parasympathetic
pupil of eye	dilates	constricts
salivary gland	inhibits secretion of saliva	stimulates secretion of saliva
lungs	dilates bronchi and bronchioles	constricts bronchi and bronchioles
arterioles to gut and smooth muscle	constricts	no effect
arterioles to brain	dilates	no effect
heart rate		
stroke volume		

a) Complete the table by filling in the spaces to suggest the effects of these two systems on heart rate and stroke volume.

Use information in the table and your own knowledge to answer the following questions.

b) In giving dental treatment, it is important that any local anaesthetic stays close to the site of its injection. Explain why dental anaesthetics usually contain adrenaline.

c) Many people suffer from motion sickness when travelling in cars. A number of drugs are used to control this and some work by inhibiting the parasympathetic system.
Explain why side-effects of such drugs may include:
(i) dryness of the mouth; (ii) blurred vision.

[AEB November 1992, Biology Paper 1, q.4]

6 The table below refers to the functions of some regions of the brain. Complete the table by inserting the correct word or words.

Region of the brain	One function
	Control of voluntary movement
Medulla	
	Control of balance and fine movement
Hypothalamus	

[Edexcel June 1998, Human Biology Module Test HB3, q.5]

Assignment

HEAD INJURIES

One in seven children in hospital surgical wards are there because they have head injuries. For the best chance of recovery, their condition must be speedily diagnosed and treated quickly and effectively.

1 Suggest some of the major causes of head injury in the UK. Which sports do you think will cause the most casualties? Anne was not wearing a cycling helmet when she fell backwards from her bike and was found unconscious. She was found by someone trained in first aid, whose first thought was, 'ABC'. Anne's breathing was poor until the first-aider released her tongue – it had been obstructing the airway.

2

a) What does ABC mean in first aid?

b) If you find a person lying on the ground with a suspected head injury, should you move them? Explain your answer.

c) How would you give mouth-to-mouth resuscitation to an unconscious person?

d) What happens when the oxygen supply to the brain is reduced?

The capillaries of the brain are much more selectively permeable than those in the rest of the body. This phenomenon is explained by saying there is a **blood–brain barrier**. Consequently, many chemicals cannot pass from the blood to the brain.

3

a) A head injury can damage the blood–brain barrier. Suggest the effects of this.

b) Anne was taken to hospital. How would damage to her skull be assessed?

Anne was found to have a skull fracture and her condition began to worsen. She had a slow pulse, rising blood pressure and slow respiration.

4 Which control centre of the brain could have been affected?

Reflexes often help to diagnose disorders of the nervous system. Anne remained unconscious and did not respond to sounds, but her basic body reflexes (including the eye's pupil reflex) still operated.

Fig 15.A1 **Anne has an urgent CT scan**

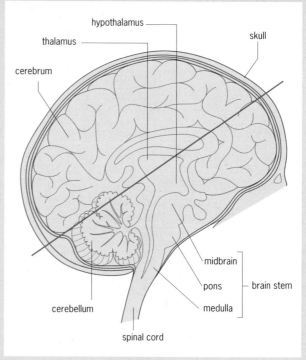

Fig 15 A2 **The diagram shows the brain 'slice' taken by the CT scan. It is illustrated in Fig 15.A4**

5

a) Describe the nerve pathways for the pupil reflex.

b) Is the knee-jerk reflex useful in determining brain function?

A common consequence of a blow to the head is a blood clot, or **haematoma**. Blood vessels rupture at the site of impact and a pool of blood accumulates, putting pressure on the surrounding brain tissue. This is called **compression**. To see if blood was collecting in or around Anne's brain, **CT (computerised tomography)** scans were

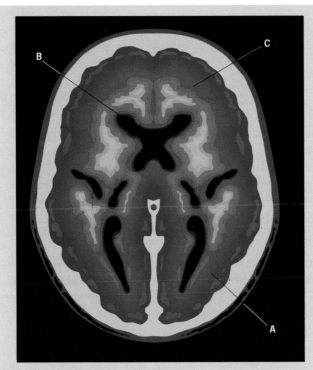

Fig 15 A3 **A CT scan of a normal brain**

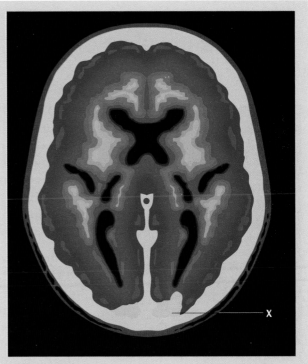

Fig 15 A4 **A CT scan of Anne's brain after the accident. Note the abnormal area marked X: this is a haematoma**

taken. The scanner detects density differences between brain, blood and bone and produces pictures of 'slices' through the brain.

6

a) Look at Fig 15.A3, a scan of a normal brain. What are **A**, **B** and **C** on the scan?

b) The pink region **X** in Fig 15.A4 is a haematoma pressing on Anne's brain. Which region of the cerebrum is this? Check with Fig 15.8. How might the brain's function be affected?

c) If the haematoma was not treated, Anne's condition would rapidly worsen because the brain stem and its structures would be under pressure. Suggest the effects of damage.

The hospital takes an electroencephalogram (EEG) of Anne's brain. The doctor diagnoses 'petit mal epilepsy', a common consequence of brain injury.

7

a) What is an EEG?

b) What is epilepsy, and how does 'petit mal' differ from 'grand mal'?

c) Explain how an EEG helps to diagnose epilepsy.

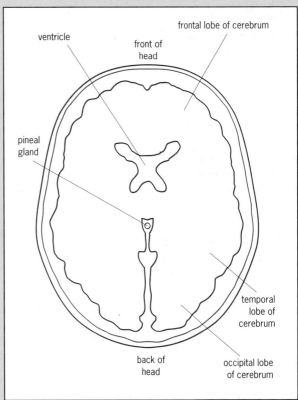

Fig 15.A5 **A simplified interpretation of the normal CT scan in Fig 15.A3**

Following an operation to remove the blood clot to her brain, Anne made a full recovery. Recent studies have shown that following a head injury, damage caused to brain cells is not instant. Instead, it is thought that injured brain cells release enzymes from their lysosomes, which slowly digest the cells. In pioneering new techniques, accident victims can have their brains cooled, and this hypothermia slows down any damaging chemical reactions until the cells have had a chance to repair themselves.

16 Hormones

Robert Wadlow made it into the record books as the tallest man who has ever lived. There might have been people who have grown taller – stories exist of men over nine feet tall – but their existence, unlike Wadlow's cannot be proved. Today, such extremes in size are rare because conditions can be diagnosed early and treated effectively

MANY SCHOOLCHILDREN know that the tallest man who ever lived was Robert Wadlow, who stood a touch under 9 feet tall and had size 36 feet. Born in Alton, Illinois, USA, Wadlow reached the height of over 7 feet at the age of 13 and continued to grow throughout his short life of 22 years. He might have continued to grow, but he died prematurely following an infected foot wound.

Waldow's condition was caused by an over-productive pituitary gland that made too much human growth hormone, or somatotropin. Today people rarely suffer such extremes because the condition is diagnosed early. Blood tests done on a particularly large (or small) child can reveal hormone imbalances. Under-production can be remedied by injections of the hormone – either the real thing or a synthetic analogue (a chemical of similar structure that has the same effect). Conversely, over-production can be treated either by reducing the size of the gland, or by administering drugs which block the action of the hormone.

Many women find the symptoms of the menopause eased by hormone replacement therapy, and some women have even been able to conceive in their 60s if they have hormone treatment. Add this to the artificial hormones taken every day by women on the pill, and you can understand why making artificial hormones is big business.

1 HORMONES ARE CHEMICAL SIGNALS

The human body communicates in two fundamentally different ways: by nerves and by chemicals. As Fig 16.2 shows, there are three main types of chemical signals:

- Locally acting chemicals such as **endorphins**, **prostaglandins** and **histamines** (Fig 16.3). These act on cells that are close to their site of production.

- **Hormones**. These are produced by special tissues called **endocrine glands**. They pass into the blood and act away from their site of production. Hormones affect **target cells** and **target organs** in the body.

- **Pheromones** are chemical messengers that allow *external* communication: simple exchanges of information *between* people. Generally, pheromones are not an important means of communication in humans, despite the claims of some newspaper reports and advertisements.

In this chapter we focus mainly on hormones. The definition of a hormone in mammals is:

A chemical messenger that is produced by cells or tissues of the endocrine system and which travels through the blood to act on target cells or target organs, before being broken down.

The main part of the chapter describes the human endocrine system, and we finish with a brief overview of prostaglandins and endorphins.

Fig 16.1 **Many aspects of sexual reproduction in humans are controlled by hormones, including gamete production, puberty, pregnancy, birth and lactation**

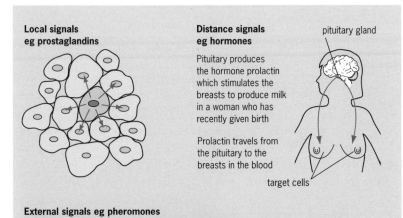

Local signals eg prostaglandins

Distance signals eg hormones

pituitary gland

Pituitary produces the hormone prolactin which stimulates the breasts to produce milk in a woman who has recently given birth

Prolactin travels from the pituitary to the breasts in the blood

target cells

External signals eg pheromones

Fig 16.2 **A summary of the different types of chemical signalling that exist within and between organisms. Local signals include neurotransmitters such as acetylcholine (see Chapter 14) and local messengers such as prostaglandins, endorphins and histamines. Hormones allow communication between different organs, via the blood. Externally-produced messengers allow communication between organisms**

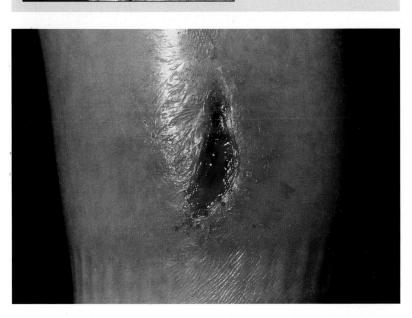

Fig 16.3 **Histamine is a local messenger which is released by injured cells. It acts on the cells around an injury like the deep cut shown above, causing blood vessels near to the injury to dilate, bringing more blood into the area. This is an important part of the inflammation response (see Chapter 34)**

2 THE HUMAN ENDOCRINE SYSTEM

Fig 16.4 provides a useful reference point for the rest of this chapter, and also for the hormones described in Chapters 10, 11 and 19.

In humans, more than a dozen tissues and organs produce hormones. Some, including the **pituitary gland**, the **thyroid gland**, the **parathyroid glands** and the **adrenal glands** are endocrine specialists: their *major* function is to secrete one or more hormones. Others, such as the pancreas, ovaries and testes, secrete hormones in addition to their other functions. Together, these **glands** make up the human **endocrine system**, shown in Fig 16.4.

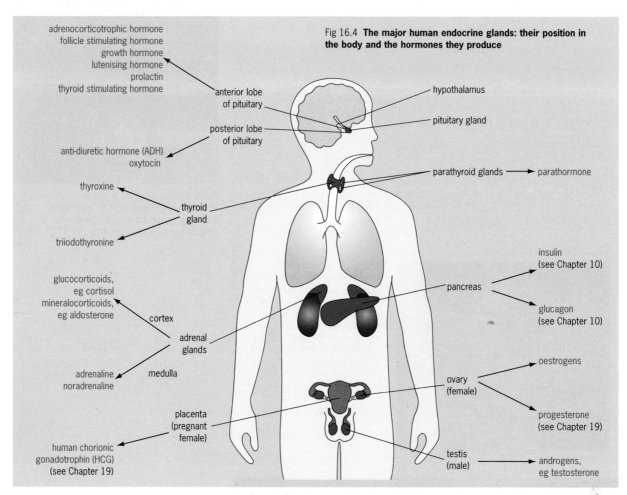

Fig 16.4 **The major human endocrine glands: their position in the body and the hormones they produce**

adrenocorticotrophic hormone
follicle stimulating hormone
growth hormone
lutenising hormone
prolactin
thyroid stimulating hormone

anterior lobe of pituitary

posterior lobe of pituitary

anti-diuretic hormone (ADH)
oxytocin

thyroxine

thyroid gland

triiodothyronine

glucocorticoids, eg cortisol
mineralocorticoids, eg aldosterone

cortex

adrenal glands

adrenaline
noradrenaline

medulla

placenta (pregnant female)

human chorionic gonadotrophin (HCG) (see Chapter 19)

hypothalamus

pituitary gland

parathyroid glands → parathormone

insulin (see Chapter 10)

pancreas

glucagon (see Chapter 10)

oestrogens

ovary (female)

progesterone (see Chapter 19)

testis (male)

androgens, eg testosterone

The endocrine system has four main functions:

- It maintains **homeostasis**, the balance of the body, by making sure the concentration of many different substances in body fluids are kept at the correct level. The control of blood sugar level, blood pH and water balance are all examples of homeostasis (see Chapters 10 to 13).

- It works with the nervous system to help the body respond to stress. The release of adrenaline in the **fight or flight** response is an example of this (see the Assignment at the end of this chapter).

- It controls the body's rate of growth.

- It controls sexual development and reproduction (see Chapter 19).

We know of more than 50 human hormones, and we divide them into two groups according to their origin: those made from *fatty acids* and those made from *amino acids*. The first group, which include **steroids** such as oestrogen and progesterone, are *lipid-soluble*.

Amino acid-based hormones are either modified amino acids, peptides or glycoproteins: they are *water-soluble* and include hormones such as insulin and adrenaline. Table 16.1 lists most of the hormones produced by the endocrine system and shows where they act and what they do. You might find it useful as a revision aid.

Table 16.1 **Human endocrine glands and their principal secretions. You do not need to learn all of this, but you should find it useful for reference as you read the rest of the chapter**

Gland	Hormone	Chemical structure	Effect
Pituitary 1 Posterior lobe	Anti-diuretic hormone (ADH)	Peptide	Reduces amount of water lost in urine. Raises blood pressure by constricting arterioles
	Oxytocin	Peptide	Contraction of smooth muscle during child-birth. Stimulates secretion of milk from mammary glands
2 Anterior lobe	Adrenocorticotrophic hormone (ACTH)	Peptide	Stimulates production and release of hormones from adrenal cortex
	Follicle stimulating hormone (FSH) also called ICSH in males	Glycoprotein	Controls the development of follicles in the ovary, and sperm cells in the testis (see Interstitial cell secreting hormone, ICSH, page 280)
	Growth hormone	Protein	Promotes growth (especially of skeleton and muscles). Affects body metabolism
	Luteinising hormone (LH)	Glycoprotein	Stimulates ovulation and formation of the corpus luteum (stimulates testosterone production in males)
	Prolactin	Protein	Stimulates milk production and release during pregnancy
	Thyroid stimulating hormone (TSH)	Glycoprotein	Stimulates growth of thyroid gland; synthesis and production of thyroid hormones
Thyroid	Thyroxine + Triiodothyronine	Iodine-containing amino acids	Thyroxine Increases rate of cell metabolism, controls aspects of growth and development and controls basal metabolic rate (BMR)
Parathyroid	Parathormone	Peptide	Raises blood calcium levels by stimulating release of calcium from bone
Pancreas (Islets of Langerhans)	Insulin (produced by the β cells)	Protein	Lowers blood glucose levels by making cell membranes more permeable to glucose, increases glycogen storage in liver
	Glucagon (produced by the α cells)	Peptide	Raises blood sugar levels by stimulating glycogen breakdown in the liver
Adrenal 1 Cortex	Glucocorticoids eg cortisol	Steroids	In response to stress, raises blood glucose.
	Mineralocorticoids eg aldosterone	Steroids	Concerned with water retention. Increases reabsorption of sodium chloride in kidneys, so important in control of blood volume and pressure
2 Medulla	Adrenaline	Modified amino acid	'Fear, flight and fight' reactions. Prepares the body for heightened activity. Mimics effects of the autonomic nervous system
	Noradrenaline	Modified amino acid	As adrenaline
Gonads 1 Ovary (follicle)	Oestrogens	Steroids	Female sex characteristics, rebuilding of uterus lining after menstruation, inhibits FSH
2 Ovary (corpus)	Progesterone	Steroids	Stimulates maturation of uterus lining, maintains pregnancy, inhibits FSH
3 Testis	Androgens (eg testosterone)	Steroids	Support sperm production. Important in the development of male secondary sexual characteristics
Placenta	Human chorionic gonadotrophin (HCG)	Steroid	Causes corpus luteum to secrete progesterone, thus maintaining pregnancy

There are two basic types of gland in the body: **exocrine glands** (eg salivary gland), whose secretions are released through a tube or **duct**; and **endocrine glands** (eg thyroid gland), whose secretions are released into the bloodstream for transport to their eventual destination.

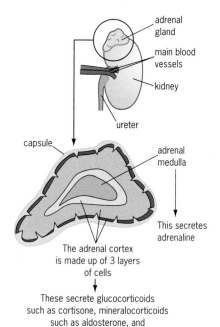

Fig 16.5 **The adrenal glands sit on top of the kidneys. They consist of the adrenal medulla, the tissue in the centre of the gland which secretes adrenaline, and an outer layer called the adrenal cortex**

Making and releasing hormones

Endocrine glands are tissues and organs that produce and release hormones. These glands have no ducts. The hormone is simply secreted into the blood and then travels to target organs or cells elsewhere in the body. A good example of an endocrine gland is the **adrenal gland**. Its structure is illustrated in Fig 16.5.

Hormones are formed inside endocrine cells. Some amino acid-based hormones, such as insulin and glucagon, are produced directly from a copy of a gene by the processes of transcription and translation (see Chapter 28). Most hormones, including the steroids, are made as an end product of a series of chemical reactions. Each reaction makes minor changes to **precursor** molecules, until, eventually, the active hormone is produced. For the steroid hormones progesterone, oestrogen and testosterone, the precursor is cholesterol (see Chapter 3).

All endocrine glands contain some of the hormone they produce, at any given time. Although the hormone thyroxine is released continuously, most are not. **Negative feedback loops** control the release of hormones from many endocrine glands so that homeostasis is maintained. This concept is covered in Chapter 10.

Thyroxine: an example of negative feedback

Metabolic rate is the rate at which all cells in the body carry out their biochemical reactions. It is a vital whole body function that must be controlled within very strict limits. The hypothalamus in the brain detects even a small decrease in metabolic rate and responds by releasing more **thyrotropin releasing hormone** (Fig 16.6). This acts on the pituitary gland, causing it to release more **thyroid stimulating hormone**. This passes to the thyroid gland which responds by secreting more **thyroxine**, the hormone which acts on individual cells to increase metabolic rate. As soon as metabolic rate gets back to normal levels, the hypothalamus responds by releasing less thyrotropin releasing hormone and, in a healthy person, homeostasis is maintained.

Fig 16.6 **Control of metabolic rate involves a negative feedback loop**

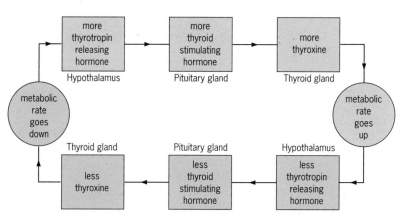

How hormones affect target cells

Hormones exert their effects in two different ways. Peptide hormones travel in the blood to all parts of the body. They do not, however, affect all cells in the body. Target cells or organs have specific proteins on their surface which act as receptor sites. The hormone fits into these sites as a key fits a lock, or like an enzyme fits its substrate (see Chapter 4). Once in place, the hormone–receptor complex brings about changes inside the cell, although the hormone itself never enters the cytoplasm. In contrast, steroid hormones pass easily into cells because they can get through the lipid bilayer. They bind to a specific receptor molecule inside the cytoplasm: this complex then causes biochemical changes inside the cell.

Peptide hormones and second messengers

Amino-acid-based hormones are water-soluble and vary in size from **thyroxine** (just two modified amino acids) to **growth hormone**, a protein made up of 190 amino acids. Because they are not lipid-soluble, they cannot get into cells through the lipid cell surface membrane. Instead, they act as **first messengers**, binding to target receptors on the outside of the cell surface membrane.

As Fig 16.7(a) shows, this binding event activates an enzyme, **adenylate cyclase** which is located on the inside surface of the membrane. The activated enzyme converts ATP into a substance called **cyclic AMP**. This is the **second messenger** which moves about inside the cell, causing biochemical changes. The final effect on cell function depends on the type of hormone and the type of target cell involved.

Steroid hormones

Steroids are small lipid-soluble molecules that can get through cell membranes easily. They enter cells and bind with **target receptors** inside the cytoplasm (Fig 16.7(b)). Cortisol, progesterone, oestrogen and testosterone all act in this way.

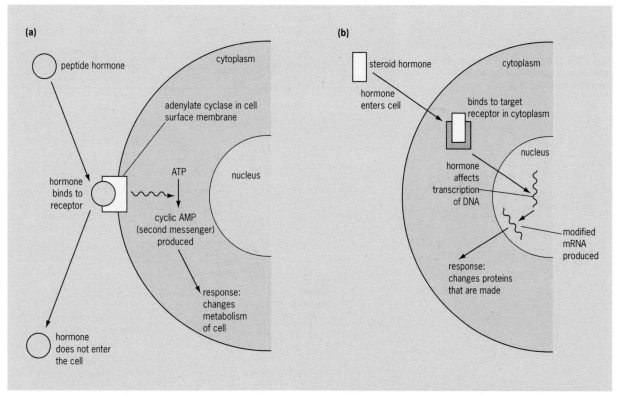

Fig 16.7(a) **Water-soluble hormones work via a second messenger which is formed inside the cell when they bind to receptors on the cell surface**

Fig 16.7(b) **When steroid hormones bind to a receptor inside the cytoplasm, they form a complex similar to an enzyme–substrate complex. This enters the nucleus of the cell and binds directly to the DNA, interfering with the cell's ability to read some of its genes. A steroid hormone can either switch on or switch off protein synthesis of particular genes. The overall effect of hormone action is different for each steroid hormone (see Table 16.1)**

3 COMPARING THE NERVOUS AND ENDOCRINE SYSTEMS

Hormones do not act independently. They work together with other hormones and also with the nervous system. The nervous system and the endocrine system both control and coordinate the function of different parts of the body and they both rely on chemical messengers, but they have obvious differences (Table 16.2). Despite these differences, we now realise that the link between the endocrine system and the nervous system in humans is more definite than once thought. The main physical link between the two systems is between the **hypothalamus**, part of the base of the brain, and the **pituitary gland**, an endocrine gland just beneath it (Fig 16.8).

Table 16.2 **Differences between the nervous system and the endocrine system in humans**

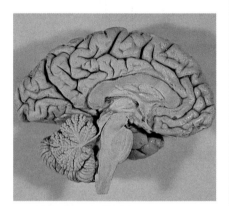

Fig 16.8(a) **The position of the hypothalamus and pituitary gland in the brain can be identified from Fig 16.8(b)**

Property	System	
	Nervous system	**Endocrine system**
Nature of signal	Nerve impulses are **electrical** signals, transfer of information across synapses, is chemical (see Chapter 14)	All hormones are **chemical** signals
Size of signal	**Frequency modulated** – determined by the frequency of nerve impulses sent along a nerve fibre, and the number of nerve fibres being stimulated	**Amplitude modulated** – determined by the concentration of hormone
Speed of signal	**Rapid**. Human nerves conduct nerve impulses at speeds ranging between 0.7 metres per second and 120 metres per second	**Usually slower**, by comparison. The release of insulin from the pancreas in response to a rise in blood sugar level takes several minutes
Effect in the body	**Localised** effect – each individual neurone links with only one or a few cells	More **general** effect – hormones can influence cells in many different parts of the body
Capacity for modification	**Can be modified** by learning from previous experience	**Cannot be modified**: no learning from previous experience

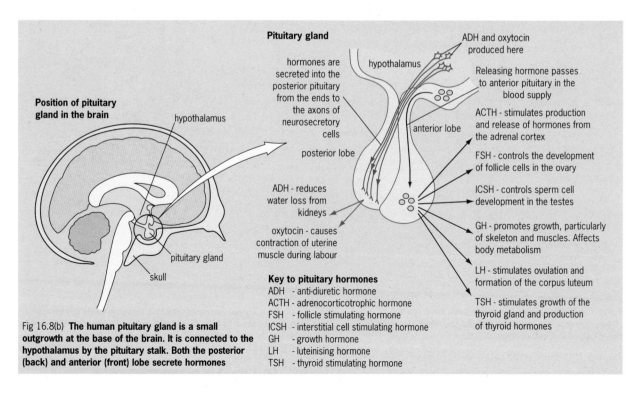

Position of pituitary gland in the brain

hypothalamus

pituitary gland

skull

Pituitary gland

hormones are secreted into the posterior pituitary from the ends to the axons of neurosecretory cells

posterior lobe

ADH - reduces water loss from kidneys

oxytocin - causes contraction of uterine muscle during labour

hypothalamus

anterior lobe

ADH and oxytocin produced here

Releasing hormone passes to anterior pituitary in the blood supply

ACTH - stimulates production and release of hormones from the adrenal cortex

FSH - controls the development of follicle cells in the ovary

ICSH - controls sperm cell development in the testes

GH - promotes growth, particularly of skeleton and muscles. Affects body metabolism

LH - stimulates ovulation and formation of the corpus luteum

TSH - stimulates growth of the thyroid gland and production of thyroid hormones

Key to pituitary hormones
ADH - anti-diuretic hormone
ACTH - adrenocorticotrophic hormone
FSH   - follicle stimulating hormone
ICSH - interstitial cell stimulating hormone
GH   - growth hormone
LH   - luteinising hormone
TSH   - thyroid stimulating hormone

Fig 16.8(b) **The human pituitary gland is a small outgrowth at the base of the brain. It is connected to the hypothalamus by the pituitary stalk. Both the posterior (back) and anterior (front) lobe secrete hormones**

The hypothalamus and pituitary: partners in communication and control

The pituitary has been called the 'master gland' because it controls the activity of most of the other endocrine glands, but it is itself actually under the control of the hypothalamus. The range of hormones produced by the pituitary and the way the hypothalamus controls pituitary function is shown in Fig 16.8(b) on the facing page.

The human pituitary gland has a **posterior lobe** and an **anterior lobe**. The **pituitary stalk** connects the posterior lobe to the brain. This direct physical link allows nervous communication between the pituitary and the hypothalamus. A rich network of blood vessels also link the hypothalamus with the anterior lobe, allowing hormonal signals to pass from the brain to the pituitary.

Anterior means 'towards the front'.
Posterior means 'towards the back'.

Neurosecretory cells and the posterior pituitary

A **neurosecretory cell** is a modified neurone that synthesises a hormone. The hormone is produced in the cell body and is then transported down the axon and released into the blood through the synapse, which terminates on a capillary. The neurosecretory cells that arise in the hypothalamus and end in the posterior pituitary, make two hormones: **anti-diuretic hormone (ADH)** and **oxytocin**. When the hypothalamus is stimulated by appropriate nerve impulses, it releases these hormones into the posterior pituitary. From here they pass into the blood and then travel around the body. ADH controls the reabsorption of water by the kidneys (see Chapter 11), oxytocin causes contraction of the uterus during childbirth (see Chapter 19).

■ See questions 1, 2 and 3.

Hormonal control of the anterior pituitary

Blood vessels that connect the hypothalamus with the anterior pituitary carry hormones and chemical messengers. These either stimulate or inhibit the release of pituitary hormones, including growth hormone prolactin, the gonadotrophins, adrenocorticotrophic hormone and thyroid stimulating hormone (see Fig 16.4).

Fig 16.4 shows which endocrine glands produce these hormones. Table 16.1 shows what each hormone does.

Nervous and hormonal control of digestive secretions

The body uses a combination of nerve communication and hormones to control the secretion of digestive juice to coincide with the presence of food in particular areas of the gut. Below is a brief summary of how this happens: for more detail see Chapter 6.

The taste or smell of food encourages the brain to signal the salivary glands and stomach to release saliva and stomach secretions. Food leaving the stomach stimulates the vagus nerve, which then triggers the release of bile and pancreatic juice. Digestive secretions can also be controlled by hormones. The hormone gastrin, produced by the stomach wall, travels in the bloodstream but exerts its effect locally, stimulating the production of both pepsinogen and hydrochloric acid.

Secretin and cholecystokinin control pancreatic and liver secretions. Both are formed by cells in the duodenal wall. Secretin causes the release of sodium hydrogencarbonate in the pancreas. In the liver, it increases the rate of bile formation. Cholecystokinin triggers the release of pancreatic enzymes such as lipase and trypsinogen.

A What would happen if the production of digestive juices was not timed to coincide with the presence of food in the gut?

The action of adrenaline

As the nervous system and the endocrine system are linked closely together, we should not be surprised that hormones can affect the way that we feel emotionally. The classic example of this is the **fear**, **fight or flight response** (see the Assignment at the end of this chapter).

At the molecular level, adrenaline binds to the outside of a target cell and activates the enzyme adenylate cyclase. This converts ATP to cyclic AMP. Cyclic AMP, is a second messenger which sets off a **cascade**, a chain of reactions that convert glycogen to glucose.

As adrenaline binds to receptor molecules on the surface of a cell, changes in the membrane cause the production of **G proteins**. For every one molecule of adrenaline that binds, 10 molecules of G protein are produced. Each of those 10 molecules of G protein catalyses the production of 10 molecules of **adenyl cyclase**. Each molecule of adenyl cyclase stimulates the production of 10 molecules of cyclic AMP. This **amplification** effect continues along a chain of reactions that eventually breaks down glycogen into glucose.

> ✔ The process of amplification seen in the enzyme cascade which results in the fear, fight or flight response, allows *one* molecule of adrenaline to stimulate the production of *millions* of glucose molecules.
>
> See question 4. ■

4 PROSTAGLANDINS

Prostaglandins are a group of hormone-like compounds. They were originally thought to be produced by the prostate gland – hence the term prostaglandins. They are not true hormones as they are not produced by endocrine glands. Most mammalian cells synthesise prostaglandins which act locally on surrounding cells.

Prostaglandins control cell metabolism, probably by modifying levels of cyclic AMP inside the cell, and are involved in a wide range of activities including blood clotting, inflammation and smooth muscle activity. Prostaglandins are extremely powerful: one billionth of a gram produces measurable effects. The prostaglandins in human semen cause muscles of the uterus and oviduct to contract during female orgasm, helping the sperm on their journey towards the ovum. Synthetic prostaglandins can be used to stimulate labour in a pregnant woman whose baby is overdue.

> ✔ Prostoglandin pessaries can be used to set off labour in a pregnant woman whose baby is overdue. However, semen is rich in prostaglandins, and some experts recommend to expectant couples the natural alternative of having frequent sex. There is a sound physiological rationale behind this, but, from a practical viewpoint, it is rather difficult advice to follow.

5 ENDORPHINS

Endorphins are one of several morphine-like chemical messengers produced in the brain (endorphin = endogenous morphine). These polypeptide molecules mimic the effects of drugs such as morphine and heroin: like prostaglandins, they are involved in a wide range of activities, but their most important role seems to be in the management of pain. It is thought that endorphins bind to pain receptors and so block the sensation of pain. Apparently long-distance runners who run until they collapse do so because abnormally high levels of endorphin block out the acute pain and discomfort.

Endorphins also seem to affect the 'pleasure centres' of the brain. Stimulation of these areas provide the intense feelings associated with orgasm. Scientists studying the chemistry of pleasure have shown that pleasurable sensations begin when the hypothalamus releases serotonin. This stimulates the release of endorphins which turns on the supply of dopamine and simultaneously turns off the supply of GABA (an amino acid which suppresses dopamine). Dopamine stimulates the pleasure centre directly.

Drugs like heroin work by mimicking endorphins. The brain of an addict stops making endorphins and so withdrawal from heroin results in a sudden lack of endorphin and a build-up of GABA, producing unpleasant withdrawal symptoms.

7 PHEROMONES AND BEHAVIOUR

Pheromones, sometimes called **ecto-hormones**, are substances that organisms release into the environment to communicate with organisms of the same species. Humans produce pheromones (see Fig 16.2). Pheromones are small volatile molecules that spread easily into the environment. They are active in very small amounts: the pheromone of the female gypsy moth causes a response in the antennae of the male moth at concentrations as low as one in a thousand million million molecules (Fig 16.9).

Pheromones are usually classified according to the type of response they produce:

- **Alarm pheromones** are produced by bees and ants when attacked. They excite other insects of the same species to swarm around the attacker.

- **Sex attractants** are released by moths, rats and possibly humans to attract members of the opposite sex. Humans do not have a particularly good sense of smell but there is some evidence that very young babies can recognise their mother by her characteristic smell. Some people also think that sexual partners can also recognise each other using their sense of smell, but there is less evidence to back this up.

- **Trail substances** are produced by ants during to show other ants where to find sources of food.

- The **queen substance** is produced by the queen bee within a hive to suppress the production of other queens.

Fig 16.9 **Sex attractants such as the pheromone bombykol are released into the air in minute quantities by female moths. The male is extremely sensitive to these pheromones, because his antennae are large and crammed with many specialist receptors. This male has about 17 000 receptors in each antenna that respond only to bombykol**

SUMMARY

After reading this chapter, you should know and understand the following:

- There are three basic types of chemical signal: **local signals** such as histamine, prostaglandins and endorphins; **hormones** such as insulin and adrenaline; external signals or **pheromones**.

- The human **endocrine system** has four main functions. It controls growth, sexual development and fear, flight and fight reactions. It maintains body homeostasis.

- Endocrine glands lack a duct: their secretions are delivered directly into the bloodstream.

- Hormone levels are usually controlled by **negative feedback loops**.

- There are two basic types of hormone: **peptide hormones** which are derived from amino acids (eg insulin) and **steroid hormones** which are made from fatty acids (eg oestrogen).

- Peptide hormones bind to receptors on the outer membrane of target cells and achieve their effect via a **second messenger** such as cyclic AMP. Steroid hormones pass straight through the cell membranes and bind to target receptors inside the cytoplasm.

- There is a close link between the nervous and endocrine systems, shown by the way in which, in the brain, the **hypothalamus** interacts with the **pituitary gland.**

- Adrenaline, the hormone which causes the classic **fear, flight and fight** response, acts at the molecular level to bring about the production of millions of glucose molecules.

- **Prostaglandins** and **endorphins** are local messengers which affect many different types of cell. Prostaglandins are best known for their effects on the female reproductive system: endorphins are chemicals which influence our perception of pleasure and pain.

QUESTIONS

1 The anterior lobe of the pituitary produces and secretes six different hormones including follicle stimulating hormone (FSH), which initiates development of gametes in ovaries and testes, and adrenocorticotrophic hormone (ACTH) which controls the secretion of some hormones by the adrenal cortex. The anterior lobe also produces thyroid stimulating hormone (TSH) which controls the formation of thyroxine by the thyroid gland. The secretion of TSH itself is controlled by thyrotrophin releasing factor (TRF) [= thyrotropin releasing hormone] in a negative feedback loop. TRF is produced by the hypothalamus.

a) Account for the different specificity of FSH and ACTH.

A neuroendocrine mechanism is involved in the control of body temperature in mammals. Some of the components of this mechanism are listed below in alphabetical order.

anterior lobe of pituitary   thermoreceptor
central nervous system   thyroid gland
cold (exteriostimulus)   thyroxine
hypothalamus   TRF
increase in metabolic rate   TSH

b) Copy and complete the flow diagram. Place each component in a box as shown, joining them with arrows as shown in the key.

> Anterior lobe of pituitary
>
> H
>
> TSH
>
> Keyed arrows:
> ······H······▶ = hormone release
> ······I······▶ = inhibition
> ······S······▶ = stimulation

[UCLES June 1995 Modular Biology: Transport, Regulation and Control Paper, q.1]

2
a) The table shows some of the hormones produced by the pituitary gland, the organ or gland which they affect, and the effect they produce on this organ or gland. Copy and complete the table.

Pituitary hormone	Target organ/gland	Effect on organ/gland
	Adrenal cortex	Secrets cortisol
		Increased secretion of thyroxin
Luteinising hormone (LH)	Ovary	

b) Somatotropic hormone is also a pituitary hormone. It plays an important part in normal growth. Suggest:
(i) how somatotropic hormone may increase growth;
(ii) how it acts to oppose the action of insulin.

[AEB 1995 Human Biology: Paper 1, q.13]

3 Fig 16.Q3 summarises the role of the hormones involved in the first part of the menstrual cycle.

a) Name hormone **A**.

b) Use the information in the diagram to explain the meaning of the term *negative feedback*.

c) Some women are infertile because they do not produce enough hormone **A**. The drug clomiphene works by lowering the concentration of oestrogen in the blood. Suggest how treatment with clomiphene might lead to conception.

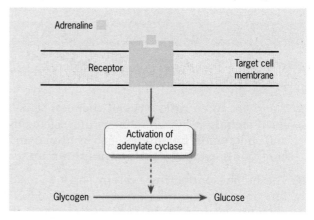

[NEAB June 1996 Biology: Process of Life Module Test, q.7]

4
a) Adrenaline increases the rate of conversion of glycogen to glucose.
(i) Under what conditions is adrenaline normally secreted?
(ii) Explain how the release of adrenaline during exercise is of benefit to the body.

b) Fig 16.Q4 shows how adrenaline affects a target cell.

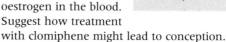

(Adenylate cyclase is also known as adenyl cyclase.)
(i) Explain how proteins are suited to their function as receptors.
(ii) Describe the precise function of adenylate cyclase.
(iii) Explain how a single molecule of adrenaline can lead to the formation of many molecules of glucose.

[AEB Summer 1998 Biology Module Paper 4, q.4]

Assignment

ADRENALINE AND STRESS

When we are nervous, familiar symptoms appear. We get butterflies in the stomach, our pulse increases, our hands become cold and clammy. This happens because our body is preparing for action by secreting **adrenaline**. This hormone:
- increases heartbeat,
- increases ventilation rate,
- causes vasoconstriction, resulting in redirection of blood from intestines and skin to brain and muscles,
- increases metabolic rate,
- dilates pupils.

1

a) Which organs secrete adrenaline?
b) What is the overall effect of adrenaline on the body?
c) Use your answer to 1b to explain why adrenaline leads to an increased blood pressure.

The effects of adrenaline are very similar to the effects of the sympathetic nervous system (see Chapter 14). For example, secretion of adrenaline speeds up the heart, as does direct stimulation by a sympathetic nerve.

The difference is that adrenaline produces a general state of readiness for action, while the sympathetic nervous system is used for fine control. If your body needs to increase the heart rate a little, but does not need the other effects of adrenaline, it does so via the sympathetic nerves.

Fig 16.A1 **Adrenaline helps this mountain biker to concentrate and compete to the best of her abilities. Her heart is pounding, blood pressure is raised and all of the capillaries in the muscles are open to maximise performance**

Fig 16.A2 **There are many stressful lifestyles: being a busy executive is just one. Stress happens when adrenaline is continually released, ready for action that is not taken. This leads to raised blood pressure which can increase the risk of heart disease or stroke**

The effects of stress

It is difficult to define stress. What is a stressful situation to some people may be sheer enjoyment to others. Certainly, some stimulation is desirable – it makes us feel more alive.

In the short term, the responses of the body to stress have a beneficial effect and leave the individual better able to cope with the crisis if and when it comes. In the long term, however, the effects can be far from beneficial. Continued secretion of adrenaline and general sympathetic stimulation bombard the body with stress chemicals. Table 16.A1 outlines some of the common symptoms of stress.

Table 16.A1 **The effects of stress on bodily functions**

Normal state	Adrenaline response	Short-term effect	Long-term effect (stress)
Brain: normal blood flow	blood flow increases	think more clearly	headaches and migraines
Muscles: normal blood flow	blood flow increases	improved performance	muscular tension and pain
Heart: normal pulse and blood pressure	output and pressure increase	improved performance	hypertension and chest pain
Intestines: normal blood flow and peristalsis	blood flow decreases, peristalsis increases	slower digestion	abdominal pain and diarrhoea
General biochemistry: normal rate of oxygen use. Glucose and lipids liberated	oxygen, glucose and lipid use increases	more energy available quickly	rapid tiredness

2

a) List some activities/sports that people enjoy because they stimulate release of adrenaline.
b) List some occupations or lifestyles that are commonly associated with stress.

3

a) Discuss or write down some ideas that might explain why there is more stress-related illness today than in previous generations.
b) Discuss or write down some strategies that people might use to minimise stress.

These men stepped on landmines while working in the fields near their village. They are learning to get around with a new prosthetic leg, but they can still feel the leg they lost.

LOSING A LIMB is a surprisingly common event. Amputation can follow the horrific injuries suffered in terrorist bomb attacks or in serious industrial or road accidents. However, the amputee wards in Britain's hospitals are more likely to be full of smokers whose circulation is no longer effective enough to keep their extremities alive.

When a person is unfortunate enough to lose an arm or a leg, you might expect that they would lose all the sensations that come from that part of the body. However, amputees can still feel pain or itching in the removed limb. This is an added misery for a person already seriously traumatised by their injury.

The phenomenon of the phantom limb arises because of the nature of the nervous system. Our brain interprets the environment by knowing where all the incoming nerves have originated. If someone stamps on your left foot, you know which foot hurts because impulses arrive via the left foot nerves. When a limb is removed, many of the nerves that connect it with the brain remain intact, and any stimulation of these nerves is interpreted by the brain as sensation in the missing limb.

1 RESPONDING TO THE ENVIRONMENT

We make sense of our surroundings by using our sense organs to gather information and relay it to the central nervous system. Table 17.1 shows the huge amount of sensory information that reaches the brain. Fortunately, we deal with most of it at a subconscious level. Just as well, otherwise we would have no time for thoughts. Overall, the brain copes with a huge range of sensory information and uses it to decide on appropriate courses of action.

In this chapter, we present an overview of how the sense organs work and then we examine the eye and the ear in more detail. You can find out more about taste and smell in Chapter 7, and touch and the skin are covered in Chapter 12.

Some important terms

Before going on to look at sensory systems in more detail, you will find it useful to become familiar with some important terms:

- A **stimulus** is a physical event, usually some form of energy change, that we can detect with our receptors. Examples include **mechanical stimuli** such as pressure, touch, and movement of air, **thermal stimuli** such as heat and cold, **light** and **chemical stimuli** such as taste, smell, and the concentration of carbon dioxide or oxygen in the blood. There are many factors in our

Fig 17.1 **We make sense of our world by using our sense organs. Try describing a visit to a restaurant, and make a note of how many times you refer to your senses. 'Nice atmosphere, cosy, comfortable. And the music, and the food, and the wine, and the company...'**

External stimuli	Internal stimuli
Sound: volume, pitch, harmonics	Blood pressure
Vision: colour, shape, intensity, movement	Solute concentration of the blood: do we need a drink?
Touch: light or heavy pressure?	Internal pain: ulcer, wind, headache?
Texture: Sandpaper or velvet?	
Temperature: hot or cold?	Tension in muscles: posture?
Pain: mild or severe?	Fatigue in muscles
Movement: direction, acceleration or deceleration	Tension in bladder or rectum
Direction of gravity	
Smell: which one of thousands of different chemicals?	
Taste: salt, sweet, sour or bitter?	

Table 17.1 **Summary of the sensory information that goes to the brain**

environment that are not stimuli – we cannot detect them because we do not have the appropriate receptors. We do not sense radio waves or high-pitched dog whistles, for example.

- A **sensation** refers to a general state of awareness of a stimulus. Aristotle first identified the five senses of hearing, sight, taste, smell and touch, but we now know that there are actually many more senses than this. The skin can experience the sensations of light touch or deep pressure, heat, cold and pain. There are also the general senses of balance and body movement. Internally, the body can sense changes in blood pressure and blood levels of chemicals; and we experience hunger and thirst when we needs food or fluids.

- **Perception** is the interpretation of stimuli by the brain. Perception of a stimulus is different from just sensing that it is there – it allows us to assess the significance of the stimulus. Many stimuli can be ignored, such as the sensations from your clothes, to allow us to concentrate on more urgent stimuli (like the information in this book).

> ✔ Perception involves reception *and* interpretation of stimulus information.

2 FEATURES OF SENSORY SYSTEMS

A sensory system must allow the following processes to take place:

- **Transduction**. This is the process by which a stimulus initiates a nerve impulse. Stimuli vary widely in their nature; they can be physical, light, heat or chemical, and so on. All, however, result in the generation of a nerve impulse (see Chapter 14). For example, when physical pressure is applied to the skin, sensory cells inside specialised receptors in the skin depolarise, and the physical stimulus is transduced, or changed, into the electrochemical event known as the action potential.

- **Transmission**. A nerve impulse is transmitted along a sensory neurone into the central nervous system (CNS) (see Chapter 15). The brain recognises the strength of a stimulus not by the size of the impulse but by its frequency. The stronger the stimulus, the higher the frequency of impulses.

- **Information processing**. The CNS processes information that it receives from several parts of the body and makes a decision about the best course of action to take. We have superb memories compared with other animals and so we are able to combine new sensory information with past experience before deciding on an appropriate course of action.

Fig 17.2 **Different species have different sense organs. People who have ever taken a dog for a walk will know that their pet lives in a different world. They can drag you from lamp-post to lamp-post, obsessed with smells that we can't even begin to detect, even if we wanted to. Their hearing is much more acute, too. Dogs can hear much higher frequencies than we can. On the other hand, their colour vision is not as good**

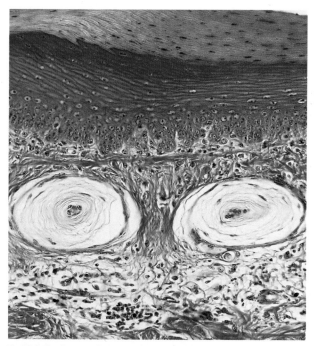

Fig 17.3 **A micrograph of two touch receptors in human skin. They are abundant at fingertips. In this touch receptor, pressure is detected by the end of the neurone at the centre of the conective-tissue capsule which is layered like an onion. When the neurone is stimulated, action potentials are propagated along the neurone. The brain interprets these impulses as pressure on that part of the body.**

Receptors

A receptor is part of the nervous system that has become adapted to receive stimuli. Some receptors are modified neurones while others are separate cells connected to neurones. Some receptors, such as those in the skin, occur alone or in small groups. Others are concentrated within a specialised sense organ such as the eye or ear.

Sensory receptors are vital. They sense the external environment and they also provide information to the CNS about the body's internal environment. Receptors that detect changes in the outside world are positioned at or near the body surface (Fig 17.3), usually at the front end. These include the specialised sense organs, the ear and eye, the smell receptors of the nose and the taste receptors of the tongue. The human head, like that of most animals, is basically a set of sense organs in front of a mass of nerves (the brain) that is capable of processing all the incoming information. In contrast, receptors that respond to internal stimulation, such as stretch receptors in blood vessels or muscles, are found deep inside the body.

Types of receptor

Receptors can be classified in three ways: by the type of stimulus they respond to, by their location within the body, or by their level of complexity (Table 17.2).

Table 17.2 **Classification of receptors**

Receptor name	Action	Example
Classified by stimulus		
Photoreceptor	responds to light	eye
Chemoreceptor	responds to chemicals	taste bud
Thermoreceptor	responds to changes in temperature	temperature receptors in skin
Mechanoreceptor	responds to mechanical (touch) stimuli	nerve fibres around hairs
Baroreceptor	responds to changes in pressure	baroreceptors in carotid artery
Classified by location		
Interoceptor	responds to stimuli within the body	baroreceptors in carotid artery
Exteroceptor	responds to stimuli outside the body	eye and ear
Proprioceptor	responds to mechanical stimuli conveying information about body position	muscle spindle
Classified by complexity		
General senses	single cells or small groups of cells	temperature receptors in skin
Special senses	complex sense organs	eye and ear

?

A Look at the cells in Fig 17.3. What type of receptor cell are they? Classify them according to stimulus, location and complexity, as in Table 17.2.

There are also two basic types of receptor, and both can be any of the classes of receptor described in Table 17.2. Simple or **primary receptors** consist of a single neurone whose axon carries a nerve impulse into the CNS. The temperature and pressure receptors in the skin are both examples of primary receptors.

Secondary receptors receive information from a cell which is not part of the nervous system and then pass the information on to a nerve cell for transmission to the CNS. Taste buds in the tongue are secondary receptors.

How do receptors work?

How do cells detect stimuli? How, for instance, can we distinguish between a light and a heavy touch? The answer lies in the fact that primary receptors have proteins in their cell surface membranes that are sensitive to the type of stimulus. Whatever the stimulus, transduction follows the same basic pattern. A receptor protein is activated by a specific stimulus. In turn, the protein opens or closes specific ion channels in the cell surface membrane. This change in ion movement causes depolarisation of the membrane; this is the **generator potential.**

If the depolarisation goes beyond the threshold level for that cell, it triggers an **action potential** (see Chapter 14). All receptors therefore act as biological transducers: they create action potentials in neurones in response to physical and chemical stimuli from the environment.

The greater the stimulation of the sense cell, the greater the depolarisation. Once the generator potential exceeds the threshold level, an action potential is produced and is transmitted along the axon. It is important to remember that all action potentials are of the same magnitude – you can't have powerful and weak nerve impulses. The body is able to tell the difference between strong and weak stimuli by the frequency of the action potential created.

Receptors are very sensitive – they have to be in order to detect small changes in the environment. Apparently, the human nose can detect ethanoic acid (acetic acid, the acid in vinegar) at a concentration of 5×10^{11} molecules per litre of air; a dog's nose, though, can detect the same smell at a concentration of about 2 million times less, 2×10^5 molecules per litre.

Body senses also amplify the incoming signal. A stimulus is often quite weak – just a few photons of light when viewing a distant star for example, or a few molecules of a chemical. But the action potential travelling from the eye to the brain, for example, has about 100 000 times as much energy as the few photons of light that triggered it.

stimulus	
↓ has effect on	⎫ Filtration and detection of stimulus
receptor protein	
↓ changes activity of	
ion channel	⎫
↓ causes build-up of	⎬ Amplification of stimulus
generator potential	⎭
↓ reaches threshold	⎫ Encoding of stimulus
action potential in sensory nerve	

17.4 **Flow chart summarising the function of receptor cells**

3 VISION

The eyes are literally our window on the world. This most sensitive and intricate of sense organs has fascinated both scientists and poets for centuries. In this section we look at the basic structure of the eye before going on to study focusing and the correction of defects. Finally, we look at the strucutre that is the key to advanced study of the eye – the retina.

B Look at Table 17.2. What type of receptors are the photoreceptors of the eye: are they exteroceptors, proprioceptors or interoceptors?

Fig 17.5 **Mammalian eyes are contained in the** orbit, **a bony socket of the skull. They are protected by the bone and also by a pad of fat at the rear of each eyeball. Other features of the eye are also protective:**
- **The eyebrows prevent sweat running from the forehead into the eye**
- **The eyelashes prevent the entry of airborne particles such as flying insects, often adding to the efficiency of the eye blink reflex**
- **Tears constantly bathe the surface of the eye removing the surface film of dust and dirt and killing bacteria**
- **The** conjunctiva, **a thin, transparent membrane which covers the cornea, also protects the eye from airborne debris**

Structure of the human eye

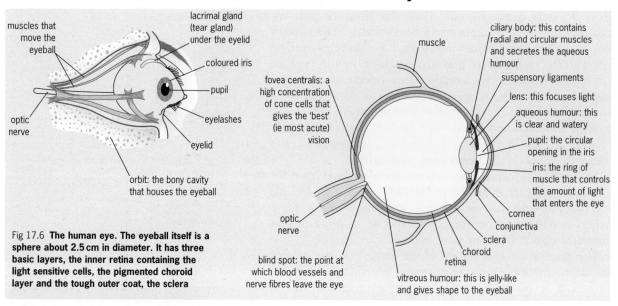

Fig 17.6 **The human eye. The eyeball itself is a sphere about 2.5 cm in diameter. It has three basic layers, the inner retina containing the light sensitive cells, the pigmented choroid layer and the tough outer coat, the sclera**

Fig 17.7 **As people get older, the lens in their eyes can become cloudy, interfering with normal vision. In this routine operation, an ultrasound probe is being used to break up a cloudy lens before a plastic lens is fitted.**

C The eye is often compared with a camera. Which part of the camera corresponds to the retina of the eye?

The eye is an important and delicate structure. The **orbits**, bony sockets in the skull, protect the eyes from physical damage, and a pad of fat behind the eyeball helps to cushion it from shocks. The other protective features of the eye are shown in Fig 17.6.

The main body of the eye is divided into two parts by a **biconvex lens**. This lens contains transparent lens fibres (long thin cells that have lost their nuclei) together with an elastic **lens capsule** made of glycoprotein. The lens is flatter at the front than at the back and is soft and slightly yellow. With age it becomes flatter, yellower, harder and less elastic. There is no blood in the lens – substances diffuse in and out of it from the surrounding fluid.

The **eyeball** has three cavities: the region between the **cornea** and the **iris**, the region between the iris and the lens, and the cavity that fills all the space behind the lens. The first two cavities are filled with a fluid called the **aqueous humour**, which is thin and watery. The third cavity, the largest (it takes up about 80 per cent of the volume of the eyeball) is filled with a thicker fluid called the **vitreous humour**. This transparent gel has the consistency of egg white and contains 99 per cent water and 1 per cent collagen fibres and hyaluronic acid. The vitreous humour preserves the spherical shape of the eyeball and helps to support the retina.

The retina

The retina is the light-sensitive layer at the back of the eye. It is a complex structure with a deep layer of light-sensitive cells called **rods** and **cones**, together with a middle layer of **bipolar neurones** which connect the rods and cones to a surface layer of **ganglion cells** (Fig 17.8). Fibres from the ganglion cells join up to form the **optic nerve**. Only the back of the retina is photosensitive. The front surface extends over the choroid and forms the inner lining of the **iris** and **ciliary body**.

The sensitive part of the retina has the area of a ten pence coin and acts as a projection screen onto which images are directed. Strangely, the nerves that come from the retina pass in front of the light sensitive cells, but these do not inerfere with vision.

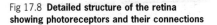

Fig 17.8 **Detailed structure of the retina showing photoreceptors and their connections**

The choroid

The choroid contains many blood vessels. Blood supplies the cells of the eye with nutrients and oxygen and removes waste. The choroid is dark-coloured due to a high concentration of the pigment **melanin** in its cells. These pigmented layers prevent internal reflection of light rays and so prevent us seeing a confused and blurred image.

At the front of the eye, the choroid expands around the edge of the lens to form the **ciliary body**. The smooth muscles in the ciliary body alter the shape of the **lens** (Fig 17.9). The lining of the ciliary body secretes aqueous humour, the fluid which fills the front of the eye.

The **iris** is also an extension of the choroid, partially covering the lens and leaving a round opening in the centre, the **pupil**. Its function is to control the amount of light entering the eye. The iris contains radial and circular muscles which can alter pupil size. Iris muscles are controlled by the **autonomic nervous system** (see Chapter 15).

The sclera

The sclera forms the 'white' of the eye. It is a tough external coat mainly of collagen fibres. Six exterior muscles attached to the sclera enable the eye to look up, down and side-to-side. The central one-sixth of the sclera is colourless and transparent, forming a clear window, the **cornea**, through which light gets into the eye.

How the eye works

The process of seeing is complex and involves five different stages:

● Light enters the eye.

● An image is focused on the retina.

● Energy in the light that makes up the image is transduced into an electrical signal.

● Nerve impulses carry information about the image into the brain.

● The brain decodes the information and perceives the image.

Let's look at each of the stages in more detail.

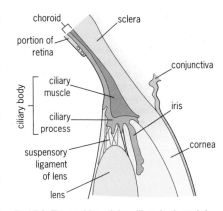

Fig 17.9 **The position of the ciliary body and the suspensory ligaments in the eye. When we focus on objects that are distant from us, the ciliary muscles relax, increasing the tension on the suspensory ligaments. This pulls the lens into a thinner and flatter shape. When we look at objects closer to us, the ciliary muscles contract, the tension on the ciliary ligaments decreases and the lens becomes fatter**

Bright light
Relaxation of radial muscles and contraction of circular muscles cause pupil to constrict

Dim light
Contraction of radial muscles and relaxation of circular muscles cause pupil to dilate

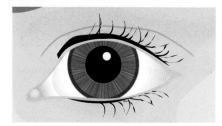

Fig 17.10 **The eye is prevented from being dazzled by a speedy reflex that makes the pupil smaller. To achieve this** constriction **of the pupil, radial muscles in the iris relax and circular muscles contract. In dim light, the opposite happens: radial muscles contract and circular muscles relax, causing the pupil to enlarge**

Fig 17.11 **The eye can focus on near and distant objects by altering the shape of the lens. This process is called accommodation**

How light enters the eye

Light enters through the cornea and then passes through the pupil, the aqueous humour, the lens and the vitreous humour, before it reaches the retina. To operate well in conditions of different light intensity, the eye must control how much light reaches the retina. It does this by changing the diameter of the pupil. In bright light, the pupil **constricts** (gets smaller) to prevent overstimulation of the retina and the perception of 'dazzle'. In dim light, the pupil **dilates** (gets wider), letting in more light.

The size of the pupil is controlled by the muscles in the iris (Fig 17.10). These are themselves controlled by the autonomic nervous system and so pupil adjustment is a **reflex response**, not under the conscious control of the brain.

Focusing light on the retina

If you do not focus a camera correctly, the picture you take is blurred. Similarly, the eye must focus light to produce a high resolution image on the retina. The eye focuses an image by **refracting** or bending light using the cornea and the lens, forming an upside down or **inverted image** on the retina (Fig 17.12).

Because of the composition and curvature of the cornea, most of the refraction of light occurs in this structure. The lens is important in the fine focusing of light onto the retina. The cornea is fixed but the lens is adjustable, allowing **accommodation**, a reflex that makes the eye focus on objects that are at different distances from the eye.

Accommodation

Generally, when we open our eyes in the morning, they are not focused on nearby objects. At rest, the ciliary muscles have relaxed, pulling the lens flat. In this state we focus on distant objects.

When we focus on an object that is only a metre or so away, the ciliary muscles of the eye contract involuntarily (Fig 17.11). This reduces the tension on the ligaments that hold the lens in place, and the lens becomes 'fatter'. The focal length changes, and the image is focused. When we then look at an object much further away, the ciliary muscles relax, again automatically. The tension on the ligaments supporting the lens increases, and the lens becomes 'thinner' allowing our vision to **accommodate**.

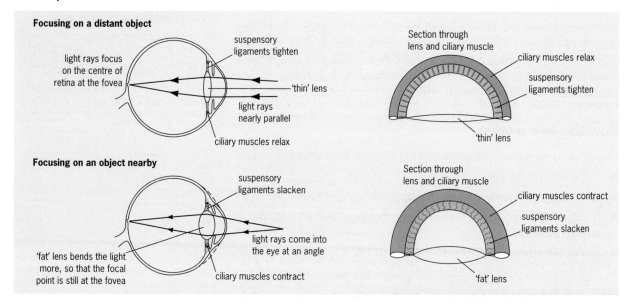

Problems with the lens system

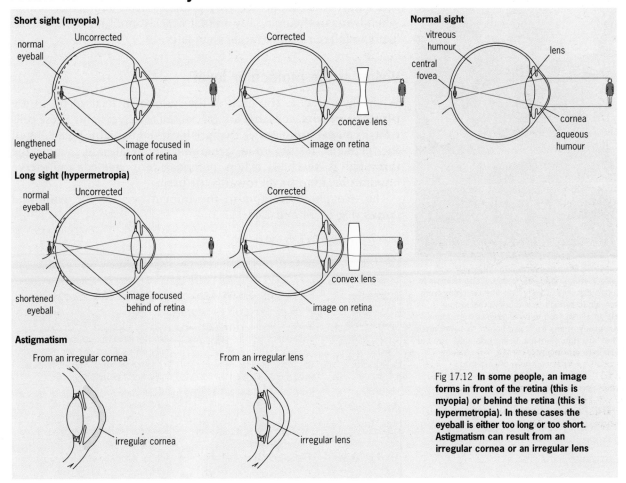

Fig 17.12 **In some people, an image forms in front of the retina (this is myopia) or behind the retina (this is hypermetropia). In these cases the eyeball is either too long or too short. Astigmatism can result from an irregular cornea or an irregular lens**

In some individuals the lens and cornea do not focus correctly (Fig 17.12). If you are short sighted, a condition known as **myopia**, you focus light from an object *in front of* the retina. This produces a blurred image. Short sight can result from an elongated eyeball or a thickened lens. It is corrected by using a diverging or concave lens.

If you are long sighted you are said to have **hypermetropia**, and you focus light *behind* the retina. Again, objects appear blurred. Long sightedness results from a shortened eyeball or a lens that is too thin, and is corrected using a converging or convex lens.

A more complicated visual defect is **astigmatism**. In astigmatism, the surface of the cornea is irregular, the object appears blurred because some of the light rays are focused and others are not (this is similar to the distortion produced by a wavy pane of glass). Astigmatism is corrected using a cylindrical lens that bends light rays in one plane only.

Binocular vision

Humans and some other mammals, notably primates (Fig 17.13) and predators such as cats have **binocular vision**. We have eyes at the front of our heads, so the field of vision seen by each eye is similar but not identical. Normally, the brain is able to compute the incoming signals so that we see only one image.

Binocular vision gives us a big advantage. When two sets of signals are decoded by the brain, we get the impression of distance, depth and of objects being three-dimensional. This **stereoscopic**

?

D Imagine you are standing at a bus stop reading a book. What happens **(a)** to the ciliary muscle, **(b)** to the lens and **(c)** to the pupil of your eye as you look up to see the bus approaching in the distance?

See question 1.

Fig 17.13 **Primates have both binocular and colour vision. Binocular vision allows animals to judge distance. This is vital when jumping from branch to branch. It is thought that colour vision allows food to be chosen with greater accuracy; many fruits use colour to signal when they are ripe. Humans, being primates, owe our vision to our ancestry**

?

E It has long been known that a deficiency of vitamin A causes the condition of night blindness. Give an explanation for this.

vision allows us and other animals to judge distance and depth accurately. It is easy to see why primates need bionocular vision; when living in the trees, any animal which cannot judge distance is likely to fall out sooner rather than later.

Vision at the molecular level

When light reaches the back of the retina it is focused on to the **photoreceptors**, cells that are specialised to detect light. These cells contain pigment molecules that are bleached by light. Bleaching of the pigment results in a generator potential and, when the threshold is reached, action potentials are created and nerve impulses begin to travel towards the brain.

The two photoreceptors in the human eye are the **rods** and **cones** (Fig 17.14 and Table 17.3).

Table 17.3 **Comparing rods and cones**

Feature	Rod cells	Cone cells
Shape	rod-shaped outer segment	cone-shaped outer segment
Connections	many rods connect with one bipolar neurone	only a single cone cell per bipolar cell
Visual acuity	low	high
Visual pigment(s)	contain rhodopsin (no colour vision)	contain three types of iodopsin responding to red, blue and green light
Frequency	120 million per retina – twenty times more common than cones	7 million per retina – twenty times less common than rods
Distribution	found evenly all over retina	found all over retina but much more concentrated in the centre, particularly the yellow spot or fovea centralis
Sensitivity	sensitive to low light intensities (used in dim light)	sensitive to high light intensities (used in bright light)
Overall function	vision in poor light	colour vision and detailed vision in bright light

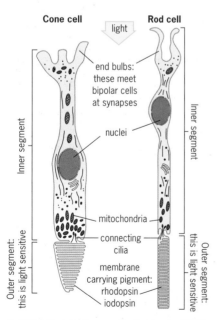

Fig 17.14 **Rods and cones: the two types of light-sensitive cell in the human eye**

Photoreception in rod cells

Rods contain a reddish-purple compound called **rhodopsin**, or **visual purple**. This consists of two joined compounds – **opsin**, a protein, and **retinal**, a light-absorbing compound derived from vitamin A. When retinal is exposed to light and absorbs light energy, it changes shape (it is actually converted into an **isomer**) and this causes the two compounds of rhodopsin to break apart. The free opsin acts as an enzyme which sets in motion a series of biochemical events which leads to the hyper-polarisation of the rod cell surface membrane, that is, it becomes more negative. This generates action potentials in the connecting nerve cells. Nerve impulses then pass to the brain for decoding. In the absence of further light stimulation, the retinal molecule goes back to its original shape and then recombines with opsin to form rhodopsin. It is worth remembering that this resynthesis process takes time.

Rhodopsin is very sensitive to light and so rods can function effectively in dim light. In bright light our rhodopsin is broken down quicker than it can be reformed, so that most of it is bleached for most of the time. In this state we are said to be **light adapted**. If we enter a dark room from a brightly lit one we cannot see anything because much of our rhodopsin is bleached. However, as we resynthesise the pigment and become **dark adapted**, things become clearer and we are able to see in dim light, even if it is only in shades of grey.

Photoreception in cone cells

The function of the cones is to see colour and detail, although they can only do this in bright light. In dim light they do not function at all. Cone cells contain the pigment **iodopsin**, a photosensitive pigment made up of **photopsin**, a different protein from that found in rods, and retinal. The events that occur in cone cells stimulated by light are basically the same as those that occur in rod cells. There are, however, three different types of cone. Each one contains a slightly different pigment and responds to a different wavelength of light. One responds to red light, one to green light and one to blue light, so allowing us to see in colour (Fig 17.15).

F Which cones are stimulated by **(a)** blue light at 490 nm, **(b)**, violet light below 440 nm? (Look at Fig 17.15.)

■ See question 2.

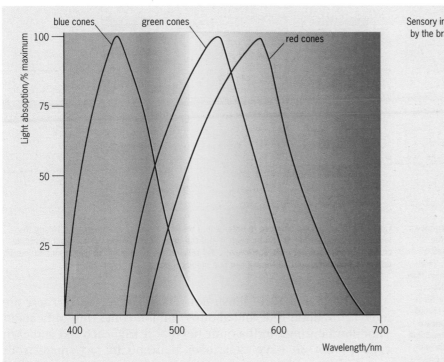

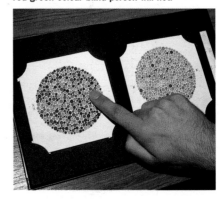

The colour that we 'see', or more accurately **perceive** (interpret in our brain), depends on which cones are stimulated. When all the cones are fully stimulated we see 'white'. When very few are stimulated we see black. Stimulation of separate types produces red, green or blue, and all colours in between are produced by combinations of different levels of stimulation of the three types together. This model of how three different types of cone can produce the range of colour vision that we experience is known as the **trichromatic theory** of colour vision.

Fig 17.15 (a) **There are three types of cone responding (optimally) to three wavelengths of light. The colour we perceive depends largely on the proportion of the stimulation of the three types. If a single type of cone is missing, the individual will not be able to distinguish between certain colours – red–green is the commonest – and is said to be colour blind.**

(b) **Colour blindness can be diagnosed by means of Ishihara tests. A person with normal vision will see the pink and red loop, while a red-green colour-blind person will not.**

Visual acuity: the ability to see detail

From what distance can you read a newspaper? Can you read the bottom line of the standard optician's eye-test board? Tests like these are a measure of visual acuity, which is defined as the ability to resolve detail. For instance, two lines such as these ══════ will obviously appear as two lines, but if you prop the book up and walk backwards, a point will be reached when they look like one grey line.

An interesting feature of visual acuity is that we only have a very narrow field of accurate vision. Focus on any particular word on

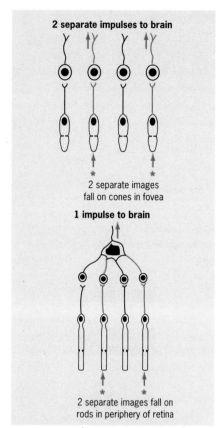

Fig 17.16 The concept of retinal convergence. Generally, the further out you go from the centre of the fovea to the edge of the retina, the greater the degree of convergence. Two separate stimuli (such as two dots or lines) may stimulate two separate cones, but not the one in between. Therefore, the brain will see these as two separate images. The same two stimuli falling on the rods will not stimulate separate neurones, because a group of rods all feed into the same neurone. In this case, the brain will perceive only one stimulus, and see only one line or dot

this page and you see it in detail, but those on either side are unclear. We see only a very small proportion of our field of vision in detail – and that is the part which is focused on the fovea, and therefore which stimulates the cone cells.

Overall, rods are more sensitive than cones. Rods can function in poor light, but they cannot perceive detail; you can't read small print in dim light. The cones have a higher visual acuity than the rods, but they need bright light. Why is there such a difference? the answer lies largely in a phenomenon known as **retinal convergence**.

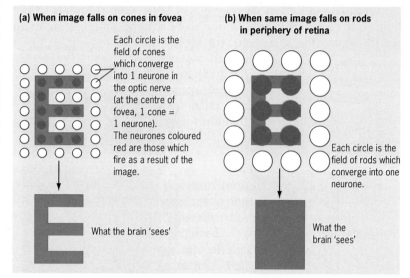

Fig 17.17 Why can we only see in detail when we look straight at something? Taking the letter E as an example, it can be seen that if the image falls on the cones it will stimulate some neurones but not others. However, if it falls on the rods it will simply stimulate lots of adjacent neurones, and all we will see is a splodge, not a clear letter E

Figs 17.16 and 17.17 illustrates retinal convergence. There are over 120 million rods and about 7 million cones, but only about 1 million neurones in the optic nerve. Clearly, each rod and each cone cannot all have their own neurone. In fact, many rods converge into one neurone, while only a few cones are associated with each neurone. In the centre of the fovea it is thought that each cone actually has its own neurone.

Transmission of nerve impulses to the brain

As you can see in Fig 17.8, the photoreceptor cells are linked to a set of nerve cells in the retina called bipolar cells. These link with a second type of nerve cell called ganglion cells. It is the axons of ganglion cells that take information from the eyes to the brain. Ganglion cells fibres are bundled together to form the **optic nerve.**

We cannot see an object whose image falls on the retina at the point where axons of ganglion cells leave to form the optic nerve. Since it contains no receptor cells, any light striking this small area is not sensed; hence its name, the **blind spot** (Fig 17.18).

See question 3. ■

Fig 17.18 To demonstrate the blind spot, hold your book at arm's length with the two symbols straight in front of your eyes. Close your left eye and concentrate on the cross with your right eye. Bring the book slowly towards you. Keep looking at the cross on the left: eventually, as the dot falls on the blind spot of your right retina, the image of the dot on the right will disappear

How the brain perceives an image

Action potentials do not differ very much from one another and so there are only two possible ways in which stimuli going into the brain can be coded. One is the *rate* at which action potentials arrive and the other is their *destination* in the brain.

The rate at which sensory neurones fire tells the brain the strength of the stimulus (see page 219): specific neurones tell the brain which part of the retina has been stimulated. The brain analyses this information to assess the nature of the original stimulus. This process, called **visual processing**, is immensely complex.

Specific areas of the cerebral cortex are associated with different sensory functions (see Chapter 15). The **visual cortex** in the **occipital lobes** deal with visual information (Fig 17.19). However, body senses must not be considered in isolation. Sensory and motor systems interact within the control regions of the CNS, and sensory information is processed at various points along the nerve pathway before it reaches the brain. In the brain it is processed further with information from other senses and stored memories.

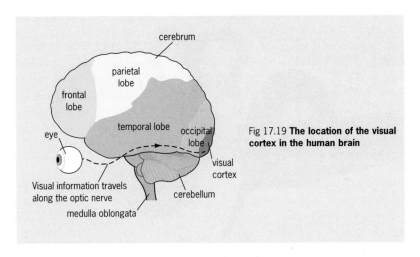

Fig 17.19 **The location of the visual cortex in the human brain**

4 HEARING AND BALANCE

The sense of hearing that allows us to detect and process sounds, and the senses of balance and proprioreception that allow us to maintain our balance and posture, are possible because of the human ear. The ear is a miniature receiver, amplifier and signal-processing system. It is divided into three parts – an outer ear, middle ear and inner ear – and is connected to the brain by the auditory nerve. The structure and function of the ear are shown in Fig 17.20.

Sound waves pass through the outer and middle ear and into the inner ear through the oval window. Pressure on the oval window squashes the fluid in the inner ear, and compression waves travel through the canals of the cochlea. The round window eventually receives these movements of fluid and dampens them down (this prevents them from being reflected back into the cochlea) by bulging into the middle ear. Sensitive hair cells in the cochlea fire off signals, sending action potentials along the auditory nerve to the brain, which analyses and processes the incoming information.

Fig 17.20 **The structure and function of the human ear**

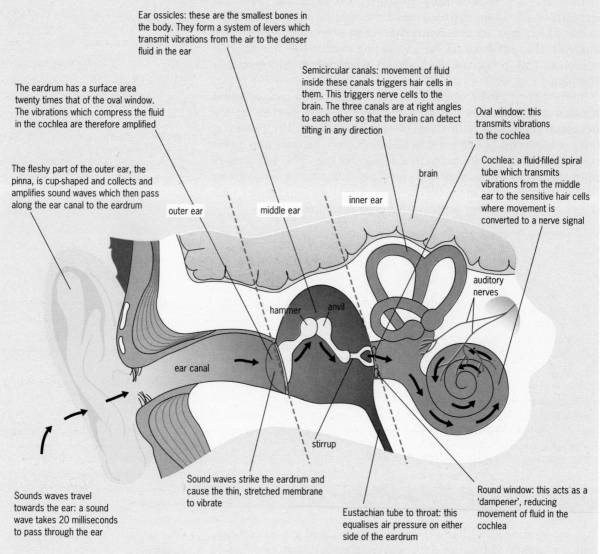

Ear ossicles: these are the smallest bones in the body. They form a system of levers which transmit vibrations from the air to the denser fluid in the ear

Semicircular canals: movement of fluid inside these canals triggers hair cells in them. This triggers nerve cells to the brain. The three canals are at right angles to each other so that the brain can detect tilting in any direction

Oval window: this transmits vibrations to the cochlea

The eardrum has a surface area twenty times that of the oval window. The vibrations which compress the fluid in the cochlea are therefore amplified

Cochlea: a fluid-filled spiral tube which transmits vibrations from the middle ear to the sensitive hair cells where movement is converted to a nerve signal

brain

The fleshy part of the outer ear, the pinna, is cup-shaped and collects and amplifies sound waves which then pass along the ear canal to the eardrum

inner ear

outer ear

middle ear

auditory nerves

hammer

anvil

ear canal

stirrup

Sounds waves travel towards the ear: a sound wave takes 20 milliseconds to pass through the ear

Sound waves strike the eardrum and cause the thin, stretched membrane to vibrate

Eustachian tube to throat: this equalises air pressure on either side of the eardrum

Round window: this acts as a 'dampener', reducing movement of fluid in the cochlea

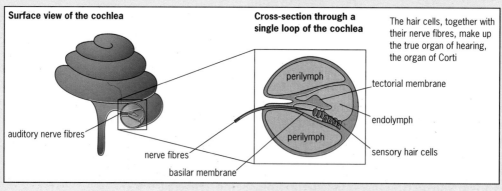

Surface view of the cochlea

Cross-section through a single loop of the cochlea

The hair cells, together with their nerve fibres, make up the true organ of hearing, the organ of Corti

auditory nerve fibres

nerve fibres

basilar membrane

perilymph

tectorial membrane

endolymph

perilymph

sensory hair cells

SUMMARY

After completing this chapter you should know that:

■ A **sensation** is a general state of awareness of a stimulus; **perception** involves the interpretation of sensory information.

■ A **receptor** is a **biological transducer**: it converts energy from one type of system (eg chemical) to another system (eg electrical). A receptor is classified according to the type of stimulus it receives (eg photoreceptor), its location (eg interoceptor) or its level of complexity.

■ The eyeball contains three outer layers, the **sclera**, **choroid** and **retina**.

■ Incoming light rays are **refracted** (bent) by the cornea and by the lens of the eye.

■ **Accommodation** is the ability to focus the eye automatically, to see near and distant objects.

■ Rod and cone cells are stimulated by the effects of light on two pigments, **rhodopsin** and **iodopsin**. Breakdown of these compounds alters the electrical properties of the cell membranes.

■ There are three types of cone cell pigment. Each responds to a different range of wavelengths of light. Humans have **trichromatic vision**: our cone cells detect three colours, blue, green, red.

■ **Stereoscopic vision** enables us to perceive depth and distance.

■ The human ear has two main functions, to detect sound and to maintain balance.

QUESTIONS

1 A number of changes occur in a human eye when it re-focuses from a distant object to a nearby object.

a) What is the term used to describe these changes?

b) Copy and complete the table to describe the state of each of the structures when looking at an object at different differences from the eye.

Structure	Looking at a distant object	Looking at a nearby object
Ciliary muscle		
Suspensory ligament		
Lens		

[AEB 1994 Biology: Specimen Paper, q. 6]

2 The diagram of Fig 17.Q2 shows a single rod from a mammalian retina.

Fig 17.Q2

a) Name the parts labelled A and B and give *one* function of each. Copy the table and write in your answers.

Part	Name	Function
A		
B		

b) Sketch the diagram and add an arrow next to the diagram to indicate the direction in which light passes through this cell.

c) State *two* ways in which vision using cones differs from vision using rods.

[ULEAC 1996 Biology Module Test B3: Specimen Paper, q. 4]

3 A person was instructed to close the left eye and stare with the right eye at a cross drawn on a plain white board. Different objects were then moved into the field of view from one side. The points where the objects were first seen and where their colours were first identified were recorded. The results are shown in Fig 17.Q3.

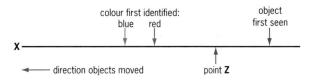

Fig 17.Q3

a) On what part of the right eye would the cross be focused?

b) Explain why an object might be seen at point **Z** but its colour would not be identifiable.

c) **(i)** Copy the diagram and add an arrow below the line showing where you would expect the colour purple to be first identified.
(ii) Explain your answer to part **(i)**.

[AEB 1995 Biology: Specimen Paper, q. 7]

18 Support and movement

It might look fun, but in zero gravity the body is under a lot of stress

ASTRONAUTS MUST OFTEN LIVE and work in zero gravity. Weightlessness is not a restful state and it can put great stress on body systems, particularly the skeletal system.

Many changes occur in zero gravity. With less work to do, body muscles begin to break down and the bones start to lose calcium at an increased rate. The number of red blood cells in the body falls and there are dramatic shifts in fluid distribution: the face becomes puffy and the legs become thinner (astronauts call this condition 'bird's legs'). The lack of gravity also affects the spine. Without the constant downward force, the spine lengthens by as much as eight millimetres. This can lead to blocked nerves and back pain. Often, astronauts also lose their touch sensitivity.

NASA scientists are busy looking at forms of treatment and exercise that might prevent bone breakdown. Such information might also have practical benefits closer to home. By studying and finding ways to slow the accelerated changes produced in space, it may be possible to develop new treatments for bone diseases such as osteoporosis which occurs when bones become brittle due to a loss of calcium.

Movement is a feature of all living things and occurs at all levels. Atoms move within molecules and cell contents move inside cytoplasm. Plants move when they tilt their leaves towards the Sun and when their flowers open and close. The human rib cage moves up and down as we breathe.

A Look at Fig 19.9. Why are the bones of the legs thicker than the corresponding bones in the arms?

B Which structure is protected by the vertebral column or backbone?

1 THE HUMAN SKELETON

The human skeleton has several functions:

- It acts as a framework that supports soft tissues.
- It allows free movement through the action of muscles across joints.
- It protects delicate organs and structures such as brain and lungs.
- It forms red blood cells in the bone marrow.
- It stores and releases minerals from bone tissue.

Fig 18.1 shows the structure and features of the human skeleton.

The structure and properties of human bone

Bone is one of the hardest tissues in the human body and is second only to cartilage in its ability to absorb stress. It is made up of bone cells called **osteocytes** which sit in a bone **matrix**. The matrix consists of inorganic matter (mainly compounds of calcium and phosphorus) interwoven with **collagen fibres** (see page 50).

Bone needs to be tough and resilient. These properties are provided by two types of bone tissue: **compact bone** and **spongy bone** (Fig 18.2). **Compact bone** is deposited in sheets called **lamellae** arranged as cylinders inside cylinders. Nerves and blood vessels run in a central canal. The compacted structure of the

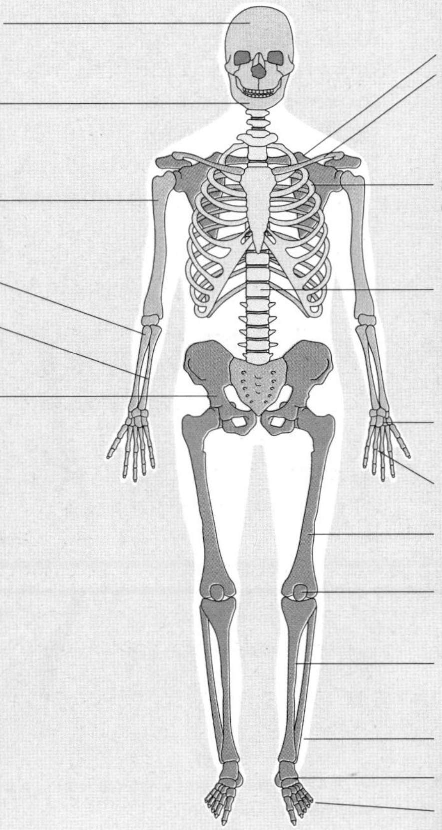

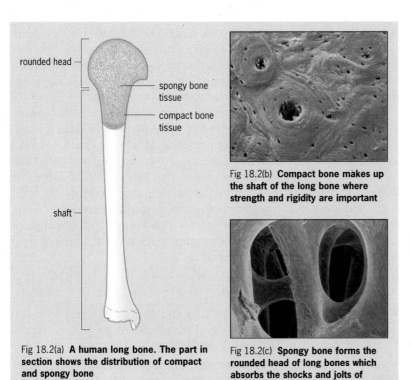

cranium (protects the brain)

mandible (lower jaw; the only movable bone in the skull, allows chewing movements)

humerus

radius

ulna

pelvic (hip) **girdle**: made up of the **pelvic bone** (itself composed of three fused bones) together with the **sacrum** at the back (fused bones at the base of the spine); a solid arrangement of bones providing stability

pectoral (shoulder) **girdle**, made up of: **scapula** (shoulder blade) and **clavicle** (collar bone); a loose arrangement of bones providing flexibility

ribs protect heart and lungs: intercostal (between the ribs) muscles aid breathing movements

vertebral column (spine): 26 individual bones held together by ligaments and separated by cartilage discs; provides support for the axis of the body, protects the nerve cord

carpals (small bones of wrist)

phalanges

femur (thigh bone)

patella (knee cap)

tibia (shin bone)

fibula

tarsals (small bones of ankle)

phalanges

Parts of the skeleton	Number of bones
Axial skeleton:	
skull	29
spine	26
rib cage	25
Total	**80**
Appendicular skeleton:	
pectoral girdle	4
pelvic girdle	2
arms	60
legs	60
Total	**126**

lamellae gives immense strength. The lamellae of **spongy bone** are arranged in a criss-cross pattern, forming a spongy honeycomb. This structure has excellent shock-absorbing properties.

Fig 18.1 **The human skeleton contains a total of 206 bones. It is divided into two parts: an** axial skeleton **which comprises the** skull **and** vertebral column **and an** appendicular skeleton **which is made up of the** limbs **and** limb girdles.

All vertebrates show skeletal modifications related to their lifestyle. Since humans are bipedal **(they walk on two legs), the hips and lower spine take most of the weight of the body and so the** pelvic girdle **(the hip) is larger and less flexible than the** pectoral girdle **(the shoulder)**

rounded head

spongy bone tissue

compact bone tissue

shaft

Fig 18.2(a) **A human long bone. The part in section shows the distribution of compact and spongy bone**

Fig 18.2(b) **Compact bone makes up the shaft of the long bone where strength and rigidity are important**

Fig 18.2(c) **Spongy bone forms the rounded head of long bones which absorbs the shocks and jolts of movement**

The human body also makes use of the supporting properties of water. The **amniotic fluid**, the fluid inside the uterus, surrounds and supports the developing fetus; and the **vitreous humour**, the jelly-like material inside the eyeball, supports the structures inside the eye.

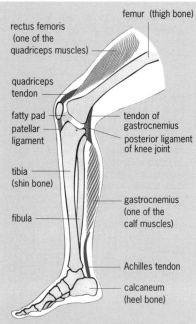

Fig 18.3(a) **A diagram of the human leg showing the position of the main ligaments (green) and tendons (blue)**

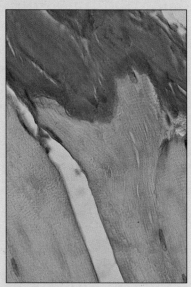

Fig 18.3(b) **This micrograph shows a slice of tendon (yellow) connected to a piece of muscle (red). Tendons and ligaments look similar but their composition is different: tendons contain mainly** collagen fibres (see page 54); ligaments are made up of fibres formed from the protein elastin

?

C Why are ligaments elastic?

D Why does the tendon need to be inelastic?

Joints

As in other vertebrates, the human skeleton is jointed. A joint is simply a place in the body where two bones meet. Most joints are **movable** but some, such as between the bones making up the **skull**, the **sacrum** (base of the spine) and the **pelvis**, are **immovable**, or **fused**. Elastic **ligaments** bind bones together, while tough inelastic **tendons** attach muscles to bone (Fig 18.3). The human knee joint is seen in Fig 18.4. Internally, the joint is lubricated by a viscous fluid, **synovial fluid**, secreted by the **synovial membrane**, the membrane that lines the joint.

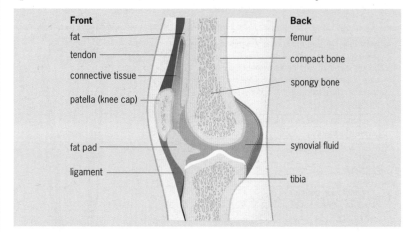

Fig 18.4 **The features of a typical joint can be seen clearly in the knee joint**

The ligaments and the synovial membrane together form a **joint capsule** which surrounds the end of the bones. Different types of joint allow for different kinds of movement (Fig 18.5).

How the human body moves

A simple action, like tapping your finger on the table, involves the skeletal system, the muscular system and the nervous system. This voluntary action begins as a stimulus in the cerebrum of the brain. Motor nerves carry impulses to muscles and cause them to contract, pulling on bones through the tough inelastic tendons.

Muscles can only pull, or contract; they cannot push. This means that muscles rarely act alone: most of the time they work in groups. Contraction of a muscle moves a bone at a joint, but a second (**antagonistic**) muscle returns the bone to its original position. Take movement in the arm, for example. The biceps muscle bends or **flexes** the arm; the triceps muscle straightens or **extends** the arm.

We all know that machines can help us to lift heavy loads using levers. The bones in our bodies also act as **levers**. The principle of leverage allows a muscle to overcome the **resistance** supplied by a weight. This happens in three ways:

● An arm bends because a force is applied between the **pivot** (the elbow) and the resistance (the weight to be lifted) (Fig 18.6(a)).

● We stand on tiptoes by pivoting the toes on the ground and using our calf muscles to raise the weight at the ankles. This is the wheelbarrow principle (Fig 18.6(b)).

● When standing or sitting, we can raise our head because muscles at the back of the neck tip the skull at the pivot at the top of the spine. This is the seesaw principle (Fig 18.6(c)).

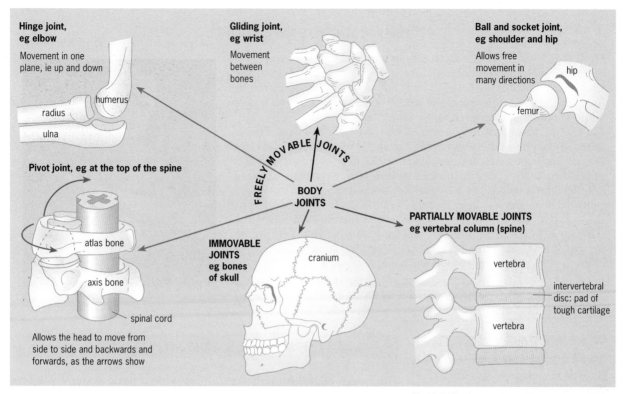

Hinge joint, eg elbow

Movement in one plane, ie up and down

humerus
radius
ulna

Gliding joint, eg wrist

Movement between bones

Ball and socket joint, eg shoulder and hip

Allows free movement in many directions

hip
femur

Pivot joint, eg at the top of the spine

FREELY MOVABLE JOINTS

BODY JOINTS

atlas bone

axis bone

spinal cord

Allows the head to move from side to side and backwards and forwards, as the arrows show

IMMOVABLE JOINTS eg bones of skull

cranium

PARTIALLY MOVABLE JOINTS eg vertebral column (spine)

vertebra

intervertebral disc: pad of tough cartilage

vertebra

Fig 18.5 **There are many different types of joint in the vertebrate skeleton. They are classified both by their shape and by their mobility**

The 'stuck finger' test demonstrates the effect of muscles working in groups (Fig 18.7). Each of your fingers has a separate tendon connecting it to muscles in the forearm. Try it. Curl up your middle finger and place the other four fingers on a hard surface as shown. Now lift up the other fingers one by one. You will find that you can lift all of your fingers except the ring finger. This is because these two fingers share the same tendon connection.

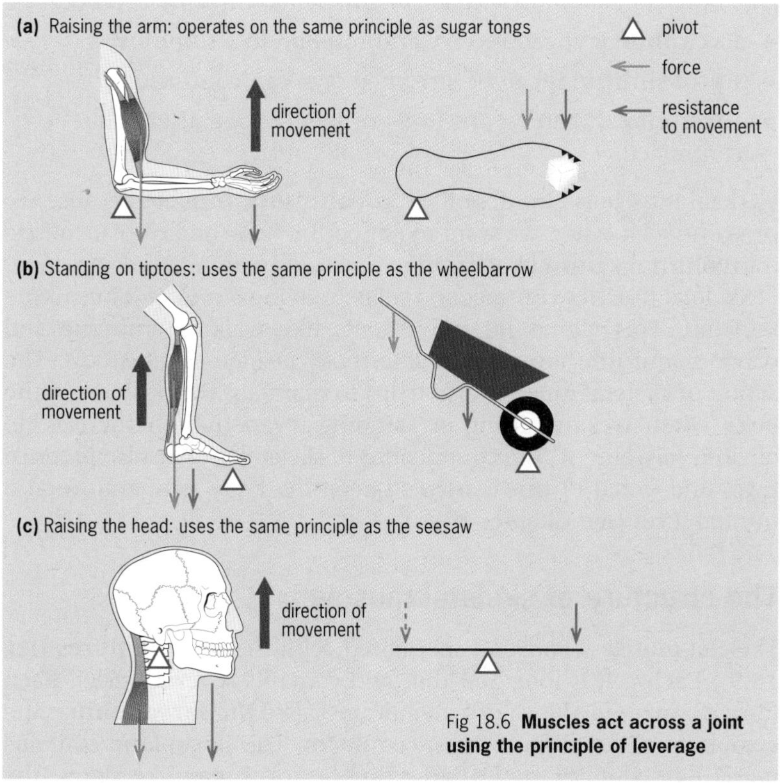

(a) Raising the arm: operates on the same principle as sugar tongs

△ pivot
← force
← resistance to movement

direction of movement

(b) Standing on tiptoes: uses the same principle as the wheelbarrow

direction of movement

(c) Raising the head: uses the same principle as the seesaw

direction of movement

Fig 18.6 **Muscles act across a joint using the principle of leverage**

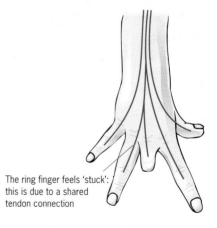

The ring finger feels 'stuck': this is due to a shared tendon connection

Fig 18.7 **The stuck finger investigation**

CORAL: A SUBSTITUTE FOR BONE?

MENDING DAMAGED BONES is a major part of a surgeon's work, yet acceptable substitutes for bone are hard to find. It is possible to use bone from another part of the patient's body but only small amounts can be used. Artificial substitutes run the risk of being rejected by the body's immune system (see Chapter 34).

A few species of coral have a porous structure similar to that of bone. When grafted into the body, the honeycomb texture of coral provides the conditions necessary for new blood vessels to grow into it, and this promotes new bone growth. In addition, coral is tough, carries no risk of infection and is unlikely to be rejected by the body.

'Liquid bone' can also be made from coral. This is based on a calcium-rich solution and a phosphate-rich solution. The surgeon mixes the compounds in a little acid before applying it, rather like toothpaste, to the site of a fracture. Within 12 minutes it has solidified, and after one hour the 'new' bone is as hard as real bone.

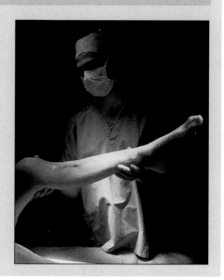

Fig 18.8 **Coral grafts are used to repair shattered limbs, backbones and jaws. This broken leg is being examined before surgery**

There are three types of muscle. **Skeletal muscles** are responsible for whole body movement. **Smooth muscle** is responsible for automatic movements such as those involved in peristalsis. **Cardiac muscle** is found only in the heart. The properties of the different muscle types are described in Chapter 2 and cardiac muscle is covered in detail in Chapter 9.

E What type of muscle fibre (skeletal, cardiac or smooth) is found in the following structures?

(a) stomach

(b) aorta

(c) biceps muscle

(d) left ventricle of heart

(e) face

---

2 MUSCLES AND MOVEMENT

The bones of the human skeleton provide a basic system of levers and joints that makes the skeleton potentially movable, but neither levers nor joints can move without muscles. Skeletal muscle (also called striped or striated muscle) provides the main source of power for human locomotion. In this section we look at the structure and function of skeletal muscle.

The properties of skeletal muscle

Like other sorts of muscle, skeletal muscle has three basic properties:

- **Excitability:** it can receive and respond to a stimulus.
- **Extensibility:** it can be stretched or it can contract.
- **Elasticity:** it can return to its original shape after it has contracted.

Skeletal muscle is also described as **voluntary muscle** because we can contract it when we want to. Smooth muscle and cardiac muscle is **involuntary muscle**: we cannot consciously control its contraction.

Skeletal muscles contract and relax, moving bones at joints in the skeleton. This allows for movements like walking, running and waving, and fine movements like those needed to use tools. The action of skeletal muscles also helps to maintain the position of the body when we are sitting or standing, even though there is no obvious movement. The contractions of skeletal muscle also produce heat, and much of this is used to keep the body temperature at a normal level (see Chapter 12).

The structure of skeletal muscle

Skeletal muscle is made of specialised cells, the **muscle fibres** (Fig 18.9). Each cell is long and thin and contains several nuclei. Each fibre is surrounded by a cell membrane called the **sarcolemma**; the cytoplasm of the fibre is the **sarcoplasm**. The sarcoplasm contains many mitochondria and a large number of thread-like fibres, the

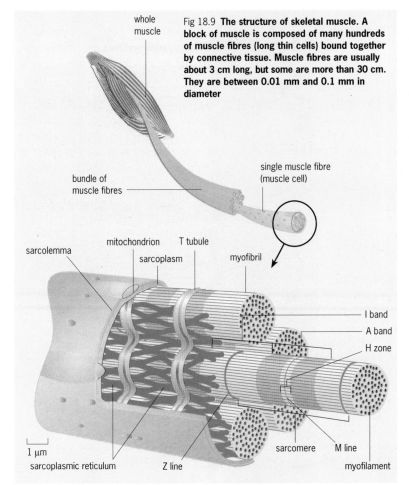

whole muscle

single muscle fibre (muscle cell)

bundle of muscle fibres

Fig 18.9 **The structure of skeletal muscle.** A block of muscle is composed of many hundreds of muscle fibres (long thin cells) bound together by connective tissue. Muscle fibres are usually about 3 cm long, but some are more than 30 cm. They are between 0.01 mm and 0.1 mm in diameter

sarcolemma

mitochondrion
sarcoplasm
T tubule
myofibril

I band
A band
H zone

1 µm
sarcoplasmic reticulum
Z line
sarcomere
M line
myofilament

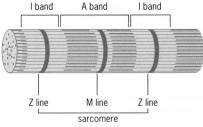

I band   A band   I band

Z line   M line   Z line
sarcomere

Fig 18.10 **The ultrastructure of the myofibril.** Each myofibril contains myosin and actin strands. Where the two overlap, the myofibril looks dark: this region is called the A band. In the regions where only actin is present, the myofibril looks lighter: this region is called the I band. At the centre of each I band is the dark Z line. At the centre of each A band is the dark M line. A bands and I bands alternate and so give skeletal muscle its striped appearance.

The sarcomere, the basic unit of muscle contraction, is the section of muscle fibre between one Z line and the next

myofibrils. These run along the length of the muscle fibre and are parallel to one another. The **sarcoplasmic reticulum** that surrounds each myofibril consists of a network of tubes that contain calcium ions. These play a major role in bringing about muscle movement (see page 272).

As Fig 18.10 shows, each **myofibril** contains many threads called **myofilaments**. Thick myofilaments are made of the large-molecule protein **myosin**, the thin myofilaments are composed of a smaller protein, **actin**. Actin myofilaments also contain two other proteins: **troponin** and **tropomyosin**.

■ See question 1.

The sliding filament theory of muscle contraction

The relationship between the ultrastructure of skeletal muscle and its ability to bring about movement is explained by the **sliding filament theory of muscle contraction** (Fig 18.11).

As we have seen, the actin and myosin filaments lie parallel to each other. When the muscle is at rest, the 'heads' that stick out from the myosin filament cannot attach to the actin filaments that are alongside them. Movement occurs when the heads move towards the actin filament, attach to it and act as hooks to pull the actin myofilaments past them. As the actin myofilaments from opposite ends of the sarcomere are pulled towards each other, the sarcomere becomes shorter. This happens along the whole length of each myofibril and causes the muscle fibre to contract.

Later, as the actin myofilaments become detached from the myosin heads, both filaments return to their original position and the muscle relaxes.

The role of calcium ions and ATP in muscle contraction

Contraction of skeletal muscle does not simply occur 'out of the blue'. It must be initiated by a stimulus in the form of a nerve impulse (see Chapter 21) which arrives at the muscle fibre from the nervous system. So when a muscle contracts and then relaxes, the following events take place (refer again to Fig 18.11):

- The brain makes a decision that the body is to move. Nerve impulses travel from the brain, along the spinal cord and through the motor nerves, towards the muscle.

- The electrical signal reaches the **neuromuscular junction**, a specialised synapse. The arriving signal causes the release of acetylcholine (see Chapter 14). This chemical is a neurotransmitter: it travels across the gap between the end of the nerve and the muscle, and binds to receptors on the **motor end plate**.

- This binding causes an electrical change in the muscle fibre which triggers the release of calcium ions from the sarcoplasmic reticulum into the myofilament.

- Calcium ions bind to troponin and tropomyosin, the proteins that are closely associated with actin filaments. This changes the three-dimensional shape of the troponin–tropomyosin–actin complex, revealing parts of the actin filament that were previously hidden. These are the active sites to which the heads of the myosin filaments will attach.

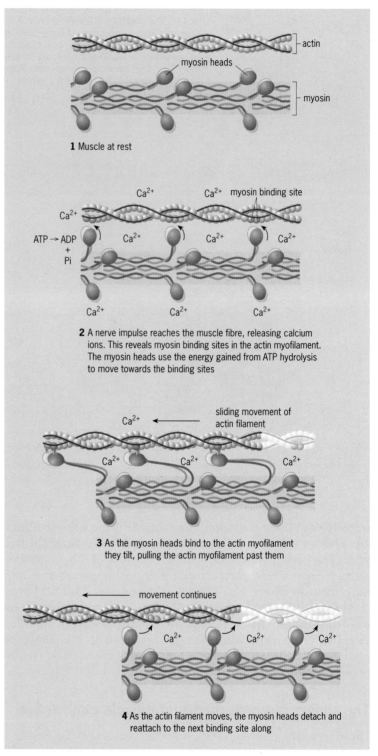

1 Muscle at rest

2 A nerve impulse reaches the muscle fibre, releasing calcium ions. This reveals myosin binding sites in the actin myofilament. The myosin heads use the energy gained from ATP hydrolysis to move towards the binding sites

3 As the myosin heads bind to the actin myofilament they tilt, pulling the actin myofilament past them

4 As the actin filament moves, the myosin heads detach and reattach to the next binding site along

Fig 18.11 **The sliding filament theory to explain muscle contraction**
■ See question 2.

- The calcium ions also act directly on myosin, activating it so that it splits ATP into ADP and inorganic phosphate, releasing energy. This energy is used to move the heads of the myosin filaments towards the newly exposed binding sites on the actin filaments.

- As cross-bridges form between the actin and myosin filaments, the myosin heads tilt, pulling the actin myofilaments past them. As the myofilaments slide, the heads detach from one site and attach to the next. As many as 100 such attachments can occur every second.

TIME OF DEATH

WHEN A BODY is discovered, a pathologist investigates and tries to estimate how long the person has been dead. This information can be important in a murder enquiry. The pathologist assesses the time of death, partly by taking the internal temperature, to see how much the body has cooled from the normal body temperature of 37 °C, and partly by looking at the state of the muscles.

When death occurs, ATP is no longer made. It is a short-lived chemical and so it runs out fairly quickly. This causes the muscles to lock into position as cross-bridges that formed between actin and myosin filaments before death can no longer be broken. This condition, **rigor mortis**, happens in all body muscles. It appears about four hours after death and lasts about 24 hours. After this time, muscle proteins are destroyed by enzymes within the cells and so rigor mortis disappears.

Fig 18.12 **All suspicious deaths are investigated by a police pathologist: one of the first things he or she does is estimate the time of death**

- As the actin myofilaments from opposite ends of the sarcomere move towards each other, the muscle fibre contracts.
- When the stimulation of the muscle fibre stops (when the brain decides the body should stop moving), the calcium ions are pumped back into the sarcoplasmic reticulum.
- As the level of calcium ions falls, troponin and tropomyosin move back to their original positions, blocking once again the active sites available for myosin attachment. The myosin heads can no longer attach to the actin filament, and both sets of filament return to their original position. When the antagonistic muscle contracts, the sarcomeres in the first muscle return to their normal length and the muscle relaxes.

Fast and slow muscle fibres

A nerve impulse is the trigger for muscle contraction, but the length of time a contraction lasts depends on how long calcium ions remain in the sarcoplasm. This time is different in different types of skeletal muscle fibres. **Fast twitch fibres** and **slow twitch fibres** are classified on the basis of their contraction times.

There are three important differences between fast and slow fibres:

- Slow twitch fibres have less sarcoplasmic reticulum than fast twitch fibres. This means that calcium ions remain in their sarcoplasm longer.
- Slow twitch fibres have more mitochondria which provide ATP for *sustained* contraction.
- Slow twitch fibres have significantly more myoglobin than fast twitch fibres. Myoglobin has a higher affinity for oxygen than the haemoglobin in blood and so is particularly efficient at extracting oxygen from the blood.

The overall result of these differences is that slow twitch fibres are responsible for sustained muscle contraction, such as that which maintains body posture, while fast twitch fibres are responsible for the shorter-acting but more powerful contractions important in locomotion (see the Assignment at the end of this chapter).

■ See question 3.

SUMMARY

After completing this chapter you should know and understand the following:

■ The skeleton has three main functions; **support, protection** and **movement.** The human skeleton is movable because different bones meet at **joints,** across which **muscles** are attached. Because muscles can pull but not push, muscles work in **antagonistic pairs.**

■ Muscles are **excitable, extensible** and **elastic. Skeletal muscle,** also called **striped, striated** or **voluntary muscle,** is responsible for human locomotion, maintaining body posture and generating body heat.

■ Skeletal muscle is made of **muscle fibres** which contain bunches of **myofibrils** which in turn contain **myofilaments.** Thick **myosin** filaments lie alongside thin **actin** filaments.

■ The **sliding filament hypothesis** explains muscle contraction at the molecular level

■ The events of muscular contraction can be summarised as follows. A nerve impulse arrives at the **neuromuscular junction.** This releases acetylcholine, which changes the ionic permeability of the **sarcolemma.** Calcium ions are released into the **myofilaments** where they bind to the tropomyosin-troponin-actin complex, and activate ATPase. Cross-bridges form between **actin** and **myosin,** the fibres slide over each other and the muscle contracts.

QUESTIONS

1 Study the electron micrograph of striated muscle tissue shown below and then answer the following questions.

Magnification:
× 13500

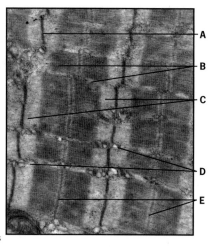

© Biophoto Associates

a) What name or term is given to the structures labelled
   (i) A,
   (ii) C,
   (iii) D.

b) **(i)** Measure, in millimetres, the distance between two successive **Z** lines. Repeat this twice at **two** other positions on the micrograph and note your three answers. Then calculate the mean value.
   (ii) Calculate the **actual** mean length of a sarcomere unit in this muscle tissue. (Show your working.)

c) Make an approximate measurement (in millimetres) of the (relaxed) length of the biceps muscle in your arm and note the result.

If a single striated muscle fibre (cell) stretches the complete length of **this** muscle, how many sarcomere units are required to reach from end to end of the muscle cell? (Show your working.)

d) **(i)** What name is given to the structures that attach your biceps muscle to the bones?
   (ii) Name the two bones onto which the biceps is **inserted.**

e) Using only the letters given on the micrograph, identify the following:
   (i) a region where cross bridges are situated;
   (ii) a region that will decrease markedly in length on contraction of the muscle;
   (iii) a region composed primarily of actin protein;
   (iv) a region composed mainly of actin and myosin proteins.

f) Name an inorganic ion you would expect to be present in relatively high concentrations in regions **B** and **D.**

[Oxford June 1993 Biology: Paper 1, q. 3]

2 The diagram below represents a longitudinal section through striated muscle.

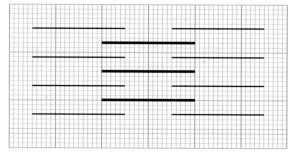

Fig 18.Q2(a)

a) Draw a similar diagram [on 2mm graph paper] to represent the appearance of the same section when the muscle has contracted.

b) Fig 18.Q2(b) represents a cross-section through this muscle.

Draw a vertical line on Fig 18.Q2(a) to show from where the cross-section might have been taken.

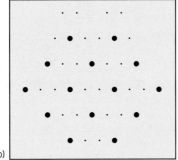

Fig 18.Q2(b)

c) Describe the part played in the contraction of striated muscle by: **(i)** calcium ions; **(ii)** the energy released from the breakdown of ATP.

[AEB 1996 Biology: Specimen Paper, q. 12]

3 The drawing below has been made from a slide of skeletal muscle tissue seen with a light microscope at a magnification of 800 times. It shows parts of two motor units.

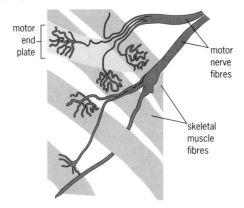

a) Use evidence from the drawing to suggest:
  (i) a meaning for the term *motor unit*;
  (ii) why all the muscle fibres shown will not necessarily contract at the same time.

b) Briefly describe the sequence of events at the muscle end plate which leads to an action potential passing along the muscle fibre.

The diagram below shows the pathways by which energy is produced for muscle contraction. Numbers **1** to **3** indicate the order in which the various pathways are called on to supply ATP as muscular effort increases.

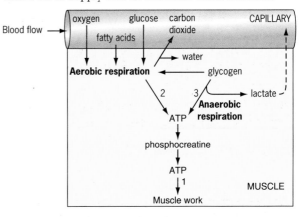

c) **(i)** What happens to the lactate produced in pathway **3**?
  (ii) Explain the part played by phosphocreatine in supplying energy to the muscle.

[AEB 1996 Human Biology: Specimen Paper 2, q. 1]

Assignment

MUSCLE FIBRES AND ATHLETICS

This Assignment follows on from the Chapter 13 Assignment **Energy systems and sport**. You may need to revise this to answer all the questions here.

The different types of muscle fibre

You many have noticed that athletes are highly specialised for their particular event. Marathon runners can run great distances without fatigue, but they lack the power to run as fast as the sprinters. In contrast, sprinters fatigue far too quickly to run long distances. These features are in part due to the type of muscle fibres they have inherited, and partly to the training they have done. Research has shown that there are two basic types of muscle fibre: **fast twitch** and **slow twitch**. Fast twitch fibres can be further subdivided, giving us the three different muscle types:

Slow twitch oxidative (type I) fibres. These contain large amount of myoglobin, many mitochondria and a dense capillary network. They split ATP slowly and therefore can only contract slowly. These fibres are red and have a high resistance to fatigue.

Fast twitch oxidative (type IIA) fibres. These show similar features to slow twitch fibres: a lot of myoglobin, many mitochondria and blood capillaries; but they are able to hydrolyse ATP far more quickly and therefore to contract rapidly. They are relatively resistant to fatigue, but not as resistant as type I fibres.

Fast twitch glycolytic (type IIB fibres). These fibres have a relatively low myoglobin content, few mitochondria and few capillaries. They contain large amounts of glycogen which provides fuel for anaerobic ATP production via the process of glycolysis. They contract rapidly but fatigue quickly, due to the build-up of lactic acid. Type IIB fibres appear whiter than the other two types.

1 Summarise the above information in the form of a table.

2 Movement of muscles requires energy which is provided by one or more of three energy systems: **(i)** the ATP/CP system, **(ii)** glycolysis and **(iii)** the aerobic system. Outline the essential features of these systems.

3
a) What is the function of myoglobin? (See page 147 for a reminder.)
b) Suggest why type IIB fibres have very little myoglobin.

Muscles are made up of motor units: a motor unit is a bundle of fibres controlled by a single motor nerve. Analysis of muscle

tissue by biopsy shows that all the muscle fibres in a single motor unit are of the same muscle type, but there can be different muscle types in different units within the same muscle. The various muscles of the body have different proportions of the three types, according to their function.

4 Why is it an advantage to have all three types of muscle fibre in the same muscle?

The distribution of the different muscle types in the body

For the rest of this Assignment we will consider the two main types of muscle fibre: slow and fast twitch.

A central principle in physiology is that structure is closely related to function and this is illustrated clearly by the distribution of the muscle types in the body. For instance, the postural muscles – such as those in the neck and back – have a high proportion of slow twitch fibres. In contrast, the muscles which bring about fast, explosive movements like running, jumping and throwing are packed with fast twitch fibres.

5
a) Where in the body would you expect the muscles to have a high proportion of fast twitch fibres?
b) Explain why slow twitch fibres are suitable for postural muscles.

The distribution of different fibre types in different athletes

The muscle fibre types of a selection of age-matched athletes were analysed, and the results are summarised in Table 19.A1. These are average values: there is considerable variation between different athletes.

Table 19.A1 **Analysis of the average proportions muscle fibre types in selected muscles in different male athletes**

Type of athlete	Approx % fast twitch	Approx % slow twitch
Marathon runner	18	82
Swimmer	25	75
Cyclist	40	60
800 m runner	52	48
Untrained person	55	45
Sprinter and jumper	62	38

6 What does 'age matched' mean?

7 Explain in general terms how the proportion of fast and slow twitch fibres is related to the nature of an athlete's chosen activity.

8 What do you think the information about untrained people tells us?

HUMAN LIFE SPAN

AS FAR AS SEX IS CONCERNED, humans are unique. We develop bodies capable of reproducing at around 12 or 13 years old – a good few years before most of us become independent from our parents.

For most female mammals mating occurs only when there is an egg to be fertilised and most ovulations result in pregnancy. The female spends her life being pregnant or lactating, or both. The ceaseless cycle of reproduction is brought to an end by death. Humans are quite different. The average woman in the UK ovulates about 400 times, but has only 2 children during her life. We take a detailed look at human reproduction in Chapter 19.

Old age is another part of life that is alien to most animals. Predators who get a bit slower catch fewer prey, get weaker and eventually starve. Ageing prey become easy targets for a predator having a lazy day.

Homo sapiens is in a unique situation. When our physical capabilities start to fail, we find less demanding things to do. Top sports people, for example, might stop performing at the highest level in their mid thirties but then they can progress into training, management or broadcasting. Eventually though, age catches up with us. We look at the human lifespan from birth onwards in Chapter 20, ending with a section on the effects of ageing.

As you read the next two chapters, you might like to reflect on the human animal; our attitudes to sex, a twenty year childhood, a retirement often longer than the working life and an ever increasing wish to prolong its own lifespan. What a strange creature indeed.

19 Human reproduction

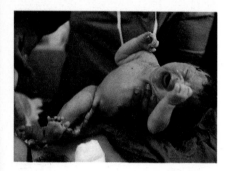

ASK ANYBODY what the purpose of sex is and, after a few funny looks, you'll probably get the answer 'to make babies'. Yet, although millions of acts of sexual intercourse take place across the globe every day, the vast majority of couples have no intention of making a baby. In practice, many of them use contraception to prevent pregnancy.

In the same way, if you ask what breasts are for, you might get the answer 'for breastfeeding'. However, when in close proximity to a topless woman, the average man probably does not spend much time thinking about how many children she could successfully breastfeed. All mammals produce milk, but humans are the only ones with permanently swollen hemispheres of fatty tissue.

Clearly, sex in humans is about more than just making babies. In fact, we seem rather obsessed with sex. Humans are one of the few species that have sex when there is no egg to be fertilised, and statistics tell us that the average human has sex several thousand times in their lifetime.

As you read through the account of human reproduction that follows, have a think about the function of sex in humans. If it is for more than basic reproduction, what is its purpose? Is it simply for pleasure? Or it is a form of 'social cement', strengthening the bond between people, encouraging them to fall in love and stay together? You decide.

> ✓ The need to reproduce, so passing genes on to the next generation to ensure the survival of the population, is one of the basic features of living organisms. In many organisms, the urge to reproduce can override the urge to live.

1 THE NEED TO REPRODUCE

None of us is immortal and so, in common with all other types of organism, we must reproduce in order for our species to survive. An intersting way of looking at the whole business of reproduction is to think of organisms as disposable containers for the genes they contain. None of us will live forever, but our genes, mixed with those from other people, might survive long after we are forgotten.

Sexual reproduction in animals

Sexual reproduction involves the fusion of **gametes**, or sex cells. Female sex cells, called **egg cells** or **ova**, fuse with the male gametes, called **spermatozoa** or **sperm**, in a process known as **fertilisation**. The resulting cell, the **zygote**, develops into a new individual.

In sexual reproduction, the genetic material of two different individuals is mixed and combined to produce an individual that is genetically different from either parent. This produces **variation** within a population (see Chapter 29). All individuals, unless they have an identical twin, are **genetically unique**: different from all other individuals in the group.

Reproduction in mammals

Mammals show a great range of reproductive strategies that reflect their circumstances, but the basic life cycle of most mammals follows a general pattern. After mating, the fertilised egg cell(s) develop inside the uterus, nourished by the placenta. The **gestation period** of a mammal, the time between fertilisation and birth, often reflects the metabolic rate of the organism. It is shorter in small short-lived mammals such as mice and longer in large long-lived mammals such as elephants.

Other factors complicate the length of gestation. For example, it is also limited by the size of the fetal skull (which must fit through the female's pelvis at birth) and the mobility of the mother. In species where the mother needs speed and agility to survive, the gestation period tends to be shorter. Birth is followed by a period when the mother gives close protection and suckles her young. The length of upbringing varies greatly: it depends on factors such as degree of maturity at birth and the amount of learning required to survive.

The onset of sexual maturity marks the change from **juvenile** to **adult**. Most female mammals conceive as soon as they become sexually mature and enter their first **oestrus** (season). From then on, throughout their reproductive life, they are either pregnant, feeding young or both. A female's reproductive life is usually brought to an end only by death. This has been the harsh fact for humans, too, until fairly recently. Even now, in societies where contraception is either unavailable or not used for cultural reasons, many women reproduce continuously.

The role of hormones in mammalian reproduction

For many animals the timing of reproduction is vital. If there is a severe variation in the seasons, animals must ensure that their offspring are born when food is plentiful and conditions are mild. **Hormones** that control and synchronise reproductive events are produced by three different organs: the **hypothalamus**, the **pituitary gland** and the **gonads** (sex organs).

The hypothalamus is a major point of contact between the nervous and **endocrine** (hormone) systems (see Chapter 16). Internal and external stimuli reach the hypothalamus, which responds by releasing hormones. They, in turn, control the rest of the endocrine system (Fig 19.1).

Day length is a classic example of an external stimulus. When the hours of daylight reach a threshold level, perhaps indicating the arrival of spring, the hypothalamus responds by releasing a hormone called a **gonadotrophin releasing factor (GnRF)**. This controls the activity of the nearby pituitary gland, stimulating it to release hormones called **gonadotrophins**.

Gonadotrophins have a direct effect on the gonads (the ovaries and testes). These respond by releasing **steroid sex hormones** (Chapter 3) such as **oestrogen** and **progesterone** and **testosterone**.

> Diploid cells contain two sets of chromosomes, haploid cells contain only one set. In animals, **somatic** (body) cells are diploid while gametes (sex cells) are haploid.

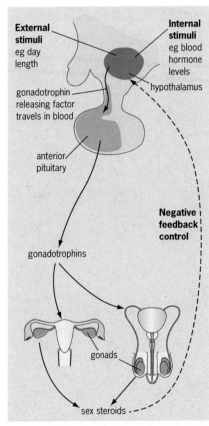

Fig 19.1 **The hormonal control of reproduction.** The hypothalamus can respond to external and internal nervous stimuli and is also sensitive to hormone levels in the blood. The hypothalamus controls the levels of several different circulating hormones using a negative feedback mechanism

> **A** The hypothalamus controls the level of sex hormones via a negative feedback mechanism. Suggest how the hypothalamus would respond to a fall in oestrogen levels.

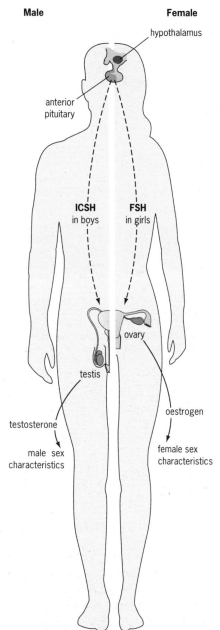

Male　　　　　　　　**Female**

hypothalamus

anterior pituitary

ICSH in boys　　FSH in girls

ovary

testis

testosterone

oestrogen

male sex characteristics

female sex characteristics

Fig 19.2 **The onset of puberty is controlled by hormones.**
　　Testosterone causes the penis, scrotum and testes to enlarge and mature. It also causes enlargement of the larynx (voice-box) which deepens the voice. Body hair is more extensive, with coarse hair appearing on the chest and face. Ironically, in later life, the same hormone leads to male-pattern baldness. A man's shoulders and chest tend to be large and there are often changes in facial structure, with the nose and chin becoming more prominent. The muscles tend to be better developed and generally the male has greater physical strength than the woman.
　　Oestrogen causes the ovaries, oviducts, uterus and vagina to mature and brings about the development of secondary sex characteristics. Breasts grow, pubic and underarm hair grows, the shape of the pelvis changes and the body fat is redistributed so that a woman tends to have wider hips and a more curved shape than a man

3 HUMAN SEXUAL REPRODUCTION

Sexual activity in humans has evolved to achieve more than just fertilisation. Although humans become sexually mature in their early to mid-teens, many societies have cultural and legal controls aimed at restricting sexual activity before the late teens. Although beyond the scope this book, it is interesting to consider whether human sexual behaviour may have evolved to strengthen the emotional bond between a couple. This 'social cement' may make it more likely that couples will stay together and so provide a stable environment for raising children.

Puberty

Puberty is the time between childhood and adulthood: it marks the process of sexual maturing. We still don't know exactly what triggers the onset of puberty, but we know that, early on, GnRF is secreted from the hypothalamus (look back to Fig 19.1). GnRF travels the short distance to the anterior pituitary, where it stimulates the release of gonadotrophins, **follicle stimulating hormone** (FSH) in females and **interstitial cell stimulating hormone (ICSH) in** males (Fig 19.2). These two hormones are chemically identical, but have different names because they have different effects in the two sexes.

In girls, FSH targets the ovaries, which it stimulates to produce **oestrogen**. This steroid hormone is responsible for many of the female sex characteristics. In addition, oestrogen stimulates the ovaries to start producing egg cells (ovulation) and this leads to the first monthly period, or **menstruation**, an event known as the **menarche**.

To start with, periods tend to be irregular and unpredictable, and may sometimes occur without ovulation. Within about a year, hormone levels have increased to the point where they stimulate the regular development of follicles. This makes periods more regular and, unfortunately for many girls, painful. Period pain is mainly due to the hormone progesterone, which causes uterine cramps.

In boys, ICSH targets the **interstitial cells** of the testes. These are embedded in the connective tissue between the seminiferous tubules – see Fig 19.3(b). ICSH stimulates these testis cells to secrete **testosterone**, the steroid hormone that stimulates development of male sex characteristics.

The age of onset of puberty

The average age for the onset of puberty is 12 to 13 in girls, 13 to 15 in boys. Interestingly, this is much earlier than in previous centuries. Two hundred years ago the average age of the menarche in girls was 16, around four years later than it is now. The reason for this is almost certainly the improvement in diet. A better diet enables us to grow faster and so reach the same stage of maturity at an earlier age. In girls, the proportion of body fat appears to be important: a girl who is a dedicated athlete (a gymnast for example) and has a high muscle:fat ratio, may find that the menarche is delayed. Also, girls who have started their periods but who then crash diet and lose a lot of weight may suddenly find their periods stop for a while.

The male reproductive system

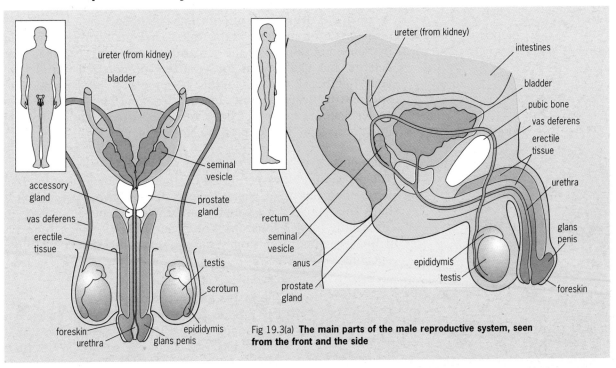

Fig 19.3(a) **The main parts of the male reproductive system, seen from the front and the side**

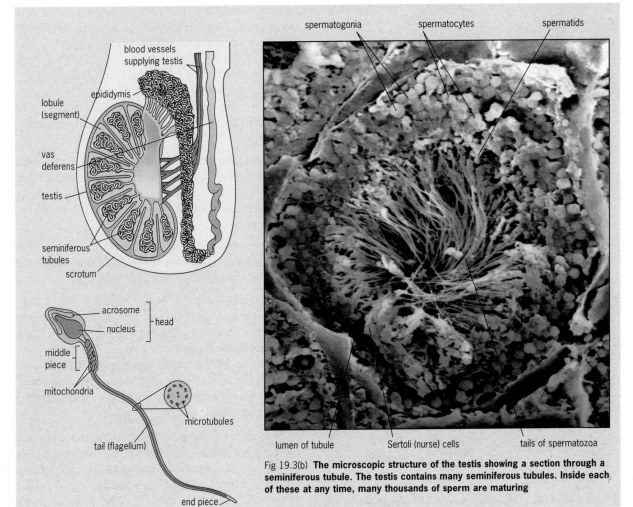

Fig 19.3(b) **The microscopic structure of the testis showing a section through a seminiferous tubule. The testis contains many seminiferous tubules. Inside each of these at any time, many thousands of sperm are maturing**

The overall structure of the male reproductive system is shown in Fig 19.3. The male system secretes testosterone, makes and stores sperm and delivers them into the female's body.

Spermatozoa are made in the testes, a pair of organs that are held in a pouch of skin called the **scrotum**. It may seem odd that such delicate and vital organs are relatively unprotected outside the body, but there is a reason for this. The process of sperm production, **spermatogenesis**, is most efficient at around 35 °C, two degrees cooler than the core of the body. Men whose testes do not descend into the scrotum cannot produce healthy sperm.

See questions 3(a) and 3(b). ■

The penis

The penis introduces sperm into the female's body so that internal fertilisation can occur. Some animals do not have a penis – most birds and reptiles, for example – and their attempts at fertilisation are a little more haphazard. The male produces semen from his genital opening and simply rubs it onto the female's genital opening.

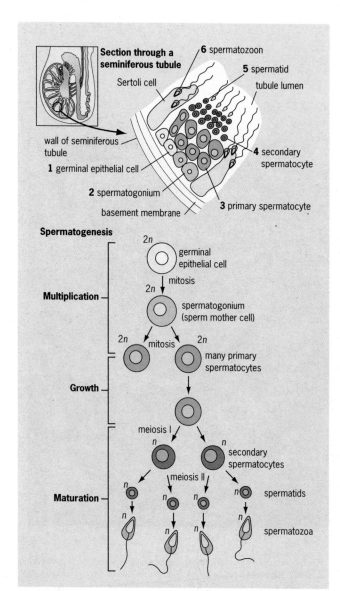

Fig 19.4 **The process of spermatogenesis. Sperm cells develop as they pass from the outer wall of the seminiferous tubule to the lumen**

The testes and spermatogenesis

The production of sperm in human males is a continuous process, beginning at puberty and continuing well into old age: men in their nineties have fathered children. Spermatogenesis centres around the process of meiosis, and occurs at a remarkable rate: over a thousand human sperm are made every second. Look again at Fig 19.3(b) and you will see that each testis is composed of a series of **lobules**. They contain **seminiferous tubules**, the structures in which sperm production takes place. This process, shown in Fig 19.4, has three main phases:

- **Multiplication.** As large numbers of sperm are needed, cells of the **germinal epithelium** divide by mitosis to produce many **spermatogonia** (sometimes called sperm mother cells).

- **Growth.** The spermatogonia grow into **primary spermatocytes**. At this stage the cells are still diploid ($2n$).

- **Maturation.** The diploid primary spermatocytes undergo meiosis. After the first division they become **secondary spermatocytes** and when meiosis is complete they have become haploid **spermatids**. In the final part of the maturation process, spermatids differentiate into the familiar spermatozoa (sperm).

Throughout their development, sperm cells are closely associated with **Sertoli** or **nurse cells**, from which they obtain nutrients. In the lumen of the seminiferous tubule – see Fig 19.3(b) – the tails of the spermatozoa are clearly visible: their heads are attached to Sertoli cells.

Sexual arousal in males

Men become sexually aroused by thinking about sex, as a result of physical stimulation or a combination of both. Nerve impulses from the brain pass down parasympathetic nerves (see Chapter 15) and cause arterioles leading to the penis to dilate. The penis receives more blood than can drain away, spongy **erectile tissue** in the shaft becomes filled with blood, and an **erection** results.

Flaccid (non-erect) penises vary greatly in size, largely depending on how much blood is retained in the spongy tissue. When erect, about 90 per cent are between 14 and 16 cm long. The end of the penis, the **glans**, is particularly sensitive, and continued stimulation from rhythmic thrusting eventually leads to a series of reflexes known as **ejaculation**. Stored spermatozoa are propelled along the **vas deferens** by powerful peristaltic waves. As they pass various accessory glands, different secretions are added to the sperm and the final ejaculate is a milky fluid called **semen** (Table 19.1).

Human males ejaculate, on average, about 5 cm³ of semen, which contains between 50 to 200 million sperm. Most sperm never get anywhere near the egg, even after unprotected sex. For most of the monthly cycle, the cervix is blocked by a plug of mucus which sperm cannot penetrate. Only at around ovulation time does the mucus consistency change, allowing sperm to pass through easily.

Table 19.1 **The constituents of semen**

Gland	Secretion	Purpose
Seminal vesicles	fructose	energy for spermatozoa
	mucus	lubrication
	protein	forms clots, which alter consistency of semen
	prostaglandins	stimulate peristalsis in the female system
Prostate	alkaline chemicals	neutralise acid in vagina
	clotting agent	clots the protein from the prostate, forming a gelatinous mass
Cowper's gland	clear fluid	cleans urethra prior to ejaculation

The female reproductive system

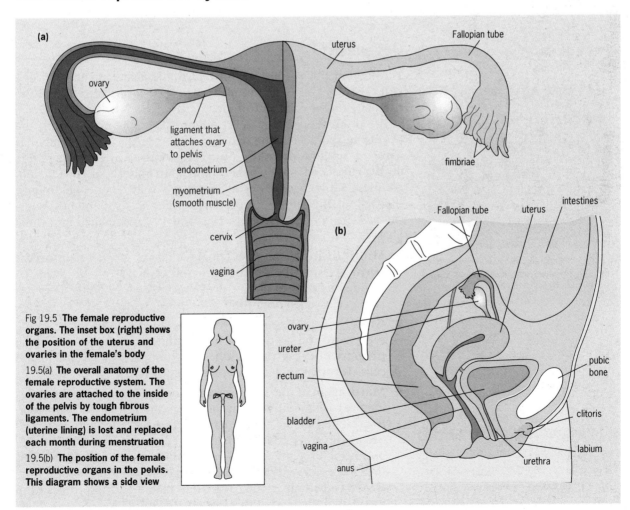

Fig 19.5 **The female reproductive organs. The inset box (right) shows the position of the uterus and ovaries in the female's body**

19.5(a) **The overall anatomy of the female reproductive system. The ovaries are attached to the inside of the pelvis by tough fibrous ligaments. The endometrium (uterine lining) is lost and replaced each month during menstruation**

19.5(b) **The position of the female reproductive organs in the pelvis. This diagram shows a side view**

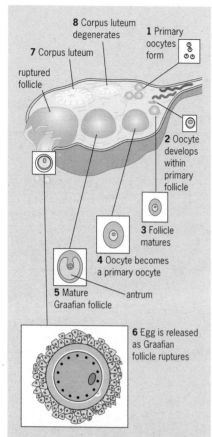

Fig 19.6(a) **Section through an ovary showing the egg cell at different stages of development**

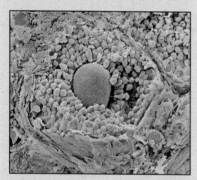

Fig 19.6(b) **Scanning electron micrograph through a primary oocyte showing an egg surrounded by follicle cells**

Meiosis produces four daughter cells. In oogenesis, one ovum is formed; the other three cells degenerate.

Fig 19.5 on the previous page shows the structure of the female reproductive system. This includes a pair of **ovaries** which are the primary female sex organs. These produce egg cells, or **ova**, and also secrete the hormones **oestrogen** and **progesterone**. Each ovary is about 3 to 4 cm across and is attached to the inside of the pelvis by a **ligament** (a tough band of connective tissue). The oviducts, or **Fallopian tubes**, connect the ovaries to the **uterus** (womb). Each tube ends in finger-like **fimbriae**, which move close to the ovary at the time of ovulation.

The uterus is a compact muscular organ that nourishes, protects and ultimately expels the fetus. The human uterus is able to expand from about the size of a small orange with a capacity of about 10 cm³ to accommodate a full-term baby. This is a five-hundred-fold increase in the capacity of the uterus. The bulk of the uterine wall is made from smooth muscle and is known as the **myometrium**. The lining of the uterus, the **endometrium**, consists of two layers. The underlying layer is a permanent **basement membrane** which produces the surface layer, the **decidua**. This layer is built up every month and shed during **menstruation**.

The **cervix**, or neck of the uterus, is a narrow muscular channel which is usually blocked by a plug of mucus. During sexual arousal the muscles of the **vagina** (a muscular tube which leads to the outside of the body) relax and glands in the vagina secrete lubricating mucus. This allows the male's penis to enter without discomfort. During childbirth, the cervix dilates to around 10 cm in diameter to allow the baby to pass through.

Oogenesis

Throughout this section we refer to the female gamete as an **egg cell**. The egg cell is surrounded by several layers of cells and the complete unit is called a **follicle**.

The production of egg cells, **oogenesis**, takes place within the ovaries of the developing female fetus. At birth, a girl already has about 2 million **primary oocytes**. Most of these degenerate during childhood and by puberty there are only about 200 000 left. Of these, only about 450 ever mature fully – one per month throughout the female's reproductive life. As in spermatogenesis, the process of oogenesis (Figs 19.6 and 19.7) is divided into three phases:

- **Multiplication**. As the female embryo grows, **primordial germ cells** in the **epithelium** (outer layer) of the ovary go through a series of mitotic divisions to produce a population of larger cells called **oogonia**.

- **Growth**. Oogonia move towards the middle of the ovary where they grow and go through further mitotic divisions to become **primary oocytes**. Each oocyte is surrounded by a layer of follicle cells. Together they form a **primary follicle**.

- **Maturation**. From puberty onwards, a few primary follicles mature each month. Usually, only one completes its development, the rest degenerate. The remaining primary follicle grows larger, becoming an **ovarian follicle**. Its cells secrete **follicular fluid**, producing droplets which join together to form a fluid that fills the space known as the **antrum**. The mature follicle, the **Graafian follicle**, is almost 1 cm in diameter. It protrudes from the wall of the ovary just before ovulation.

Inside the developing follicle, the oocyte begins its first meiotic division. There is no need for more than one egg cell, so the second set of chromosomes formed at meiosis I is discarded, passing into a small cell (with very little cytoplasm) known as the **first polar body**. This appears to have no function, but it often completes the meiotic division, producing two similar cells; both later break down. After meiosis I, the egg cell is known as a **secondary oocyte**. It then begins the second meiotic division but gets no further than metaphase. The division is completed only if the egg cell is fertilised.

When fertilisation occurs, meiosis II is completed and the egg cell becomes the mature **ovum**. This produces another 'spare' set of chromosomes, the **second polar body**, a cell that also degenerates.

Ovulation

On around day 14 of the human female's menstrual cycle, an egg cell (a secondary oocyte) is released from the ovary in a process called **ovulation**. Pressure in the antrum builds up and ruptures the Graafian follicle, forcing the egg cell out. The egg cell is released into the body cavity, but very few get lost since the fimbriae of the Fallopian tubes hover close to the ovaries, and the ciliated lining of the tubes creates a current that gently sucks in the egg cells.

Remarkably, a woman can become pregnant even if she has lost an ovary on one side and a Fallopian tube on the other. This suggests that the functioning fimbria actively seeks out the productive ovary by moving right across to the other side of the woman's body.

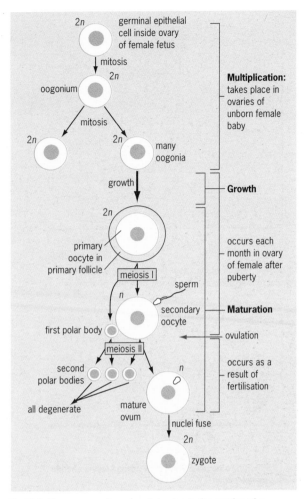

Fig 19.7 **The process of oogenesis begins in the ovaries of a developing fetus but is not completed unless the egg is fertilised**

THE MENOPAUSE AND HORMONE REPLACEMENT THERAPY

THE MENOPAUSE IS a natural event that occurs when the ovaries stop working. For most women, this happens between the ages of 45 and 54. This time is often difficult and traumatic. The lowered levels of oestrogen and progesterone are directly responsible for the unpleasant symptoms of the menopause:

- Circulatory problems such as hot flushes and night sweats.

- Psychological problems, such as depression, anxiety and insomnia.

- Skeletal problems. Oestrogen inhibits reabsorption of bone, and after the menopause, bone loss can be as great as 7 per cent per year. This condition, **osteoporosis**, affects the spongy bone particularly and the sufferer is more likely to break a bone.

- Oestrogen is also thought to give women some protection against some types of heart disease. Women below menopausal age are less likely than men to have heart disease, but afterwards they catch up.

Hormone replacement therapy can help to reverse many of these symptoms and effects. The basic idea behind HRT is simple: to restore hormone levels to those of the early follicular phase of the menstrual cycle using tablets, implants or transdermal patches.

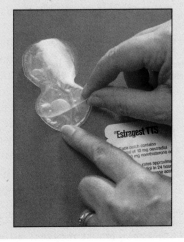

Fig 19.8 **One method of administering the hormones in HRT: a transdermal patch (a patch attached to the skin) that releases controlled amounts into the blood-stream. The hormones used in HRT are extracted from the urine of pregnant mares or made artificially**

See questions 1(b) and 7.

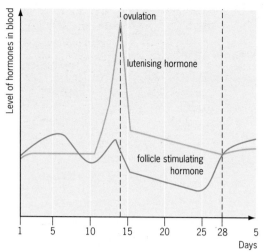

Fig 19.9(a) **How levels in luteinising hormone and follicle simulating hormone change during the menstrual cycle**

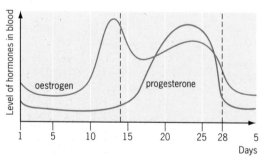

Fig 19.9(b) **How oestrogen and progesterone levels change during menstruation**

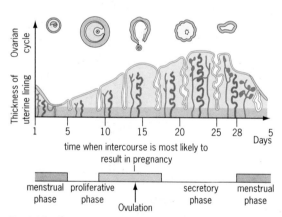

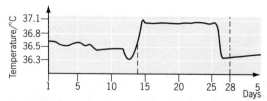

Fig 19.9(c) **Stages of follicle development and changes in the thickness of the endometrium. Note that pregnancy is more likely to occur after unprotected sex between days 9 and 17, in a cycle where ovulation occurs on day 14**

Fig 19.9(d) **A woman's body temperature rises just after ovulation. In couples finding it difficult to conceive, taking daily temperature measurements is a good way to pinpoint ovulation**

The human female menstrual cycle

The average length of the human menstrual cycle is 28 days, but variations from 24 to 35 days are normal, and greater variations are not uncommon. The cycle is divided into four phases, Fig 19.9(c):

- The **proliferative phase** in which the endometrium regenerates.
- The **ovulation phase** in which the ovum (egg cell) is released.
- The **secretory phase** in which the endometrium secretes nutrients in preparation for implantation.
- The **menstrual phase** in which the endometrium is shed from the body.

By convention, the first day of the period (the most obvious event) is called day 1 of the menstrual cycle. Fig 19.9 shows the timing of the phases listed above. It is important to relate the events of the menstrual cycle to the hormonal changes shown. You will need to refer to these diagrams as you read the following text.

The proliferative phase

On day 2 of the cycle, the pituitary gland releases **follicle stimulating hormone** (**FSH**). This stimulates the development of several ovarian follicles. At around day 6, one of the follicles dominates and begins to secrete oestrogen. The others degenerate. The remaining follicle develops into a **Graafian follicle** and continues to secrete oestrogen until day 14 of the cycle. Consequently, blood oestrogen levels rise. Oestrogen causes **proliferation** (growth) of the endometrium to replace the layer lost during the previous menstruation. After 14 days the repair is complete.

The ovulation phase

On day 12 to 13, the blood oestrogen level reaches a threshold level which triggers the release of **luteinising hormone** (**LH**), from the anterior pituitary gland. The rapid increase in LH levels triggers ovulation around day 14. The ovum lives for only 24 to 36 hours. During this time it moves only a few centimetres from the ovary.

The secretory phase

Luteinising hormone has a second effect: it causes the Graafian follicle to develop into a **corpus luteum** ('yellow body'). The name comes from the yellow appearance of the secretory cells that develop inside the 'remains' of the Graafian follicle. The corpus luteum secretes oestrogen and progesterone. Progesterone causes spiral-shaped blood vessels to grow into the endometrium. This thickened lining begins to secrete nutrients and mucus to prepare for an embryo to be implanted. During this phase the high levels of progesterone inhibit the production of FSH. As long as progesterone levels are high, the endometrium is maintained and no new follicles are stimulated. The 'contraceptive pill' takes advantage of this inhibition (see Feature box on page 291).

If the egg cell is not fertilised, the corpus luteum lasts for about 10 to 12 days and then degenerates, ceasing to secrete progesterone. This is a key event because the inhibition of FSH is lifted. The endometrium is no longer protected and the cycle can start again.

The menstrual phase

The drop in progesterone and oestrogen levels causes the uterine capillaries to rupture, and the endometrium is lost from the body through the cervix, together with some blood. Renewed secretion of FSH begins around day 2, and the cycle begins again.

The events of pregnancy

If the egg cell is fertilised, the menstrual cycle is interrupted and the female's body changes in response to the events of pregnancy.

Fertilisation

Fertilisation is a complex sequence of events that begins when the sperm reaches the egg cell (oocyte). The events of fertilisation are shown in Fig 19.10.

It may also help you to look back at Fig 19.7. The secondary oocyte is surrounded by several layers of follicle cells, the **corona radiata**, and a layer of glycoprotein, the **zona pellucida**. Before a sperm can penetrate the layers surrounding the oocyte, it undergoes a process called **capacitation**. The actual mechanism is poorly understood, but it seems to set off the **acrosome reaction**, which allows the sperm head to enter the oocyte. The **acrosome** is a bag of digestive enzymes on the tip of the sperm head (see Fig 19.10). During the acrosome reaction the bag splits, releasing the enzymes, which digest a pathway through any remaining follicle cells and the zona pellucida.

As soon as the outer membrane of the first sperm penetrates the cell surface membrane of the oocyte, a rapid reaction occurs. Many **cortical granules** fuse with the zona pellucida, forming a **fertilisation membrane**. This reaction starts at the point of entry of sperm head and spreads rapidly over the surface of the oocyte, preventing entry of other sperm. So, only one sperm enters the diploid secondary oocyte, even though many reach it at the same time. Entry of the sperm nucleus triggers the completion of meiosis II in the female nucleus, leading to the formation of the second polar body and a haploid mature ovum.

Almost immediately afterwards, a spindle forms and the paternal and maternal chromosomes come together, forming a diploid zygote. Within 12 hours, the first mitotic division takes place. Cell division is now rapid, forming a bundle of cells called a **morula** (Latin for the blackberry it resembles). As divisions continue, this becomes a **blastocyst** and moves slowly along the Fallopian tube through the action of cilia, which create a steady current of fluid towards the uterus (see Fig 19.11).

It takes around 6 or 7 days for the embryo to complete the journey down the Fallopian tube. When it arrives at the uterus on day 21 of the cycle, the lining must be in just the right condition to accept it. For a successful pregnancy there must be exact timing between the preparation of the endometrium and the development of the embryo.

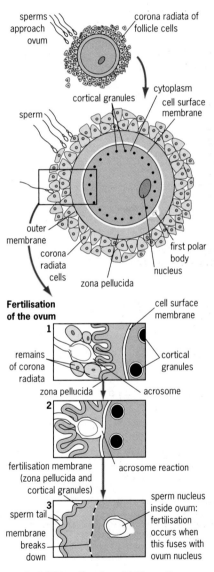

?

B Many women do not have a 28 day cycle. Although the second half of the cycle (ovulation to menstruation) is usually 14 days, the first half can vary considerably. If a woman's cycle is 35 days, on which day is she likely to ovulate?

■ See questions 1(a), 6 and 7.

Fig 19.10 **With a diameter of 100 μm, the egg cell is the largest cell in the human body. Most of the volume is taken up by inert food material that will fuel the embryo during its first few cell divisions. A spermatozoon must pass through any remaining follicle cells (the corona radiata) and the zona pellucida, before it can fertilise the egg cell. The head of the sperm enters the body of the egg cell, but the tail remains outside**

■ See questions 3(b) and 3(c).

Implantation

Pregnancy begins not with fertilisation but when the embryo **implants** in the wall of the uterus. This happens about a week after fertilisation and is not always successfully completed. Many women trying to conceive may have a 'near miss', when an egg is fertilised but fails to implant.

Implantation begins when the blastocyst makes contact with the endometrium (Fig 19.11), usually on the back wall. The outer layer of the blastocyst, the **trophoblast**, causes an **inflammatory-type response** (normally a response to damage) and this causes an outgrowth of the endometrium at this point. The placenta develops where the trophoblast and the endometrium interact.

Fig 19.11 **Following fertilisation, the embryo undergoes several mitotic divisions as it passes slowly down the Fallopian tube. When the embryo reaches the uterus, it implants in the endometrium**

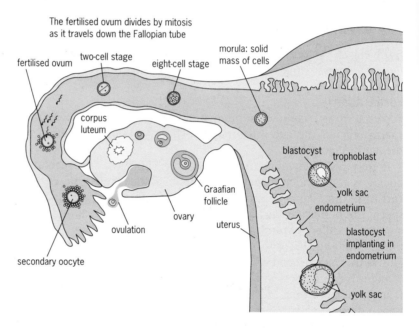

If implantation is successful, the embryo begins to secrete **human chorionic gonadotrophin (HCG)**. This hormone forces the corpus luteum in the ovary to continue to secrete progesterone, thereby maintaining the endometrium and inhibiting FSH production.

The **chorion**, one of the membranes that later grows and surrounds the embryo, develops **villi** (projections) that burrow into the endometrium. These are thought to break down the mother's blood vessels, causing the chorionic villi to become bathed in maternal blood.

See questions 3 and 4. ■

The placenta

The placenta is a temporary organ that allows the blood systems of the fetus and the mother to come into close contact, without actually mixing (Fig 19.14). The placenta allows nutrients and oxygen to pass to the fetus from the mother, and allows metabolic waste back into the mother's blood (see Table 19.2).

The placenta is an organ adapted to maximise the exchange of materials so, as you would expect, it has a large surface area provided by the **chorionic villi**. A close look at the cells of the villi shows that the membranes are folded into microvilli and also contain many mitochondria: these two features maximise the processes of diffusion and active transport (see Chapter 5). There are many small vesicles in the cells of the villi, suggesting that substances are being absorbed by pinocytosis (see Chapter 5).

Table 19.2 **Exchange of materials across the placenta**

Mother to fetus	Fetus to mother
oxygen	carbon dioxide
glucose	urea
amino acids	other waste products
lipids, fatty acids and glycerol	
vitamins	
ions; Na, Cl, K, Ca, Fe	
alcohol, nicotine, many drugs	
viruses	
antibodies	

THE PREGNANCY TEST

BY THE TIME A PREGNANT woman would have been due for her next period, the embryo will have implanted and would be starting to secrete HCG. Enough of this hormone passes into the urine for it to be detected by a modern pregnancy testing kit. This kit makes use of monoclonal antibodies (Chapter 34), which are specific for HCG and which are attached to a coloured chemical.

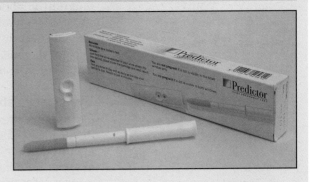

Fig 19.12 **A home pregnancy test kit**

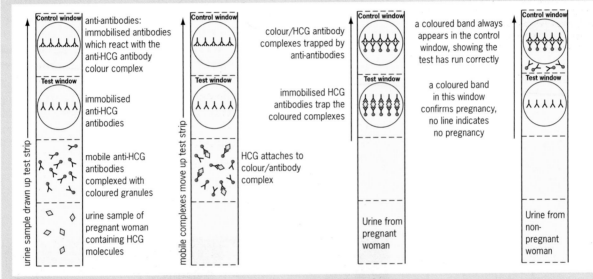

anti-antibodies: immobilised antibodies which react with the anti-HCG antibody colour complex

immobilised anti-HCG antibodies

mobile anti-HCG antibodies complexed with coloured granules

urine sample of pregnant woman containing HCG molecules

urine sample drawn up test strip

colour/HCG antibody complexes trapped by anti-antibodies

immobilised HCG antibodies trap the coloured complexes

HCG attaches to colour/antibody complex

mobile complexes move up test strip

a coloured band always appears in the control window, showing the test has run correctly

a coloured band in this window confirms pregnancy, no line indicates no pregnancy

Urine from pregnant woman

Urine from non-pregnant woman

Fig 19.13 **How a home pregnancy test kit is used.**

(a) **Urine containing HCG is drawn up the test strip by capillary action**

(b) **HCG attached to the colour/antibody complex moves up the test strip**

(c) **In the urine of a pregnant woman, HCG attached to the colour/antibody complexes moves up to dock with the immobilised (fixed) HCG antibodies in the test window. All the complexes are stopped at one place, forming a visible coloured line and confirming pregnancy**

(d) **For the urine of a non-pregnant woman, the colour/antibody complexes have no attached HCG, so all the complexes move past the immobilised HCG antibodies in the test window and attach to the anti-antibodies in the control window showing the test has worked**

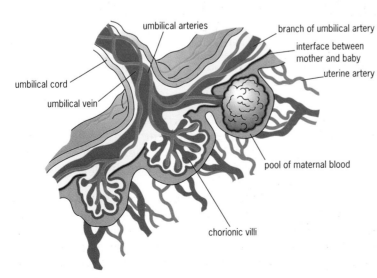

umbilical arteries

branch of umbilical artery

interface between mother and baby

uterine artery

umbilical cord

umbilical vein

pool of maternal blood

chorionic villi

Fig 19.14 **The fine structure of the placenta. The capillaries within the chorionic villi are bathed in pools (lacunae) of maternal blood. The placenta is a large disc shaped organ which is usually attached to the back wall of the uterus. At birth, contraction of the uterine muscle causes the chorionic villi to split away from the endometrium and the placenta is delivered after the baby: hence its common name of afterbirth. Most mammals eat the placenta after the birth of their young. It is an important source of nourishment at a time of great need**

See question 4. ■

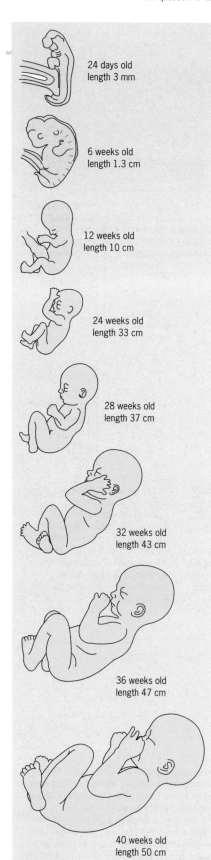

The placenta is an important **endocrine organ**. It secretes the hormones that maintain pregnancy, taking over from the corpus luteum at about 12 weeks. The placenta secretes progesterone (which maintains the endometrium), oestrogen (which inhibits the ovulatory cycle), human chorionic gonadotrophin (see the Feature box on the opposite page) and **human placental lactogen** (which stimulates the development of the mammary glands).

Pregnancy, labour and birth

In humans, pregnancy lasts for an average of 40 weeks. Some of the main stages of growth of a baby are shown in Fig 19.15.

From around the twelfth week of pregnancy, progesterone secreted by the placenta inhibits uterine contractions. The level of progesterone rises steadily until just before birth, usually around 38 weeks after conception, when it starts to fall dramatically. This lifts the inhibition of uterine contractions. The mother's anterior pituitary begins to secrete **oxytocin** and the placenta secretes prostaglandins, two hormones that actively promote contractions. Oxytocin stimulates uterine contractions at about 40 weeks. The resulting tension in the muscle and pressure on the cervix are stimuli that bring about further secretion of oxytocin, causing more powerful contractions.

There are three stages of labour:

- Stage 1. The cervix dilates (opens) to a diameter of 10 cm.
- Stage 2. The fetus is pushed out of the uterus.
- Stage 3. The placenta and umbilical cord are expelled.

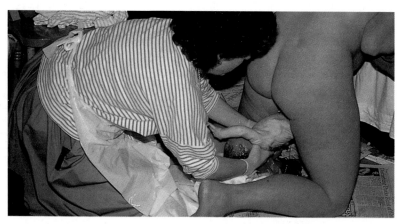

Fig 19.16 **Childbirth is one of the most intense human experiences. Childbirth used to be a time of great danger to both mother and child, but advances in medicine have greatly reduced the risk. Today, in the developed world, 99 out of every 100 babies born survive beyond their first birthday (two hundred years ago, only 54 out of every 100 lived)**

Lactation

All mammals produce milk from specialised **mammary glands**. Humans are unique in having permanent breasts but these are normally composed only of fatty tissue. Throughout pregnancy, however, oestrogen and progesterone stimulate the development of milk-producing tissue.

After birth, the first fluid released from the breasts is called **colostrum**. This contains no fat and very little sugar, but has important antibodies that 'lend' the baby immunity until it has time to develop its own. After about three to four days, normal milk is produced.

24 days old
length 3 mm

6 weeks old
length 1.3 cm

12 weeks old
length 10 cm

24 weeks old
length 33 cm

28 weeks old
length 37 cm

32 weeks old
length 43 cm

36 weeks old
length 47 cm

40 weeks old
length 50 cm

Fig 19.15 **Stages in the growth of a baby in the uterus**

Milk is produced constantly and, although a certain amount of leaking from time to time is usual, milk flow only happens when the baby suckles (Fig 19.17). The stimulus of sucking at the nipple causes the posterior pituitary gland to secrete oxytocin. This hormone travels in the blood and causes contraction of the muscular **myo-epithelial cells** surrounding the milk glands or alveoli. This squeezes the milk out of the alveoli, through the milk ducts and into the infant's mouth. This mechanism, called the **let-down reflex**, takes several seconds to take effect but is quite powerful. If the unsuspecting infant lets go of the nipple, it can get a jet of milk in the eye.

Throughout the period of lactation the pituitary continues to secrete **prolactin**, a hormone that maintains the milk ducts and, to some extent, inhibits ovulation. This inhibition is called **lactational anoestrus** and reduces the risk of conception so soon after birth. However, it is not always effective and breastfeeding cannot be relied on as a contraceptive.

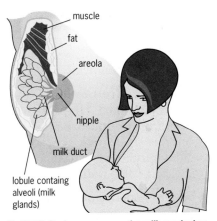

Fig 19.17 **During pregnancy the milk-producing ducts grow, substantially increasing the size of the breasts. When the new baby suckles, the physical stimulation leads to a series of hormonal events that increase the amount of milk that is made**

■ See questions 5 and 6.

CONTRACEPTION

THE TWENTIETH CENTURY has seen a revolution in 'family planning'. People can regulate the number and spacing of their children by using one or more of the wide range of effective methods of contraception now available (Table 19.3).

Table 19.3 **The efficiency of the commoner methods of contraception**

Method	Mode of action	Failure rate (pregnancies per 100 women/year)
The pill	prevents ovulation	0–3
IUD (coil)	prevents implantation	0.5–6
Condom	prevents sperm from reaching cervix	3–20
Diaphragm + spermicidal gel	prevents sperm from reaching cervix, gel kills sperm	3–25
Spermicidal gel alone	kills sperm	3–30
Coitus interruptus	penis withdrawn before ejaculation	10–40
The rhythm method	abstain from sex near to time of ovulation	15–35
Vaginal douche	washes sperm out of vagina	80
Male sterilisation (vasectomy)	vas deferens cut so that semen contains no spermatozoa	0–0.15
Female sterilisation (tubal ligation)	prevents egg from passing into uterus	0–0.05
Nothing	no contraception at all – a control for the other methods	85

Hormones and contraception

'The pill'

The pill – or, to use its full name, the **combined oral contraceptive pill** – has been used widely since the 1960s. It contains a mixture of artificially produced oestrogen and progesterone. The pill is effective because it raises the blood hormone levels to those encountered during the secretion phase of the menstrual cycle, or those of early pregnancy. The high level of progesterone inhibits the secretion of FSH, preventing the development of new follicles and therefore egg cells. The high progesterone levels also inhibit menstruation, so the pill is usually taken for 21 days and then not taken for 7 days. This allows menstruation to take place but does not leave enough time for any follicles to develop. Strictly speaking, this is not normal menstruation but a 'withdrawal bleed' caused by the drop in progesterone.

The 'mini-pill'

This variation of the pill contains low doses of hormones, not enough to prevent ovulation, but enough to change the consistency of the mucus at the cervix. Instead of becoming permeable to sperm at ovulation, the mini-pill causes continued secretion of the thick, non-ovulatory mucus, which prevents sperm getting through.

Injections and implants

Injections and implants both contain progestogen, an artificial form of progesterone. Injections are intra-muscular – usually in the buttock – after which the hormone is slowly released into the bloodstream. Progestogen acts as a contraceptive by inhibiting FSH, although the cervical mucus is also thickened.

Implants, such as Norplant, come in the form of six small tubes that are surgically placed under the skin of the upper arm. The progestogen is released slowly and lasts for five years. In this case, the contraceptive effect is mainly due to an alteration of the thickness of the mucus at the cervix, although there can be some inhibition of FSH.

SUMMARY

After studying this chapter, you should know and understand the following:

■ **Sexual reproduction** involves two sexes which produce **haploid gametes**. Males produce **spermatozoa** by **spermatogenesis**, females make **ova** (egg cells) by **oogenesis**. Both processes involve a special type of cell division called **meiosis**.

■ At fertilisation, a new **diploid zygote** is formed. The zygote, which is genetically unique, grows and develops by a series of mitotic divisions.

■ The timing of reproductive events in humans is controlled by hormones. The **hypothalamus** controls the **pituitary gland**, which releases **gonadotrophins**. These stimulate the **gonads** to secrete steroids: **oestrogen**, **progesterone** and **testosterone**.

■ The human menstrual cycle lasts, on average, for 28 days. The cycle begins with **menstruation**, which occurs as the next ovum (egg cell) is

prepared inside the ovary. By day 14 a new endometrium has grown. Ovulation takes place on day 14 and the egg cell is then available to be fertilised. If the egg cell is not fertilised the endometrium breaks down and the cycle repeats.

■ In humans, the egg cell is fertilised in the Fallopian tube. The **zygote** begins mitotic divisions, becoming a **morula** and then a **blastocyst** as it moves along the tube towards the uterus. When the blastocyst has implanted in the endometrium, it is known as an **embryo**.

■ The developing embryo (which is called a **fetus** after about 8 weeks) is nourished via the placenta, a temporary organ that does the job of the fetal lungs, intestines and excretory system.

■ Birth is brought about by a series of hormonal changes that begin uterine contractions. In mammals, the infant is nourished by milk produced by the mother.

QUESTIONS

1

a)   (i) Describe the changes that occur in the uterus endometrium during a normal menstrual cycle.
  (ii) Explain how progesterone is involved in the control of these changes in the endometrium.

b)   (i) Explain what is meant by hormone replacement therapy (HRT), and
  (ii) describe the circumstances in which it might be used.

[UCLES Autumn 1995 Modular Biology: Growth, Development and Reproduction Paper, q. 1]

2

a) Copy and complete the table which refers to hormones controlling the mammalian oestrous cycle.

Hormone	Site of production	Effect
FSH	anterior lobe of pituitary gland	
	ovary	repair of uterine lining
	anterior lobe of pituitary gland	ovulation
progesterone		maintenance of uterine lining

b) In some mammals changes in day length (amount of daylight) stimulate the release of reproductive hormones and the onset of a breeding season.

Outline how the nervous system is involved in detecting changes in daylength and co-ordinating the release of reproductive hormones.

[NEAB June 1995 Modular Biology: Processes of Life Paper, q. 7]

3

Fig 19.Q3 shows the structure of a human sperm.

a)   (i) Name the parts **A** to **F**.
  (ii) Apart from **B** to **E**, sperm contain few other cytoplasmic organelles. Suggest why this is so.

b) Outline the main events that take place from when a sperm first reaches an egg until a zygote is formed.

c) Suggest why very large numbers of sperm are produced yet only one is involved in fertilisation.

Fig 19.Q3

[UCLES 1996 Modular Biology: Growth, Development and Reproduction, Specimen Paper, q. 2]

4 The graph in Fig 19.Q4 shows the concentration of the hormones progesterone and human chorionic gonadotrophin (HCG) in the blood during the early stages of a human pregnancy.

a) Describe how the progesterone curve would differ if pregnancy had not occurred.

b) Name the site of secretion of progesterone during **(i)** the period shown in the graph; **(ii)** the last 3 months of the pregnancy.

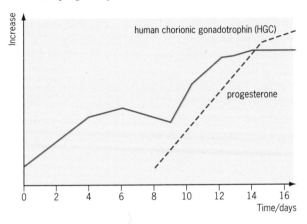

Fig 19.Q4

c) **(i)** Suggest the main function of human chorionic gonadotrophin in early pregnancy.
(ii) Give evidence from the graph to support your answer to (c)(i).

[AEB 1996 Biology: Specimen Paper 1, q. 13]

5 **When the Pill gets under your skin**

When news of a new contraceptive method reached British women in 1993, family planning organisations were flooded with enquiries. The contraceptive is an implant called Norplant. Not everyone, however, is thrilled by Norplant. Organisations around the world are concerned that it might be used as a method of social control.

Norplant delivers in a new way. It consists of six capsules, each 34 millimetres long. Each capsule contains 38 milligrams of a synthetic progesterone hormone. This hormone thickens the mucus produced by the cervix (neck) of the uterus. It also inhibits the production of LH (luteinising hormone).

A health worker inserts the contraceptive capsules through an incision on the inside of the upper arm and they remain under the skin for five years, steadily releasing progesterone into the blood-stream.

(Adapted from an article in *New Scientist*)

a) Suggest how the thickening of the mucus produced by the neck of the uterus might help to prevent contraception.

b) **(i)** Describe the role of LH in the menstrual cycle and then explain how the inhibition of LH production prevents contraception.
(ii) Hormone levels are often affected by negative feedback processes. Explain as full as you can what is meant by negative feedback.

c) Suggest and explain the advantages and disadvantages of using Norplant as a contraceptive.

[NEAB Feb 1995 Modular Biology: Life Processes Paper, q. 8]

6 Fig 19.Q6 summarises the role of the hormones involved in the first part of the menstrual cycle.

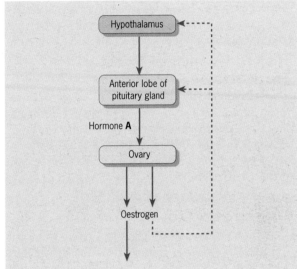

Fig 19.Q6

a) Name hormone **A**.

b) Use the information in the diagram to explain the meaning of the term *negative feedback*.

c) Some women are infertile because they do not produce enough hormone **A**. The drug clomiphene works by lowering the concentration of oestrogen in the blood. Suggest how treatment with clomiphene might lead to conception.

[NEAB June 1996 Biology: Processes of Life Module Test, q.7]

7

a) Describe the roles of hormones in pre-menstrual tension, hormone replacement therapy and the menopause.

b) Discuss the ethical issues raised by abortion.

[UCLES November 1997 Sciences: Development and Reproduction, Section B, q.1]

Assignment

INFERTILITY TREATMENT

Nine out of ten couples who use no contraception and actively try to conceive, are successful within a year. But, one in ten couples may face the distressing problem of infertility. Many are 'sub-fertile' rather than infertile, and can be helped by the various methods described in this Assignment. However, for a few, nothing works.

In this exercise we look at the possible causes of infertility and at some of the treatments. When you have read both and worked through the questions, you should be in a position to prescribe the best course of treatment for a particular couple.

1 Causes of infertility

Doctors accept that a couple may need infertility treatment if they have been trying to conceive for at least a year without any success. The first objective is to establish whether either partner has an obvious problem which is preventing conception.

Female infertility may be caused by:

● Blocked Fallopian tubes.

● Altered hormone levels leading to a failure to ovulate or implant. Failure to ovulate, known as **anovulatory infertility**, is usually due to a failure to secrete the right balance of hormones.

● Cervical mucus that halts, repels or kills sperm.

1

a) How would a specialist find out if a woman's Fallopian tubes were blocked?
b) Why might blocked Fallopian tubes cause infertility?
c) Which hormones combine to cause ovulation?
d) What should happen to the cervical mucus around the time of ovulation?

Male infertility may be caused by:

● A low sperm count. Samples which are found to have fewer than 20 million sperm per cm^3 are said to be abnormally low.

● Production of large numbers of abnormal sperm (more than 4 per cent).

● Production of antibodies that make the sperm stick together.

2 Men who wear tight-fitting clothes, or who spend a lot of time in a hot bath, have sometimes been found to have a low sperm count. Suggest a reason for this.

Treatment methods: ovulation induction

A woman may not ovulate because the balance of hormones in her body is abnormal. To restore the hormone levels and initiate follicle development and ovulation, she can have treatment with artificial gonadotrophins or drugs that stimulate the natural secretion of gonadotrophins. One such drug, clomiphene, works by increasing FSH secretion.

After such treatment, the response of the ovaries can be followed by ultrasound, which shows how many follicles are developing in each ovary. A follicle is considered to be ready for ovulation when it reaches 17 mm in diameter. The endometrium is also checked – it should be at least 8 mm thick at this time.

When a ripe follicle is detected, ovulation can be stimulated artificially by injecting human chorionic gonadotrophin (HCG), a hormone that has a similar effect to LH. Ovulation should occur after about 36 to 48 hours, and so intercourse should be timed to coincide with this.

3

a) Why will increased FSH secretion help a woman who isn't ovulating regularly?
b) What is a potential problem with treatment which stimulates ovulation? Hint: this may not be a problem if the couple want a large family.

Treatment methods: in vitro fertilisation (IVF)

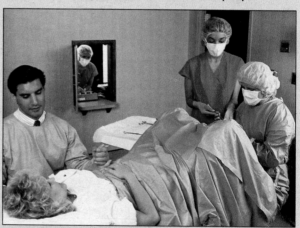

Fig 19.A1 **A woman undergoing IVF treatment must endure weeks of hormone treatments, followed by the uncomfortable process of egg cell collection. Up to about 20 egg cells may be recovered. Her egg cells are fertilised by her partner's sperm 'in vitro' and, when the tiny embryos have developed, two of these are put back into her uterus. IVF has, at best, about a 30 per cent success rate**

In vitro (= 'in glass') fertilisation is what most people think of as 'test-tube baby' treatment. In IVF, the egg cells and sperm are taken from the couple and fertilised in a dish. The process of fertilisation normally takes 12 to 15 hours and after this the new embryos start to develop. Cell division is taken as a sign that fertilisation has been successful. Then, the tiny embryos (no more than balls of 8 to 16 cells at this stage – see Fig 19.A2) are placed into the uterus.

4 How often does an ovary normally produce an ovum?

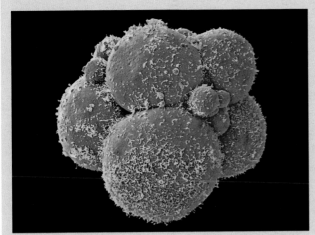

Fig 19.A2 **A human embryo at the 8-cell stage. The eight large cells (red) are covered with microvilli (yellow). The smaller cells with degenerate**

Treatment methods: gamete intra-Fallopian transfer (GIFT)

GIFT involves stimulating the ovaries and collecting the ova in much the same way as in IVF treatment. The important difference is that in GIFT the egg cells are mixed with sperm and immediately introduced into the Fallopian tubes (Fig 19.A3), without waiting to see if fertilisation occurs. The advantage of this procedure – which would seem to be less controlled than IVF – is a 5 per cent better success rate. This is possibly because the Fallopian tube is the natural site for fertilisation: it may secrete chemicals which stimulate the process.

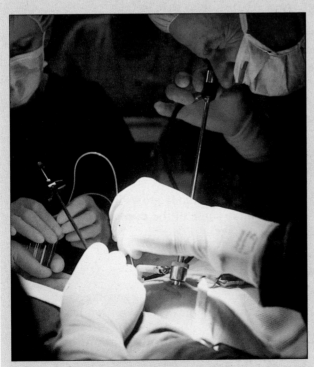

Fig 19.A3 **A mixture of egg cells and sperm is being inserted into the woman's Fallopian tube through a small insertion**

5 Many specialists feel that counselling couples is an important part of IVF and GIFT treatment. Think of two reasons why these treatments might cause anxiety.

Treatment methods: intra-uterine insemination (IUI)

In around 20 per cent of infertility cases, there seems to be no problem with either partner. Surprisingly, in some of these cases, intra-uterine insemination proves successful. The basic idea behind IUI is to introduce the partner's semen into the uterus.

The steps in IUI are as follows:

● Follicle development is stimulated and monitored.

● Treatment to induce ovulation is given.

● A fresh sperm sample is introduced into the uterus.

6 What advantage does IUI give over trying to conceive naturally?

After working through the Assignment so far, you should now be able to tackle the following question:

7 Copy and complete the following table, matching up the causes of infertility to suitable treatments. Remember that there may be more than one suitable treatment.

Your options are:

OI – ovulation induction

IVF – in vitro fertilisation

GIFT – gamete intra-Fallopian transfer

IUI – intra-uterine insemination

Problem	Possible treatments
Blocked Fallopian tubes	
Hostile cervical mucus	
Low sperm count	
Woman has antibodies against partner's sperm	
Many abnormal sperm	

8 A programme of infertility treatment can put a great strain on a relationship. Suggest reasons for this.

9 Hormone replacement therapy can be used to allow post-menopausal women to become pregnant. A donated egg can be fertilised by sperm from the woman's partner, and can lead to women as old as 60 bearing children. Discuss the advantages and disadvantages of this technology.

20 Growth and ageing

Life is a sexually transmitted fatal condition. As the proportion of elderly people increases in many populations in the western world, there are serious implications for our societies. Older people suffer more from degenerative diseases such as Alzheimer's and will need to be cared for, putting an increasing burden on health care resources

IN THE TIME IT TAKES to read this sentence, fifty million of your cells will have died. This is an old biological 'fact', but serves as a stark reminder of our mortality. The brain, like all the other organs in the body, degenerates as we get older. In most people this results in a gradual loss of memory. Many older people struggle with their short-term memory – they forget everyday things – but have very clear and lucid memories of what happened in their childhood.

Alzheimer's disease accelerates memory loss and causes a decline in many other mental abilities. Eventually, people with Alzheimer's might not remember who they are or who the people around them are, and this can lead to them to behave aggressively or even violently. Few things in life can be as distressing as watching a close relative decline in this way.

The causes of Alzheimer's disease are not yet properly understood. One suggestion that might account for the symptoms is the accumulation of aluminium in the brain tissue. Post-mortem examinations show that the brains of those who have died from Alzheimer's contain significantly more aluminium than normal. Also, kidney patients who can't excrete aluminium and therefore build up higher concentrations of aluminium ions in their blood are more likely to suffer from Alzheimer's disease.

However, this evidence is circumstantial and it is not possible to say whether there is a clear causal link between aluminium and Alzheimer's disease. Aluminium deposits might just as easily be an effect caused by some other underlying problem. Further research is in progress and the evidence will need to be analysed carefully to find the answer.

1 GETTING BIGGER AND GETTING OLDER

In this chapter we look at how the body grows from birth until the late teens, and then we study some of the events that occur as the body starts to decline in old age.

What is growth?

We all know that when something grows, it gets bigger. But to a biologist, growth is a lot more complex. This is a useful definition:

Growth is an increase in size brought about by the addition of more body tissue.

Growth therefore excludes temporary changes such as those that result from water retention in the cells and tissues.

We say that human growth is **diffuse**: it occurs all over the body, not just at one growing point. A child grows to adult size because all the organs grow. A 10-year-old is not only taller than a 3-year-old but has a larger heart, liver, pancreas and so on. Not all of these organs, however, grow at the same rate. In the first part of this chapter we look at patterns of growth and the way in which they are controlled by internal factors such as hormones and by external influences such as nutrition.

Ageing in humans

Many animals die long before they reach their full lifespan (Fig 20.1). Humans are unusual since they often go on living on for years after they have lost their ability to reproduce. A longer life, however, has its disadvantages. Organs degenerate and systems become less efficient. In the second part of this chapter we look at some of the changes that affect the body as it gets older.

Fig 20.1 **In the wild, mammals like the Grant's gazelle rarely reach old age. Their usual fate is to be killed and eaten by carnivores or to die of disease long before this. Modern medicine, however, has enabled humans to grow old and 'wear out'. Some of the same changes that we experience as we age are seen in the animals we protect, such as our pets**

2 MEASURING GROWTH

As people grow, not only do they become taller but they also become broader, and the distance between their front and their back increases: growth takes place in all three dimensions. How can we measure these changes to give an accurate picture of overall growth? We commonly use three basic criteria: increase in **body mass**, increase in **height** and increase in **supine length**.

Body mass

Body mass is quick and easy to measure. It is useful because it enables us to measure total body size. Body mass changes slightly, even from day to day and hour to hour – we weigh more when we have just eaten a large meal or when we have a full bladder, for example – but these fluctuations are minor. Larger gains in weight occur during childhood, and we use standardised charts so that children's size can be plotted to make sure that any potential problems in growth are spotted early (Fig 20.2).

Women gain weight as they go through pregnancy and, as we get older, we can gain weight just because we eat a bit too much and are a bit less active. Homeostatic mechanisms try to keep our weight stable (see Chapter 10), but we can override this control – all too easily, judging by the epidemic of obesity currently sweeping the developed countries.

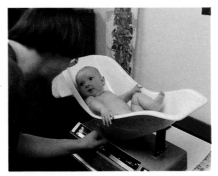

Fig 20.2 **We monitor the way babies grow by weighing them regularly**

Height

Like body mass, height is easy to measure. The person being measured stands straight with heels flat on the ground and facing forwards. A horizontal bar is moved down until it touches the top of the head and the height is read off from a suitably positioned scale (Fig 20.3). Height also gives a good indication of growth over time (Fig 20.4).

Using height as the only measure of body size has its disadvantages. It cannot, for example, give information about the changes in the width of the shoulders or hips, changes that are just as important as increases in height. And it's no good for assessing the growth of babies who are too young to stand.

Fig 20.3 **Medical checks usually involve the measurement of both body mass and height. The body mass index, the ratio of body mass (in kilograms) to the square of the height (in metres) gives important information about a person's health and fitness. A BMI of 20–25 is ideal, over 25 is getting a bit overweight, and over 30 is classed as clinically obese. This ratio is not reliable in very short or tall people, or in children and adolescents who are still growing**

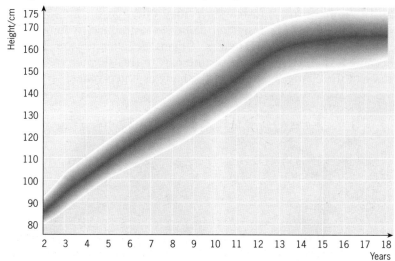

Fig 20.4 **Typical height curves for girls from the ages of 2 to 18 years. The spread of values is the result of large studies that measured the growth patterns in thousands of children. The centre line is the average curve; the shaded areas show the normal range of height at each age**

Supine length

You can only measure a person's height if they are old enough to stand. In very young children, length is measured instead. The infant is placed as flat on its back as possible. The ankles are gently pulled to stretch the child and straighten its legs. The feet are turned up vertically. The **supine length** can now be measured. Adults are, in fact, taller lying down than standing up because gravity causes a slight compression of the skeleton when we are standing. The supine length of an adult is usually between one and three centimetres greater than his or her standing height.

Techniques to assess average body size and growth

In 1759, a Frenchman, Count Philibert de Montbeillard, decided to measure the height of his newborn son. He took measurements every six months from birth until the boy was 18 years old. Collecting data like this by measuring a single individual or group of people on many occasions forms what we now call a **longitudinal study**. It gives a very accurate picture of growth, but it does have one big disadvantage. Unless, like the Count, you only study your own children, it is often difficult to maintain contact with the people involved in the study for the length of time required to produce meaningful results.

An alternative approach is to use a **cross-sectional study**. Suppose we wanted to use this approach to investigate the growth of human males between 10 and 18 years of age. We would select different age groups – 10-year-olds, 11-year-olds and so on. Then we would measure a large group of boys in each age group. Each individual would be measured once only. From the average height of all the members in each age group we could plot a graph to show the overall pattern of growth.

These two approaches can lead to rather different results. Look at the graph in Fig 20.5.

A Look at Fig 20.5. What do you think the peaks in the blue curves represent?

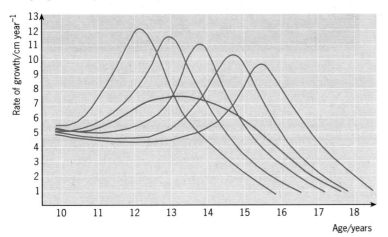

Fig 20.5 **The blue curves show the rates of growth of five different boys obtained from a longitudinal study. These curves differ in shape from the red curve. This curve was obtained from a cross-sectional study**

The blue curves show the rate of growth in centimetres per year for five boys. This was a longitudinal study in which the height of each boy was measured at regular intervals over the full period of time. The red curve shows what happens when you calculate and plot the average height of all the individuals in each age group. This is what you would do in a cross-sectional study. Note how this curve has been smoothed out. The characteristic increase in growth rate around puberty, clearly visible for each individual, has almost disappeared.

Representing growth

Imagine you are making a journey by train. If you know the *distance* you have travelled, you can work out where you are. To know the distance, you need to know the *speed* of the train. Since speed varies, you would need to know the speed during different parts of the journey.

Look at the graphs in Fig 20.6 showing the growth of de Montbeillard's son.

Fig 20.6(a) shows his height in centimetres over the whole period. It is sometimes called a **cumulative growth curve** because it shows his total height at a specific age. We can think of this as like the total distance that a train travels during the whole journey.

Fig 20.6(b) shows **growth rate**. This is how you calculate growth rate: subtract the height at a particular time from the height at some time later, and divide the result by the time interval. Here, growth rate is expressed in centimetres per year. (Growth rate is like the speed of the train.)

What we must remember is that the cumulative growth curve will only go up or remain stationary. Children get taller; they don't shrink. Growth *rate*, on the other hand, can decrease as well as increase.

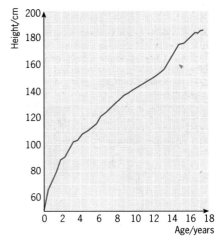

(a) **Cumulative growth curve**

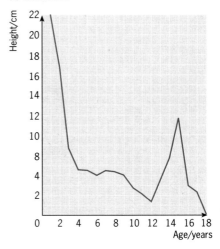

(b) **Growth rate**

Fig 20.6 **The growth of de Montbeillard's son: (a) is a cumulative growth curve showing changes in his total height; (b) is a graph showing his growth rate**

3 LOOKING AT PATTERNS OF GROWTH

Fig 20.7 **These two organisms have different patterns of growth. An elephant's growth is limited and allometric, while a crocodile shows unlimited, isometric growth**

Different organisms show differences in their patterns of growth. A crocodile is about 15 centimetres long when it hatches from its egg and can grow to a length of over five metres. It increases in length because all parts of it grow at more or less the same rate. As a result, an adult crocodile has very similar body proportions to a young crocodile. Its pattern of growth is described as **isometric**. Growth in a mammal follows a different pattern. The various parts of the body grow at different rates. Mammalian growth is therefore described as **allometric**.

Different organisms also show different growing periods (Fig 20.7). Some, like the elephant, reach maturity, then stop growing. Their pattern of growth is described as **limited.** Others – the crocodile, for example – continue to grow, often slowly, throughout their lives. These organisms show **unlimited** growth.

?

B Would you expect the ratio of tail length to total body length to change as a crocodile grows? Give a reason for your answer.

The pattern of human growth

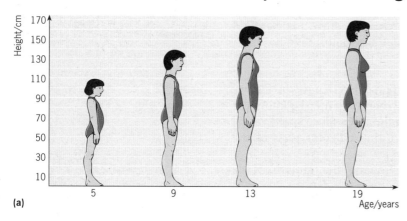

(a)

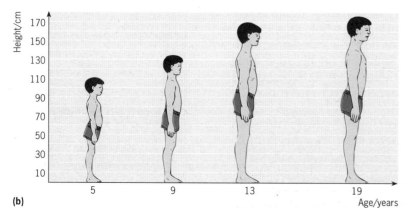

(b)

We shall now look at the pattern of human growth in a little more detail. Look at Fig 20.8.

You can see that human growth is clearly limited. A maximum height is reached. Although the diagrams suggest that growth stops during adolescence, there is a very small amount of growth later – no more than 2 or 3 millimetres. Until the age of about 30, the length of the spine increases as small amounts of bone are added to the vertebrae. For practical purposes, however, we usually say that in Europe and North America, growth stops at around the age of 15.5 years for the average girl and 17.5 years for the average boy.

Fig 20.8 **Growth in height of** (a) **a typical human female and** (b) **a typical human male**

The pattern of growth can also be described as allometric and the diagram shows us that there is a change in body proportions. Although organs such as the kidney and liver have a growth curve similar to that of the body as a whole, there are some notable exceptions to this pattern and these are shown in the graph in Fig 20.9.

The brain and skull grow very rapidly after birth, reaching 90 per cent of adult size soon after the age of six. The lymphatic system also develops early. It has a very different growth curve from the rest of the body, reaching full size well before puberty and then decreasing. Finally, the reproductive organs grow very slowly in childhood. Maximum growth of these organs occurs during puberty.

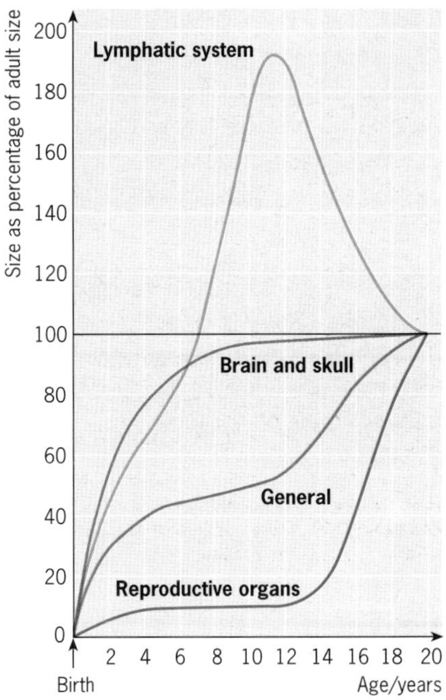

The control of human growth

Human growth is complex. It involves, directly or indirectly, all the systems of the body, and takes place over a long period of time. In addition, different organs and parts of the body grow and mature at different rates.

Growth obviously has to be coordinated. This is achieved by hormones. Before you read this section, it would be helpful to look back at Chapter 16 and remind yourself of the ways in which hormones can affect the activities of their target cells. There are at least twelve different hormones that have a direct effect on growth at different times during a person's life. The functions of some of these are summarised in Fig 20.10.

Fig 20.9 **Growth curves of different parts of the human body compared with the overall growth curve**

?

C Suggest a biological advantage for the pattern of growth of **(a)** the brain, **(b)** the reproductive organs.

D At what age is a child half its adult height?

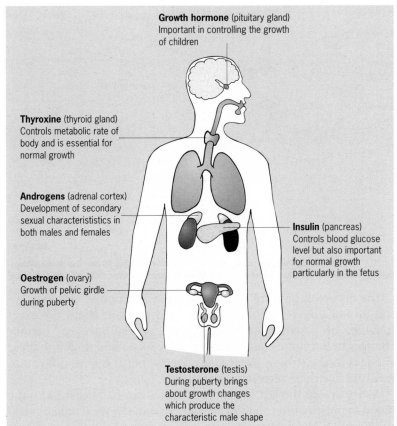

Fig 20.10 **The main endocrine glands associated with human growth**

Growth hormone

Growth hormone is produced by the anterior lobe of the pituitary gland. It has a rather unusual property for a mammalian hormone in that it is species specific. Hormones such as insulin can be extracted

E Before growth hormone was produced by genetic engineering, what would have been the only source to use for treating human growth disorders?

F A blood sample is collected and found to have a very low concentration of a particular hormone. Suggest three different ways of explaining this.

from other animals and will work in humans, but not growth hormone. Only human growth hormone works in humans.

Growth hormone controls the growth of children, though it is not necessary for fetal growth, and we do not know if it has a function in adults. Like other hormones produced by the pituitary gland, it is secreted in pulses. For most of each day, the concentration of growth hormone in the blood is so low that it can hardly be detected, but several times each day, levels rise dramatically. This does not mean that people grow in spurts that correspond with these peaks in hormone concentration. What actually happens is that growth hormone stimulates the release of another hormone from the liver. It is this second hormone which is responsible for stimulating growth: it acts on the cartilage cells at the ends of bones and on muscle cells where it stimulates protein synthesis.

Thyroxine

Thyroxine, produced by the thyroid gland, is also essential for normal growth. As well as controlling basal metabolic rate, it stimulates the process of growth in general, but also has a specific effect on protein synthesis and growth of the skeleton.

The reproductive hormones and their role in puberty

Puberty is the period of life when the sex organs grow and start to produce gametes. Secondary sexual characteristics develop, such as the distribution of body hair and the differences in body shape that distinguish males from females (Fig 20.11).

Fig 20.11(a) **Not a job for life: at puberty, the male voice breaks – suddenly gets deeper – as the size of the larynx increases**

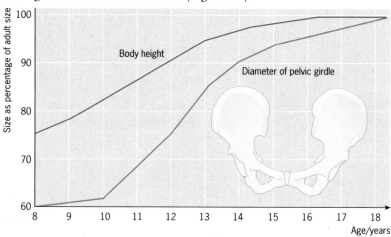

Fig 20.11(b) **During birth, the baby passes through the mother's pelvic girdle. This graph shows that the increase in the diameter of the pelvic girdle of females occurs relatively late in puberty**

The events of puberty are controlled by hormones, although we still do not understand the full story. To find out about the exact function of a particular hormone we would have to monitor changes in hormone concentration in the blood and then link these changes to the various aspects of puberty; but it is very difficult to measure hormone secretion accurately.

As we saw in the last section, hormones can be released into the blood in pulses rather than in a steady stream. So we would have to collect small blood samples from a large enough sample of people over a long period before we could know what was happening. Another problem is that many of the hormones involved in controlling puberty are broken down very rapidly.

The hormonal changes that control the onset of puberty are described in detail on page 280.

4 AGEING

As we pointed out at the start of this chapter, few wild animals reach old age. There are many reasons for this but predation and disease are obviously important. In humans, the situation is very different. Because of improvements in the standards of living in many parts of the world over the past 200 years, more people live to a ripe old age. In developed countries, we are already experiencing the effects a 'top heavy' population. In the next few years, the same population changes will occur in the less well developed nations. Fig 20.12 shows how, in China, India, Colombia, Bolivia and Kenya, the percentage of people over sixty is likely to increase between now and the year 2020.

Old age affects individuals and brings about changes in cell structure and organ function. An ageing population also affects society. An increase in the number of people living to old age places a huge strain on budgets for health and social care – one of the great challenges of the 21st century. In this chapter we concentrate on the biological aspects of ageing.

G It is expected that the total number of old people in Kenya will rise considerably, even though their percentage in the population will change little. Explain why.

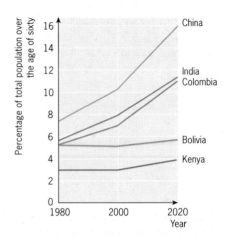

Fig 20.12 **The proportion of old people in the populations of many less developed countries is likely to rise during the first 20 years of the twenty-first century**

WHY DO WE AGE?

GRAB THE SKIN on the back of your hand, pinch it and let it go. If you are young, it will snap back immediately. Try it (gently) with an older person and the pinched skin stays there for a while. One of the features of ageing is that our tissues lose their elasticity, causing that classic symptom of advancing years, wrinkles. Lines appear as our skin goes baggy instead of being stretched tightly over our bodies. Our hair falls out, bones soften, muscles weaken, spine curves, arteries harden, gums recede... and we may even start to enjoy gardening or making marmalade.

So what causes all this? Scientists have been pondering for years. There are no definite answers, but there are two aspects to consider. Do we age because we have a genetic clock ticking away inside us, determining how long we live – the **deterministic** view; or are the cumulative effects of things in our environment to blame – the **non-deterministic** view? The answer is almost certainly that it is a complex interaction of both. As we get older, we lose the ability to repair the damage caused by various internal and external factors.

A clue to the role played by our genes comes from the simple observation that all babies are born young. This seems obvious, but begs a fascinating question. How can it be that a baby born to a 14-year-old mother, and therefore made with a 14-year-old egg, is born at the same age as one born to a 50-year-old mother, with a much 'older' egg? It seems that there is a genetic clock inside the egg that doesn't start ticking until fertilisation.

Recent research into the ageing process of cells has focused on the ends of the chromosomes – stretches of DNA known as **telomeres**. When we are conceived, our telomeres are full length: the cells of a newborn baby have the potential to divide 80 or 90 times. Each time a cell divides, the telomeres shorten (like the way that lead in a pencil wears down when you write with it). Cells seem to 'know' how old they are because they can 'count' the number of divisions they have made, and when the telomeres have been shortened to a critical length, the cell is no longer able to divide.

When cells can't divide, damaged tissues can't regenerate, and we begin to wear out. But as we say in the opener to Chapter 28, we might in future be able to control the shortening of the telomeres, and therefore control the genetic cause of ageing. We might even be able to extend life well beyond 100 – which would have huge socio-economic consequences.

But how does the *environment* cause ageing? There are certainly external factors: exposure to sunlight, for example, accelerates the process of DNA damage. But internal factors are also very significant. Our internal environment can be very hostile. Key molecules such as **free radicals**, reactive chemicals which react rapidly with anything around them, are capable of damaging DNA and proteins in a random fashion. One of the most damaging free radicals is the oxygen radical, •O (a single oxygen atom with an unpaired electron). A small amount of all the O_2 oxygen in the air we breathe is split into free radicals. It is therefore always present in the body, and our metabolic defences cannot repair all the damage it causes. Even without external factors such as radiation, our vital molecules are damaged in time, and ageing is inevitable.

Cell structure, organ function and ageing

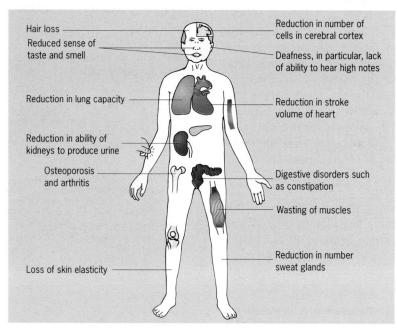

Hair loss

Reduced sense of taste and smell

Reduction in lung capacity

Reduction in ability of kidneys to produce urine

Osteoporosis and arthritis

Loss of skin elasticity

Reduction in number of cells in cerebral cortex

Deafness, in particular, lack of ability to hear high notes

Reduction in stroke volume of heart

Digestive disorders such as constipation

Wasting of muscles

Reduction in number sweat glands

Fig 20.13 **Some of the many effects of ageing. It is important to remember that different individuals can be affected to different extents**

Biologists who study the effects of ageing in humans would like to be able to define what is 'normal'. Illness and incapacity are so common among old people that complete freedom from disease is extremely rare. In addition, there is a lot of individual variation.

Let's look at the breathing system, for example: we could to say that there is a 50 per cent reduction in lung capacity by the time a person reaches the age of eighty. But, with so much variation, some individuals will be almost unaffected, while others will have such a severe reduction in lung capacity that they are barely able to move.

Fig 20.13 shows some of the ways old age affects organs and systems.

Fig 20.14 **Charlie Chaplin, star of many silent films, was not unusual in becoming a father when he was in his seventies. In contrast, few women over the age of fifty have had children without hormone treatment**

Ageing and the reproductive system

Fertility in men gradually declines with age. A man's general state of health usually determines how old he is when he loses the ability to father children. Some healthy men father children in old age (Fig 20.14).

In women, however, the inability to produce children is a more marked event, usually occurring around the age of fifty. This is known as the **menopause**. In the four or five years before the menopause, as the ovaries gradually fail, there is a progressive decline in the number of menstrual cycles in which an ovum is produced.

A reduction in the amount of oestrogen that the ovaries produce is the cause of many of the unpleasant symptoms of the menopause. Hormone replacement therapy (HRT) involves supplying extra oestrogen and can be used to control these symptoms (see page 285). Fig 20.15 summarises the hormonal events associated with the menopause.

Ageing bones and joints

As we get older, the skeleton suffers from a lifetime of wear and tear. The physical stresses on the body throughout life lead to two important medical conditions: **osteoarthritis** and **osteoporosis**.

Osteoarthritis

Osteoarthritis is one of the commonest conditions associated with old age (Fig 20.16). X-ray evidence suggests that nearly 80 per cent of those over 65 years have some arthritic damage to their joints. Thankfully, in most of them, this is not severe enough to cause painful symptoms.

Many different factors cause osteoarthritis to develop, but in most cases, there is unnatural stress on the joint concerned. Surveys have shown, for example, that the physical work carried out by farmers makes them very prone to osteoarthritis of the hips; the elbows are

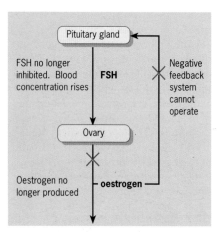

Pituitary gland

FSH no longer inhibited. Blood concentration rises

FSH

Negative feedback system cannot operate

Ovary

Oestrogen no longer produced

oestrogen

Fig 20.15 **Some of the hormonal changes that accompany the menopause**

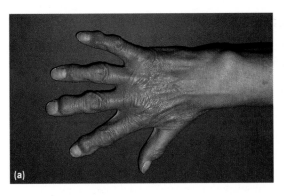

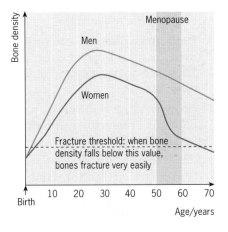

Fig 20.16 **Osteoarthritis is not a single disease. It describes the various conditions that result from joint failure. Joints commonly affected are** (a) **those in the hands, and** (b) **in the knees as shown by this X-ray image, and in the hips.** (c) **If the hips are severely affected, surgery might be required to replace the hip joint with an artificial one**

affected in people who operate pneumatic drills; and those who are overweight suffer problems with their knees. Whatever the actual cause, this stress leads to damage to cartilage and its loss at the **articular surfaces** of bones that form the joints (Fig 20.17). In addition, there is an increase in bone-forming cells called **osteoblasts**. New bone may be laid down, deforming the shape of the joint.

Although we cannot reverse the changes that accompany the development of osteoarthritis, we can do a lot to ease the symptoms. Anti-inflammatory drugs form the first line of treatment and the affected person can fit rubber heels to their shoes to reduce jarring of the joint as they walk.

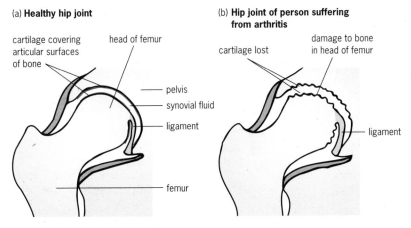

Fig 20.17 (a) **In a healthy hip joint, cartilage covers the articulating surfaces of the bones which form the joint.** (b) **In the hip joint of a person suffering from osteoarthritis, this cartilage has been lost**

Osteoporosis

This is another condition in which the effects become increasingly marked with age. Look at Fig 20.18.

You can see that as you get older, up to the age of about 20, bone gets denser. In other words, a given volume of bone will have a greater mass. After 20, there is a gradual decrease in density. In men, this loss of bone is not usually a matter for concern. In women, however, bone density is always less than in men, and it falls further during the menopause. The result is that many olderwomen suffer from osteoporosis, a loss of bone mass. Their bones lack strength, and very little stress can fracture them.

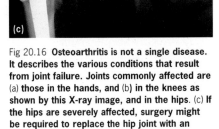

Fig 20.18 **The effect of age on bone density**

USING X-RAYS TO SEE FURTHER

WE NORMALLY ASSOCIATE X-rays of bones with fractures, but a hospital X-ray department investigates many conditions that affect the skeleton. Fig 20.19 shows some computer records from the radiography department of a large hospital.

DIAGNOSIS	AGE ON ADMIT	SEX
FRACTURE OF ANKLE	17	M
FRACTURE OF NECK OF FEMUR	17	M
ACQUIRED DEFORMITIES OF TOE	17	M
INTERNAL DERANGEMENT OF KNEE	17	M

Fig 20.19 **A hospital record of patients receiving X-rays**

The pie charts in Fig 20.20 show the age on admission of males and females different conditions.

- Some onditions are the the result of gender-specific behaviour. Young boys tend to be rougher in their play than girls of the same age and so they break their arms and leg more often. It is also likely that older women referred to the hospital with mis-shapen toes insisted on wearing tight fashionable shoes when they were younger.
- Congenital conditions are present at birth. Children with these conditions are X-rayed more often.
- Osteoarthritis is a disease of the elderly and is a major reason for X-ray investigations in the 70–79 age group.
- Older women tend to break bones often because their sense of balance has declined, but osteoporosis is also a major factor. Fractures of the neck of the femur are particularly common and many elderly women receive hip replacements.

Fig 20.20 **Pie charts drawn from data like that in Fig 20.19(b)**

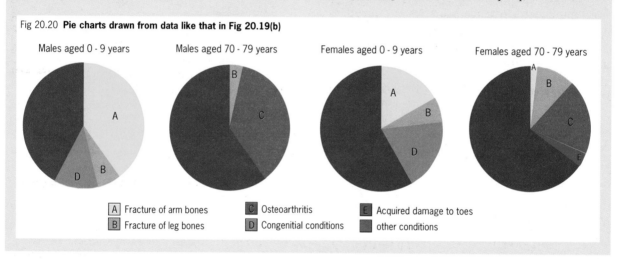

Males aged 0 - 9 years

Males aged 70 - 79 years

Females aged 0 - 9 years

Females aged 70 - 79 years

A	Fracture of arm bones	C	Osteoarthritis	E	Acquired damage to toes
B	Fracture of leg bones	D	Congenitial conditions		other conditions

SUMMARY

After reading this chapter, you should know and understand the following:

■ **Body mass**, **height** and **supine length** can be used to measure body size and body growth. There are advantages and disadvantages with all three criteria.

■ Growth curves may be plotted using data collected from either **longitudinal** or **cross-sectional** studies. They can show cumulative growth or growth rate.

■ Human growth is **allometric** and **limited**.

■ Growth is controlled by hormones. These include **growth hormone** and **thyroxine**.

■ **Puberty** is the period of life when the sex organs grow and start to produce gametes. The events of puberty are initiated and controlled by changes in hormone concentrations.

■ Ageing brings about changes in cell structure and organ function.

■ Fertility in men gradually declines with age. In women, however, the inability to produce children is a associated with the menopause.

■ **Osteoarthritis** and **osteoporosis** are conditions that commonly affect the skeletal system of older people.

QUESTIONS

Trait	Mean differences			
	Identical twins		Non-identical twins reared together(52 pairs)	Siblings of the same sex reared together (52 pairs)
	reared together (50 pairs)	reared apart (19 pairs)		
Height/cm	1.7	1.8	4.4	4.5
Mass/kg	1.9	4.5	4.6	2.1
Head length/mm	2.9	2.2	6.2	not available
Head width/mm	2.8	2.9	4.2	not available
Intelligence quotient	5.9	8.2	9.9	9.8

1 The table above shows data from studies of twins and their siblings (brothers and sisters).

a) Identical twins are sometimes called 'monozygotic twins'. Suggest why.

b) Using the information in the table,
   (i) give **one** trait, the difference in which was largely due to environmental factors. Explain the evidence of your answer.
   (ii) give **one** trait, the difference in which was largely due to environmental factors. Explain the evidence for your answer.

[NEAB June 1998 Biology: Social Biology Module Test, q.2]

2 The graph of Fig 20.Q2 shows the changes in lean body mass (protein and bone) and fat content of the body with age in men and women.

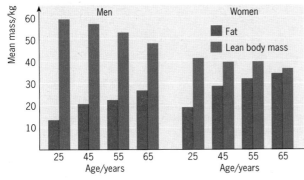

Fig 20.Q2

a) Suggest **one** explanation for each of the following:
   (i) the change in lean body mass with age;
   (ii) the change in fat content with age;
   (iii) the difference in the proportion of lean body mass and fat between a 25-year-old man and a 25-year-old woman.

b) Describe the effects of ageing on the functioning of a **named** organ or system of the human body.

[NEAB February 1998 Biology: Social Biology Module Test, q.6]

3 The flow chart of Fig 20.Q3 summarises the hormonal events that occur during the first 14 days of the human menstrual cycle.

a) Name
   (i) gland A,
   (ii) organ B,
   (iii) hormone C,
   (iv) hormone D.

b) With the aid of the flow chart, explain what is meant by negative feedback.

c) The menopause is the time when the menstrual cycle ceases. In women who have just undergone the menopause, hormone C is found in very high concentrations. Explain why this is so.

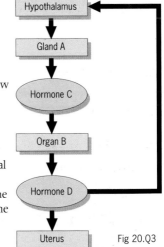

Fig 20.Q3

[NEAB June 1998 Biology: Social Biology Module Test, q.3]

4 Fig 20.Q4 shows a section through a human hip joint.

a) Describe how the structure of this joint would be affected by osteoarthritis.

b) The table shows the percentage of women of different age groups reporting to a large hospital with fracture of the femur.

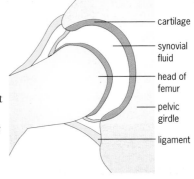

Age group/years	Percentage of women in age group reporting with fracture of the femur
20–29	0
30–39	0
40–49	0.4
50–59	1.1
60–69	2.6
70–79	7.4

Apart from an increased likelihood of falling with age, suggest an explanation for the trend shown by the figures in this table.

[AEB June 1997 Biology: Human Life Span Module Test, q.2]

Assignment

WHY DO WE GROW OLD?

Why do some people live longer than others? In this Assignment we look at the evidence that supports the idea that longevity is controlled by genes.Studying twins is one way of finding out more about how genes control variable characteristics. First we need to look at the ways in which twins are formed. Fig 20.A1 shows the formation of **monozygotic** twins and **dizygotic** twins.

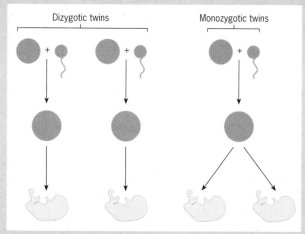

Fig 20.A1 **Monozygotic twins result from the splitting of a single fertilised egg cell or zygote. Dizygotic twins develop from two separate zygotes**

1

a) Use Fig 20.A1 to explain why monozygotic twins are also called identical twins.

b) Explain why it is not really accurate to describe dizygotic twins as non-identical.

In one investigation concerned with ageing, many pairs of monozygotic and dizygotic twins were studied. Records were kept of the age of the first twin when he/she died and the number of months later that the second twin died. The data were collected only on pairs of twins in which the first sibling was at least 60 years old when he/she died. The results of this investigation are summarised in Table 20.A1.

Table 20.A1

	Monozygotic		Dizygotic		
	Male	Female	Same sex		Different sex
			Male	Female	
Average difference in length of life between members of twin pairs/months	29.4	47.6	89.1	61.3	126.6

2

a) Suggest why the data were collected only from twins in which the first sibling was at least 60 years old when he/ she died.

b) Evaluate the evidence from the table that the length of a person's life is controlled by their genes.

c) Women generally live longer than men. How does this explain the figures for the difference in length of life between members of dizygotic twin pairs of the same sex and of different sex?

If individual cells are removed from an animal and cultured, most do not live forever. Similarly, most cells from different human tissues grow in a culture for only a limited amount of time. Then they degenerate and die.

In one investigation, cells were isolated from fetal skin and cultured in a suitable container with oxygen and a supply of nutrients. They grew rapidly until they had formed a layer over the inner surface of the container. At this stage they stopped dividing. The culture was then divided and placed in two new containers. The cells started growing again until, once more, they formed complete layers over the inner surfaces of the containers. The culture was divided in half repeatedly until the cells stopped dividing. The results are shown in the graph in Fig 20.A2.

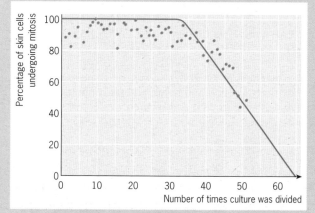

Fig 20.A2 **In this graph the percentage of fetal skin cells estimated to have undergone mitosis has been plotted against the number of times the culture was divided**

3

a) Look at the information about the skin in Chapter 12 and suggest the type of cell that could have been used in this investigation. Give a reason for your choice.

b) Use the information from the graph to suggest a hypothesis to explain why humans do not live forever.

c) Assuming your hypothesis is correct, predict the shape of the curve you would expect if the initial sample of skin cells had been taken from a young adult.

An inherited conditions in which the process of ageing is more rapid than normal is Hutchinson–Gilford syndrome. It is controlled by a recessive allele. Sufferers show many visible signs of old age even when quite young. Fetal skin cells containing two alleles for Hutchinson–Gilford syndrome will have about 10 culture cycles before they die.

THE WORLD WE LIVE IN

WE ALL LIVE in a constantly changing world. Most of the minor fluctuations in our environment are predictable: day turns into night; tides come in and go out twice a day; the weather changes with the seasons. Geology has shown that much greater changes have occurred during the past; ice ages have come and gone and there have been times of global warming. Living things, including humans, have to adapt to their changing environment, or they do not survive.

This intricate and complex relationship between us, other living things and the environment is at the centre of the branch of biology known as ecology. Ecology is a vast topic that can include an in-depth investigation of a nettle bed and a detailed survey of the sexual preferences of an obscure species of beetle. However, some aspects of ecology are very relevant to health and a study of human physiology. Ecology can mean studying the ecosystem that exists in your mouth and it also encompasses the whole-planet issues that affect us all, such as global warming and ozone depletion.

In this section, we begin in Chapter 21 with some of the basic concepts of ecology. In the following chapter we look in detail at the flow of energy in ecosystems. Later, in Chapter 23 we examine the importance of recycling, and finally we consider the effects of the most resource-depleting of population explosions – that of human beings.

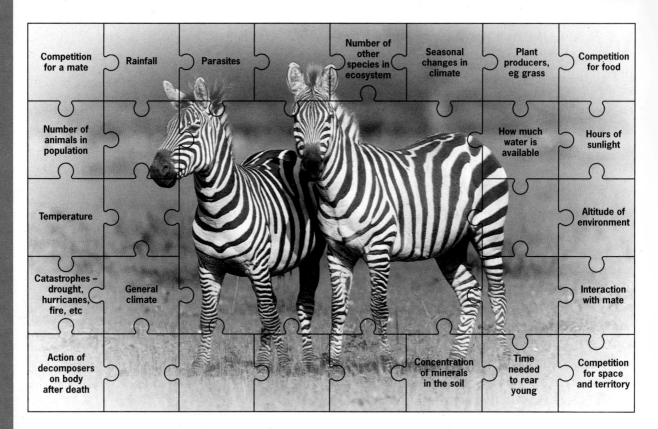

Competition for a mate	Rainfall	Parasites	Number of other species in ecosystem	Seasonal changes in climate	Plant producers, eg grass	Competition for food
Number of animals in population					How much water is available	Hours of sunlight
Temperature						Altitude of environment
Catastrophes – drought, hurricanes, fire, etc	General climate					Interaction with mate
Action of decomposers on body after death				Concentration of minerals in the soil	Time needed to rear young	Competition for space and territory

Fig 21.1 **Ecology is the study of whole organisms: we learn how an organism interacts with individuals of its own species and other species, and with the physical environment that surrounds it. The jigsaw pieces show just a few of the factors that influence the lives of these zebras in the savannah grasslands of Africa**

The word ecology is loosely used to mean environmental concern or environmental conservation. These topics are important, and they depend heavily on knowledge gained through ecology, but they are not the same thing. We define ecology as:
the scientific investigation of living organisms in their natural surroundings.

1 THE SCIENCE OF ECOLOGY

Ecology is a branch of biology. The word ecology was first used by Ernst Haeckel in the 1860s. It comes from two Greek words: *oikos* meaning home and *logos* meaning understanding. His definition of ecology was 'the knowledge of the sum of the relations of organisms to the surrounding outer world, to organic and inorganic conditions of existence'. Put more simply, ecology is the study of organisms in their natural surroundings. It looks at how they are adapted to their environment and how they interact with both the living and non-living world around them.

Each organism and each physical feature of an environment is separate, but they all interact and interlock, forming a complex system that is a bit like a three-dimensional jigsaw puzzle: ecologists study different parts of the puzzle and then try to work out how it is put together (Fig 21.1).

Ecology and other branches of science

Ecology is closely related to two other branches of science: **environmental biology** and **environmental science**. The first of these looks more generally at living things in the environment. The second uses the biology, geology and the physical sciences to help us to explore our environment.

It is also impossible to study ecology without bearing in mind the genetics, evolution and behaviour of the organisms at the centre of a study. Ecological genetics and behavioural ecology are beyond the scope of this book but it is important to remember that each organism you study in the context of its environment has developed to fit that environment through the process of natural selection (see Chapter 30).

Practical ecology

Fig 21.2 **Safety and field trips. Ecologists conduct studies outside and must consider safety when working near water or in isolated areas.**

Here, a student is identifying organisms from a pond, and two students are measuring the rate of water flow in a stream

As in other branches of science, progress in ecology depends heavily on scientific investigation. Looking back at our definition of ecology, it should come as no surprise that many of these studies are done outside. Field studies involve a great deal of observation and can include experimental studies, thought these are more often carried outt in the laboratory where they can be more strictly controlled. Ecologists use these three basic approaches together to gain a complete picture of the organism or the environment that is being investigated.

The exciting thing about ecology is that it is still a relatively young science. In just over 100 years of ecological study, we have barely begun to even scratch the surface. Many environments of the world remain unexplored and there are many organisms that we know of, but whose habits and living conditions are known only sketchily, or are a complete mystery. Unlike biochemistry or physiology, ecology offers an A-level student the chance to do original practical work, and perhaps to make an important scientific discovery.

Fig 21.3 **Habitats change frequently. Porlock used to have a mile-wide strip of farmland between it and the sea. In 1996, the shingle barrier that formed the sea defences was breached and a salt marsh with salt water lakes has developed rapidly. This is a new habitat that now supports birds such as egrets. Coastal areas are subject to many changes like this, creating new areas for observing the processes of colonisation and adaptation**

WHY DO HUMAN BIOLOGISTS NEED TO STUDY ECOLOGY?

YOU MIGHT BE wondering what a section on ecology is doing in a human biology textbook. Ecology is sometimes thought of as a study of nature – lots of field trips and wading around in mud looking at insects. But ecology is much more wide ranging than that.

The quality of our environment and the way we interact with the outside world are major factors that determine whether people are healthy or face serious health problems. Substances that contaminate our environment in water, food, air and soil contribute to the untimely death of millions of people every year.

The problem is worst in the developing world where 4 million babies and young children die from diarrhoea every year because of dirty water and bad food. Over 1 million people die from malaria each year and over 267 million others suffer recurrent symptoms. Treating these illnesses with medicines is only of limited use: to really tackle them effectively, it is crucial to find out what is causing them and take steps to reduce people's risk. That doesn't mean just finding out what bacteria, virus or parasite is causing the disease directly. It means identifying the environmental factors that are allowing the disease to spread.

Many third world countries, for example, have seen huge increases in cases of malaria with the more widespread availability of plastic 'disposable' containers. When these are dumped, they provide the perfect trap for rainwater and then form a wonderful stagnant pool in which the anopheles mosquito can lay its eggs. Recognising this as a problem has helped to prevent more cases of malaria than could have been treated using millions of pounds' worth of drugs.

Fig 21.4 **We have a two-way relationship with our environment, and our actions can have unexpected consequences. These people in the Philippines live by recycling rubbish. The plastic bags around them look harmless enough but they could be lethal**

2 TERMS AND CONCEPTS IN ECOLOGY

Fig 21.5 **The biosphere is the part of the Earth's surface that supports life. It extends from the bottom of the deepest oceans into the part of the Earth's atmosphere that contains breathable air: in total a vertical distance of between 30 and 40 kilometres.**

It covers most of the hydrosphere, the surface layers of the lithosphere and some of the atmosphere. The hydrosphere comprises the Earth's bodies of water – oceans, seas and lakes. The lithosphere is the layer of soil and rock that forms the Earth's crust. The atmosphere is the blanket of gases that envelop the planet, maintaining a mean temperature of about 7 °C. This temperature is crucial to life because it allows most of the water on Earth to exist as a liquid (see Appendix 1)

The relationships between an organism and the physical features of its environment, and between all the other organisms that live with it, are incredibly complex. Any description of them uses terms that you have probably not come across before. So, before getting into detail about any particular aspect of ecology, the next section gives an overview of some basic terms and concepts.

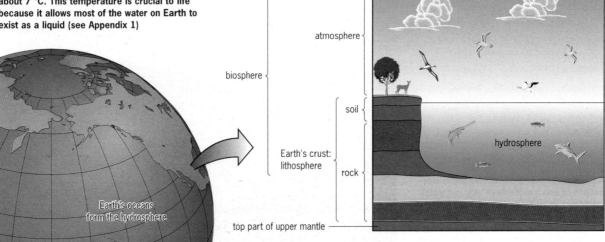

The biosphere and its division into biomes

All the living organisms that we currently know of live on Earth. But not all parts of the Earth support life. The part that does forms a sort of 'skin', the **biosphere,** which is shown in Fig 21.5.

The biosphere is made up of lots of different areas that have very different environmental conditions. We call each of these fairly broad areas **biomes**. The major biomes of the world are listed in Table 21.1 and some of these are shown in Fig 21.6. Many of the best-studied biomes are the **terrestrial biomes** (those that exist on land) but there are also **aquatic biomes** (those that exist in water).

?

A Look at Table 21.1. Which biomes do you think exist in Britain?

Table 21.1 **The major biomes of the world**

Biome	Type	Major features
Savannah	terrestrial	Tropical grassland with few trees. Example: the Serengeti in Africa
Temperate grassland	terrestrial	Grassland with hot summers and cold winters. Some broad-leaved plants. Example: the prairies in the US
Desert	terrestrial	Hot and dry. A few highly specialised plants and animals. Example: the Kalahari desert in Africa
Tundra	terrestrial	Cold for most of the year: very short growing season. No trees: some specialised plants. Example: northern Canada
Tropical rainforest	terrestrial	Tropical climate. Lush vegetation. Great diversity of animal and plant life. Example: forests of equatorial Africa
Temperate rainforest	terrestrial	Cool and wet. Tallest trees in the world. Example: redwood forests of North America
Temperate deciduous forest	terrestrial	Hot summers and cold winters. Dominated by broad-leaved deciduous trees. Example: oak woodlands of central and western Europe
Lakes and ponds	aquatic: freshwater	Large bodies of standing freshwater. Example: Great Lakes of North America
Streams and rivers	aquatic: freshwater	Flowing freshwater. Example: the Amazon
Marine rocky shore	aquatic: sea water	The border between the land and the sea. Most rocky shores show **zonation**: there are characteristic bands of different environmental conditions that lead to colonisation by very different plants and animals
Coral reef	aquatic: sea water	The tropical rainforests of the seas. Reefs form in warm, shallow waters and form a biome that contains an incredible diversity of living organisms

Fig 21.6 **The biomes of the world show great extremes of environmental conditions**

(a) **Arctic tundra**

(b) **Desert**

(d) **Rainforest**

(c) **Coral reef**

The ecosystem concept

The **ecosystem** is the basic functional unit of ecology. It is a single working unit that consists of a group of interrelated organisms and their physical environment (Fig 21.7). We can divide any one of the biomes described above into many different ecosystems.

When we study a particular organism in its ecosystem, we look at the physical features that might affect it, such as rainfall, soil type, temperature and so on. These physical or **abiotic** features help to determine what range and type and numbers of living organisms live inside the ecosystem. Of course, we also look at the living or **biotic** part of the ecosystem, to see how our chosen organism relates to the other organisms present. This is particularly important if we are looking at the impact of human activities.

On page 316, we look at the abiotic and biotic features that affect the distribution of organisms in an ecosystem in much more detail.

Fig 21.7 **A schematic diagram showing the essential features of an ecosystem**

| Ecosystem | = | **Producers** (green plants) | + | **Consumers** (herbivores, omnivores, carnivores) | + | **Decomposers** (mainly bacteria and fungi) | + | **Non-living components** (eg weather, soil composition) |

Levels of organisation in an ecosystem

Individual living organisms may look randomly arranged in an ecosystem, but they are organised into recognisable units. Fig 21.8 provides an overview of organisation in an ecosystem.

From ecosystem to individual

We can study the human body and narrow down our investigation from a body system to an important organ, to the level of the individual cell. In a similar way, we can focus in on the components of an ecosystem. Each one contains a **community** of organisms, which consist of many different **populations**. Each population is made up of many **individuals**. It is important to define these basic features of an ecosystem:

- A community is a collection of groups of organisms from different species that live in close association in the same ecosystem. Organisms in a community interrelate (see Chapter 24).

- A population is a group of individuals of the same species that live in a particular area at any one moment in time. We can look at a population of rabbits or a population of beech trees. Different populations interact in an ecosystem, forming a community.

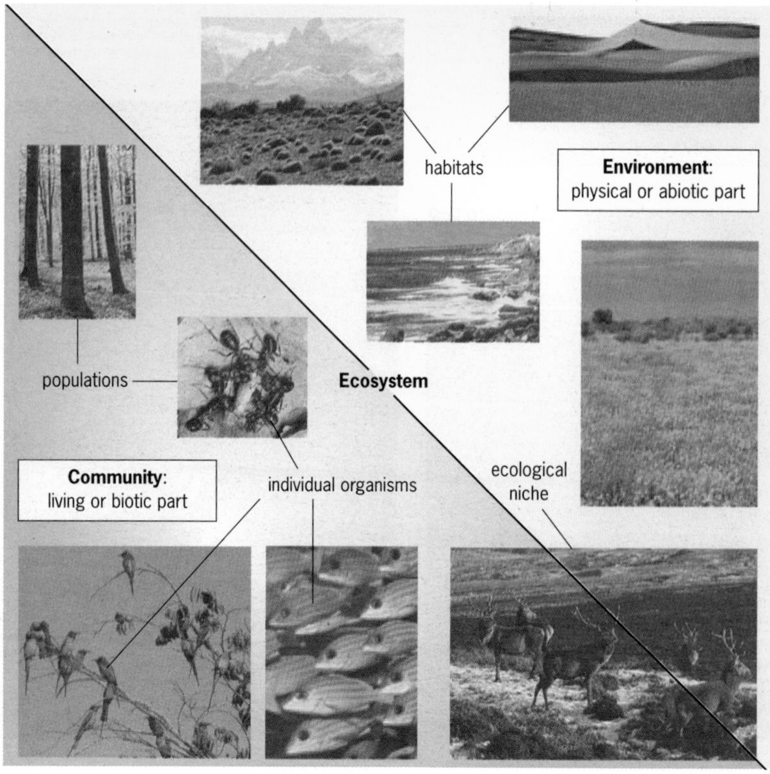

habitats

Environment: physical or abiotic part

populations

Ecosystem

Community: living or biotic part

individual organisms

ecological niche

Fig 21.8 **An ecosystem has two basic parts: the abiotic part and the biotic part. Within both of these, there are further levels of organisation**

● An individual is a single organism within a population. Some populations contain organisms that are genetically identical. Most populations of bacteria, for example, are identical because they have developed as a result of asexual reproduction. But many populations contain individuals that are the result of sexual reproduction: these individuals show genetic variation (Fig 21.9 and see Chapter 30).

Environment, habitat and ecological niche

The divisions listed above relate primarily to the *living* components of an ecosystem. We can also sub-divide an ecosystem with reference to its *non-living* components. The term **environment** describes the overall physical surroundings that occur in a biome or an ecosystem (although it is often used more generally: people speak of the environment of the Earth).

Within the environment of a single ecosystem are **habitats**, individual areas in which particular groups or individual organisms live. The habitat of any organism is its normal home. Some organisms have only one habitat, for example, the giant panda lives only in bamboo forest. Others, like humans, live in many different ecosystems and biomes and so have many different habitats.

Another important term is **ecological niche**. To describe an organism's niche, we need to show how that organism relates to the physical and biological components of its surroundings. It is not just *where* it lives, it is also *how* it lives and *what* it does.

Fig 21.9 **This individual stoat is part of a larger population. Like other members of the group, she is genetically unique. She has evolved to become well adapted to her woodland environment and, if her particular genes allow her to live a long life, she will pass her genes on to several new members of the next generation**

?

C What is the difference between an organism's habitat and its ecological niche?

3 THE BASIC FEATURES OF AN ECOSYSTEM

Let's consider an example of a real ecosystem. Fig 21.10 shows the main features of a familiar aquatic ecosystem, the garden pond. This is simplified (a real pond contains more organisms), but you can see that an ecosystem is made up of physical or abiotic components such as water, mineral nutrients, soil and rock, and living or biotic components such as fish, snails, insects and water plants.

Think about one of the insects in the pond. It can only survive and live there if the physical conditions in the pond suit it. So the pH and the temperature of the water must be within a suitable

Fig 21.10 **A schematic diagram showing a garden pond as an ecosystem. Every living organism in the pond is affected by abiotic factors such as temperature, light levels and mineral availability, and by biotic components – other living organisms**

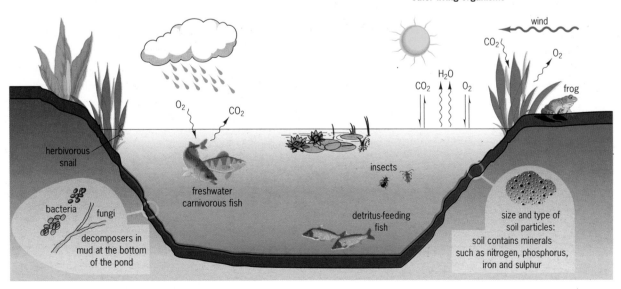

herbivorous snail

bacteria   fungi

decomposers in mud at the bottom of the pond

freshwater carnivorous fish

detritus-feeding fish

insects

size and type of soil particles:
soil contains minerals such as nitrogen, phosphorus, iron and sulphur

frog

wind

CO_2 O_2

H_2O

CO_2 O_2

O_2 CO_2

range. The insect also needs to interact with other organisms: it needs to eat smaller organisms or to feed on the sap of plants; and it needs to be with other insects of the same species so that it can find a mate and reproduce. And there is a 'grey area': some of its physical needs, dissolved oxygen for instance, are only met if the insect lives with organisms that release oxygen as a product of photosynthesis.

The abiotic components of an ecosystem

The abiotic components of an ecosystem give it a physical form, providing places for organisms to live. Physical conditions help to determine where each organism survives best: plants adapted to dry, almost desert conditions will not thrive if a river changes its course and the environment becomes boggy and wet. The nature of the physical environment in any ecosystem depends on:

- the rock and soil types present,
- the type of landscape: mountain range, flood plain etc,
- the position on Earth: altitude, latitude etc,
- the climate and weather, including light, temperature, water availability, wind and water movements,
- the potential for catastrophe: fire, flood etc.

> In a real ecosystem the abiotic and biotic factors all act together to determine the environmental conditions for each organism. Both interact to constantly change the conditions in the ecosystem. How this happens depends on the ecosystem and the physical conditions. The Feature box below gives a thought-provoking example.

A MODERN-DAY PLAGUE

THE ABIOTIC AND BIOTIC parts of any ecosystem interact in many different ways, some subtle and some not so subtle. Australia provides one of the most dramatic examples of how disrupting the delicate balance of an ecosystem can devastate both the environment and the populations of living organisms that depend on it.

Fig 21.11 **A rabbit suffering from the symptoms of myxomatosis, the deadly viral disease that wiped out most of the rabbit population. Most rabbits today are resistant to the virus**

Two hundred years ago, Australia was hot, dry, and covered predominantly by grasslands, with some areas of forest. When European settlers arrived they started to clear the forests and brought in two alien species of mammal: sheep and rabbits. The rabbits in particular multiplied extremely rapidly in conditions that suited them very well, and were soon eating the grass faster than it could grow.

Ten years after just 24 rabbits were introduced, the wild population had grown to millions and started destroying crops and wild vegetation as well as any grass that was left. They out-competed native mammal species, driving many of them to extinction. This plague continued until the late 1950s, when farmers introduced the myxomatosis virus. This killed over 99 per cent of the rabbits, with a few surviving because they were naturally resistant. Today, Australia is in danger of being overrun by another rabbit plague.

Some abiotic factors, such as the position of an ecosystem on Earth, remain unchanged as time passes. Others, such as the landscape and soil type, change very slowly. Although the basic climate of an ecosystem is usually pretty constant, you can get daily and often hourly changes in weather. Catastrophes such as fires and floods are sudden and short-lived, but they can produce severe, long-lasting effects (Fig 21.12).

Biotic components of an ecosystem

The biotic environment that surrounds an organism in an ecosystem results from the activities of all the other organisms living there. Complex relationships called **intraspecific interactions** occur between organisms of the same species. **Interspecific interactions** happen between organisms from different species (Table 21.1).

Fig 21.12 **In some ecosystems, catastrophic events are necessary. Some species of plant, such as** Banksia **from Western Australia (above) reproduce only after a serious fire (below).** Banksia **seeds are protected by closed capsules that can be broken open only by the heat of the flames: without exposure to fire, the seeds cannot germinate**

Table 21.1 **Types of interactions that occur between organisms in an ecosystem**

Activity	Type of interaction	What happens
Reproduction	Intraspecific	Location of, selection of and competition for a mate
Caring for young	Intraspecific	Both parents, one parent or, more rarely, older siblings feed, protect and shelter the young
Social behaviour	Intraspecific	Animals cooperate to find food and to defend themselves against competitors or predators (Fig 21.13)
Competition	Intraspecific	Organisms compete for resources such as food, space, light, water and mineral nutrients
Reproduction	Interspecific	Animal vectors pollinate some species of plant and help to disperse the fruits and seeds of others (Fig 21.14)
Caring for young	Interspecific	Rare, but some species rear the young of another species
Mutualism	Interspecific	Two organisms both benefit from a long-term association
Parasitism	Interspecific	Individuals from one species use another species to provide food and shelter whilst giving nothing in return. Many hosts actually suffer from the interaction: many parasites cause disease
Predation	Interspecific	All animals obtain food by eating another organism
Protection	Interspecific	Many species make use of other species to hide from predators. Some copy the markings of other species to mimic warning signals that predators avoid
Competition	Interspecific	Organisms in a community compete for resources such as food, space, light, water and mineral nutrients
Defence	Interspecific	Organisms have developed various strategies to defend themselves against predators, parasites and competitors

4 AN OVERVIEW OF THIS SECTION OF THE BOOK

In this section of the book, we provide an introduction to ecology at advanced level. with an emphasis on what is important for human biologists to know. In the next chapter there is a thorough overview of the importance of energy and a description of the biochemical processes of respiration and photosynthesis. In Chapter 23 we look at recycling in the natural world, with detail on the carbon cycle and nitrogen cycle. Finally, in the last chapter in this section, we look at how humans affect their environment and look at some of the big issues that seem likely to threaten our environment in the future.

Fig 21.13 **Meercat sentries keep a look out for predators while other members of the group search for food and look after the young**

Fig 21.14 **Black-chinned humming birds pollinate the penstemon as they drink its nectar**

22 Energy in living systems

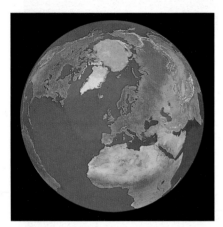

An understanding of energy is of fundamental importance in biology, from the study of individual molecules and reactions to global issues such as the use of fossil fuels, pollution, damage to the ozone layer and the greenhouse effect

ALL LIVING ORGANISMS depend on energy. Most of them live by the energy that comes from sunlight. Studying energy and how it flows through organisms and ecosystems is valuable for many topics in human biology.

Part I of this chapter gives an overview of some of the key concepts that you will need to understand in order to study energy flow in living systems. Part II gives a fairly detailed account of photosynthesis. This topic is crucial to understanding how energy flows through ecosystems and it is a core topic on most human biology syllabuses. Part III concludes the chapter with the process of cell respiration. Again we have included details of the biochemical pathways involved in this vital life process, but the information is presented in a step-wise format, so that you can build up your understanding.

With all three topics, remember that you should concentrate on understanding the relevance and overall importance of the process, not the details of individual biochemical reactions and pathways.

PART 1: OVERVIEW OF KEY CONCEPTS

1 SOLAR ENERGY, PHOTOSYNTHESIS AND RESPIRATION

This planet, and all the organisms on it, must have energy and almost all of it is supplied by our nearest star, the Sun. A huge amount of sunlight reaches the surface of the Earth but only about 2 per cent of it is actually absorbed by plants. The rest of the energy heats up the land, air and water, preventing the planet from freezing.

In **photosynthesis,** plants use radiant energy from the Sun to convert simple inorganic substances (mainly carbon dioxide and water) into larger organic molecules (glucose, starch, lipids and proteins). The plant tissues built from the products of photosynthesis form food for organisms that cannot make their own, such as animals, fungi and most bacteria. Some of these feed other organisms. Ultimately, therefore, the products of photosynthesis provide the energy which most living things need to carry out the processes of life. This energy is made available to organisms through the process of **cell respiration**.

We look at the biochemistry of photosynthesis and cell respiration later in this chapter, but first we revise some key concepts.

Fig 22.1 **Photosynthesis by blue-green bacteria was responsible for the appearance of oxygen in the Earth's atmosphere 2 billion years ago, and photosynthesis by plants maintains the oxygen level at 20 per cent of the atmosphere today**

2 WHY DO WE NEED ENERGY?

Humans need a constant supply of energy to maintain their life processes:

- **Growth and repair** of cells and tissues. Energy is required for the biochemical reactions that build large organic molecules from simpler ones. For example, energy is needed to build proteins from amino acids.

- **Active transport**. Energy is required to move some substances in or out of cells. The transport of amino acids from the small intestine into the blood is achieved by active transport. Active transport often takes place against a diffusion gradient (see Chapter 5) and it allows the body to control its internal environment more efficiently.

- **Movement.** All movement requires energy, and human movement occurs on several levels:

  inside cells, eg chromosomes separating

  whole cells, eg sperm swimming (see Fig 22.2)

  tissues, eg muscles contracting

  whole organs, eg a heart beating

  parts or whole organisms, eg talking, walking

- **Temperature control**. Humans are warm-blooded animals, **homiotherms**, and we use around 70 per cent of the energy from respiration to maintain the body at a constant 37°C.

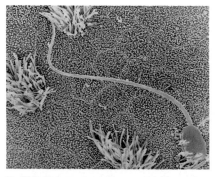

Fig 22.2 **Movement requires energy. The sperm cell shown here is on the ciliated epithelium of the Fallopian tube. To swim far enough up into the Fallopian tubeto reach an egg, a sperm cell needs enough energy to enable it to swim over 7500 times its own length – the equivalent of a 10 kilometre swim for an adult human**

3 HOW DO LIVING ORGANISMS GET THE ENERGY THEY NEED?

In order to respire and so release energy, organisms need a supply of food. Food can be obtained in two ways:

1 Some organisms can make their own food. These organisms are called **autotrophs** (meaning 'self-feeders') and they use energy from the surroundings to make the materials they need. There are two kinds of autotroph, the **photoautotroph** and the **chemoautotroph**. Photoautotrophs, such as green plants, photosynthesise, using sunlight as an energy source. Chemoautotrophs, all of which are bacteria, use energy made available from chemical reactions other than those involved in photosynthesis. Some bacteria involved in nitrogen recycling are chemoautotrophs (see Chapter 23).

2 Some organisms need to obtain ready-made food because they cannot make their own. Such organisms are called **heterotrophs**. Heterotrophs must obtain their food, either directly or indirectly, from organisms that carry out photosynthesis. They either eat their food (ingesting it into a gut, as most animals do) or digest the food first and then absorb the nutrients, as decomposers, mainly fungi and bacteria, do. See Fig 22.3.

Human nutrition is covered in more detail in Chapter 6.

Fig 22.3 **All of these organisms have something in common – they are heterotrophs. They cannot make their own food and must take it, ready made, from an outside source. As you can see, there are several different ways of doing this**

4 ENERGY FLOW: WHERE DOES ENERGY COME FROM, WHERE DOES IT GO?

Fig 22.4 **The basic idea of energy flow. At the start of any food chain there must be an organism capable of making food – usually a green plant. From there, the energy is passed along the chain within food molecules. Some of this energy is eventually used to make the cells, tissues and organs of heterotrophic organisms**

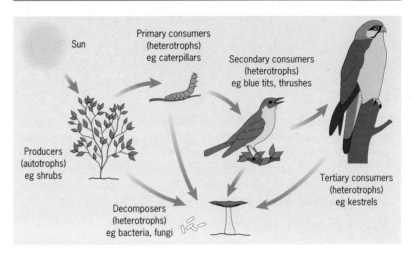

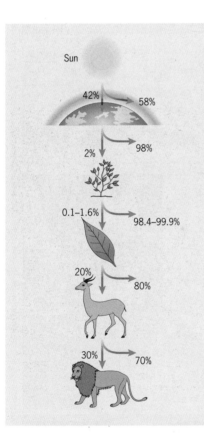

Of the light energy from the Sun that reaches the Earth, only about 42 per cent actually gets through to the surface of the planet. The rest is either reflected by the atmosphere, or is absorbed by it, a process that warms the atmosphere.

Of the light energy reaching the Earth's surface, about 2 per cent is absorbed by plants. The rest goes to produce the heat which warms up the land and the oceans.

Of the energy absorbed by plants, only 0.1–1.6 per cent is incorporated into plant tissue – and even some of this is lost as heat because the plant itself must respire.

Of the plant tissue eaten by herbivores, about 20 per cent (at most) will become incorporated into the tissues of the animal. Most of the rest is lost as heat as the animal respires.

Of the animal tissue eaten by a carnivore a maximum of 30 per cent becomes incorporated into the body of the carnivore. Most of the rest is lost as heat.

Clearly, as so little energy is transferred at each point in a food chain, there is relatively little energy available for the final consumers, usually large carnivores. For this reason, such animals are quite rare.

Think of your own diet. You consume large amounts of food but very little of it is incorporated into your body. Compare your own body mass to an estimate of the mass of all the food you have ever eaten: you can see that a lot of energy must have been lost as heat. This heat is used to keep your body temperature at a constant 37 °C. So much is needed because you are constantly losing heat to the air, your clothes and the objects you touch.

?

A Give three reasons why we need energy when we are asleep.

?

B Heterotrophs must eat autotrophs; true or false? Explain.

5 ENERGY TRANSFER IN ECOSYSTEMS

In photosynthesis, green plants and some algae and bacteria use the energy from sunlight to power chemical reactions that combine simple inorganic molecules to form complex organic compounds. We say that the autotrophs in an ecosystem are **producers**: the products that they make form the food that ultimately feeds all other organisms in the ecosystem.

The heterotrophs in an ecosystem meet their need for energy by feeding on other organisms. We say that they are the **consumers**. Generally, **primary consumers** are herbivores: they obtain their energy by eating producers. **Secondary consumers** are carnivores or scavengers: they eat primary consumers. **Tertiary consumers** are carnivores that prey on other meat-eaters. This chain of dependence is usually called a **food chain** (Fig 22.5). The distinct levels of each chain are called **trophic levels**.

Humans are **omnivores**: we eat a mixture of plant and animal food. An omnivore can be a primary, secondary and even a tertiary consumer all at the same time.

?

C Look at Fig 22.5. What do the arrows in a food chain represent?

Fig 22.5 **A simple flow chart summarising the overall structure of a food chain that has four trophic levels**

Food chain

As one organism eats another, there is a transfer of energy from the bodies of the producers (or consumers) into the bodies of consumers through the trophic levels of a food chain. Energy transfer occurs in one direction only: it is not recycled, and much of the energy escapes from the ecosystem as heat (see Fig 23.1).

Another major group of heterotrophs in an ecosystem are the **decomposers**. These are mainly bacteria and fungi. They break down the bodies of dead animals and plants, usually by extracellular digestion (Fig 22.6) and absorb the soluble products. Transfer of energy from producers and consumers to decomposers is also one-way, with a high proportion eventually lost as heat. The process of decomposition allows mineral nutrients to be cycled in the ecosystem (see page 349).

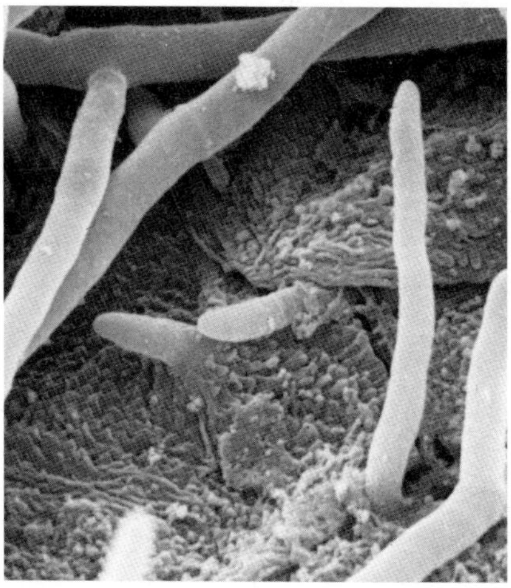

Food chains and food webs

An example of a food chain is shown in Fig 22.7. Most animals have a varied diet. A food chain is therefore only part of a much larger feeding picture: the **food web**. Take our top freshwater carnivore, the pike. It does not feed on stickleback alone but also on roach and insects such as dragonfly nymphs and pond skaters.

Most rivers and lakes include a wide variety of producers, ranging from the microscopic plankton to the larger pondweeds, rushes and flowering plants.

Fig 22.6 **Fungi grow into their food (in this case, human skin), secreting enzymes which digest the food externally. The soluble products of digestion are then absorbed through the walls of the hyphae**

Fig 22.7 **An example of a freshwater food chain. Each chain has five trophic levels (most food chains have fewer trophic levels than this)**

Freshwater food chain

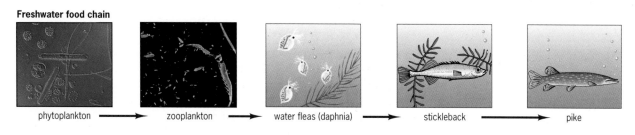

phytoplankton ⟶ zooplankton ⟶ water fleas (daphnia) ⟶ stickleback ⟶ pike

FOOD WEBS AND PESTICIDES

IN DEVELOPING a new pesticide, scientists look for a chemical which kills only the pest that we want to get rid of, harms no other species, breaks down into something harmless after a short time, does not lead to the development of resistance, and is cheap. This is a pretty tall order: no pesticide yet developed is perfect, but modern pesticides are much safer than some of those developed and used in the 1950s and 1960s.

DDT (dichlorodiphenyltriflouroethane) was used as a pesticide in many countries after World War II. It breaks down slowly, remaining in the environment for between two and five years.

During the 1950s and 1960s it became obvious that DDT was affecting whole food webs. Fig 22.8 shows what happened when DDT was transferred up the trophic levels of a food chain in an ecosystem in the US. DDT is not excreted and concentrates in the fatty tissues of organisms. So, although the concentration of DDT in zooplankton, the primary consumers in the chain, was only 0.04 parts per million, its concentration in the tissues of a top carnivore such as the osprey or bald eagle was 25 parts per million. The bodies of the large birds of prey converted the DDT into a substance that made their eggshells very fragile. As a result, very few birds managed to breed and their numbers fell quickly.

In 1972, DDT was banned and since then, the numbers of the large predatory birds have recovered. However, there are still problems because of the illegal use of DDT.

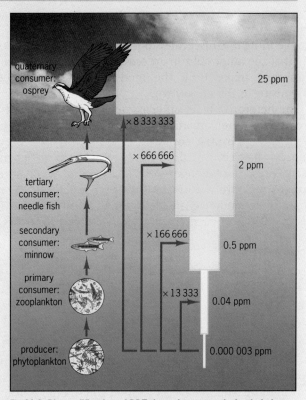

Fig 22.8 **Bioamplification of DDT through an aquatic food chain. DDT accumulates in the fatty tissues and cannot be excreted. The fat of the osprey contains over 8 million times more DDT than the water at the bottom of the food chain in which the producers live**

Pyramids of numbers

Food chains and food webs tell us a great deal about the feeding relationships that occur between organisms in an ecosystem, but they are **qualitative** rather than **quantitative**. That means that we know *which* organisms are part of the chain or web, but we have no idea of the exact *numbers* of organisms involved at each trophic level.

Clearly, numbers are very important. There are many more individual producers than primary consumers, and, as you carry on along a food chain, the numbers of organisms at each trophic level continue to decrease. At the same time, the body size of individual organisms usually increases. It is difficult, if not impossible, to find out exactly how many producers are eaten by a primary consumer and so on, but we use a pyramid of numbers to show the general idea (Fig 22.9).

Fig 22.9 **A simple pyramid of numbers. The width of each box represents the relative number of organisms present in each trophic level at any one particular time. In this example, many grass plants are needed to sustain all the antelope that one individual lion needs to eat to survive**

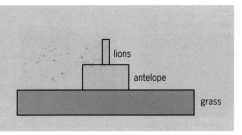

Two odd things can happen when you use pyramids of numbers. Sometimes the base of the pyramid is very narrow, indicating that there are fewer producers than primary consumers. This seems not to make sense, but these examples arise when a few large plants such as trees produce food for thousands of tiny plant-feeders, such as aphids (Fig 22.10(a)). In pyramids of numbers that represent food chains that have a population of parasites at the top, the upper level can be much larger than the level below it. This is because many parasites can feed on one host (Fig 22.10(b)). We say that these pyramids are **inverted pyramids**.

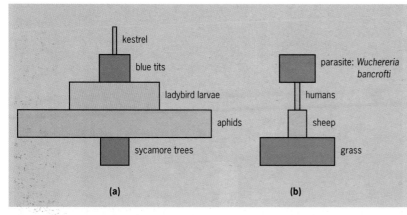

(a) (b)

Fig 22.10 **Two examples of inverted pyramids of numbers. Neither is completely inverted; they simply do not conform to the expected pyramidal shape because one of the levels is distorted**

(a) **One sycamore tree can feed many thousands of aphids, making the first trophic level of this pyramid of numbers much smaller than the second**

(b) **This pyramid is complicated by the top trophic level which involves a parasite. Many parasites can infect one host.** *Wuchereria bancrofti* **is a nematode worm which causes the disease elephantiasis in humans (see Chapter 34)**

Pyramids of biomass

Ecologists often use pyramids of **biomass** (Fig 22.11(a)), to avoid the problem of inverted pyramids of numbers. Pyramids of biomass show the mass of all the organisms at each trophic level – dry mass is the most useful measure. This is obviously extremely difficult: obtaining the dry mass of organisms means killing them, drying them and measuring the remains. Not surprisingly, very few pyramids of biomass have ever been determined in this way.

However, using samples of organisms and their wet mass allows ecologists to make estimates and devise models. But, even this type of pyramid can be inverted, as Fig 22.11(b)) shows.

■ See questions 1 and 2.

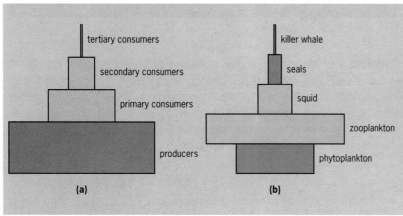

(a) (b)

Fig 22.11(a) **A generalised pyramid of biomass for a terrestrial ecosystem**

(b) **In the waters of the Antarctic, the mass of zooplankton is five times that of the phytoplankton. If we draw up a pyramid of biomass for one of the food chains, we get an inverted pyramid. The zooplankton eat the phytoplankton so quickly that the latter never get the chance to attain a large biomass. Instead, phytoplankton have a high turn-over rate, reproducing very quickly. So the biomass of organisms produced each year is large, although the number alive at any one time is often less than the number of consumers**

To obtain an even better model of an ecosystem, one which avoids inverted pyramids altogether, we must look at the way energy is transferred between the different trophic levels.

Pyramids of energy

Fig 22.12 shows a typical pyramid of energy. It shows the amount of energy that is transferred along the food chain and how much is lost as heat at each level. The more levels in the chain, the less energy there is available at the top. Consequently, very few food chains have more than four or five levels. It also explains why the top carnivores such as the big cats, tigers and large whales are the first to become endangered when their ecosystems come under pressure from humans.

Fig 22.12 **A pyramid of energy for an Antarctic food chain similar to the one shown as a pyramid of biomass in** Fig 22.11(b). **As you go up the pyramid, there is a 90 per cent loss of energy at each trophic level. This energy, of course, does not disappear: it is 'lost' as heat. For a more detailed investigation of energy transfer, see the Assignment for this chapter**

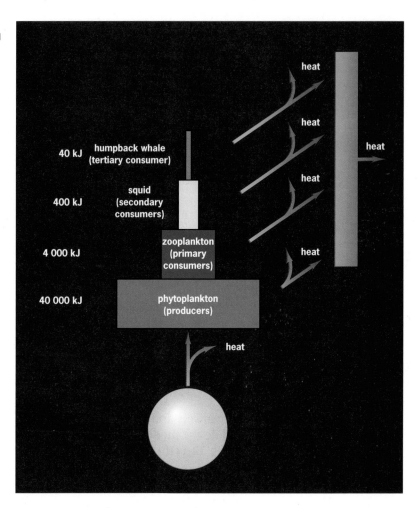

Efficiency of energy transfer between different trophic levels

As we saw earlier in this chapter, the energy transfer between organisms at different trophic levels of a food chain is never anything like 100 per cent efficient. In fact, less than 30 per cent of the energy available ever reaches the organisms in the next level. There are three main reasons for this:

● Not all the organisms in one trophic level are eaten by organisms from the next level. Many die and provide food for decomposers.

● Not all food that is eaten is digested and not all digested food is absorbed – some is lost as faeces.

● Most of the energy gained from the food that is absorbed is lost through the processes of respiration and excretion (production of *metabolic* waste).

D List the foods you ate at your last meal. Sketch out the food chains which produced them and, for each food chain, say which trophic level you, the consumer, represent.

PART II: PHOTOSYNTHESIS

6 THE OVERALL PROCESS OF PHOTOSYNTHESIS

Most plants make their own food by photosynthesis. A plant uses light from the Sun as an energy source to convert carbon dioxide, hydrogen and water into sugars. The plant either respires these simple sugars to gain energy or uses them as raw materials to build more complex molecules such as starches and cellulose.

Photosynthesis is the source of energy for virtually all living organisms.

The overall process of photosynthesis can be summarised as:

$$\textbf{carbon dioxide + water} \xrightarrow{\text{light}} \textbf{sugar + oxygen}$$

Since the sugar produced is usually glucose, the equation is often written:

$$\mathbf{6CO_2 + 6H_2O \longrightarrow C_6H_{12}O_6 + 6O_2}$$

In this section we look at the biochemistry of photosynthesis. Before you start on the details, look at the summary in Fig 22.13. This should help you to see the process as a whole.

The site of photosynthesis

Fig 22.14 shows the site of photosynthesis in a plant. Photosynthesis occurs in the green tissues. These are green because they contain the pigment, **chlorophyll**. Leaves have evolved to become efficient traps for light energy and they are the major site of photosynthesis.

The chemical reactions of photosynthesis take place in **chloroplasts**. Chloroplasts are plant cell organelles that are between 3 and 6 μm in diameter. The structure of chloroplasts was covered briefly in Chapter 1, but we study it in detail, to see exactly where the reactions of photosynthesis take place.

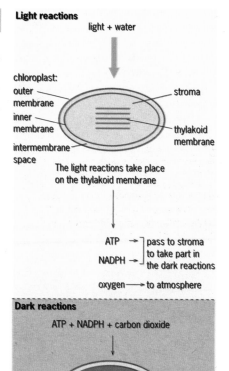

Light reactions

light + water

chloroplast:
outer membrane
inner membrane
intermembrane space
stroma
thylakoid membrane

The light reactions take place on the thylakoid membrane

ATP → ⎤ pass to stroma
NADPH → ⎦ to take part in the dark reactions

oxygen → to atmosphere

Dark reactions

ATP + NADPH + carbon dioxide

The dark reactions take place in the stroma

sugar

Fig 22.13 **Photosynthesis is divided into two main stages. In the light reaction, ATP and NADPH are generated (see page 326). In the dark reaction, ATP and NADPH are used to make sugars (see page 329)**

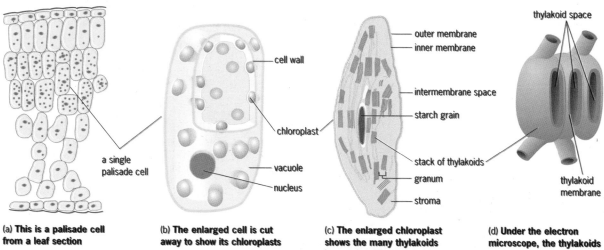

Fig 22.14 **These diagrams show the site of photosynthesis in the leaf of a plant**

(a) **This is a palisade cell from a leaf section**

a single palisade cell

(b) **The enlarged cell is cut away to show its chloroplasts**

cell wall
chloroplast
vacuole
nucleus

(c) **The enlarged chloroplast shows the many thylakoids it contains. The thylakoid membrane is crucial in photosythesis**

outer membrane
inner membrane
intermembrane space
starch grain
stack of thylakoids
granum
stroma

(d) **Under the electron microscope, the thylakoids appear to be separate but they are actually connected**

thylakoid space
thylakoid membrane

As Fig 22.14(c) shows, the chloroplast has two membranes that surround a space, the **stroma**. The stroma contains various enzymes, ribosomes, RNA and DNA. In chloroplasts the inner membrane connects with an elaborate arrangement, the **thylakoids**. These look like hollow discs and are stacked in groups or **grana**. The space inside a thylakoid is connected with every other thylakoid, forming a continuous internal compartment, the **thylakoid space**. Most of the key reactions in photosynthesis occur on or across the thylakoid membrane, the membrane between the stroma and the thylakoid space. This contains chlorophyll and other pigments of photosynthesis.

7 THE BIOCHEMISTRY OF PHOTOSYNTHESIS

Photosynthesis has two main stages (see Fig 22.13):

- **The light-dependent reaction**. Light energy becomes trapped in chlorophyll molecules. This energy is used to make ATP and reduced NADP, and oxygen is released.

- **The light-independent reaction**. ATP and reduced NADP are used to convert carbon dioxide into carbohydrate.

The light-dependent reaction

In the **light-dependent reaction** of photosynthesis, energy from sunlight excites electrons in chlorophyll molecules. The electrons pass on their energy to carrier molecules. These enzymes make ATP and reduced NADP, used in the light-independent reaction of photosynthesis to convert carbon dioxide to sugar.

The light-dependent reaction occurs in the thylakoid membrane system of the chloroplast. Two chlorophyll molecules at the heart of the **antenna complex** shown in Fig 22.15 collect the energy from all of the light that falls on the antenna.

Fig 22.15 **Each antenna contains an array of chlorophyll molecules. This increases the area over which light energy can be trapped. The antenna acts like a funnel. Energy is transferred from the outer chlorophyll molecules to a special pair of chlorophyll molecules near the centre of the antenna. Some antennae contain carotenoids, pigments which collect light of wavelengths outside the range for chlorophyll. Carotenoids hand their energy on to the chlorophyll**

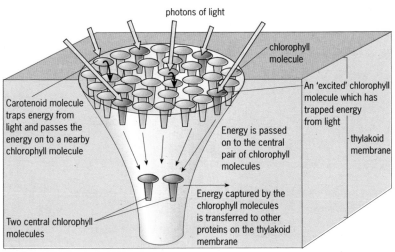

photons of light

chlorophyll molecule

An 'excited' chlorophyll molecule which has trapped energy from light

thylakoid membrane

Carotenoid molecule traps energy from light and passes the energy on to a nearby chlorophyll molecule

Energy is passed on to the central pair of chlorophyll molecules

Two central chlorophyll molecules

Energy captured by the chlorophyll molecules is transferred to other proteins on the thylakoid membrane

ANTENNA COMPLEX

This energy is passed on to a **photochemical reaction centre**. Together, the antenna complex and the photochemical reaction centre form a system called a **photosystem.** Two types of photosystem, PSI and PSII, work together in photosynthesis.

The central pair of chlorophyll molecules in the antenna complex captures light energy, and an electron is excited. It moves immediately to the **photochemical reaction centre**, a series of proteins in the thylakoid membrane that act as electron carriers. Each protein accepts an electron and then hands it on to the next.

When a molecule gains an electron it is **reduced**; when it loses an electron it is **oxidised**. So, since the electron moves from one electron carrier to the next, we describe the series of reactions as reduction and oxidation (**redox**) reactions (see page 332). In the light-dependent stage of photosynthesis, these reactions eventually lead to the production of ATP and reduced NADP – both needed later by the reactions of the light-independent stage.

Let's follow the path of electrons that leave the antenna complex of photosystem II (PSII).

Synthesis of ATP and reduced NADP

Fig 22.16 shows the arrangement of electron carrier proteins in the thylakoid membrane. Refer to this diagram as you read on, so that you can follow the path of the electrons.

(a) Next to the antenna complex at PSII is a special enzyme that splits water. As each electron 'escapes' from the chlorophyll to the photochemical reaction centre, another electron takes its place.

The replacement electrons are taken from water, by the water-splitting enzyme. When four electrons have been replaced, hydrogen ions (protons, H^+) are formed and a molecule of oxygen is released:

$$2H_2O \xrightarrow{\text{energy of 4 photons}} 4H^+ + 4e^- + O_2$$

Fig 22.16 **The arrangement of the thylakoid membrane proteins which are involved in photosynthesis. From left to right, you can follow the sequence of events in the light-dependent phase of photosynthesis**

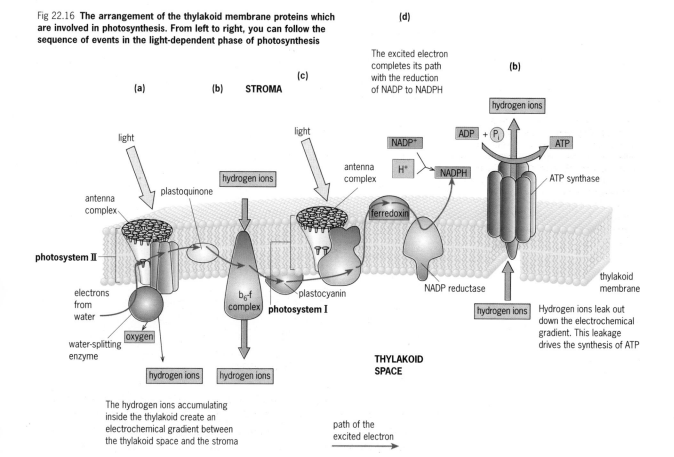

(b) Meanwhile, the escaped electrons pass down the series of electron carrier proteins. Energy released during this process results in inorganic phosphate being added to ADP to give ATP:

$$ADP + P_i \rightarrow ATP$$

(c) While ATP is being produced, the electrons pass from the electron carriers to photosystem I (PSI). PSI operates in a similar way to PSII: the chlorophyll molecules at the heart of the antenna complex trap light energy, releasing excited electrons to a photochemical reaction centre.

The difference between PSI and PSII is the energy level of the electrons in their chlorophyll molecules. The electrons which pass from PSII to PSI still have some energy left over from their first boost in PSII. So, when light energy is trapped by the antenna complex in PSI, the electrons coming into the chlorophyll molecules are excited to a level higher than they reached in PSII.

(d) These electrons combine with protons from the stroma to convert NADP to reduced NADP:

$$NADP^+ + 2H^+ + 2e^- \rightarrow reduced\ NADP + H^+$$

This whole process is called **photophosphorylation:**

- *phosphorylation* because ADP is phosphorylated to ATP,
- *photo*phosphorylation because the process is driven by light,

Overall, PSII produces ATP, and PSI produces reduced NADP. The electrons to produce reduced NADP come first from PSII, so reduced NADP and ATP production are linked. Just over one molecule of ATP is produced for every molecule of reduced NADP. Both chemicals are needed for the light-independent reactions of photosynthesis.

CHLOROPHYLL, PLANKTON AND GLOBAL WARMING

PHYTOPLANKTON are tiny organisms that live near to the surface of the oceans. For years, scientists have wondered about the very uneven distribution of these organisms, even in areas that are rich in nutrients. It seemed that something vital was missing and one suggestion was iron. Some scientists suggested that if iron were added, phytoplankton might grow better and so soak up more carbon dioxide to help slow global warming. But how could this hypothesis be tested?

Laboratory experiments were tried but, with so many variables to control, it could only be properly tested in the sea. In principle it sounds simple: just add iron and see if you get more plankton. But how can you see where the iron that you put into the sea goes, and how do you measure the phytoplankton?

Scientists came up with two solutions. Vast amounts of iron sulphate were poured into the sea together with small amounts of radioactive sulphur hexafluoride. Oceanographers could track the radioactivity over a wide area and so could follow the iron.

Secondly, scientists surveyed the plankton from the air. If you illuminate chloroplasts strongly, they **fluoresce**. By firing a laser at the sea and measuring chlorophyll fluorescence, scientists could estimate the amount of phytoplankton at a particular spot. After adding iron, they recorded a very rapid increase in the numbers of phytoplankton within just two days. But the effect was very short-lived. Either the iron dropped quickly to depths too low for plankton to be able to use, or animal plankton also grew rapidly to feast on the phytoplankton that suddenly became available.

Fig 22.17 **A bloom of marine phytoplankton**

FEEDING THE WORLD: PRODUCTIVITY IN ECOSYSTEMS

THE RATE AT WHICH the producers in an ecosystem capture light energy to build biomass is known as the **gross primary productivity** of that ecosystem. This is a theoretical maximum. In reality, all producers use energy for respiration and other life processes. When we subtract the energy that they use to live, taking it away from the energy that they capture, we find the **net primary productivity** of the ecosystem. This is the amount of energy that is available to the consumers of the system.

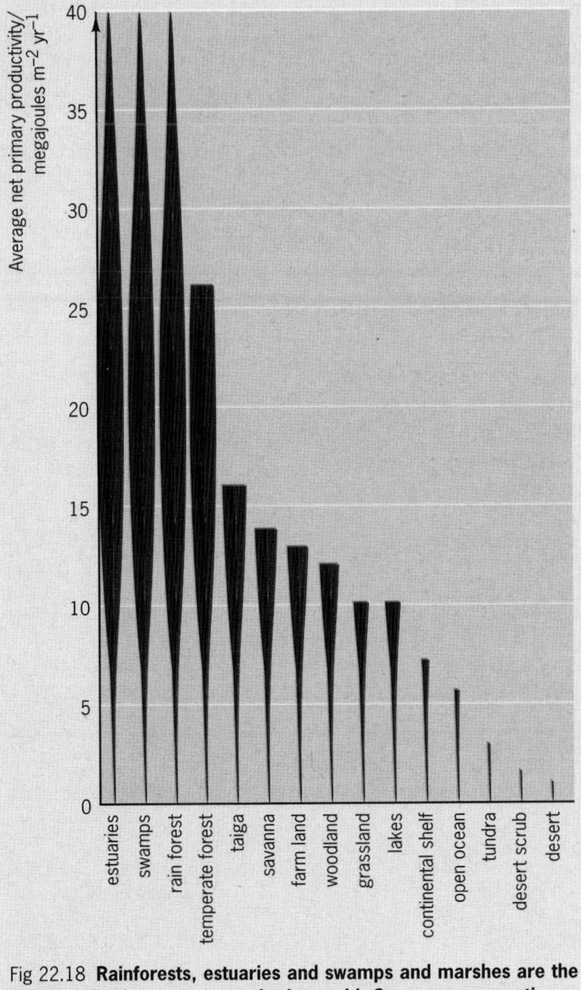

net primary productivity = gross primary productivity – energy used by producers to live

Different ecosystems of the world have different primary productivities. As Fig 22.18 shows, an ecosystem which is part of tropical rainforest is one of the most productive in the world. A major reason for this is that tropical rainforests grow around the Equator of the Earth, where more light energy is available than further north or south. Ecosystems which have high levels of water and mineral nutrients, such as an estuary or a marsh swamp, are many times more productive than deserts which are dry and severely lacking in nutrients.

The primary productivity of an ecosystem limits the number and variety of consumers that can survive there. This explains why tropical rainforest ecosystems support more diverse animal populations than other parts of the world. As rainforests are cut down to make room for expanding human populations, the potential of the area to produce enough food to support its populations decreases. As less energy is available in food webs and food chains, the numbers of animals at the higher levels are likely to fall. More species are likely to become endangered and, eventually, extinct.

Many scientists think that the conflict between the need to use the land for humans to live on and the need to leave enough land to produce food will reach a crisis in many areas of the world in the next 50 years. The famines that are tragically common in some areas may well become more of a feature of life in many other parts of the world, as primary productivity cannot keep pace with the growth of human populations.

The scale of this problem is so large that it is difficult for any individual country to tackle alone. However, many countries world-wide are now working together to try to ensure that the worst outcome does not happen.

Fig 22.18 **Rainforests, estuaries and swamps and marshes are the most productive ecosystems in the world. Open oceans, arctic tundra and deserts are the least productive**

Carbon fixation – the light-independent reaction

In the light-independent reactions of photosynthesis, the ATP and reduced NADP made in the light dependent reactions are the source of energy and reducing power to convert carbon dioxide to sugar. This takes place in the stroma. The important reaction in carbon fixing is when carbon dioxide is joined to the 5-carbon compound **ribulose bisphosphate (RuBP)**. This reaction is shown in Fig 22.19.

Sugars are too unreactive to take part in many biochemical reactions: they need to be **phosphorylated** first. This simply means that a phosphate group is added on. Ribulose bisphosphate is a sugar with two phosphate groups. The enzyme **ribulose bisphosphate carboxylase** (often abbreviated to **rubisco**) catalyses the addition of carbon dioxide to form a very short-lived compound with six carbon atoms. It decays rapidly (within a tiny fraction of a second)

■ See question 3.

Fig 22.19 **The key chemical reaction of the dark reactions: carbon dioxide joins with a 5-carbon compound. This is the *main carbon-fixing reaction*. Each circle represents a group of atoms containing carbon. Rubisco, the enzyme for this reaction is rather slow and so there is a lot of it in the stroma of the chloroplast. Rubisco makes up about 50 per cent of all the protein in a chloroplast**

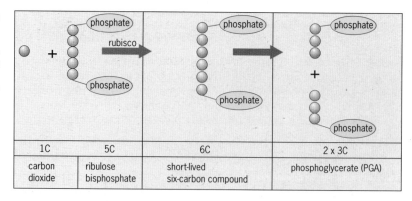

1C	5C	6C	2 x 3C
carbon dioxide	ribulose bisphosphate	short-lived six-carbon compound	phosphoglycerate (PGA)

to two 3-carbon compounds. The following summarises the carbon fixation reactions:

$$\text{ribulose bisphosphate} + CO_2 \rightarrow 2(\text{glycerate 3-phosphate})$$

At this point, new carbon molecules enter a cyclical series of chemical reactions called the **Calvin cycle** (Fig 22.20).

Fig 22.20 **The Calvin cycle: in each 'turn' of this cycle another carbon is added. It shows (from the top) that 3 atoms of carbon in carbon dioxide are fixed using 6 molecules of ATP and 6 hydrogens from NADPH, and that (at the bottom) one molecule of glycerate 3-phosphate is produced**

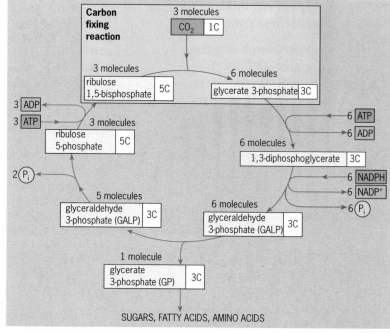

SUGARS, FATTY ACIDS, AMINO ACIDS

E Use Fig 22.20 to complete the following summary reaction for photosynthesis:

$3CO_2 + 9ATP + 6NADPH + \text{water} \rightarrow$
____ + ____ + $8P_i$ + ____

F What is the first sugar phosphate produced as a result of photosynthesis?

leaf mesophyll cell

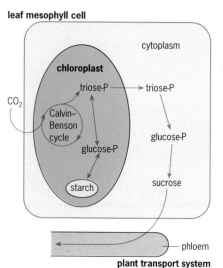

plant transport system

The **glycerate 3-phosphate** (GP) formed in the first carbon-fixing reaction is not a sugar. It needs energy and hydrogen to convert it to a 3-carbon sugar phosphate, **glyceraldehyde 3-phosphate** (GALP). Some of the GALP is converted to other sugars and some used to synthesise more ribulose bisphosphate. For every turn of the cycle, 3 molecules of ATP and 2 molecules of reduced NADP are needed to fix each CO_2 molecule into glucose.

The export of sugar

Most of the GALP is exported from the stroma of the chloroplast to the cytoplasm of the cell (Fig 22.21). Here it is converted to glucose and then other sugars, fatty acids or amino acids. Once in the cytoplasm, sugars are metabolised or exported to other parts of the plant.

Fig 22.21 **A summary of the main routes taken by the 3-carbon sugars formed in the light-independent reaction of photosynthesis. Phosphorylated 3-carbon sugars (eg GALP) leave the chloroplast. In the cytoplasm they are converted into glucose and other complex molecules**

8 WHAT LIMITS THE RATE OF PHOTOSYNTHESIS?

There are three main **limiting factors** to photosynthesis:

- light intensity
- concentration of carbon dioxide
- temperature

Light intensity

Photosynthesis cannot take place in the dark. As light intensity increases, the number of photons available to be caught by the antennae increases. The more light, the more ATP and reduced NADP is produced. But, as Fig 22.22 shows, increasing the light intensity does not increase the rate of photosynthesis indefinitely. At very high light intensities, other factors limit photosynthesis.

Carbon dioxide concentration

Photosynthesis needs only two raw materials: carbon dioxide and water. There is plenty of water in any living cell, but carbon dioxide has to diffuse in from outside the plant. The carbon dioxide concentration of normal air is low, at 0.03 per cent, and there are several barriers on the route to the chloroplast. The availability of carbon dioxide can easily limit the rate of photosynthesis.

Temperature

When neither light nor carbon dioxide is limiting, the rate of photosynthesis is limited only by the rate at which enzyme reactions can take place. The initial photochemical processes are not affected much by changes in temperature but the enzyme reactions of the Calvin cycle are temperature dependent.

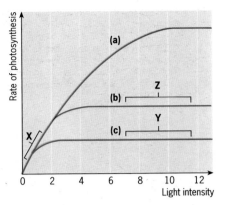

Fig 22.22 **The effect of light intensity on the rate of photosynthesis:**

(a) **the effect of light intensity at 25 °C and 0.4 per cent carbon dioxide,**

(b) **the effect of light intensity at 15 °C and 0.4 per cent carbon dioxide,**

(c) **the effect of light intensity at 25 °C and 0.01 per cent carbon dioxide**

?

G In Fig 22.22, what is the main factor limiting photosynthesis in each of the zones marked X, Y and Z?

■ See question 4.

PART III RESPIRATION

9 RESPIRATION: A VITAL LIFE PROCESS

Humans like all living organisms obtain the energy they need from **respiration**. This chemical process breaks down the simple food molecules such as glucose that are produced by digestion, absorbed into the blood and then transported round the body.

We mainly respire **aerobically** (using oxygen). Oxygen is absorbed by lungs is distributed around the body by the **circulatory system**. Only when food molecules and oxygen are together inside cells can the complex process of aerobic cell respiration begin.

Some important definitions

The terms respiration and breathing have different meanings. We use the following definitions.

Breathing is the mechanical process that supplies oxygen to the body to drive respiration, and that removes the carbon dioxide produced.

Respiration is the complex series of reactions that occurs in all living cells, which releases the energy in food and makes it available to the organism.

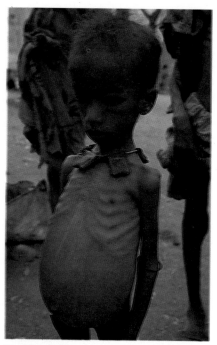

Fig 22.23 **Children who are very short of food for long periods of time often have swollen abdomens. They are suffering from kwashiorkor. Their bodies swell because fluid builds up in the tissues instead of draining back into the blood, a problem caused by low levels of protein in the blood**

Organisms that carry out aerobic respiration need to take in oxygen, and they produce carbon dioxide, a waste product that needs to be removed. Larger organisms have increased their capacity for gas exchange by developing specialised organs that have large surface areas (see Chapter 8).

Most organisms respire aerobically. Aerobic respiration releases a relatively large amount of energy and gives organisms a survival advantage in conditions where oxygen is plentiful.

Aerobic respiration is respiration that requires oxygen. Anaerobic respiration is respiration without oxygen.

Some organisms, mainly bacteria, can only respire anaerobically. A few more, yeast for example, can turn to anaerobic respiration when there is no oxygen. Some human tissues (eg muscle during strenuous exercise) can respire anaerobically.

10 THE BIOCHEMISTRY OF AEROBIC RESPIRATION

Aerobic respiration describes the cell process that requires oxygen to release energy from all types of food molecules. Humans respire mainly sugars with a small percentage of amino acids and fatty acids but, when the need arises, such as during starvation, the balance can change (Fig 22.23).

To make a very complex topic easier to study, we shall concentrate first on the aerobic respiration of only one type of food molecule: glucose. The aerobic respiration of other food molecules, and anaerobic respiration, are dealt with later (see pages 338 and 341).

Redox reactions

Redox is short for **red**uction and **ox**idation, two chemical processes that often occur together in the same reaction.

Oxidation reactions may involve the addition of oxygen or the removal of hydrogen from molecules, but the important underlying rule is that a molecule that is oxidised *loses* electrons.

Conversely, reduction reactions involve a *gain* of electrons. As electrons carry energy, a molecule that has been reduced, and that therefore has gained one or more electrons, will usually carry more energy than the oxidised form of the same molecule.

Reduction and oxidation reactions usually occur together, because if one substance loses electrons another must gain them (Fig 22.24).

The concept of redox reactions is a vital one in biology. The molecules that make up living organisms are produced, directly or indirectly, by photo-synthesis. This process reduces carbon dioxide to form organic molecules (such as sugars) that contain energy. These can later be oxidised in the process of cell respiration to release energy. Lipids contain more hydrogen than carbohydrates: they have more C—H bonds than C—O bonds. For this

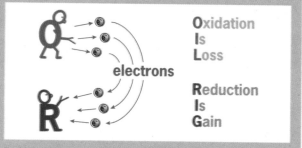

Fig 22.24 **Remember – OIL RIG: oxidation is loss, reduction is gain**

reason, lipids are said to be more highly reduced chemicals than carbohydrates, and more energy can be released by the oxidation of their many C—H bonds. Because they contain more energy per gram than other substances, lipids are ideal storage compounds (see Chapter 3).

Respiration is the release of energy from food molecules by progressive oxidation.

In respiration, glucose is oxidised to form carbon dioxide, and oxygen is reduced to form water. Most of the ATP created during this process comes from the final stage: a series of redox reactions that occur on mitochondrial membranes.

For any chemical reaction to take place, energy is required to break existing molecular bonds. The process of forming new bonds can either require or release energy. For there to be a *net release of free energy* (the whole point of respiration), the products of respiration must be at a lower energy level than the reactants.

The basic equation we use to sum up respiration is:

$$C_6H_{12}O_6 + 6O_2 \rightarrow 6CO_2 + 6H_2O + ENERGY$$

The total energy contained in the reactants (the 6 molecules of glucose and the 6 molecules of oxygen on the left) is greater than the total energy contained in the products (the 6 molecules of carbon dioxide and the 6 molecules of water on the right). The difference is the energy released by respiration.

In reality, glucose does not react directly with oxygen. Respiration is a sequence of many different reactions. Together, these can produce up to 36 molecules of a chemical called ATP per molecule of glucose:

$$C_6H_{12}O_6 + 6O_2 \rightarrow 6CO_2 + 6H_2O + 36ATP$$

The large amount of energy available from this number of ATP molecules can do useful work in the cell. The steps involved in the production of ATP depend heavily on **redox reactions**. These are reactions that involve oxidation of one reactant and reduction of another (see the box on redox reactions on the opposite page).

Respiration or combustion?

It is often said that we 'burn up' food in respiration but this is not really true. The energy in glucose *could* be released by setting fire to it (Fig 22.25). The glucose would be oxidised to carbon dioxide and water, but the reaction would be rapid and uncontrolled.

During the process of cell respiration, glucose and other organic molecules are broken down in small stages, some of which release energy. This step-by-step breakdown (Fig 22.26) is more gradual than combustion.

Fig 22.25 **The energy in food can be released by combustion, but almost all of it is lost as heat, raising the temperature of the surroundings to levels that could not be tolerated inside any living cell**

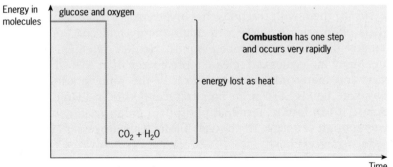

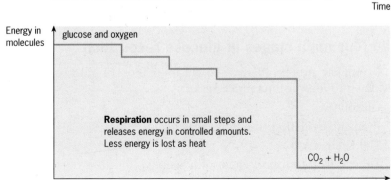

The entire process of respiration is controlled by enzymes that transfer energy in food molecules to ATP, a substance that is able to power cell processes by supplying on-the-spot, instant and usable energy in controlled amounts.

Fig 22.26 **The difference between combustion and respiration. Although the difference in energy between the reactants (glucose and oxygen) and the products (carbon dioxide and water) is the same in both cases, the way in which the energy is released is totally different**

All about ATP

ATP stands for **adenosine triphosphate**. ATP is a relatively small, soluble organic molecule that consists of a base (adenine), a sugar (ribose) and three inorganic phosphate (P_i) groups (Fig 22.27). ATP has a high **free energy of hydrolysis**, which means that when ATP is hydrolysed, a relatively large amount of energy is released. In hydrolysis, ATP reacts with water, producing ADP and inorganic phosphate (P_i).

Fig 22.27 **The hydrolysis ('water-splitting') of ATP. Under the control of the enzyme ATPase, the terminal phosphate group of ATP is removed and combines with water. This reaction releases free energy which can be used to drive energy-requiring reactions such as muscular contraction**

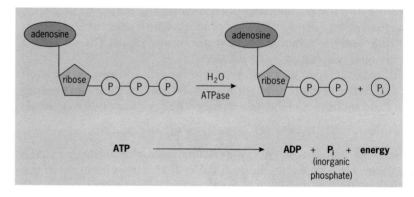

Because of its solubility and small size, ATP can be transported rapidly around cells and so can supply energy where it is needed (Fig 22.28). Metabolically active cells, such as those in muscle and in secretory tissue, transport and break down many ATP molecules.

How does ATP release energy?

When ATP loses a phosphate group to become ADP, adenosine *di*phosphate, the reaction *releases* energy. This can be used to do useful work in the cell. The same amount of energy is released when ADP loses another phosphate group to become AMP (adenosine *mono*phosphate). Less energy is released when the last phosphate group is lost.

Many of the reactions in cells *require* energy. This energy is supplied by *coupling* them with reactions that involve ATP breakdown.

It is often said that ATP is a 'high energy molecule' or that is contains 'high energy phosphate bonds'. Neither statement is really true. Many molecules can release more energy than ATP when they react, and many others contain exactly the same phosphate-to-phosphate bonds. ATP is an important molecule in living systems because it can lose its terminal phosphate group *readily*, releasing just enough energy to power biological processes without producing excess heat.

$$ATP \rightarrow ADP + P_i + energy \ (30.6 \ kJ \ mol^{-1})$$

The four main stages in glucose respiration

The complete process of the aerobic respiration of glucose can be divided into four distinct processes:

- Glycolysis
- Pyruvate oxidation (the 'link' reaction)
- The Krebs cycle
- The electron transport chain of reactions

Fig 22.29 shows how the four processes are connected.

Fig 22.28 **For explosive bursts of effort, such as a short sprint, most of the energy comes from ATP already present in the muscles. During strenuous exercise, when the muscles quickly use up all their ATP, energy is supplied by a back-up chemical called creatine phosphate (CP). Since CP has a higher free energy of hydrolysis than ATP, its breakdown releases enough energy to make more ATP instantly and so allows exercise to continue (see Chapter 13)**

Glycolysis

In this first stage a glucose molecule is converted, by a series of enzyme-controlled reactions, into 2 molecules of **pyruvate**, a 3-carbon compound. The process yields relatively little energy, only producing 2 molecules of ATP). But this stage does not require oxygen and it takes relatively little time to complete, so it can provide immediate energy.

Glycolysis takes place in the cytosol, the fluid part of the cytoplasm. Overall, per glucose molecule, glycolysis produces:

● 2 molecules of ATP. Four are produced but two are used up in glycolysis

● 2 molecules of reduced NAD. This reduced coenzyme later feeds electrons into the electron transport chain

● 2 molecules of pyruvate, which enters the 'link' reaction if oxygen is available

Pyruvate oxidation: the 'link' reaction

The pyruvate oxidation reaction links glycolysis with the Krebs cycle. It is often thought of as part of the Krebs cycle, but we consider it as a separate reaction.

In the presence of oxygen, pyruvate (the 3C molecule produced by glycolysis) moves from the cytosol to the mitochondrial matrix where it is oxidised to **acetate** (a 2C molecule), producing carbon dioxide as a by-product. This reaction also produces 2 molecules of reduced NAD. The acetate is picked up by a carrier molecule, **coenzyme A**, and **acetyl coenzyme A** is formed.

Overall, from one molecule of glucose, the reaction is:

$$2 \text{ pyruvate} + 2\text{NAD}^+ + 2\text{H}_2\text{O} \rightarrow 2 \text{ acetate} + 2\text{NADH} + 2\text{H}^+ + 2\text{H}_2\text{O}$$

The Krebs cycle

The **Krebs cycle** is a series of reactions named after Sir Hans Krebs, the biochemist who first worked out the details. The main purpose of the Krebs cycle is to provide a continuous supply of electrons to feed into the electron transport chain.

The details of the Krebs cycle, which takes place in the matrix of the mitochondria, can be seen in Fig 22.30 on the next page. Overall, per turn, the Krebs cycle produces:

● 3 molecules of reduced NAD (NADH)

● 1 molecule of reduced (FADH$_2$)

● 1 molecule of ATP (by substrate level phosphorylation)

● 2 molecules of arbon dioxide

● 1 molecule of oxaloacetate (to allow the cycle to continue)

The Krebs cycle 'turns' twice for every glucose molecule, so, per glucose, 6 molecules of NADH and 2 molecules of FADH$_2$ are produced. NADH and FADH$_2$ are particularly important because they carry the electrons that power the next stage of glucose respiration.

The electron transport chain

This final phase of respiration produces the largest number of ATP molecules. The electron transport chain can be thought of as the 'pay day' for all the 'hard work' that has been done in the earlier parts of the process.

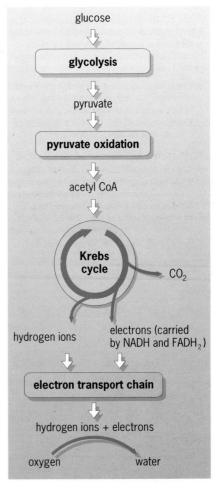

Fig 22.29 **The overall process of cell respiration, showing the order of the four main stages**

H The function of a kidney tubule cell is to move substances by active transport. Suggest why these cells are packed with mitochondria?

I To move our muscles, we need energy to be available 'on demand'. Why do we use ATP as an immediate energy source, instead of glucose?

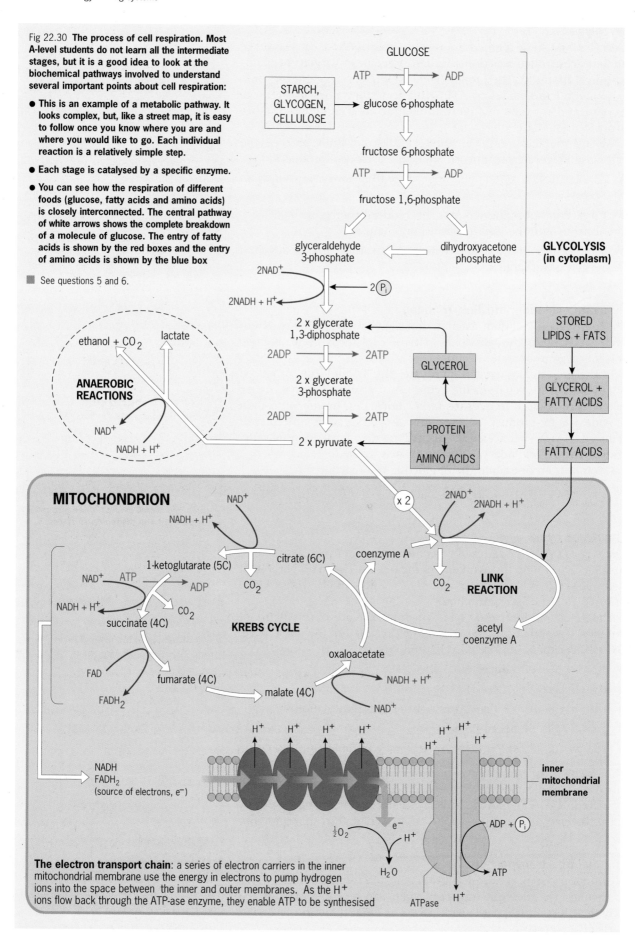

Fig 22.30 **The process of cell respiration.** Most A-level students do not learn all the intermediate stages, but it is a good idea to look at the biochemical pathways involved to understand several important points about cell respiration:

● This is an example of a metabolic pathway. It looks complex, but, like a street map, it is easy to follow once you know where you are and where you would like to go. Each individual reaction is a relatively simple step.

● Each stage is catalysed by a specific enzyme.

● You can see how the respiration of different foods (glucose, fatty acids and amino acids) is closely interconnected. The central pathway of white arrows shows the complete breakdown of a molecule of glucose. The entry of fatty acids is shown by the red boxes and the entry of amino acids is shown by the blue box

■ See questions 5 and 6.

The electron transport chain consists of a series of carrier molecules (Fig 22.31). These proteins first accept an electron (thereby becoming reduced) and then lose it again (so becoming oxidised). At each of these transfers the electrons lose some energy that can be used to power the active transport of hydrogen ions across the inner mitochondrial membrane. This results in a high concentration of hydrogen ions in the outer mitochondrial space and a low concentration in the inner mitochondrial space.

?

J Muscle cells and cells that manufacture and secrete large amounts of hormones have more mitochondria than, say, a skin cell. Suggest why.

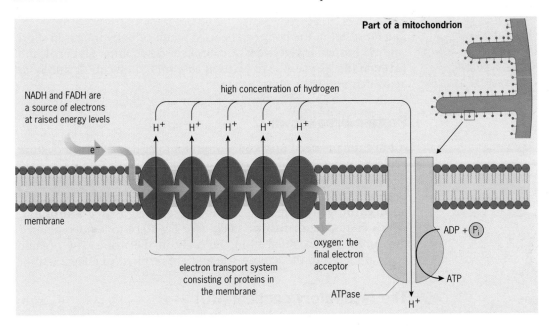

Fig 22.31
The electrons and hydrogen ions made during the first stages of respiration are finally used to synthesise ATP

Because of the difference in concentration, hydrogen ions leak back into the inner compartment. Their only route is through the middle of the stalked granules – the ATPase enzymes. As the stream of hydrogen ions flows down the concentration gradient, enough energy is released to form new molecules of ATP.

The end-product of the electron transport chain is spare electrons and hydrogen ions that combine with oxygen to form water. Although this reaction is at the end of the process, it is a key one: it is why most organisms need oxygen. If there is no oxygen to mop up the electrons and hydrogen ions from the electron transport chain, the pathway cannot be completed. The intermediate compounds of glucose respiration build up and the cell loses its ability to make most of the ATP it needs.

✔

This model of ATP synthesis by oxidative phosphorylation is called the **chemiosmotic hypothesis** and was first proposed by Peter Mitchell in 1961. Mitchell received the Nobel Prize for Chemistry in 1978.

NAD AND FAD

NAD (**nicotinamide adenine dinucleotide**) and FAD (**flavine adenine dinucleotide**) are **coenzymes**, organic compounds that catalyse reactions. As their name implies, coenzymes work closely with enzymes. Unlike enzymes, coenzymes are not proteins. NAD and FAD carry electrons from the **electron donors** in glycolysis, pyruvate oxidation and the Krebs cycle, to **electron acceptors** in the electron transport chain that takes place in the mitochondrial membrane.

When NAD and FAD accept electrons, they are reduced, becoming NADH and $FADH_2$. In this form

they are called **reduced coenzymes**. Reduction of a coenzyme always requires association with a **dehydrogenase enzyme**, an enzyme that removes hydrogen from other molecules. The hydrogen removed by the dehydrogenase enzyme is split into an electron and a hydrogen ion. NAD (or FAD) accepts the electrons produced and the hydrogen ions play an important part in the electron transport chain.

In the human body, NAD is synthesised from vitamin B_3 (nicotinic acid) and FAD is made from vitamin B_2 (riboflavin): this explains why both vitamins are essential components of our diet.

11 OTHER FUELS IN RESPIRATION

So far, we have studied the biochemical pathway of respiration which uses glucose as a **respiratory substrate**. But many organic molecules can be fed in the same central pathway to produce ATP. Look back at Fig 22.30 before you read on.

Lipid breakdown

The fat stores in the body can be used when carbohydrate is in short supply. Stored triglycerides are broken down into **glycerol** and **fatty acids**. Glycerol can be used as a fuel in glycolysis and fatty acids can enter the Krebs cycle.

✔

Lipids can fuel aerobic exercise but not glycolysis. Aerobic exercise is therefore a good way to burn excess fat (see Assignment on page 346).

Protein breakdown

Adult humans need up to 60 g of protein per day. We cannot store protein, so excess **amino acids**, the building blocks of proteins, are degraded in the liver by the process of **deamination**. Enzymes in the liver separate the amine group from the rest of the molecule, leaving an **organic acid**. The amine group is released as free **ammonia**, which enters the **ornithine cycle** (see Fig 10.6) to be incorporated into **urea** and excreted from the body in the urine. The organic acids are fed into the Krebs cycle and are respired (see Fig 22.30).

The respiratory quotient (RQ)

The respiratory quotient, or RQ, is a ratio of gas exchange, worked out by comparing carbon dioxide production with oxygen uptake. The RQ can be used to determine the type of food being respired. The basic equation for the respiration of glucose is:

$$C_6H_{12}O_6 + 6O_2 \rightarrow 6CO_2 + 6H_2O + \text{energy}$$

shows that, for every glucose molecule respired, 6 molecules of oxygen are required and 6 molecules of CO_2 are produced. Equal numbers of gas molecules occupy the same volume and so the volume of oxygen taken up by an organism is equal to the volume of CO_2 produced.

The RQ is worked out using the simple formula:

$$RQ = \frac{\text{volume of } CO_2 \text{ given out}}{\text{volume of } O_2 \text{ taken in}}$$

For the glucose equation, the RQ is: $\frac{6}{6} = 1$.

Using the respiratory quotient

If we measure the volumes of oxygen and carbon dioxide that are exchanged by an organism and find that the RQ is 1, we can infer that the organism is respiring mainly carbohydrate. Different substrates produce different RQ values:

- Carbohydrate respiration gives an RQ of 1.
- Protein respiration gives an RQ of 0.8–0.9.
- Lipid respiration gives an RQ of 0.7.
- Anaerobic fermentation gives an RQ of infinity.

?

K (a) Why is there a range of values for the respiratory quotient of proteins?

(b) Why is the respiratory quotient for fermentation infinity?

The problem with trying to work out what substrate is being respired by looking at RQ values is that many organisms, including humans, respire more than one substrate. In humans, carbohydrate is the main fuel for respiration, but small amounts of fat and protein are also respired. The RQ figure obtained is therefore variable.

Using a respirometer

The rate of aerobic respiration can be estimated by measuring either how much oxygen is taken up or how much heat is produced. Gas exchange, including oxygen uptake, is measured using a sealed chamber called a **respirometer**.

A simple respirometer

A simple respirometer is shown in Fig 22.32. Organisms placed inside exchange gases and so alter the composition of the air around them.

To measure gas exchange due to a respiring organism, we could remove the air after a fixed time and analyse it. A simpler method is to place a carbon dioxide absorber such as sodium hydroxide in the respirometer chamber. Over a set time, a volume of oxygen is used up. Carbon dioxide is produced but is absorbed by the sodium hydroxide. So, overall, the volume of gas falls by the volume of oxygen used. The gas pressure inside the chamber falls and coloured liquid is drawn along the tube towards the chamber. The volume of oxygen used up equals the volume of the liquid moved in the fixed time. The Example below shows how the respiration rate is calculated from readings taken from the tube.

Fig 22.32 **A simple respirometer**

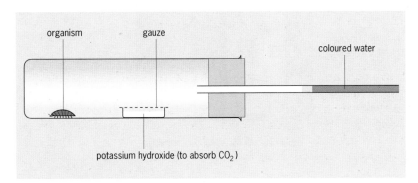

organism gauze

coloured water

potassium hydroxide (to absorb CO_2)

EXAMPLE

Q

a) When measuring the gas exchange of an organism with a simple respirometer, how do you calculate the volume of oxygen used?

b) In what units do you express the rate of respiration?

A

a) The position of the liquid along the tube is recorded at the start and end of the set time and the distance the liquid moves is calculated. The volume of liquid (equal to the change in oxygen volume) is a cylinder inside the tube.

Therefore:

volume of oxygen used up (in cm³) $= \pi \times r^2 \times h$

where r = radius or diameter/2 (in cm) and h = distance moved by the liquid (in cm).

b) The rate of respiration is given as the volume of oxygen used per organism (or per unit mass, eg per gram) per unit time; for example, as cubic centimetres of oxygen per gram per hour. So:

rate of respiration = x cm³ O_2 g^{-1} h^{-1}

A more complex respirometer

The simple respirometer shows the principles of respirometry, but has these limitations:

- Changes in temperature or atmospheric pressure make gases expand or contract. These changes affect the distance the fluid moves, so the measured distance is not just due to the change in oxygen volume.

- Sodium hydroxide might alter the composition of the gases in the chamber.

- It is difficult to restart the experiment without taking the apparatus apart.

- The volume change calculated using the diameter of the tube might be inaccurate.

To avoid these problems, a more sophisticated respirometer has been developed. This is shown in Fig 22.33. This apparatus works on the same principle as the simple respirometer, but the syringe allows the volume change to be measured directly. After a set time, the syringe can be pulled up until the coloured water goes back to its starting point. This way, you can find the exact volume of oxygen used without having to work out the volume of a cylinder. It also allows the experimenter to restart the experiment at the same time.

Fig 22.33 **A more complex respirometer. The left tube contains the organisms while the right tube acts as a control. In this way, any changes in fluid level which are not due to the organisms, such as changes in room temperature and pressure, are balanced out by the control tube. The experiment can be set up again using the syringe to level up the coloured water, which will also accurately measure the changes in volume due to the respiring organisms**

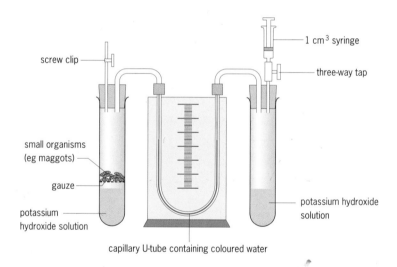

13 RESPIRATION, ENZYMES AND TEMPERATURE

Like all metabolic reactions, respiration is controlled by enzymes. The rate of an enzyme-catalysed reaction will approximately double with every 10 °C rise in temperature between 4 and 40 °C. So, with a temperature increase of 10 °C, much more ATP becomes available to power other metabolic processes.

A dramatic example is seen in cold-blooded or **poikilothermic** animals such as reptiles, which can experience wide variations in body temperature during a 24-hour period (Fig 22.34). Their rate of respiration, and therefore their metabolic rate, is roughly twice as fast at 30 °C than it is at 20 °C. Such organisms are only active when their body temperature allows them to react and move quickly.

Fig 22.34 **Reptiles are** poikilotherms: **they cannot control their body temperature in the way that mammals and birds do. This lizard must absorb sunlight to raise its body temperature to the level that allows its enzymes to function at their optimum efficiency**

14 ANAEROBIC RESPIRATION

Anaerobic respiration does not require oxygen, so it can take place in oxygen-poor environments such as stagnant water or deep soil. It can also occur in parts of an aerobic organism, such as ourselves, that are starved of oxygen,. For example, anaerobic respiration occurs in the muscles during strenuous exercise.

In anaerobic respiration, glucose is broken down into pyruvate but, because no oxygen is available, the pyruvate cannot be broken down any further. Instead of entering the link reaction and the Krebs cycle, the pyruvate is converted to a waste product. The nature of the waste product depends on the organism. Fig 22.35 shows the waste products of anaerobic respiration in different organisms.

Generally speaking, animals and some bacteria convert pyruvate into **lactate** by a simple reduction reaction. In contrast, plants and fungi such as yeast convert pyruvate to **ethanol** (alcohol) and carbon dioxide.

In humans, vigorous exercise leads to anaerobic respiration in the muscles and the resulting build-up of lactate causes fatigue (see Chapter 13.

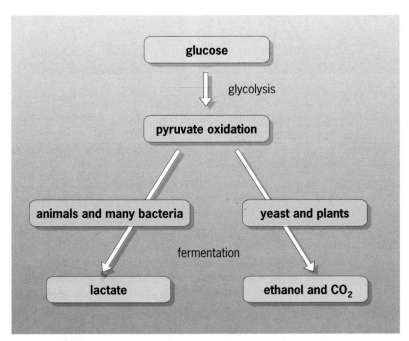

Fig 22.35 **The main problem with anaerobic respiration is the build-up of NADH and the shortage of NAD⁺. To solve this, H⁺ is added to pyruvate to make lactate or ethanol and carbon dioxide.** Fermentation **is glycolysis plus the reduction of pyruvate to either lactate or ethanol and carbon dioxide**

?

L What is the difference between fermentation and glycolysis?

Anaerobic respiration in microorganisms is generally known as **fermentation**. As Fig 22.36 shows, this process can give rise to many products that are useful to humans. Bacteria that produce lactic acid are used in the manufacture of dairy products such as yoghurt. The tangy taste is due to the high concentration of this organic acid. In addition to lactic acid and ethanol, other microorganisms can make solvents such as propanone (acetone) and butanol.

Alcoholic fermentation

Anaerobic respiration in yeast (Fig 22.37) is also known as **alcoholic fermentation**. This process was discovered by accident

Fig 22.36 **Fermentation in different microorganisms can be used to make a variety of useful products**

Left: A vat of curds and whey in cheese making

Below: During fermentation, yeast produces carbon dioxide which makes bread dough rise

Above: Making red wine: yeast produces alcohol from the sugar in grapes

Right: Alcohol made from the fermentation of sugar cane is used as fuel in countries of South America

Fig 22.37 **Yeast is a single-celled fungus that usually respires aerobically. When deprived of oxygen, it switches to anaerobic respiration: sugars are not completely broken down and the by-products are ethanol (alcohol) and carbon dioxide**

and alcoholic drinks such as wine and beer were made for many centuries before the science behind the process was known. For fermentation, yeast simply needs a source of carbohydrate, anaerobic conditions and a suitable temperature. So wine can be made from many organic materials, such as peaches, elderberries or even potato peelings, as well as grapes (Fig 22.38).

Fermentation stops when the yeast becomes poisoned by its own waste, the alcohol. This is why most alcoholic drinks made by fermentation only are no stronger than about 14 per cent ABV (alcohol by volume). Spirits such as whisky and gin, which have a higher alcohol content, are **distilled** from a fermented mixture.

Since alcohol is an intermediate product in the breakdown of sugar, it contains energy that can be released by respiration. Many drinks, particularly beer, are also rich in carbohydrates that have not been turned to alcohol. Heavy drinkers get fat because they drink large amounts as well as eating meals that, by themselves, provide enough calories to satisfy the body's energy demands.

Fig 22.38 **Yeast occurs naturally on the surface of many fruits. For instance, yeast is partly responsible for the bloom on a grape. Fruit stored in anaerobic conditions will therefore ferment and produce liquids with 'interesting' effects!**

SUMMARY

By the end of this chapter, you should know and understand the following:

■ Energy transfer through organisms in an ecosystem is one-way.

■ Green plants are **autotrophs**, organisms which can make sugars from carbon dioxide using energy. Organisms which cannot make their own food are called **heterotrophs**. Not all autotrophs use light as a source of energy, water as a source of hydrogen or carbon dioxide as a source of carbon.

■ In a **food chain**, energy is passed from producers to primary consumers to secondary and tertiary consumers. Each level of the food chain is called a trophic level. In most ecosystems, many food chains are linked together in a food web.

■ There are three main types of **ecological pyramid**: pyramids of numbers, pyramids of biomass and pyramids of energy. Pyramids of energy show that over 90 per cent of the energy transferred to the next trophic level is lost as heat.

■ **Photosynthesis** occurs in the green tissues of a plant. Leaves have become specially adapted and trap energy from sunlight efficiently.

■ Inside the plant cell, the reactions of photosynthesis occur in the chloroplast. Most of the enzymes of the light-dependent reactions are embedded in the thylakoid membrane. The light-independent reactions take place in the stroma.

■ In the **light-dependent reactions**, energy from sunlight is trapped by chlorophyll molecules. This energy is used to boost electrons to a higher energy level. The extra energy in these electrons drives a series of redox reactions that produce ATP and reduced NADP.

■ The **light-independent reactions** do not need light. The enzyme rubisco catalyses the fixation of carbon in carbon dioxide by ribulose bisphosphate to produce a 3-carbon compound, which is converted into a 3-carbon sugar using ATP and reduced NADP.

■ The sugars produced by photosynthesis can be converted into other sugars.

■ Photosynthesis is affected by various factors, but particularly three limiting factors, light intensity, carbon dioxide concentration and temperature.

■ **Respiration** is the release of energy from food. The energy released is transferred to ATP, a chemical that can provide the cell with energy in small, instant, controllable amounts.

■ Respiration consists of many reactions. Each reaction is controlled by a different enzyme. Some of the reactions occur in the cytoplasm and some occur inside the mitochondrion.

■ Complete aerobic respiration of glucose has four stages: glycolysis, pyruvate oxidation, the reactions of the Krebs cycle and the reactions of the electron transport chain.

■ **Glycolysis**: Glucose is broken down into 2 molecules of pyruvate. There is a net gain of 2 molecules of ATP. This happens in the cytoplasm of cells.

■ **Pyruvate oxidation** (the 'link' reaction): In the presence of oxygen, pyruvate enters the mitochondrion and is converted into acetyl coenzyme A.

■ The **Krebs cycle**: A series of reactions, fuelled by acetyl coenzyme A, which produces 2 molecules of ATP and many electrons and hydrogen ions.

■ The **electron transport chain**: A series of redox reactions, fuelled by the electrons and protons from the preceding three processes, as supplied by the coenzymes NAD and FAD. Electron transport releases enough energy to synthesise most of the ATP produced by aerobic respiration.

■ Most food chemicals – carbohydrates, lipids and proteins – can be respired to provide energy.

■ The rate of respiration can be measured as a **respiratory quotient**, the ratio of carbon dioxide evolved to oxygen absorbed. It can be calculated from measurements made using a respirometer.

■ Most organisms respire **aerobically** (in the presence of oxygen). The substrate (eg glucose) is broken down completely to produce carbon dioxide, water and a lot of ATP.

■ Some organisms and some tissues respire **anaerobically**. In the absence of oxygen, substrates are only partially broken down, leading to waste products such as lactate (in animals and bacteria) or ethanol and carbon dioxide (in plants and yeast).

QUESTIONS

1 In a study of an oak tree the following numbers of organisms were obtained at each trophic level.

Trophic level	Number of organisms
producer	1
primary consumer	260 000
secondary consumer	40
tertiary consumer	3

a) Sketch a pyramid of biomass to represent this food chain.

b) Suggest **two** reasons why there is such a large difference in the numbers of primary and secondary consumers.

c) The Venus flytrap is a green plant that catches and digests insects. It lives in wet, boggy places where mineral ions are scarce because they are washed out of the soil.
    (i) At which trophic level or levels would you put the Venus flytrap? Give a reason for your answer.
    (ii) Describe the difference between the way in which the Venus flytrap obtains its nitrogen and the way in which other plants normally obtain this element.

[NEAB June 1995 Biology: Module Test, Section A, q.5]

2 The graph of Fig 22.Q2 shows the relation between the productivity of phytoplankton (producers) and zooplankton (primary consumers) in a large lake.

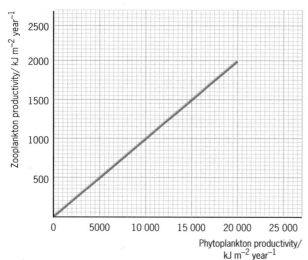

Fig 22.Q2

a) Use the graph to calculate the proportion of energy in phytoplankton which goes directly to the zooplankton. Show how you arrive at your answer.

b) Give **two** different fates of the energy that does not go directly to the zooplankton from the phytoplankton.

[AEB June 1994, Biology: Paper 1, q.14]

3 Read through the following account of photosynthesis, then write it out and add at the dashed lines the most appropriate word or words to complete the account.

Photosynthesis is a type of __ nutrition, involving the synthesis of organic molecules from inorganic materials. The process involves two types of reactions, light-dependent and light-independent.

In the light-dependent reactions, light energy is absorbed by chlorophyll molecules located on the __ of the chloroplasts; __ and __ are produced and oxygen gas is given off as a by-product.

In the light-independent reactions, __ accepts molecules of carbon dioxide, which together with the products of the light-dependent reactions, results in the formation of __ . This compound can be converted to a hexose sugar or used to regenerate the carbon dioxide acceptor molecule.

[ULEAC 1996 Biology: Specimen Paper 2, q.1]

4 An investigation was carried out into the effect of carbon dioxide concentration and light intensity on the productivity of lettuces in a glasshouse. The productivity was determined by measuring the rate of carbon dioxide fixation in milligrams per dm^2 leaf area per hour

Experiments were carried out at three different light intenities, 0.05, 0.25 and 0.45 (arbitary units), the highest approximating to full sunlight. A constant temperature of 22°C was maintained throughout.

The results are given in the table that follows.

Carbon dioxide concentration/ppm	Productivity at different light intensities/mg $dm^{-2} h^{-1}$		
	At 0.05 units light intensity	At 0.25 units light intensity	At 0.45 units light intensity
300	12	25	27
500	14	30	36
700	15	35	42
900	15	37	46
1100	15	37	47
1300	12	31	46

a) For the experiment at 0.25 units light intensity, described and comment on the effect on the productivity of the lettuces of increasing carbon dioxide concentration in the ranges **(i)** 300 to 900 ppm, and **(ii)** 900 to 1300 ppm.

b) Explain why the carbon dioxide concentration affects the productivity of plants.

c) State why the temperature should be kept constant during this experiment.

d) Suggest why, even with artificial lighting, glasshouse crops generally need to have more carbon dioxide added when temperatures are low, than when temperatures are high.

[ULEAC Jan 1994 Biology: Specimen Synoptic Paper, q.5]

5 The diagram below shows some of the stages in cell respiration.

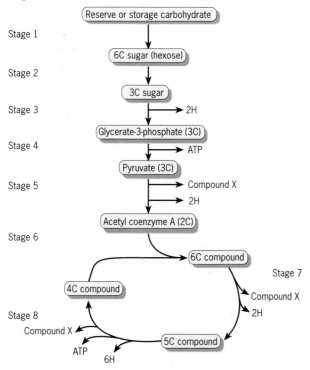

Stage 1

Reserve or storage carbohydrate

6C sugar (hexose)

Stage 2

3C sugar

Stage 3 → 2H

Glycerate-3-phosphate (3C)

Stage 4 → ATP

Pyruvate (3C)

Stage 5 → Compound X
→ 2H

Acetyl coenzyme A (2C)

Stage 6

6C compound

Stage 7
→ Compound X
→ 2H

4C compound

Stage 8
Compound X ←
ATP
6H

5C compound

a) What respiratory substrate would be used in a liver cell?

b) State in which part of a cell Stage 6 occurs.

c) Identify compound X, removed at stages 5, 7 and 8.

d) Describe what happens to the hydrogen atoms removed at stages 3, 5, 7 and 8.

[ULEAC 1996 Biology/Human Biology: Specimen Paper, q.3]

6 The diagram summarises the process of cellular respiration.

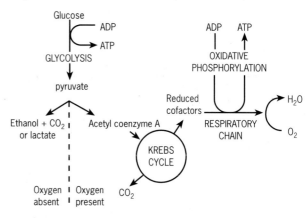

Glucose
ADP
ATP
GLYCOLYSIS

pyruvate

Ethanol + CO_2 or lactate

Acetyl coenzyme A

ADP   ATP
OXIDATIVE
PHOSPHORYLATION

Reduced cofactors

RESPIRATORY CHAIN

H_2O
O_2

KREBS CYCLE

CO_2

Oxygen absent | Oxygen present

a) **(i)** Why is glycolysis sometimes referred to as '*the common pathway*'?
(ii) Where in the cell does glycolysis take place?

b) What is the significance of the Krebs cycle being a cyclical process?

c) **(i)** What is the importance of oxidative phosphorylation?
(ii) Where exactly in the cell does it take place?

[AEB 1996 Biology: Specimen Paper, q.3]

Assignment

ARE PLANTS GOOD AT PHOTOSYNTHESIS?

One way to measure photosynthetic efficiency is to look at the proportion of light that plants can convert into stored energy in food.

Let us start with the energy in a glucose molecule. We can base our calculations on a mole of glucose molecules. A mole is the amount of a substance equal to its relative molecular mass in grams. A mole contains 6×10^{23} particles. When a mole of glucose is burned it releases about 2820 kJ.

1 How much does a mole of glucose weigh?

In the first half of this century, scientists believed that 24 photons of light were needed to synthesise one mole of glucose. Let us see if they were right. You have met the following summary reaction for photosynthesis:

$$6CO_2 + 6H_2O \rightarrow C_6H_{12}O_6 + 6O_2$$

Look back at Fig 9.7 and the text which describes it.

2
a) (i) How many electrons are needed to release one molecule of oxygen at PSII?
   (ii) How many electrons are needed to release 6 molecules of oxygen?
b) If one photon of light is needed for each electron excited, then how many photons would be needed to generate 6 molecules of oxygen at PSII?

But there is a complication to consider. The chloroplast has two photosystems which do not act independently of one another. Each electron released from PSII ends up at PSI, where another photon is absorbed.

3 Taking into account both photosystems, what is the minimum number of photons required to produce one molecule of glucose?

The latest research shows that the figure is closer to 54 photons per glucose molecule, but perhaps the difference you find may be caused by problems in making accurate measurements.

Scientists have compared the amount of energy that would be trapped in a glucose molecule with the energy in the photons needed to produce it. The actual efficiency, the **yield**, for most crop plants is below 1 per cent.

4 List as many factors as you can which could affect the efficiency of photosynthesis.

Crop researchers have estimated the energy in joules of light falling on a crop and compared it with the yield observed at harvest time. The UK receives sunlight at 100 joules per metre squared per second ($100 \text{ J m}^{-2} \text{ s}^{-1}$) in a 24-hour period.

5
a) Calculate the average energy input per hectare per year. (1 hectare = 10 000 m^2)
b) Assume the energy value of the crop of potatoes shown in Fig 22.A1 is 17.85 kilojoules per gram dry weight (17.85 kJ g^{-1}). If light were converted to chemical energy with an efficiency of 5 per cent, then what is the maximum crop (dry weight) you could expect in tonnes per hectare?
   (1 tonne = 1000 kg)

Fig 22.A1 **Potato crop in Norfolk**

6 Copy and complete the following table:

Yield	% conversion of light to chemical energy	Total dry weight /tonnes hectare^{-1}
Possible maximum	5	
Record UK crop		22
Average		3

7 The values of the record and average conversion efficiencies are very low. Find out the strategies that scientists and farmers are using to improve on them.

Assignment

WHY ARE BIG FIERCE ANIMALS SO RARE?

We normally think of large predators like lions or killer whales as being the most successful organism in their ecosystem because they can take the food that they need without worrying about predators themselves. But if you look at their numbers relative to the numbers of other animals further down the food chain, it is clear that these top carnivores are really quite rare. In this Assignment, we see why.

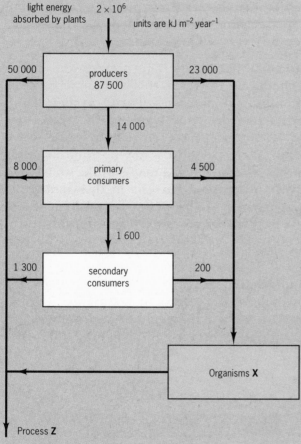

Fig 22.A2 **In this schematic diagram, the producers are the grass plants which make up most of the vegetation. The primary consumers are deer and the secondary consumers are lions**

1 Look at Fig 22.A2. This shows an energy flow diagram for a food chain in the Serengeti.
a) What is process **Z**?
b) What are organisms **X**?
c) How many trophic levels are there in the food chain?
d) Calculate the percentage efficiency with which light energy is converted into food energy in the producers.
e) What percentage of the light energy absorbed by plants is eventually available for secondary productivity in the lion?

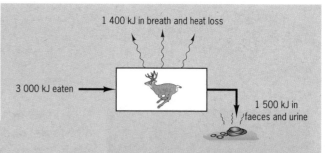

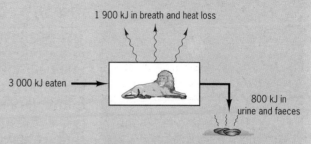

Fig 22.A3 **A comparison of the proportion of energy lost as heat and in faeces and urine in a deer and a lion which eat 3000 kJ of food**

2 Look at Fig 22.A3.
a) For every 3000 kJ of food eaten, what amount of energy is available to the deer for growth and repair?
b) What amount is available to the lion for this purpose?
c) Try to suggest why the deer loses a higher proportion of energy in faeces and urine than the lion.
d) Why should the lion use up more energy in respiration than the deer?
e) Suggest why the proportion of energy available for secondary productivity is about three times greater in the lion than in the deer.

3 From the information you have gained so far, why are large carnivores so rare?

Discussion questions

4 Do you think that the fact that humans are omnivores has anything to do with their evolutionary success?

5 Some people hold the view that, if humans were all vegetarian, the world food shortage would be largely solved. Do you agree or disagree, and why?

6 Several species of fish, such as trout and turbot, can be farmed. These fish are carnivores. Do you think this is an efficient way of producing food? Discuss the reasons for your answer.

23 Recycling in living systems

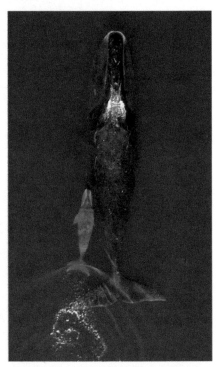

Bacteria that are a normal part of this whale's stomach contents can break down crude oil fractions and PCBs. If these bacteria can be introduced into the stomachs of other species such as dolphins and seals, this could guard them against the worst effects of some pollutants

ECOSYSTEMS ARE sophisticated recycling 'factories' in which elements and simple molecules are continually extracted from 'waste' chemicals and then used to synthesise new chemicals.

Toxic pollutants that are released into the environment by industrial processes cause problems if they cannot be mopped up or broken down, and recycled by living organisms. Pesticides, some oil components and PCBs (polychlorinated biphenyls) are particularly persistent and can build up to dangerous concentrations as they move through the trophic levels of food chains and webs.

One solution to toxic pollutants is to prevent the use or release of such chemicals. However, this is not practicable as many important industrial processes depend on them. Until recently, scientists had drawn a blank in the search for microorganisms that might have enzyme systems weird enough to break such chemicals down into harmless products. But progress is now being made. In the mid-1990s, American scientists discovered bacteria in the stomach of the bowhead whale which can break down naphthalene and anthracene, two persistent and cancer-causing fractions of oil. A couple of the other species found there prefer to feast on PCBs.

Not only could this finding explain why bowhead whales are particularly resistant to the effects of pollutants; it could lead to the identification and mass culture of bacteria which can help clear up the effects of oil spills and industrial contamination.

1 RECYCLING IS NOT NEW

In the last years of the twentieth century, recycling and reusing 'waste' has become a normal part of life. However, we humans cannot claim to have thought of it first. Every ecosystem that has ever existed has practised the same principle. Water molecules are constantly cycling through the environment, evaporating from the surface of the Earth, condensing as clouds, falling again as rain and then cycling through various living organisms. When you turn on the tap, many of the water molecules that come out have been through the kidneys of many people, via water treatment systems. Carbon, nitrogen and other elements also cycle through the environment.

The supply of these vital chemical elements is limited, so they are used again and again. Autotrophs like plants build them into large,

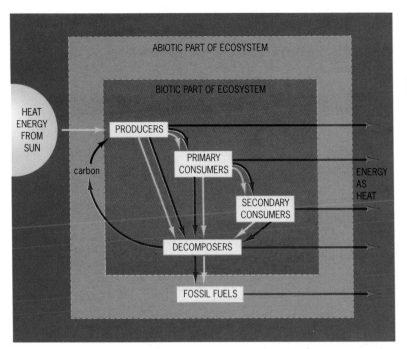

Fig 23.1 **The one-way transfer of energy (yellow arrows) and the cyclical movement of mineral nutrients (for example carbon) in an ecosystem**

Autotrophs make their own food using an external energy source (usually sunlight) and a simple inorganic supply of carbon (usually carbon dioxide).

Photoautotrophs include green plants, algae and bacteria which make their own food by photosynthesis.

Chemoautotrophs are all bacteria: they also make their own food but they may use sources of carbon dioxide and energy that are different from those used by photosynthetic organisms. See Chapter 22.

Heterotrophs cannot make their own food: they must take in food molecules from their surroundings. They use an **organic carbon source** (carbon that was once in living things). Because of this they are totally dependent upon organisms like green plants that can make their own food. See Chapter 22.

■ See question 1(b).

complex molecules, and heterotrophs break the molecules down again, respiring them to gain energy.

The very simple model of an ecosystem in Fig 23.1 illustrates how elements cycle through the compartments of an ecosystem. These processes are all driven by energy from the Sun. As you can see from the diagram, energy is transferred through different organisms in the ecosystem: it is never recycled.

3 MINERAL NUTRIENT CYCLING

All organisms need a supply of energy to carry out the processes of life, but they also need to obtain the mineral nutrients (carbon, nitrogen, phosphorus etc) which make up the complex chemicals in their bodies. Unlike energy, mineral nutrients are recycled over and over again, passing between organisms in the same ecosystem, and also other ecosystems of the world. So the carbon, oxygen and nitrogen atoms in our bodies could have been part of the soil a year ago. They may have made up the proteins of some long extinct dinosaur. And one popular anecdote says that, during the course of our lives, each of us breathes out six carbon atoms that were once part of Napoleon Bonaparte.

Producers accumulate mineral nutrients. These then pass along the food chain and are finally released back into the abiotic part of an ecosystem by decomposers. Decomposers replenish supplies of these elements in the soil and in water, and they also help to return them to the atmosphere.

Why living organisms need mineral nutrients

As we saw in Chapter 3, the chemicals of life, carbohydrates, proteins, fats and nucleic acids, contain mainly carbon, hydrogen and oxygen. Proteins also always contain nitrogen and often sulphur, and nucleic acids contain phosphorus. In addition to these elements, living things need small amounts of calcium, sodium, potassium, iron, copper, chlorine, magnesium and other trace elements. (Table 6.5 gives a list of minerals that the human body needs.)

In this chapter we concentrate on the cycling of carbon and nitrogen.

Cycling of individual mineral nutrients

Mineral nutrients occur in four basic compartments in an ecosystem:

- The organic compartment: living organisms and their debris (faeces etc).
- As available nutrients that are held on the surfaces of clay particles in soil or in solution, where they can be taken up by plants and microorganisms.
- As nutrients that are temporarily unavailable because they are bound up in soil and rocks.
- In the atmosphere, as gases that occur in the air.

The carbon cycle

Life on Earth is based on carbon compounds (see Appendix 2, page 555). Carbon circulates between the abiotic and biotic parts of the environment, as shown in Fig 23.2. Autotrophs capture carbon during photosynthesis, and all living organisms return it to the atmosphere as they respire and also when they die and decompose.

The role of microorganisms in the carbon cycle

By looking at Fig 23.2 or Fig 23.3 you should be able to see that carbon would still cycle through the biosphere for some time if decomposers were removed. However, because of their general role

Fig 23.2 **The carbon cycle. Photosynthetic organisms fix about 80 billion tonnes of carbon each year. But this represents less than 1 per cent of the total carbon in the biosphere.**

Carbon can also be locked up for long periods of time in limestone rocks, inorganic solutions (containing soluble carbonates) and fossil fuels such as coal: these reservoirs of carbon are often called carbon sinks. When volcanoes erupt, they return carbon dioxide to the carbon cycle

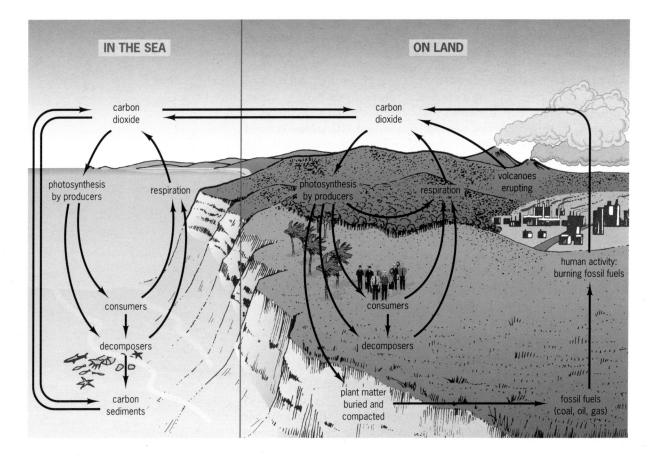

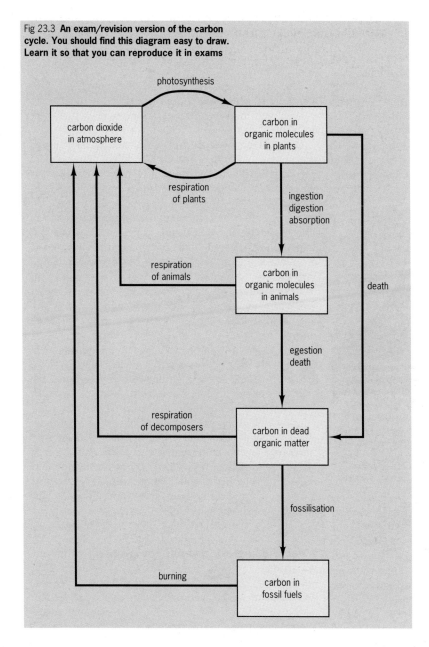

Fig 23.3 **An exam/revision version of the carbon cycle. You should find this diagram easy to draw. Learn it so that you can reproduce it in exams**

in decomposition, decomposers play a key role in the carbon cycle. Bacteria and fungi feed by secreting digestive enzymes onto organic material to digest it. Some of the 'freed' mineral nutrients are absorbed by these microorganisms. Others escape into the soil.

How humans influence the carbon cycle

Humans influence the carbon cycle by two activities not practised by other species. Firstly, humans burn fossil fuels such as coal, oil and gas. This has been going on in a big way since the start of the Industrial Revolution in Europe. Burning such fuels is the only way that their 'locked' carbon is released back into the atmosphere.

The second activity is cutting down forests, particularly rainforests in recent times, to make room for farms, fields and homes. This increases the amount of carbon dioxide in the atmosphere because it reduces the total number of photosynthetic organisms, the 'trappers' of CO_2. The possible effects of these activities are discussed in the Feature box on the next page.

CARBON DIOXIDE, THE GREENHOUSE EFFECT AND GLOBAL WARMING

CARBON DIOXIDE IS one of the main **greenhouse gases**. Together with other gases, it reduces the heat that can escape from the atmosphere (Fig 23.4). This effect is often called the **greenhouse effect**. Although pollution could be increasing the greenhouse effect, this phenomenon is not *caused* by pollution: it is a natural effect due to the presence of an atmosphere. In fact, having an atmosphere is one of the reasons that life developed on Earth: it allows surface temperatures to be much more stable than they would otherwise be.

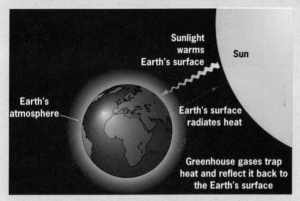

Fig 23.4 **The greenhouse effect. Energy from the Sun is either transmitted, reflected or trapped by the atmosphere. Energy radiated back into space is at longer wavelengths (in the infrared region) than the energy arriving from the Sun, and the consequence is that more heat enters than can escape.**
Greenhouse gases (carbon dioxide, methane, nitrous oxides, chlorofluorocarbons), together with water vapour in clouds, contribute to the 'blanketing' effect of the atmosphere

Accurate measurements of atmospheric carbon dioxide concentrations began in the mid 1950s, in Hawaii and at the South Pole. These sites were chosen because they are far away from any major sources of pollution. The curve, shown in Fig 23.5 clearly shows an annual cycle of peaks and troughs corresponding to summer and winter. Superimposed on this is a small but steady rise in overall CO_2 levels. Levels of other greenhouse gases such as methane, nitrogen oxides and chlorofluorocarbons (CFCs) have also increased during the same period.

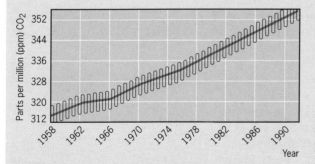

Fig 23.5 **The rising level of carbon dioxide in the atmosphere. Data was collected at the Mauna Loa observatory in Hawaii. There is good evidence to suggest that changes in the average temperature at the surface of the Earth are closely linked to changes in the levels of greenhouse gases in the atmosphere**

What are the likely effects of global warming?

Many scientists think that an increase in carbon dioxide in the atmosphere will cause global warming. They believe that the average temperatures on the Earth's surface will increase, causing widespread disturbances in world climate.

Using all the evidence and data that is available at the moment, a computer model has been developed to predict what might happen in the future. This projection shows that the mean surface temperature might increase.

It is important to remember that any projection is an informed guess: we cannot say exactly what will happen. Let us assume that nothing is done to stop the amount of greenhouse gases that are being released into the atmosphere by human activity. The projection indicates that, at best, there might be a mean temperature increase of only 1.5 °C by the year 2050. At worst, there could be a temperature rise of more than 5 °C.

The effect of any global warming that does occur will not be felt evenly around the globe. There will be greater warming at the high latitudes (at the poles, for example) and at middle latitudes than at the equator. The Northern Hemisphere will warm more than the Southern Hemisphere, because it has more land mass (land heats up more quickly than water). More areas of the world would have extreme heatwaves and more forest and grassland fires. The average sea level would rise, perhaps by two to four centimetres every 10 years. Many low-lying areas could become flooded and maybe even submerged.

What is being done about the problem of global warming?

Though scientists debate the extent and exact effects of global warming, they generally agree that we should be taking steps to cut the emissions of greenhouse gases. This means reducing the amount of fossil fuels that we burn: we could make a big difference by improving the efficiency of heating systems and by reusing 'waste' heat from industry. Wherever possible, we should use renewable energy sources such as wind, water and solar power. Although many people object on safety grounds, making more use of nuclear energy would also help to reduce emissions of greenhouse gases.

In addition, we should limit the destruction of forests and use methods of agriculture that do not irreversibly damage the land. Of course, all of this is cancelled out if the human population continues to increase, so efforts should also be made to stabilise the world's population, to prevent further pressure on energy resources.

Many of these measures are being put into practice, and if people become more aware of this and other environmental problems, international cooperation might soon start to offset the effects of global warming.

The nitrogen cycle

Like carbon, nitrogen is vital to organisms for the formation of proteins, nucleic acids and their products. Although nitrogen is all around us (it makes up 79 per cent of the air we breathe), most living organisms cannot access this supply. Nitrogen gas is unreactive because the two atoms which make up the N_2 molecule are bound together by a strong triple bond. Nitrogen can, however, be fixed by microorganisms, and then it cycles through ecosystems, as shown in Fig 23.6.

Fig 23.6 **The nitrogen cycle. Bacteria play an important role in the movement of nitrogen in the biosphere**

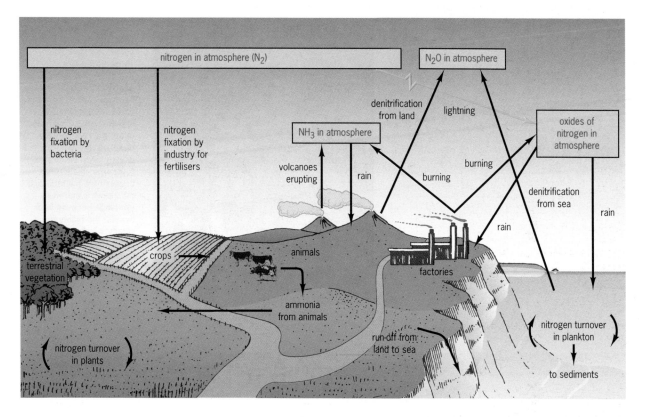

Nitrogen-rich waste from animals (compounds of ammonia, for example) is converted to nitrates by **nitrifying bacteria**. These microorganisms also fix nitrogen by converting atmospheric nitrogen into useful products. In terrestrial ecosystems, nitrogen is fixed by free-living nitrifying bacteria in soil and by **symbiotic bacteria** that live in nodules, the swellings on the roots of legumes (plants like peas and beans). In aquatic ecosystems, some blue-green bacteria fix nitrogen. Nitrogen returns to the atmosphere through the action of **denitrifying bacteria**.

These processes, in more detail, are:

● **Nitrogen fixation.** This is carried out mainly by soil bacteria such as *Azotobacter*. Some bacteria, such as *Rhizobium* are symbiotic: they live in the root nodules of some plants such as peas and beans (Fig 23.8).

● **Ammonification.** This is the breakdown of organic nitrogen such as proteins and urea into ammonia. It is also known as **saprotrophic decay** and is usually thought of as 'rotting', or decomposition. Bacteria and fungi carry out most of the ammonification in the nitrogen cycle.

● **Nitrification**. Nitrifying bacteria such as *Nitrosomonas* oxidise ammonia to nitrite (NO_2^-). Other bacteria, such as *Nitrobacter*

> ✔
> Decomposers stop organic material accumulating in an ecosystem. They also ensure that essential nutrients, such as nitrates, sulphates and phosphates, are released back into the soil.

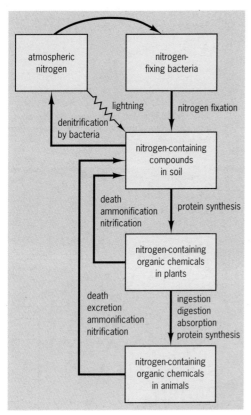

Fig 23.7 **Copiable diagram of the nitrogen cycle for exams**

See question 2. ■

oxidise nitrite to nitrate (NO_3^-). These oxidation reactions release energy which the bacteria use to make food (they are chemoautotrophs).

● **Denitrification**. Bacteria such as *Pseudomonas denitrificans* obtain energy by using the nitrite or nitrate ion as an electron acceptor for the oxidation of organic compounds. They live in conditions with low oxygen levels such as waterlogged soils, and can be important in reducing nitrate levels further.

Fig 23.8 **Legumes such as peas and beans have nodules on their roots. These house nitrogen-fixing species of the bacterium *Rhizobium*. The bacteria convert nitrogen gas from the atmosphere into ammonia. This passes out of the root nodules, dissolves in soil water and is then taken up by plant roots. The relationship between the bacteria and the plants is therefore symbiotic because both benefit: the plants get a steady supply of nitrogen and the bacteria get a home and a constant supply of sugars from the plant**

SUMMARY

After reading this chapter you should know and understand the following:

■ Energy transfer through organisms in an ecosystem is one-way. Elements such as carbon and nitrogen cycle between living organisms and the abiotic environment.

■ **Decomposers** break down the dead bodies of producers and consumers and play an important role in mineral cycling. Most decomposers are bacteria and fungi.

■ The **carbon** and **nitrogen cycles** show how these elements circulate between the biotic and abiotic parts of an ecosystem. You should appreciate the overall cycle and be able to draw a clear summary diagram.

■ Carbon dioxide is a **greenhouse gas**: together with other gases in the atmosphere, it acts as a heat insulator. The presence of carbon dioxide in the atmosphere helps to maintain temperatures on Earth that are warm enough to sustain life.

■ Human activities have led to an increased concentration of carbon dioxide in the atmosphere. Many scientists believe that this is causing a general increase in average temperatures throughout the world, a phenomenon commonly called **global warming**.

■ Microorganisms play an important role in the nitrogen cycle. Different species of bacteria carry out the processes of **nitrogen fixation** the conversion of atmospheric nitrogen into useful products), **ammonification** (the breakdown of organic nitrogen such as proteins and urea into ammonia, **nitrification** (the oxidation of ammonia to nitrite and the conversion of nitrite to nitrate) and **denitrification** (use the nitrite or nitrate ion as an electron acceptor for the oxidation of organic compounds.

QUESTIONS

Some of the questions at the end of this chapter contain material that is covered in Chapter 22. These questions are marked with an asterisk *.

*1 a) In an investigation into seashore ecology, a group of students estimated the biomass of the producers, primary consumers and secondary consumers in three areas of a rocky shore. The results are shown in Fig 22.Q1(a).

    (i) Suggest how many students may have obtained the data shown

    (ii) Estimate the biomass of the primary consumers in Area 1.

    (iii) In which of the Areas, 1 or 2, does the primary consumer make most efficient use of the producer? Explain your answer.

    (iv) In Area 3 the biomass of the primary consumers is greater than that of the producers. Suggest how a pyramid of biomass of this shape is possible.

Fig 23.Q1(a)

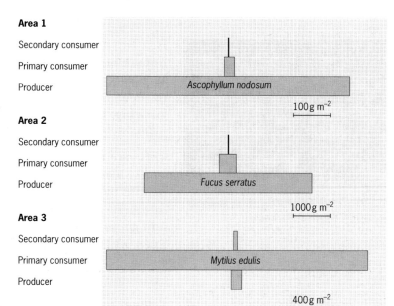

b) The diagram shown in Fig 22.Q1(b) represents how energy flows through each of the communities investigated.

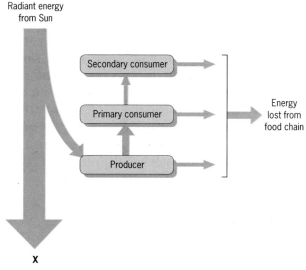

    (i) Suggest two reasons why the energy labelled X does not enter the food chain.

    (ii) Explain two ways in which energy can be lost from the food chain.

[AEB June 1997 Human Biology: Paper 2, q.4]

2 The diagram below shows stages in the nitrogen cycle.

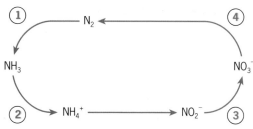

a) What name is given to each of the following stages:
    (i) 1
    (ii) 2 and 3
    (iii) 4

b) How are plants affected by stages 1 to 4

c) In what ways do green plants contribute to the nitrogen cycle?

d) State three ways in which human activity may affect the nitrogen cycle.

[UCLES June 1997 Structured Science Scheme: Unit B3 Ecology and Environmental Physiology Paper, q.1]

24 Populations

Gatherings of people like this are common today but could not even have been imagined a hundred thousand years ago

FOR HUNDREDS OF THOUSANDS of years, the population of human beings on Earth grew at a steady rate. But in the last 200 years, the world's population has increased from 1 billion to over 5 billion people. World population is currently growing by 80 to 90 million people each year and will probably pass the 6 billion mark in early 1999.

This rate of growth is projected to continue for the next 30 years, leading to a global population of around 8.5 billion people by 2030. The rate of increase and the sheer massive numbers of human beings that this population explosion will generate are unprecedented in the history of the planet. Even with our constantly developing technology and ability to adapt, making the Earth's finite resources stretch to accommodate everyone will be a huge challenge.

Trying to predict what will happen to the population further in the future is difficult because it depends on so many factors. If the population carries on increasing at the current rate, a century from now there could be as many as 15 billion people.

But this is the worst case scenario: United Nations projections assume that family planning will become widespread all over the world and that birth rates everywhere will fall. Even taking into account that life expectancy would continue to increase as medicine continues to advance, the UN estimates that population growth will slow down during the early part of the next century and will stabilise around the year 2100. We return to the issue of human population growth throughout this chapter.

1 HOW POPULATIONS GROW

Populations are changing constantly. Imagine a new population starting in a particular area with a handful of individuals and ideal environmental conditions. A good example would be a pair of healthy rabbits introduced onto a small island which had plenty of edible vegetation and no predators. The population would grow slowly at first: this is phase A in the graph shown in Fig 24.1.

After a few months, the population would start to grow very rapidly and would double at regular intervals. This type of growth is called **exponential growth**, and is illustrated as phase B in the same graph. As more rabbits were born, they would then breed and there would be a huge rabbit population in a couple of years.

At some point, the population would be prevented from increasing further by various **limiting factors** in the environment – phase C. When food, space and water supply are stretched to their limit,

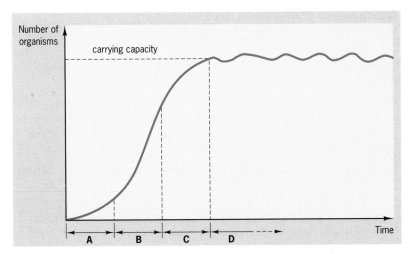

Fig 24.1 **This graph shows an idealised curve of population growth. As the curve A to C looks like a flattened S, it is called an** S-shaped or sigmoid curve. **The text describes phases A to D in detail**

The **biotic potential** is the maximum rate at which a population can increase in size when there are no limits on its growth.

The **environmental resistance** which affects a population is made up from all the factors which act together to limit its size.

When the biotic potential and the environmental resistance reach a balance, the size of the population levels off: it reaches its **carrying capacity**.

more rabbits would die and they might breed less successfully. (If possible, they would leave the environment.) Because there is less food to go around, they would eventually eat the plants faster than the plants could grow.

Most populations stabilise at some point and the growth curve becomes flat. The number of individuals in the population at this point is the **carrying capacity** of the environment – phase D. Sometimes there is a sudden increase in numbers and a population overshoots its carrying capacity (Fig 24.2). The environment cannot provide the resources for this number of organisms and a **population crash** usually follows. The population dies back below the level of the carrying capacity, and then takes time to stabilise. If the overshoot has damaged the environment, the population might stabilise at a different level.

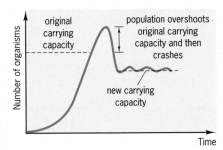

The two growth curves described in Figs 24.1 and 24.2 occur in laboratory situations but you will not find them in real ecosystems. Fig 24.4 shows the three main types of growth curve that occur in natural conditions:

Fig 24.2 **A population crash follows a sudden increase in numbers which exceeds the carrying capacity. If the environment is damaged by the temporary overshoot, there may be a new, reduced carrying capacity**

- **Stable**: the population size remains at roughly the same level over a long time. A study of a population of trees in a well-established woodland would produce this sort of curve.

- **Irruptive**: the population size is basically stable, with the occasional dramatic increase followed by a population crash. This type of population growth occurs in algae. A population is mostly stable but, in hot weather, when water is contaminated by phosphate or nitrate fertilisers, there is an **algal bloom** (Fig 24.3).

- **Cyclic**: the population increases and decreases in a regular cycle of growth and die-back.

Fig 24.3 **An algal bloom on a dyke**

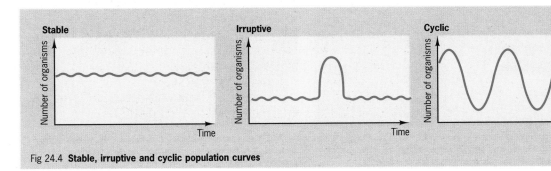

Fig 24.4 **Stable, irruptive and cyclic population curves**

What limits population size?

In an ecosystem, **abiotic** (physical) factors in the environment can limit population size. When we look at the whole range of abiotic factors that affect population size, we can classify them into three general categories:

- Limiting factors that are *always present*. These include space and light. A cliff that supports a population of kittiwakes, for example, is a fixed size and can accommodate a fixed number of birds (Fig 24.5).

- Limiting factors that *vary predictably over time*. These include changes in temperature and weather patterns in different seasons which lead to population movement or a change in behaviour among different organisms. As winter approaches, birds which live in an ecosystem in the summer might migrate to a warmer place (Fig 24.6). Small mammals such as dormice hibernate.

- Limiting factors that are *unpredictable*. Some unpredictable limiting factors are **density dependent**: they cause a greater effect if there is a high population than if the population is quite small. Sudden changes in the availability of water and nutrients, or a localised natural disaster, for example, can devastate an overcrowded population but leave a widely dispersed population relatively untouched.

- Other factors are **density independent**: they can devastate a population, whatever its size. Fires, floods, severe frosts and other catastrophes, for example, might destroy part or the whole of an ecosystem without warning (Fig 24.7).

 See questions 2 and 3.

Fig 24.5 **The size of the ledges is an unchanging limiting factor. The kittiwakes can get nesting sites only by displacing other birds**

Fig 24.6 **Arctic terns are the world's champion travellers. In their search for food and the right breeding conditions, they range from deep inside the Arctic Circle, over the Atlantic to the pack ice of Antarctica**

Fig 24.7 **Catastrophes, including those caused by human activity such as oil spills, have a density-independent effect on populations**

HUMAN POPULATION GROWTH: IMPLICATIONS FOR THE NEXT CENTURY

HUMAN POPULATION GROWTH shows the classic J-shaped curve of exponential growth. If you compare the graph shown in Fig 24.8 with the one in Fig 24.1, you can see that the J-curve is the first part of the classic S-curve.

As in any exponential growth curve, the population starts off increasing slowly: from the early origins of human beings to the last 200 years or so, the curve remained fairly flat. However, we are now in the period where the curve is becoming extremely steep, with the human population of the world doubling about every 50 years or so. At some point, the curve will level off and the population will stabilise. It is difficult to know when this will happen and what forces will bring it about.

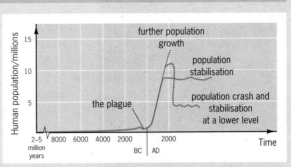

Fig 24.8 **Growth in the world human population. The dotted lines show three possible predictions of what will happen to the human population in the next two hundred years or so. It is inevitable that in some way the population will stabilise at some point in the future: the two main worries are when and how it will happen**

3 INTERACTIONS BETWEEN ORGANISMS

Every living organism in an ecosystem affects the others. Some of these interactions are **intraspecific**: they take place between organisms of the same species. Others are **interspecific**: they take place between organisms of different species.

Intraspecific interactions

Relationships between organisms of the same species in an ecosystem fall into three main categories:

- Reproduction and care of the young.
- Social behaviour.
- Competition for resources.

We consider these in more detail below.

Reproduction and care of the young

Organisms of the same species interact at many different stages of sexual reproduction. In some species the relationship is brief and minimal: a wind-pollinated plant simply releases pollen into the air and accepts pollen from another plant of the same species. In others, the association is complex and long term. For example, swans, like humans, find a mate and stay together for many years, sometimes for life (Fig 24.9). Both parents feed and care for the young and, to some extent, provide companionship and help for each other.

In both examples, the level of interaction achieves the aim of reproduction: producing enough surviving offspring to maintain the population level in the ecosystem.

In many animal species, reproduction involves interactions with more than just one other member of the same species. Many animals compete with each other for a suitable mate: for example, male red deer fight fiercely for the control of groups of females. Animals that are more social, such as termites and meercats, provide joint care for each others' young. Some adults in the group, not necessarily just the parents, guard, transport and protect the young, while the others go off to search for food.

Fig 24.9 **The female swan carries her young, while the male displays to deter intruders**

Fig 24.10 **There is evidence that bees can communicate the position of good pollen sources by elaborate dancing displays to other bees in the hive. Here, a worker bee with full pollen sacs is dancing to indicate the amount, distance and direction of its source of pollen**

Social behaviour

Many animals are social: they live in groups rather than as a collection of individuals that just happen to live in the same place. Group living has its advantages and disadvantages. Animals that live in groups are usually less likely to be eaten by predators than solitary animals. There seem to be several reasons for this:

● A large number of potential prey in one place can confuse a predator – the predator cannot decide which one to attack.

● Different individuals in a large group of animals can perform different tasks: some can look for food, others care for the young and some can act as look-outs for predators. Large groups therefore spot approaching predators much sooner than animals on their own and so can take more successful evading action.

● A large number of animals are more successful than just a few at defending against an attacking predator. If a predator threatens, elephants group together, forming a circle to protect their young, which are herded into the centre of the group.

Animals that live in groups also stand more chance of hunting or finding food successfully (Fig 24.10, and the Feature box below). Hyenas, for example, hunt in packs and can bring down a deer that is much larger than one hyena would be prepared to tackle. Other group associations can help to protect animals from the cold as they huddle together.

COSTS AND BENEFITS OF SOCIAL BEHAVIOUR

PRAIRIE DOGS are rabbit-sized rodents with small ears and short legs (Fig 24.11). They live in large groups in the plains of North America digging a complex of tunnels up to 30 metres long. There are two species of prairie dog, the black-tailed and white-tailed forms and each community or 'town' may contain up to a thousand animals.

Prairie dogs graze above ground during the day when there are many predators around such as coyotes, badgers and hawks. They have 'sentries' sitting upright outside the burrows keeping a watchful eye for predators. The appearance of a predator provokes a series of whistling barks from the sentries.

A cost–benefit analysis has been carried out on the social behaviour of the prairie dog. An anti-predator response was tested by dragging a stuffed badger through a colony. In the larger groups of prairie dog, the predator was detected sooner. Black-tailed prairie dogs live in larger groups than their white-tailed cousins. Consequently the black-tails can afford to spend significantly less time scanning their surroundings for predators (35 per cent of their average daily time budget compared with 45 per cent in the white-tails). Less time being vigilant means more time to feed.

So why don't white-tailed prairie dogs live in large groups as well? One answer is that living in large groups has costs as well as benefits. Within either species, individuals in large groups fight more often than those in small groups. Also disease-spreading fleas are more common in large groups (3.3 fleas per black-tailed burrow, 0.5 fleas per white-tailed burrow).

The evolution of group behaviour therefore depends on balancing advantages against disadvantages. Other factors may also be involved – black-tailed prairie dogs live in open plains, for example, and, since there are many predators around, it is safer for them to live in larger groups.

Fig 24.11 **A prairie dog keeping guard as sentry**

Human social behaviour

If children see pictures of small children's faces and adult faces they prefer the adult face. A change in preference occurs at 12 to 14 years of age in girls when they express a preference for baby pictures. Boys display the same trend two years later. If we have to sit too close to a stranger in a library, we tend to use physical 'barriers' such as books or bags to keep the stranger at a distance. If this person gets even closer, we often move seats.

These are examples of the types of behaviour studied in **human ethology** – the biology of human behaviour. It asks the question: Why do we behave the way we do?

Ethologists view human behaviour in the same way as that of other animals: they think it is adaptive and has evolved to suit its purpose. In the two examples above, there is strong evidence to link children's behaviour with child rearing and adult behaviour with protection.

One goal of human ethologists is to look for patterns of behaviour shared by all peoples of the world. These patterns are seen in behaviour that involves body movements, rather than speech (Fig 24.12). Gestures such as smiling, raising the eyebrows, and behaviours such as grooming and hugging seem to be common to all human societies.

We all think that we have a free choice in our behaviour, but do we really? We may be under the control of our genes or be shaped by our surroundings. There is evidence for both points of view.

Fig 24.12 **Human speech is often thought of as our major way of communicating. However, feelings are often expressed by the look on our faces and by our general 'body language'**

Competition for resources

Most of the disadvantages of group living arise because animals have to compete with each other for vital resources. Hyenas may make a bigger kill when they hunt in a pack, but the meat has to fill more stomachs. If food becomes scarce, rivalry within the pack becomes intense and some may need to find a new habitat to survive. Animals of the same species also compete for mates, shelter, water and social position.

Interspecific interactions

A full list of the interactions that can occur between individuals of different species in an ecosystem is given in Chapter 26, Table 26.1. In this chapter we will look in detail at four of these:

● Predator–prey relationships.
● Parasitism and disease.
● Mutualism and commensalism.
● Competing for resources.

Predator–prey relationships

A basic definition of a predator is an organisms which eats another organism to obtain energy and nutrients. This definition includes animals that feed on plants.

The relationship between predator and prey is based on a constant battle. The actual 'catching and eating' part of the battle is only a small part of the story. The more serious conflict involves evolution.

All species evolve by the processes of genetic mutation and natural selection (see Chapter 30): prey species are constantly evolving and becoming better able to cope with predators; predator

Symbiosis comes from a Greek word that means 'to live together'. We use it to describe a close physical interaction between two organisms from different species. There are three types of symbiosis:

Parasitism: one of the partners benefits, the other is harmed.

Commensalism: one partner benefits and the other is unaffected.

Mutualism: both partners benefit from the association.

Fig 24.13 **Bats have evolved sophisticated sonar systems to detect flying insects with great accuracy. Some insects, however, have evolved the ability to detect bat sonar signals and respond by closing their wings and dropping out of the sky. This makes it much harder for the bat to catch them and they gain an important survival advantage over other insect species**

species are constantly evolving to out-manoeuvre their prey (Fig 24.13). Sometimes, either the predator or prey loses the long-term battle. If the prey evolves to completely avoid a predator, and that predator has no alternative food supply, the predator species can be driven to extinction. If a predator out-evolves its prey, it will do well in the short term. But, if the prey species becomes extinct, the predator then needs to find an alternative food supply to survive.

In general, predators thrive when they develop adaptations that allow them to eat well and make the most of their food. Herbivores, animals that eat plants, often have specialised mouthparts and digestive system to gain the most nourishment from the plants available to them. Many herbivores are ruminants: their digestive system has four stomachs and they chew their food twice, once after eating it and once after regurgitating it.

Prey species continue to survive because they have evolved defence mechanisms against predators. In animals, these include camouflage, spines and barbs and chemical deterrents (Fig 24.14). Many, particularly insects, produce foul tasting or poisonous chemicals and usually advertise this by the colours and patterns on their bodies and wings.

Although plants cannot run away from grazing animals, they are not totally defenceless. Their defences include thorns, tough tissues and thick waxy cuticles. Many plants have also developed **secondary compounds**, chemicals that have no metabolic role but which put off potential grazers. Examples include tannins in oak leaves and cardiac glycosides in foxglove plants.

Fig 24.14 **Animals use many different techniques to avoid capture by predators**

Right: The camouflage of this iguana matches the colours of the trunk and lichens in a Peruvian rainforest

Lower left: The deadly sting of a scorpion deters predators as well as being useful for hunting food

Lower right: When under threat, a porcupine runs backwards at the enemy with quills raised. They can fall off, and an embedded quill can cause a septic wound that would kill a lion

Parasitism and disease

Predators usually kill their prey: parasites have a different strategy because they depend on their prey, the **host** (a different species), staying alive. They live and reproduce using the resources of the host organism that is damaged in the process. This is why many parasites cause disease in humans and other animals (Table 24.2). The more successful parasites cause less damage to their host. The tapeworm, for example, can survive in the human digestive system for many years, causing relatively minor damage in the affected person.

Table 24.2 **Some major parasitic diseases and the parasites that cause them**

Disease	Parasite	Symptoms and effects
Malaria	*Plasmodium* (4 different species of this protoctist parasite cause malaria)	Regular cycles of fever: disease transmitted to people by the anopheles mosquito (see Feature box)
Sleeping sickness	*Trypanosoma* species (also a protoctist parasite)	Transmitted to people by insects such as the tsetse fly which bite infected animals
Leishmaniasis (also called kala-azar and oriental sore)	*Leishmania* species (another protoctist parasite)	Transmitted to humans via sandflies
Schistosomiasis (also called bilharzia)	*Schistosoma* (blood fluke)	Flukes live in water (their intermediate host is a water snail) and infect humans who bathe
Tapeworm infection	*Taenia solium* (tapeworm)	Parasite lives in intestine: passed to humans in infected and undercooked meat
Toxoplasmosis	*Toxoplasma* species (protoctist parasite)	Causes weakness and fever – passed from mother to baby across placenta. Can be transmitted by domestic cats

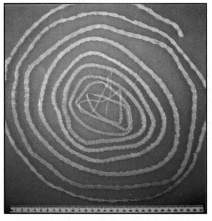

Fig 24.15 **Above: The beef tapeworm. The head is at the thin end, attached to many hundreds of segments. They break off and pass out of the host (cattle), to infect another victim. The ruler is 30 cm long.**

Below: At the head end (the scolex) are four suckers and a ring of hooks that attach the tapeworm to the gut wall of its host

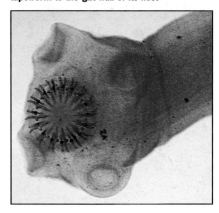

Parasites show very specialised features which fit them to their lifestyle (Fig 24.15). All parasites show some general adaptations:

● They have ways of getting into the body of the host. Fungi which parasitise plants produce the enzyme **cellulase**, which breaks down plant cell walls. The schistosome (blood fluke) which causes bilharzia can bore through human skin.

● They have structures that enable them to attach to their host. Tapeworms have a ring of hooks called a **scolex** which they insert into the wall of the host's intestine.

● They have lost the organ systems that they no longer need. Tapeworms and other gut parasites have, for example, lost their own digestive system because they can absorb already digested food across their body surface. We call this loss of structure and function **parasitic degeneration**.

● They protect themselves against the internal defences of the host. Trypanosomes, which cause sleeping sickness in humans, produce an outer layer which varies constantly in its structure. Just as the host's body starts to make antibodies (see Chapter 34) to one type of coating, the trypanosome produces a different coat which the antibodies cannot recognise.

● They often have complex life cycles which allow them to infect new hosts. See the Feature box on malaria on the next page.

● They show great **fecundity**, or capacity for reproduction. Many tapeworm segments break off every day and pass out of the body in the faeces. Each segment can contain hundreds or even thousands of eggs.

A Organisms can interact intraspecifically or interspecifically. What do these two terms mean?

THE FIGHT AGAINST MALARIA

MALARIA IS A DISEASE caused by a parasite. Four species of protoctist of the genus *Plasmodium* are responsible for 200 to 300 million cases of malaria every year, worldwide. The disease is common in the tropical and subtropical regions of the world – the home of 40 per cent of the world's population. The death rate is appalling, somewhere between one million and five million people each year, many of them children under five.

Malaria is extremely difficult to control for several reasons. Firstly, it has a complex life cycle (Fig 24.16). Any four of the *Plasmodium* species that cause it can be passed on to human hosts by any one of 60 species of the anopheles mosquito. Draining swamps and marshes and using insecticides to spray mosquito breeding areas was temporarily successful and reduced the spread of malaria during the 1950s and 1960s. However, not every species of carrier mosquito was wiped out. Many of them have developed resistance to the early insecticides and are now much more difficult to control.

Secondly, the parasite itself is a tricky customer. It produces symptoms which get worse and then better in cycles. Anti-malarial drugs target some but not all stages in the life cycle (Fig 24.17). The sufferer therefore needs to take the drugs over a long period of time, or the parasite 'hides' from their effects. The common anti-malaria drugs have been used so widely that they have put *selection pressure* on the parasite and it has now developed resistance to many of them.

Although new drugs are being developed all the time (some of the most recent are based on extracts of rainforest plants), workers now concentrate on prevention. Any stagnant water that can be eliminated is removed and mosquito nets in houses are dipped in a safe insecticide such as permethrin. Biological methods of control of the mosquito larvae are being used – fish such as guppies are being bred to eat them. Vegetation around houses is cleared, and trees are being planted in marsh areas to destroy the wet breeding grounds of the mosquito.

Fig 24.16 **The life cycle of the malarial parasite, *Plasmodium vivax*. The anopheles mosquito is the first host, humans are the second host. People with malaria get regular cycles of illness because merozoites are continually released. When the parasite is 'hiding' in the red blood cells, symptoms are not too bad: it is the sudden rush of foreign proteins in the bloodstream that leads to the bouts of fever that are typical of malaria**

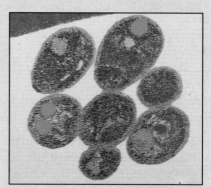

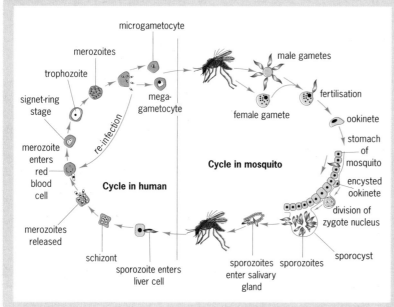

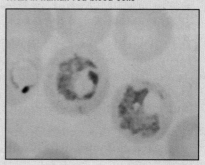

Fig 24.17 **The merozoite stage (above) and signet ring stage (below) of *Plasmodium vivax* in human red blood cells**

Mutualism and commensalism

The term **mutualism** is used to describe a relationship between two organisms from different species that receive *mutual* benefit from the association. Neither is harmed and both do better by being together than they would alone. The African bird, the oxpecker, for example, cleans the skin of the rhinoceros, removing all the parasites such as ticks and mites. The bird eats the insects, gaining the advantage of the nutrients. Legumes, with their root nodules filled with bacteria (see Chapter 23), are an example of a mutualistic association.

Two organisms which are **commensal** live together and, while one of the partners benefits, the other does not. The relationship between clownfish and sea anemones is a good example of commensalism (Fig 24.18). The fish hide from predators in the mass of the anemone's stinging tentacles. Although the anemone is unharmed, it does not seem to benefit in any way.

Competing for resources

Species can compete in different ways. Some organisms stop the organisms of other species from getting near a particular resource. Wrens, for example, defend food plants in their territory. In other situations, two species have access to the resource, but one is better at exploiting it.

Some species have evolved to coexist, sharing the resources that are available in an ecosystem. This is termed **symbiosis.** Some plants, for example, have a mesh of roots close to the surface of the soil to trap water from short showers of rain. Other neighbouring plants have long roots which take advantage of deeper groundwater. Both species are using the same resource, but they are sharing it in a way that allows both to benefit. This is called **resource partitioning** and it also occurs between animal species.

Fig 24.18 **There are 26 species of clownfish which all associate with sea anemones. This colourful fish hides from predators among the venom-producing tentacles of the sea anemone**

4 A WIDER LOOK AT COMMUNITIES IN AN ECOSYSTEM

We have seen that ecosystems are dynamic systems and that the physical parts of an ecosystem, such as the weather, affect and interact with the organisms that live there. In this section we look at how an ecosystem forms.

Imagine a completely physical environment such as the bare rock and ash left behind after a volcano has erupted and then become quiet. After the ground cools and water reenters the environment, some plants start to grow there. After a time, other species of plant establish themselves and eventually, probably after many years, woodland forms (Fig 24.19). This process is called **succession**.

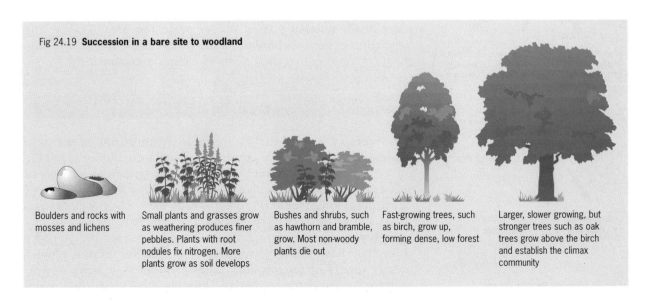

Fig 24.19 **Succession in a bare site to woodland**

Boulders and rocks with mosses and lichens

Small plants and grasses grow as weathering produces finer pebbles. Plants with root nodules fix nitrogen. More plants grow as soil develops

Bushes and shrubs, such as hawthorn and bramble, grow. Most non-woody plants die out

Fast-growing trees, such as birch, grow up, forming dense, low forest

Larger, slower growing, but stronger trees such as oak trees grow above the birch and establish the climax community

Succession also describes the more minor changes in the vegetation of an ecosystem which occur after a less catastrophic environmental change. For example, after a small patch of trees in woodland die after a series of hot, dry summers, the land becomes repopulated by the same sorts of trees very quickly because their seeds are already present in fertile ground.

To distinguish between these two types of situation, we say that the process that results in new vegetation on a bare site is **primary succession.** This takes time and the final appearance of the community depends on which plant spores and seeds come into the area from neighbouring ecosystems. The process which repopulates a previously well vegetated area after a minor environmental change is called **secondary succession**.

Strictly, succession is used only to describe the development of and changes in *plant* communities. In any ecosystem, however, the plant communities help to form the conditions which enable other classes of organism (animals, bacteria, fungi etc) to colonise the environment.

A developing ecosystem

In order to understand the stages that occur in succession, think of an environment consisting just of bare rock. Colonisation occurs in many different stages called **seres**. The rocks first provide a home for spiders which feed on insects in the area.

Succession begins as mosses and lichens start to grow on the bare rocks. These first plant species are called **pioneer species**. Over many years, perhaps thousands of years, the environment is battered by rain and wind and other climatic events, and the surface of the rock is broken down into smaller fragments. The mosses and lichens help soil to form in cracks and crevices, and eventually a few small broad-leaved plants start to grow.

As the soil becomes enriched by organic matter from the action of decomposers on dead plants, more species of plant can grow, including grasses and then plants with root nodules that have nitrogen-fixing bacteria. As these plants start to thrive, a few larger plants, such as shrubs, begin to grow.

After this, things speed up because shrubs provide roosting and nesting sites for birds which then supply many more seeds in their droppings. Pioneer tree species that need high levels of light come into the area next. As woodland becomes more established, other, more shade-tolerating trees can colonise the area. Finally, the vegetation of the area becomes more or less stable: we describe the plant community that then exists as a **climax community**.

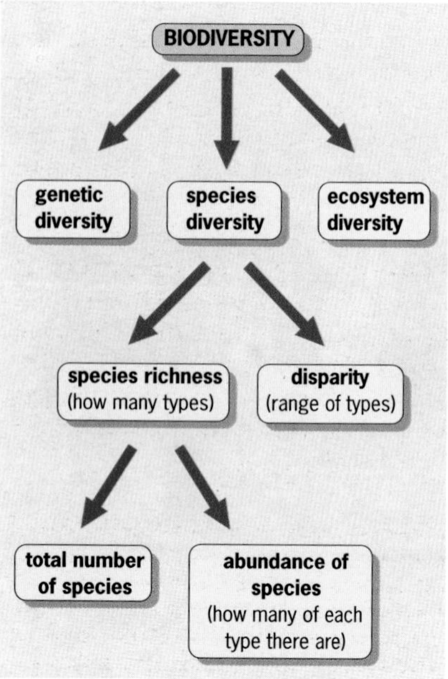

BIODIVERSITY

Biodiversity, a word that is the popular contraction of the term **biological diversity**, is used generally to refer to the variety of life on Earth. More specifically, we use it to describe the *numbers* of species present in a particular habitat or to show the *range* of types of organism found there (Fig 24.20). Take the sea bed, for example. It has a large number of species but it is largely made up of a few basic types: annelid worms, molluscs and echinoderms. A rocky shore may have the same number of species but there is often a greater variety of animals and plants: arthropods, annelid worms, molluscs, echinoderms, cnidarians, fish, birds and seaweeds. We say

Fig 24.20 ■ **What biodiversity means**

that the shore has a greater **species diversity**: it has a greater species richness (number of different species) and a greater species disparity (range of different species types), when compared to the sea bed.

Diversity can be used to indicate the 'biological health' of a particular habitat. A slow increase in plant species diversity is normal for stable habitats such as hedgerows. If a habitat suddenly begins to lose its animal or plant types, ecologists become worried and search for causes (a pollution incident, for example).

You might also have heard of the term **genetic diversity**. Different species are obviously genetically different from each other, but there is also genetic variation within a species (no two people look identical, for example).

A third form is **ecosystem diversity**. The environment is an important factor that determines the number and types of species present in an ecosystem. At its simplest, ecosystem diversity is described as the number of different types of habitat in a given area.

Fig 24.21 **The tropical rainforests of the world encompass a massive diversity of all forms of life**

Why biodiversity is important

When a species becomes extinct, many other species can be affected: no type of living organism exists independently. No one can really predict the long-term effects of repeated extinctions. Organisms have, of course, always become extinct, because of evolutionary pressures. What concerns us today is the rate at which species are disappearing because of the impact of human populations.

Although the diverse world of animals and plants provides us with a massive resource of food, we actually use only 30 species or so, to provide 90 per cent of all the food we eat world-wide. Plants have already given us many useful medicines (Fig 24.22 and see Feature box below) and there are undoubtedly more compounds still to be found. In addition, many different animals and plants stabilise soil and so minimise erosion, support fisheries, provide income through tourism and provide beauty and interest.

Fig 24.22 **Aspirin, the most commonly used drug in the world, was originally extracted from the bark of the willow**

MEDICINES FROM THE RAINFOREST

THE WEATHER CONDITIONS in the tropical rainforests around the Equator have encouraged the evolution of areas of high plant and animal diversity. The indigenous human populations who live and work in the rainforests need to make a living, and working for developers who sell timber and other products to the developed world is an obvious means of doing this. The consequences for the environment are serious. Some scientists estimate that three species of living things are being made extinct every hour, because the rainforest is being destroyed.

But what difference can a few plants make? Possibly the difference between life and death for many people.

Screening plants from the rainforest has shown that they contain many compounds that protect them against predators and parasites. These compounds might also be useful in human medicine. Since 1990, at least four compounds have been identified that might provide drugs for use against the virus which causes AIDS (Acquired Immuno-Deficiency Syndrome).

The work of developing medicines needs sophisticated chemical and pharmaceutical techniques, but it cannot be done unless plants that are collected for screening can be identified accurately and consistently. Speed is also important. If destruction continues at its present rate, the rainforests will have been completely wiped out by the end of the next century.

Fig 24.23 Scientists who travel to remote areas of rainforest to search for and identify plants which may be useful in medicine are called ethnobotanists. They work with local traditional healers to narrow down the choice of plants to study. Samples are sent back to laboratories for testing. Research programmes are now being set up to share the profits from successful compounds, so that indigenous people can benefit from the discovery, instead of suffering exploitation

HOW OLD IS MY HEDGE?

AS YOU NEXT WALK DOWN a country lane, take a good look at the hedges that you pass. Many of these structures date from the Middle Ages. They were planted by medieval farmers as boundaries to mark off their land and to keep in livestock. They probably planted shrubs like hawthorn a few feet apart. The gaps atthe base of these plants then allowed waves of colonisers.

First of all, fast growing, non-woody annual plants like typical garden weeds would have thrived: chickweed, groundsel and shepherd's purse for example. Then, non-woody perennials such as bluebells, primroses, stinging nettles, dandelions and thistles may have become established. These plants can produce food reserves to survive the winter. Eventually, woody perennials, shrubs and small trees such as elder, holly, and even some of the larger forest trees could have invaded the hedge.

Fig 24.24 **In theory, it is possible to date a hedge by looking at the range, or** diversity **of plants that are in it. The more diverse the community of the hedge, the older it is**

Measuring biodiversity

Estimates for the total number of species that still exist range from 1 million to 50 million. We base our estimates on the knowledge we already have of numbers and types of species found in particular habitats and on the rate at which our knowledge is increasing. Between 1978 and 1987, 367 new vertebrates, 173 new annelids and 7222 new species of insect were identified.

Why do we need to measure biodiversity?

There is more to measuring biodiversity than just being able to impress your friends by knowing how many species of living things there are on Earth. We need to measure the diversity:

- To compare the diversity of the same habitat over time, to assess for example, whether pollution is damaging that habitat.

- To compare two different habitats at the same time, to find out which is the more diverse.

- To check whether a new habitat is being colonised by the number of species that it should. This would be important in the case of a polluted river or lake that had been 'cleaned up' and then repopulated with living organisms.

See question 1. ■

Putting a number on biodiversity – the Simpson diversity index

Being able to put a number on the diversity of a habitat makes the life of an ecologist much easier. A **diversity index** allows us to estimate the variety of living things in a particular area. However, we must make sure that the value we use accurately reflects all aspects of diversity. We could, for example, simply count the number of different species present:

Stream 1 (100 animals) 16 species

Stream 2 (100 animals) 10 species

The conclusion would be that stream 1 was more diverse. But there is a problem: this method gives an idea of *species richness*, but it does not reflect *how many* of each species are present. In stream 1, there may be 85 animals of one species and just one of each of the others. In stream 2, there may be 10 animals of each species. Ecologists would say that stream 2 was the more diverse habitat, and so our first conclusion would be wrong.

To avoid this problem, we can use the Simpson index. Its formula takes into account that diversity depends on the number of individuals of each species present and the species richness (the number of different species present). The higher the value of the index, the greater the variety of living organisms found in the area.

$$D = \frac{N(N-1)}{\sum n(n-1)}$$

where D is the Simpson diversity index, N is the total number of individuals of all species found, n is the total number of organisms of a particular species found, and $\sum$ means 'the sum of'.

B The Simpson index was used to calculate the diversity of an oak wood and a conifer plantation. The values obtained were 19.6 for the oak wood and 12.04 for the conifer plantation. Which habitat is the more diverse?

EXAMPLE

Q Assume that you have studied a particular habitat and you have recorded the types of 20 animals you have found. The different types are represented by letters of the alphabet, and the record shows five species:

A A A A B B A C C B B A B B A D C C D E

What is the diversity of this habitat?

A Putting the figures into the equation for the Simpson index:

$$D = \frac{20 \times 19}{(7 \times 6) + (6 \times 5) + (4 \times 3) + (2 \times 1) + (1 \times 0)} = \frac{380}{86} = 4.42$$

SUMMARY

By the end of this chapter you should know and understand the following:

■ A population grows **exponentially** in ideal conditions. Growth slows down because of **limiting factors** in the environment. This produces an S-shaped or **sigmoid** growth curve.

■ When the size of the population becomes stable, we say that this is the **carrying capacity** of the environment. Sometimes, population growth exceeds the carrying capacity and a **population crash** follows. The population may then settle at a new carrying capacity.

■ Three main types of growth curve occur in nature: **stable**, **irruptive** and **cyclic**.

■ Some limiting factors which limit population growth are **density dependent**: their effects are worse in a large, closely packed population. Others are **density independent**: their effects do not depend on the population distribution or size.

■ **Interspecific interactions** occur between organisms of the same species; **Intraspecific interactions** occur between organisms of different species. You should know some examples of each type.

■ The living part of an ecosystem develops by the process of **succession**. If the area has never been colonised before, a newly formed volcanic island, for example, the process is called **primary succession**. If the area was previously covered with vegetation, a forest site that suffers a severe fire, for example, **secondary succession** occurs.

■ There are three types of **biological diversity: species diversity, genetic diversity** and **ecosystem diversity.**

■ Measuring biodiversity is important because it allows us to monitor a habitat over time, to compare two different habitats and to find out if a new habitat is being colonised by the correct number and type of species. This is vital if we are to assess the impact of our 'clean-up' efforts after environmental tragedies such as oil spills.

QUESTIONS

1 Various types of trap are used to sample the animals which live in grassland. The animals are released later.

a) Fig 24.Q1(a) shows a *Longworth* trap which is used to catch small mammals. Explain why this type of trap should be checked at least twice each day.

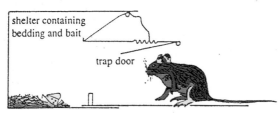

Fig 24.Q1(a)

b) Twenty *Longworth* traps were used over a period of ten days to sample the small mammals living in the grassland. Table 24.Q1 shows the results which were obtained.

Table 24.Q1

Mammal	Number trapped	% of total
Field vole	180	50
Woodmouse	108	
Bank vole	45	
Others	27	
Total	360	

  (i) Copy and complete Table 24.Q1.
  (ii) Copy and complete the pie graph of Fig 24.Q2(b) to show the species composition of the small mammal population.

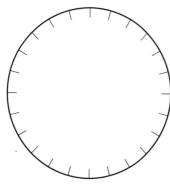

Fig 24.Q2(b)

c) Fig Fig 24.Q2(c) shows a pitfall trap which is used to catch invertebrate animals.

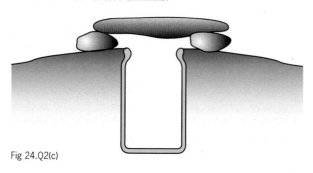

Fig 24.Q2(c)

Suggest **two** reasons why the results from this type of trap might give a false impression of the number and types of invertebrate animals which live in the grassland.

[NEAB: February 1996 AS Biology: Advanced Ecology, Section A, q.1]

2 Moose colonised Isle Royale, an island in Lake Michigan in Canada, in the early 1900s. They are large herbivores; no predators of the moose live on Isle Royale. Table 24.Q2 shows the estimated population size of moose between 1915 and 1960.

Year	Estimated size of moose population
1915	200
1917	300
1921	1000
1925	2000
1928	2500
1930	3000
1934	400
1943	170
1947	600
1950	500
1960	600

Table 24.Q2

a) Describe the pattern shown in the results.

b) Give **two** density dependent factors which might account for the pattern shown in the results between 1915 and 1943. Explain how each factor may have had its effect.

[NEAB February 1996 AS Biology: Advanced Ecology, Section A, q.2]

3 A group of students investigated the distribution of plants bordering a small stream in a salt marsh. Fig 24.Q3 shows their results.

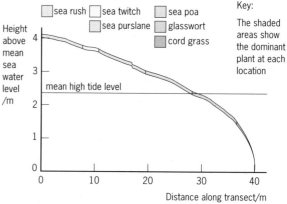

Fig 24.Q3

a) Describe the techniques which the students could use to obtain the data shown in the diagram.

b) The distribution of plants in this habitat is governed mainly by abiotic factors.

  Suggest **two** abiotic factors that could restrict the distribution of sea rush and cord grass in this habitat and explain how each factor would have its effect

[NEAB February 1996 AS Biology: Advanced Ecology, Section A, q.3]

Assignment

AN UNCERTAIN FUTURE

In Chapter 30 we look at how humans evolved from their primate ancestors. In this Assignment, we investigate what has happened to the human population over the last 12 000 years and we raise some questions about what will happen in the next 12 000.

We start our story at the start of the 'Agricultural revolution' in about 10 000 BC, when there were around 10 million people on Earth. People had changed from being hunter gatherers to being farmers. With this new settled lifestyle, there was a dramatic increase in the birth rate and a decline in the death rate.

a) Suggest why nomadic travellers might have had fewer children than farmers living in permanent settlements.

b) Why would life expectancy be lower in hunter-gatherer societies?

By the time of the First Egyptian Dynasty (3000 BC) the global human population had increased from about 10 million people to 100 million. By the height of the Roman Empire and the birth of Jesus Christ, there were 300 million. Civilisations flourished in three main regions of the world: Southwest Asia, China and the American continent.

2 What properties do you think these areas might have had that enabled them to support large numbers of people?

Table 24.A1 shows how population continued to increase until 1804, when it reached the 1 billion milestone.

Table 24.A1 **Population trends to the first billion humans**

Year	0	1000	1250	1500	1750	1800
Population (in billions)	0.30	0.31	0.40	0.50	0.79	1.00

3 Plot the figures given in Table 24.A1 as a line graph. Extend the time axis to the year 2100 AD.

a) Estimate how many people would have been alive in the year 2000 if the rate of growth had remained constant after 1800 AD.

b) Current estimates put the world population at 6 billion in the year 2000. Why is your estimate less than this?

c) If population trends continue like this, in which year will the number of people on Earth reach 11.6 billion, the estimated maximum population that the planet's known resources can support?

Fig 24.A1 **The suffering caused by starvation, famine and overcrowding seems likely to continue as the human population continues to increase exponentially**

Staggeringly, adding yet another billion people after 1800 took only 123 years, and another billion was added 33 years after that. Since 1960 we have been in the middle of an incredible population explosion with another billion people being added every 12 years.

4 Experts think that population growth will slow down during the next century and stablise at around 11 billion people. What factors are likely to be responsible?

Most of the Earth's land surface is now inhabited by humans. Table 24.A1 shows the 15 most populated countries today and gives an estimate of how their individual populations will grow in the next 50 years.

Table 24.A2 **Human population, in millions, of the 15 most populated countries in the world with estimates of their population in the year 2050 AD**

2000		2050	
Country	Population	Country	Population
China	1232	India	1530
India	945	China	1514
USA	269	Pakistan	357
Indonesia	200	USA	348
Brazil	161	Nigeria	339
Russia	148	Indonesia	318
Pakistan	140	Brazil	243
Japan	125	Bangladesh	218
Bangladesh	120	Ethiopia	213
Nigeria	115	Iran	170

6 Explain the different reasons why China, Indonesia and the USA all have a lower place in 2050 compared to the year 2000.

7 Imagine that it is the year 2100. You have been asked to write a book telling the history of the previous century. The population of Earth stands at 3 billion people. Explain in about one thousand words what happened.

25 Human activity and the environment

MANY PROBLEMS IN the environment happen when people use a resource for today, hoping that everything will be okay tomorrow. In the Philippines sea-horse fishing is very important to the poor fishermen and their families: they can sell sea-horses to traders who then sell them on to the lucrative Chinese medicine market in Hong Kong and elsewhere. Without the sea-horses, many of the people would not be able to feed their families.

As fishing continues, the sea-horses have started to disappear. Fewer sea-horses means that the fishermen catch fewer and earn less and so start to catch smaller and younger sea-horses to supplement their income. The population is further depleted and a vicious circle is set up.

In a ground-breaking environmental project, scientists have studied the life-cycle and behaviour of sea-horses and are working with local fishermen to try to change fishing habits, so that the sea-horse population recovers.

Rather than take a standard environmentalist hard-line view and try to persuade the villagers to stop fishing altogether to protect all the sea-horses, the team recognises the conflict between human need and the needs of the environment. They have shown the fishermen that an enforced 'safe area' will allow sea-horses to breed without threat, and have taught them to avoid catching pregnant male sea-horses. In this way, the scientists have proved to the local community that they can help to restore sea-horse numbers. Not only will the sea-horse population be saved for the future, the fishermen will get back their source of income and will be free from the worries of starvation.

Sea-horses are unique in the animal kingdom: the male not the female gets pregnant and gives birth

1 WHY DO HUMANS AFFECT THEIR WORLD SO MUCH?

We humans are a unique species: we live in most of the ecosystems of the world, adapting our habitats to suit us (Fig 25.1), rather than having to adapt to the environment as most other species do. And, unlike other species, our behaviour affects the environment on a global as well as a local level.

The main human activity that affects the world environment is farming. Close behind is industry and mining for fuels such as coal, oil and gas. The more technologically advanced the society, the

Fig 25.1 **We think of the British countryside as 'natural' but, in fact, it is almost entirely artificially made. Only a few areas in Scotland, Wales and Northern England remain in their 'original' form – all farmland and most forests are the result of human influence on the environment**

greater the overall environmental impact of each individual person. So, for example, a child born in Britain into a fairly well-off family may live for 80 years or more. In that time the amount of resources they use is enormous: think of the furniture, electrical appliances, cars, fuel, food, clothing, etc used up by someone like this. At the other extreme, a child born in a poor family in India or South America might live only 20 years (Fig 25.2). They could live in poorly built housing, or be homeless and forced to live on the streets, probably never get quite enough to eat, and would certainly never use a car or any significant amount of fuel.

In this chapter, we take a brief look at some of the complex issues involved in controlling the effect people have on the ecology of the Earth. There is room only to give an overview of some of the problems and potential solutions.

Fig 25.2 **In many areas of the world, extreme poverty, homelessness and malnutrition affect millions of people**

2 FARMING AND ITS EFFECTS ON THE ENVIRONMENT

It is very common to hear environmentalists complain that farming is destroying hedgerows and wildlife habitats, that woodlands and forests are being cut down to make new farmland and that it should all be stopped. In addition, farmers are often portrayed as hard-line business men and women who put profit before anything else. There is a clear conflict here, but the issue is not as simple as these points of view suggest.

We must consider five basic facts:

● People need to have food to eat: staple foods need to be cheap.

● There are more people on Earth to feed than ever before.

● Farmers need to make a living.

● Farming needs land.

● Some farming techniques are damaging the environment so badly that the capacity of the land to produce food is being reduced.

The key to solving the problem of food production is to develop farming methods that are efficient enough to provide the amount of food that we need in the short term, but which protect the environment to ensure that food production levels can be maintained, or even increased, in the future.

Food production on a global scale

On a global scale, food production has increased during the last 40 years (Fig 25.3). Until the early 1980s, this increase matched the increase in world population, so that, overall, more people had enough to eat than ever before. There were still, of course, countries and areas where food supply did not meet demand and many millions of people were malnourished or starving.

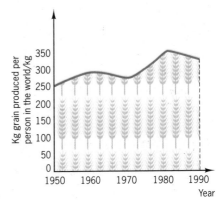

Fig 25.3 **The amount of grain produced per person in the world between 1950 and 1990**

Since 1984, population growth has increased faster than food production. Part of the reason for this has been the damaging effects of farming practices which produce food in the short term, but lead to long-term environmental damage, particularly soil erosion and desertification (see page 380). Very poor countries where food production has fallen since the mid-eighties have tried to compensate by importing food from Europe and the USA where more food is produced than is needed by the local population. However, this only increases the debt of the poor countries, leading to more poverty and, in the long term, more malnutrition and starvation.

Many scientists believe that soil erosion, water shortages and pollution from excessive use of pesticides and fertilisers will lead to a crisis in food production during the twenty-first century. Only a change in farming techniques might prevent this.

The concept of sustainable agriculture

The problem of providing food for an increasing world population without reducing the capacity of the land to produce food is a difficult one, but it has some solutions. By using farming techniques which are sustainable, rather than successful in the short term only, many areas of the world should be able to continue food production at the right level for the foreseeable future.

Moves towards sustainable agriculture would include:

- Using new varieties of crop plants, perhaps produced by genetic engineering, which can survive and give good yields in poor soils, dry conditions and without the need for expensive pesticides and fertilisers (Fig 25.4; see also Chapter 31).

- Making use of non-traditional food sources – more plants that have nitrogen-fixing bacteria in their roots (these need no fertiliser), insects (many cultures of the world eat insects and their larvae as protein-rich delicacies) and unusual animals. In Europe, for example, ostrich farming is becoming more widespread.

?

A What is the different between a sustainable resource and an unsustainable one? Think of an example of each.

Fig 25.4 **This scientist in the Philippines is testing the disease resistance of genetically modified varieties of rice**

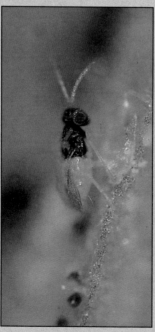

Fig 25.5 **Until recently the only biological control for whitefly, a common plague on greenhouse-grown tomatoes (left) was *Encarsia formosa* (centre), the parasitic wasp which eats only adult flies, leaving larvae to develop and so cause a recurring problem. Gardeners and commercial growers can now obtain the beetle *Delphastus pusillus* (above) which kills larvae as well as adult flies, so providing much more effective and long-lasting control**

- Reducing the amount of fish caught in the seas to levels at which the sizes of the populations do not continue to decrease. At the same time, the number of fish farms could increase.

- Encouraging and rewarding farming methods which do not lead to soil erosion and other environmental problems.

- Using biological pest control (Fig 25.5), rather than artificial pesticides which can persist in the environment.

- Encouraging farmers in poor countries to grow a selection of crops that can feed local people, rather than cash crops such as tobacco and coffee for export.

- Increasing financial and technological aid to poor countries to allow them to set up more sustainable methods of farming and food production.

3 HUMAN ACTIVITIES AND POLLUTION

(a)

(b)

(c)

(d)

Fig 25.6 **Examples of how we humans pollute our environment.**

Fig 25.6(a) **The outlet pipe in the background is discharging raw sewage which is being washed up onto this beach**

Fig 39.6(b) **Effluent from the chemical works is polluting the river which supplies it with vital water**

Fig 25.6(c) **This landfill site will eventually be levelled and should look more pleasant, but methane production under the surface might still cause problems**

Fig 25.6(d) **Pollutants in smoke from industrial plants add to the acid content of rain**

Pollution occurs when substances are released into the environment in harmful amounts, usually as a direct result of human activity (Fig 25.6). Most pollution results from excessively high concentrations of substances, but there are exceptions.

Some pollutants are harmless substances already found in the natural environment. They cause problems when they are produced in large quantities by human activity; examples include carbon dioxide and organic waste (sewage). Other pollutants produced by human activity are compounds that are known to be toxic, such as pesticides. These often remain in the environment for a long time because there is no living organism to break them down. A third type of pollutant is a chemical that pass all the safety tests, but then has totally unexpected effects, as chlorofluorocarbons have done. Pollutants can affect the air, the sea, waterways and the land.

Fig 25.7 **Unpolluted rain has a pH of about 5.6 because of the CO_2 that occurs naturally in the atmosphere: rainwater in most industrialised countries has a pH of between 4 and 4.5. Some cities in the US are bathed in acid fog with a pH as low as 2.3 which is 1000 times as acidic as normal rain, and the same acidity as lemon juice**

Fig 25.8 **As they are carried in the air, oxides of sulphur and nitrogen form secondary pollutants such as nitric acid vapour, droplets of sulphuric acid and particles of nitrates and sulphates. These chemicals reach the surface of the Earth in two forms: *wet*, as acid rain, snow or cloud vapour; or *dry* as acidic particles**

Fig 25.9 **These trees in the Czech Republic have been killed by acid rain that has resulted from emissions of sulphur dioxide and oxides of nitrogen given off by heavy industry all over Europe**

Air pollution: acid rain

To many people, walking through a high Swiss meadow in summer during a shower of rain, sounds idyllic. Peace and quiet, fresh mountain air and refreshing rain. But what if the rain has the pH of vinegar or lemon juice (Fig 25.7)?

In Switzerland, Norway, Sweden and other parts of Scandinavia, acid rain falls on some of the most unspoilt natural landscape in the world. It starts off, as far as 1000 kilometres away, as emissions of sulphur and nitrogen oxides released from the more highly industrial areas of Europe. Although there are some natural sources of these gases, by far the largest amount is generated by burning fossil fuels: the petrol used to power road vehicles and the oil, coal and gas used in power stations (Fig 25.8).

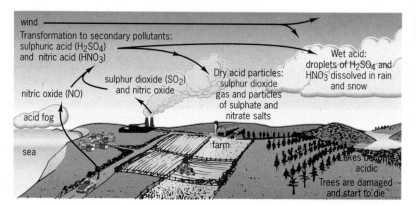

The effects of acid rain on the environment are still causing arguments. It is quite difficult to get good evidence to demonstrate a definite link between the occurrence of acid rain and the effects that ecologists and environmental scientists have noticed. For example, many European and North American forests have been damaged during the last 30 years, especially at high altitudes and at the edges of some forests which have large areas of dying trees (Fig 25.9). In Britain the pattern of loss of lichen species seems to follow the distribution of sulphur dioxide pollution. In fact, lichens are so sensitive to this sort of pollution that they are used as pollution indicators. Other possible effects of acid rain are listed in Table 25.1.

Many countries, including Britain, have now recognised acid rain as a serious problem and have reduced their emissions of sulphur dioxide in the last 25 years.

Possible effect of acid rain	How it might happen
Fish deaths in Scandinavian lakes and elsewhere	Indirect mercury poisoning: more acidic water is thought to convert inorganic mercury compounds in lake sediments into compounds which are soluble in the fatty tissues of fish
	Indirect aluminium poisoning: in acidic conditions, more aluminium ions are released from the soil and washed into lakes. Fish respond by making excess mucus which clogs their gills with fatal results
Damage to statues and buildings	Chemical reactions between components of acid rain and rock
Overstimulation of plant growth	Acid rain falling on the ground means excess nitrogen. The plants then use other nutrients faster, reducing soil fertility
Thought to make human respiratory problems, such as asthma, worse	Irritation of surface membranes in lungs and bronchi

Table 25.1 **Possible effects of acid rain**

Air pollution: the ozone layer

The effects of acid rain occur hundreds of miles away from the source of the pollutants that cause it. This is bad enough, but other pollutants affect the whole planet. Examples include the effect of greenhouse gases on global warming (see Chapter 23) and ozone depletion.

Ozone, O_3 is a form of oxygen: its molecules contain 3 oxygen atoms, rather than the 2 that occur in molecules of atmospheric oxygen. Both gases have been a major feature of the Earth's atmosphere for the last 450 million years. Oxygen is vital to life: it is the fuel for aerobic cell respiration. The role of ozone is less direct: it forms a sort of sunscreen high up in the atmosphere, shielding the biosphere from the harmful effects of ultraviolet light that beams down from the Sun.

Chlorine- and bromine-containing compounds, mainly chloro-fluorocarbons (CFCs), that we released into the atmosphere during the 20th century seem to be causing a measurable thinning of the ozone layer (Fig 25.10). This could cause serious problems in the future.

Ozone (O_3) is present in the atmosphere as a whole in very small amounts, a few parts per million at the most. However, it is not evenly spread. It occurs as a definite layer in the stratosphere, 15–50 kilometres above the Earth's surface. Low level ozone occurs at ground level when bright sunlight acts on pollutants produced by heavy traffic – we see this as photochemical smog in large cities in high summer.

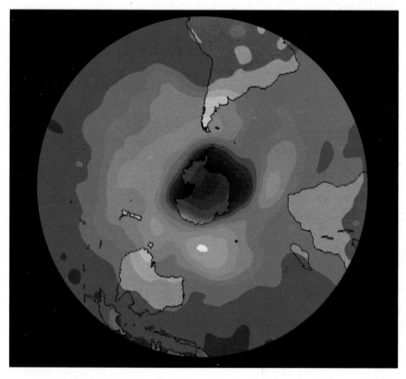

Fig 25.10 **In the mid-1980s, scientists discovered something unexpected: sunlight on the Antarctic in spring causing immense destruction of ozone and a huge hole appearing in that part of the ozone layer. This 'ozone hole' has been noticed every year since, and appears to be growing**

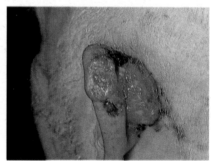

Fig 25.11 **In countries such as Australia, New Zealand, South Africa and Chile, which are protected only by a very thin layer of ozone for many months of the year, there have already been many more cases of skin cancer than ever before. Children in Australian schools are required by law to wear hats and blocking creams whenever they go outside**

Originally, CFCs were thought to be 'ideal' chemicals: they were non-flammable, non-toxic, non-corrosive, cheap to make and could be used as coolants in fridges, propellants in aerosol spray cans, and in the manufacture of the packing material Styrofoam. However, in the mid-1970s, scientists showed that CFCs rise slowly into the stratosphere and are then converted by the action of ultraviolet light into chlorine atoms. These accelerate the breakdown of ozone into O_2 and O. CFC molecules can stay in the stratosphere for about 100 years, and each one can destroy hundreds of thousands of molecules of ozone.

CFCs have now been phased out in most countries of the world, but their effects continue. Even if no more CFCs were put into the air from tomorrow, the ozone layer would still take about 100 years to recover from the worst of the damage. Less ozone in the stratosphere allows more of the harmful ultraviolet radiation from sunlight to reach the Earth's surface. Scientists predict that this will lead to a huge increase in skin cancers and cataracts (Fig 25.11).

Brands of products which carry the phrase 'CFC-free' may do so as a marketing ploy to persuade the consumer that they are more 'environmentally worthy' than other products. In fact, by law, no aerosols sold in Britain contain CFCs.

Water pollution: getting into hot water

Fig 25.12 **As well as causing the death of fish directly, by depriving them of oxygen, warm water can also encourage the growth of parasites which would otherwise not be a problem**

When you think of water pollution, you probably think of chemical effluent, outflows of raw sewage or huge oil spills. While these undoubtedly are examples of pollution, probably the most overlooked cause of damage to inland aquatic ecosystems is the warm water released as a by-product of many industrial processes. Fish and water-living invertebrates are killed by warm water, not because they are scalded directly, but because warm water can carry much less dissolved oxygen than cool water and they 'suffocate' (Fig 25.12).

Water pollution and sewage

Contamination of rivers, lakes and seas by sewage causes two main problems. Firstly, it can introduce potentially dangerous pathogens (organisms that can cause disease) into the environment (see Fig 25.6(a)). This is a particular problem if the water is drunk by animals or people. The second problem is that sewage and other organic wastes are decomposed by the action of aerobic bacteria. These use large quantities of oxygen, leaving less for other organisms living in the water. Sewage contamination can lead to **eutrophication** (see Fig 24.3).

Biological oxygen demand

It is easy to find out the oxygen requirements of water from a particular site by taking a sample of the water and measuring its oxygen content. After keeping it sealed and in the dark for 5 days at 20 °C, the oxygen content is measured again. The rate at which the oxygen has been used up is the **biological oxygen demand (BOD)**. The BOD of unpolluted river water is about 3 mg O_2 per litre (dm^3) of water, per day. Raw sewage uses over *100 times* more oxygen in the same time (Table 25.2).

Although pollution by sewage is a major cause of eutrophication, there are others. If large amounts of nitrate- and phosphate-rich fertilisers are put onto farmland, the excess can run off and contaminate local rivers and lakes. Industrial processes which pollute local waterways with large amounts of concentrated sugars or other organic waste products can also cause the BOD of the water to increase significantly.

Find out more about eutrophication in the Assignment at the end of this chapter.

> ✔
>
> Organic pollution consists of sewage, a mixture of faeces and urine. It contains salts, urea (a nitrogen-containing compound – see page 160), dead gut-lining cells, undigested food and living bacteria.

> ✔
>
> The term **eutrophic** means over-fertile. So, a eutrophic river or lake contains more nitrates and phosphates than normal, and abnormal growth, particularly of algae, can occur.
>
> The opposite of eutrophic is **oligotrophic**: this means under-fertile. So an oligotrophic river cannot sustain much life.

Table 25.2 **Typical BOD values per day of different types of water**

Water type	BOD value per day
Clean river water	3 mg dm^{-3}
Water from polluted stream	10 mg dm^{-3}
Domestic sewage, untreated	250–350 mg dm^{-3}

Contamination of water by oil and detergents

The names Exxon Valdez, Braer and Sea Empress may sound familiar, even if you can't quite remember that they are oil tankers that have famously run aground causing huge oil spills (Fig 25.13). These pollution disasters make headline news but, in 1993, a study by Friends of the Earth estimated that over 1000 times the amount of oil carried by the Exxon Valdez (which polluted the coast of Alaska in 1989) is lost in the US each year – not through national catastrophes, but through the normal operations of washing tankers and releasing the oily water, from pipeline and storage tank leaks and from accidental loss from offshore oil wells.

The effects of oil spills on ocean ecosystems depend on:

- The amount of oil released.
- How far away from the shore the spill happens.
- The weather conditions, water temperature and speed of ocean currents.
- The type of oil released: an area affected by a spill of refined oil takes over three times longer to recover than a similar area that has a crude oil spill.

In the clean-up operation following a large spill, oil is dispersed by detergents. The aim is to spread the oil molecules to limit damage to the local environment. Unfortunately, many detergents are extremely toxic and lead to many sea-bird deaths.

25.13 **This cormorant was caught in the oil spill which happened at Manobier beach, Wales in February 1996 after the tanker Sea Empress ran aground. Birds covered in oil cannot swim very easily and sometimes tire and drown. Those which end up on the beach try desperately to clean their feathers, ingesting and inhaling large amounts of oil which damages their digestive system and often causes pneumonia**

Spoiling the land

People can pollute areas of land directly by dumping toxic waste, by industrial processes such as mining and by simple neglect: leaving non-biodegradable rubbish behind after a picnic at a local beauty spot, for example. However, human activity, particularly farming, is indirectly causing two main problems: **deforestation** and **desertification**. Both have far-reaching global effects.

Deforestation

Deforestation has been practised by humans for centuries. As human population and technology have increased, so has the need to use wood. (The discovery of fire has had an enormous impact.) Many European forests were cleared by various different settlers from about the fourteenth century onwards. Today, the human population is larger than ever before and the process of deforestation has now spread to all parts of the world, particularly the tropical rainforests.

Deforestation is a process which results from the conflict between immediate human need and the necessity to protect the environment from serious and long-term damage. Tropical hardwoods are prized as building materials throughout the world, and other types of wood are used to make the thousands of tons of paper that we use every day. Many people in the poor countries of the world depend on wood or charcoal for fuel, and wood, twigs, crop residues and grass are used for fuel in many of the 'better off' countries such as China, Brazil and Egypt. As well as cutting down forests to use the wood they provide, deforestation is also carried on to clear land for farming. People have cleared large areas of forest for plantations of other trees (rubber and palm oil), and for cattle ranches and enormous wheat fields, in order to fulfil the needs of the human population for more raw materials and more food.

The consequences of deforestation are potentially serious. Locally, clearing the land of trees and plants leads to increased soil erosion (see below). Globally, reducing the biomass of productive trees may contribute to the greenhouse effect (see page 352) as less carbon dioxide is used in photosynthesis. Also, loss of rainforest, one of the most diverse ecosystems of the world, is likely to affect humans and other organisms in ways we cannot predict, although we already know that many of the plants found in rainforests are useful sources of powerful drugs (see page 367). Balancing the needs of people with those of the environment is difficult.

Desertification

Soil is an important abiotic factor that determines the distribution of organisms in an ecosystem: few plants can grow without the anchorage of soil or the water and minerals it supplies to them. When the soil is lost, a process called **soil erosion** occurs. This is a worldwide problem.

Soil erosion is a natural process: wind and water movements move surface debris and topsoil from one place to another. In ecosystems that are undisturbed by human activity, healthy plant roots 'bind' the soil, and soil is lost and made at about the same rate. Problems occur when the soil is removed faster than it is formed: this can happen because of farming, deforestation, building, over-grazing by animals, physical damage by off-road vehicles or fire and some kinds of mining.

Farming is probably the worst culprit. Topsoil is eroding on about a third of the world's cropland. Each year, the land must feed 90 million more people with about 25 billion tonnes less topsoil. In Africa, soil erosion is now 20 times faster than it was 30 years ago. The cattle-producing areas are now severely affected by desertification (Fig 25.14). But, there are solutions to soil erosion and desertification, and many of these are starting to be put into practice.

Many countries world-wide are now working together to try to reduce deforestation:

People are being encouraged to manage forests sustainably. This means using different techniques to make sure we have the wood we need without destroying large areas of forest.

People are being persuaded to use paper substitutes.

There is growing investment in rainforest communities to find new medicines from plants.

Eco-tourism is being promoted. If people pay to see wildlife in its natural surroundings, this contributes to the income of an area and makes it less likely that habitats are destroyed.

Fig 25.14 **Areas of severe desertification occur all over the world. Desertification is defined as the process which reduces the productive potential of hot dry areas by 10 per cent or more. Moderate desertification is a 10–25 per cent drop in productivity: severe desertification is a 25–50 per cent drop. At its most severe, desertification actually creates deserts**

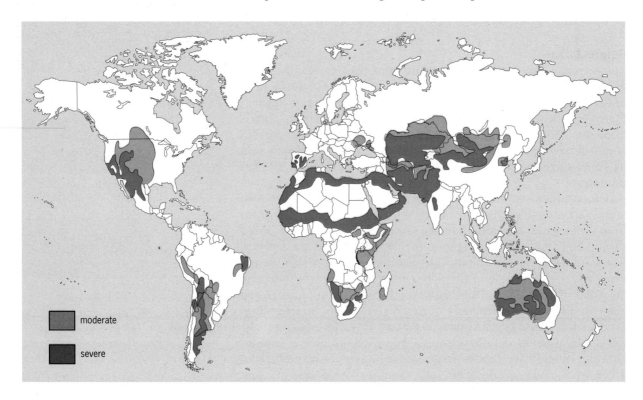

moderate

severe

SUMMARY

After reading this chapter you should know and understand the following:

■ Humans influence the environment of the Earth on a *global* as well as a *local* level.

■ Farming is the major human activity that causes environmental damage.

■ People need to farm to feed the increasing human population but many farming techniques cause long-term damage to the land, lowering its capacity to grow food in the future.

■ To avoid this problem it is important that farmers start to use **sustainable** methods of agriculture.

■ Many human activities cause **pollution**. Pollutants can affect the air, the sea and rivers and lakes, and the land.

■ Acid rain, use of CFCs and other substances which destroy the ozone layer, and increased production of greenhouse gases all contribute to air pollution.

■ Water can be contaminated by organic pollution. This can lead to a sequence of events called **eutrophication**. Heat, oil and other substances can also pollute waterways.

■ When forests are cleared to make new farmland, fewer trees are available to photosynthesise, we lose species diversity and the land becomes vulnerable to **erosion**.

■ Severe soil erosion can lead to **desertification**. Many sustainable methods of agriculture try to minimise the effects of soil erosion.

QUESTIONS

1 The map shows a small area in the Midlands. The sites indicated with letters **A** to **C** are sources of pollution.

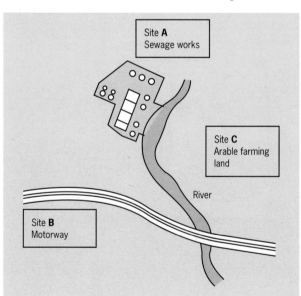

Choose any **two** of sites **A** to **C**. For these sites, suggest the main pollutants associated with **each** and explain how these pollutants affect living organisms and ecosystems.

[NEAB June 1995 Biology: Ecology Module Test, q.8]

2 One cause of acid rain is the release of sulphur dioxide into the atmosphere.

a) Describe the principal way in which sulphur dioxide is released into the atmosphere.

b) Describe **two** specific effects of acid rain on living organisms.

c) The map (Fig 25.Q2) shows the percentage of trees which have been damaged by acid rain in different European countries.

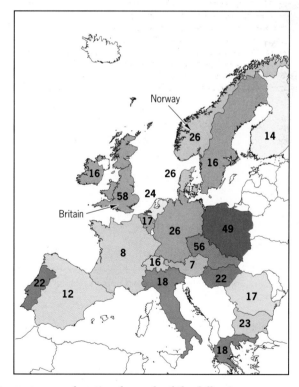

Suggest an explanation for each of the following:
  (i) the high percentage of damaged trees in Britain;
  (ii) the fact that Norway has a high percentage of damaged trees even though it is sparsely populated.

[NEAB February 1996 Biology: Ecology Module Test, q.6]

Assignment

JOURNEY ALONG A RIVER

As Fig 25.A1 shows, rivers tend to be relatively unpolluted near to their source. As they flow on towards the sea they are contaminated with organic pollution and industrial effluent, creating increased levels of pollution and eutrophication.

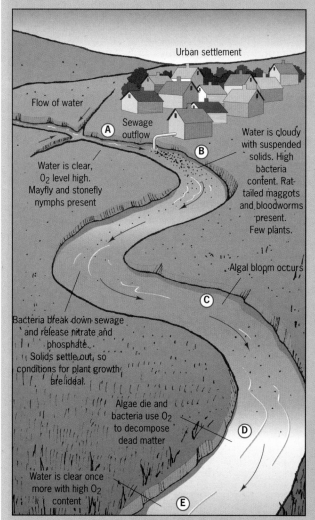

Fig 25.A1 **This diagram shows the different conditions found along a river**

It is often difficult to tell exactly how polluted the water is at any point, just by looking at it. We need to sample the water and sample the organisms that live there to find out whether the water is safe.

There are two easy methods which allow us to find out which organisms live in a particular part of the river. The first is to do a kick sample. The number and diversity of organisms found are used to give the stretch of waterway a diversity index (see page 368).

1 What sort of diversity index would you expect to find in:
a) a stretch of unpolluted river, such as point **A** in Fig 25.A1?
b) a stretch of polluted river, such as point **B** in Fig 25.A1?

2
a) What factor is likely to be the most important in determining whether water can support a large or a small number of species?
b) What method would you use to measure this factor?

3 Initially, the water at point **B** is turbid (cloudy and murky). From your knowledge of the nature of sewage, explain why few animals or plants can grow in this zone of the river.

4 What do you think the BOD value may be for the water in this part of the river?

5 A student who was feeling particularly brave did a kick sample in this area and found very few species. They were the classic indicator species of heavy organic pollution: rat-tailed maggots, bloodworms and tubifex worms – see Fig 25.A2.

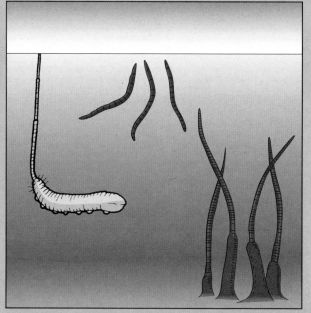

Fig 25.A2 **The larva of a rat-tailed maggot, some bloodworms and tubifex worms. The tails of the tubifex worms protrude from tubes that the worms make using mucus and debris**

a) What is an indicator species?
b) Explain why each of the three species shown in Fig 25.A2 is able to tolerate low oxygen levels.
c) When the water was tested in the laboratory, large numbers of *E. coli* bacteria were found. Using this and the other evidence, what would you conclude about the sewage outfall? (See page 5 for an extra clue.)

More sophisticated investigations of the water further below the sewage outlet produced the data shown in Fig 25.A3.

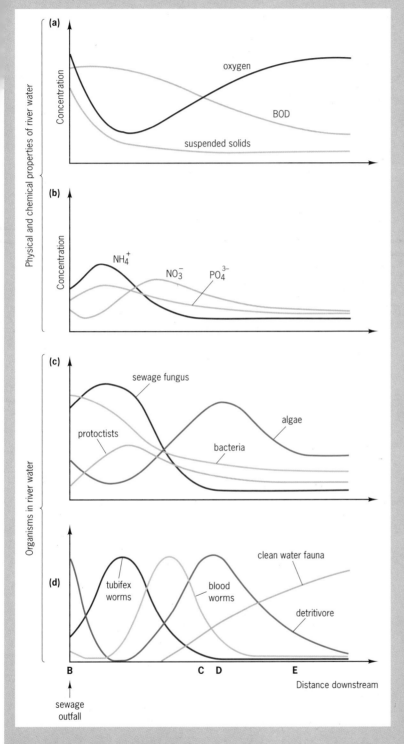

Fig 25.A3 **Changes in the river after a sewage outfall**

6 Look at Fig 25.A3(b).
a) Use your knowledge of the nitrogen cycle to explain why there is first a peak in the levels of dissolved NH_4^+, followed by a peak in NO_3^-.
b) The population of sewage fungus falls to zero, just before point **C**. What does this tell you about the metabolism of the sewage fungus?
c) Why does the level of bacteria not fall to zero?

7 At point **C**, the water begins to look like pea soup – conditions have become perfect for the growth of algae. Contrary to what you might first think, the problem here is that the water is too fertile – it has become eutrophic.
a) Use the graphs shown in Fig 25.A3 (a) and (b) to list three factors which are responsible for the algal bloom.
b) Explain why the algal bloom is harmful to the normal clean-water organisms.
c) Why do the populations of the different invertebrates present in the river water peak at different times?

8 Look at Fig 25.A3(d).
a) Find out the characteristics of tubifex worms and blood worms that allow them to flourish in the river between points **B** and **E**.
b) How do the detritivores help to change the water conditions between points **C** and **E**? Find out the names of some common detritivores that might be found in this sort of river.

Once the algae have absorbed the available nutrients, they start to die. Once again, bacterial activity becomes a problem, as decomposers rot the dead algae and use up large amounts of oxygen. After this, if there is no more organic discharge, the water further downstream begins to recover. Eventually, the water becomes clean once more.

Research and discussion questions

Find out about the basic principles behind sewage treatment. What major health problems does effective sewage treatment prevent?

Find out what the current laws are in the UK which control the discharge of sewage into rivers and into the sea. Do you think they are strict enough?

GENES, GENETICS AND EVOLUTION

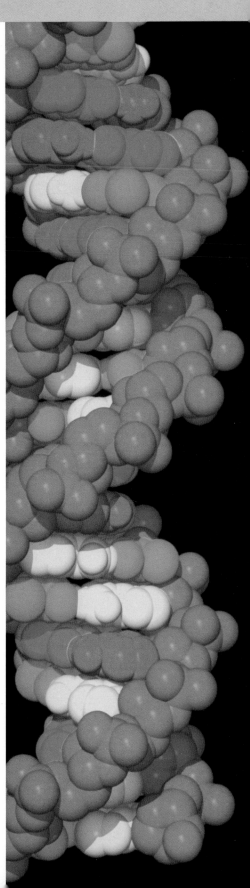

WHY ARE WE all similar? Why are we all different? These are two of the fundamental questions on which the science of genetics is based. When the sperm from your father met the egg from your mother, a unique individual was created, different from anyone who had gone before or will ever exist again.

But, would it surprise you to learn that you have quite a lot of genes in common with bacteria, carrots, pine trees and vampire bats? Certain essential genes – such as those which code for the enzymes of respiration – are very similar in all organisms as they are vital to this basic life process. Such genes must have been copied faithfully, generation after generation, for the last three billion years or so. It is also intriguing to realise that, although almost all organisms die young before reproducing, not a single one of our ancestors did. Their DNA must have had what it takes to be passed on.

In this section we begin by outlining some basic concepts in genetics in Chapter 26. In Chapter 27 we study the cell cycle and in Chapter 28 go on to look in more detail at the DNA molecule: how it codes for the manufacture of proteins, and how it can be copied faithfully, time after time. Chapter 29 focuses on inheritance and particularly on the work of Gregor Mendel. Then, in Chapter 30, we look at the mechanisms of evolution, and consider human evolution briefly, before ending the section in Chapter 31 with an overview of modern genetics. We see, above all, how the DNA technology revolution is likely to have an increasing effect on all of our lives.

26 Genetics: the basics

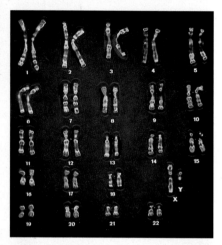

Fig 26.1 **A complete set of human chromosomes. Each body cell contains two sets of 23 chromosomes and each chromosome contains up to 4000 genes. The total amount of DNA in a cell is called the genome. The human genome consists of about 60 000 genes**

In organisms which reproduce sexually, a set of genes from the male combines with one from the female to form a unique individual (see Chapter 19).

1 WHAT ARE GENES?

Genes are the 'instructions' that control everything that happens inside a cell. Physically, genes are different sized stretches of DNA that code for the manufacture of proteins such as enzymes. Enzymes organise the chemistry that goes on in our bodies and are behind all the life processes, including growth, repair and reproduction. Throughout our lives, the genes remain in the nuclei, directing the activities of each of our cells.

Children grow and develop according to the genes they have inherited from their parents. It's a harsh fact, but nothing lives forever: the inevitable fate of all organisms, often sooner rather than later, is death. However, although we die, our genes live on in our offspring.

2 STUDYING GENETICS WITH THIS BOOK

Studying genetics means learning about DNA and what it does. In this section, we divide our study of this remarkable molecule into five areas:

- The **cell cycle** is the study of how cells divide. Discovering just how a single fertilised egg can develop into an organism as complex as a human is one of the central challenges of biology. When the control of cell division goes wrong, it can lead to cancer. Find out more about mitosis and meiosis, the two types of cell division, in Chapter 27.

- **Molecular biology** is the branch of science which deals with all aspects of the DNA molecule: its structure and function.

  How are genes used to make products? Why are some genes active when others aren't? How can DNA copy itself? How do mutations occur? This is covered in Chapter 28.

- **Genetics** is the study of inheritance: the way in which genetic information is passed on from one generation to the next.

  Why are only some genes passed on? Why are some characteristics always shown while others are hidden? This is covered in Chapter 29.

- **Evolution** looks at how species change and develop with time.

  What is the mechanism of this change? How long does the process take? Can we observe evolution in action? What do we know about the evolution of humans. This is covered in Chapter 30.

- **Genetic engineering** is the popular name for the technology which manipulates the DNA molecule.

  How can we change DNA? How can we transfer DNA between organisms? What are the benefits of genetic engineering? Can we cure genetic disease? Should we pursue this technology at all? This is covered in Chapter 31.

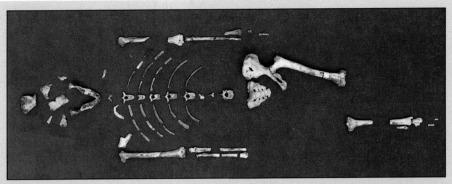

Extracting DNA from fossil remains such as this Australopithecus could give us some clues when piecing together the story of human evolution (see pages 458–462)

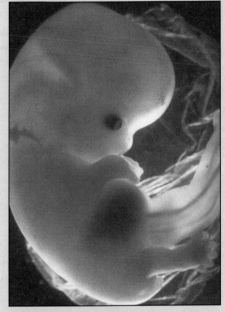

A human embryo at 7weeks after conception. We have developed the technology to clone animals as advanced as humans, but this area of genetics is an ethical minefield

Wheat has been bred selectively to produce a greater yield. Many new types are easier to harvest and are also resistant to disease

This fermenter houses genetically engineered bacteria which contain the human growth hormone gene. Inside the fermenter the bacteria multiply and produce large amounts of human growth hormone

Fig 26.2 **The applications of our knowledge of DNA**

Male and female sperm can be separated according to the amount of DNA they contain. The sperm are stored in liquid nitrogen until needed. Farm animals can be artificially inseminated to produce offspring of the required sex

3 THE POTENTIAL OF GENETICS

Few scientific subjects make the headlines as often as advances in genetics and DNA technology. We are witnessing an accelerating revolution which is likely to have far-reaching effects. Soon, we will have mapped the complete human genome and perhaps that of other species too.

The implications of this are enormous:

- Two out of every three people die for reasons connected to their genes. Although only a small percentage inherit a lethal genetic disease, many more inherit a *tendency* to develop a condition such as heart disease, premature senility or cancer.

  An in-depth knowledge of the human genome will make it possible for doctors to screen babies before or soon after birth for genetic abnormalities and tendencies to develop such diseases. But this raises ethical and moral questions, not just scientific ones: would you want to know just how likely you are to develop cancer in middle age? Would you want anyone else to know – life insurance companies, potential employers, friends or partners? Who would have access to your DNA files?

- We can isolate genes for useful products, such as insulin and growth hormone, and place those genes into another organism. This is **recombinant DNA technology**. Bacteria already produce human insulin and other products on a large industrial scale, and it is also possible to transfer such genes into higher animals such as sheep so that they produce milk containing the required protein.

- We can isolate faulty genes, such as those which cause cystic fibrosis, and replace them with multiple copies of healthy genes. Although still at the experimental stage, this **gene therapy** holds much promise for the future.

- We could be able to halt some of the genetic causes of ageing, and allow more people to live even longer.

Only the bravest scientists would dare to speculate about where our genetic knowledge might eventually take us. It could be to a greatly improved world where suffering is reduced while quality of life and life expectancy are improved. Genetic engineering and agricultural techniques might give us food-producing species that can live in the harshest of conditions, so providing food in areas of most need.

But can we be this optimistic? There are still many things we don't know about living organisms and it is impossible to predict exactly the effects of interfering with an organism's DNA. Already, companies are trying to patent genetic material and its products, leading to hot debate about whether or not such natural products can be owned, like a brand name or an invention. Will genetic advances be used solely to make money, and therefore be available only to those who can pay?

The ethical implications of some very recent developments also need to be considered. The first mammal has now been cloned, but, at present, it is inconceivable that research in this country could ever move towards cloning humans. The controls and safeguards would not permit it. Such safeguards will need to be updated constantly. As you study genetics, keep an open mind.

4 WHAT GOES ON IN THE NUCLEUS?

The nucleus of every human cell contains DNA (Fig 26.3). Each nucleus contains a set number of long, elaborately coiled DNA molecules which condense into chromosomes. Genes, the individual instructions, are dotted along the chromosomes.

In an adult organism most cells are not dividing: the chromosomes are uncoiled, and active genes are producing proteins to control the activities of the cells. Not all genes are active at the same time: controlling which genes are active and which are not enables different cells to carry out different functions. For instance, all cells in the human body contain two copies of the insulin gene. Only in certain specialised cells in the pancreas, however, are the genes switched on and used to make insulin.

Glossary of genetic terms

Like many branches of science, genetics comes with its own jargon, seemingly designed to make some very straightforward ideas inaccessible to the average person. You will need to know the following terms:

Diploid: cells or organisms containing two sets of genes. Human body cells, for instance, are all diploid.

Haploid: cells or organisms containing one set of genes. Human sex cells (egg cells and sperm) are haploid.

Gene: a length of DNA which codes for a particular polypeptide. Some proteins consist of more than one polypeptide; these are coded for by more than one gene.

Allele: one form of a gene. There may be two or more alleles of any particular gene. Diploid organisms contain two alleles, one on each pair of chromosomes. For example, in pea plants, the gene for flower colour may have a red allele and a white allele.

Genotype: the genetic make-up of an individual.

Phenotype: the physical expression of the genotype. Many alleles are expressed, or used, and these contribute to the characteristics of the individual. Other genes – recessive alleles in the presence of a dominant version – are not expressed and so remain hidden.

Dominant: the allele which, if present, is expresssed. If the red allele in the pea plant is dominant, the pea has a red flower, even if it has a white allele as well.

Recessive: the allele which is expressed only in the absence of the dominant. In the pea, if the white allele is recessive, the plant only has white flowers if it has two white alleles.

Homozygous: when both alleles of a particular gene are the same. In the pea example, an individual would be homozygous if it has two red alleles or two white alleles.

Heterozygous: the two alleles of a gene are different. In a pea plant, a heterozygous individual could have a red allele and a white allele.

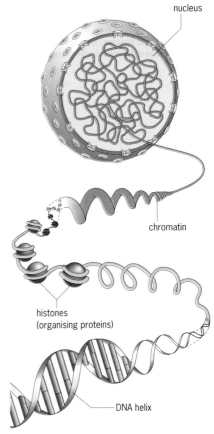

nucleus

chromatin

histones
(organising proteins)

DNA helix

Fig 26.3 **Most of the DNA in the cell is locked away in the nucleus where, in effect, it forms a reference library. The genes, which act as instructions for all cell functions, must be passed on when the organism reproduces**

Chromosomes condense and become visible during cell division, as we see in more detail in Chapter 27.

Normal cell division – **mitosis** – duplicates the DNA and then divides it accurately into two so that the two new cells formed are genetically identical.

Meiosis is the type of cell division which produces haploid cells such as eggs and sperm. This process halves the amount of genetic material that goes into each new cell and shuffles it at the same time so that no two sex cells are the same.

Fig 26.5(a) **Height in human males is an example of continuous variation. Most men have values around the middle, giving a bell-shaped curve known as a** normal distribution

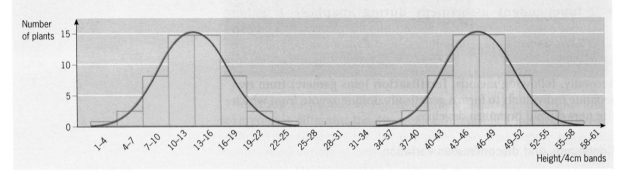

Fig 26.5(b) **Height in pea plants is an example of discontinuous variation. Individual plants are either dwarf or tall. Although there is some variation in both categories, there is no overlap: individuals fall into one category or the other**

6 A BRIEF HISTORY OF GENETICS

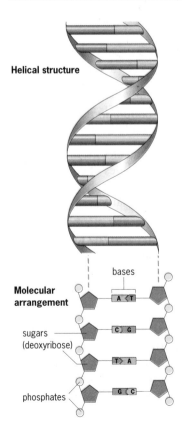

Helical structure

Molecular arrangement

bases

A ⟨ T

C ⟩ G

T ⟩ A

G ⟨ C

sugars (deoxyribose)

phosphates

The history of genetics gives an insight into the changing nature of scientific progress. Early scientific discoveries and breakthroughs were few and far between, often made by individuals who observed and studied nature and put forward radical theories to explain what they had seen. Mendel (page 425) and Darwin (page 444) are classic examples.

More recent discoveries have been made by teams of dedicated people working together, using the scientific method and latest technology to build on knowledge accumulated by others. Table 26.2 outlines some of the landmarks in this most dynamic of scientific areas.

Fig 26.6 **The discovery of the structure and function of DNA is one of the greatest scientific discoveries of the twentieth century. The DNA molecule stores genetic information and, with the help of enzymes, makes exact copies of itself, time after time**

Year	Scientist(s)	Discovery
1858	Charles Darwin, Alfred Russel Wallace	Jointly announced their theory of evolution by natural selection: 'Survival of the fittest'
1859	Charles Darwin	Published his book *The origin of species*
1866	Gregor Mendel	Published his laws of genetics following studies on pea plants. His findings were ignored
1900	Hugo de Vries, Carl Correns, Eric von Tshermak	Discovered meiosis, providing an explanation for Mendel's laws: his work is rediscovered by all three and gains recognition
1905	Nettie Stephens, Edmund Wilson	Independently discovered the principles of sex determination: XX = female, XY = male
1910	Thomas Hunt Morgan	Proposed the theories of sex linkage, mutation, linkage and chromosome maps following work on *Drosophila* (fruit fly)
1928	Fred Griffiths	Proposed that some 'transforming principle' had changed a harmless strain of bacteria into a lethal one. The hunt for DNA began
1941	Beadle, Tatum	Irradiated the bread mould *Neurospora*, producing mutations which suggested that genes code for enzymes
1944	Oswald Avery	Purified the 'transforming principle' in Griffiths' experiment, showing it to be nucleic acid (DNA)
1950	Erwin Chargaff	Discovered that the base pairing ratios in DNA were always the same, whatever the organism (Chargaff's principles: A=T, C≡G)
1951	Rosalind Franklin	Obtained high-quality X-ray diffraction studies of DNA, showing that it has a helical structure
1952	Hershey, Chase	Used bacteriophages to show that DNA, not protein, is the material of heredity
1953	Crick, Watson	Built on the work of Chargaff and Franklin to work out the three-dimensional structure of DNA
1958	Meselson, Stahl	Used radioisotopes of nitrogen to prove the semi-conservative mechanism of DNA replication
1958	Arthur Kornberg	Purified DNA polymerase from *E. coli* and used it to make DNA from nucleotides in a test tube
1961	Jacob, Monod	Propose a mechanism for switching genes 'on' or 'off': the operon hypothesis of metabolic control
1966	Marshall Nirenberg, Gobind Kharana	Cracked the genetic code: particular triplets of bases on DNA code – via mRNA – for the 20 amino acids
1970	Smith, Wilcox	Isolated the first restriction enzyme, *Hindll*, which can cut DNA
1972	Paul Berg, Herb Boyer	Produced the first recombinant DNA molecules
1973	Annie Chang, Stanley Cohen	Showed that DNA could be inserted and cloned inside a bacterium
1977	Fred Sanger	Developed a method for sequencing DNA
1977		The first Genetic engineering company (Genentech) is founded, using recombinant DNA to make pharmaceuticals
1983	James Gusella	Located the gene responsible for Huntington's chorea on chromosome 4 (see Fig 30.1)
1984	Cary Mullis	Developed the polymerase chain reaction (PCR) in which minute samples of DNA can be copied, prior to analysis
1984	Alec Jeffreys	Developed DNA profiling (or 'fingerprinting')
1988		The Human Genome Project began to map the human DNA sequence
1989	Francis Collins	Identified the gene (CFTR) responsible for cystic fibrosis on chromosome 7
1990	French Anderson	First attempts at gene therapy: T-cells of a 4-year-old girl were exposed to viruses containing working copies of her defective gene. After treatment, her immune system began working again
1994		Genetically engineered 'Flavr savr' tomatoes went on sale
1995		Transgenic sheep made to express human genes in their mammary glands, so that they produce milk containing valuable pharmaceuticals
1997	Roslin Institute, Edinburgh	First mammal cloned. Dolly the sheep fuels controvesy about genetic research
1998		Dolly the cloned sheep gives birth to lambs conceived normally.

Table 26.2 **Some landmark discoveries in genetics. In many of these, the scientists mentioned by name led a much larger team of scientists who worked towards the discovery described**

IN A HEALTHY human adult, cells divide only when they should. Some cells, such as those that line the gut, are replaced at a remarkable rate. Other cells, such as muscle cells, live longer and need to be replaced far less often. Occasionally, the systems that control cell division break down, and a cell that should be stable divides uncontrollably. Soon, a mass of tissue called a tumour forms. Tumours can be either benign or malignant.

A benign tumour is not cancerous. Cells divide within a small, confined area and the growth is often surrounded by a membrane. New cells form in the centre of the mass. This type of tumour can be dangerous if it presses on important organs, or if it grows very large. It is usually treated by surgery. Because the tumour cells are confined and do not invade other tissues, recovery and survival rates are good.

A malignant tumour is commonly known as cancer. The tumour grows at the edges, spreading out and invading the surrounding tissues (supposedly like a crab – hence the name cancer). This type of tumour is much more dangerous. Vital tissues and organs can be destroyed quickly and this can lead to death. Even if the tumour is removed from the body, actively dividing cells may have already broken off and set up secondary growths elsewhere in the body.

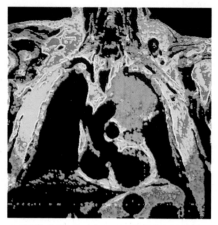

Highly sophisticated imaging techniques can be used to diagnose tumours. This CT (computer tomography) scan shows a large, spherical pink tumour in the person's upper left lung (the heart is in the centre)

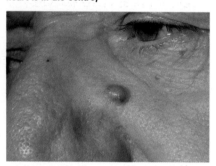

Skin cancer is common, especially in countries such as Australia, where fair-skinned people are exposed to strong sunlight

1 SOME BASIC PRINCIPLES

An understanding of cell division is of great importance in the study of human biology. It is the process by which we grow and develop. As adults, our body uses the main type of cell division for maintenance and repair. When control of cell division is lost, the result can be cancer (see Chapter 35). Another type of cell division halves the chromosome number in eggs and sperm (the sex cells, or **gametes**) so that when fertilisation occurs, the resulting zygote has the normal double number.

Humans, like all animals, are eukaryotic organisms: in eukaryotic cells, DNA is organised into chromosomes and is enclosed within a double nuclear membrane. Before going on to study the mechanism of cell division in eukaryotic cells, familiarise yourself with some of the basic terms and concepts by looking at Table 27.1.

An introduction to cell division

In humans, the nucleus of each **body cell**, or **somatic** cell, contains two sets of chromosomes and so has two sets of genes. Body cells are described as **diploid** (from Greek, meaning 'double number')(Fig 27.1).

Table 27.1 **Some basic terms important in cell division**

Term or feature	What it is or what it means
Gene	A length of DNA that codes for the production of a particular polypeptide or protein
Genome	The name given to the full set of genes in a cell. The human genome consists of between 50 000 to 100 000 genes (the exact number is not yet known)
Diploid	Diploid cells contain two versions of every gene
Haploid	Haploid cells contain one version of every gene
Chromosome	A long single molecule of DNA, organised around proteins called histones. The largest human chromosomes contain about 4000 genes. Chromosomes exist in cells all the time but they can be seen only during cell division, when they condense and separate
Homologous chromosomes	Chromosomes exist in pairs – humans have 23 pairs. Homologous chromosomes have the same genes at the same positions, but not necessarily the same versions of each gene
Chromatid	During cell division, the DNA of a cell is replicated (copied). When the chromosomes condense, they therefore appear as double structures: each unseparated chromosome within such a pair is called a chromatid
Sister chromatid	When two chromatids are genetically identical (as in mitosis), they are called sister chromatids
Bivalent	A bivalent is a pair of homologous chromosomes that line up together, as they do during meiosis
Locus	The position of a gene on a chromosome
Transcription	The process in which a molecule of mRNA is assembled on an active gene. mRNA thus becomes a mobile copy of the gene
Translation	The process of converting the code on the mRNA into a protein. This is achieved by protein synthesis: amino acids are joined in a particular order to make a protein such as an enzyme

A 'DNA makes RNA makes proteins'. Explain this statement.

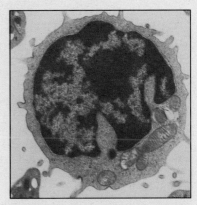

Fig 27.1(a) **A human white cell (a lymphocyte). The nucleus of this cell contains all the genes necessary to make a whole new organism, a clone of the person from whom this cell was taken. Although the technology exists to clone humans, the area is fraught with huge moral, ethical and legal problems**

Fig 27.1(b) **In 1996, Dolly the sheep was the first ever mammal to be cloned from an adult (body) cell**

Mitosis is the process of normal cell division. In mitosis, the chromosomes are copied and then divided equally between the two new daughter cells. So each mitotic division produces two cells, both diploid and each with exactly the same genes as the parent cell.

Some cells in the human body are not diploid. Gametes, the sperm and eggs, contain only one copy of each gene as they have only one set of chromosomes. These cells are haploid and are produced by a special type of cell division called **meiosis**.

A male and female gamete join together at fertilisation, to form a new diploid cell called a **zygote**. This one cell divides by mitosis to produce a complete new organism (Fig 27.2). Reproduction and growth are covered in Chapters 19 and 20.

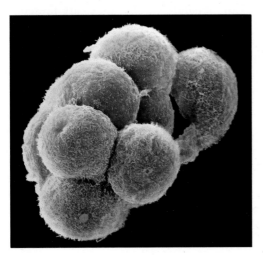

Fig 27.2 **This human embryo is 4 days old, but the fertilised egg has already divided mitotically 4 times, producing a cluster of 16 cells. The cells will soon specialise to form the tissues and organs of the body**

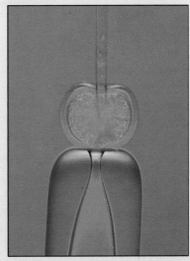

Fig 27.1(c) **A micropipette puts the nucleus from an adult cell into a non-fertilised egg cell that has had its own nucleus removed**

CELL TURNOVER

AS CHILDREN, we grow because the cells in our bodies divide mitotically, and cell production outnumbers cell death. When we reach our adult size, our cell population stays constant. The rate at which cells die and are replaced is known as **cell turnover**.

Research has shown that different tissues and organs have very different turnover rates. Brain cells, for instance, are not usually replaced, and after our twenties there is a slow but steady decline in number. This is not normally something to worry about – you won't run out. Other cells have a low turnover. Liver cells, for example, might divide only once every one to two years although, unlike the brain, they can regenerate if some cells are damaged or destroyed. Most of the high turnover cells are epithelial cells (Fig 27.3 and Fig 27.4). These are the cells of the body that most frequently give rise to tumours. Cells of the bone marrow (Chapter 34) and cells in the testes (Fig 27.5) also divide very frequently.

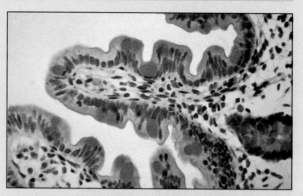

Fig 27.4 **Intestinal epithelial cells have a very rapid turnover: the entire gut lining is replaced every couple of days. Much of the content of these cells is digested and reabsorbed in an efficient recycling system, but even so, gut cells form a significant proportion of faecal matter**

Fig 27.3 **Scanning electron micrograph of human epidermis. Dead skin cells are constantly being lost and replaced by mitosis. A large proportion of household dust consists of human skin cells**

Fig 27.5 **The production of sperm is a remarkably intense process. The average healthy human male produces about 1000 new sperm every second. The process of sperm production,** spermatogenesis, **which involves meiosis, is described in Chapter 19**

2 THE CELL NUCLEUS

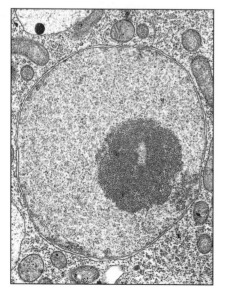

The DNA of a cell is contained in the nucleus. In a cell that is not actively dividing, DNA exists as **chromatin** (see Fig 3.35). This has a granular appearance (Fig 27.6). Separate chromosomes are present, but the DNA is so spread out that we cannot tell one from another.

When a cell is about to divide, the chromosomes condense and separate. They become visible under the light microscope as dark, rod-like structures. Each pair of chromosomes is known as a **homologous pair**. This means that they both contain genes at the same positions, or **loci**. You received one chromosome of each pair from your mother and one from your father. So, although homologous pairs contain the same genes, they do not necessarily carry the same *versions* of each gene. This is the key to understanding why organisms vary. Alternative forms of genes are called **alleles** (see Chapters 26 and 29).

Fig 27.6 **In a non-dividing cell, the chromosomes are spread out as chromatin and appear as pale granules. In this form, the DNA is used as a template to make mRNA in the process of transcription. mRNA passes to the cytoplasm and is itself used as a template for protein production in the process of translation. In this cell, the nucleolus is also clearly visible: it is the dark sphere inside the nucleus**

The appearance, number and arrangement of chromosomes in the nucleus is called the **karyotype**. The human karyotype consists of 23 pairs of chromosomes (Fig 27.7). All diploid human cells contain 23 pairs of chromosomes and this is often given the notation $2n = 46$. This means that a diploid cell has 46 chromosomes (a haploid cell has 23). Other organisms have different numbers of chromosomes (Table 27.2).

?

B Which notation would you use to describe the number of chromosomes in human sperm?

Table 27.2 **The number of chromosomes in the cells of some plants and animals, compared to humans. Note that there is no connection between chromosome number and the complexity of an organism. Some ferns have hundreds of chromosomes**

Species	Number of chromosomes
Penicillin mould (Penicillium notatum)	5
Broad bean (Vicia faba)	12
Lettuce (Lactuca sativa)	18
Yeast (Saccharomyces cerevisiae)	34
Cat (Felis cattus)	38
Human (Homo sapiens)	46
Potato (Solanum tuberosum)	48
Chimpanzee (Pan troglodytes)	48
Horse (Equus caballus)	64
Chicken (Gallus gallus)	78
Dog (Canis familiaris)	78

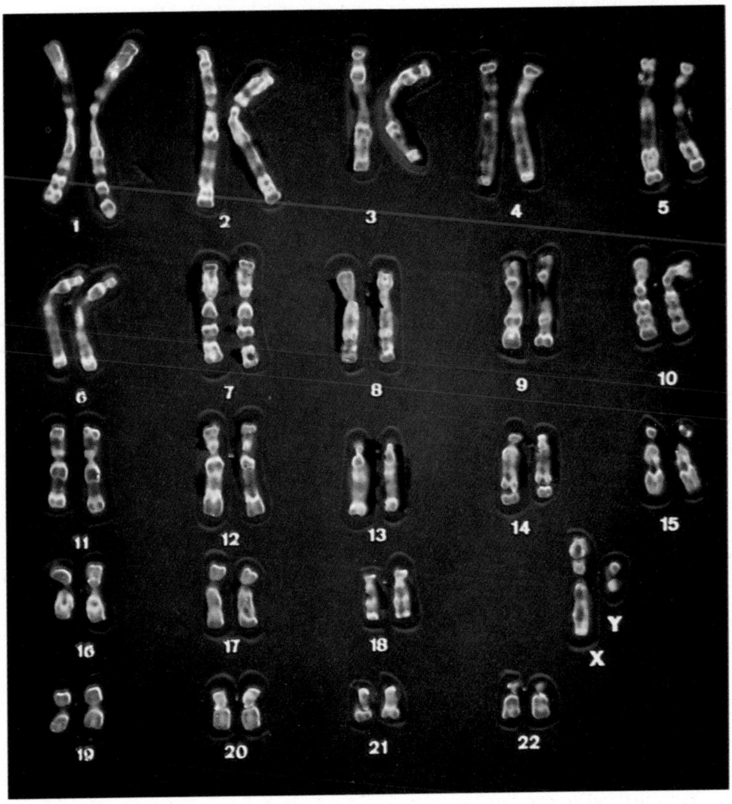

Fig 27.7 **The 23 pairs of chromosomes in the human karyotype. Karyotypes are used to detect chromosomal abnormalities and to perform sex tests on athletes. The presence of XX confirms the athlete is female, the presence of XY shows the athlete is male**

3 THE STAGES OF THE CELL CYCLE

The cell cycle is the complete sequence of events in the life of an individual diploid cell. The cycle starts and ends with cell division and consists of the stages of mitosis plus **interphase**, the interval between divisions in which the cell carries out its normal functions. Fig 27.8 shows the main stages of the life cycle of a cell.

Cells that do not divide, such as those in muscle and nerve tissue, are always in interphase: this is the normal state for a functioning cell. In interphase the DNA in the nucleus is unwound, and active genes are read to produce proteins such as enzymes.

A new cell has three options:

- It can remain stable, in interphase for many months or years. Brain and other nerve cells rarely divide, if ever.

- It can undergo mitosis within a short period of time. Skin cells and cells than line the gut all have a very rapid turnover (see the Feature box on Cell Turnover on the opposite page).

- It can undergo meiosis. Specialised germ cells in the ovary and testes include meiosis in their cell cycle and produce egg cells or sperm.

Fig 27.8 **The four stages of the cell cycle: the three stages of interphase and the stage of active division in which mitosis occurs**

G_1 – growth of cytoplasm and organelles
S – synthesis of DNA
G_2 – second growth phase
M – mitotic phase (nucleus and cytoplasm divide to produce two diploid cells)

Most non-dividing cells, eg brain, are in the G_0 phase

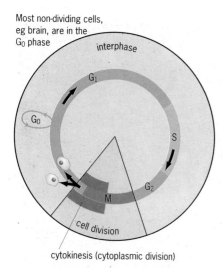

cytokinesis (cytoplasmic division)

?

C Name the three main stages of interphase.

D Without DNA replication, what would happen to the genetic material each time cell division took place?

✓

Exam hint – Remember the mnemonic:
IPMAT – **I**nterphase, **P**rophase, **M**etaphase, **A**naphase, **T**elophase.

Interphase

We start looking at the cell cycle at the point when a cell has just divided. The cell is now in interphase. Interphase has three distinct stages: G_1, **S** and G_2.

G_1 – the first growth phase

Just after it has been produced by division of its parent, a new cell is in the early part of the first **g**rowth phase, G_1, sometimes described as G_0. Cells that do not divide remain at this point in the cell cycle because they do not need to go further: they never replicate their DNA. 'New' cells are relatively small, with a full-sized nucleus but relatively little cytoplasm.

During G_1, protein synthesis starts and the volume of cytoplasm and the number of organelles increase rapidly. The process is actually quite complicated, not least because some organelles such as mitochondria (which have their own DNA) divide independently of the cell nucleus. In later G_1, the cell takes on more 'normal' proportions: the nucleus begins to look smaller as the surrounding cytoplasm increases in volume.

The S phase

The **S** phase – DNA **s**ynthesis phase – follows G_1. In this phase, the cell's DNA **replicates** (copies itself). Predictably, a cell only enters the S phase if it is going to divide. The point at which DNA replication starts is called the **restriction point**. After this, the cell becomes **restricted**, or locked into an automatic sequence that moves inevitably on to cell division.

G_2 – the second growth phase

Before the actual mitotic cell division (see below), the cell enters a second, shorter **g**rowth phase, G_2, in which the proteins necessary for cell division are synthesised.

Mitosis

Mitosis is a continuous sequence of events but, for clarity, they are divided into four distinct stages: **prophase**, **metaphase**, **anaphase** and **telophase**. The stages of mitosis are illustrated in Fig 27.9. You will need to refer to this constantly as you read the text.

The movement of chromosomes during cell division is controlled by microtubules (page 19), which form the **spindle**. In a non-dividing cell the microtubules are found as two bundles of fibres, the **centrioles**, in an area of cytoplasm known as the **centrosome**, or **microtubule organising centre (MTOC)**. During cell division the centrioles move to opposite sides of the nucleus, from where they form the spindle.

Prophase

Chromatin begins to condense into chromosomes: Fig 27.9(a). At this stage, each chromosome has replicated itself and now consists of two identical **chromatids**. These are known as **sister chromatids** and are joined by a **centromere**. Unless there has been a mutation (a fault in the DNA replication), these sister chromatids are genetically identical. The subsequent stages of mitosis organise and split the pairs of chromatids, so that one chromatid of each pair goes into each daughter cell.

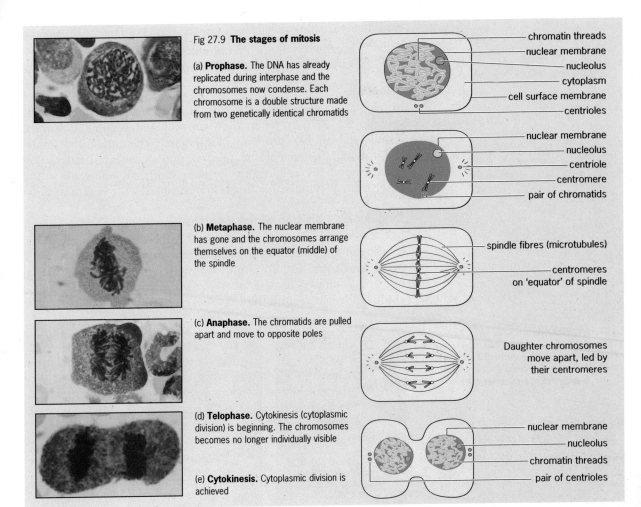

Fig 27.9 **The stages of mitosis**

(a) **Prophase.** The DNA has already replicated during interphase and the chromosomes now condense. Each chromosome is a double structure made from two genetically identical chromatids

- chromatin threads
- nuclear membrane
- nucleolus
- cytoplasm
- cell surface membrane
- centrioles

- nuclear membrane
- nucleolus
- centriole
- centromere
- pair of chromatids

(b) **Metaphase.** The nuclear membrane has gone and the chromosomes arrange themselves on the equator (middle) of the spindle

- spindle fibres (microtubules)
- centromeres on 'equator' of spindle

(c) **Anaphase.** The chromatids are pulled apart and move to opposite poles

Daughter chromosomes move apart, led by their centromeres

(d) **Telophase.** Cytokinesis (cytoplasmic division) is beginning. The chromosomes becomes no longer individually visible

- nuclear membrane
- nucleolus
- chromatin threads
- pair of centrioles

(e) **Cytokinesis.** Cytoplasmic division is achieved

As the chromosomes condense, other changes occur in the cell:

- The nucleolus begins to break down.
- The centrioles move to opposite sides of the nucleus.
- The centrioles begin to assemble the spindle.
- The nucleolus disappears.
- The nuclear membrane begins to break up.

From prophase onwards, most 'normal' cell activity, such as protein synthesis and secretion, is halted until division is over.

Metaphase

The beginning of metaphase (meta = middle) is marked by the disappearance of the nuclear membrane, which breaks down into separate vesicles, moves into the surrounding cytoplasm and joins with the endoplasmic reticulum: Fig 27.9(b).

The spindle (Fig 27.10) becomes fully developed and fills the space that was occupied by the nucleus. Then the most obvious event of metaphase happens: the chromatid pairs attach themselves to individual spindle fibres and align themselves on the equator of the spindle.

Anaphase

At the start of anaphase (ana = apart), as Figs 27.9(c) and 27.10 show, the chromatids are pulled apart by movements of the spindle fibres: sister chromatids are pulled to opposite poles. The newly separated

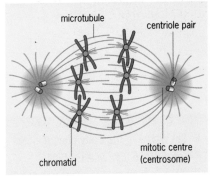

- microtubule
- centriole pair
- chromatid
- mitotic centre (centrosome)

Fig 27.10 **The spindle is a cradle of microtubule fibres which organise the chromosomes during cell division**

chromatids are now called chromosomes, and are single structures. If you watch a film of mitosis (highly recommended for learning purposes), you will see that anaphase is the most obvious event.

Telophase

In telophase (telo = final) the chromosomes reach the poles of the spindle: Fig 27.9(d). They then unravel and become indistinct, forming the familiar chromatin of interphase. The nuclear membrane re-forms and the nucleolus reappears. Soon afterwards, transcription resumes and the cell restarts protein synthesis, endocytosis and other normal cytoplasmic functions. This marks the end of mitosis.

Cytokinesis – division of the cytoplasm

In the events of mitosis just described, we have looked only at the splitting of the nucleus of the cell. Splitting of the cell itself is called **cytokinesis** and usually begins during anaphase. In most animal cells, microtubules form a furrow in a ring around the cell. These gradually constrict until the cells separate, as shown in Fig 27.11.

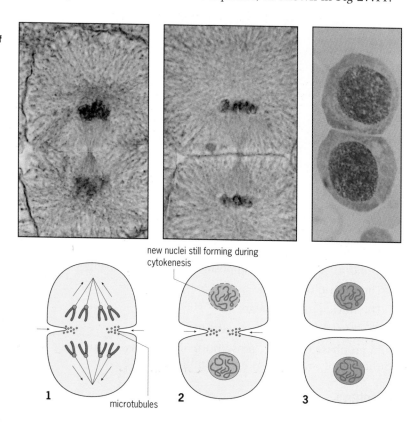

See questions 1 and 2. ■

Fig 27.11 **Cytokinesis in animal cells. As the events of mitosis come to an end, movement of microtubules causes a constriction around the centre of the cell. Eventually the cytoplasm divides, leaving two new daughter cells**

new nuclei still forming during cytokenesis

1   microtubules   2   3

Meiosis: making the gametes

Meiosis is the type of cell division that halves the genetic material in cells. For this reason, it is known as a **reduction division**. In humans, as in all animals, meiosis makes haploid gametes (sex cells).

Meiosis is a remarkable process that produces haploid gametes *and* shuffles the genes, so that each gamete produced is genetically different. This is why children born to the same parents are usually not identical: they are produced by the fusion of a genetically unique sperm and egg. Identical twins are an exception because they originate from the same fertilised egg.

E (a) Where in the body does meiosis occur in **(i)** women and **(ii)** men?

(b) Why is meiosis sometimes referred to as reduction division?

The stages of meiosis

Meiosis has two separate divisions. Both divisions have stages that are given the same names as in mitosis, but they have a number to denote whether they refer to the first or second division, for example anaphase I, prophase II.

It is important to remember here that a normal diploid cell contains two copies of each chromosome, one from the 'mother', and one from the 'father'. Just before meiosis, the cell's DNA becomes organised into chromosomes and replicates to form pairs of chromatids in preparation for division, just as it does before mitosis. So at the first stage of meiosis, the cell contains 92 (2 × 46) chromatids. The 23 chromosomes originally from the mother have been duplicated, making 23 identical pairs of chromatids, as have the 23 chromosomes originally from the father.

An overview of meiosis is shown in Fig 27.12. There is a basic difference between meiosis and mitosis: in meiosis, the homologous chromosomes pair up, in mitosis this does not happen. During the first part of meiosis, each homologous pair can swap sections of DNA, so that each pair becomes 'genetically mixed'.

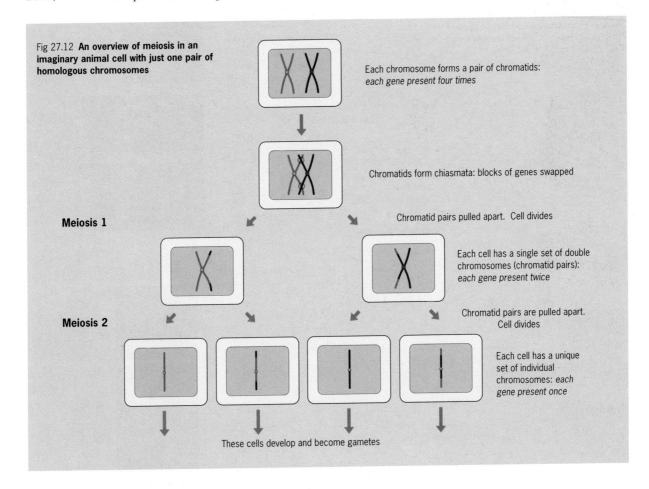

Fig 27.12 **An overview of meiosis in an imaginary animal cell with just one pair of homologous chromosomes**

Each chromosome forms a pair of chromatids: *each gene present four times*

Chromatids form chiasmata: blocks of genes swapped

Chromatid pairs pulled apart. Cell divides

Meiosis 1

Each cell has a single set of double chromosomes (chromatid pairs): *each gene present twice*

Chromatid pairs are pulled apart. Cell divides

Meiosis 2

Each cell has a unique set of individual chromosomes: *each gene present once*

These cells develop and become gametes

The individual stages of meiosis are shown in Fig 27.13. In the first meiotic division (prophase I to telophase I), one of each homologous pair of 'genetically mixed' chromosomes passes into each new cell. In the second meiotic division, the individual chromatids are pulled apart so that one chromatid (now a chromosome) goes into each daughter cell. The end result of meiosis is four haploid cells, each containing a single set of chromosomes. Each cell is genetically unique.

Fig 27.13 **The stages of meiosis in an imaginary animal cell showing just one of the pairs of chromosomes**

Stage of meiosis First division	What is happening	What it looks like
Interphase	Just before meiosis, DNA replicates, so cells which contained two copies of each chromosome now have four. Chromosomes not yet visible	
Early prophase I	Chromosomes become visible. Centromeres move to opposite sides of cell	
Mid prophase I	Each homologous pair of chromosomes comes together to form a bivalent	
Late prophase I	Each chromosome in a bivalent forms two chromatids. Genetic mixing occurs: chiasmata, the points of cross-over, are visible	
Metaphase I	The bivalents arrange themselves on the equator of the spindle	
Anaphase I	The chromatid pairs from each homologous chromosome split apart and move to opposite poles of the cell	
Telophase I	Cytokinesis begins, two new cells form, each has two copies of each chromosome. These chromosomes are genetically different from those in the original cell	
Interphase	A resting time (length varies between cell types)	

Second division

Prophase II	A new spindle forms, at right angles to the first	
Metaphase II	Chromosomes, each of which is a pair of chromatids, align themselves on the equator of the spindle	
Anaphase II	Chromatids are pulled apart to form two chromosomes that then move to opposite poles of the cell	
Telophase II	Cytokinesis begins. Four haploid cells, each with only a single chromosome, have been formed. Each chromosome is genetically different	

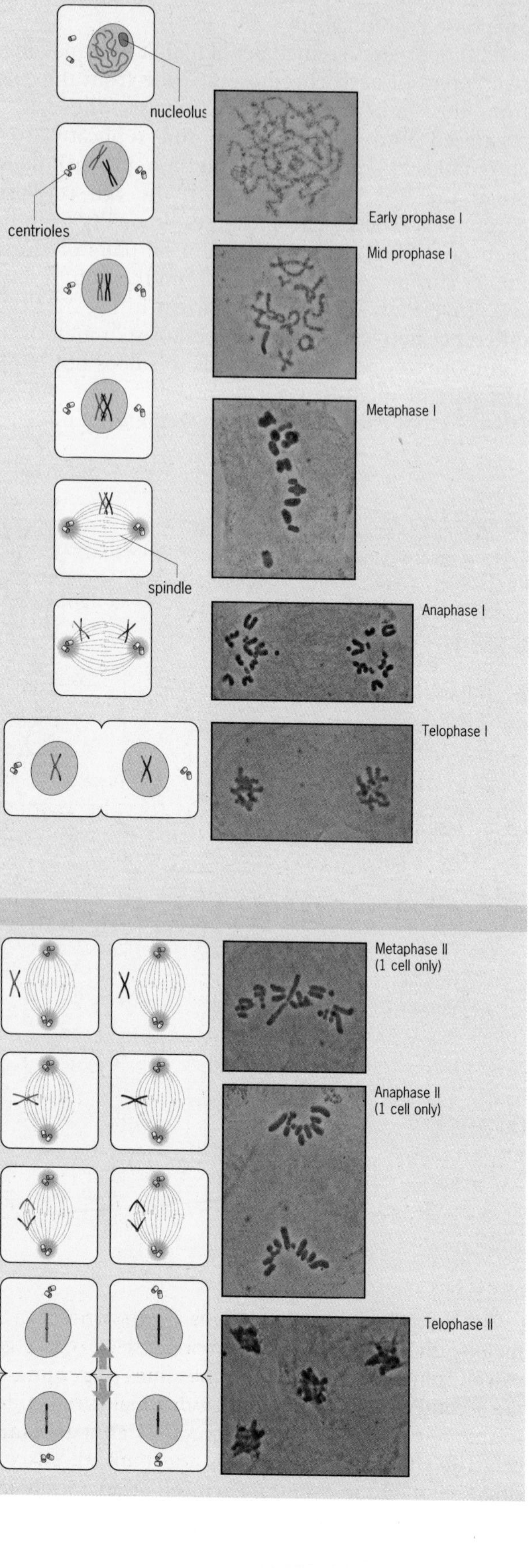

Prophase 1

Prophase 1 is more complex than the corresponding phase in mitosis, and can be subdivided into early, middle and late.

Early prophase I starts when the chromosomes condense and the nucleolus disappears.

In **mid prophase I**, the homologous chromosomes, one from each parent, pair up. Each pair forms a **bivalent**. This does not happen in mitosis. When a chromosome pair is exactly aligned, it is said to be at **synapsis**. Next, each chromosome in a pair divides into two chromatids, giving four chromatids per bivalent.

In **late prophase I**, **recombination** – or cross-over – takes place (Fig 27.14). One (or both) of the chromatids of the two homologous chromosomes breaks off at certain points and fuses with a chromatid of the other chromosome in the bivalent, forming joints called **chiasmata** (the singular, *chiasma*, means 'crosspiece'). This process ensures that blocks of genes are swapped between maternal and paternal chromosomes. The position of chiasma formation varies, even within the same species, and this produces a large variety of new gene combinations.

Prophase ends with the chromatids of each bivalent (pair of homologous chromosomes) entwined and joined by chiasmata.

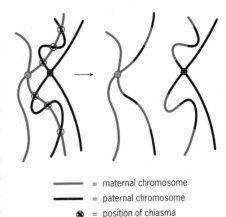

————— = maternal chromosome
————— = paternal chromosome
⊛ = position of chiasma

Fig 27.14 **Chiasmata form between homologous chromosomes during late prophase I of meiosis. At these points, parts of the maternal chromosome separate and join with the paternal chromosome, and vice versa. This produces new chromosomes that are genetically different from each other, and from both the maternal and paternal chromosomes from which they are derived**

Metaphase 1

In metaphase 1, the nuclear membrane disappears and the spindle is fully developed. The bivalents move to the equator of the spindle in the same way as individual chromosomes do during the matching phase in mitosis.

Anaphase I

The chromatid pairs of a bivalent are pulled apart (Fig 27.15) because of the action of spindle fibres, a process that separates the entwined chromatids. At the end of anaphase 1, the chromatid pair from one of the original homologous chromosomes is positioned at one pole of the cell and the chromatid pair from the other homologous chromosome is at the other pole. Either the maternal or the paternal chromosome can pass into either cell. This is the process that allows **independent assortment of alleles** (see Chapter 29). It is another reason why meiosis increases variation.

Occasionally, this phase is not completed successfully and a pair chromosomes fails to separate. The result is that both homologous chromosomes pass into one daughter cell, the other receiving neither. This situation can lead to conditions such as Down's syndrome.

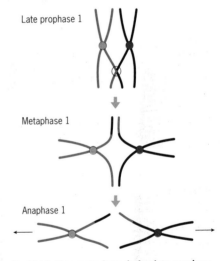

Late prophase 1

Metaphase 1

Anaphase 1

Fig 27.15 **Chiasmata form during late prophase I and cross-over takes place. The newly formed chromatid pairs separate during metaphase I and then the two pairs move to opposite poles of the cell during anaphase I**

Telophase I

The spindle disappears and the nuclear envelope re-forms around the two sets of chromosomes. At the same time cytokinesis separates the cytoplasm, forming two daughter cells, ready for the second meiotic division.

Interphase

The length of the resting interphase between the two meiotic divisions varies widely. It is sometimes short or even non-existent. If there is no interphase, the chromosomes remain condensed and the cell passes straight from telophase 1 into prophase II. In human females, however, ova may remain in interphase for decades. Basic egg cells are made in a girl before birth, but they do not complete meiosis until just before ovulation, anything up to fifty years later.

At the end of meiosis 1, there are two cells, each containing two copies of each chromosome on a chromatid pair. The second meiotic division separates the chromatids, so that each daughter cell formed is haploid (has one set of single chromosomes).

The second meiotic division

Prophase II

For each chromosome, the chromatid pair attaches itself to the new spindle, which forms at right angles to the first.

Metaphase II

Each chromatid pair lines up on the equator of the spindle.

Anaphase II

The chromatids are pulled apart to form two chromosomes that move to opposite poles.

Telophase II

The spindle disappears, the nuclear membrane re-forms, chromosomes unravel and cytokinesis produces two separate cells.

The end product of meiosis

Meiosis produces four genetically different haploid cells, known as a **tetrad**. Genetic variation has been produced in three ways:

- The homologous chromosome pairs originate in different organisms, one maternal and one paternal, and so are genetically different.
- Blocks of genes are swapped between the chromatids of homologous chromosomes as the chiasmata form during prophase I.
- Each daughter cell can receive a copy of either chromosome from a pair, and each copy may have undergone cross-over and have different genes from the other three. This is called indpendent assortment and is covered in more detail in Chapter 29.

Learning meiosis

Meiosis is a complex process that many people find difficult to learn. Here are some hints that you may find useful. First, divide your study of meiosis into three sections by looking at three questions:

What is the point of meiosis?

It makes gametes by shuffling the genes to produce haploid cells that are all genetically different from either of the parents. Each fertilisation (combination of an egg and a sperm) produces an individual that is genetically unique.

What do each of the meiotic divisions achieve?

Start by learning an overview of each division: Fig 27.13 is a good place to start. In the first meiotic division, homologous chromosomes pair up, divide into chromatids, swap blocks of genes and then separate. In the second division the individual chromatids separate. The end result is four genetically different haploid cells from one diploid original.

What are the stages of the two divisions?

Once you are confident that you can answer the first two questions, you can put some flesh on the bones. Describe one stage at a time and then draw what you have just described. Try using different colours for paternal and maternal chromosomes. Alternatively, you could make some chromosomes out of modelling clay and work through the meiotic sequence yourself. Many students find this a valuable exercise.

F How many bivalents are formed in a human cell undergoing meiosis?

The importance of genetic variation should become apparent as you study evolution. Find out more in Chapter 30.

See questions 3 to 7. ■

DOWN'S SYNDROME

IN THE UNITED KINGDOM, two children in every 1000 are born with Down's syndrome (Fig 27.16(a)), a genetic condition that arises from a fault in meiosis. Individuals with this condition have an extra chromosome 21 (Fig 27.16(b)) and usually have physical and mental disabilities. They often have a small mouth with a normal-sized tongue (making eating and speech difficult), reduced resistance to disease and heart abnormalities. Approximately one in three Down's syndrome children do not survive to their twelfth birthday.

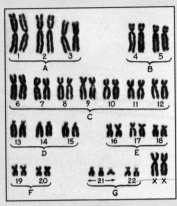

Fig 27.16(a) **A child with Down's syndrome**

■ See question 8.

Fig 27.16(b) **The complement of chromosomes (the karyotype) of a person with Down's syndrome, showing the extra chromosome 21**

The reason for the extra chromosome is usually a failure of the chromosome 21 pair to separate during anaphase 1 of meiosis (see Fig 27.13). Both chromosomes 21 pass into one daughter cell, and the other gets no copy at all. If a gamete with no chromosome 21 forms a zygote, the embryo fails to develop. But, if the gamete containing both chromosomes is fertilised, the resulting baby later develops Down's syndrome.

One reason for the non-separation may be the length of time the egg spends in the ovary. In females, egg cells do not complete their meiotic division until they are 'selected' for ovulation (see Chapter 19). The longer the eggs remain in the ovary, the 'stickier' the chromosomes become and the greater the probability that they will not separate during anaphase I. This explains why the chance of having a child with Down's syndrome increases as women get older. Women below the age of 25 have less than a 1 in 2000 chance, but this rises to 1 in 50 for women over 45.

The age of the father is also important: recent studies have shown that in a significant minority of cases the extra chromosome comes from the father, that is, a normal ovum is fertilised by a sperm containing two chromosomes 21s.

SUMMARY

After reading this chapter, you should know and understand the following:

■ Almost all the cells in the human body are **diploid**: they contain a double set of chromosomes. The only exceptions are the gametes: sex cells – egg cells and sperm – are **haploid** and contain only one set of chromosomes.

■ The cell cycle consists of four stages: G_1, in which the cell grows; **S**, in which the DNA is doubled; G_2, a second growth phase, and finally **mitosis** or **meiosis**. G_1, S and G_2 are collectively known as **interphase**.

■ Mitosis is normal cell division, the process by which we grow and develop. One diploid cell divides to produce two genetically identical diploid cells.

■ You can remember the stages of mitosis using the nmemonic IPMAT: Interphase, Prophase, Metaphase, Anaphase, Telophase.

■ Non-dividing cells are said to be in interphase. There are no chromosomes visible and the active genes in the DNA are being used to direct the activities of the cell. If the cell is going to divide, the cell prepares itself during interphase and the DNA is replicated.

■ In prophase, the chromosomes condense and become visible. The nuclear membrane disappears and the **spindle** forms. The chromosomes are double: two identical chromatids are joined at the **centromere.**

■ In metaphase, the chromosomes arrange themselves in the middle of the spindle (remember, meta = middle).

■ In anaphase, the double chromosomes are pulled apart (remember, ana = apart).

■ In telophase, the nuclear membrane re-forms around the two new sets of chromosomes and the cytoplasm divides forming two new, genetically identical cells.

■ Meiosis is a special type of cell division which produces sex cells or gametes. It is also known as a 'reduction division'. One diploid cell will divide meiotically to produce four genetically different haploid cells.

QUESTIONS

1 The drawing of Fig 27.Q1 shows animal cells in different stages of mitosis

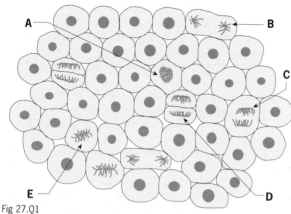

Fig 27.Q1

a) Using only the letters in the diagram, list the cells in the correct sequence, beginning with the cell in the earliest stage.

b) **(i)** Describe what is happening to the genetic material in cell **B**.
   (ii) How are the events you have described in **(i)** being brought about?

c) Explain why the cells produced by mitosis are genetically identical.

[AEB 1994 Human Biology: Paper 1, q.9]

2 The graph in Fig 27.Q2 shows the movement of chromosomes during mitosis. Curve **A** shows the mean distance between the centromeres of the chromosomes and the corresponding pole of the spindle.

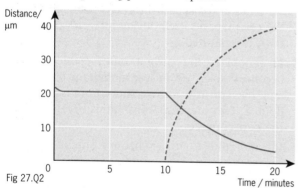

Fig 27.Q2

a) What is represented by curve **B**?

b) **(i)** At what time did anaphase begin?
   (ii) Explain how **one** piece of evidence from the graph supports your answer.

[AEB 1996 Biology: Specimen Paper, q.5]

3 Fig 27.Q3 shows chromosomes in a cell undergoing meiosis.

a) Name **(i)** the stage of meiosis shown in Fig 27.Q3;
   (ii) the process that has occurred at A.

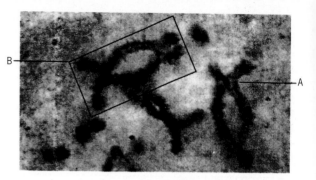

Fig 27.Q3

b) Make a labelled drawing to interpret what has happened to the bivalent labelled **B**.

c) Explain how meiosis gives rise to variation.

[UCLES June 1994 Modular Biology: Applications Paper, q.1]

4 Fig 27.Q4 shows an animal cell in prophase of mitosis. The letters represent alleles of a gene that codes for eye colour, and a gene that codes for body colour. Allele **R** gives red eyes, and is dominant to allele **r**, which gives white eyes. Allele **B** gives black body, and is dominant to allele **b**, which gives grey body.

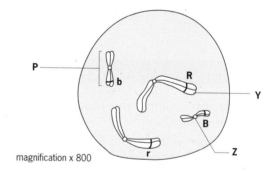

magnification x 800

Fig 27.Q4

a) Draw the diagram, and on it, **(i)** name structures **Y** and **Z**; **(ii)** shade with pencil **two** homologous chromosomes.

b) Calculate the actual length of chromosome **P**. Express your answer in micrometres (μm). Show your working.

c) What will be the genotypes of the cells produced as a result of mitosis of the cell in Fig 27.Q4?

d) **(i)** If the cell in Fig 27.Q4 were to undergo **meiosis** to produce gametes, state the genotypes of the gametes formed.
   (ii) If gametes formed from several cells like the one in Fig 27.Q4 were to fuse randomly, state the phenotypic ratios expected in the resulting offspring.

[UCLES March 1995 Modular Biology: Foundation Module Paper, q.2]

5

a) Explain what is meant by a *homologous pair of chromosomes*.

Fig 27.Q5 is a simplified diagram of all of the chromosomes in a cell undergoing meiosis in a mosquito.

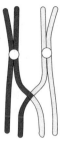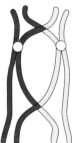

Fig 27.Q5

b) With reference to Fig 27.Q5,
  (i) name the stage of meiosis shown;
  (ii) state the number of chromosomes that will be found in each daughter nucleus when meiosis has been completed.

c) State three ways in which meiosis could lead to variation in the daughter cells produced.

d) Describe the significance of mitosis in growth.

[UCLES June 1998 Sciences: Biology Foundation, q.4]

6 Sketch Fig 27.Q6(a), which shows the chromosomes in the nucleus of a cell.

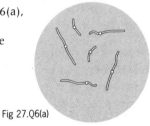

Fig 27.Q6(a)

a) **(i)** On your copy of Fig 27.Q6(a), shade in **one pair** of homologous chromosomes.
  (ii) Copy and complete Fig 27.Q6(b) to show the arrangement of these chromosomes at metaphase 1 and metaphase 2 of meiosis. Use the asterisks as the poles of the spindle.

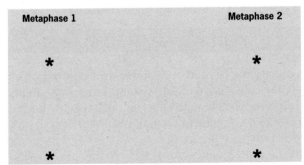

Fig 27.Q6(b)

  (iii) Draw a diagram which will show the chromosome content of a cell which is the product of a meiotic division of the cell shown in Fig 27.Q6(a).

[UCLES February 1997 Sciences: Biology Foundation, q.3]

7 Fig 27.Q7 shows the changes in the amount of DNA (arbitrary values) in the spermatogonia of a diploid organism as the spermatogonia undergo mitotic nuclear division, and the amount of DNA in the primary and secondary spermatocytes as they undergo meiosis.

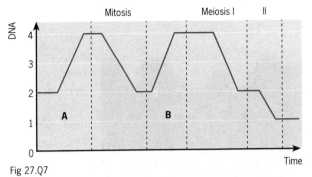

Fig 27.Q7

a) Explain what process accounts for the change in DNA quantity during periods A and B in the diagram.

b) Notice that the amount of DNA doubles before mitosis and meiosis. Does the number of chromosomes double? Explain your answer.

c) If the diploid chromosome number in this organism is 8 and the haploid number is 4, what is the chromosome number of cells:
  (i) at the end of mitotic division?
  (ii) at the end of meiotic division?

d) The DNA in the cells at the end of mitosis contains 56% G–C base pairs. What are the percentages of the four bases (adenine, guanine, cytosine, thymine) in this DNA?

[OCSEB June 1997 Structured Science Scheme: Unit B2 Genetics and Evolution, q.4]

8 Most cells in the human male and female contain 46 chromosomes.

Gametes (ova and sperm) contain 23 chromosomes.

a) How many copies of each gene are present in an ovum or a sperm of a normal individual?

b) Some humans have an unusual chromosome number. State how the normal chromosome number is altered in a **named** human condition.

c) What is the chromosomal difference between human males and females?

d) **(i)** Red blood cells are continuously replaced by cell division. What type of nuclear division precedes this cell division?
  (ii) What are the chromosomal and genetic effects of this type of nuclear division?
  (iii) What nuclear events must take place before a nucleus divides, so that both daughter cells are genetically normal?

[OCSEB January 1998 Structured Science Scheme: Unit B2 Genetics and Evolution, q.1]

28 Genes and chromosomes

Imagine the implications of prolonging your life well past 100. How long would you work for? When would you retire? Could any government afford pensions and healthcare bills? What would happen to societies in which 6,7 or even 8 generations of the same family were alive together?

WHY DO WE AGE? Until a few years ago, scientists and doctors thought that our bodies simply wore out like machines. They now think that many aspects of ageing are genetically determined. Organisms develop and grow by cell division, and the same process happens in adults to repair and maintain their bodies. It seems likely that each of our cells has a genetic timer which counts how many times a cell divides: after a maximum number of divisions, the cell dies. But what sets this limit?

The latest research is focusing on the ends of chromosomes, the telomeres. Every time a cell divides, the telomeres shorten, and when they disappear completely, the cell is unable to divide further and its repair systems break down. Researchers have been intrigued to find that eggs and sperm are 'immortal' – they have an enzyme, telomerase, which ensures that the embryo's telomeres are complete. This is necessary to make sure that the genetic clock that produces a baby is set to zero. It might seem obvious, but babies need to be born with 'new' cells so that they have their full life expectancy, regardless of the age of their parents.

The discovery of telomerase opens up the exciting, if ethically difficult possibility that we may be able to manipulate the ageing process. If we can use telomerase to repair the telomeres of normal body cells, it may be possible to extend potential life expectancy to 150 years, or more. The social and economic consequences of this are massive, and are bound to be the cause of much heated debate in years to come.

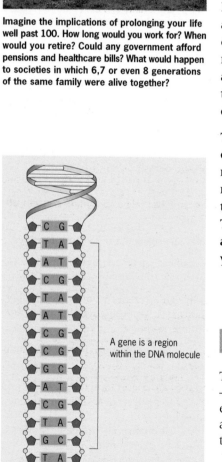

A gene is a region within the DNA molecule

1 WHAT IS A GENE?

The nucleus of a eukaryotic cell has a set number of chromosomes – every human body cell has 46. Each chromosome is a single, elaborately-coiled DNA molecule which has individual genes dotted along its length (Fig. 28.1). A gene is a length of DNA that codes for the synthesis of one polypeptide.

We used to think that a gene contained all the information needed to build a complete protein, such as an enzyme. However,

Fig 28.1 **Genes are sequences in the DNA molecule. The sequence of bases shown here represents one very short gene. In the genetic code, a group of three bases codes for one amino acid, so a protein consisting of 500 amino acids requires a gene of at least 1500 base pairs long, probably more. When unravelled, chromosomes are incredibly long – each one consists of up to 4000 genes and these make up only about 10 per cent of its total length (the rest is non-coding DNA)**

many proteins consist of more than one polypeptide chain, so these molecules are coded for by more than one gene. For instance, haemoglobin molecules consist of four polypeptide chains, two alpha and two beta, so it takes two genes to make a complete haemoglobin molecule.

Genes in action:
the central concept of molecular biology

Can we sum up the way genes work in one sentence? It is a tall order, but the underlying theme, or **central dogma** of genetics is:

> **The DNA of a gene codes for the production of messenger RNA which, in turn, codes for the production of a polypeptide** (Fig 28.2).

In this chapter we look at how genes work, and at how they are copied to allow cell division, reproduction and growth.

After Crick and Watson worked out the structure of DNA in 1953, two vital questions remained:

- How does information get from the DNA in the nucleus to the site of protein synthesis in the cytoplasm?

- How does the base sequence of the DNA translate into the amino acid sequence in a polypeptide?

To answer the first question, Crick, Brener and Monod developed the **messenger hypothesis**. This states that a specific molecule, named **messenger RNA** (mRNA), is copied directly from the DNA sequence of the gene. mRNA is therefore a mobile copy of a gene. It can move from the nucleus to the cytoplasm. Here, ribosomes use mRNA as a template (pattern) to assemble amino acids in the correct order to make the required polypeptide.

Crick also answered the second question: he proposed that an 'adaptor' molecule existed which had two specific binding sites. At one side, the molecule fitted a specific DNA base sequence and at the other, a specific amino acid. This molecule was later discovered and named **transfer RNA** (tRNA) (see Fig 28.5).

2 THE GENETIC CODE

Along the DNA molecule the base sequence is continuous, but it is read in blocks of three. Each three, or triplet, is called a **codon.** Each codon codes for a particular amino acid. A particular sequence of bases therefore codes for a specific amino acid sequence. When the amino acids are assembled into a polypeptide, they interact to twist and bend the chain into its final shape. When the polypeptide chain is complete, it forms all or part of a protein which performs a vital role in the organism.

You will have noticed that there are two strands of bases in the DNA molecule. One strand, called the **template strand** or **sense strand**, contains the all-important genetic code for any particular gene. The corresponding part of the other strand is simply there to stabilise the molecule. However, in the same DNA molecule, different genes are present on different sides and the enzymes which transcribe DNA can use either side as the sense strand.

✔ The genome is defined as 'all the DNA sequences in an organism'. It is a complete set of genes, together with the non-coding DNA in between. The human genome is thought to consist of over 3 billion base pairs. Remarkably, the entire genome is present in every one of our body cells.

✔ The position of a gene on a chromosome is known as its **locus** (plural: loci).

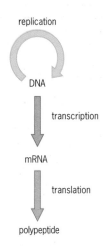

Fig 28.2 **The central concept of molecular biology. The base sequence on a particular gene is copied, or transcribed, to produce a messenger RNA template. This mRNA template passes out of the nucleus and moves to the ribosomes where it is read, or translated, into a polypeptide**

✔ Retroviruses such as the Human Immunodeficiency Virus (HIV) are an exception to the central dogma. Retroviruses contain RNA from which they make a DNA copy using the enzyme **reverse transcriptase**.

? **A** Name three structural differences between DNA and RNA molecules.

✔ The base pairings are always the same. In DNA, A bonds with T and G with C. In RNA, T is replaced by U, so A bonds with U, and G with C (see page 52).

?

B The RNA polymerase enzyme moves along a DNA sequence reading ATACGCTAT.

(a) What is the corresponding sequence on the RNA molecule?

(b) How many amino acids are encoded in this sequence?

Transcription

DNA can be thought of as a permanent reference library: it does not leave the nucleus and so cannot be used directly for protein synthesis. Instead, the genetic code for a gene is transferred, or **transcribed,** onto a smaller RNA molecule, **messenger RNA**, which can then pass to the ribosomes to act as a template for protein synthesis. In effect, mRNA is a mobile copy of a gene.

Before transcription can begin, the DNA helix must unwind and the two halves of the molecule must come apart, so exposing the base sequence. This process begins when the DNA is 'unzipped' by specific enzymes and the enzyme **RNA polymerase** attaches to the DNA molecule at the **initiation site**. This is a base sequence at one

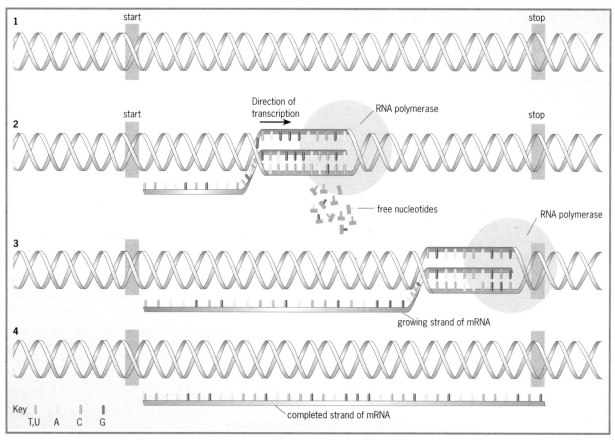

Key ‖ ‖ ‖ ‖
   T,U A C G

completed strand of mRNA

Fig 28.3(a) **This diagram shows the process of transcription. The enzyme RNA polymerase is a big molecule, (not shown to scale) and covers about 50 base pairs at once. Beginning at start, the initiation site, the enzyme moves along the template or sense strand of DNA, using free nucleotides from the cytoplasm to make messenger RNA**

Fig 28.3(b) **This electronmicrograph shows part of two DNA strands. Each part contains a gene. The 'branches' are many lengths of mRNA that are being transcribed from the gene.**

Each length of mRNA is moving from left to right, attached to the DNA by an RNA polymerase molecule. RNA polymerase travels along the DNA with its length of mRNA, and adds nucleotides to the mRNA. When transcription of an mRNA molecule is complete, as happens at the bottom right of this micrograph, the mRNA strand detaches from the DNA strand and exits the nucleus

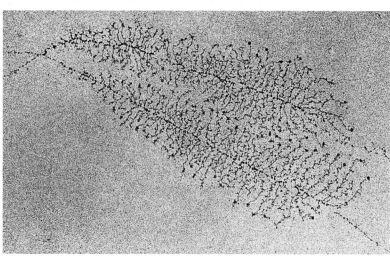

end of the gene which effectively says: Start here. Once in place, the enzyme moves along the gene (Fig 28.3(a)), assembling the messenger RNA molecule by adding the matching nucleotides, one at a time.

Once formed, the mRNA copy begins to peel away from the DNA Fig 28.3(b) shows many mRNA copies of the same gene being made at the same time. When RNA polymerase has passed a particular region of DNA, the double helix rewinds. When the whole gene has been transcribed, the complete mRNA molecule leaves the nucleus.

Translation

Translation occurs when the base sequence on the mRNA is used to synthesise a polypeptide. The process is called translation because the information contained in the gene – and delivered by the mRNA – is *translated* into a polypeptide.

Protein synthesis is one of the major activities of cells, and it is highly likely that you are already familiar with many of the organelles and chemicals involved. The roles of individual organelles in the process are listed in Table 28.1. See also Chapter 1 and Chapter 3 for background information.

C How do mammals ensure that a regular supply of amino acids reaches all of their cells?

D Is protein synthesis an anabolic or a catabolic reaction? (Hint, it's a vital part of body building.)

Table 28.1 **The main organelles and chemicals involved in protein synthesis**

Organelle/molecule	Role in protein synthesis
Nucleus	Houses the DNA
Nucleolus	Manufactures the ribosomes
Ribosome	Site of protein synthesis – where amino acid assembly occurs
Endoplasmic reticulum	Isolates, stores and transports polypeptides
Golgi apparatus	Modifies and packages polypeptides
DNA	Stores the genetic information
Messenger RNA	Is a mobile copy of a gene on DNA
Transfer RNA	Each brings a specific amino acid to the ribosomes

Translation occurs on **ribosomes**, structures which hold all the components together as an amino acid chain is made. The mRNA strand passes out of the nucleus and attaches to the ribosome. At the point of attachment, two tRNA molecules deliver the required amino acids and hold them in position so that they can be added to the growing polypeptide.

The mechanism of translation is shown in detail in Fig 28.4. The process continues until an mRNA **stop codon** – which has no tRNA molecule – moves into the ribosome's first binding site. At this point all the components separate, releasing the completed polypeptide.

After translation, polypeptides are processed according to their final destination:

- Those which are to be exported from the cell, such as digestive enzymes, are threaded through pores in the endoplasmic reticulum to accumulate on the inside. Here they are processed and packaged before being secreted (see pages 13 and 15).

- Polypeptides that will form membrane proteins follow the same route as those for export, but they remain on the cell surface membrane rather than being released.

- Polypeptides that will be used inside the cell, such as those that form haemoglobin, remain free in the cytoplasm.

Genes code for the manufacture of polypeptides. Fig 3.19 on page 45 shows the importance of protein in the human body.

Many people find protein synthesis a difficult topic to learn. A valuable exercise is to make the different components – either from modelling clay or just paper – and work through the events shown in Fig 28.4(b).

You will need to make a ribosome, an mRNA molecule with several codons written on it, and enough tRNA molecules and amino acids to match to all the codons on the mRNA.

Fig 28.4(a) **The ingredients needed for protein synthesis**

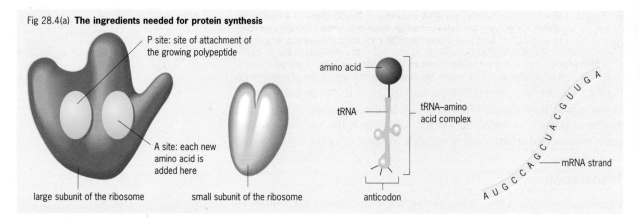

P site: site of attachment of the growing polypeptide

A site: each new amino acid is added here

large subunit of the ribosome

small subunit of the ribosome

amino acid

tRNA

tRNA–amino acid complex

anticodon

mRNA strand

AUGCCAGCUACGUUGA

Fig 28.4(b) **The process of protein synthesis**

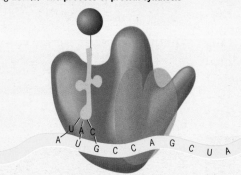

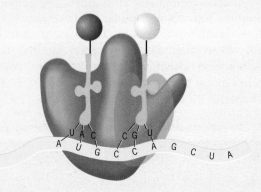

(1) The two subunits of the ribosome come together. The mRNA strand binds to the ribosome. Then an amino acid–tRNA complex with an anticodon complementary to the first codon on the mRNA, binds to the P site

(2) In the next step, an amino acid–tRNA complex with an anticodon complementary to the next codon on the mRNA strand binds to the A site

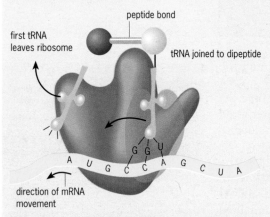

peptide bond

first tRNA leaves ribosome

tRNA joined to dipeptide

direction of mRNA movement

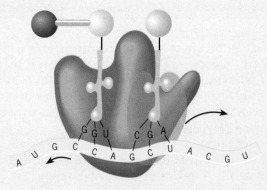

(3) A peptide bond forms between the two amino acids which have been brought together on the ribosome. The bond between the first amino acid and the first tRNA is broken and the tRNA molecule leaves the ribosome. The ribosome moves along the mRNA strand and the second tRNA molecule, now joined to a dipeptide, shifts across from the A site to the P site

(4) A third amino acid–tRNA complex with an anticodon that matches the third codon on the mRNA strands comes into the A site

The sequence of events from **(2)** to **(3)** are repeated until the entire length of the mRNA strand has been translated. The polypeptide is then complete

The role of transfer RNA

Transfer RNA molecules are smaller than mRNA molecules, containing only about 75 to 80 nucleotides. Each consists of a single strand of nucleic acid folded back on itself to form a 'clover-leaf' shape (Fig 28.5). Transfer RNA molecules bring specific amino acids from the cytoplasm to the ribosome so that they attach to the growing polypeptide.

Fig 28.5 **Two ways of representing the tRNA molecule.** (a) **The 3D shape of the molecule, stabilised by hydrogen bonds as the single nucleotide chain folds back on itself.** (b) **A schematic diagram: each type of transfer RNA transports a specific amino acid to the ribosomes, the site of protein synthesis. At one end of the molecule is the anticodon which attaches to the corresponding codon on the messenger RNA. At the other end of the molecule is the amino acid attachment site**

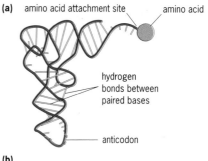

(a) amino acid attachment site — amino acid — hydrogen bonds between paired bases — anticodon

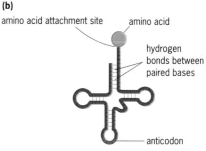

(b) amino acid attachment site — amino acid — hydrogen bonds between paired bases — anticodon

The process by which an amino acid binds to a tRNA molecule is controlled by an enzyme. The process also involves the splitting of ATP. This is important because ATP gives the tRNA–amino acid complex enough energy to form a peptide bond when the amino acid is added to the growing polypeptide.

At one end of the tRNA molecule is the **anticodon**, a three-letter base sequence which matches the codon on the mRNA molecule. At the other end is a particular amino acid. So, for instance, if the codon on the mRNA reads AUG, which codes for the amino acid methionine, it needs a tRNA molecule with an anticodon of UAC which arrives at the ribosome carrying a methionine molecule.

There are 64 different codons, but three of them code for 'Stop translating'. When such a codon arrives at the ribosome, no tRNA is needed. This means that tRNA molecules must match with 61 different codons. There are only 20 different amino acids, so some amino acids are translated from more than one codon. For example, the codons GGG, GGA, GGC and GGU all code for the amino acid glycine.

Like messenger RNA, transfer RNA is also made by transcription. In eukaryote DNA there are many genes whose sole function is to make tRNA. In any cell, most genes occur only once, but there are hundreds of genes that code for tRNA synthesis, to provide the vast numbers of these work-horse molecules that are needed. That tRNA genes do not code for proteins is another exception to the central dogma.

Polysomes and the rate of translation

We have seen that the mRNA molecule is a long single strand of nucleotides, and that the code is translated into a polypeptide at the point of contact with a ribosome. To achieve protein synthesis at a reasonable speed, ribosomes occur in clusters, called **polysomes** or **polyribosomes**, which all translate a different bit of the mRNA at the same time (Fig 28.6).

?

E Why must several different tRNA molecules carry the same type of amino acid?

▨ See questions 1 and 2.

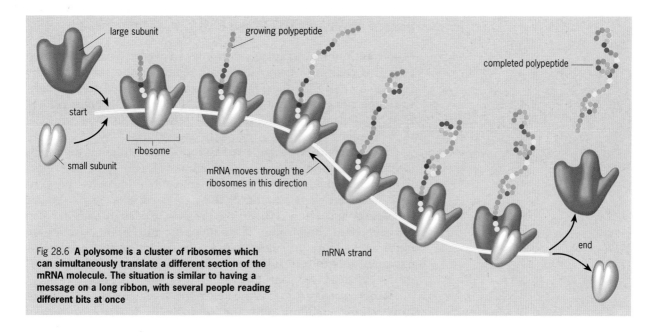

Fig 28.6 **A polysome is a cluster of ribosomes which can simultaneously translate a different section of the mRNA molecule. The situation is similar to having a message on a long ribbon, with several people reading different bits at once**

large subunit — growing polypeptide — completed polypeptide — start — ribosome — small subunit — mRNA moves through the ribosomes in this direction — mRNA strand — end

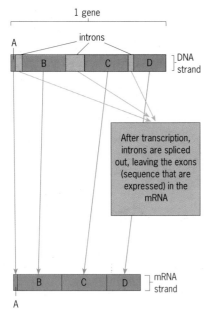

Fig 28.7 **Even in a gene, there are non-coding lengths of DNA called** introns. Only the exons **are encoded into the mRNA molecule, and therefore expressed**

3 A JOURNEY ALONG A CHROMOSOME

If one single chromosome were unravelled, like pulling the wool out of a piece of knitting, we would be left with one long DNA strand. Although this is one single molecule, it is several *centimetres* long and contains up to four thousand genes.

How do you go about investigating such a molecule? One way is to work backwards; if you can isolate the mRNA molecules which are in the cytoplasm, you can trace their origin in the DNA.

A clever technique called **nucleic acid hybridisation** is used to pinpoint the origin of the mRNA, giving the location of the gene itself. In this technique, DNA strands are denatured: when heated to 87 °C, hydrogen bonds which hold the two strands together are broken, and the individual strands separate. If messenger RNA strands from the cytoplasm are added to the mixture, they stick to the region of DNA on which they were originally made.

Studies using this technique have unexpectedly revealed that there is some non-coding DNA in genes (Fig 28.7). Most eukaryote genes contain regions, called **introns**, which do not find their way into the mRNA molecule at all. The parts of the gene which *are* expressed in the mRNA are called **exons** because they are the **ex**pressed regions of the gene.

Generally, a gene consists of a DNA sequence which is present only once. The non-coding sequences, however, tend to consist of repeated or 'stuttered' sequences. Humans all have the same repeated sequence, but in different individuals it is repeated a different number of times and in different places. This is the basis of **DNA profiling** – see the Assignment in Chapter 3.

The function of non-coding DNA is not known. It is probably not useless: much of the repeated DNA in a chromosome occurs around the region of its centromere and may have an important role in maintaining the physical stability of the chromosome as a whole.

The DNA could be like a page of writing in a book. If the letters were strung together continuously, we would find it difficult to read: the gaps between words help us to make sense of the sentences.

4 THE CONTROL OF GENE EXPRESSION

Most multicellular organisms start life as a zygote: a single fertilised cell. This cell has a complete set of genes. As the cell divides by mitosis, the DNA is copied faithfully, so each new cell also contains the same complete set of genes. How, then, do cells differentiate? How, for example, do some animal cells specialise into muscle, nerve or skin?

The answer lies in **gene expression** – how different genes are 'switched on', or **expressed**, in different cells (Fig 28.8).

At any one time, the average cell is probably using only 1 per cent of its available genes. Some of the expressed genes are used for 'housekeeping' – they code for the proteins needed by all cells, such as the enzymes involved in respiration. In contrast, other genes are expressed only in specialised cells. The genes for making insulin, for example, are expressed only in the β cells of the islets of Langerhans in the pancreas (see page 158).

Selective activation of genes is the key to the development of a complex multicellular organism and, not surprisingly, it is the subject of

Fig 28.8 **A spectacular example of the power of gene expression is seen in the life cycle of some insects such as this peacock butterfly. The genes present in the fertilised egg are still present in the adult, but a different set are activated to give rise to the caterpillar and to control its metamorphosis into a pupa and then an adult – which looks like a completely different organism**

intense research. If we could understand how and why some genes are activated while others remain dormant, we could unlock the mystery of embryonic development – how a single fertilised egg grows and differentiates into something as complex as a human being. Unfortunately, this interesting area is very complicated and is beyond the scope of this book.

5 DNA REPLICATION

Whenever a cell divides, the DNA must be copied. Otherwise, there would be no reproduction and no growth. When cells divide, each new daughter cell must have a complete set of genes. **Replication**, the mechanism of DNA copying, is therefore of fundamental importance.

Just three years after Crick and Watson published their model of DNA structure in the journal *Nature*, their theory about DNA replication was confirmed by Arthur Kornberg. He put DNA into a mixture containing all four **nucleotides** (each of the four bases attached to a phosphate and a sugar), together with the enzyme DNA polymerase, and he showed that the DNA could replicate without any other factors present.

In the mixture, each DNA strand unwinds and acts as a template for the construction of a new strand. The exposed strand acts as a template on which the free nucleotides arrange themselves in exactly the same sequence as the intact strand they replace. This model is called **semi-conservative replication** because each of the resulting strands of DNA contains one strand from the original DNA and one newly synthesised strand: half has been conserved and half is new.

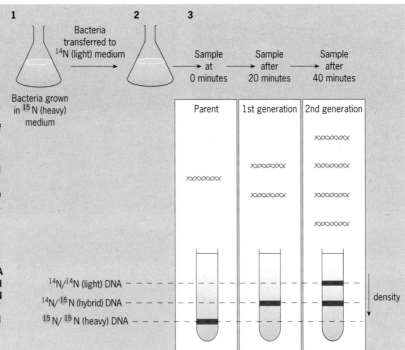

Fig 28.9 Meselson and Stahl's experiment to support the semi-conservative replication model of DNA replication

1 To start with, many generations of *E. coli* bacteria are grown in a medium containing nucleotides labelled with the heavy nitrogen radioisotope, ^{15}N . Eventually, almost all of the bacterial DNA contains ^{15}N, and so is denser than normal DNA. A control culture of the same bacterium is grown with 'normal' ^{14}N nucleotides for comparison

2 The bacteria grown in ^{15}N are transferred to ^{14}N medium. After 0, 20 and 40 minutes, samples of the DNA are extracted, placed in a solution of caesium chloride and centrifuged. This separates out the DNA according to its density and allowed Meselson and Stahl to distinguish between DNA containing ^{15}N and ^{14}N.

3 The results confirm that, initially, all the DNA contained ^{15}N. After 20 minutes the DNA had replicated, producing two strands of $^{15}N/^{14}N$ hybrid DNA. After 40 minutes, the second generation strands consisted of half $^{15}N/^{14}N$ hybrids and half all-new ^{14}N strands

?

F Predict the outcome of the third generation of DNA molecules in Meselson and Stahl's experiment: Of the eight DNA strands produced, how many would be hybrids and how many would be all 14N strands?

Fig 28.10(a) When DNA replicates, the double helix unwinds and each exposed strand acts as a template for the synthesis of a new strand. This is a simplified diagram showing how free nucleotides are added to produce two new DNA helices

Fig 28.10(b) DNA replication is actually more complicated and several enzymes other than DNA polymerase are involved. DNA helicase unwinds the parent helix and then this is 'held open' by DNA binding proteins so that the DNA polymerase can gain access to the parent strand

Strong support for the semi-conservative model of DNA replication came from the work of Meselson and Stahl (Fig 28.9). By growing bacteria with bases containing the radioisotope ^{15}N (sometimes called *heavy nitrogen*) and then following its progress, they showed that each new DNA strand contains half the original strand.

The mechanism of semi-conservative replication

The basic mechanism of DNA replication is shown in Fig 28.10(a). DNA is copied before the cell divides (see Chapter 27) and the basic idea is simple: the helix unwinds and each strand acts as a template for the manufacture of a new, identical strand.

In practice, DNA replication is rather more complex than it appears, and is a good illustration of the importance of enzymes (Fig 28.10(b)). Two **helicases** use energy from ATP to separate the DNA strands. Next, **DNA binding proteins** attach to keep the strands separate. **DNA polymerase** enzymes then move along

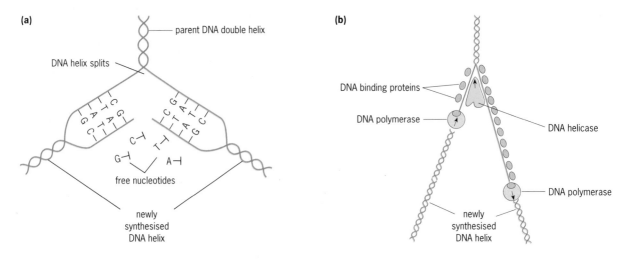

(a)

parent DNA double helix

DNA helix splits

free nucleotides

newly synthesised DNA helix

(b)

DNA binding proteins

DNA polymerase

DNA helicase

DNA polymerase

newly synthesised DNA helix

the exposed strands, synthesising the new strands, often in short segments. Finally, **DNA ligase** enzymes join the segments together, completing the new DNA strands.

■ See question 4.

DNA proofreading

Mistakes in DNA replication occur regularly. The chances of the wrong base being added are between 1 in 10^8 and 1 in 10^{12}. This may seem low, but so many nucleotides are added that this rate means approximately one mistake occurs for every ten genes that are copied. Clearly, such a rate would cause the death of most organisms. This disaster is prevented by the polymerase enzymes themselves. They can detect when the wrong base has been inserted, and pause to correct the mistake. In this way 99.9 per cent of mistakes are corrected. Uncorrected mistakes give rise to **mutations**.

There are many different DNA repair mechanisms in cells to put right the genetic damage which inevitably accumulates during the life of an organism. The condition xeroderma pigmentosum illustrates the importance of these mechanisms: an affected person is liable to get skin cancer because their cells cannot repair genetic damage caused by exposure to ultraviolet light (Fig 28.11).

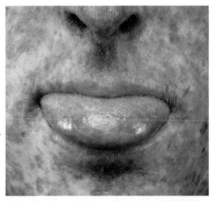

Fig 28.11 **Xeroderma pigmentosum is a rare genetic disease characterised by dry, freckled skin which is extremely sensitive to sunlight**

HOW DO YOU COMPARE THE DNA OF DIFFERENT ORGANISMS?

THE TECHNIQUE OF DNA hybridisation (Fig 28.12) can be used to compare the similarity of the DNA of different species, and is proving to be a useful tool in the search to piece together evolutionary trees.

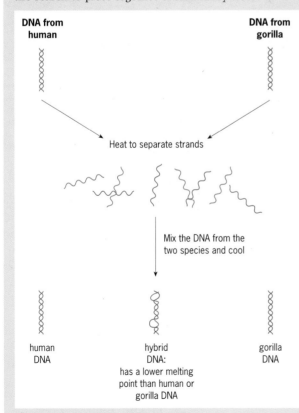

Fig 28.12 **The principle of DNA hybridisation**

DNA is much more stable than protein because it is held together by regular hydrogen bonds along the whole of its length. However, when DNA is heated to about 87 °C, known as the melting point of DNA, the hydrogen bonds break and the two halves of the molecule separate. When the temperature drops below 87 °C, the two strands join up: they **re-anneal**.

This is the basic process of DNA hybridisation:

DNA from two different species is cut into small fragments and heated to 87 °C so that the two strands separate. The separated strands from each species are then mixed together. As the temperature drops the DNA strands re-anneal. Some of the molecules formed will be hybrids: they contain one strand from each species.

The temperature at which the two hybrid strands re-anneal tells us about the degree of similarity between them. When two strands from the *same* species re-anneal, they do so at 87 °C. Closely related DNA re-anneals at temperatures slightly lower but close to 87 °C. This indicates that they have many base sequences in common and so form many hydrogen bonds along the length of the molecule. In contrast, DNA from more distantly related species must cool to much lower temperatures before they can anneal as there are fewer hydrogen bonds to hold the strands together.

For example, human DNA can be hybridised with gorilla DNA or rabbit DNA. The melting point for DNA from all three single species is 87 °C. The melting point for the human/rabbit DNA is lower than for the human/gorilla DNA. This shows that humans and gorillas have more DNA sequences in common than humans and rabbits.

■ See question 3.

Mutations

A mutation is a spontaneous change in the genetic material of an organism (Fig 28.13(b)). It usually occurs when DNA is copied and cells divide. Generally, there are two types of mutation:

● a gene mutation,

● a chromosome mutation.

Both of these may occur in **somatic cells** (normal body cells) or **germ cells** (sex cells).

Somatic cell mutations are not passed on to the next generation. Generally, organisms accumulate somatic mutations as they get older, and this is thought to be one of the causes of cancer (see Chapter 35). In contrast, when a mutation arises in an egg or sperm cell which then goes on to become a zygote, the mutation is passed on to all the cells of the new individual. Germ cell mutations alter the *genome* of an organism.

Fig 28.13(a) **To many people, this is the image conjured up by the word mutant. But in biology, a mutation is a fault in the copying of genetic material and is usually lethal or makes no difference to the organism. Only in rare cases will mutation give rise to new variation, and it is sure to be less spectacular than science fiction writers would have us believe**

Fig 28.13(b) **An example of a real-life mutation. A chance mutation in a single gene of the speckled moth *Biston betularia* produced a fortunate change from the normal speckled coloration, seen in the moth on the left, to the melanic form seen on the right. In sooty areas, this new form was better camouflaged and so had a better chance of avoiding predators. More of the black moths passed on their genes to the next generation, making the melanic form dominant in industrialised areas**

Gene mutations

A gene mutation is a change in the base sequence of DNA. We have already seen that when DNA replicates, two identical strands are formed. Occasionally, however, a fault can occur. In a typical gene of, say, 1400 base pairs, a change in just one of the base pairs has the potential to make the whole gene useless, and can prove lethal to the organism. A change in a single base is known as a **point mutation**.

How can such a tiny change be so disastrous? It is because a change to any base will alter a codon so that it will probably code for a different amino acid. In turn, this 'wrong' amino acid can affect the way the polypeptide chain folds, so changing the shape of the whole protein molecule. It might then be unable to function. (Remember: many genes code for enzymes or parts of enzymes whose proper functioning relies on precise shape.) Table 28.2 shows the different types of gene mutation that can occur, using words instead of codons. In all cases, the original meaning of the message is lost.

From Table 28.2 we can see that the two least damaging types of mutation are substitutions and inversions, because these alter only one of the seven amino acids. In some cases, if the new amino acid occurs in a non-vital part of the chain, or has similar properties to the correct one, then the mutation might not actually matter at all.

In contrast, deletions and additions are potentially much more damaging because they cause **frame shifts**: only one base may be added or lost, but this causes the whole sequence to shift along, altering all the codes. In this case, the protein made is completely different from the original and is unlikely to function as it should. The Assignment at the end of this chapter deals with sickle cell anaemia, a disease caused by a fault in just one base.

How common are gene mutations?

The rate at which genes mutate varies between species and between genes at different loci. At a rough estimate, there is an average mutation rate of about 10^{-5} per locus per generation, meaning 1 in 100 000. So, in a population of one million individuals, you could expect ten to have inherited a mutation at any particular gene. The rate of gene mutation (and chromosome mutation) can be greatly increased by environmental factors, known as **mutagens**, such as ultraviolet, X-rays, alpha and beta radiation, and chemicals such as mustard gas and cigarette smoke.

Chromosome mutations

During mitosis (normal cell division) chromosomes are copied, condensed and pulled apart. The end result should be that each daughter cell receives a complete set of perfect chromosomes. But faults can occur. A chromosome mutation occurs when the structure of a whole chromosome – or set of chromosomes – is altered in some way.

Fig 28.14 shows the four basic types of chromosome mutation: deletion, duplication, inversion and reciprocal translation.

Table 28.2 **Different types of gene mutation**

Type of gene mutation	Effect on code (using words rather than codons)
Normal code	THE FAT OLD CAT SAW THE DOG
Addition (frame shift)	THE EFA TOL DCA TSA WTH EDO G
Deletion (frame shift)	THE ATO LDC ATS AWT HED OG
Substitution	THE FAT OLD BAT SAW THE DOG
Duplication	THE FAT TFA TOL DCA TSA WTH EDO G
Inversion	THE TAF OLD CAT SAW THE DOG

Fig 28.14 **The main types of chromosome mutations**

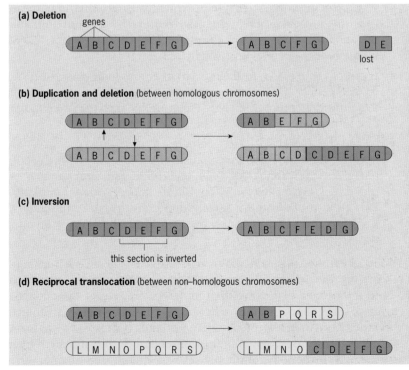

(a) Deletion
genes

(b) Duplication and deletion (between homologous chromosomes)

(c) Inversion
this section is inverted

(d) Reciprocal translocation (between non–homologous chromosomes)

(a) Deletion occurs when a chromosome splits and a fragment is lost as the parts rejoin. In this case, a whole set of genes is lost, and so this is often fatal

(b) A duplication occurs when a section of chromosome is copied twice. This can occur when pairs of homologous chromosomes break at different places and then reconnect to the wrong partners. Thus, one of the partners would have two copies of a particular sequence, but the other would have neither. This can have serious consequences

(c) An inversion results when a broken segment is reinserted in the wrong order. Any genes contained in this region will be transcribed and translated backwards, and will therefore not make the correct proteins

(d) A reciprocal translocation occurs when two non-homologous chromosomes exchange segments. This makes the chromosome pairs of unequal size, and this can cause problems in meiosis, with daughter cells receiving the wrong chromosomes

While gene mutations normally affect just one gene, chromosome mutations involve the disruption of whole blocks of genes. For this reason, many chromosomal mutations give rise to **syndromes**, which are complex sets of symptoms with a single underlying cause.

In addition to the structural mutations outlined above, another common failure is **non-disjunction**. This occurs when pairs of chromosome fail to separate during anaphase I of meiosis. Some gametes get both chromosomes, while others receive neither. Downs syndrome, for example, occurs when a gamete containing two copies of chromosome 21 joins with a normal gamete containing one, giving the affected individual three copies of chromosome 21 (see page 405).

Polyploidy

Occasionally, a mutation causes a whole set of chromosomes to be changed. For instance, a fault in meiosis might result in some gametes having two sets of chromosomes while others receive none. When such a diploid gamete joins with a normal one, the result is a **triploid** zygote – one with three sets of chromosomes. In another case, a fault in mitosis after fertilisation could result in the chromosomes doubling but not separating, leaving four sets in a single **tetraploid** cell. Neither of these embryos could develop.

The possession of multiple sets of chromosomes is known as **polyploidy**. This phenomenon is very common in flowering plants and has played a vital role in their evolution. Botanists estimate that over half of the world's flowering plant species are polyploid, including some of the world's most economically important crops: sugar beet, tomatoes and tobacco.

Why mutations are important

A mutation can have one of several effects:

- It can be lethal: an organism with a lethal mutation cannot survive and might not even develop beyond the zygote. Genes code for vital proteins such as enzymes, so many gene mutations mean that the protein is the wrong shape, and so cannot function metabolically. Chromosome mutations which involve the loss of whole blocks of genes are usually lethal.

- It can have no effect. It might occur in a non-coding part of DNA, or in a gene which is not expressed. Alternatively, it can result in a different amino acid being incorporated into the protein, but one which does not alter its ability to function.

- It can be beneficial. Occasionally, a mutation produces an improvement in phenotype. Statistically, this is a very rare event, but given the number of DNA replications and the number of individual organisms, it is bound to happen sooner or later. Beneficial mutations are hugely important because, ultimately, they are the source of all variation. Natural selection favours these 'improved' mutants (see Chapter 30).

Fig 28.13(b) on page 418 shows an example of a beneficial mutation. Following the Industrial Revolution, a mutation in the peppered moth produced a black (melanic) form that was better camouflaged against sooty backgrounds. The black moth had an advantage because it was hard for predators to see and so survived to pass on its genes to the next generation. Consequently, the mutated gene spread and the genome of the species was altered (see Chapter 30).

Meiosis is also known as reduction division. See Chapter 27 for more detailed information about this important process.

G Polyploid cells can be induced experimentally by adding the chemical colchicine, a substance extracted from the crocus. Colchicine inhibits the formation of the spindle during cell division. Suggest why colchicine results in polyploid cells.

SUMMARY

■ A chromosome is a large, densely staining body consisting of a single, supercoiled DNA strand. The genes along a chromosome are found at particular points called **loci**.

■ A **gene** is a length of DNA that codes for the synthesis of a particular polypeptide or protein. Only about 10 per cent of a chromosome consists of genes; the rest is non-coding DNA.

■ The genetic code is the sequence of bases contained in the DNA molecule. The code is read in groups of three bases, called **codons**. Each codon codes for a particular amino acid.

■ The central dogma (principle) of molecular biology is that genes are read by **messenger RNA** which in turn moves out into the cytoplasm where it codes for the assembly of a protein.

■ At the ribosome, messenger RNA attracts **transfer RNA** molecules with the right amino acids attached. The amino acids are held together so that peptide bonds form; in this way the protein is assembled.

■ DNA replication occurs when the two strands separate and the enzyme **DNA polymerase** assembles two new strands on the originals. This is called semi-conservative replication as each old strand is *conserved* as half of the new DNA.

■ A fault in DNA replication leads to a **gene mutation**. The base sequence is altered so that the gene no longer codes for the correct protein. Chromosome mutations can result in whole blocks of genes being lost, or translated backwards for example.

■ Most mutations are harmful, some have no effect, but a few can be beneficial. Beneficial mutations might give the organism an advantage and in this way mutations give rise to variation and hence drive evolution.

QUESTIONS

1 The DNA coding system contains the information for the production of polypeptides by a cell.

a) In the DNA coding system, describe:
  (i) the form in which the message is transmitted from the nucleus to the cytoplasm;
  (ii) precisely where in the cell the message is translated.

b) Explain why different alleles of the same gene produce similar, but not identical, polypeptides.

[AEB June 1996 Human Biology: Paper 1, q.5]

2 The diagram in Fig 28.Q2 shows part of a messenger RNA (mRNA) molecule.

U	A	C	C	G	A	C	C	U	U	A	A

Fig 28.Q2

a) **(i)** How many codons are shown in this section of mRNA?
  (ii) What is specified by a sequence of codons in an mRNA molecule?

b) A tRNA molecule carries a complementary base sequence for a particular codon.
  (i) write the complementary sequence for the first codon in the mRNA sequence given above.
  (ii) Describe the role of tRNA molecules in the process of protein synthesis.

[ULEAC June 1996 Biology and Human Biology Module Test, q.5]

3 DNA hybridisation is a method of comparing DNA from different species. In this technique purified DNA samples from two different species are heated to separate the strands, then mixed and allowed to cool together. As the separated strands cool they re-combine into double helices. Some of the new double helices are hybrid: that is, they consist of one strand from each species. This process is shown in Fig 28.Q3.

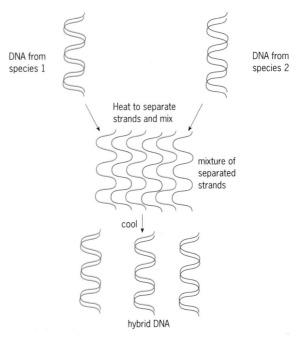

DNA from species 1

DNA from species 2

Heat to separate strands and mix

mixture of separated strands

cool

hybrid DNA

Fig 28.Q3(a)

a) The strands in hybrid DNA separate at a lower temperature than those in DNA from a single species. The more closely related the two species, the lower the difference between the separation temperatures of hybrid DNA and single-species DNA.

    (i) Suggest why the strands in hybrid DNA separate at a lower temperature than those in DNA from a single species.

    (ii) Suggest why the separation temperature of hybrid DNA from distantly related species is lower than that of hybrid DNA from closely related species.

b) The table below shows the difference between the separation temperatures of hybrid DNA and single-species DNA for a number of pairs of primate species or groups. Each figure is a mean of many tests using DNA from different individuals.

Sources of hybrid DNA	Difference in separation temperature/°C
Human/chimpanzee	1.8
Gorilla/chimpanzee	2.2
Human/gorilla	2.3
Human/orang-outan	3.6
Gorilla/orang-outan	3.6
Chimpanzee/orang-outan	3.6
Gibbon/other apes	4.8
Old World monkeys/apes	7.2

It is assumed that the difference in separation temperature is directly proportional to the time since the evolutionary lines of the two groups diverged.

    (i) The evolutionary lines of Old World monkeys and apes are thought to have diverged 30 million years ago. Use this figure to calculate the time represented by a difference of 1 °C in separation temperature of hybrid DNA. Show your working.

    (ii) Calculate how long ago the evolutionary lines of humans and chimpanzees diverged, according to DNA hybridisation studies. Show your working.

c) The diagram of Fig 28.Q3(b) is an incomplete family tree of primate evolution based on the table in part **b)**.

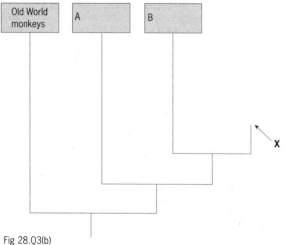

Fig 28.Q3(b)

    (i) Fill in the names of the missing species or groups in boxes A and B.

    (ii) Complete the family tree from the branching point marked X.

d) The relationship and divergence times calculated using DNA hybridisation data have been the subject of considerable debate and disagreement among students of primate evolution. In particular, it has been claimed that the difference in separation temperature between hybrid DNA and single-species DNA may not be directly proportional to the time since the evolutionary lines of the two groups diverged.

    (i) Give *two* possible reasons for doubting that the difference in separation temperature between hybrid DNA and single-species DNA is directly proportional to the time since the evolutionary lines of the two groups diverged.

    (ii) Give *two* alternative sources of evidence concerning relationships and divergence times in primate evolution.

[ULEAC June 1994 Human Biology: Paper 1, q.13]

4 In 1958, Meselson and Stahl published the results of an experiment which provided strong evidence that cells produce new DNA by a process of semi-conservative replication.

a) Why is replication of DNA described as semi-conservative?
Meselson and Stahl's experiment is outlined in Fig 28.Q4 (^{15}N is a heavy isotope of nitrogen).

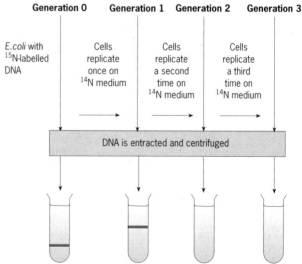

Fig 28.Q4

b) Which component of the DNA was labelled with the ^{15}N?

c) Explain why centrifugation separates the DNA labelled with different isotopes of nitrogen.

d) Copy 28.Q4 and draw in the results you would expect for generations 2 and 3.

[NEAB February 1996 Modular Biology: Continuity of Life Paper, q.4]

Assignment

SICKLE CELL ANAEMIA

In addition to the topics covered in this chapter, this Assignment assumes that you know some simple genetics (monohybrid inheritance) and evolution (natural selection).

A small minority of people, particularly those of African origin, inherit sickle cell anaemia (SCA), a condition in which their haemoglobin acts abnormally. SCA is a genetic disease: it results from a gene mutation which is inherited. Carriers of the sickle cell allele are usually perfectly healthy, and in some circumstances even have an advantage over people who possess two normal alleles. However, individuals with two sickle cell alleles have the disease.

The faulty allele makes haemoglobin S instead of the normal haemoglobin A. Problems arise when oxygen tensions are low, as they are in actively respiring tissues. At low oxygen tension, abnormal haemoglobin tends to come out of solution and crystallise, distorting the red blood cells into crescent shapes (Fig 13.A1). This has two main consequences. Firstly, sickle cells tend to form clots which can block blood vessels. Secondly, the spleen destroys abnormal sickle cells at a much greater rate than normal, causing anaemia.

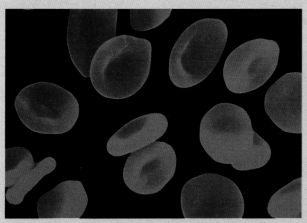

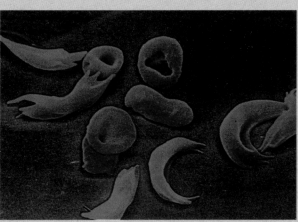

1 What is the function of haemoglobin?

Haemoglobin molecules consist of four polypeptide chains: two α-globins and two β-globins. Research shows that sickle cell anaemia results from a fault in just one base in the β-globin gene. Here is the base sequence of the first seven codons of this gene:

Normal base sequence GUA–CAU–UUA–ACU–CCU–GAA–GAG

Sickle cell sequence   GUA–CAU–UUA–ACU–CCU–GUA–GAG

2
a) How can you tell that these codons are from mRNA and not DNA?
b) Which codon has mutated?
c) What is the name given to this type of mutation?
d) Work out the amino acid sequence for both base sequences (you will need to refer to a table that shows which base codes for which amino acid).
e) Explain in general terms how this slight change can produce abnormal haemoglobin.
f) Explain why other types of gene mutation can have a much greater effect on the final protein than the example seen here.

The inheritance of sickle cell anaemia

Generally, people have two copies of each gene, and this is the case with sickle cell anaemia. The two alleles involved are:

Hb^A = coding for normal haemoglobin
Hb^S = coding for sickle cell haemoglobin

So the three possible genotypes are:
$Hb^A Hb^A$ Normal
$Hb^A Hb^S$ Carriers of the disease but with few, if any, symptoms
$Hb^S Hb^S$ Have the disease

3 Sickle cell anaemia is described as an autosomal, recessive disease.
a) What do the terms **(i)** autosomal and **(ii)** recessive mean?
b) Explain why carriers are usually healthy.
c) If two carriers have children, what is the probability that the first child will have the disease?

The sickle cell gene is far commoner than you would expect for one which is selected against. The reason for this is that the carriers usually show a higher resistance to malaria than normal.

4
a) What is meant by the phrase 'selected against'?
b) Explain why the frequency of the sickle cell allele is higher than you would expect in some areas.

Fig 28.A1 **Above left: Normal red blood cells.
Below left: Abnormal red blood cells from someone with sickle cell anaemia. Haemoglobin S tends to polymerise, forming crystal-like structures which distort the red blood cells**

29 How genes are inherited

Woody Guthrie, the famous American folk singer, suffered from Huntington's disease. His son, Arlo Guthrie was born before his father knew he had a disease that could be passed on to his children. Luckily, Arlo is unaffected: he has not inherited his father's faulty gene.

The degeneration caused by Huntington's disease can be revealed by a brain scan. A protein called huntingtin, produced by a faulty gene that was discovered in 1993, damages nerve cells in some areas of the brain, including the cerebral cortex. The gene persists because affected people tend to reproduce before symptoms appear

ABOUT EIGHT PEOPLE in every 100 000 are born with the genetic disorder Huntington's disease. This condition, which involves premature degeneration of the brain, is caused by a single faulty gene on chromosome four. The slow decline into dementia that develops ultimately, is fatal.

Diagnosis is difficult because early symptoms are vague: sufferers can have bursts of temper, become more clumsy than usual or have difficulty remembering things. Also, the symptoms do not usually become apparent until the sufferer is in their thirties or forties, usually after they have passed the gene on to the next generation.

Until recently, people with a family history of the disease knew that they were at risk, but they could never be sure whether they were actually carrying the gene. This led to a dreadful dilemma: should they go ahead and have children and risk passing the gene on, or should they decide to remain childless, only later to experience the anguish of finding out that they were free of Huntington's disease when it was too late to try for a baby.

Today, tests are available which can tell people whether they are carriers of the Huntington's gene. A sample of DNA can even show if an unborn baby carries the disease. This does not make the decisions involved any easier, but it does give affected families accurate information on which to base those decisions.

✔ Some important terms and concepts relating to inheritance are covered in Chapter 26. If you are new to genetics, read that chapter before you start to study this one.

✔ A gene is a length of DNA which acts as a template for the production of messenger RNA. mRNA forms a mobile copy of a gene, travelling from the cell's nucleus to the cytoplasm. Ribosomes use this template to synthesise a specific amino acid chain. Complete proteins are made from one or more of these chains.

1 MENDELIAN INHERITANCE

As Fig 29.1 shows, there is often a marked family resemblance between parents and their children and between siblings. However, all new-born babies, except identical twins, are genetically unique. Every new-born baby has around 60 000 genes and because humans – like most animals – are diploid organisms, a baby has two versions of each gene. It inherits one version from its mother and one version from its father.

It is difficult to imagine what happens when 60 000 different genes are mixed. Not surprisingly, although a vast amount of genetics research is underway, we are still a long way from completely understanding what happens when those mixed genes are passed on to the next generation. Some genes are activated while others remain hidden, some master genes switch on whole blocks of other genes.

How can we begin to make sense of it all? Amazingly, the man who first worked out the underlying rules did so without any knowledge of genes, chromosomes, DNA or cell division. Known

Fig 29.1 **The tendency for certain features to be inherited is shown clearly in this family. However, no two people ever inherit exactly the same set of genes unless they are identical twins**

Fig 29.2 **Gregor Mendel (1822–1884). Mendel's discoveries, like those of many scientists, were due to a combination of hard work, good scientific method, a touch of genius and a significant amount of luck**

as the father of genetics, **Gregor Mendel** (Fig 29.2) worked with pea plants. These are much simpler organisms than humans and provide a good starting point to understand some of the basic rules of **Mendelian inheritance**.

The work of Gregor Mendel

Gregor Mendel, a Czech monk, was the first to work out the basic laws which govern the inheritance of genes. Before Mendel's work became known, it was widely thought that inheritance was due to 'blending' of different features. So, for example, a tall person married to a short person would produce children who would eventually grow to an intermediate height. Mendel showed it was wrong to assume that characteristics blended and demonstrated how individual characteristics are passed down to the next generation.

For years, Mendel bred and studied the edible pea plant (Fig 29.3). He chose these plants because they had easily observable features, they were easy to cultivate, they had rapid life cycles and their pollination could be controlled. His breeding experiments centred on the inheritance of a few features such as plant height, flower colour and pea shape/appearance.

Mendel was lucky because he chose features that were controlled by single genes. In addition, pea plants tend to fertilise themselves and produce pure breeding homozygous individuals. Had Mendel worked with more complex features or a different species, he might not have been able to work out the underlying laws of inheritance.

Mendel conducted his experiments using a good scientific method. He controlled his experimental conditions carefully, ensuring that plants were pollinated only by the pollen he transferred. He made careful observations and then repeated experiments as many times as was practicable, keeping meticulous records. Mendel was trained in mathematics and he applied statistical methods to his results in a way that was very advanced for the time.

His genius really showed in the way he interpreted his results. Mendel proposed the **particulate theory**, in which he stated that there are individual units of heredity which cannot be diluted, only

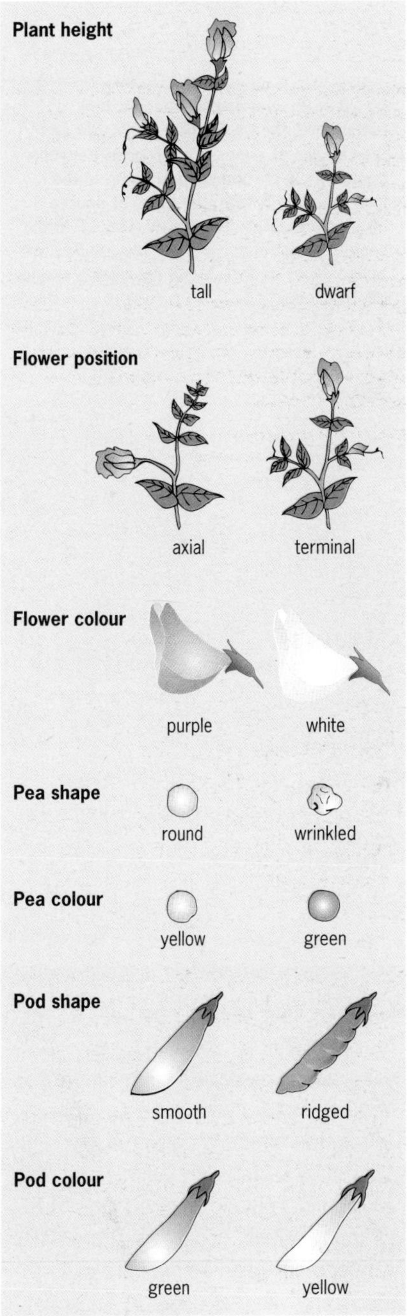

Plant height

tall   dwarf

Flower position

axial   terminal

Flower colour

purple   white

Pea shape

round   wrinkled

Pea colour

yellow   green

Pod shape

smooth   ridged

Pod colour

green   yellow

Fig 29.3 **The features of pea plants studied by Mendel. Pea plants tend to self fertilise, producing true-breeding strains. For example, white-flowered plants contain only white-flowered alleles: they are not 'hiding' any purple alleles and so always produce white flowers**

hidden. Mendel came up with the idea that each pea plant has two factors for each character, one inherited from each parent. Only one unit can be present in the gamete and so be passed on to the next generation. Mendel's 'particles' are, of course, genes.

In 1866, Mendel published a detailed account of his findings, but their significance was not appreciated by the scientific community. In 1884, Mendel died without ever receiving true recognition for his work. In 1900, his work re-emerged thanks to the independent work of three other European scientists – see Table 26.2 on page 393. In the time between 1866 and 1900, the process of meiosis has been discovered and this provided a mechanism which would explain Mendel's observations. All three scientists cited Mendel's original work, which had suggested the basic mechanism of inheritance.

Meiosis and Mendel's findings

Mendel's findings can be summarised into two laws, both of which are explained by the process of meiosis (Fig 29.4).

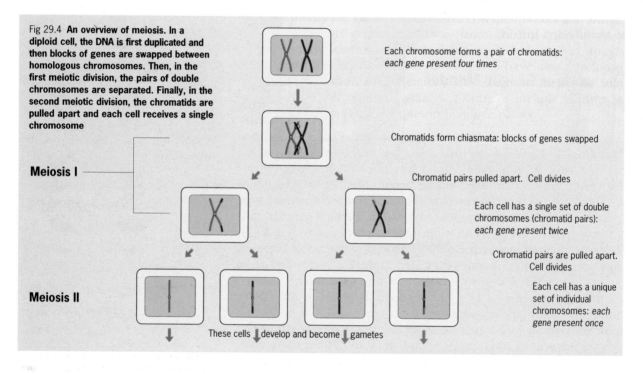

Fig 29.4 **An overview of meiosis. In a diploid cell, the DNA is first duplicated and then blocks of genes are swapped between homologous chromosomes. Then, in the first meiotic division, the pairs of double chromosomes are separated. Finally, in the second meiotic division, the chromatids are pulled apart and each cell receives a single chromosome**

Meiosis I

Meiosis II

Each chromosome forms a pair of chromatids: *each gene present four times*

Chromatids form chiasmata: blocks of genes swapped

Chromatid pairs pulled apart. Cell divides

Each cell has a single set of double chromosomes (chromatid pairs): *each gene present twice*

Chromatid pairs are pulled apart. Cell divides

Each cell has a unique set of individual chromosomes: *each gene present once*

These cells ↓ develop and become ↓ gametes

● Mendel's **law of segregation** states:

An organism's characteristics are determined by internal factors which occur in pairs, only one of which can be present in a single gamete.

In other words an organism has two versions of each gene, but only one of these version is passed on to the offspring. This is because meiosis separates homologous pairs of chromosomes, so that only one of each pair passes into the gamete.

● Mendel's **law of independent assortment** states:

Each of a pair of contrasted characteristics can be combined with each of another pair.

Again, this is explained by meiosis; any one of a pair of chromosomes can pass into the gamete with any member of another pair. The significance of this is seen in dihybrid (two-gene) inheritance.

✔

Mendel's second law is perhaps easier to understand by considering an example. If you have a plant with red or white flowers and round or wrinkled seeds, the offspring can have any combination of flowers and seeds; red and round, red and wrinkled, etc. The offspring will not necessarily have the same combination as the parent.

Monohybrid inheritance: how single genes are passed on

Monohybrid (single gene) inheritance concerns the inheritance of different alleles (usually two) of a single gene. Like Mendel, we start with the pea plant. Peas have several easily observable features which are controlled by single genes (see Fig 29.3).

For example, pea plants have one gene for height. The height gene has two alleles: tall (T) and dwarf (t). Pea plants are diploid and so have two alleles. There are therefore three possible genotypes:

TT – homozygous for T (homo = same)

Tt – heterozygous (hetero = different)

tt – homozygous for t

Consider what happens when a homozygous tall plant is crossed with a homozygous dwarf plant (Fig 29.5). All the gametes from the tall plant contain a T allele and all those from the dwarf plant have a t allele. These combine at fertilisation to give offspring all with the genotype Tt. The **first generation**, known as the **F1**, are all tall, because tallness is dominant. However, although they look identical to the tall parent plant, they are different in one very important respect: they are heterozygous and not homozygous.

If two of these heterozygous plants are crossed, half of the gametes from each parent are T and half are t, giving us four possible genotypes in the **second generation**, the **F2**:

25% are   TT
25% are   tT
25% are   Tt
25% are   tt

The first three genotypes give tall plants (the T allele is present) but a quarter of the plants are dwarf (they are homozygous for the t allele).

It is important to realise that chance plays a very important part in genetics. If we took four F2 pea plants we might expect to get three tall plants and one dwarf, but it is highly likely that we would get something else, such as four and none or two and two. The greater the number of offspring, the higher the chance that the numbers reflect the expected ratio. This is why Mendel had to carry out *hundreds* of crosses before he became convinced of the underlying ratio – see the Assignment at the end of this chapter.

The test cross

It is often important to know whether an individual displaying a dominant trait is homozygous or heterozygous for a particular allele. For example, how can we tell if a tall pea plant is pure breeding, or if it is carrying a hidden dwarf allele? To find out we do a **test cross**. This involves crossing the tall plant with a homozygous dwarf plant (tt). If the tall plant is homozygous (TT), all the offspring produced are tall. However, if it is a heterozygote (Tt), about half the offspring are tall and half are dwarf.

In genetics, different alleles are often denoted by letters, such as Aa. Usually, the capital letter stands for the dominant allele. As a tip, if you are choosing letters to represent alleles, chose those whose capital cannot be confused with the lower case version. So, Rr is better than Ss. This is very important in exams, where your handwriting may be worse than usual.

Homozygous individuals are said to be **true breeding**. This means that they will always produce the same phenotype of offspring because they are not hiding a recessive allele.

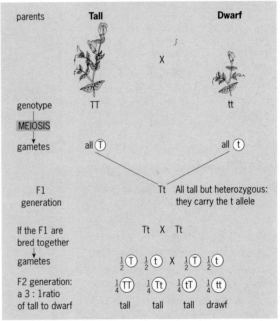

Fig 29.5 **Monohybrid inheritance in pea plants. Pure breeding tall and dwarf pea plants are homozygous (TT and tt). When crossed, the plants of the first (F1) generation are all tall but they are heterozygous (Tt), carrying the allele for dwarfness. If these heterozygous individuals are selfed (bred together) the dwarf form reappears in the next (F2) generation in a ratio of 1 : 3 (ie 1 in 4). We say that the heterozygotes do not breed true**

A In pea plants, yellow seeds are dominant to green. If a heterozygous yellow-seeded plant were crossed with a green-seeded plant, what ratio would you expect in the F1 generation?

B Why is a homozygous dwarf plant used in a test cross? Why not use a homozygous tall plant?

Monohybrid inheritance in humans

Clear-cut examples of monohybrid inheritance in humans are relatively rare, and often involve genetic disease where people inherit one or more faulty alleles. Genetic diseases are often recessive because faulty alleles which fail to make an important protein can be masked by normal ones which function properly. Table 29.1 shows examples of monohybrid inheritance in humans.

In contrast, some genetic disorders such as Huntington's disease (see the Opener of this chapter) are caused by dominant alleles. The alleles concerned code for a product which actively causes damage; the symptoms are not due to an allele not doing its job. Such alleles are dominant because the presence of a normal allele cannot mask the symptoms.

Table 29.1 **Examples of monohybrid inheritance in humans**

Traits	Features
Dominant traits	
Huntington's disease	See Opener
Freckles	
Dimple in chin	
Recessive traits	
Sickle cell anaemia	Haemoglobin polymerises, distorting red blood cells into a sickle shape. This leads to circulatory blockages and anaemia. See the Assignment on page 423.
Attached ear lobes	
Phenylketonuria (PKU)	Inability to process the amino acid phenylalanine. Protein intake must be monitored to prevent build-up of the amino acid to harmful concentrations. Can be fatal if not treated.
Haemophilia (also sex linked – see below)	Inability to produce a critical blood-clotting factor, carried on the X chromosome. Haemophilia is now treated by supplying the affected individuals with the clotting factor.
Albinism	Iinability to make pigment melanin.
Rhesus blood group	Presence (Rh+) or absence (Rh-) of rhesus protein on red blood cells.
Cystic fibrosis	Excessive mucus production, especially in lungs and pancreas. Breathing and digestion are affected and sufferers are very susceptible to lung infections.
Galactosaemia	Inability to convert galactose to glucose in the liver. All lactose must be avoided to prevent fatal levels of galactose from accumulating.
Tay-Sachs disease	Inability to produce critical lipases, which results in fatty accumulations in the brain.
Lactose intolerance	Inability to breakdown the disaccharide lactose to glucose and galactose. Leads to vomiting, diarrhoea, flatulence. Avoiding lactose prevents symptoms.

✔ The ability to roll the tongue is often cited as a simple example of inheritance in humans. However, although people seem to fall neatly into one of two categories – you can either roll your tongue or you can't – there is considerable debate about whether or not this skill can be learned, and whether it is controlled by one gene or more.

Inbreeding and outbreeding

Everyone agrees that brothers and sisters or other closely related people should not have children. In most human societies such **inbreeding** is illegal for social, moral and biological reasons.

Biologically, inbreeding is undesirable because it reduces genetic variation in the offspring. Inbreeding promotes homozygosity: inbred individuals are homozygous for many more alleles than individuals produced by outbreeding. So, faulty recessive alleles are more likely to be paired up, resulting in a much higher proportion of genetic defects and disease.

In practice, inbreeding tends to happen in small, isolated human and animal populations, or as a consequence of artificial selection. Pedigree dog and cat breeding is a good example of this. To maintain the most desirable features, champion animals are often bred with near relatives. Analysis of the pedigree certificates of the champions

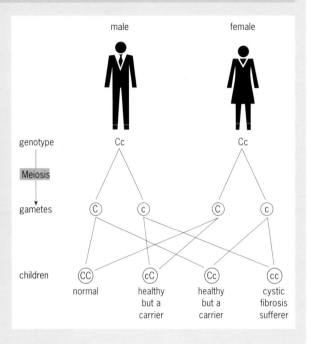

CYSTIC FIBROSIS: AN EXAMPLE OF MONOHYBRID INHERITANCE IN HUMANS

THE DISEASE CYSTIC FIBROSIS, covered in detail in Chapter 36, is caused by the mutation of a single allele. A normal allele (C) makes a membrane protein essential for the proper functioning of certain epithelial cells. The mutated allele (c) does not code for a functional protein.

Most people have two healthy alleles: their genotype is CC. In the UK population, however, around one person in 25 carries a faulty allele. These carriers (Cc) are perfectly healthy because they also possess a normal allele which makes the working protein.

However, in one couple in 625 (1 in 25 × 1 in 25) both partners are carriers. There is a one in four chance that any child they have could inherit both faulty alleles and therefore be a cystic fibrosis sufferer. As one cystic fibrosis child in four is born to one couple in 625, approximately one child in every 2500 has this condition.

Fig 29.6 **The inheritance of cystic fibrosis. If both parents are carriers, there is a 25 per cent chance that any child will be a cystic fibrosis sufferer. There is also an equal chance that a child will inherit no faulty alleles and that his/her descendants will be completely free from the disease. There is a 50 per cent chance of the child being a carrier of cystic fibrosis**

in a particular breed will often show that the same individuals crop up again and again. The consequence of this artificial selection is a relatively high proportion of genetic defects.

Outbreeding (the breeding of unrelated individuals) is more desirable as it produces more new genetic combinations and greatly reduces the chance that faulty genes will be expressed. Many organisms have developed mechanisms which prevent inbreeding (Fig 29.7). If pedigree animals are cross-bred, rather than inbred (cross a Siamese cat with a Persian, for example), you often get a fitter, healthier animal. We say that such offspring show **hybrid vigour**, and cross-breeding is a favourite tool of plant breeders who find that hybrids often grow faster and show better disease resistance.

Fig 29.7 **When woodlice lay a batch of eggs, the offspring are either all male or all female. This means that brothers and sisters cannot mate, thus reducing the chances of inbreeding**

Dihybrid inheritance: how two genes are passed on

What happens when a tall, purple-flowered pea plant is crossed with a dwarf, white-flowered individual? Do you always get tall plants with purple flowers, or can the features re-combine, producing tall, white-flowered plants, and dwarf, purple-flowered ones. The simple answer is yes, you can get these recombinants, thanks to the process of meiosis.

The inheritance of two separate genes is called **dihybrid inheritance**. If, as is usual, the two genes are on separate chromosomes, Mendel's second law (page 426) applies: each of a pair of contrasted characteristics can be combined with each of another pair. So, an organism with a genotype of AaBb can produce gametes of AB, Ab, aB or ab. In other words, one allele from each pair passes into the gamete, and all four combinations are possible. This is due to **independent assortment** during meiosis (Fig 29.8).

?

C There are approximately 700 000 live births every year in the UK. How many of these babies would you expect to have cystic fibrosis?

D Suggest why there is less inbreeding in most human societies today than in previous centuries.

Fig 29.10 **When two genes are close together on the same chromosome they are almost always inherited together: they are linked. The chance of being separated by crossover depends on the distance between them**

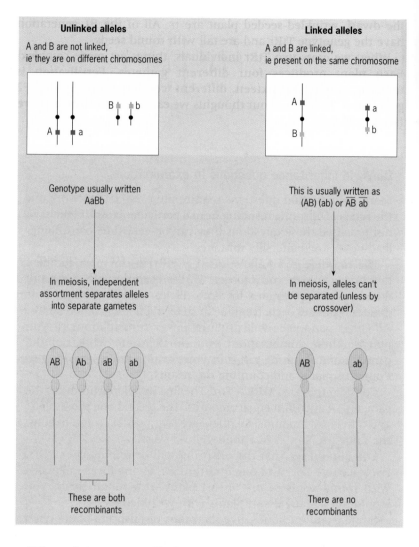

When the F1 are **selfed** (bred together) you would expect a 9:3:3:1 ratio, but this does not happen. The F2 generation are mainly like the parents (purple and elongated or red and round) but there are a few recombinants (red and elongated or purple and round). This is because the genes for flower colour and pollen shape are linked (present on the same chromosome) and so are inherited together. Any recombinants we get must therefore be due to crossing over.

Consider the set of results shown in Table 29.2.

Table 29.2 **An example of a phenotypic ratio which indicates linkage**

Phenotype	Approx. expected numbers if no linkage (9:3:3:1)	Observed numbers in F2 generation
Purple, elongated	405	336
Purple, rounded	135	31
Red, elongated	135	28
Red, rounded	45	325
Total	720	720

The total number of offspring is 720 of which 59 (31 + 28) are recombinants. Eight per cent of the offspring result from crossing over and so, by convention, we can say that the **loci** for flower colour and pollen shape are 8 units apart on the chromosome. If the percentage of recombinants was higher, we would know that the loci were further apart on the chromosome.

The term **locus** is used to describe the position of a gene on a chromosome.

When genes are not linked, we expect them to be separated 50 per cent of the time because there is a 50 per cent chance that alleles on separate chromosomes will be inherited together. When looking for linkage, geneticists look for a frequency of recombinants significantly less than 50 per cent.

■ See question 1.

Chromosome maps

The frequency with which linked genes are separated by crossover can tell us a lot about the relative positions, or loci, of these genes on the chromosome. Two alleles that are close together are rarely separated by crossing over, while those located on opposite ends of the chromosome are separated almost every time. Analysis of crossover frequency can be used to work out the relative positions of alleles and so make chromosome maps (Fig 29.11).

Genes Y and Z have a crossover frequency of 10 per cent and so are 10 units apart. A third cross between Y and X determines the relative positions of X, Y and Z. If the order of genes is XYZ (as shown), the crossover frequency between X and Y is 13 per cent. If the order is XZY, the crossover frequency is 33 per cent.

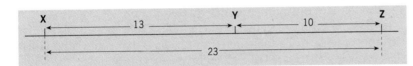

Fig 29.11 **The position of genes on chromosomes can be worked out by interpreting crossover frequencies. Consider three genes, X, Y and Z. Genes X and Z are found to have a crossover frequency of 23 per cent, so they are 23 units apart on the chromosome**

EXAMPLE

An A-level genetics question involving linkage

This example is a modified question from an UCLES Modular Biology examination in Applications of Genetics.

The mosquito, *Aedes aegypti*, has a spotted and a spotless form on a grey or yellow body. The dominant allele is spotless, S, and the recessive allele is spotted, s. The allele for grey body G, is dominant to the allele for yellow body, g. A cross was made between homozygous spotless, grey bodied mosquitoes and spotted, yellow-bodied mosquitoes. The F1 individuals were then crossed with the double recessive strain and the numbers of the resulting phenotypes were counted. The results are shown below.

Phenotype	Number
Spotless, grey bodied	442
Spotless, yellow bodied	458
Spotless, yellow bodied	46
Spotted, grey bodied	54

Q State the genotype and the phenotype of the F1 using the above symbols.

A The question is asking for the genotype of the first generation. You know that the parents are homozygous spotless, grey (genotype: SSGG) and spotted, yellow (genotype: ssgg). The genotype of the F1 must be SsGg, giving a phenotype of spotless, grey bodied.

Q Using a genetic diagram, explain fully the results shown in the table.

A The table below shows the result of a cross between an F1 individual, SsGg, and a double recessive which must be ssgg.

So we have:

Parents with genotype	Spotless, grey × Spotted, yellow SsGg ssgg			
Gametes	SG Sg sG sg × all sg			
Expected ratio	25% SsGg	25% Ssgg	25% ssGg	25 % ssgg
Phenotypes	Spotless, grey	Spotless, yellow	Spotted, grey	Spotted, yellow
	1 :	1 :	1 :	1

You should see that the expected results are not the same as those given in the table of results on the left, so there must be an extra complication. The key feature from the table is that there are more of the parental phenotypes than expected, and fewer recombinants. This should scream **LINKAGE** at you!

To produce a really thorough answer you should also work out the percentage of recombinants.

The total number of crosses in the table is 1000, of which 100 (46 + 54) are recombinants. This leaves you with the simple calculation to show that 10 per cent of the mosquitoes are recombinants, and so the two genes are 10 units apart on the chromosome.

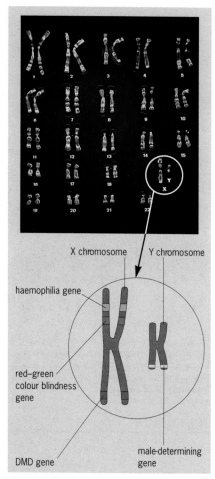

X chromosome   Y chromosome

haemophilia gene

red–green
colour blindness
gene

DMD gene

male-determining
gene

Fig 29.12 **The human karyotype showing 22 pairs of autosomes and one pair of sex chromosomes. This is a male: females have two identical X chromosomes. Note that the Y is much smaller than the X, and carries fewer genes. The Y chromosome does, however, carry the male determining gene**

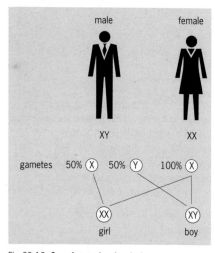

Fig 29.13 **Sex determination in humans. Half of the sperm carry an X chromosome, half carry a Y**

See question 2. ■

3 SEX DETERMINATION

What decides whether a baby will be a boy or a girl?

In humans, sex determination depends on the inheritance of special **sex chromosomes**. Every human cell contains 23 pairs of chromosomes: 22 pairs of **autosomes** and one pair of sex chromosomes (Fig 29.12). Females have two large X chromosomes and so we say they are the **homogametic sex**. Males have one X chromosome and one smaller Y chromosome, and are therefore the **heterogametic sex**.

In the female, all eggs receive an X chromosome during meiosis (Fig 29.13). However, meiosis in the male produces sperm, half of which receive a Y chromosome and half an X. So as the sperm swim for the female's egg it's a real race: if a Y sperm fertilises the female's egg, the offspring is male. If an X sperm wins, the baby will be a female.

In humans, about 114 boys are conceived for every 100 girls. We do not yet fully understand the reason for this. Interestingly, however, more male embryos die and so at birth the proportions are down to 106 to 100. By puberty the numbers are equal and in old age the females outnumber males by two to one.

In 1990, a group of genetic researchers demonstrated that the Y chromosome carries a male determining gene that codes for a **testis determining factor** (**TDF**). All embryos are female unless the active TDF imposes maleness upon it. When a male and female embryo share the same blood supply, it is possible that the TDF can produce hormones that can force an XY *and* an XX embryo develop as a male. This is an extremely rare situation in humans but it is common in cattle.

4 SEX-LINKED INHERITANCE

Genes carried on the sex chromosomes are said to be **sex-linked**. Human females have two X chromosomes which, like the autosomes, carry two alleles of every gene. Females therefore have two sets of sex-linked alleles. In males, however, the Y chromosome is smaller and cannot mirror all of the genes on the X chromosome, so males have only one set of most sex-linked alleles. This is why males suffer from the effects of X-linked genetic diseases more often than females. There are no known Y-linked diseases or conditions, probably because the Y chromosome carries so few genes.

A classic example of a sex-linked trait transmitted by the X chromosome is **haemophilia**. People suffering from this disease do not make factor VIII, an essential component in the complex chain reaction of blood clotting (see page 519). In addition to the problems caused if they injure themselves, haemophiliacs suffer from internal bleeding as a result of normal activity. Bleeding at the joints during even light exercise is a particular problem. However, haemophiliacs can usually live a full and active life by having regular injections of factor VIII.

Haemophilia is caused by a sex-linked recessive allele. Females have a pair of alleles but males possess only one. So, if a male inherits the haemophilia allele, he has the disease since he cannot possess another healthy allele to mask its effect.

Using the following notation:

X^h = the haemophilia allele on the X chromosome

X^H = the healthy allele on the X chromosome

Y = the Y chromosome which does not carry either allele

we can see that the possible genotypes are:

$X^H X^H$ = healthy female

$X^H X^h$ = healthy, carrier female

$X^H Y$ = healthy male

$X^h Y$ = haemophiliac male

We can study the inheritance of haemophilia using a **pedigree diagram**. The word pedigree means 'of known ancestry' and applies as much to humans as it does to domestic animals. The pedigree of the Royal Family can be traced back to 1066 and shows that Queen Victoria was a carrier of the haemophilia allele and passed it on to four of her nine children (Fig 29.14). Other sex linked traits in humans include Duchenne muscular dystrophy (DMD, Figs 29.12 and 29.15) and red–green colour blindness (Fig 29.16).

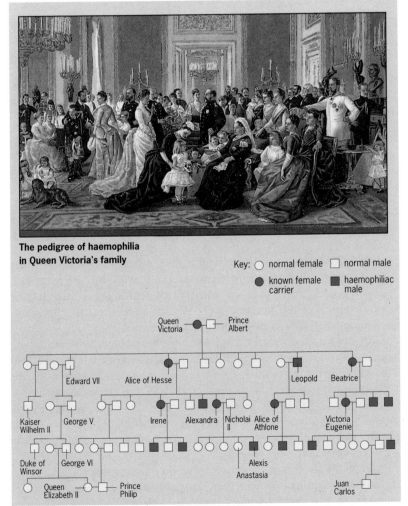

The pedigree of haemophilia in Queen Victoria's family

Key: ○ normal female □ normal male ● known female carrier ■ haemophiliac male

Fig 29.14 **Queen Victoria (1819–1901) had nine children and passed the haemophilia allele on to two daughters and one son. Her eldest son, Edward VII, did not inherit the faulty allele and so the disease is not carried by his descendants, the present-day British Royal Family**

Fig 29.15 **Duchenne muscular dystrophy affects mainly boys. The body muscles are gradually replaced by fibrous tissue and become weak, usually confining sufferers to a wheelchair by the age of 10. Few live beyond 20. One male child in 4000 is born with this disease, sometimes caused by a spontaneous mutation, so the mother may not even be a carrier. Gene therapy holds new hope for the future**

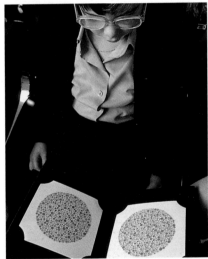

Fig 29.16 **A person with red/green colour blindness cannot distinguish between red, green, orange and yellow. On these Ishihara test charts, a person with normal vision would see numbers but a sufferer would see only a random pattern of dots. This condition is caused by a defective allele which codes for the one of the three groups of light-sensitive cone cells in the retina. Eight per cent of males but only 0.4 per cent of females suffer from colour blindness**

5 MULTIPLE ALLELES

Sometimes there are more than two alleles for a particular gene. A classic example which shows multiple alleles of the same gene is the ABO blood group in humans.

The entire human population can be classified into A, B, AB or O depending on the proteins they carry on the membrane of their red blood cells. The ABO system is controlled by one gene (I) with three alleles, I^A, I^B and I^O. The I^A allele codes for A proteins, I^B for B proteins and I^O for no relevant proteins. I^A and I^B are codominant over I^O. Each individual inherits two alleles which combine to produce the blood group as shown in Table 29.3.

Blood group testing can sometimes be used to disprove (but not prove) parentage. Table 29.4 shows, for instance, that a man of blood group AB must pass on either an I^A or an I^B allele, and so cannot possibly be the father of a child with blood group O. DNA fingerprinting, of course, provides a much more accurate (if more expensive) test which can say with a great deal of certainty who **is** the father. The Assignment in Chapter 3 looks at the process of DNA fingerprinting.

Table 29.3 **The genetics of the ABO blood group system**

Genotype	Phenotype (blood group)
$I^A I^A$ or $I^A I^O$	A
$I^B I^B$ or $I^B I^O$	B
$I^A I^B$	AB
$I^O I^O$	O

Table 29.4 **Possible blood groups of children born to parents of particular blood groups**

			Paternal blood group			
			A	**B**	**AB**	**O**
		Genotypes	$I^A I^A$ or $I^A I^O$	$I^B I^B$ or $I^B I^O$	$I^A I^B$	$I^O I^O$
	A	$I^A I^A$ or $I^A I^O$	A, O	A, B, AB, O	A, B, AB	A, O
Maternal blood group	**B**	$I^B I^B$ or $I^B I^O$	A, B, AB, O	B, O	A, B, AB	B, O
	AB	$I^A I^B$	A, B, AB	A, B, AB	A, B, AB	A, B
	O	$I^O I^O$	A, O	B, O	A, B	O

G If a child has the blood group B and his mother is blood group O, can his father be a man with blood group A? Say why.

6 POLYGENIC INHERITANCE

So far in this chapter, we have been concerned with **discontinuous inheritance**: all-or-nothing features controlled by single genes. Many features, however, are **continuous** and are controlled by several genes which act together. We describe the inheritance of such characteristics as **polygenic inheritance**.

Skin colour in humans is an example of polygenic inheritance. The depth of skin colour depends on the amount of melanin present. This is determined by at least three genes. As an example of this type of inheritance at work, imagine that the alleles A, B and C contribute cumulatively to the overall amount of melanin but the alleles a, b and c do not. So, someone with the genotype AABBCC would have the darkest skin, while the genotype aabbcc would confer a very pale skin colour.

As Fig 29.17 shows, a cross between two purely homozygous individuals would result in an intermediate (heterozygous) F1 generation, but the F2 generation would show a wide variation in skin colour depth. The graph shows a **normal distribution**: most individuals have skin colours in the mid-range. This range is one of the characteristic features of continuous variation (see page 391).

The inheritance of height in humans is polygenic.

H Non-identical twins are born to a couple who both have the genotype AaBbCc. Is it possible for one baby to be born with white skin and the other to have black skin? Explain your answer.

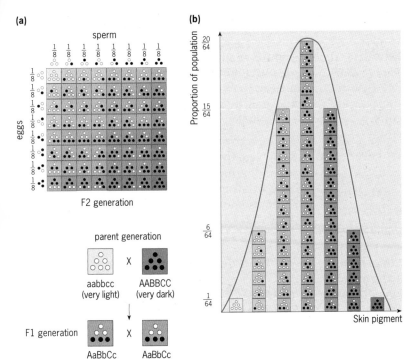

Fig 29.17 **Human skin colour is an example of polygenic inheritance which shows a normal distribution in the F2 generation**

7 EPISTASIS (WHEN GENES INTERACT)

Up to now, we have discussed the inheritance and effect of genes which act on their own. In practice, however, many genes interact with each other: the presence of one allele often affects the expression of others. We call this interaction **epistasis**. A classic example of epistatic gene interaction is coat colour in mice. The natural coloration of wild mice is called **agouti** and is produced from banded hairs (Fig 29.18).

Two genes are involved, each with a dominant and a recessive allele. We will represent them by the notation Aa and Bb, where A and B are dominant over a and b. The allele A codes for the ability to produce hair pigment: AA and Aa mice have pigmented hairs but all aa individuals are albinos. The B allele codes for the ability to make hair with graduated coloration: BB and Bb mice have graduated hair, bb mice have hair which is all one colour.

Only mice which have both dominant alleles A and B show the agouti coloration. These can have the genotype AABB, AaBB, AABb or AaBb. Mice which are aa cannot make pigment, so, whether they are BB, Bb or bb, no bands are visible. Mice which can make pigments but not banded hair (genotypes AAbb or Aabb) are plain black.

■ See question 5.

Fig 29.18 **The base of the hair of a mouse with agouti coloration is dark brown/black fading to lighter brown towards the tip**

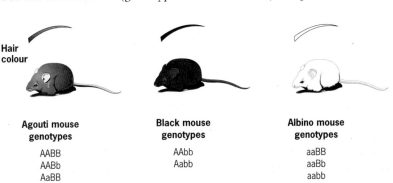

Agouti mouse genotypes	Black mouse genotypes	Albino mouse genotypes
AABB	AAbb	aaBB
AABb	Aabb	aaBb
AaBB		aabb
AaBb		

?

I Two agouti mice, genotypes AaBb, are bred together. What phenotypic ratio would you expect in the next generation?

SUMMARY

■ Humans, like most organisms, are **diploid**; each cell carries two versions, or **alleles**, of each **gene**. If both alleles are the same, the organism is **homozygous** for that allele. If the pair are different, the organism is **heterozygous**.

■ During **meiosis**, only one of the alleles passes into the **gamete**. This is Mendel's first law.

■ **Monohybrid inheritance** is the study of one gene. A cross of AA × aa gives all Aa in the **first generation (F1)**. A cross of Aa × Aa gives a 3:1 **phenotypic ratio** in the **second generation (F2)**.

■ **Dihybrid inheritance** involves two separate genes, usually on separate chromosomes. Homozygous individuals AABB × aabb produce offspring which are all AaBb in the F1 generation. However, due to **independent assortment** during meiosis, AaBb individuals produce gametes of AB, Ab, aB and ab. If bred together, AaBb individuals produce sixteen different genotypes, and four different phenotypes with a ratio of 9:3:3:1.

■ Genes present on the same chromosome are said to be **linked**. Such genes are inherited together unless separated by crossover in prophase I of meiosis. The greater the distance between the genes on the chromosome, the more often they are inherited separately.

■ Humans have 23 pairs of chromosomes: 22 pairs of **autosomes** and one pair of **sex chromosomes**. The X chromosome is larger than the Y and carries more genes.

■ The sex of a baby is determined at the moment of conception. Because gametes are produced by meiosis, half of the sperm carry a Y chromosome and half carry an X. If a Y sperm fertilises the egg, a male results; X gives a female.

■ Genes carried on the sex chromosomes are said to be **sex-linked**. Females have two complete sets of sex-linked genes, but males have only one copy of all but the few genes carried on the Y chromosome.

■ Some diseases such as haemophilia are sex-linked; caused by defective genes on the X chromosome. Females have two alleles and so can be carriers. Males have only one allele and so cannot be carriers: they are either healthy or they have the disease.

■ Some genes have more than two alleles. The human ABO blood group is an example of such a **multiple allele**. In this case, one gene has three alleles, A, B and O.

■ Many characteristics, such as height in humans, are controlled by many different genes. We describe such characteristics as **polygenic**; they often produce a range of values. Human height is a good example of a polygenic characteristic.

■ The presence of one allele can affect the expression of an allele of a different gene. This type of gene interaction is called **epistasis**. For example, one allele may give an organism the ability to make pigment, while another allele determines the colours and/or the distribution of that pigment.

QUESTIONS

1 In maize, the allele for coloured grain, **C**, is dominant to the allele for colourless, **c**, and the allele for rounded grain, **R**, is dominant to the allele for wrinkled grain, **r**.

a) Explain what is meant by: **(i)** a gene; **(ii)** an allele.

b) Maize plants heterozygous for both characteristics were crossed. The following results were obtained:
83 coloured, rounded
33 coloured, wrinkled
29 colourless, rounded
8 colourless wrinkled
Suggest why the results are not an exact 9:3:3:1 ratio.

c) Some of the colourless rounded grains would have the genotype **ccRr** and some would be **ccRR**.

Explain the genetic crosses necessary to distinguish between these genotypes.

[NEAB Feb 1996 Modular Biology: Continuity of Life Paper, q.2]

2 The diagram of Fig 29.Q2 shows the inheritance of one type of myopia in three generations of a human family.

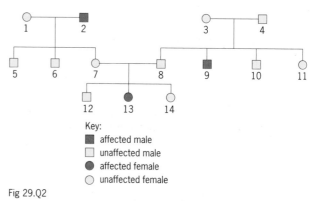

Key:
■ affected male
□ unaffected male
● affected female
○ unaffected female

Fig 29.Q2

a) Give **one** piece of evidence from the diagram which suggests:

    (i) that the myopia is determined by a recessive allele;

    (ii) why it is **not** likely that myopia is a sex-linked condition.

b) Using **A** for the dominant allele and **a** for the recessive allele, give the possible genotype or genotypes for: **(i)** individual 7; **(ii)** individual 12.

[NEAB Feb 1996 Modular Biology: Continuity of Life Paper, q.7]

3 Pure-breeding pea plants with grey seed coats and tall stems were crossed with plants with white seed coats and short stems. The offspring (F1 generation) all had grey seed coats and tall stems.

a) State suitable symbols for the alleles for colour of seed coat and height of stem.

b) Explain, by means of a genetic diagram, the ratios of genotypes and phenotypes you would expect if the F1 generation were self-fertilised.

c) Explain how you could find out the genotype of a plant with grey seed coats and tall stems.

[UCLES Spring 1996, Modular Biology: Central Concepts in Biology Specimen Paper, Section A, q.1]

4 Fig 29.Q4 shows part of a family tree in which the inherited condition of phenylketonuria occurs.

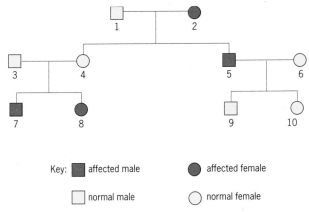

Key:
- ■ affected male
- ● affected female
- □ normal male
- ○ normal female

Fig 29.Q4

a) Identify and explain **one** piece of evidence from this family tree to show that the allele for phenylketonuria is recessive to the allele for the normal condition.

b) Giving a reason for your answer in each case, identify **one** individual who must be: **(i)** heterozygous; **(ii)** homozygous.

c) If individual 10 married a man who was heterozygous for the gene, what is the probability that their first child would be affected?

[AEB 1996 Human Biology: Specimen Paper 1, q.7]

5 Fig 29.Q5 shows four different comb shapes in chickens, resulting from the interaction of two unlinked gene loci, **R/r** and **P/p**. The genotypes of each comb shape are given. A dash (–) in the genotype indicates that either the dominant or the recessive allele may be present.

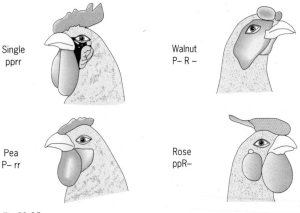

Single
pprr

Walnut
P– R –

Pea
P– rr

Rose
ppR–

Fig 29.Q5

a) Single-combed chickens were crossed with pure-breeding walnut-combed birds. The F1 were then bred together to produce the F2 generation. Draw genetic diagrams to show the genotypes and phenotypes of the F1 and F2 generations.

    NB Take care that **P** and **p** cannot be confused in your answer.

b) Explain how varieties of chickens with very different plumage or comb shapes can be produced.

    The interaction of these two unlinked gene loci produces four separate phenotypes and is an example of discontinuous variation.

c)   **(i)** Suggest **one** phenotypic characteristic in chickens which shows **continuous** variation.

        (ii) Explain the genetic basis of continuous variation.

[UCLES Spring 1996 Modular Biology: Genetics and its Applications Module Specimen Paper, Section A, q.1]

6 Familial hypophosphataemia is a sex-linked condition caused by a dominant allele on the x chromosome. The pedigree of a family with this condition is shown in Fig 29.Q6.

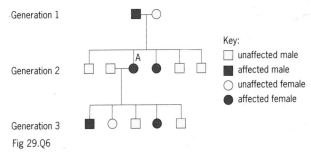

Key:
- □ unaffected male
- ■ affected male
- ○ unaffected female
- ● affected female

Fig 29.Q6

a) What is meant by the term *sex linkage*?

b) Explain why none of the males in generation 2 suffers from hypophosphataemia.

c) If person A were to have another son, what is the probability that this son would suffer from hypophosphataemia? Give a reason for your answer.

[ULEAC June 1994 Human Biology: Paper 3, q.3]

Assignment

THE CHI-SQUARED TEST

The chi-squared test is a simple statistical method that scientists use to tell if their results are significant or due to chance. For instance, if you were to roll a dice, you would have a one in six chance of getting any particular number. If you rolled the dice six times, you would probably not get each number once; you may get three twos or some other combination. However, the more times you roll the dice, the greater the probability that all six numbers will be represented evenly.

If you are playing a game with someone who gets a higher proportion of sixes than normal, you might suspect them of playing with a loaded dice. Without investigating the dice itself, how could you decide whether that person was just lucky or that the difference was due to cheating? You could use the chi-squared test.

The nature of science and the null hypothesis

Most people accept that there is a link between smoking and lung cancer but, in science, it is virtually impossible to prove anything absolutely. What you can do is gather support for your **hypothesis**.

You could do some research into lung cancer victims and divide them into smokers and non-smokers. You could then use the chi squared test to decide if there was a significant difference in the number of smokers who developed lung cancer compared with the non-smokers.

To use this statistical test you must approach the problem from a specific angle, and this involves developing a **null hypothesis**.

In the lung cancer example, you would use the hypothesis: 'There is **no** difference in the incidence of lung cancer between smokers and non-smokers'. You could then perform the chi-squared test (or some more sophisticated statistics) and come up with the conclusion that 'there is a very low probability that these results are due to chance' in which case you can reject the null hypothesis and support the idea that there *is* a link between smoking and lung cancer.

1 Write a null hypothesis that you might use to investigate the following questions:

a) What is the effect of sunbathing on the incidence of skin cancer?

b) What is the effect of cooking on the vitamin C content of vegetables?

The test itself

The basic idea behind the chi-squared test is that you compare the observed results with those you would expect. The extent to which the results vary is sometimes described as 'goodness of fit'. This means do the actual results fit the results that we had expected?

The formula for the chi-squared test is:

$$X^2 = \sum \frac{(O - E)^2}{E}$$

In this formula
O = the observed result,
E = the expected result and
Σ = the sum of

In practice it is best to use a table like the one below to progressively work out all the steps needed by the formula. In this worked example, a geneticist performed one of Mendel's classic pea experiments and got the following results.

Yellow, round	318
Yellow, wrinkled	103
Green, round	106
Green, wrinkled	33
Total	560

The geneticist wanted to know whether these results did actually reflect the expected 9:3:3:1 ratio, or whether they deviated so significantly that other forces were at work, and that the 9:3:3:1 hypothesis was in doubt. She did the following calculations:

Category	Hypothesis	Observed	Expected	$O - E$	$(O-E)^2$	$\frac{(O-E)^2}{E}$
Yellow, round	9	318	315	3	9	0.028
Yellow, wrinkled	3	103	105	−2	4	0.038
Green, round	3	106	105	1	1	0.009
Green, wrinkled	1	33	35	−2	4	0.114
Total		560				= 0.189

So the value of chi squared is 0.189. What next? The hard work is now done and we only need one more piece of information; the number of **degrees of freedom**, which is a measure of the spread of the data. The number of degrees of freedom is always one less than the number of categories of information. So in this case, it is four categories minus one which equals three degrees of freedom.

We now look up the value of 0.189 with three degrees of freedom on a table of chi-squared values. The following is an extract from the chi-squared tables:

Table 29.A1 **Extract from chi-squared tables**

	Probability (eg 0.5 = 50%)									
	0.99	**0.98**	**0.95**	**0.9**	**0.7**	**0.5**	**0.3**	**0.2**	**0.1**	**0.02**
1 D of F	.00016	.0063	.0039	.0158	.148	.455	1.07	1.64	2.71	5.41
2 D of F	.0201	.0404	.103	.211	.713	1.39	2.41	3.22	4.60	7.82
3 D of F	.115	.185	.352	.584	1.42	2.37	3.66	4.64	6.25	9.84

From Table 29.A1 we can see that our value of 0.189 lies between the 0.98 and 0.95 column. From this we can say that the probability that our results are significant is between 95 and 98 per cent. Put another way, it tells us that the probability that these results are due to chance is between 2 and 5 per cent. By convention, scientists say take the cut-off point as 5 per cent. If the test shows that there is a less than 5 per cent probability that your results are due to chance, you can reject your null hypothesis.

$\mathcal{2}$ Use the chi-squared formula to work out the following:

a) A student tosses a coin one hundred times. She gets 64 tails and 36 heads. Is the coin biased or are the results due to chance?

b) A geneticist managed to get his hands on a breeding pair of incredibly rare Matabili dung beetles. The male had green eyes and a blue abdomen. The female had pink eyes and an orange abdomen. When he crossed them he got the following results:

Table 32.A2

Phenotype	Numbers of offspring in F1
Green eyes, blue abdomen	563
Pink eyes, blue abdomen	190
Green eyes, orange abdomen	183
Pink eyes, orange abdomen	64
Total	1000

Do these results conform to the expected 9:3:3:1 ratio?

c) Apart from their diet, these beetles look like being ideal organisms for the study of genetics. What features make an organism suitable for the study of inheritance in the laboratory?

d) Much progress in the field of genetics has been made using the fruit fly **Drosophila**. Find out some of the features that can be studied using *Drosophila*.

Chi-squared in examinations

If you have worked your way through the above examples, you may have concluded that chi squared involves a lot of arithmetic and is time consuming. The examiners have come to the same conclusion and therefore it is unlikely that you will be set a full investigation to work through from scratch: they want to test your biology, not your maths. However, this test is a very useful tool and examiners will want you to appreciate its uses, and possibly to do some simple calculations.

30 Evolution

Adam and Eve. Were they really the start of the human race?

HUMANS ARE A VERY strange species. We stand on two legs, naked apart from hair on the top of our heads and at a few strategic places. We are covered in sweat glands, we have a flat face and a long neck. Males have very large genitals compared to body size, whilst females have permanently swollen breasts. An alien zoologist looking at humans for the first time might be surprised that we dominate the Earth.

In the absence of any survival equipment, such as sharp claws or teeth, great speed or protective armour, why have we been so successful? The answer lies, of course, in our brains. Given our body size, we have the largest brain of any animal. Coupled with dextrous hands, our brains allow us to use language, cooperate and make tools to a far greater extent than any other species.

But where did humans come from? Until fairly recently, the answer was clear. 'And the Lord God formed man of the dust of the ground, and breathed into his nostrils the breath of life; and man became a living soul.' (Genesis 2:7) Women followed soon after: 'And the rib, which the Lord God had taken from man, made he a woman, and brought her unto the man' (Genesis 2:22)

Until relatively recently most people believed this, or another explanation based on religious beliefs. Some still do, but Charles Darwin's theories on evolution changed fundamental scientific thinking. It is clear that humans evolved from an ape-like ancestor.

1 WHERE DO LIVING THINGS COME FROM?

To most human eyes, the living world seems fairly static. Like produces like: elephants give birth to baby elephants, frogs make frogs. It seems that no new species appear. There are literally millions of different species on Earth. Where did they come from?

Theories on the origin of species

Ideas about the origins of species – including ourselves – have been central themes in philosophical and religious thought for centuries. Until the nineteenth century, virtually everybody in the Christian world looked to the Bible. The book of Genesis provided the answers: all animals and plants were created at the same time, by the great Creator. Even the eminent taxonomist Linnaeus, who in 1742 published his system for classifying all living plant species

Fig 30.1 **The discovery in the nineteenth century of fossils such as this *Triceratops* fuelled the idea that the living world was not static. Different layers of rock held different sets of organisms and this led to the acceptance that the living world was not fixed: plants and animals could change with time**

known at the time, said 'there are as many species as God created in the beginning'. Other religions had different ideas.

The idea that species were not fixed, and that the infinite variety of living things had developed through a slow process of evolution, seems to have arisen first in early Greek philosophy and reappeared from time to time throughout the following centuries. It failed to gain lasting acceptance because no explanation could be found to show how the process might have come about. In addition, in some cultures, any 'non-religious' thought was strongly discouraged, even on penalty of death.

A catalyst in the development of the theory of evolution was the discovery of fossils (Fig 30.1). These were clearly the remains of extinct creatures. Why were they no longer alive?

One Christian explanation said that extinct species were the victims of a great global catastrophe, and that a new set of organisms had been created to replace them. It was even suggested that only the passengers aboard Noah's Ark survived a great global flood. When it was pointed out that different rocks contained different types of plants and animals, it seemed that there would have had to be many such floods, and so the **catastrophe theory** gradually lost credibility.

An early attempt to explain the mechanism of evolution came from the French naturalist Lamarck, who had made extensive studies of different life forms. He was convinced of the process of evolution, but the mechanism he proposed in 1809 to explain it was wrong. Lamarck supposed that organisms adapt to their environment by developing new structures and losing old ones. Such differences, *acquired during the animal's life*, were then passed on to the next generation. As an example he cited the webbed feet of water birds. Lamarck suggested that, in an effort to swim, the birds extended their toes and so stretched the skin between them. The stretched condition was then inherited and this process was repeated until it produced a fully-webbed foot. With no knowledge of genes, chromosomes or DNA, this idea of the **inheritance of acquired characteristics** was perfectly reasonable for the time.

?

A What was wrong with Lamarck's theory of evolution?

CHARLES DARWIN

CHARLES DARWIN (Fig 30.2) was the son of a doctor. After studying medicine at Edinburgh he changed his mind and prepared for a career in the Church. However, despite going to Cambridge to study theology (religion), Darwin retained his interest in biology and geology.

Darwin was offered the post of naturalist on the HMS Beagle, embarking on a scientific survey of South American waters. It was observations he made

Fig 30.2 **Charles Darwin, 1809 to 1892**

during this trip that first stimulated the young Darwin into contemplating the origins of species. Not only did he become convinced of the process of evolution, but he developed an idea about how it might happen. This trip was not only the turning point in Darwin's life; it was also the start of the greatest ever revolution in our perception of the living world, and of our place within it.

On returning home, Darwin buried himself in his studies, determined to prove or disprove his ideas about the evolution of species. Like Lamarck, Darwin developed his theories with no knowledge of genes, chromosomes or DNA. Mendel's work with peas (see page 425) was going on during Darwin's lifetime but it remained hidden in an obscure journal.

Even though he managed to build up an overwhelming case for evolution, Darwin was reluctant to publish his theories, knowing the conflict they would cause with the Church. For more than 20 years Darwin carried on gathering evidence to support his theories. He was finally pushed into publication after exchanging letters with **Alfred Russel Wallace**: he too had come to the same conclusion about evolution. They agreed to make a joint announcement, and the Darwin–Wallace paper was presented to the Linnaean Society of London in July 1858.

In November 1859, Darwin published his revolutionary work, ***The Origin of Species by Means of Natural Selection***. Few books before or since have caused quite such a controversy. Although Church leaders largely remained quiet, many people bitterly attacked it because it questioned the theory of creation. Darwin was called 'the most dangerous man in England'.

A working definition of evolution is 'a change in the genetic composition of a population over time'. It has also been defined as 'a gradual process in which new species develop from pre-existing ones'. But as we shall see in this chapter, populations can evolve without changing into new species.

2 CHARLES DARWIN AND THE THEORY OF EVOLUTION

In a book whose title was shortened to *The Origin of Species*, Charles Darwin included these four ideas in his theory of evolution:

● **The living world is changing**, not static.

● **The evolutionary process is usually gradual**, not a series of jumps such as catastrophes and re-creations.

Fig 30.3 **This cartoon first appeared in Punch in 1861. It shows how poorly people of the time understood Darwin's theories**

THE LION OF THE SEASON.

ACCURATELY ESTIMATING THE AGE OF THE EARTH is central to the theory of evolution. In 1650 James Usher (then the Archbishop of Armagh) announced that his study of Hebrew literature had led him to the conclusion that the date of creation was the year 4004 BC. Dr Lightfoot, vicar of the University church at Cambridge, took it a step further and proclaimed that the date of creation was 9 a.m. on 23 October, 4004 BC. Baron George Cuvier later stated that the Earth must be much older, more like 70 000 years.

Faced with the large array of different fossils, nineteenth century geologists decided that the Earth was much older still. In 1830, Charles Lyell suggested that the true age of the Earth would turn out to be millions, rather than thousands of years. He was the first to propose that old rocks could be uncovered or brought to the Earth's surface (by volcanic activity, for example), so revealing the remains of long-dead species.

This century, ultra-sensitive dating techniques which measure the decay of radioactive isotopes have put the age of the Earth at about 4 600 000 000 (4.6 billion) years. This is an important breakthrough because, although our brains cannot comprehend such a time scale, it fits in well with the idea of progressive evolution. The Assignment at the end of this chapter should help you to put this geological time scale into perspective.

● **The common ancestor theory**. This suggested that closely related species evolved from one basic ancestor. Darwin said that man and apes evolved from a common ancestor, but the popular press took this to mean that humans had recently evolved from apes, leading to cartoons such as the one in Fig 30.3.

● **The theory of natural selection** as a mechanism to explain evolution.

Natural selection: 'survival of the fittest'

Darwin's observations of the living world led to three basic conclusions:

● Generally, organisms produce far more offspring than can possibly survive.

● Living things are locked in a struggle for survival. They compete for food, space, mates etc. In short, they are trying to eat and not be eaten, so that they can reproduce.

● Individuals of the same species are rarely identical: they show **variation**.

There was nothing particularly original about these observations, or in the idea that organisms evolved. What Darwin did, however, was to put all these factors together and suggest **natural selection** as the mechanism for evolutionary change.

Natural selection is commonly simplified to 'survival of the fittest'. Biologists define fitness as the ability of an organism to survive and reproduce. The fittest organisms survive to produce more offspring than less fit ones. In this way, the characteristic of a population can gradually change from generation to generation.

To illustrate the principle, imagine a population of blackbirds in a woodland. Like most organisms, they reproduce sexually and there is variation in the population. Some have better reflexes than others, forage for food more successfully, others have a better immune system. Normally, there is competition for resources such as food, mates and nesting sites. In this situation, the fittest birds survive and produce more offspring than others. The fitter birds may gather more food and so rear larger families than less fit individuals (Fig 30.4). Or they may rear two clutches of eggs in the time it takes others to produce one. The vital point here is that the *fittest individuals pass their genes on to more of the next generation.*

The cornerstone of Darwin's theory is that the mechanism of evolution is natural selection – this is known as **Darwinism**. The application of twentieth century knowledge about chromosomes, genes and DNA to Darwin's theories is called **neo-Darwinism**.

Fig 30.4 **Blackbirds which can gather the most food tend to raise the greatest number of offspring, and so pass on their genes to more of the next generation**

Fitness in biology is a measure or reproductive success. So, a fit flowering plant produces and disperses more seeds than its competitors. Fit fungi are more successful because they produce more spores.

Fig 30.5 **When the European rabbit is introduced into countries where it is not endemic, it tends to out-compete the native herbivores. With few predators around, the rabbit population can reach plague proportions. When this has happened in the UK and Australia, farmers faced ruin. The deadly myxoma virus was introduced into the rabbit population, killing 99 per cent. The remaining 1 per cent were naturally resistant to the virus and are the ancestors of today's, largely resistant, rabbit population**

The severity of an organism's circumstances is called the **selection pressure**. The greater the selection pressure, the faster the evolutionary process. When organisms have a short life cycle, the process can be very swift indeed. In the 1950s farmers in the UK and Australia took the drastic step of introducing a deadly disease, myxomatosis, into the ever-expanding rabbit population (Fig 30.5). This virus killed over 99 per cent of the rabbits: this is selection pressure at a massive level. Only the rabbits whose genes gave them resistance to myxomatosis were able to survive and pass on their genes to the next generation.

Natural selection in action

Many people tend to think that evolution is something that happened in the past and that the world of today is the finished product. But this is far from the truth: evolutionary forces are still at work and natural selection is continually changing the characteristics of populations, often with great speed. Consider the following examples:

- The widespread use of antibiotics such as penicillin rapidly led to the development of resistant strains of bacteria. Penicillin works by inhibiting one of the enzymes which the bacteria need in order to manufacture cell walls. However, many resistant strains have a slightly different enzyme, one which is unaffected by penicillin.

- The use of warfarin as a rat poison has led to the development of warfarin-resistant rats. Warfarin is an anti-coagulant which kills rats by inducing **haemorrhage** (internal bleeding). Resistant rats have a modified enzyme in their blood-clotting system which allows their blood to clot in the presence of warfarin.

- Copper tolerance in the grass *Agrostis tenuis*. Copper is a metabolic poison which prevents many plants from growing on copper-polluted land (Fig 30.6). However, resistant plants have evolved. These can transport copper out of their cells, so that it accumulates in the cell wall and does not interfere with the plant's metabolism. (In non-polluted areas, the normal type of grass out-competes the copper-resistant strain.)

The organisms in these examples have a short life cycle and high reproductive capacity, and so can respond quickly to the selection pressure placed on them. The effects of selection pressure on organisms which live longer and have a lower reproductive capacity take much longer to become apparent.

?

B Although we have waged war against species such as locusts, weevils and beetles, we have never succeeded in making any pest species extinct. On the other hand, many larger species such as rhinos, whales and giant pandas are facing extinction, despite our efforts to save them. Give two reasons for this difference.

Fig 30.6 **This disused copper mine at Parys Mountain in Anglesey (near right) is devoid of life. Copper is a metabolic poison which kills virtually all plants. The grass *Agrostis tenuis* (far right) is one of the few that has evolved copper-resistance**

ARTIFICIAL SELECTION

HUMANS HAVE PRACTISED artificial selection, or selective breeding, for thousands of years. We haven't needed a scientific understanding of genetics to do this: animals or plants with the most desirable features are simply crossed or mated. For example, if cows with the highest milk yield are impregnated by the healthiest and fastest-growing bulls, many of the next generation are big, healthy cows which produce large quantities of milk. Those which do not have the desired features are not used for breeding. In this way, *artificial selection* rather than natural selection is used to bring about great changes in the phenotype in a short time: animals and plants can be 'domesticated' in just a few generations.

Selective breeding is still important in many different areas of food production:

● Increasing the growth rate and the meat, milk or egg production of livestock.

● Increasing the disease resistance of crop plants (increasing the disease resistance of animals by selective breeding is not as reliable).

● Changing the muscle to fat ratio of livestock, so that more lean meat is produced.

● Increasing the yield and nutritive value of crop plants such as wheat.

● Increasing the tolerance of plants and animals to drought, heat or pollutants.

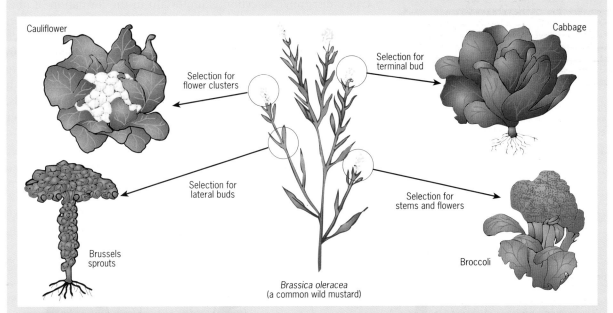

Fig 30.7 **Many familiar vegetable types have arisen from the artificial selection of the wild mustard, *Brassica oleracea*. In the original wild population, plants show variation in the size and shape of leaves, stems, terminal buds, lateral buds and flower clusters. Selection for individual features rapidly produces the familiar crops you see here**

Fig 30.8 **Like all pedigree animals, this champion Abyssinian cat is the product of many generations of artificial selection. Breeders consider the wedge-shaped face a desirable characteristic and so have concentrated on this feature. Breeding animals in this way can lead to problems, and some people strongly disagree with breeding animals for superficial 'perfection'**

Different types of natural selection

From what you have learned about natural selection, it would be tempting to assume that it is just a mechanism for change. However, this is not always the case: it can be a means of keeping things static. Many types, crocodiles and sharks for example, have not changed significantly for millions of years.

There are basically three different types of natural selection:

● **Stabilising selection** (Fig 30.9(a)). In a favourable and unchanging environment, stabilising selection can lead to a

EVOLUTION ON ISLANDS

AS DARWIN FOUND OUT ON HIS VOYAGE, islands are particularly interesting places to study evolution. Young, volcanic islands such as the Galapagos (Fig 30.12) are often colonised by plants whose seeds have been blown there by the wind, and by flying animals which have been driven off course by storms. In later years, of course, colonisation by non-flying animals (such as rats and cats) is likely to be a result of human activity.

The evolution of birds on islands can be very spectacular. Free from predators, birds often lose the power of flight and sometimes grow very large. Before the Maori settlers arrived from Polynesia, New Zealand was a bird-dominated ecosystem. Huge flightless birds, **moas**, some as tall as 3.5 metres, evolved to fill the niches normally occupied by mammals. They were easy prey for the Maoris, however, who hunted them to extinction within a few centuries. Ironically, the only survivor of the moa family, the kiwi, has become the country's national emblem.

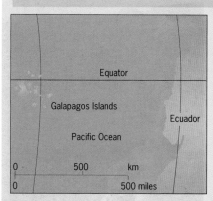

Fig 30.12 **The Galapagos Islands are an archipelago situated near the Equator in the Pacific Ocean, about 600 miles west of Ecuador**

Comparative anatomy

The anatomy (body plan) of plants and animals can reveal a lot about evolutionary relationships. Fig 30.13 shows one of the classic examples of comparative anatomy, the **pentadactyl limb** (five-fingered limb). This structure appears in various forms throughout the vertebrates: the feet of reptiles and amphibians, the human hand, the bat's wing, the seal's flipper, for example.

This basic five-fingered plan has been modified in a variety of ways to suit the animal's mode of locomotion. Structures with a common origin but with a different function are said to be **homologous**. So, the wing of a bat is homologous to a human hand. The possession of the pentadactyl limb in so many different organisms is persuasive evidence that these animals have evolved from a common ancestor.

Fig 30.13 **The vertebrate pentadactyl limb provides strong evidence for evolution. In all these examples, the basic bone layout is the same (in the diagram, the same type of bone is given the same colour in each animal) but the limbs have become modified in different ways according to the way the animal moves**

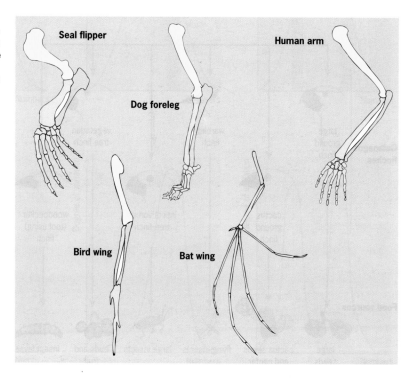

Fig 30.14 **Convergent evolution occurs when unrelated species solve the same evolutionary problem in a similar way**

(a) **Cacti, which are native only to the Americas, have evolved to cope with desert living. The leaves have become spines which lose no water and protect the plant. The fleshy stem expands and stores water and has taken over the job of photosynthesis**

(b) **You might guess that this euphorbia from Africa was a cactus, but you would be wrong. The two plants have evolved in different continents and are unrelated; they look very similar because they have developed the same features to survive desert conditions**

(c) (left) **The flipper of the whale and the fin of the shark** (right) **are a classic example of convergent evolution. The two animals are completely unrelated and the internal structure of forelimbs is totally different. However, they are both the same shape because of the job they have to perform**

In contrast, totally *unrelated* structures can become adapted for the same function. The wings of insects and the wings of birds, for example, have a totally different anatomy and origin in the embryo, but are similar in shape because they perform the same job. Such structures are said to be **analogous**. The process by which unrelated species evolve to resemble each other is called **convergent evolution** (Fig 30.14). A classic example of this is seen in seals (mammals), penguins (birds) and fish. Due to their need to swim efficiently, all have evolved the same streamlined shape. The forelimbs of the seal and the wings of the penguin have both evolved to resemble fins.

?

D Are the following pairs of structures analogous or homologous?

(a) Trout's fin and dolphin's flipper.

(b) Horse's leg and bat's wing.

Embryology

Studying the development of an embryo reveals some interesting evolutionary secrets. The scientist Ernst Haeckl said 'ontogeny recapitulates phylogeny' which means that the evolutionary development of a species (its **phylogeny**) is mirrored by the changes of each individual from birth to maturity (its **ontogeny**).

Fig 30.15 **Four developmental stages in embryos of fish, turtles, chicken and humans. You can see that the early stages at the top are very similar in all four. For example, in the drawings in the second row, all four embryos have branchial arches, the 'folded' areas just under the head at the front. Only in the later stages do differences become obvious**

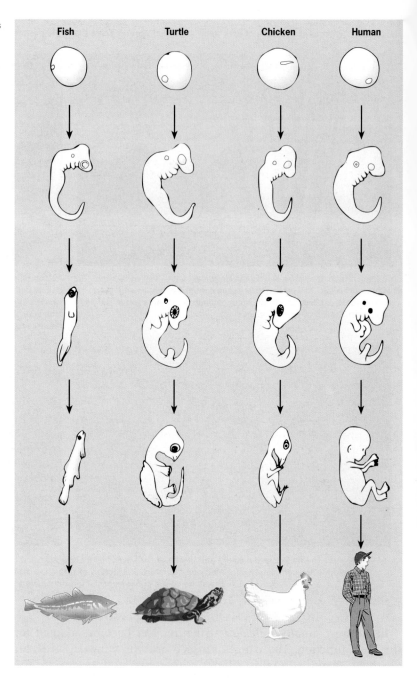

A number of vertebrate embryos follow a very similar path from fertilisation through several stages of early development – so similar that it is often difficult to tell them apart (Fig 30.15). For example, all vertebrate embryos develop branchial arches; these are relics of the gills possessed by aquatic ancestors. However, although the study of the developing embryo provides further evidence for evolution, it does not give us much information about the actual evolutionary process.

Comparative biochemistry

The biochemistry of different species shows some striking similarities. Many chemicals such as nucleic acids, ATP and cytochromes, organelles such as ribosomes, and pathways such as respiration, are almost universal. This is strong evidence that all living things have a common ancestry.

4 HOW A NEW SPECIES EVOLVES

We have looked at natural selection and seen that it can lead to stability or change. In this section we examine how natural selection can lead to the formation of a new species.

What is a species?

We instinctively think that we know what the word 'species' means, but an accurate definition is difficult, if not impossible. This is a common definition:

A species is a population or group of populations of similar organisms which are able to interbreed and produce fertile offspring.

This would seem to be reasonable; a horse and a donkey can mate to produce a mule (Fig 30.16), which is sterile, so the horse and the donkey can be said to belong to different species. However, this definition runs into trouble. There are many closely-related species, wolves, coyotes and domestic dogs, for example, which can interbreed to produce perfectly fertile offspring. The only reason why the wolves and coyotes remain distinct is that the hybrids do not compete as successfully as the pure strains.

The difficulty we have in defining the term species stems from the very nature of the process of evolution. New species evolve from pre-existing ones, and this is usually a slow transition which may take many generations and thousands of years. There are many living species which are still in the process of separating from each other. The wolf and the coyote almost certainly evolved from a common ancestor and for all practical purposes are separate species (Fig 30.17). If left to evolve naturally over the next few thousand years, the two species may well accumulate enough genetic differences to prevent interbreeding.

So, we can modify the definition to:

A species is a collection of recognisable organisms which shares a unique evolutionary history and is held together by cohesive forces of reproduction, development and ecology.

Although more of a mouthful, it is a more satisfactory description because it recognises that all members of a species are usually (but not always) recognisable, and that they share the same 'biology', occupying the same niche in the ecosystem and having particular social structures, courtship, mating habits etc.

Fig 30.16 **And nothing to look forward to either. The mule is the product of a mating between a donkey and a horse, two species that have different numbers of chromosomes. It cannot reproduce because it is sterile. Because of its stamina, the mule has been used as a beast of burden for thousands of years**

Fig 30.17 **The coyote (*Canis latrans* – below left) and the wolf (*Canis lupus* – below right) are classed as different species, but they are capable of interbreeding and producing fertile offspring**

The development of new species

Speciation, the development of new species, happens when different populations of the same species evolve along different lines. **Populations** are groups of interbreeding individuals of the same species occupying a particular geographical region. Examples of populations may be the water fleas in a pond, all the oaks in a forest or the elephants in a game reserve.

It is thought that speciation usually happens in the following way:

● Part of a population becomes isolated in some way that prevents them from breeding with the rest of the population. Very often, the breakaway population will, by chance, have a different genetic constitution from the original (see genetic drift – page 457).

● The two populations experience different environmental conditions, so that natural selection acts in different ways, causing the genetic make-up of the two populations to differ.

● Eventually, genetic differences accumulate to a point where individuals from separate populations that come together again can no longer interbreed.

It is worth mentioning that new species can also arise from a hybridisation between two different species. This particularly applies to plants, and plant breeders take advantage of this ability to produce new crop varieties.

There are two basic methods of speciation:

● **Allopatric speciation** (allo = different, patris = country), where the two populations are physically separated.

● **Sympatric speciation**, where the two populations are reproductively isolated in the same environment.

How populations become isolated

Usually, species evolve when populations become isolated by physical barriers such as stretches of water or mountain ranges. This is known as **geographic isolation** (Fig 30.18).

Fig 30.18 **The principle of geographic isolation**

(a) The original population of mice has a 50:50 distribution of the alleles A and a, ie A is as common as a. A river changes its course and cuts off a few individuals. In the new, isolated population the allele frequency, by pure chance, is 4:10. This change in allele frequency is called the founder effect. The fewer the individuals in the breakaway founder population, the more dramatic the change in allele frequency is likely to be

(b) The founder population starts to grow, but they are in different conditions from the original population. The alleles A and a were neutral up to now, but then a water-borne disease spreads through the new population, which are confined to marsh land. The aa genotype confers resistance on those who carry it. By natural selection, the A allele is selected against, and so the a allele becomes even more frequent. In this way, the genetic make-up of the two populations diverges

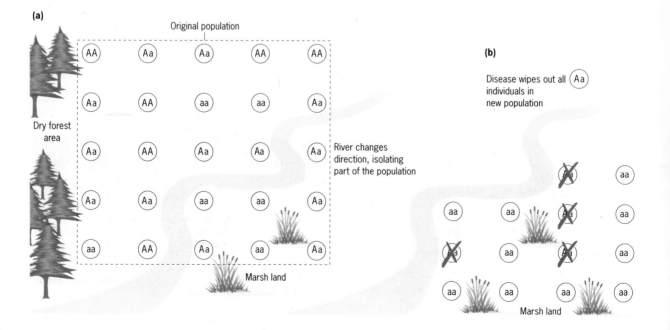

Evidence suggests that geographical isolation is the most common way in which populations become reproductively separated. However, speciation can happen even when two populations overlap. For instance, two populations of frogs may have a different mating call, and the females may respond only to calls of males in 'their' population. Table 33.1 summarises the different isolating mechanisms.

Mechanism	How it works
Prezygotic mechanisms, ie gametes don't come into contact	
Geographic	Two populations do not come into contact due to some physical barrier, eg mountain range, water
Temporal	Different timing, eg flowering times or mating seasons do not coincide
Behavioural	Individuals of other populations are not acceptable as mates, eg wrong courtship, mating call, or pheromones
Mechanical	Structural differences prevent mating, eg the shapes of the genitals in insects
Postzygotic mechanisms, ie even if mating takes place	
Hybrid non-viability	Hybrid zygotes fail to develop to sexual maturity
Hybrid sterility	Hybrids do not produce functional gametes
Hybrid weakness	Hybrids have reduced survival or reproductive rates

Table 30.1 **A summary of isolating mechanisms**

5 EVOLUTION IN ACTION

The process of evolution is generally slow, and it is no surprise that for centuries people thought that the living world was static. However, there are situations where you can see the process in action. Organisms with short life spans and high reproductive capacities, such as bacteria or insects, can evolve in a very short time indeed – see page 457.

Also of great interest to evolutionary biologists is a **polymorphism**, a situation where two or more different forms, **morphs**, of a species exist in a population. Each morph confers a different survival advantage in different situations. Polymorphisms such as that of wing colour in the peppered moth (below) allow us to see how natural selection leads to changes in allele frequency, and so we can observe evolution in action.

The peppered moth

Fig 30.19 shows the original speckled form of the peppered moth *Biston betularia*. This moth was common in the UK at the time of the industrial revolution. In areas of heavy industry, many buildings and trees became covered in soot. The speckled moth lost its camouflage in these areas and so was eaten more often by its predators.

Luckily for the moth, a mutation occurred around this time that produced a black moth known as the **melanic form**. In industrialised areas the melanic moth had a selective advantage. However, in 'cleaner' areas the reverse was true: the speckled moth thrived and the melanic form was rarely seen. Both morphs belong to the same species, but their survival depends largely on the environment.

How can we study evolution?

If evolution is so slow, how can we observe it happening? An important principle here is that evolution involves a *change in allele frequency*. For instance, when myxomatosis was introduced into the rabbit population, the alleles for resistance were selected for, and so

Fig 30.19 **The two forms of the peppered moth, normal (speckled) and melanic (black). Against the background of the bark of a tree shown in this photograph, the two speckled moths on the right are almost invisible while the melanic form on the left is far more likely to be seen and caught by a predator. However, it becomes invisible on a black surface such as a sooty wall, and so is more likely to survive in an industrial environment**

they increased in frequency. In any particular situation, if we can show that allele frequencies are changing, we know that evolutionary forces are at work.

Evolutionary studies are usually based on interbreeding populations, which are known as **demes** or **Mendelian populations**. The sum total of all the genes circulating within such a population is called the **gene pool**.

Populations and allele frequency

When the gene pool remains more or less the same, a population is not evolving. Natural selection may be acting, but it is often a force for stability rather than change. However, if the gene pool of a population is changing, natural selection is acting as a force for change and this is evolution in action.

To show how we measure allele frequency, imagine a population of 500 rats. In the population are two alleles for coat colour: black (B) is dominant over brown (b). The rats are diploid – each individual has two alleles – so the population contains 1000 coat colour alleles. If 600 of the alleles are B, then we can say that the frequency of B is 60 per cent, or 0.6 (statisticians prefer decimals). As there are only two alleles, it follows that the frequency of b must be 40 per cent or 0.4. (Remember that the sum of the allele frequencies must be 100%, or 1.)

If one gene is dominant to the other – as in this case – the symbol p is used for the frequency of the dominant allele B, and q is used for the frequency of b. It is clear that $p + q$ must equal 1.

The values of p and q change when a population evolves. To find out if this is happening we use the **Hardy–Weinberg principle**.

$$p + q = 1$$

Frequency of dominant allele + frequency of recessive allele = total number of alleles of that gene.

The Hardy–Weinberg principle

The principle, named after the British mathematician G.H. Hardy and the German biologist W. Weinberg, states that there will be no change in the frequency of alleles in a population so long as the following conditions are met:

● The population is very large, diploid and reproduces sexually.

● Mating is totally random – there is no tendency for certain genotypes to mate together.

● All genotypes are equally fertile, so there is no natural selection.

● There is no emigration or immigration.

● No mutations occur.

We can use the Hardy–Weinberg rule to estimate allele frequencies using just the information about the frequency of a particular genotype. Consider the example of albinism in humans. One in 10 000 humans are born albinos – they have two copies of a recessive gene which means they cannot make the pigment which gives colour to hair, skin and eyes (Fig 30.20). The frequency of the allele for albinism in the population is difficult to estimate because carriers of the allele are not albinos. So, how can we find the frequency of the albinism allele?

If we call the normal allele A and the albinism allele a, most of the population are AA, some are carriers: Aa or aA, and a small number are albinos: aa. Using the Hardy–Weinberg formula we can estimate the percentage of the population who carry the a allele. Remember

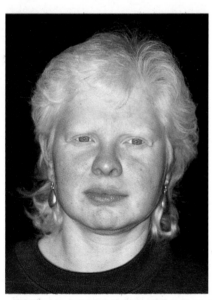

Fig 30.20 **This person is an albino. The white hair, pink eyes and pale skin are the result of an inability to make the pigment melanin**

that p and q represent the frequency of the A and a alleles respectively. The Hardy–Weinberg equation:

$$p^2 + 2pq + q^2 = 1$$

Put into words, it says:

The frequency of AA individuals, added to the frequencies of Aa, aA and aa, must equal 1.

We also know that $p + q = 1$, and we can use these two equations to work out the proportion of heterozygotes in the population, and any other frequencies involved:

1 in 10 000 is a frequency of 0.01%, or 0.0001 expressed as a decimal.

1 in 10 000 have the genotype aa, so the frequency of aa; qq or q^2, has the value of 0.0001.

So $q = \sqrt{0.0001} = 0.01$.

If $q = 0.01$, then p must be $1 - q$, which is 0.99.

So, in other words, 99 per cent or 0.99 of the alleles in the human population are A.

From this we can calculate that the number of carriers of the albino allele $2pq$ is:

$$2pq = 2 \times 0.99 \times 0.01 = 0.0198 \approx 0.2.$$

So, almost 2 per cent of the population carry the albinism allele.   ■ See question 4.

When allele frequencies change

The Hardy–Weinberg principle states that allele frequencies remain constant as long as certain conditions apply. It follows that when these conditions are not met, the allele frequencies can change and the population can evolve. Consider what happens when the conditions listed on page 456 are not met.

- **The population is small**. If the population is small, chance plays a large part in determining which alleles pass to the next generation. The smaller the population, the greater the probability that the genes of one generation will not be accurately represented in the next. The name given to a change in allele frequency caused by chance is **genetic drift**. These are two important causes of genetic drift: **bottlenecks**, when a population declines and then recovers; and the **founder effect**, when a population starts from just a few individuals (see below). As Figs 30.18 and 30.21 show, this can result in a drastic change in the allele frequency.

- **Mutations occur**. Mutations are ultimately the origin of all genetic variation (see Chapter 28). They are rare, and most are harmful or neutral. Occasionally, however, a mutation gives an organism a survival advantage, or the environment changes so that previously harmful/neutral mutations become advantageous. For example, a mutation in a bacterial gene which leads to the production of a modified enzyme could make the bacterium resistant to an antibiotic. That individual would survive, even in the presence of the antibiotic, and pass its resistance gene on to the next generation.

- **Individuals move in or out of populations**, that is, immigration or emigration. If the new individuals breed, this movement may lead to gene flow. When a few individuals move into a new environment and start a population, we see the founder effect. Like the beads in Fig 30.21, the individuals in the founder population are probably not representative of the original population. A group of six human survivors on a desert island, for example, might have two red-haired people and two blondes. A population founded by these people would have a much higher percentage of red-heads and blondes than the population in their country of origin. In the most extreme case, a new population can be founded by one individual, such as a self-fertilising plant or a pregnant female animal.

● **Mating is non-random**. In many cases, individuals choose mates with a particular genotype. In this case, mating is not random, it is **assortive**. Such assortive mating can change allele frequency.

● **Natural selection occurs**. When there is competition for resources – and this is usually the case in an ecosystem – natural selection acts and only the most successful individuals pass on their genes.

Fig 30.21 An illustration of genetic drift. In the original 'population' there are 1000 beads; half red, half blue. If ten beads are extracted at random, there is a high probability that there won't be five of each. If these ten 'organisms' then become the founders of a new population, the proportion of colour (the allele frequency) will be significantly changed, in this example to 7:3

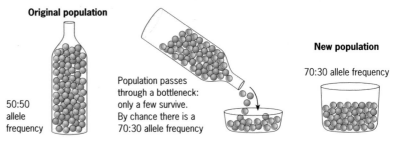

Original population

50:50 allele frequency

Population passes through a bottleneck: only a few survive. By chance there is a 70:30 allele frequency

New population

70:30 allele frequency

6 EVOLUTION: OUR PLACE IN THE GREAT SCHEME OF THINGS

There are millions of different species on Earth, and we have given all the ones we know about a Latin name. The gorilla is simply *Gorilla gorilla*, the chimpanzee is *Pan troglodytes* while we have given ourselves the name *Homo sapiens*, which means 'thinking man'. A Latin name tells us both the **genus** and the **species**. The generic name has a capital latter while the specific name does not. When written, Latin names should always be in italics or underlined.

We classify organisms according to the similarities between them. Understanding relationships between species allows us to piece together the story of life on Earth. Classification is therefore a bit like a huge family tree over a massive timescale. Just like an address has a house number, a road, a town etc, we classify organisms into progressively narrower groups. Table 30.2 shows this 'address', which we more correctly call a **taxonomic hierarchy**. Taxonomy is the science of naming and classifying organisms and the term hierarchy reflects the layered structure of the family tree of living organisms that consists of groups within groups.

Latin names are important. If you call an organism by its Latin name, any biologist in the world, whatever language they speak, will know what you are talking about. The bird known as the sparrow in America is a completely different species from the one we are familiar with in Britain, but confusion can be avoided as the two birds have completely different Latin names.

Organisms do not evolve according to a set pattern, just to make the taxomist's life easier. A detailed study of a particular organism might show that it falls neatly into Kingdom, Phylum, Class, Order, Family, Genus and Species. Or it might not. This is why you sometimes find extra groups such as sub-phyla, super-families or sub-species.

Humans and their close relatives

Humans vary in size, shape, colour, blood groups and in thousands of other ways, but we are all one species, *Homo sapiens*. *Homo* is a genus, a set of species that are grouped together because of their close similarity. There are no living species considered to be close enough to humans to be put in our genus, so we are the only

Table 30.2 **The taxonomic hierarchy: how we classify human beings**

Taxonomic group	Humans	Explanation
Kingdom	Animalia	There are five kingdoms, animals, plants, fungi, protoctista and bacteria
Phylum	Vertebrata or Chordata	We are animals with backbones
Class	Mammalia	There are five classes, fish, amphibians, reptiles, birds and mammals. The defining features of mammals are fur/hair and milk production
Order	Primates	Primates are adapted for life in the trees. Key primate features are covered in the main text
Family	Hominidae	Man-like creatures – includes several extinct species
Genus	*Homo*	No other living species is sufficiently like humans to be included in this genus
Species	*sapiens*	We are the thinking species, modern man

?

E Think up a mnemonic (memory aid) to help you learn the sequence Kingdom, Phylum, Class, Order, Family, Genus, Species: KPCOFGS.

✔

Taxonomic groups do not overlap. For instance, there are no animals that are both reptiles and amphibians. An animal must belong to one class or the other; for example, a toad is an amphibian and an iguana is a reptile. Neither animal can be part reptile and part amphibian.

species described as *Homo*. However, there have been human ancestors that belonged to the genus *Homo*.

Humans are also the only species in the family *Hominidae*. However, the basic body plan shared by humans and the great apes is so similar that biologists have put us all in the super-family of *Hominoidae*. The word **hominoid** describes any ape-like or human-like creature.

Human evolution

So how did humans evolve? It is widely accepted that human evolved in Africa, from the same ancestors that gave rise to the great apes; chimps (Fig 30.22), gorillas and orang-utans. We did not, however, evolve from these species, as is often thought. Piecing together a complete evolutionary picture is difficult. If you took all the remains that have ever been found of ancestral humans, they would all fit into an average-sized living room. However, we do have some vital fossil evidence that has allowed us to piece together a likely **lineage**, or line of development (see the Feature box on the next page).

The first primates

The first primates to evolve appeared about 70 million years ago. We owe a lot of our human features to our primate ancestry. The key features of primates are:

- Flexible limbs with gripping hands. Hands and feet have **opposable digits** – thumbs that can oppose all the other digits, giving an excellent grip.

- Finger nails rather than claws. This allows some protection, but doesn't interfere with the sensitive fingertips. Primate hands are highly dextrous. This makes it easier to obtain food and it allows more effective grooming. It has also helped us develop the ability to make tools.

- Flat faces with forward-facing eyes. This allows stereoscopic vision, giving judgement of distance, an essential skill for a life in the trees. Primates also have colour vision, which is believed to help them find ripe fruit.

- Large brains. Primate brains have large areas that process information from the eyes, ears, mouth and skin.

- Socially, primates tend to live in family groups, giving birth to single offspring that are dependent on their parent for a long time.

Fig 30.22 **Chimpanzees show the classic ape characteristics; a sloping face with protruding jaw and heavy brow ridges. It is debatable whether the chimp or the gorilla is man's closest relative – the evidence is conflicting**

MAKING SENSE OF FOSSILS

THIS PHOTOGRAPH shows the fossil remains of Lucy, a young girl who lived approximately 3.5 million years ago.

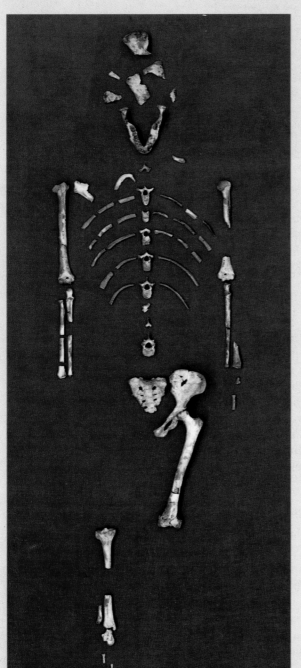

Fig 30.23 **This is Lucy, the most complete specimen of** *Australopithecus afarensis* **ever found. She was discovered in 1974 in what is now Ethiopia, and named after the Beatles song 'Lucy in the sky with diamonds' that the archaeologists Don Johanson and Tom Gray were playing at the time of the discovery.**

Overall, the great skill in piecing together fossil evidence comes from knowing where to look, and in recognising hominid remains when you see them. There are incredibly skilled people who can look at a pile of rubble and spot a tooth, jaw fragment or vertebrae of a primitive human, without mistaking it for modern ape or human remains. Who knows what remains are yet to be discovered?

Looking at it can illustrate quite well how fossils are able to tell us a lot more than they appear to at first glance.

How can we tell this fossil is a 12-year-old girl?

You can sex a human, or humanoid, skeleton from the shape of the pelvis. Females have a wider pelvis, with a larger gap to allow a baby's head to pass through. Size and bone development indicates age.

How do we know the age of these remains?

Fossils are formed in sedimentary rocks. When this girl died, her skeleton was covered with layers of silt and and sand in conditions that protected it from the normal processes of decay. The soft tissues such as the skin, muscles and internal organs decay, but the bones undergo a chemical reaction that turns them into rock. In time, the sediments bury the fossils deeper and deeper. We can estimate the age of the skeleton by comparing this layer with other strata, but for a more accurate measurement we can use **potassium/argon dating** – see the Assignment at the end of this chapter.

What did she look like when she was alive?

From the face we can see some obvious features: the sloping face, protruding jaw and prominent eyebrow ridges; she was obviously not a modern human. We can reconstruct the soft tissues in the same way that forensic artists do with the skull of a murder victim. If you know the muscle structure and the average depth of the soft tissues at strategic points on the skull, you can use clay, and a lot of skill, to reconstruct the face.

How large was her brain?

This is easy to do – you just look at the space left inside the skull. You can estimate the volume by filling the cranium with water or sand and then emptying into a measuring cylinder. You can even make a latex cast of the brain cavity and measure its volume more accurately by displacement.

What did she eat?

There are several clues about diet. Generally, the more vegetation in the diet, the stronger the muscles which pull the jaw from side to side, and the flatter the molars. This girl has teeth very similar to ours, so she probably had a varied, omnivorous diet. Sometimes microscopic examination of the scratches on the enamel can tell us a lot about the diet of a fossilised human.

How do we know that she walked upright?

Firstly, the hole where the spinal cord goes into the skull (the foramen magnum: Latin for 'big hole') is under the skull so that the head pivots on the spinal column. Also, the leg bones are thicker than those of the arm, and the pelvic girdle is stronger than the shoulder girdle. This implies that the weight was placed mainly on the legs rather than evenly distributed.

Assignment

THE HISTORY OF THE WORLD

Rocks, and the fossils they contain, can now be dated by **radiometric dating techniques**, which measure levels of certain isotopes. We can use this technique to construct a geological time scale – a biological history of the world. This Assignment outlines the process of dating using radioactive isotopes and puts the history of life on Earth into perspective.

Dating rocks and fossils

The principle behind radiometric dating is simple: some elements exist as **radioactive isotopes** which decay at a predictable rate. Measurement of this decay allows us to estimate the age of the rock.

For example, when volcanoes erupt, the lava contains radioactive potassium, ^{40}K. This decays steadily into argon (^{40}Ar). The half-life of ^{40}K is 1.3 billion (1.3×10^9) years. So, in 1.3 billion years, half of the potassium decays. Half of the remaining potassium will decay in the next 1.3 billion years, and so on. By measuring the ratio of radioactive potassium to argon, scientists can estimate the age of the rock strata to the nearest 50 000 years.

In many places, layers of volcanic lava alternate with sedimentary rocks which contain fossils. If we can date the lava, we can also estimate the age of the fossils. Some fossils of primitive bacterium-like organisms are thought to be around 3 500 000 000 years old.

a) What does half-life mean?
b) What is the half-life of ^{40}K?
c) What fraction of the original ^{40}K will be left after 3.9 billion years?

Carbon dating

This process can be used to date samples which still contain some organic material. Cosmic radiation is constantly bombarding the atmosphere, turning $^{12}CO_2$ into $^{14}CO_2$. This becomes fixed into organic molecules by the process of photosynthesis and then passes up the food chain. When an organism dies, the ^{14}C decays into ^{14}N, with a half-life of 5730 years. Carbon dating can be used to measure the age of any carbon-containing material up to 50 000 years old.

2 A human skull is found to have only one sixteenth of the ^{14}C it would have has when 'new'. Estimate its age.

The Earth is currently estimated to be about 4 600 000 000 years old. This time scale is impossible to comprehend, but if we condense this huge span into one year we can begin to put it into perspective.

The Earth calendar

Date	Event
1 Jan	Earth forms
26 Feb	Oldest rocks – nothing was permanent before this time
23 March	First life – bacteria like organisms obtain energy from organic molecules (food) by the process of glycolysis
April–May	Organic food runs out; photosynthesis evolves, makes food and releases oxygen as a by-product
30 June	Oxygen atmosphere fully developed. Aerobic respiration evolves some time after this
17 Sept	First eukaryotes
19 October	First coordinated multicellular colonies (like sponges)
5 Nov	First invertebrates (worms and jellyfish)
15 Nov	Rapid increase in invertebrate diversity
20 Nov	First fish
21 Nov	First coral reefs
29 Nov	First colonisation of land – vascular plants appear
4 Dec	First amphibians
6 Dec	First trees
7 Dec	First insects
8 Dec	First reptiles
17 Dec	First mammals
19 Dec	First birds and dinosaurs
21 Dec	First flowering plants
25 Dec	Last of the dinosaurs
26 Dec	First primates
31 Dec	
14.00	First upright bipedal hominids
21.00	*Homo erectus*
22.45	Recent Ice Age starts
23.00	Evidence that primitive humans can cook food
23.48	Neanderthal man appears
23.54	*Homo sapiens* appears
23.55	Australian aborigines settle
23.58.49	Agriculture invented
23.59.28	Beginning of recorded history
23.59.29	Pyramids built
23.59.41	Buddha born
23.59.45	Christ born
23.59.53	Battle of Hastings
23.59.57	Start of the scientific method
23.59.58	Australia 'discovered' by Cook. The Industrial Revolution
23.59.59	Most of the knowledge contained in this book is discovered

3
a) In our one-year scale, how many seconds represents 1000 years?
b) How long in actual time is represented by one minute, one hour and one day?
c) Use your answer to **(a)** to estimate how long the dinosaurs lived for.

4 The rise of humans as the dominant species has been meteoric.
a) What features of humans account for our success.
b) What dangers does this success bring?

5 Write down what you feel to be the four most significant advances made by humans. These could be inventions or social and cultural advances. If you are working in a group, use your own list to provide a discussion.

technology

Tracey, a transgenic sheep at PPL Therapeutics in Edinburgh

THIS IS TRACEY, THE FIRST TRANSGENIC SHEEP. Although Tracey is a normal, healthy animal, she has a remarkable talent: each litre of her milk contains about 35 grams of the protein human alpha-1-antitrypsin, a potential treatment for lung diseases such as emphysema and cystic fibrosis.

Transgenic organisms have DNA that has been modified by geneticists. In Tracey's case, the human gene which codes for alpha-1-antitrypsin was inserted into her DNA when she was an embryo. The gene was 'adopted' by her cells and is now expressed in her milk-producing glands.

Like many biological molecules, alpha-1-antitrypsin is far too complex to synthesise in a laboratory, and extracting the protein from human tissue is not a possibility – the tissue is just not available in the quantity required for medical use. Although all Tracey needs to make it is a diet of grass, her milk is worth several thousand pounds per litre! In the future we might have whole herds of transgenic organisms making life-saving medical treatments.

1 THE NEW GENETICS

Genetic engineers manipulate DNA to allow individual genes to be transferred from one organism to another. Often the two organisms are from completely different species.

Genetics seeks to explain the mechanisms of inheritance and evolution and, like many other 'pure' sciences, it can be applied to solve practical problems. In the 'new genetics' that has developed rapidly in recent years, humans are taking a much more active role in determining how genes pass between individual organisms. Not only can we study DNA, we can manipulate it and use it to our advantage. This branch of genetics is called **genetic engineering, or DNA technology**.

Transferring genes between organisms offers many exciting prospects. In agriculture, for example, genes from nitrogen-fixing bacteria could be put into plants which would then fix their own nitrogen and so make some of their own fertiliser. This could greatly improve the efficiency of food production. Crop plants given genes which code for disease resistance or tolerance to pollutants could be grown on previously inhospitable land, such as industrial land or desert margins. DNA technology also has many potential applications in medicine.

The genome of an organism, its entire collection of DNA sequences, can now be explored and mapped. Theoretically, any individual gene can be located, cut out, replaced or inserted into another organism. Of course, this sort of research is regulated very carefully, and the ethical, moral and social consequences need to be taken into account (Fig 31.1).

We are in the process of mapping the entire genome of some species. The bacterium *Escherichia coli* – a prokaryote – has already been sequenced, and so has a simple eukaryote, *Saccharomyces cerevisiae* (baker's yeast). The Human Genome Project, a far more ambitious project, is due to finish in 2005 (see the Feature box on this page).

Once we know what individual human genes do and where to find them, the possibilities are endless:

● We can take individual genes and make multiple copies: this is **gene cloning**.

● We can replace defective genes with healthy ones: this is **gene therapy**.

● We can extract genes which code for useful products and insert them into other organisms: this is **recombinant DNA technology**. Organisms produced in this way are called **transgenic organisms**. More genes which code for human products could be transferred into microorganisms for mass production. Products already made in this way include insulin, somatostatin, alpha-1-antitrypsin, human growth hormone and blood clotting substances such as factor VIII.

In this chapter we focus on the techniques that allow gene manipulation, and we look briefly at some of the possibilities for the future and at some of the ethical implications of genetic technology.

Fig 31.1 **'You were so keen to see what you could do, you didn't stop to think about whether or not you should.'**
  This is a scene and a quote from the film **Jurassic Park**. It is quite unlikely that we will ever be able to recreate dinosaurs, but the quote reflects an important point: all genetic research has the potential for abuse and we should think very carefully about its consequences

2 GENE HUNTING

The first problem that genetic engineers tackle is finding the gene they are interested in. Every human body cell contains 46 chromosomes, each having several thousand genes. So, how, for example, do you go about finding the insulin gene that you want to insert into a bacterium? Let's look at three different approaches.

✔ DNA is common to all organisms, and it always has the same basic structure. This allows us to combine genes from organisms as diverse as humans and bacteria.

A: Isolate the gene from the rest of the genome

This has been compared with looking for a straw needle in a haystack. However, there are several sophisticated techniques that can be used to map chromosomes. For instance, if you know part of the base sequence in the required gene you can make a **genetic probe**. This is a single-stranded piece of matching DNA labelled with a radioactive or fluorescent marker. When mixed with the **genomic DNA** (DNA from the genome) – in the right conditions – the probe seeks out the complementary base sequence and binds to it, showing you exactly where the gene is.

THE HUMAN GENOME PROJECT

THE 1990S SAW THE BEGINNING OF the most ambitious project ever to be undertaken in molecular biology: the Human Genome Project (HGP). The aim of this project is simple: to map out the sequence of DNA that makes up all the human genes. It is a phenomenal task. The sequence of chemicals involved is estimated to be 3.2 billion base pairs long.

Critics of the project look to the cost: estimates are currently running at anything between 1 and 10 US dollars per base pair. The money could be better spent, they argue, on projects that could give a much faster return on the investment. The completion date for the project is the year 2005. Despite the cost and the critics, most scientists think that the potential benefits of such knowledge are enormous. If we know where individual genes can be found, and we know the DNA sequence of those genes in a normal healthy individual, we should be better equipped to diagnose and treat all sorts of diseases, not only those currently considered to be genetic diseases.

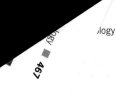
In the long term, of course, having a complete library of human genes will make this a lot easier. The ultimate goal is to know where each gene is located, what it does, how it interacts with other genes and what goes wrong in cases of genetic disease.

B: Use messenger RNA

If you can isolate the mRNA molecule that is a copy of the gene that you are trying to find, you can use it as a genetic probe as described above. You can also use it to make a DNA copy using the enzyme reverse transcriptase. This means essentially that you make an 'artificial' gene. For example, you could find the mRNA that codes for insulin in the cells where the insulin gene is expressed – the cells in the islets of Langerhans. DNA made from mRNA is called **complementary DNA**, or **cDNA**. Artificial genes are shorter than those in the genome because they contain no introns (Chapter 28).

> **?**
>
> **A** Suggest what the enzyme reverse transcriptase does.

C: Work backwards from the protein

If you work out the amino acid sequence of the desired protein – a relatively easy process – you can make a piece of DNA which codes for it. It would obviously be a tedious process to make a large gene, but now there is equipment which can synthesise artificial DNA quickly and easily. The end product of this process is also a short, artificial gene made of cDNA.

3 MANIPULATING DNA USING ENZYMES

Once you have located the gene you want, the next step is to use it. Enzymes are the tools of the trade here.

Restriction enzymes

Table 31.1 **The target sites of some restriction enzymes**

Enzyme	Bacterial origin	Recognition site
EcoRI	E. coli	G|AATTC CTTAA|G
HindIII	H. influenzae	A|AGCTT TTCGA|A
Bam H1	B. amyloliquefaciens	G|GATCC CCTAG|G

Restriction enzymes can be thought of as 'molecular scissors' – they cut DNA strands at specific points (Table 31.1). More properly called **restriction endonucleases**, they are made by bacteria in response to attack by viruses called bacteriophages. Their name reflects their function: they *restrict* damage by chopping bacteriophage DNA into smaller, non-infectious fragments, and they make their cuts *inside* the nucleic acid molecules.

There are many different endonuclease enzymes, produced originally by different species of bacteria. Each enzyme cuts DNA at a different base sequence, a point known as the **recognition site**. The enzymes make staggered cuts in the DNA (Fig 31.2), commonly called **sticky ends**. These can combine with complementary sticky ends, and this allows lengths of DNA that have been removed from one organism to be **spliced** – inserted – into the DNA of another. EcoR1, for example, is a common endonuclease isolated from the bacterium *Escherichia coli*. This is a very popular enzyme with genetic engineers because it is readily available. It is also reliable: it cuts DNA at its recognition site with precise accuracy, showing very little **star activity** (cutting DNA at sites other than the recognition site).

You might be asking, 'If restriction enzymes cut DNA, why is the bacterium's own DNA not damaged?' The answer lies in a process

> ✓
>
> The prefix *endo-* means *inside*; the prefix *exo-* means *outside*. An *endonuclease* cuts up a DNA strand by making cuts along the whole length of the molecule. In contrast, an *exonuclease* cuts DNA by chopping off the nucleotides at the ends of the molecule.

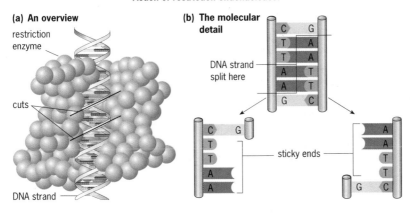

Action of restriction endonuclease:

(a) An overview

restriction enzyme

cuts

DNA strand

(b) The molecular detail

DNA strand split here

sticky ends

Fig 31.2 **Endonuclease enzymes are wonderful tools for the genetic engineer because not only do they cut DNA, but they make *staggered* cuts, leaving 'sticky' ends which make re-joining DNA strands much easier. Endonucleases cut at the sugar–phosphate bonds in the DNA molecule, leaving only the hydrogen bonds intact. At room temperature and above, these bonds are not enough to hold the molecule together, and so it splits, producing a 'sticky end'**

?

B Different restriction enzymes have differently sized recognition sites. Those in Table 31.1 are described as 'six cutters' because their recognition sites are six bases long, but four or five cutters are also common. Why do you think four cutters produce a larger number of fragments than five cutters?

called **methylation**: the bacteria add a methyl ($-CH_3$) group to the sequences in their own DNA that could act as recognition sites. This prevents the restriction enzymes doing any damage, presumably because the methylated region no longer fits into the active site of the enzyme.

DNA ligases

The sticky ends of the DNA strands produced by the action of restriction enzymes are joined together by enzymes called **DNA ligases** (ligate means *to join*). The normal function of a DNA ligase enzyme is to join strands of DNA during replication (Chapter 31). They are also involved in DNA repair. Genetic engineers use them as 'molecular glue' and they are essential partners for restriction enzymes. An example is **T4 DNA ligase**, which is made by the bacterium *E. coli*.

THE POLYMERASE CHAIN REACTION (PCR)

THE FIRST RECOMBINANT DNA molecules were produced in the early 1970s when bacteria were made to copy foreign DNA along with their own. However, in 1983 an American called Cary Mullis cloned DNA without bacteria using the relevant enzymes. This process, the **polymerase chain reaction**, is gene cloning in a test-tube. It allows DNA to be **amplified** (copied many times).

The idea behind PCR is very simple (Fig 31.3). You mix together your original piece of DNA with the enzyme DNA polymerase in a solution of nucleotides. Next, add some **primers**, short pieces of DNA which act as signals to the enzymes, effectively saying 'start copying here'. You then heat the DNA so that it denatures into two strands. The thermostable DNA polymerase gets to work and produces two identical strands of DNA (see Chapter 28 for the detailed mechanism of DNA replication).

In a second cycle of reactions the two strands become four, and so on. Typically, the cycle is repeated about 20 times and within an hour you have millions of copies of the original piece of DNA.

This technique has many applications. Forensic scientists use it to amplify tiny DNA samples from spots of blood or hair roots, to obtain enough material for forensic analysis by DNA profiling. PCR also allows archaeologists to study small samples of DNA from historical material such as the preserved bodies found in peat bogs or ice graves.

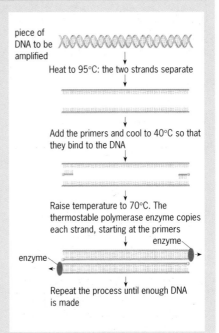

piece of DNA to be amplified

Heat to 95°C: the two strands separate

Add the primers and cool to 40°C so that they bind to the DNA

Raise temperature to 70°C. The thermostable polymerase enzyme copies each strand, starting at the primers

enzyme

enzyme

Repeat the process until enough DNA is made

Fig 31.3 **The polymerase chain reaction**

Enzymes that modify DNA

Restriction and ligase enzymes can cut and splice DNA, but there are many other enzymes which are useful tools for the genetic engineer. Some of these are shown in Table 31.2.

Table 31.2 **Some enzymes commonly used in genetic engineering**

Enzyme	Function
Exonucleases	Remove terminal base from DNA – useful in analysis of base sequences.
DNase	Makes random cuts in DNA: chopping into varying-sized fragments.
Kinases	Add groups such as phosphates to DNA: this helps with labelling and analysis, particularly when phosphates are made with ^{32}P.
Polymerases	Synthesise nucleic acid molecules from nucleotides. This is essential for the polymerase chain reaction (PCR) – see Feature box. Reverse transcriptase is a polymerase which makes DNA from RNA. It occurs naturally in retroviruses such as the human immunodeficiency virus, HIV.

4 CLONING GENES

Armed with restriction enzymes and ligases, genetic engineers can create recombinant DNA molecules. However, once a recombinant DNA fragment has been made in vitro ('in glass'), it usually needs to be copied repeatedly so that enough molecules are available for analysis or for commercial use.

Although DNA fragments can be multiplied in a test-tube (see the Feature box on PCR on the previous page), it is often much easier to put the DNA into organisms such as bacteria which will then adopt the new DNA as their own, and copy it for you. This is called **gene cloning**. The basic procedure is outlined in Fig 31.4. As an added bonus, the bacteria can also express the gene, making large amounts of the product.

The word *protocol* is often used in genetic engineering. It means 'standard procedure' and so a cloning protocol would be a standard method for cloning a gene.

See questions 1, 2 and 3. ■

Fig 31.4 **The principles of recombinant DNA technology. A human gene that codes for a useful product, such as insulin, is transferred into a bacterium which then multiplies to form a large population. The bacteria all express the human gene, and large amounts of the gene product can be recovered**

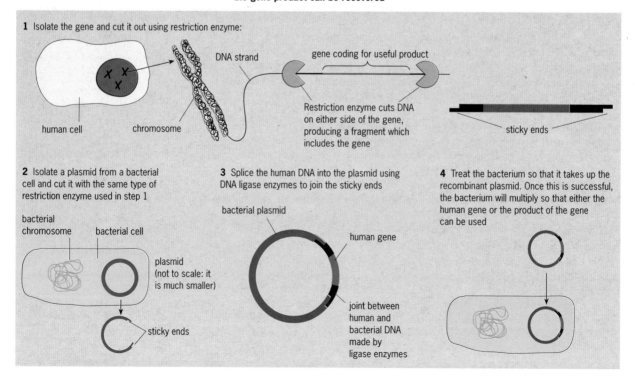

1 Isolate the gene and cut it out using restriction enzyme:

human cell — chromosome — DNA strand — gene coding for useful product — Restriction enzyme cuts DNA on either side of the gene, producing a fragment which includes the gene — sticky ends

2 Isolate a plasmid from a bacterial cell and cut it with the same type of restriction enzyme used in step 1

bacterial chromosome — bacterial cell — plasmid (not to scale: it is much smaller) — sticky ends

3 Splice the human DNA into the plasmid using DNA ligase enzymes to join the sticky ends

bacterial plasmid — human gene — joint between human and bacterial DNA made by ligase enzymes

4 Treat the bacterium so that it takes up the recombinant plasmid. Once this is successful, the bacterium will multiply so that either the human gene or the product of the gene can be used

USING VECTORS TO TRANSFER GENES

HUMAN GENES are often transferred into bacteria or yeast because these microorganisms multiply quickly to form a huge population that expresses the gene and makes large quantities of the gene product. This can then be extracted and purified. The microorganisms are grown in a **fermenter**, a giant vat with ideal conditions for growth (Fig 31.5).

Fig 31.5 **Transgenic bacteria or yeasts are cultured in large cylindrical sterile fermenters, although the process is not strictly speaking fermentation. These containers were originally used for making wine and the name has stuck**

Putting genes into bacteria

Putting a human gene into another organism is not an easy task. The first step is to attach the gene to a **vector**, or carrier. One such vector is a **plasmid**, a tiny circular piece of DNA that occurs in bacteria (see Chapter 1). Like restriction enzymes and ligases, plasmids can be bought commercially.

The plasmid is cut using the same endonuclease enzyme used to cut the human DNA to obtain the gene. Using the same endonuclease is important – the DNA must be cut at the *same base sequence* to produce complementary sticky ends that will join up. The gene and the plasmid are mixed together and then a DNA ligase is added to join up the sticky ends. The DNA molecule produced is circular, like the original plasmid.

In the second step, the genetic engineer must induce the bacterium to accept the plasmid. One technique involves soaking the bacteria in ice-cold calcium chloride and then incubating them at 42 °C for 2 minutes. Nobody seems to know exactly why this works, but it does. Bacteria which have accepted the plasmid now contain recombinant DNA, and are, by definition, transgenic organisms.

The conversion process is not very reliable: for every bacterium that takes up the recombinant plasmid, about 40 000 do not. So how do you tell which ones are recombinant? A clever trick here is to insert two genes into the plasmid; the one you want to use and one that makes the recombinant bacteria easy to detect. For instance, you can add a gene that confers antibiotic resistance and then culture the bacteria on agar plates containing the antibiotic. Only the bacteria with the modified plasmid will survive and grow.

An alternative is to add a second gene that codes for an enzyme that metabolises a coloured substrate. When bacteria are grown on agar plates made with the substrate, colonies that have taken up the plasmid are a different colour from colonies of non-recombinant bacteria.

Once you have identified colonies of recombinant bacteria, you can start pure cultures. Transgenic bacteria are often described as 'sick' because they multiply more slowly than normal bacteria: the population doubles every 30 minutes instead of every 20 minutes. Within ten hours, one transgenic bacterium can produce over a million copies of itself, each one containing a working clone of the original gene.

Modified bacteriophages also make good vectors. They reproduce by inserting their own DNA into the DNA of a host bacterium. Bacteriophages can transfer larger amounts of DNA than plasmids, but, at present, their use is limited.

Putting genes into eukaryotic cells

Usually, a gene is transferred into a different organism so that it can be expressed, making greater volumes of product. This happens only when the gene is able to use the mRNA, ribosomes, and Golgi apparatus etc, for protein synthesis. Some eukaryotic genes are not expressed effectively by prokaryotic cells, and so must be transferred into another eukaryotic cell. Yeast, a single-celled fungus, is often used.

Producing recombinant eukaryotic cells is more difficult than producing recombinant bacteria. The cell walls of fungal cells are a major barrier. But they can be digested away with suitable enzymes to form a **protoplast, a** cell without a cell wall. In this 'naked' state, the yeast cell will accept plasmids.

Other methods of transferring DNA into eukaryotic cells include the following techniques.

Microprojectiles. It is possible to shoot DNA into host cells. Tiny pellets of metal are coated with DNA and fired at high speed at the target cells. Remarkably, some of the cells recover and accept the foreign DNA.

Electroporation. This involves exposing the host cells to rapid, brief bursts of electricity to create temporary gaps in the cell surface membrane, through which foreign DNA can enter.

Liposomes. In this more subtle method, DNA is inserted into liposomes – spheres formed from a lipid bilayer. The liposomes fuse with the surface membrane of the host cell and introduce the DNA into the cytoplasm.

Calcium phosphate precipitation. Plasmids are mixed with calcium phosphate. As this precipitates, the grains that form contain DNA. These grains can enter cells by endocytosis (see page 83).

DNA injection. DNA can be inserted directly into a cell using a very fine pipette called a micromanipulator that avoids the inevitable hand tremor (see Fig 31.6). Tracey the transgenic sheep (see Opener) was created when the gene for alpha-1-antitrypsin was manually injected into the cells of a sheep embryo. This method leaves much to chance: many embryos have to be treated before one takes up the foreign DNA.

CLONING WHOLE ORGANISMS

IN 1997, WHEN SCIENTISTS SUCCEEDED in cloning a sheep, there was a lot of fuss in the media. Was it ethically and morally right to clone animals as advanced as mammals? Would this lead to scientists attempting to clone humans?

The definition of a clone is 'a genetically identical copy' and the term can apply to individual strands of DNA, individual cells or whole organisms. Cloning is, in fact, a common natural process: all organisms made by asexual reproduction are clones of their parent, unless there is a mutation. Even humans make clones – they are more commonly called identical twins.

There is much commercial interest in the cloning process because it can produce exact copies of organisms with desirable characteristics. If, for example, you have a 'perfect' apple tree which produces large amounts of delicious, healthy, blemish-free apples, you can clone it rather than risk losing its valuable genotype in the lottery of sexual reproduction.

Cloning plants is easy, and can be done with relatively little equipment in schools and colleges. Cloning mammals, however, is a lot more difficult but technically feasible. Eggs and sperm or fertilised embryos are collected from prize specimens of cattle, for example. Cells can be pulled apart when the embryo is at the eight or sixteen cell stage and each fragment can then develop into a complete embryo. Each embryo is a clone of the original. The process can be repeated many times and then the cloned embryos can be introduced into the uteruses of normal cows. After the normal gestation period, several very average animals give birth to a herd of prize specimens.

There is no doubt that is it *possible* to clone humans, and this has led to some alarm. Much of the concern is justified and, in the UK, the Human Fertilisation and Embryology Authority has banned experiments on human cloning. However, some people have been worried unnecessarily by media reports based on ignorance and misunderstanding. A human clone would be, in every respect, a twin of the original person. He or she would still be a unique human being, with an individual personality. Your experiences and memories, as well as your genes, make you into the person you are. The misconception that people can have their cells frozen so that they can be cloned after their death to 'live again' is complete nonsense.

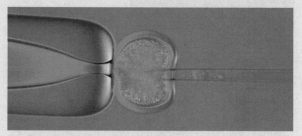

Fig 31.6 **Injecting DNA directly into a sheep cell. The cell is held steady by gentle suction from the pipette on the left, while the even finer pipette on the right penetrates the cell surface membrane. This technique is used to produce transgenic and cloned animals**

Fig 31.7 **This tomato has been genetically modified to improve the quality of the fruit and its processing characteristics.**

The normal ripening process in a tomato is accompanied by softening of the tomato fruit, making it susceptible to disease and damage during harvesting. By targeting and reducing the expression of the one of the tomato's ripening genes, polygalacturonase, a fruit has been produced which ripens normally but softens more slowly. The genetically modified fresh tomatoes delivered for processing are of greater quality and, because their 'softening' enzyme has been deactivated, they can be used to make a thicker tomato puree or firmer tomato dice

5 GENETIC ENGINEERING AND FOOD

As we saw in the introduction to this chapter, genetic techniques have an enormous potential to improve plant crops to feed our ever-expanding population. So far, genetic engineers have concentrated on producing the following improvements to some of the world's most important crop plants:

- **Tolerance to herbicides**. Adding genes to make plants resistant to herbicides allows us to kill weeks without affecting crop plants.

- **Resistance to insect pests**. Genes that code for products which are lethal to insects are added to the plant's genome.

- **Resistance to viral disease**. Genes for resistance to important diseases such as mosaic virus can now be built-in.

- **Improvements in crop quality**. Plants such as oilseed rape can be modified so that they make other, more useful oils that can be used as detergents, fuel oils and other non-food products. Other crops, such as tomatoes, are being modified to make them easier to store and process (Fig 31.7).

- **Manipulation of growth rate**. Plant growth rate is genetically determined. If we can learn how to transfer the genetic clock from a faster growing species, such as the poplar, into the oak, a normally slow-growing tree, we might obtain a renewable source of valuable hardwood.

SUMMARY

■ Genetic engineers isolate, cut out and transfer genes between organisms. Genetic engineering technology has many applications, including making human products such as insulin on a large scale.

■ Strands of DNA are cut using **restriction endonuclease enzymes**, which always cut DNA at a particular base sequence. Strands of DNA can be joined by using **DNA ligase** enzymes.

■ In order to transfer DNA into a living cell such as a bacterium, a **vector** (carrier) is needed. Common vectors are **plasmids** (circles of DNA found in bacteria) or viruses.

■ Once the foreign DNA is inside the host cell, it is adopted by the host. Genes in the foreign DNA are expressed using the host cell's synthetic machinery (ribosomes, mRNA etc). In addition, the foreign DNA is **cloned** (copied) every time the host cell divides.

■ **Gene therapy** is a technique which replaces faulty genes with healthy ones. It provides great hope for sufferers of genetic diseases such as cystic fibrosis.

■ The **polymerase chain reaction** (PCR) allows samples of DNA to be cloned in a test-tube using the necessary enzymes, instead of inside a host cell.

QUESTIONS

1 Read the following passage on gene technology (genetic engineering), then replace the dashed lines with the most appropriate word or words to complete the text.

The isolation of specific genes during a genetic engineering process involves forming eukaryotic DNA fragments. These fragments are formed using enzymes which make staggered cuts in the DNA within specific base sequences. This leaves a single-stranded 'sticky end' at each end. The same enzyme is used to open up a circular loop of bacterial DNA which acts as a ___ for the eukaryotic DNA. The complementary sticky ends of the bacterial DNA are joined to the DNA fragment using another enzyme called ___ . DNA fragments can also be made from ___ template. Reverse transcriptase is used to produce a single strand of DNA and the enzyme ___ catalyses the formation of a double helix. Finally new DNA is introduced into host ___ cells. These can then be cloned on an industrial scale and large amounts of protein harvested. An example of a protein currently manufactured using this technique is ___ .

[ULEAC 1996 Biology/Human Biology, Specimen Module Paper, q.4]

2 The Colorado beetle is a pest of potato crops. A soil bacterium, *Bacillus thuringiensis*, produces a substance called Bt which kills Colorado beetles but is harmless to humans. Scientist have isolated the gene for Bt production from bacteria and inserted it into potato plants so that the plant produces Bt in its leaf tissues.

a) **(i)** What is a gene?
(ii) Suggest how the gene for Bt production could be isolated from the bacteria and inserted into cells of the potato plant.

b) Bt can also be used as a spray. Colorado beetles may be killed if they ingest potato leaves which have been sprayed with Bt.

Suggest and explain **one** reason why using Bt-producing potato plants might increase the rate of evolution of Bt-resistance in the beetles compared with using Bt as a spray.

[NEAB February 1997 Modular Biology: BY2 Continuity of Life, q.3]

3

a) Suggest why it is important that there are many different types of restriction endonuclease enzymes available to genetic engineers.

People suffering from haemophilia need treatment with the blood-clotting protein, factor VIII, because they are genetically unable to produce their own.
Factor VIII used to be extracted from human blood but a genetically engineered kind has just been made available. An artificial version of the human gene has been produced which codes for the same sequence of 2338 amino acids that is present in natural factor VIII. This is inserted into hamster kidney cells which are grown in large fermenters. The cells secrete large amounts of factor VIII into the surrounding fluid.

b) **(i)** Why was it necessary to know the amino acid sequence of factor VIII before an artificial version of the human gene could be produced?
(ii) What is the minimum number of nucleotides which must be present in the gene for factor VIII?
(iii) Suggest **one** advantage of the use of genetically enegineered factor VIII over that extracted from blood.

[NEAB February 1995 Modular Biology: BY2 Continuity of Life, q.4]

Assignment

CYSTIC FIBROSIS AND GENE THERAPY

Cystic fibrosis (CF) is a genetic disease that is due to one defective allele (see Chapter 31). About one person in 25 carries the defective allele, but they also carry the normal allele and do not suffer from the disease.

1 Why do heterozygotes not suffer from the disease?

2 A couple are both carriers for cystic fibrosis.
a) What is the probability that their first child will have cystic fibrosis?
b) What is the probability that their second child will have cystic fibrosis?

The normal allele codes for a protein called **cystic fibrosis transmembrane regulator** (CFTR). This essential membrane protein in epithelial cells transports chloride ions out of the cells and into mucus. Normally, when chloride ions are secreted, sodium ions follow, and this decreases the water potential of the epithelial mucus. Water follows outwards by osmosis, making normal, watery mucus which can be moved by the cilia (tiny hairs) lining the airways.

Cystic fibrosis sufferers make a protein which differs in just one of its 1480 amino acids. Although slight, this fault prevents CFTR from functioning normally. Chloride ions and sodium ions cannot be secreted, and so the mucus becomes much thicker than normal. Dead epithelial cells also accumulate in the mucus, adding to the general congestion of the airways.

3 Explain how a fault in just one amino acid can have such a drastic effect.

Sticky mucus is also a real problem in the pancreas. The mucus blocks the pancreatic duct, preventing the secretion of pancreatic juice.

4 Suggest why cystic fibrosis sufferers have to take tablets containing digestive enzymes.

Treatment of CF includes regular physiotherapy in which the chest is patted to dislodge the mucus. Even so, infections are common and most CF sufferers have to take a variety of antibiotics according to the infection they have at the time.

Gene therapy for cystic fibrosis

Now, one exciting possibility is that the faulty genes can be replaced by healthy ones. In 1989 the cystic fibrosis allele was located on chromosome 7. The base sequence of the healthy gene was then compared to the defective allele, and the nature of the fault was narrowed down at the molecular level. This opened up the exciting possibility that if healthy genes could somehow be introduced into the epithelial cells, they might be expressed and so make the correct membrane protein, solving the problem for a time.

The basic steps are as follows:

1. The CFTR gene is isolated, and cut out.
2. The gene is cloned many times.
3. The genes are encapsulated, either by putting them into liposomes (spheres made from lipid) or viruses.
4. The gene particles are inhaled, so that they can pass into the epithelial cells of the lung to be incorporated into the DNA of the cells.

Once in place, if all goes well, the healthy genes are transcribed, making the correct protein.

5
a) What is used to cut the CFTR gene out of chromosome 7?
b) Name two different methods of cloning the CFTR gene.
c) Suggest why the pancreatic genes would be more difficult to replace.
d) Suggest why repeat treatments at regular intervals would be necessary.

Germ-line gene therapy

This area of research is an ethical minefield. So far in this Assignment we have looked at body cell gene therapy – the replacement of the faulty allele in certain cells of the body. However, the eggs or sperm of the sufferer would still carry the defective alleles. In **germ-line gene therapy,** the original alleles in the zygote would be changed. The individual would grow and develop with healthy alleles in all cells and all of the individual's children would inherit the healthy allele.

Germ-line gene therapy sounds attractive but there are problems, so much so that this area of research is banned in the UK. Ethically, there is the familiar 'where do you stop?' problem. A new treatment for cystic fibrosis or haemophilia would be desirable, but having the technology might allow for all sorts of temptations. Our knowledge of the human genome could reveal, for example, genes which code for intelligence, height or skin colour. Tampering with these characteristics would lead to obvious problems.

Biologically, germ-line gene therapy is potentially dangerous because we know very little about how genes function in the embryo. Tampering with genes in the zygote could have effects which might become apparent only in later life.

6 In discussion groups, prepare a case which you could present either for, or against, germ-line gene therapy.

HEALTH AND DISEASE

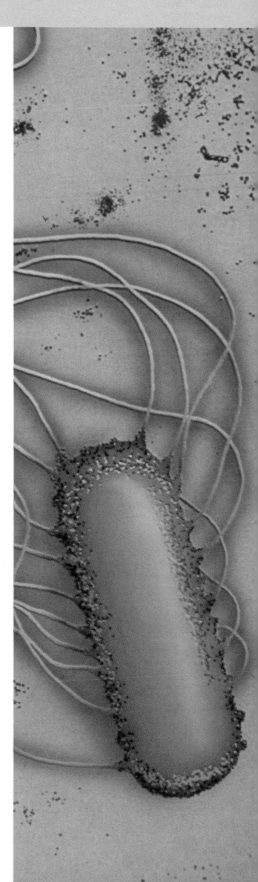

HUMANS ARE A SELF-AWARE SPECIES and, sooner or later, most of us are conscious of ill health – not a happy thought when it is ours, or that of people close to us. At the same time, as a topic for study, illness is fascinating. Just think of how popular medical dramas and documentaries are.

Until this century, the commonest causes of death were the 'catchable' diseases – those carried by bacteria, viruses and other organisms too small to detect. Then the microscope showed that diseases such as dysentery and tuberculosis were caused by infective agents whose spread could be worked out and then controlled. Later came antibiotics, vaccines and other sophisticated medicines.

Nowadays, the commonest causes of death in the western world are all non-communicable, caused in part by our lifestyle and in part by the fact that modern medicine prevents most of us from dying from diseases that are communicable. In this final section of the book, we look first at infectious diseases, and then at the ways in which the body defends itself against these potentially lethal invaders. We progress to the biggest killers, the lifestyle diseases, before finishing with genetic diseases and some thoughts for the future.

Can we look forward to medical science solving all our health problems? In all probability, we will simply swap one set of problems for another. What is certain is that the proportion of elderly people in the population will continue to increase, as will the demand by all age groups for even more health care.

In the UK, to best use limited health service funds, hard decisions have to be made. Early diagnosis and better public health awareness – taking responsibility for our own health – could help to reduce the demand for much costly corrective treatment. To what extent future generations will act on advice and lead a healthy lifestyle remains to be seen.

32 Health and disease

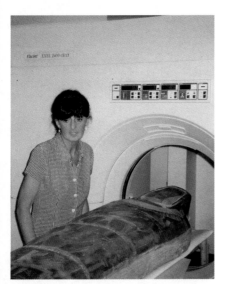

When the mummified bodies of ancient Egyptians are examined using modern technology, we can find out a lot about diseases that affected them. Some bodies show very clear signs of infectious diseases such as leprosy and tuberculosis

DJEDMAATESANKH was an ancient Egyptian princess who died when she was about 35 years old. She was considered an important person, and so her body was mummified. Her attendants removed her internal organs and wrapped her body in cloth that had been soaked with resin.

A few years ago, nearly 3000 years after her death, she finally had a post-mortem. Canadian scientists used modern medical techniques to examine her preserved body remains. They built up a three-dimensional picture of her head using an X-ray technique called computerised axial tomography, or CAT scanning.

The pictures that emerged were fascinating. They revealed that one of her upper teeth was loose and underneath it was a hole about two centimetres in diameter. So the researchers concluded that she probably had a large abscess in one of her teeth.

Toothache must have been a feature of everyday life for the ancient Egyptians. The abscess could have led to blood poisoning or meningitis, and it is likely that such an infection killed her.

Fig 32.1 **Fit does not necessarily mean healthy.** Studies carried out on athletes suggest that the gruelling programme of training that they must undertake to reach the high standard of the runners in this picture actually damages their health. They are much more likely than the average person to suffer from viral infections. This is possibly because their immune systems are not as effective

1 DEFINING HEALTH AND DISEASE

We start this chapter by looking at what health and disease mean. Then, in the main part of the chapter, we look at the causes of human disease: infections, our lifestyle and our genes. Finally, we see how doctors use modern techniques to make an accurate diagnosis before they begin to treat people in their care.

What do we mean by healthy?

We consider someone healthy if he or she is not suffering from a disease. But the World Health Organisation recognises health as more than this. Its definition of health is more useful:

Health is a state of complete physical, mental and social well being.

This definition recognises the importance of mental as well as physical health. However, agreeing a definition of 'a state of complete well being' is virtually impossible. We often encounter problems like this when we attempt to give biological definitions to words that we use in everyday language.

We also need to consider a person who has lost an arm in an accident. Are they healthy? What about someone with Down's syndrome? A definition of health must be flexible enough to include special cases such as these. However, to be realistic it should not be as broad as the WHO definition, which excludes most of us.

Without defining it formally, we can say that health:

- is more than a lack of disease,
- describes a state in which the body's organs and processes work properly,
- takes into account particular genetic and environmental factors.

What do we mean by disease?

Defining disease is more straightforward. Diseases arise when the body's functions are disturbed or when its structures are altered. Diseases can be defined as **communicable** if they are caused by **infectious agents** and can be passed from one person to another. **Non-communicable diseases** are caused by a broad range of environmental and genetic factors. It is important to remember that the development of a particular disease and the way it responds to treatment varies from person to person.

2 THE MAIN CAUSES OF DISEASE

Communicable disease

Communicable or infectious diseases are those that can pass from one person to another. These diseases are caused by microorganisms such as bacteria and viruses (Figs 32.3, 32.3 and 32.4). Although common in many developing countries, infectious disease is much rarer in modern Europe because of effective treatments such as vaccines and antibiotics.

Communicable diseases: infections

Many communicable diseases are caused by microorganisms.

Fig 32.2 **Smallpox has now been totally eradicated. In 1967, when the World Health Organisation began its campaign to eliminate this dreadful viral disease, there were ten million cases spread through thirty different countries. In 1977, the last ever case was recorded in Somalia**

DIPHTHERIA.   SCROFULA.   CHOLERA.

FATHER THAMES INTRODUCING HIS OFFSPRING TO THE FAIR CITY OF LONDON.
(*A Design for a Fresco in the New Houses of Parliament.*)

Fig 32.3 **This Punch cartoon shows three diseases caused by bacteria: diphtheria, scrofula (a form of tuberculosis) and cholera. Between them, they affected large numbers of people in London and other parts of the United Kingdom in the mid-nineteenth century**

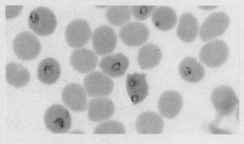

Fig 32.4 **The malarial parasite is a small one-celled organism belonging to the Kingdom Protoctista. It is carried by the anopheles mosquito and invades human red blood cells. World-wide, malaria kills about 2.7 million people every year**

Non-communicable disease

Non-communicable diseases seem to be connected with our lifestyle. Cancer, for example, is a disease that can affect many different parts of the body. This disease clearly does not have a single cause. Smoking tobacco is the major cause of lung cancer, and too much sunbathing can lead to skin cancer. People who inherit certain genes can be at an increased risk from some cancers, particularly those that affect the colon or the breast. Viruses seem to be involved in the development of cervical cancer.

?

A Suggest why communicable diseases are not as common in Europe as they were two hundred years ago.

Lifestyle diseases

Lifestyle diseases are those which result from the conditions in which we live.

Fig 32.5 **Heart disease is the greatest single cause of death in England and Wales. It accounts for nearly one in three of all deaths. Many environmental factors are linked to a greater risk of heart disease. These include stress, smoking, a high-fat diet and lack of exercise**

Genetic diseases

It has been estimated that approximately 2 per cent of the UK population have some form of genetic disease.

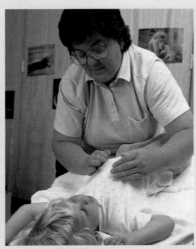

Fig 32.6 **Cystic fibrosis is caused by a single gene. It is a recessive characteristic. Individuals who inherit two defective alleles cannot make a protein responsible for transporting chloride ions across cell membranes. As a result, they produce extremely thick mucus which affects the functioning of organs such as the lungs. Physiotherapy helps to disperse this mucus**

Degenerative diseases

Age brings about gradual decline in the efficiency of many of the systems in the human body.

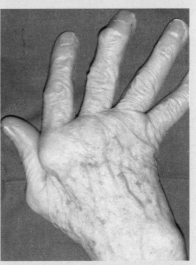

Fig 32.7 **Osteoarthritis is a painful condition which affects the joints of many older people. The cartilage at joints is lost and this may result in damage to the underlying bone, so that the sufferer has difficulty in moving the joints**

How disease-causing factors can influence each other

Cystic fibrosis is an inherited disease. It is due to a fault in a single gene. The normal allele of the gene makes a protein responsible for transporting chloride ions across cell membranes. If a person inherits one faulty allele, they are fine because they have a normal allele that still makes the protein. People develop the disease only if they inherit two faulty alleles, which makes cystic fibrosis a **recessive characteristic**.

People with no normal allele cannot make any functional transport protein and, as a result, produce extremely thick mucus. This affects organs such as the lungs and the pancreas. So cystic fibrosis has a genetic cause. At the same time, people who inherit the condition are likely to suffer from chest infections caused by microorganisms which become trapped in the mucus in the lungs.

Malaria is a severe disease which affects many people living in the tropics. It is caused by a single-celled parasite that is transmitted from the blood of one person to that of another by anopheles mosquitoes. People who can afford preventive medicines, mosquito nets over beds and insect repellants are less likely to be bitten by mosquitoes than the poorer members of the community who cannot pay for such protection.

In addition, people who live in areas where malaria is common often have genes which cause them to make slightly different sorts of haemoglobin that confer some protection from malaria.

So, with malaria, there is a link between lifestyle, infection and genetics. Whether or not a particular person catches malaria and then dies from it depends on all three factors.

INVESTIGATING THE CAUSE OF DISEASE: ASTHMA IN BARCELONA

BARCELONA IS an industrial port in Spain. Doctors throughout the area noticed that on some days large numbers of people were admitted to hospital suffering from asthma attacks, and wondered what might be causing the problem.

The investigators knew how many people were being admitted to hospital with asthma every day. They used the figures for the month between 18 November and 18 December 1984 (Fig 32.8), and mapped the places where the attacks of asthma started (Fig 32.9).

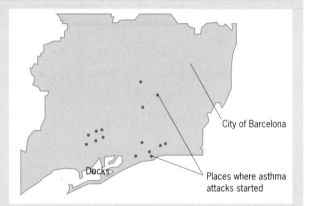

Fig 32.9 **Distribution of the asthma attacks. It was first thought that the attacks would be related to the industrial areas of the city, but the data suggested that the problem was connected with the docks**

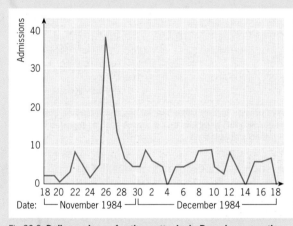

Fig 32.8 **Daily numbers of asthma attacks in Barcelona over the period of a month**

They discovered that people had more asthma attacks on the days when soya beans were being unloaded. Further, a detailed case study showed that people whose asthma attacks happened on soya bean unloading days were much more likely to be allergic to soya beans than people admitted on days when there was no unloading.

Such studies can give people the information they need to avoid contact with substances that might cause an allergic reaction.

3 PATTERNS OF DISEASE

The earliest people were hunter–gatherers who lived in small isolated groups. What did they die of?

Even if they had a reliable food supply, they still had to search for food whatever the weather. Some of the very old and the very young might have died of the effects of extreme cold or heat, and fighting between rival groups could have accounted for a proportion of deaths. However, the most likely cause of death was infectious disease. Not the massive epidemics of measles, plague or smallpox that killed so many later in history, but chronic (long-lasting) diseases such as tuberculosis, malaria and leprosy.

Why was this? The simple answer is that there weren't enough people to sustain an epidemic of a short-lived infection such as

An **endemic** disease is one that is always present in the population.

An **epidemic** is an outbreak of disease which spreads rapidly through a population, affecting a large number of people.

measles. Today, there are a lot more people and infectious diseases can spread more easily. In Africa, for example, measles is **endemic** – on any one day, someone, somewhere in Africa has the measles. If the people who are infected have contact with many other people who have no immunity to measles, a large proportion of the population can become infected all at the same time. This is an **epidemic**.

The typical hunter gatherer society probably consisted of no more than a couple of hundred people and they probably had little contact with outsiders. Diseases like measles would not have been able to spread easily, and would have never become a problem (Fig 32.10).

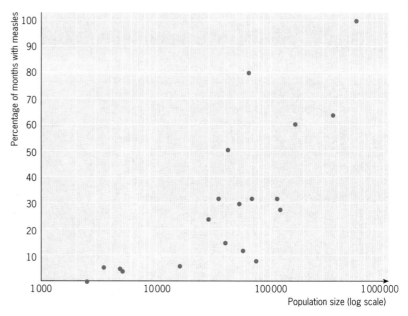

Fig 32.10 **Measles is always present in very large populations, but it is often absent from smaller populations. Evidence like this suggests that diseases such as measles were unknown in early hunter–gatherer populations**

B Use the graph in Fig 32.10 to estimate the minimum size of the population in which cases of measles are always present.

Chronic diseases such as tuberculosis, malaria and leprosy were much more of a problem. These conditions affect the body for a long period of time and infection could have been passed on even when contact between different groups of people did not happen very often.

When people began to domesticate animals and cultivate crops human societies changed enormously. The land could support far more people and communities grew larger. As settled agriculture developed, towns and cities came were built, and trade brought increasing contact between different groups of people. From that time, we have been literally plagued by infectious diseases.

Agriculture influenced disease in two other ways:

● The cultivation of crops led to changes to the environment. For example, simple irrigation systems that made more water available for crops provided the perfect breeding grounds for the larval forms of the parasites that cause malaria and schistosomiasis (see Chapter 33).

● Closer contacts between humans and their domestic animals allowed new diseases to emerge, such as smallpox, plague and influenza.

● Trading links allowed diseases to spread rapidly.

Influenza, an example of a communicable disease

Outbreaks of influenza or flu occur every year in the UK. The disease is caused by a virus that affects the epithelial cells that line the nose and throat (Fig 32.11). After a short period of incubation, flu causes headache, fever and aching muscles and joints. With rest, these symptoms usually disappear after a week or so. In a few cases, secondary infections develop and these can cause life-threatening conditions such as pneumonia. The elderly are particularly vulnerable as their immune systems are less efficient. They are most at risk if they have lung problems already.

Flu epidemics are also bad for the economy, since so many people lose time at work.

C Suggest how a knowledge of the way in which the influenza virus is spread could be used to limit the spread of a flu epidemic.

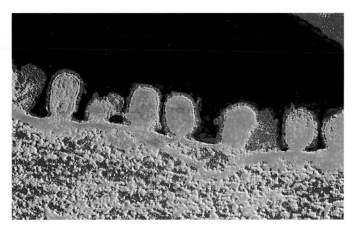

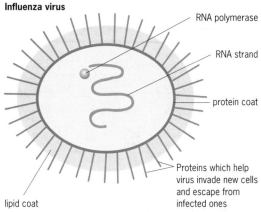

Fig 32.11 **The electron micrograph shows influenza viruses leaving an infected cell. The drawing shows the main features of an influenza virus**

Influenza virus

RNA polymerase

RNA strand

protein coat

Proteins which help virus invade new cells and escape from infected ones

lipid coat

When we breathe out, sneeze or cough, tiny droplets of mucus and saliva fly into the air. If you have flu, these droplets contain flu viruses and this is how they are passed on to other people. Small droplets can remain suspended in the air in poorly ventilated crowded buildings for ages, so it is not surprising that flu epidemics often occur during the winter months.

A global disease

The influenza virus has different forms. Two of these, **Type A** and **Type B,** are associated with major epidemics.

Influenza epidemics are most often caused by Type A viruses. Type B is not as common but outbreaks of Type B influenza occur on a global scale. Such large scale epidemics are known as **pandemics**.

One pandemic broke out in Europe in 1556 and lasted until 1560. Historians estimate that during this period almost 20 per cent of the entire population of the United Kingdom died of influenza. There was a more recent pandemic in the years 1918–1919 (Fig 32.12). Worldwide, almost 20 million people died, more than died in the entire First World War that immediately preceded it.

Protection against influenza

Why is it that you can have a single attack of measles and then be immune for life, but you can get influenza many times?

A person who has had influenza becomes immune to the strain of virus that caused it. However, the influenza virus has a very high rate of mutation. This means that different strains arise, each with slight differences in their outer protein coat. The body might be immune to the strain responsible for an outbreak of influenza one year, but this immunity does not protect against other strains that cause later outbreaks of flu.

Fig 32.12 **Spanish flu, as it was known, killed nearly 20 million people in the period immediately after the First World War. There was little doubt that the speed at which the disease spread from country to country and continent to continent was partly due to troops returning home**

D Explain why the 1918–1919 influenza pandemic spread much more rapidly than the pandemic which occurred in 1556.

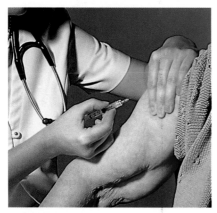

Fig 32.13 **Vaccination against influenza is offered to the most vulnerable members of the community**

Flu vaccines are available. They are usually given to those at risk such as elderly people (Fig 32.13) and those with asthma and heart conditions. The actual vaccine produced is effective only against the two or three strains of the virus common that year. It is necessary to have another injection with a different vaccine the following year since new strains will probably have evolved.

4 IDENTIFYING A DISEASE

Looking for the obvious

Before a doctor can treat a disease effectively, he or she must identify the condition that is making the patient ill. Someone usually goes to the doctor with a collection of symptoms. For example, a woman finds that she gets tired very easily. Her husband thinks she looks rather pale and she notices that she gets out of breath when she exerts herself. The doctor observes the patient and listens to her describing these symptoms, and suspects that she is suffering from anaemia.

Anaemia is not a disease itself, but it has a cause. It could be due to:

- a shortage of iron in the diet; iron is an important part of haemoglobin molecules.

- loss of blood; this might result from heavy periods or perhaps an ulcer that bleeds constantly.

- destruction of red blood cells caused by sickle-cell anaemia, an inherited condition. An affected person has an abnormal form of haemoglobin that transports oxygen with reduced efficiency. The red blood cells of someone with this condition have a shorter life-span than normal red blood cells.

The doctor looks for clinical signs that might show which one of these is most likely. Only when an accurate diagnosis has been made, can treatment begin.

> Symptoms are indications of a disease that are noticed by the patient.
>
> Signs are indications of a disease that are picked up by a doctor but not necessarily by the patient.

Identifying pancreatitis

The pancreas is an important organ located just below the stomach (Fig 32.14). It has two main functions. It produces the hormones insulin and glucagon that regulate blood glucose concentration, and it secretes digestive enzymes. The enzymes include trypsin which breaks down proteins, lipase which digests fats, and amylase which hydrolyses starch to maltose.

Pancreatitis is a severe inflammation of the pancreas, caused when the enzymes that normally break down food in the intestine start to break down and destroy the tissue of the pancreas itself. It can be **acute** (when it comes on suddenly) or it can be **chronic** (long term).

Pancreatitis can be difficult to identify. The pancreas is in the first loop of the small intestine and is close to the liver and stomach, so pain in this area could be caused by either pancreatitis, liver disease, stomach ulcers or gallstones. A correct diagnosis is vital, and this is where biochemical tests come in handy.

Liver:
hepatitis

Stomach:
gastric ulcer

Gall bladder:
gallstones

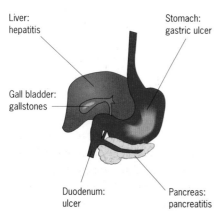

Duodenum:
ulcer

Pancreas:
pancreatitis

Fig 32.14 **The pancreas is situated in the first loop of the small intestine and near to a number of other important organs. Diseases that affect any of these organs could give similar symptoms**

If the pancreas tissues are damaged rapidly, the enzymes in the cells escape into the blood. Tests involve checking whether lipase and amylase are in a blood sample, and if they are, it is a good indication that the patient has acute pancreatitis.

If the condition is long term, however, the pancreas tissue is more slowly and progressively destroyed, and the ability to secrete enzymes declines gradually. This is chronic pancreatitis and is identified by measuring the amounts of enzymes in the patient's faeces. Recently, researchers have found that chronic pancreatitis is common in people who have a mutation in the gene that is faulty in cystic fibrosis. Although their mutation causes a less severe set of symptoms, it might soon be possible to develop a genetic screening technique to identify people at risk from developing pancreatitis later in life.

E Trypsin is an enzyme secreted by the pancreas. Explain why:

(a) you would expect to find trypsin in the faeces of a healthy person;

(b) the amount of trypsin in the faeces will decrease in a person with chronic pancreatitis.

Phenylketonuria: an inherited enzyme deficiency

Phenylketonuria is an inherited condition that affects approximately 1 in every 10 000 people. All new-born babies in the UK are tested for phenylketonuria. In the heel or **Guthrie test**, the midwife takes a blood sample from the heel of the baby and tests the sample for the presence of phenylalanine, an amino acid.

In a healthy baby, phenylalanine is converted into tyrosine by the enzyme phenylalanine hydroxylase (Fig 32.15). This enzyme is missing in a baby with phenylketonuria, and phenylalanine and other metabolic substances made from it build up in the baby's blood. These substances interfere with brain development, so it is very important that the condition is identified as soon as possible after birth.

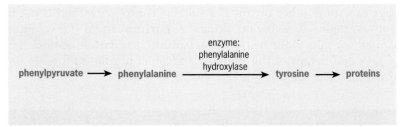

Fig 32.15 **Part of the biochemical pathway by which phenylalanine is converted to other substances in the body. Each of the steps in this biochemical pathway is controlled by a different enzyme**

Suppose the enzyme phenylalanine hydroxylase were missing. Let's look at the possible effects on the biochemical pathway, and how we could use these effects to show that the patient had phenylketonuria.

- First, there would be no enzyme. So we could do a direct test for the enzyme and show that it was absent from the patient's blood.

- If phenylalanine hydroxylase were missing, its substrate, phenylalanine, would not be converted into tyrosine, so the amount of phenylalanine should be higher than normal, and the amount of tyrosine should be lower or absent. We could test for raised phenylalanine levels and reduced tyrosine levels.

- Finally, the biochemical pathway from phenylalanine to phenylpyruvate might be affected. There is more phenylalanine present and, since it cannot be converted to tyrosine, it might be converted to phenylpyruvate instead. We could test for an increase in phenylpyruvate.

F Suggest how phenylketonuria might be treated by altering the amounts of phenylalanine and tyrosine in the diet.

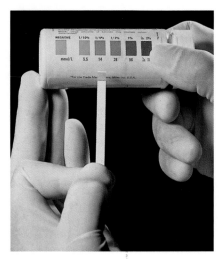

Fig 32.16 **The Clinistix test**

G Suggest two advantages of diabetics being able to test a urine sample for glucose with Clinistix rather than with Benedict's solution.

Diabetes and testing for glucose

Diabetics are unable to control the concentration of glucose in their blood (see Chapter 10). The blood glucose concentration rises and glucose appears in the urine. A hundred years ago, the condition was always fatal. Now, as long as it is correctly diagnosed, patients can be treated successfully and lead normal lives.

Monitoring the amount of glucose in the blood and urine of a diabetic patient is an essential part of this treatment. Early methods relied on chemical tests like Benedict's test in which glucose is used as a reducing agent. More modern methods make use of enzymes. Enzymes are highly specific and sensitive. This makes them ideal for analysing small samples. Today, diabetics use Clinistix™ to test for the presence of glucose in urine (see Fig 32.16 and page 71 in Chapter 4).

Looking inside the body

Sometimes, the only way to make an accurate diagnosis is to look inside the body. Cutting it open is one way, but there are always risks with surgery. The risks are rare, but it is safer to use **non-invasive** methods where possible. Several techniques now produce images of what is going on inside the body. The images sometimes look very strange to the untrained eye, but they can be interpreted by a skilled technician or doctor.

X-rays

X-rays have a much shorter wavelength than visible light. They penetrate different tissues in the body to different extents. Although they are most often used to look at bones and teeth (Fig 32.17), X-rays can show other organs if a suitable contrast medium is used. For example, if someone drinks a **barium meal** (a solution containing salts of the heavy metal barium), the barium absorbs the X-rays and details of the stomach and small intestine show up on the final image of X-rays taken shortly afterwards.

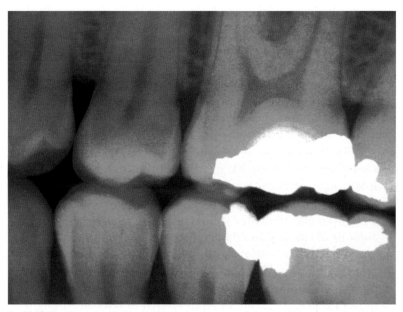

Fig 32.17 **X-ray images enable dentists to locate areas of decay which may not be obvious. The white area is a filling.**

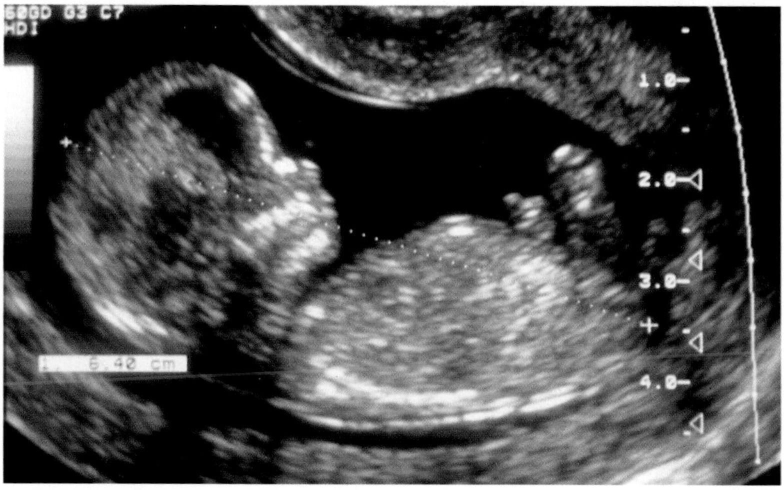

Fig 32.18 **An ultrasound image of a human fetus**

?

H Explain which of the above methods would be most suitable for:

(a) examining the ovaries of a woman undergoing fertility treatment in order to see the developing follicles;

(b) examining the inside of a knee joint before performing a cartilage operation.

The disadvantage of X-rays is that they damage DNA. For this reason, X-rays are not used to examine the reproductive organs or to scan the developing baby inside a pregnant woman, and the people who use X-ray equipment have to take safety precautions.

Ultrasound

Ultrasound is high frequency sound, so high that it cannot be heard by the human ear. An ultrasound beam passes into the body and bounces off dense tissue. The reflected waves produce an image that shows internal organs in great detail. Unlike X-rays, it does not damage DNA and so can be used to check the development of a baby in the uterus (Fig 32.18).

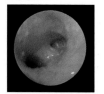

Fig 32.19 **An image of the inside of a bronchus in the respiratory tract taken with an endoscope**

Endoscopy

An endoscope is a long tube with a light at one end and a means of displaying an image at the other. It can be used to examine the inside of hollow body parts such as the oesophagus and stomach or the trachea and bronchioles (Fig 32.19). Many modern endoscopes rely on fibre optics. This means that they are small in diameter and flexible. Endoscopes are used for a variety of purposes such as inspecting the external ear and eardrum for signs of damage or infection, or for looking at the inside of the gut.

SUMMARY

After reading this chapter you should know and understand the following:

■ **Health** is difficult to define, but it is more than just an absence of disease.

■ **Communicable** or infectious diseases may be spread from person to person and are caused by **microorganisms**.

■ **Non-communicable** diseases have many causes. They can result from our **lifestyle** or from our **genes**, or they can occur as we age as a consequence of the process of **degeneration**.

■ Infectious diseases became increasingly important as agriculture developed and humans began to live in large settled communities.

■ Diseases are characterised by distinct **signs** and **symptoms**.

■ Diagnosis may be confirmed by a range of **biochemical tests**. Many of these rely on **enzymes**.

■ **X-rays**, **ultrasound** and **endoscopy** allow us to investigate the internal structure and functioning of the body.

Assignment

THE CHILDREN IN THE CEMETERY

Today, Lyme Regis is a small, pleasant seaside town on the Dorset coast. It is a popular tourist spot set in a bay in a region noted for its attractive scenery.

In Victorian times, however, it was very different. There was considerable poverty and overcrowding in the lower part of the town along local river banks. Just over a hundred years ago, the medical officer of health wrote that it was common for dwellings with just one living room and one bedroom to house a family of ten. In these conditions, communicable diseases spread very rapidly.

Fig 32.A1 shows the seasonal pattern of deaths from communicable diseases typical of towns like Lyme Regis in the middle of the nineteenth century.

1 Suggest an explanation for:

a) the pattern of deaths from respiratory diseases;

b) the pattern of deaths from diseases affecting the gut.

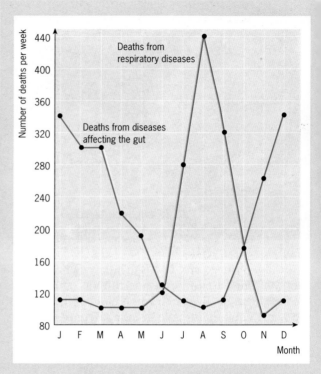

Fig 32.A1 **Graph showing the seasonal pattern of deaths from communicable diseases of the respiratory tract and gut**

Fig 32.A2 shows the results of an investigation of the burials in Lyme Regis Cemetery in the period 1856 to 1900. The upper line on the graph shows the total number of burials per year; and the lower line shows the total number of burials of children aged under 16 years.

2

a) Explain what is meant by a disease epidemic.

b) What is the evidence from the graph that epidemics were responsible for many deaths in the period shown on the graph?

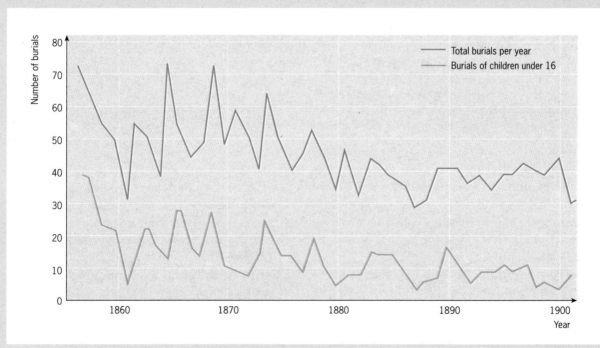

Fig 32.A2 **Burials in Lyme Regis Cemetery during 1856–1900**

Important diseases which produced epidemics in Lyme Regis during the nineteenth century included whooping cough, measles and smallpox.

3 Suggest why these diseases are no longer a major cause of death in the United Kingdom.

4

a) Copy and complete the table below to show the percentage of the total number of burials that were children under 16.

Decade	Number of burials		Percentage of total burials that were under 16
	Total	Under 16	
1866	539	224	
1876	519	145	
1886	402	105	
1896	373	75	
1916	313	39	
1936	308	21	
1966	301	7	

b) Plot the data in the table as a suitable graph to show how the percentage of total burials that were children under the age of 16 years changed over the period shown.

c) Use the information in this Assignment to suggest why there was a change in percentage of total burials that were children under the age of 16 over the period shown.

Cholera is a bacterial disease that produces severe vomiting and diarrhoea. Unless treated promptly, the patient becomes severely dehydrated and death can occur within 24 hours of the symptoms first appearing. Epidemics of cholera caused large numbers of deaths in nineteenth century Britain. After one outbreak which occurred in Portsmouth, a local doctor wrote:

'At present the part of Portsmouth called Portsea was one large cesspool, for there could not be less than 16 000 cesspools daily permitting 30 000 gallons of urine to penetrate into the soil. Just reflect on the character of the well-water of a district which became mixed every year with 365 times 30 000 gallons of urine, to say nothing of a host of other abominations!'

5

a) Use this information to suggest how the epidemic of cholera spread through Portsmouth.

b) Suggest why cholera is no longer a major cause of death in the United Kingdom.

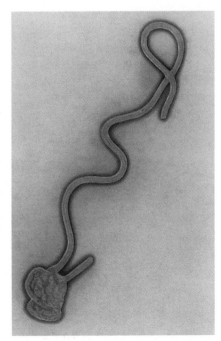

An electron micrograph showing an ebola virus particle. Ebola virus particularly targets liver cells and macrophages: massive destruction of the liver is a hallmark feature of infection

WE ARE USED TO HEARING about new treatments, new vaccines and new medical advances on an almost daily basis. But it comes as something of a shock when a new disease appears. Everyone knows about AIDS, but other new diseases have appeared within the last 30 years. Ebola virus, for example, first emerged in Zaire and Sudan in 1976. Over 500 cases were reported and infection proved very dangerous: almost 88 per cent of victims died in Zaire and 53 per cent perished in Sudan. Ebola haemorrhagic fever occurred again in 1978, in virtually the same place in Sudan, but only three other cases have been reported since.

The 'shock, horror, gasp' factor of ebola fever has driven journalists into a frenzy. Reports, articles and even films predicted the mayhem that would follow if Ebola virus hit the USA, even thought the number of people who have ever died from Ebola is tiny. But important questions need to be answered: Where has this virus been hiding? Why did it suddenly emerge to cause disease? Could there be a large epidemic?

The animal reservoir for ebola virus seems to be a species of monkey that lives in areas of dense forest that are rarely visited by people. Yes, a few places like this still exist, and as people start to go into them, they come into contact with such monkeys. The virus is transmitted and, what was a harmless organism in the monkey, is deadly in its new host.

In Zaire and Sudan, ebola virus spread quickly between people because of close contact. The centre of the epidemic in Zaire involved a missionary hospital where needles and syringes were re-used without sterilisation. Most of the staff of that hospital died and there were a few cases involving people taking care of the sick or preparing bodies for burial, but the epidemic was self-limiting. Because of its relatively short incubation period and the severe symptoms it causes, it is unlikely ever to cause a world-wide epidemic.

1 MICROORGANISMS AND DISEASE

Disease-causing microorganisms have been responsible for enormous numbers of human deaths. They have frequently altered the course of history. The species that probably had the greatest effect on human populations was *Yersinia pestis*, the bacterium responsible for the great outbreaks of plague such as the Black Death and the Great Plague. The Black Death swept through Europe in the middle of the fourteenth century, killing almost a quarter of the entire population. The Great Plague of 1665 had a devastating effect on the population of London and other British cities and towns.

Plague is not a disease confined to the rat-infested slums of the Middle Ages. It still exists today. Every year there are about fifteen cases in the western United States, and the disease is still able to terrify. An outbreak of plague in India in 1994 brought headline fears in many newspapers of a new epidemic. But it didn't happen: only a handful of people died.

Why is it that a disease which once killed literally millions of people can now be readily controlled? The story of the decline of plague reflects much of the story of infectious disease that forms the basis of the next two chapters. In this chapter we look at the different organisms that cause infectious disease. In Chapter 34 we look at the ways in which the immune system deals with infectious disease.

What makes microorganisms dangerous?

Microorganisms are very common in and on the human body. There are more bacteria living on your skin than there are people in the world. Most bacteria cause you no harm; the few that do are called **pathogens**. It is important to find out which microorganisms are responsible for which diseases for, until the infection has been identified, it is impossible to give effective treatment.

The nineteenth century biologist Robert Koch put forward a set of basic principles for identifying a pathogen: these principles are known as **Koch's postulates**.

- The microorganism must always be present when the disease is present. It should not be present if the disease is absent.

- It should be possible to isolate the microorganism from an infected host and grow it in culture.

- When cultured microorganisms are introduced into a healthy host, the disease should develop.

- It should be possible to isolate the microorganism from the new host.

By following these principles, researchers have identified the microorganisms responsible for many diseases. But there are problems in applying Koch's ideas especially when humans are involved. For example, it would be unethical to introduce a suspected pathogen to humans to see whether or not they developed a disease. In addition, some people fail to show disease symptoms after being infected. They might be resistant to that disease or they might be **carriers** that can pass on the disease to others, even when they appear disease-free themselves (Fig 33.2).

Fig 33.1 **Part of the stained glass window in the church at Eyam in Derbyshire. It tells the story of how plague was brought to the village in a bundle of cloth from London. This bundle was opened by the tailor George Viccars who became the first of the 267 out of the total of 350 people in the village to die from plague**

?

A Suggest why it might have been difficult to use Robert Koch's ideas to prove that typhoid fever was caused by *Salmonella typhimurium*.

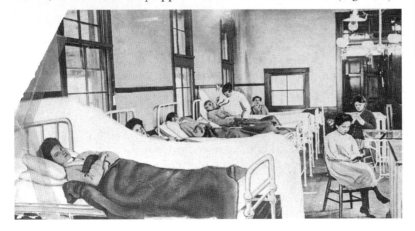

Fig 33.2 **Typhoid fever is a severe disease of the digestive system caused by the bacterium** *Salmonella typhimurium*. **In carriers, people who are infected but show no signs of the disease, the bacteria living in the gut pass out of the body in the faeces. Typhoid Mary, who lived in New York earlier this century, was a carrier of** *Salmonella typhimurium*. **She gained a bit of a bad name for herself when, unwittingly, she shared her infection with many of the people she came into contact with. While she remained the picture of health, they sickened and died**

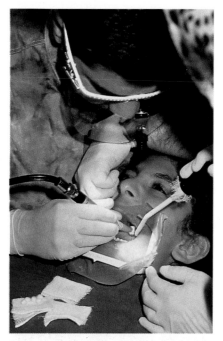

Fig 33.3 **During dental treatment, the patient is at risk from infection through the wounds in the gums, but so is the dentist. The patient can pass on airborne viral infections and there is a slight risk of blood-to-blood transmission**

How do microorganisms get into the host?

Your body provides an ideal environment for pathogens to multiply rapidly. It's warm and wet and there are abundant supplies of nutrients. Fortunately for you, your skin prevents most of them taking advantage. The outer layer of skin consists of dead, dry cells made of an indigestible protein called **keratin**. Microorganisms find it difficult to penetrate this layer to enter the body. They are more likely to use the slightly easier entry route offered by the mucous membranes of the breathing system, the gut, the urinary system or the reproductive system. But the body has various strategies to make things as difficult as possible. The mechanisms involved are explained on page 510.

In order to cause disease, pathogens must be transmitted from one person to another. Many of them cannot survive for very long outside the body. The main transmission routes are:

- through the air in droplets of mucus or saliva,
- in contaminated food or water,
- by sexual contact with another person,
- through breaks in the skin.

B Study the photograph of the dentist at work in Fig 33.3. What precautions are being taken to avoid transmission of any infection: **(a)** from the dentist to the patient? **(b)** from the patient to the dentist?

C Everyone gets colds, but measles, mumps and diphtheria are now very rare in the UK. Can you guess why this is? (Clue on page 516.)

Airborne infections

Many microorganisms are transmitted through the air. These include influenza, colds and measles, which are all caused by viruses; and the bacterial diseases tuberculosis, diphtheria and meningitis. When an infected person coughs, sneezes or even just breathes out, tiny droplets of mucus and saliva are expelled (Fig 33.4). These droplets carry microorganisms with them. The larger ones fall to the ground rapidly, but the smaller ones can remain suspended in the air for long periods of time. In poorly ventilated, crowded conditions, diseases such as influenza can spread very rapidly.

Fig 33.4 **The power of a sneeze. When someone sneezes, the droplets that explode from the mouth and nose start their journey at over 150 kilometres per hour**

Contaminated food and water

Every year over 5 million people die from infections that cause diarrhoea. Many victims are small children. In Bangladesh three out of five children under five die from infant diarrhoea. The microorganisms responsible spread in water or food contaminated by human faeces.

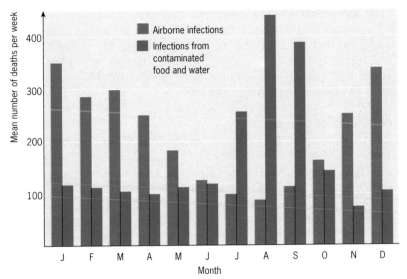

D Suggest why, in the period shown in Fig 33.5:

(a) the highest number of deaths from airborne infections occurred during the winter months;

(b) the highest number of deaths from contaminated food and water occurred during the summer months.

Direct contact

Direct contact between the skin of one person and that of another is unlikely to produce infection. Sexual contact is another matter, however. Microorganisms that spread like this include the human immunodeficiency virus (HIV), the virus that causes genital herpes, the bacterium responsible for gonorrhoea, the spirochaete that causes syphilis, and *Candida*, the yeast that causes thrush.

Through the skin

The outer layer of the skin consists of tough cells that protect the body very effectively against infection. But this layer can be broken, and cuts and grazes form an entry point for bacteria. For example, tetanus is a serious disease that affects the nervous system. The bacteria that cause tetanus are common in soil and can enter the body through puncture wounds and animal bites.

Mosquitoes, ticks and lice all feed by puncturing the skin. In doing this they transmit microorganisms directly into the blood. One of the most important insect-borne diseases is malaria, which is transmitted by anopheles mosquitoes (Fig 33.6).

The difference between infection and disease

Infection is not the same as having the disease. Before the disease develops, there is an **incubation period** in which the pathogen multiplies in the cells and tissues of the host. Think about what happens when you have been in a room with someone who is suffering from a cold. The infected person might sneeze and expel tiny droplets of mucus and saliva carrying the cold viruses on them. You inhale these particles and they get into your throat. There they attach to the epithelial cells in the mucous membrane. Like all cells in the body, these cells have cell surface membranes that contain many different protein molecules. The type of proteins present depends on the cell type, and on the genetics of the individual. Only if the right protein is present in the cell surface membrane, does the virus attach.

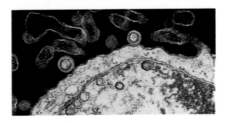

Fig 33.6 **An anopheles mosquito feeding on human blood. 2.7 million people die each year from malaria**

E The proteins of the cell surface membrane are determined by genes. Use this information to explain why some people are susceptible to particular diseases while others are not.

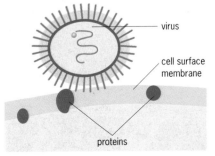

Fig 33.7 **A pathogen such as the virus shown in the diagram attaches only to a cell that has the right protein molecules in its cell surface membrane. The fact that proteins are genetically controlled helps to explain why some people are susceptible to some infections, while others fight them off easily. The photo shows two herpes viruses** (blue) **attached to a cell surface membrane.**

The virus then enters the cell and begins a cycle of reproduction and colonisation of new cells. But infection is not entirely one-sided. The body has a complex immune system (see Chapter 34). This might prevent the symptoms of the disease from ever developing. Even if some illness does follow, the immune system works to allow us to recover quickly.

The effects of disease

Infectious disease has two main effects. Firstly, the pathogen can damage the cells of the body directly. The human immunodeficiency virus, HIV, gives rise to the symptoms of AIDS mainly as a result of destroying large numbers of a type of white blood cell. As a result, the immune system stops working effectively and this allows other infections or certain types of tumour to develop. Similarly, some of the symptoms of malaria result from a loss of red blood cells or the blocking of capillaries that supply particular organs in the body.

Alternatively, the pathogen can cause the release of substances that injure the body. Many bacteria, for example, produce **toxins**. **Exotoxins** are released by the bacteria as they grow. **Endotoxins,** which form part of the bacterial cell wall, escape into the body when the bacterium dies and the cell wall breaks down.

Diphtheria is a good example of a bacterium that causes disease because it produces a toxin. It was a dangerous bacterial infection until about 40 years ago, when it used to kill many children. Since then, antibiotics and an effective vaccine have virtually eliminated it.

Diphtheria has an incubation period of two to six days, and then a sore throat and fever develop. A membrane forms across the throat and makes it difficult to breathe. As the bacteria multiply, they release a toxin into the blood. This toxin inhibits protein synthesis and results in damage to many parts of the body, particularly to the heart and the nerves. Interestingly, this toxin is produced only in bacteria of the species *Corynebacterium diphtheriae* that are themselves infected by a bacteriophage. The toxin is actually produced by one of the bacteriophage genes (see Fig 3.13).

?

F Treatment for diphtheria involves the use of an antibiotic such as penicillin and an antitoxin. Explain why it is necessary to use both an antibiotic and an antitoxin.

2 INFECTIONS CAUSED BY BACTERIA

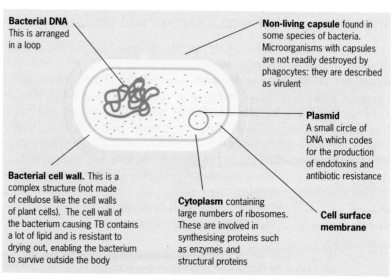

Bacterial DNA
This is arranged in a loop

Non-living capsule found in some species of bacteria. Microorganisms with capsules are not readily destroyed by phagocytes: they are described as virulent

Plasmid
A small circle of DNA which codes for the production of endotoxins and antibiotic resistance

Bacterial cell wall. This is a complex structure (not made of cellulose like the cell walls of plant cells). The cell wall of the bacterium causing TB contains a lot of lipid and is resistant to drying out, enabling the bacterium to survive outside the body

Cytoplasm containing large numbers of ribosomes. These are involved in synthesising proteins such as enzymes and structural proteins

Cell surface membrane

Fig 33.8 **Some of the strategies that bacteria use to invade the human body and cause disease**

Bacteria are **prokaryotes** (see Chapter 1). Each bacterial cell is a mass of cytoplasm enclosed by a cell surface membrane and a cell wall. Bacterial cells have no nucleus and their DNA is in the form of a loop in the cytoplasm. There are none of the membrane-bounded organelles such as mitochondria, endoplasmic reticulum and lysosomes that are found in eukaryotic cells. Fig 33.8 shows a typical bacterium, and how it causes disease

Bacteria don't need much to grow and reproduce. Give a single bacterial cell carbohydrate or some other source of carbon, a supply of nitrogen and a few mineral salts, and it can divide to produce a population of millions. It is easy to mix all these substances together in the lab to produce an artificial medium on which to culture bacteria. It's even easier to provide them with some leftover food in the home. In either case, the bacteria start to increase rapidly. A typical population curve is shown in Fig 33.9. The actual rate of growth depends on environmental conditions. Low temperatures or a shortage of nutrients, for example, limit the rate of growth.

?

G Explain how storing food in a refrigerator will affect the population growth curve of bacteria on a piece of chicken.

Lag phase	When introduced to a medium, bacteria take time before they start to increase in number. They need to produce the enzymes necessary to digest the nutrients in the medium. This requires activation of genes and the synthesis of proteins.
Exponential phase	Bacteria reproduce by binary fission. Once the cell has grown and reached a certain critical size, it divides in two. In this way, the population grows very rapidly. In optimum conditions, it may double in as little as 20 minutes. During the exponential stage, there are plenty of nutrients available, and toxic waste products have not accumulated to the extent that they slow down bacterial growth.
Stationary phase	Population growth begins to slow down and finally stops. There is a balance between the number of new bacterial cells being added to the population and the number dying. Nutrients are beginning to be in short supply and there may be a build-up of waste products which will inhibit the growth and reproduction of the bacteria.
Decline phase	Conditions in the culture become increasingly limiting. Nutrient supplies continue to fall and waste products to build up. Bacterial cells are unable to survive in these conditions and die in increasing numbers.

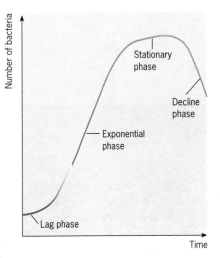

Fig 33.9 **Right: A population growth curve for the bacterium *Escherichia coli* growing on an artificial medium. Bacterial food poisoning results from eating foods that have been contaminated with certain types of bacteria or with the toxins that they have produced. Food that gives rise to this type of food poisoning has often been stored under conditions in which bacteria can multiply rapidly**

Tuberculosis

Tuberculosis (TB) is a disease caused by an infection with the bacterium *Mycobacterium tuberculosis*. Globally, each year, there are 3.8 million cases of TB, and 2.6 million of them prove fatal.

In the United Kingdom and other more developed countries, the incidence of tuberculosis fell dramatically during the twentieth century. Many factors contributed to this decline. There was a general improvement in living standards during this time, and advances in medicine such as the discovery of effective antibiotics and vaccines have made tuberculosis treatable and preventable. However, there are now signs that the number of cases is rising again (Fig 33.10). What has gone wrong?

Part of the problem is that drug-resistant strains of mycobacteria have evolved. Surveys in selected sites in Asia show high levels of multiple-drug-resistant TB that cannot be treated with the most powerful anti-TB drugs. The number of people being immunised has decreased and large numbers of people have become more susceptible because they have AIDS and the decreased immunity

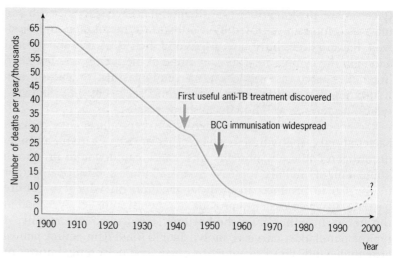

Fig 33.10 **The graph shows the change in incidence of tuberculosis during the twentieth century. Six countries account for over 50 per cent of the current TB epidemic. According to WHO estimates, 4.5 million of the 8 million new cases that occur each year are in India, China, Bangladesh, Pakistan, Indonesia and the Philippines. The problem in these countries is largely due to three factors: multidrug resistant TB (MDR-TB), HIV and the economic crisis**

that goes with it. Experts predict a further increase in the number of TB cases because of higher rates of TB transmission to HIV-positive people. By the end of the century, HIV will be a contributory factor in three-quarters of a million new TB cases world-wide.

Table 33.1 shows some of the groups of people in the UK who are at greater risk of tuberculosis.

Table 33.1 **Tuberculosis risk groups in the 1990s in the UK**

Risk group	Reasons for greater likelihood of developing tuberculosis
AIDS patients	AIDS patients have immune systems that are less effective at combating infection. Tuberculosis is more common in AIDS patients.
Recent immigrants	Recent UK immigrants, especially from the Indian subcontinent, are less likely to have been immunised against tuberculosis. They are also more likely to have had an earlier infection.
Vagrants	Misuse of alcohol and other drugs is associated with an increased risk of tuberculosis. Once diagnosed, it is difficult to ensure the necessary long term treatment.

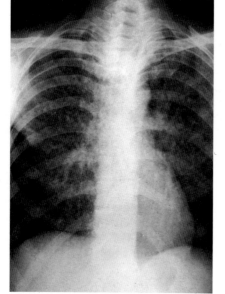

Fig 33.11 **In this chest X-ray, the smoky white areas are where the bacteria that cause tuberculosis have damaged the lung tissue**

Signs and symptoms of TB

Tuberculosis can affect many of the organs in the body. In almost 80 per cent of cases it affects the lungs. Someone with primary tuberculosis has a fever, tends to lose weight and often has a persistent cough. They might cough up blood because the bacteria destroy the lung tissue: a chest X-ray shows this damage as a dark shadow (Fig 33.11). The condition can be confirmed by identifying the bacteria in a sputum sample or in a small piece of lung tissue removed with the help of an endoscope.

Treatment and control of TB

There is no completely effective vaccine against TB of the lungs. BCG vaccine (Bacillus Calmette Guérin) was developed in 1921. It is very useful in preventing certain types of TB including the more severe types that can affect babies during the first year of life. The vaccine, which is a live but **attenuated**, or altered, strain of *Mycobacterium tuberculosis*, is given at or soon after birth to 85 per cent of babies world-wide. The vaccine provides cross-protection against leprosy, but is unlikely to protect adults against TB, and some people question why children in the UK are still given the BCG vaccine at the age of 13.

The disease itself can be treated successfully with antibiotics. Combinations of drugs are used as this discourages the development of strains that are resistant to particular antibiotics. Since tuberculosis is infectious, health workers trace all the people who have been in close contact with patients. These contacts can then be screened for the disease and treated if necessary.

In 1998, a group of researchers announced that they had decoded the entire genome of the tuberculosis bacterium. Knowing the sequence marks a new phase in the battle against one of the most successful pathogens that affects humans. The sequence contains 4 411 529 base pairs, which combine to form over 4,000 individual genes. The genome is already yielding information that could prove invaluable to future research on anti-TB drugs and vaccines.

Salmonellosis: signs and symptoms

There are many different species of bacteria that belong to the genus *Salmonella*. Only about ten of them cause food poisoning, known as **salmonellosis**. When food contaminated with the bacteria is eaten, the bacteria pass into the intestine. Other bacteria that cause food poisoning do not multiply once they are inside the body – they cause illness as they release toxins. *Salmonella* species actually enter the cells lining the small intestine and multiply actively. As the population increases, some bacteria die and release endotoxin. This is the toxin that causes the typical symptoms of food poisoning. Bad personal hygiene and poor food preparation skills can spread *Salmonella* food poisoning (Fig 33.12).

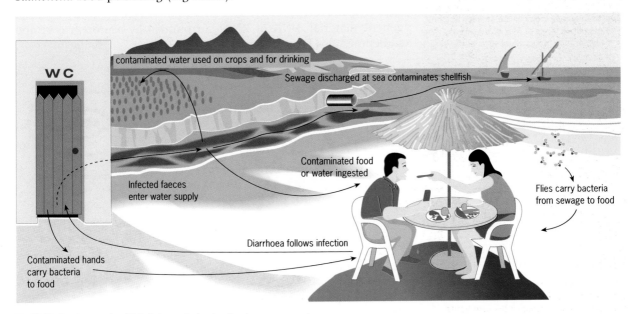

Fig 33.12 **Some ways in which *Salmonella* food poisoning can spread**

The endotoxin released as the bacteria die irritates the lining of the intestine, and causes the unpleasant and typical symptoms of gastroenteritis, nausea, vomiting, diarrhoea and abdominal pain.

Typhoid fever

Typhoid fever is caused by another species of *Salmonella*. It is a much more severe disease than salmonellosis. The patient often has a long-lasting fever, becomes delirious, and the spleen becomes inflamed. The wall of the intestine can be so badly damaged that it bleeds.

H The symptoms of salmonellosis usually take between 12 and 72 hours to develop. The symptoms of other forms of food poisoning develop much more quickly. Explain why.

See question 4.

If you want to avoid salmonellosis or other food poisoning, avoid buying and eating food after the recommended date. Store perishable food in a cool place or, preferably, in a refrigerator. Do not re-freeze food that has already been frozen and thawed once. Avoid the risk of contamination by maintaining good personal hygiene and by keeping raw and cooked foods separate from each other. Make sure that cooking and re-heating is thorough at sufficiently high temperatures.

Fig 33.13 In the blue box, six different viruses have been drawn to the same scale as the two bacterial cells *E. coli* and *Chlamydia*, and an animal cell.

Viruses are generally much smaller than bacteria, but there is some overlap. *Herpes simplex*, one of the larger viruses, is similar in size to *Chlamydia*, one of the smaller bacteria. The magnified views of four of the viruses show the appearance of individual virus particles

Treatment and control of salmonellosis and typhoid fever

Most cases of food poisoning are left to run their course. Sometimes, loss of body fluid may be severe, particularly in infants. Then, applying oral rehydration therapy, ORT (see Chapter 11) or using an intravenous drip will replace body fluids. Antibiotics may be used to treat typhoid fever.

The best way of dealing with *Salmonella* or any other organism that causes food poisoning or gastro-enteritis is to avoid getting it in the first place: careful attention to basic food hygiene removes much of the risk.

3 INFECTIONS CAUSED BY VIRUSES

It is difficult to decide whether viruses are living organisms at all. They do not show several of the characteristics that define a living organism; for example, they don't respire or feed or multiply. But they are more than just a collection of very complex chemicals. Their origins are also rather uncertain. Some biologists think they are stray bits of DNA that have 'escaped' from the genomes of higher plants and animals.

Viruses vary considerably in size but they are all very small (Fig 33.13). The virus responsible for polio, for example, is one of the smallest and is only 20 to 30 nm in diameter. The herpes virus is much larger; its diameter is about 250 nm. In addition to their small size, viruses have a very simple structure. They consist of a piece of genetic material, DNA or RNA, surrounded by a protein coat and sometimes also by a membrane.

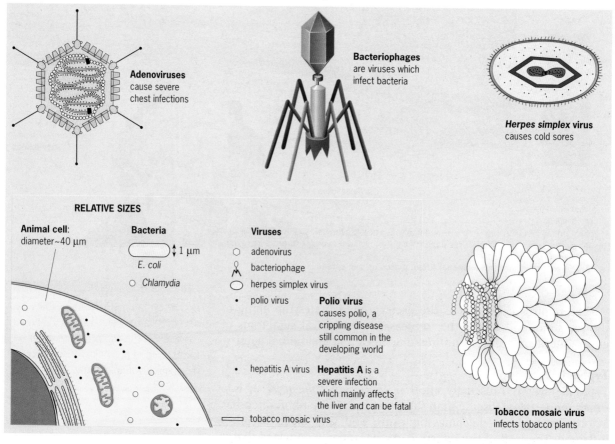

Adenoviruses cause severe chest infections

Bacteriophages are viruses which infect bacteria

Herpes simplex virus causes cold sores

RELATIVE SIZES

Animal cell: diameter ~40 μm

Bacteria
↕ 1 μm
E. coli
○ *Chlamydia*

Viruses
○ adenovirus
🜕 bacteriophage
◯ herpes simplex virus
• polio virus

Polio virus causes polio, a crippling disease still common in the developing world

• hepatitis A virus **Hepatitis A** is a severe infection which mainly affects the liver and can be fatal

▭ tobacco mosaic virus

Tobacco mosaic virus infects tobacco plants

In order to multiply, a virus must take over the host cell. It uses the cell's organelles and biochemical processes to make more virus particles. Viruses therefore have a great capacity to harm the cells and tissues of their host, and they cause a huge range of diseases including influenza and AIDS.

AIDS

AIDS, short for acquired immunodeficiency syndrome, is characterised by the destruction of vital cells in the immune system. It is caused by the human immunodeficiency virus (see page 505). The disease was first recognised in the early 1980s and the virus itself was identified in 1986. It has spread dramatically (Fig 33.14). The World Health Organisation estimates that up to 10 million people are now infected.

HIV is a retrovirus: it carries the enzyme reverse transcriptase. Once the virus infects a cell, often a T lymphocyte, this enzyme allows the cell to make virus DNA from virus RNA. The virus DNA is then inserted into the cell's own DNA. Here it acts as a gene. It might do nothing for a long time but eventually it will cause the production of more virus RNA. This will result in thousands of new viruses that burst out through the cell surface membrane and infect other cells. This life cycle is summarised in Fig 33.15.

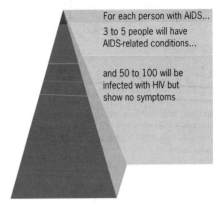

For each person with AIDS... 3 to 5 people will have AIDS-related conditions...

and 50 to 100 will be infected with HIV but show no symptoms

Fig 33.14 **The HIV pyramid**

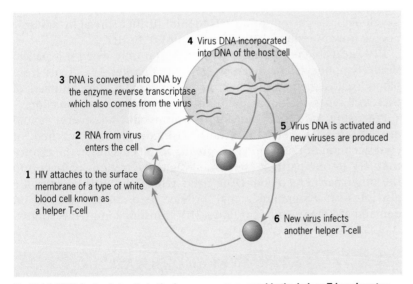

4 Virus DNA incorporated into DNA of the host cell

3 RNA is converted into DNA by the enzyme reverse transcriptase which also comes from the virus

2 RNA from virus enters the cell

1 HIV attaches to the surface membrane of a type of white blood cell known as a helper T-cell

5 Virus DNA is activated and new viruses are produced

6 New virus infects another helper T-cell

Fig 33.15 **HIV infects vital cells in the immune system, notably the helper T lymphocytes. This diagram shows how a single virus can infect a host cell and multiply to produce many thousands of new viruses**

HIV is often transmitted from one person to another during sex, but the virus can also spread when drug users share dirty needles, when infected blood is transfused, or when contaminated blood is used to make blood products. When HIV enters the body, it can take many years – maybe 10 or 15 years – before AIDS develops. A very small number of people are known to have been HIV positive since the mid-1980s and have not so far developed the syndrome. Perhaps they never will, and AIDS researchers are trying to find out why. This might provide really important clues about how AIDS can be held in check in other people.

Signs and symptoms

Table 33.2 shows the main stages in the development of AIDS from first infection with HIV.

Table 33.2 **The stages in the development of AIDS**

Stage	Number of helper T lymphocytes per mm³ of blood	Main features of stage
A	over 500	No symptoms but patient has antibodies to HIV in the blood.
B	200–499	Patient usually has fever or diarrhoea which lasts for a month or more. He or she might have infections such as thrush, caused by *Candida*, a type of yeast.
C	less than 200	Likely to have one or more of the following diseases known to be related to AIDS: Tuberculosis Karposi's sarcoma, a form of cancer in which large black or brown tumours appear on the skin (Fig 33.16). Pneumonia

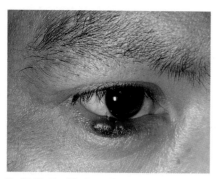

Fig 33.16 **The black swelling shows that this person has Karposi's sarcoma, a form of cancer, which affects blood vessels in the skin. Karposi's sarcoma is much more common in patients with AIDS than it is in the rest of the population**

Treatment and control

Drugs are available for the treatment of AIDS, and they are getting more effective all the time. However, they are expensive and can only slow the progress of the disease – there is no cure. An effective vaccine is thought to be a long way in the future because HIV can change its surface proteins and evade host immune responses. Gene therapy might provide some hope (see Chapter 36) but, until it is developed, the current priority is to limit further spread by advising people how to reduce their risk of exposure to HIV.

HIV is capable of infecting another person only when it is passed inside lymphocytes or macrophages (see Chapter 34). Apart from blood itself, semen and vaginal secretions contain large numbers of these cell types, making body fluids high-risk sources of infection.

Practising safe sex – using condoms during every sexual encounter – is very important. Using a barrier method of contraception not only prevents pregnancy, it cuts down the chance that HIV in one partner's semen or vaginal secretions will get into the other partner's blood. Drug users who share dirty needles are also at risk. Programmes that provide free, sterile needles on demand are designed to cut down HIV transmission by this route.

See questions 3 and 5. ■

4 PARASITES AND DISEASE

Bacteria and viruses are microorganisms; they are too small to be seen with the naked eye. However, some larger organisms also cause disease. These are the protoctists and the parasitic worms. In Europe, they are relatively rare, but on a global scale they have an enormous impact. Some examples of these larger parasites are shown in Fig 33.17. The roundworm *Ascaris lumbricoides* is the subject of the Assignment at the end of this chapter.

It is all too easy to concentrate on the horrors of parasitic infections. As biologists we should not forget that parasites as a group provide one of the best examples of **adaptation**.

A parasite is an organism that lives in or on a host organism. It gains a nutritional advantage from this relationship while the host suffers a disadvantage.

It might seem that this is a very easy way of making a living, but

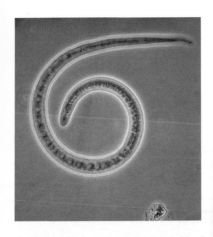

Fig 33.17 **Some common human parasites. Over half the world's population is infected by at least one of them**

a) **Roundworm**

(b) **Tapeworm: head end**

(c) **Bilharzia: male (fat) and female (thin) flukes**

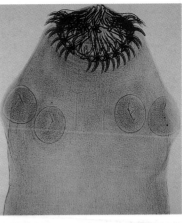

(d) **Malaria: thread-like sporozoites in blood**

(e) **Hookworm**

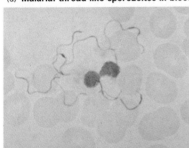

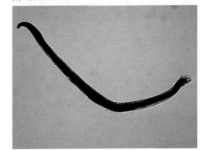

think about the problems a parasitic worm faces, living in your intestine. There are obvious advantages in lying in a soup of pre-digested food. There is no need to produce digestive enzymes, a complex nervous system is unnecessary because food does not need to be searched for, and a locomotory system is almost a disadvantage since the worm might move to less favourable areas of the gut.

On the other hand, there are disadvantages. Parasites are made of the same complex macromolecules as the food that you eat, so they must avoid being digested and must try to stay put despite the constant contractions of the gut muscles. It's also difficult to move from the intestine of one person into the intestine of another. Because of this, life cycles can be very complex and many involve intermediate hosts. Since the chances of completing the life cycle and reaching a new host are small, many parasites produce vast numbers of offspring.

?

I What particular problems are likely to be encountered by a parasitic blood fluke living in a vein?

Malaria

Malaria is caused by four species of the protoctist *Plasmodium* – *P. vivax*, *P. falciparum*, *P. ovale* and *P. malariae*. They are transmitted by the bite of the *Anopheles* mosquito. The life cycle of *Plasmodium* and an introduction to malaria is on page 364.

During the 1960s, malaria was successfully eradicated from the warmer areas of southern Europe and the United States, and its incidence was reduced significantly in many countries in the tropics. But the disease is now on the increase again. The WHO estimates that each year 270 million new malaria infections occur world-wide, along with 110 million cases of illness and 2.7 million deaths. Malaria is endemic in at least 100 countries of the tropical and subtropical regions of the world, there are an astounding two billion people at risk of malaria infection, and malaria is the cause of a quarter of all childhood deaths in Africa.

There are several reasons why the disease is so widespread: the cost of maintaining control programmes is high, insecticide

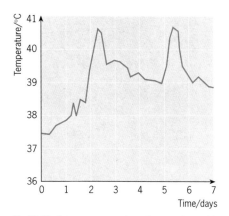

Fig 33.18 **A temperature chart for a patient with malaria. Note the peaks that occur at regular intervals. This cyclic pattern of fever is characteristic of malaria**

resistance has become more common in the *Anopheles* mosquito that transmits malaria, and the malarial parasite itself has developed drug resistance.

Signs and symptoms

One of the most characteristic features of malaria is the pattern of fever (Fig 33.18), with temperature peaks at regular intervals. These correspond to a new generation of parasites bursting out of the red blood cells, destroying them and causing anaemia. The parasites can block capillaries, limiting the blood supply to organs in the body: the brain, digestive system and kidneys can all be affected.

Treatment and control

One way to combat malaria is to break the cycle of infection. This can be done by killing the mosquitoes with insecticides clearing the pools of stagnant water in which the larvae live and preventing the insects from biting humans. Where possible, the risk can be reduced by sleeping under mosquito nets (Fig 33.19), wearing long-sleeved shirts and trousers after dark when mosquitoes are most active, and using mosquito repellents. But all these measures are too expensive for many local people. Also, they are not certain to work.

Fig 33.19 **The use of mosquito nets soaked in insecticide can be very effective in controlling malaria. Unfortunately, the cost may be beyond the resources of people living in malarial areas**

Fig 33.20 **The map shows that malaria is confined to tropical and subtropical regions. Chloroquine has been used very successfully in the control of malaria but resistance to chloroquine is now a major problem**

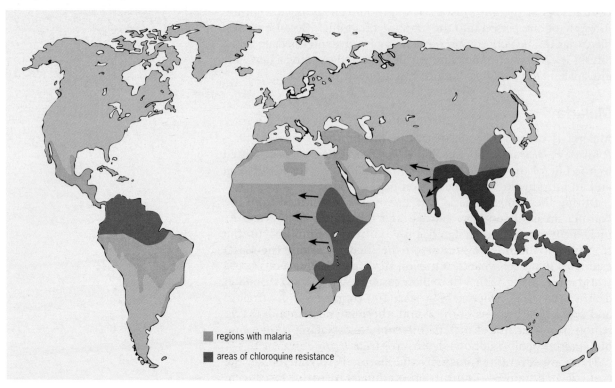

regions with malaria

areas of chloroquine resistance

Anti-malaria drugs

The anti-fever properties of the bitter bark of the tree *Cinchona ledgeriana* were known in Peru before the fifteenth century. **Quinine**, the active ingredient in the bark, was first isolated in 1820. **Chloroquine** is now the most commonly used antimalaria drug in Africa. It lowers fever and reduces the number of parasites in the body. Resistance to chloroquine, however, is widespread and has been rising steadily since the early 1980s (Fig 33.20). For this reason, it is used in combination with other drugs. Research continues into resistance-reversing drugs, but so far no useful drugs have emerged.

Drugs based on artemisinin are important treatments for severe malaria. Both are derived from the Chinese plant *Artemesia annua* (or quinghaosu). They are effective, fast-acting, but relatively expensive drugs.

Artemether or quinine are the antimalarial drugs most suitable for the treatment of severe malaria in children.

An infusion of qinghaosu (*Artemesia annua*) has been used for at least the last 2000 years in China. Its active ingredient artemisinin has only recently been scientifically identified.

■ See question 6.

WILL THERE EVER BE A VACCINE FOR MALARIA?

DESPITE DECADES of research effort, no one has yet managed to develop a vaccine that is effective against a protoctist such as *Plasmodium*. Several potential malarial vaccines have been developed, but many have run into problems. The vaccine developed by Manuel Patarroyo in Colombia has been trialled in people and seemed to be the answer everyone was waiting for. However, further trials showed that the vaccine was not protective, and the arguments about it have raged ever since. (See the Assignment on page 523.)

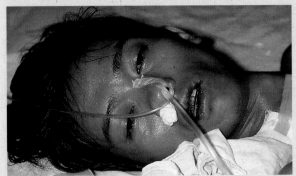

Fig 33.21 **An effective malarial vaccine would cut the misery of malaria infection for millions of people**

Currently, molecular biologists are using two main tactics to find a suitable vaccine. Some are trying to identify antigens that appear on the parasite's surface at various stages of its life cycle. The idea is to put several *Plasmodium* genes into the harmless *Vaccinia* virus, and then introduce it as a **recombinant vaccine** into the body. Once inside, it divides a few times and sets up a good immune response to *Plasmodium*. Some of these recombinant vaccines look promising and may be trialled in humans very soon.

The other approach is to develop a DNA vaccine for malaria. The aim is to isolate the *Plasmodium* genes that provoke an immune response during infection, and to put them into a plasmid. This is a ring of DNA obtained from bacteria. The DNA itself is introduced into the body where it uses the human cell machinery to make the proteins coded for in the genes. The person's immune system detects these proteins, and makes an immune response. It is this response that could combat the real infection if the person is later exposed to *Plasmodium*. The first early trials of a malarial DNA vaccine were announced in 1998, but, even if things go well from now on, it could be at least five years before we will know if they are going to work.

Schistosomiasis

Schistosomiasis, also known as **bilharzia**, results from infection by one of three species of a blood fluke called *Schistosoma*. The three species are: *Schistosoma mansoni* (Africa and South America), *Schistosoma haematobium* (Africa and the Middle East), and *Schistosoma japonicum* (Asia). Like many other parasitic worms, these flukes have a very complex life cycle (Fig 33.22).

Schistosomiasis is generally believed to rank second after malaria as the most important parasitic disease in the world. A total of 500–600 million people are at risk of infection and more than 200 million people in 74 countries succumb to the disease. Of these, 800 000 die each year and 120 million others have severe symptoms.

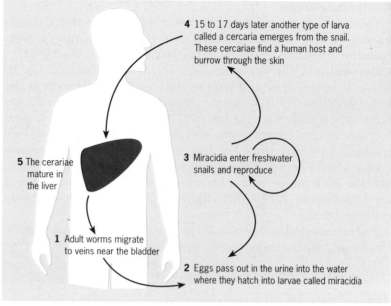

4 15 to 17 days later another type of larva called a cercaria emerges from the snail. These cercariae find a human host and burrow through the skin

5 The cerariae mature in the liver

3 Miracidia enter freshwater snails and reproduce

1 Adult worms migrate to veins near the bladder

2 Eggs pass out in the urine into the water where they hatch into larvae called miracidia

Fig 33.22 **Schistosomiasis and its cycle of infection. Part of the life cycle takes place in humans and part in water snails. Unfortunately the snails are very difficult to kill. In the photo the male schistosome is blue and the female is white.**

Fig 33.23 **(Left) The best way to control the disease is therefore for people to avoid the freshwater areas that form the snail's breeding ground. This is good in theory, but if all supplies of freshwater are affected, where can the local people bathe and wash clothes? Until the poverty and poor living conditions of many of the areas where schisosomiasis is rife are tackled, people have no choice but to expose themselves to repeated infection**

Signs and symptoms

The body responds to an infection with schistosomes by trying to 'wall off' the eggs that are produced by the worms within the veins. This scar tissue can cause the liver to enlarge or can lead to blockages in the bladder and kidney. The severity of the damage is directly related to the numbers of parasites in the body. Since the parasite does not reproduce inside the human host, clinical signs do not appear until the number of worms in the body reaches a critical level.

Treatment and control

The main drug used to treat schistosomiasis is highly effective and has few side effects. However, it does not prevent reinfection – only continued, repeated treatments will keep the disease under control – and it is expensive. This creates an impossible situation because the people worst affected by the disease are the ones least able to afford treatment. Increased use of drugs also increases the likelihood that drug resistant strains will evolve.

Tackling the snail that carries the schistosome is also difficult. The chemicals used to kill the snails are thought to be harmful to the environment. Furthermore, the increased use of irrigation world-wide has aided the spread of schistosomiasis by expanding the habitat of the freshwater snail (Fig 33.23).

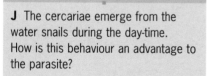

J The cercariae emerge from the water snails during the day-time. How is this behaviour an advantage to the parasite?

See questions 1 and 2. ■

How parasites are adapted for their lifestyle

The parasites that cause malaria and schistosomiasis illustrate how parasites become very specialised as they have adapted to living in the blood system of their human host. The two parasites and their adaptations are compared in Table 33.3.

Table 33.3 **Adaptations to a parasitic lifestyle:** *Plasmodium* and *Schistosoma* compared

Adaptation	*Plasmodium*	*Schistosoma*
Attaching to the host	The malarial parasite spends much of its adult life inside the cells of its host. Attachment is not a problem.	Adult worms live in the veins around the bladder. Because they are small they could easily be washed away by the blood. Suckers help them attach to the walls of the veins.
Resisting host defence mechanisms	The blood is a dangerous place for any parasite because of the white blood cells of the immune system (see Chapter 34). Within an hour of entering the body, malarial parasites are inside liver cells. Here they hide from the immune system. The antigens on the surface of a malarial parasite change frequently and the immune system cannot keep up.	Adult worms have a remarkable way of avoiding the host's immune system: they disguise themselves by covering up with molecules from the host's red blood cells.
Reproduction and reproductive organs	There are many reproductive stages in the life cycle of the malarial parasite. The parasites reproduce in the mosquito, but this insect is small compared to a human. Once in their main host, malaria parasites also reproduce in the liver and in the blood cells, causing a rapid increase in parasitic load.	The life cycle of this worm depends very heavily on chance. A person must urinate into fresh water; the miracidia larvae that hatch from the eggs must find a snail host and the cercariae must encounter a human shortly after emerging from the snail. The production of large numbers of offspring is therefore the best way of ensuring survival. Each egg laid by an adult worm can produce up to 200 000 cercariae.
The life cycle	Usually, blood from one person does not mix with that of someone else. So malaria parasites cannot depend on person-to-person contact for transmission. A second host is used. In malaria, the female *Anopheles* mosquito is the organism. When she bites someone with malaria, she takes in infected red blood cells. These parasites complete their life cycle in her body, finally migrating to her salivary glands. When she bites someone else, she injects a small amount of infected saliva, and malaria is passed on.	Part of the life cycle of *Schistosoma* is spent inside its human host; part is spent inside a freshwater snail. In many parts of the tropics where schistosomiasis is endemic, humans spend much of their time around areas of fresh water. They use it for washing clothes, irrigating crops, watering animals and bathing. This increases the chances of the parasite being passed from one person to another.
Reduction of body systems	The malaria parasite spends most of its life inside one of its host's red blood cells. It does not have to move to find food, and the medium in which it lives has the same water potential as its own cytoplasm. It has therefore lost the capacity to locomote and to regulate cell water content – features of many free-living single-celled organisms.	Schistosomes do not have a complex nervous system because conditions they find in the blood are remarkably constant. They cannot move either: a locomotory system is also unnecessary since the parasites' survival depends on it being anchored in the host.

?

K What is the advantage to the malarial parasite in living in a medium that has the same water potential as its own cytoplasm?

SUMMARY

After reading this chapter you should know and understand the following:

■ Robert Koch produced a series of principles that enabled biologists to identify the microorganism that causes a specific disease. These principles are known as **Koch's postulates**.

■ In order to cause disease, a pathogen must gain access to the body. This involves penetrating the body's defences.

■ **Transmission** of pathogens can take place through the air, in contaminated food or water, by sexual contact or through the skin.

■ For infection to cause disease, the pathogen must multiply inside the host. Disease can result from **direct damage** to the cells and tissues of the host or from indirect damage cause by **toxins** produced by the infecting microorganism.

■ Some species of bacteria and viruses are pathogenic. So are some species of larger organisms such as protoctists and worms. These are **parasites** and they show many **adaptations** to their way of life.

QUESTIONS

1 *Schistosoma mansoni* is a platyhelminth parasite. Part of its life cycle is spent in a human host and part in a freshwater snail. The life cycle is summarised in Fig 33.Q1.

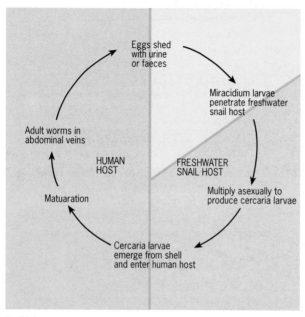

Fig 33.Q1

Suggest how each of the following features is an adaptation to the animal's way of life.

a) Adult worms respire anaerobically.

b) Hatching of the eggs is stimulated by a less negative solute potential.

c) Cercaria larvae show positive thermotaxis.

[AEB June 1992, Biology: Biology Paper 1, q.4]

2 Tapeworms are platyhelminth parasites which are swallowed by their hosts as cysts. Under appropriate conditions the cysts hatch into mature tapeworms. The graph of Fig 33.Q2 shows the effects of various treatments on the hatching of tapeworm cysts.

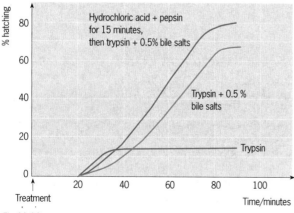

Fig 33.Q2

a) Suggest one control that should have been set up and explain why it was necessary.

b) **(i)** In which part of the gut would you expect this tapeworm to hatch from its cyst?
   (ii) Give evidence from the graph to support your answer.

[AEB Nov 1991, Biology Paper 2, q.11]

3

a) Influenza starts with a tickling in the throat, watery eyes and the occasional sneeze. Give **three** symptoms of influenza which may develop over the next few days.

b) Explain how a knowledge of its method of transmission can be used to reduce the incidence of influenza.

c) The drawings of Fig 33.3 show changes to the protein coat in the descendants of one strain of the influenza virus between 1968 and 1985. The shaded areas are 'spikes' on the protein coat of the virus.

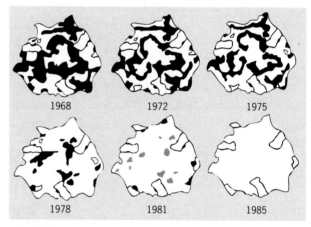

Fig 33.Q3

   (i) Explain briefly how these changes to the protein coat of the virus have occurred.
   (ii) Using information from the drawings and your own knowledge of the workings of the immune system, suggest why some people caught this strain of the influenza more than once between 1972 and 1985.

[NEAB February 1996 Biology: Health and Disease, q.3]

4 One species of *Salmonella* causes food poisoning by producing toxins in the human intestine. Symptoms, such as vomiting and diarrhoea, usually occur within 24 hours of eating contaminated food. Another species of *Salmonella* causes typhoid fever. The symptoms of typhoid fever do not develop for 10 to 14 days.

a) Suggest why the interval between infection and symptoms appearing is much longer for typhoid fever than for salmonella food poisoning.

b) Suggest why antibiotics are used to treat typhoid fever, but are not normally used to treat salmonella food poisoning.

c) Describe how cooked chicken may be a cause of salmonella food poisoning when served cold as part of a buffet meal.

[NEAB June 1998 Biology: Health and Disease, q.6]

5 Fig 33.Q5 shows the structure of a human immunodeficiency virus (HIV).

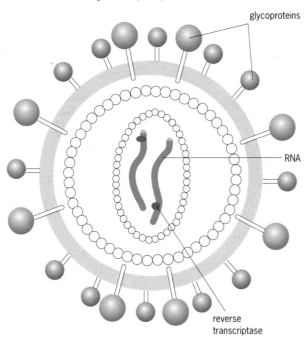

Fig 33.Q5

a) The virus attaches tot he cell surface membrane of a helper T-lymphocyte. The viral RNA and the reverse transcriptase then enter the lymphocyte. What is the function of each of the following: **(i)** the glycoproteins; **(ii)** the reverse transcriptase?

b) One form of pneumonia is normally very rare. It is much more common in people who are infected with HIV. Explain why.

c) Viruses such as the influenza virus and HIV have very high mutation rates. Suggest why this makes it difficult to produce a vaccine against them.

[AEB Human Biology 2000 Specimen Paper: Module 3, q.3]

6 Fig 33.Q6 shows the life cycle of the malarial parasite, *Plasmodium falciparum*.

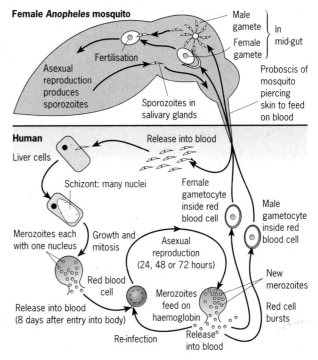

Fig 33.Q6

a) **(i)** What part does the female mosquito play in the transmission of the parasite?
(ii) From the diagram, what is the minimum time which must elapse after injection before a human could infect another mosquito?

b) Suggest one stage in the life cycle when the parasite would be vulnerable to attach by human antibodies. Give a reason for your answer.

c) The drugs quinine and chloroquine have been used to treat malaria by killing the parasites in the red blood cells. These drugs are now less effective than they used to be. Suggest why.

d) Recently, a drug called artemisinin has been developed which has proved effective against the malarial parasite. This drug reacts with iron to release a highly reactive iron oxide which is extremely toxic. Suggest why artemisinin is effective against the malarial parasite.

[AEB 1997 Human Biology Paper 1, q.13]

Assignment

ROUNDWORMS

Ascaris lumbricoides is a large roundworm that lives as a parasite in the human intestine. A few roundworms in the gut of an otherwise healthy person have relatively little effect. But if that person is already suffering from malnutrition, it's a different story. The parasite absorbs most of the little food that is eaten, making the malnutrition much worse.

Roundworms also cause problems when they are present in large numbers. The mass of worms in the intestine can block the passage of waste through the colon, causing severe pain. If the worm infection is not treated, the intestine can rupture and the affected person is then seriously ill and can die without emergency medical attention.

Female worms lay eggs that can contaminate food or that can be spread by hand–mouth contact. Adult male worms have an average length of 15 cm, while females can be half as long again.

Fig 33.A1 shows an adult female roundworm and a cross-section through its body.

Fig 33.A1 **When a roundworm infection is treated, many worms can be expelled. This photograph shows the number of worms present in a person with a relatively light infestation**

1 Look at Figs 33.A1 and 33.A2 and suggest **two** features of *Ascaris lumbricoides* that represent adaptations to its life as a parasite in the human intestine.

Ascaris lumbricoides is very common in Nigeria, a large country with a population of over 150 million. Many of them live in rural communities.

We cannot estimate exactly what proportion of the population is infected with roundworms: different studies have produced wildly different figures that range from 10.8 per cent to 96.3 per cent.

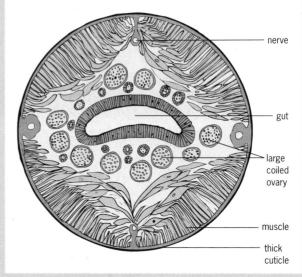

Adult female

Cross section

nerve

gut

large coiled ovary

muscle

thick cuticle

Fig 33.A1 *Ascaris lumbricoides*

2 Suggest why it is difficult to estimate accurately the proportion of the population affected by roundworms.

The studies that have been carried out suggest that the proportion of people affected differs according to the group of people that you are studying. Table 33.A1 shows the results of five different surveys.

Table 33.A1

Place	Urban/rural community	Percentage of sample infected with roundworms
Ibadan	urban	48.1
Ewekoro	urban	52
Benin City	urban	19.5
Oyo State villages	rural	71.4
Lagos State villages	rural	74.2

3

a) Why is it not possible to compare the urban and rural results shown in Table 33.A1 by taking averages for the urban and rural studies?

b) Suggest why the percentages of infected people in the urban and the rural areas are different.

Table 33.A2 shows the relationship between the father's occupation and the percentage of children infected with roundworms.

Table 33.A2

Father's occupation	Percentage of children infected with roundworms
professional	9.5
clerical officer	26.7
semi-skilled worker	20.2
trader	32.1
unemployed	51.4

4 Explain the relationship between the fathers' occupation and the percentage of children who are infected with roundworms.

A study was carried out to see how effective treatment for roundworm infections was in primary schoolchildren. Each child was treated three times and, after each treatment, the total number of worms passed out in the faeces was counted (Fig 33.A3).

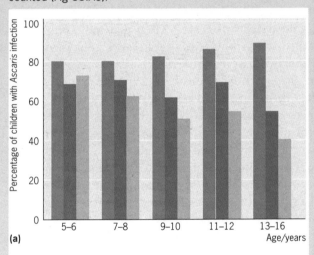

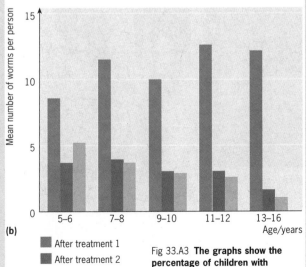

After treatment 1
After treatment 2
After treatment 3

Fig 33.A3 **The graphs show the percentage of children with** *Ascaris lumbricoides*

5

a) Describe the effects of treatment on the children in the 5–6 age group.

b) How does age affect the success of the treatment?

Elephantiasis

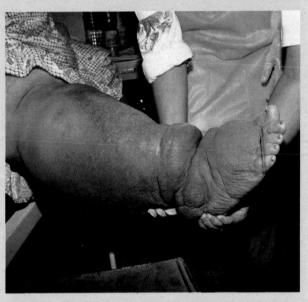

Fig 33.A4 **The swelling disease elephantiasis is caused by an accumulation of fluid due to an infection of the nematode worm** *Wuchereria bancrofti*. **The worm blocks lymph vessels and so stops lymph draining away**

The effects of the swelling disease elephantiasis is shown in Fig 33.A4. The condition is caused by the nematode worm *Wuchereria bancrofti*. The worm blocks lymph vessels. Infection usually sets in and fluid accumulates until the affected limb is several times its normal size.

6

a) Explain why blockages in the lymph vessels of the legs would cause fluid to build up.

b) How do you think the work that causes elephantiasis gets into its human host?

For research and discussion

7 Several other worms cause disease in the developing world. Schistosomes and liver flukes are just two others. Find out how many people in the world are affected by these four parasites. Why are they common in the developing world but virtually unheard of in a temperate developed country like the UK? Are there parasitic worms that cause problems in more affluent areas of the world?

IN 1995, A MAJOR MEDICAL JOURNAL reported the tragic case of a teenage girl who died within 20 minutes of taking a health supplement containing 'royal jelly', the substance made by queen bees. Sudden death in young people usually makes headlines, particularly when it seems to be related to a trivial event such as taking a normally harmless tablet. But similar deaths have also occurred as a result of a bee sting.

The victims all have in common a very rare and severe allergy. Just as some of us suffer from hay fever in the summer because we overreact to pollen in the air, some people overreact to the 'foreign' proteins produced by bees.

Normally, when someone is stung by a bee, the result is a painful red swelling which disappears again in a couple of days. In someone who is allergic to bee protein, the effects can be catastrophic. As soon as the poison gets into their bloodstream, many of their white cells release large amounts of histamine and other chemicals which cause severe inflammation. Fluid builds up in the tissues and all of the smooth muscles contract. The whole body is affected and both blood volume and blood pressure quickly plummet. Sometimes, the airways in the lungs narrow so much that the affected person can no longer breathe. These symptoms, collectively known as anaphylactic shock, can happen very quickly and can be fatal unless immediate treatment with adrenaline and antihistamines is available.

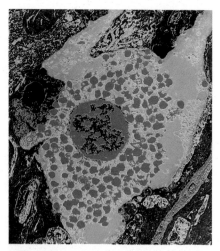

A false colour electron micrograph of a mast cell, a type of white cell involved in allergic reactions. The small red vesicles contain histamine, the chemical which produces many of the symptoms of allergy. Many people take antihistamines to dampen down their allergy to such things as pollen, house dust or animals

Fig 34.1(a) **Below: this child has smallpox, a viral disease. Once a widespread killer disease, the last recorded case of smallpox occurred in October 1977. In 1980 the World Health Organisation declared that the disease had officially been eradicated – stamped out completely. This is the first time humans have managed to do this**

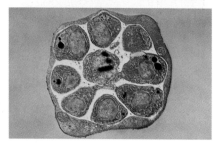

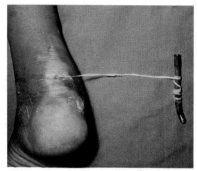

Fig 34.1(b) **Above: a guinea worm being removed from an ankle joint. This parasitic worm can reach a length of 120 cm (4 feet) and is caught by bathing in infected water**

Fig 34.1(c) **Left: a single-celled protoctist called *Plasmodium* causes malaria. After being injected into the bloodstream by mosquito bite, the parasites invade red blood cells where they multiply, hidden from the host's immune system. Nine malaria parasites are seen in this red blood cell**

Fig 34 1(d) **This man has leprosy, a bacterial disease that affects the skin, nerves and mucous membranes. With modern treatment, the deformities seen here should not develop**

1 INTRODUCING IMMUNITY

We are all surrounded by bacteria, viruses, fungi and other organisms that are capable of invading our bodies and causing disease (Fig 34.1). We are able to overcome infections by these **pathogens** (disease-causing organisms) because we have an **immune system**. This is a complex system involving many different cells and tissues which allows us to develop **immunity** – resistance to infections.

Common pathogens of humans include bacteria, fungi, viruses, protoctists, such as *Plasmodium* which causes malaria, and parasitic animals such as schistosomes that cause bilharzia (see page 501). A pathogenic organism is able to:

- breach the physical barriers of the body and enter tissues or cells;
- resist the efforts of the immune system to destroy it, long enough to multiply inside the host's body;
- get out of one host and into another;
- damage the host's tissues – either directly, or indirectly by means of **toxins** (poisons) that it releases. Some bacterial **exotoxins**, such as the one produced by *E. coli* 0507 (which caused an epidemic of gastroenteritis in Scotland in 1996), are very powerful and can be fatal.

The easiest way to start understanding the immune system is to look at its overall functions, rather than concentrating on the individual parts. Fig 34.2 summarises the main lines of defence that an organism comes up against when it tries to infect a healthy person. We look at each of these in more detail in the sections that follow.

A What conditions inside our bodies make it ideal for the growth of microorganisms?

Pathogens are disease-causing organisms. Such diseases are **communicable** or **infectious**: they can pass from one person to another. Other diseases, such as diabetes and cancer are **non-communicable**.

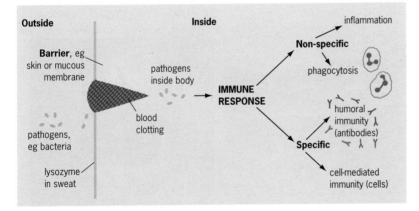

Fig 34.2 **An overview of the body's defences: non-specific responses are general responses to damage. They include inflammation and phagocytosis of debris. Specific responses are targeted against individual types of microorganism**

Our study of the immune system covers:

- The ability of the immune system to distinguish between invading organisms and the tissues of its own body. This concept of **self** and **non-self** is important in determining whether the immune system keeps the body healthy, or whether it is overcome by infection.
- Problems of the immune system. In the Feature box on page 513, we see how the body can react against its own tissues, causing **autoimmune disease**.
- How medicine today can manipulate the immune system to provide life-saving treatments such as vaccines, blood transfusions, organ and tissue transplants and specific cancer therapy.

2 THE BODY'S BARRIERS TO INFECTION

One of the most obvious ways to avoid infection is to stop potential pathogens getting into the body in the first place. The four main strategies that the body uses are shown in Fig 34.3 and are summarised below:

● **Mechanical** defence. **Nasal hairs** filter the air that is drawn into the upper airways. **Cilia**, which line the airways, sweep bacteria and other particles away from the lungs (Fig 34.3(a)).

● **Physical** defence. The skin (Fig 34.3(b)), made from **stratified squamous epithelium** (see page 28 in Chapter 2), forms a tough, impermeable barrier which normally keeps out bacteria and viruses. The mucous membranes that line the entry points to the body such as the nose, eyes (Fig 34.3(c)), mouth, airways, genital openings and anus produce fluids (see below) and/or sticky mucus. These fluids trap microorganisms and stop them attacking the cells underneath.

● **Chemical** defence. Fluids such as sweat, saliva and tears contain chemicals that create harsh environments for microorganisms. Sweat contains lactic acid and the enzyme lysozyme, both of which slow down bacterial growth. Stomach acid kills many microorganisms that manage to get that far. When we are injured, blood clots at the injury site, sealing the breach to prevent entry of bacteria (see the next section).

● **Biological** defence. Normally, a vast number of non-pathogenic bacteria live on the skin and mucous membranes. These do not harm the body but they out-compete pathogenic bacteria (Fig 34.3(d)), preventing them from gaining a foothold from which to launch a full-scale infection.

Fig 34.3 **Some of the body's barriers to infection**

Fig 34.3(a) **Light micrograph of a section through the human trachea. Household dust contains human skin cells, pollen, bacteria and fungal spores. Cilia in the respiratory system are able to filter out much of this before it reaches the alveoli, sweeping it upwards to minimise the chances of infection**

Fig 34.3(b) **The outer layer of (red) skin consists of flat, dead cells that consist mainly of the protein keratin. This forms a tough barrier which is impermeable to micro-organisms**

Fig 34.3(c) **You can see the pink mucous membranes of the eye when the eyelid is pulled down. These membranes also line the intestines and the respiratory, reproductive and urinary tracts. Cells and glands at these sites produce secretions that contain chemicals which kill bacteria**

Fig 34.3(d) **The skin and mucous membranes are covered in microorganisms such as these hyphae of _Trichophyton_, a fungus which causes athlete's foot. Normally there is a balance of harmless organisms which keep the pathogenic ones in check**

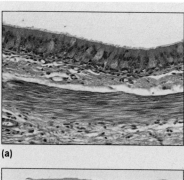
(a)

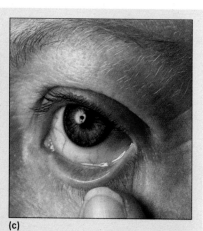
(c)

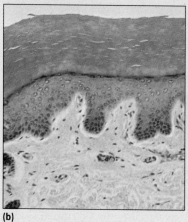
(b)

(d)

3 BLOOD CELLS AND DEFENCE AGAINST DISEASE

Blood clotting

Whenever blood vessels break, blood leaks out and clots (Figs 34.4 to 34.6). We can see this happening when we cut ourselves, but blood can also clot deep inside the body. Clotting enables the body to avoid blood loss and, at the surface, to prevent infection. It is important that blood clots only when it should, because when a blood clot, a **thrombus**, blocks a vital blood vessel, the consequences can be fatal (Chapter 35).

Fig 34.5 outlines the major steps involved in the control of blood clotting. When blood vessels are injured or burst, a **cascade reaction** is initiated. The activation of one molecule leads to the activation of many more. In the final steps of the process, an inactive enzyme, prothrombin, is converted to active thrombin. This form of the enzyme converts soluble fibrin into insoluble fibrin. The fibrin fibres form a mesh that traps red blood cells and forms a clot (Fig 34.6).

If any of the factors in the cascade are missing, the blood cannot clot. This occurs in the condition haemophilia, a sex-linked genetic disease in which the sufferer (usually a male) cannot make factor VIII. The genetics of haemophilia are discussed in Chapter 30.

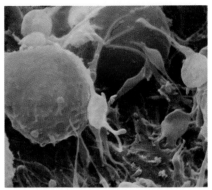

Fig 34.4 **This photograph shows platelets (yellow) surrounding a red blood cell and a small white cell (mauve). Platelets are vital to the blood clotting process. Each one is packed with the enzymes and chemicals needed for the cascade of reactions that results in a clot. When they come into contact with damaged blood vessels, platelets become sticky, attracting and activating other platelets**

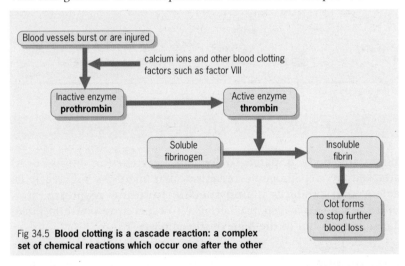

Fig 34.5 **Blood clotting is a cascade reaction: a complex set of chemical reactions which occur one after the other**

The process of **thrombosis** causes blood to clot inside vessels. The clot itself – a **thrombus** – can be fatal if it prevents blood from reaching vital tissues such as heart muscle. A **coronary thrombosis** is a common cause of heart attack while a thrombosis in the brain can cause a **stroke**.

C The saliva of several species of leech contains an anticoagulant. What are anticoagulants and why does the leech need them? Could 'leech spit' have any medical uses?

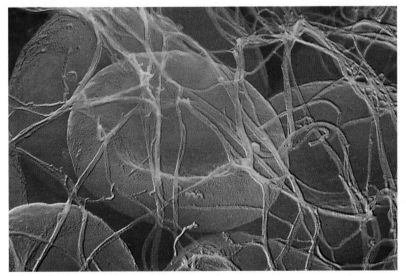

Fig 34.6 **A scanning electron micrograph of a blood clot. You can see red blood cells and platelets trapped in a fine mesh of fibrin. Blood clotting is a complex process that is set off when blood vessel walls are damaged and finishes when soluble fibrinogen is converted into an insoluble fibrin mesh**

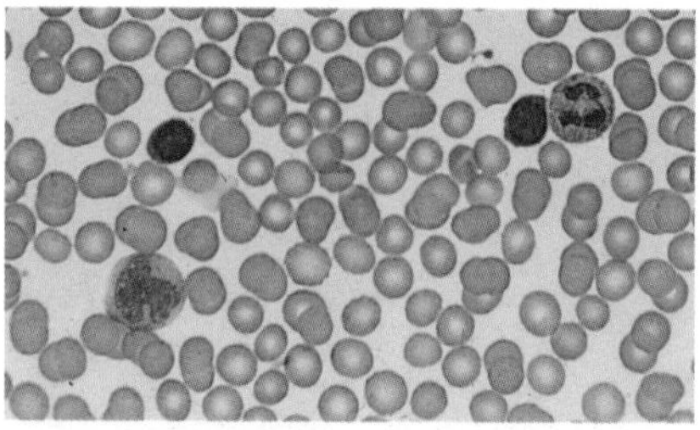

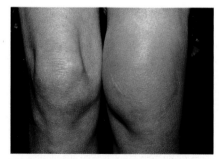

Fig 34.7 **In this blood smear, the white cells stand out because their nuclei are stained. The numerous red cells, which have no nuclei, do not take up the purple stain**

?

D Use Table 34.1 to identify the white cells shown in Fig 34.6.

✔

White cells spend only about 10 per cent of their time in the blood, and so cannot really be called white *blood* cells. Leucocyte is the general name for all white cells.

Classification of white cells

Fig 34.7 is a blood smear showing some white cells – **leucocytes** – along with many red blood cells. (The red cells outnumber the white by about 700 to 1). White cells are made in the bone marrow and are found throughout the body. They can move and are able to squeeze between cells, passing freely in and out of the circulation. Individual types of white cell are classified according to their appearance, origin or function (Table 34.1).

Table 34.1 **Different types of white cell. Neutrophils, eosinophils and basophils have a granular cytoplasm and are therefore called granulocytes. The other types are called agranulocytes**

Cell type	Diagram	How to recognise them	Relative abundance/ %	Function
Neutrophils		Lobed nucleus	57	Phagocytosis
Lymphocytes		Large round nucleus; little cytoplasm	33	Specific immunity. B-cells make antibodies. T cells involved in cell-mediated immunity
Monocytes		Large, kidney-shaped nucleus	6	Phagocytosis. Monocytes develop into macrophages: general 'rubbish collecting' cells
Eosinophils		Stain red with eosin	3.5	Associated with allergy
Basophils	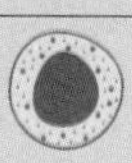	Stain with basic dye	0.5	Release chemicals such as histamine that are responsible for inflammation

4 THE NON-SPECIFIC IMMUNE RESPONSE

When the body is damaged by cuts, scratches or burns, or is attacked by a pathogenic organism that manages to breach its defences, it produces a **non-specific immune response**. It is called a *non-specific* response because it occurs in response to tissue damage itself, not to the *cause* of the damage.

Inflammation

Inflammation is a rapid reaction to tissue damage. Whether it is in response to a cut, insect bite or a heavy blow such as a sport injury, the classic signs of inflammation are always the same:

● **Redness**: blood vessels dilate, increasing blood flow to the area.

● **Heat**: also caused by the extra blood flow.

● **Swelling**: the extra blood forces more tissue fluid into damaged tissues.

● **Pain**: swollen tissues press on receptors and nerves. Also, chemicals produced by cells in the area stimulate the nerves.

Inflammation (Fig 34.8) is triggered by damaged cells. Ruptured cells and some white cells (**mast cells** and **basophils**) release 'alarm' chemicals such as **histamine**. These substances dilate blood vessels and the increased blood flow leads to the classic signs of inflammation. The 'alarm' chemicals also attract white cells that remove bacteria and debris by **phagocytosis**.

Fig 34.8 **The familiar effects of inflammation: reddened skin, swollen tissues and tenderness**

Why inflammation is useful

Inflammation prevents the spread of infection and speeds up the healing process. It also provides a way of telling the rest of the immune system what is going on. When microorganisms are phagocytosed, fragments of their cells (particularly molecules that were originally on their surface) are processed by the phagocytes. Some of these surface molecules, which we call **antigens**, allow the specific immune system to recognise and remember the type of microorganism that has tried to invade the body.

Antigens stimulate the specific immune system to produce cells and chemicals that bind specifically to that antigen, and to no others. Find out more about specific immunity in Section 5.

> An **antigen** is a molecule or part of a molecule, for example, a protein, which is detected by the specific immune system as 'foreign': not part of the host's body. The immune system responds to an antigen by producing a very specific protein, an **antibody**, which reacts with that antigen, and that antigen only.

Phagocytosis

Neutrophils are the commonest type of white cell. Together with monocytes, they are known as **phagocytes** because of their ability to 'eat' pathogens by phagocytosis (Fig 34.9). In this process, the white cell engulfs the pathogen, takes it into a vacuole inside its cytoplasm and then digests it with **lytic** enzymes (see Chapter 5).

Fig 34.9 **This phagocyte (a neutrophil) is engulfing a speck of dust. These white cells occur in the bloodstream and throughout the tissues. Each phagocyte (which is spherical when not active) can ingest between five and 25 bacteria before the toxic products of breakdown kill the cell. At sites of infection there may be many dead, liquefied white cells which, together with dead bacteria and other debris, form** pus

---

AUTOIMMUNE DISEASE: WHEN THE BODY ATTACKS ITS OWN TISSUES

IN PEOPLE who have autoimmune disease, the mechanism that enables the immune system to tell what is *self* and what is *non-self* breaks down. T and B cells (types of white cell) begin to attack the body's own cells and tissues. This is the underlying cause of **multiple sclerosis, insulin dependent diabetes, myasthenia gravis** and **rheumatoid arthritis**.

In multiple sclerosis, T cells attack the myelin sheath around nerves. This severely limits nerve function, resulting in loss of movement and sometimes blindness.

In insulin dependent diabetes, the body makes anitbodies that destroy the β cells in the islets of Langerhans in the pancreas. This means that the pancreas becomes unable to produce insulin and the affected person can no longer control their blood glucose.

In myasthenia gravis, the body makes antibodies that attack the motor end plates, the specialised synapses that connect motor nerves to muscles. If the motor end plates become damaged, the muscles cannot contract. The first symptom of this disease, which affects one in 30 000 people (mainly female), is rapid tiring during exertion. The muscles become progressively more unresponsive and the affected person can have difficulty breathing.

People with rheumatoid arthritis suffer from swollen and deformed joints (Fig 34.10).

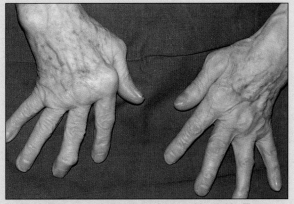

Fig 34.10 **This person suffers from rheumatoid arthritis. The cartilage at the joints has been attacked by the immune system, causing swollen and painful joints**

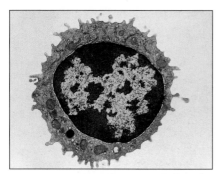

Fig 34.11 **A T lymphocyte. T and B cells are identical: they have a characteristically large round nucleus and relatively little cytoplasm. The difference between them is their function: B cells secrete antibodies while T cells are involved in cell mediated immunity. They are also produced by different parts of the body: B cells are produced by the bone marrow, T cells mature in the thymus gland**

5 SPECIFIC IMMUNITY

The specific immune system protects the body from 'invasion' by microorganisms and parasites and also makes sure that the body's defences do not turn on its own tissues. The specific immune response is made up from two different systems that cooperate closely:

● **Humoral immunity**, also called antibody-mediated immunity, involves only chemicals: no cells are directly involved. The chemicals, called antibodies, attack bacteria and viruses before they get inside body cells. They also react with toxins and other soluble 'foreign' proteins. Antibodies are produced by white cells called **B lymphocytes**, or **B cells**.

● **Cell-mediated immunity**, as the name suggests, involves cells which attack 'foreign' organisms directly. Activated **T lymphocytes**, or **T cells** (Fig 34.11), kill some microorganisms, but they mostly attack infected body cells. Cell-mediated immunity is used by the body to deal with multicellular parasites, fungi, cancer cells and rather unhelpfully, tissue transplants.

Fig 34.12 summarises the differences between these two systems and Fig 34.13 shows how antigens from pathogens stimulate antibody production by B cells.

Fig 34.12 **Lymphocytes are made in the bone marrow by stem cells. Two distinct types of lymphocytes develop, T-lymphocytes and B-lymphocytes (T cells and B cells). Both T- and B cells migrate from the bone marrow to the lymph glands, but the T cells go via the Thymus while the B cells go straight from the Bone marrow**

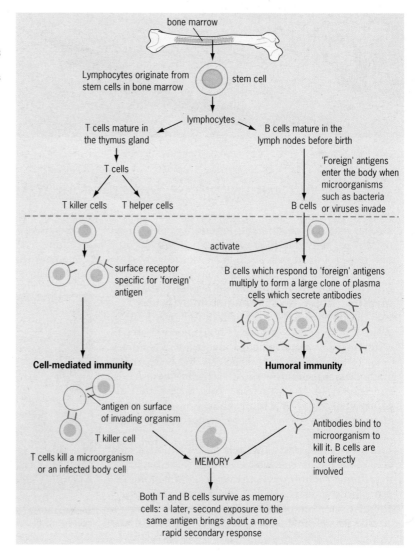

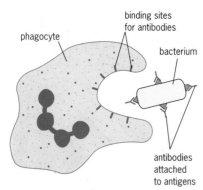

Fig 34.13 **Pathogens such as bacteria are covered in antigens which stimulate B cells to produce antibodies. Many antibodies coat the pathogen, labelling it as foreign to stimulate attack by phagocytes**

B cells and humoral immunity

B cells are produced in the **bone marrow** and are distributed throughout the body in the **lymph nodes**. B cells respond to the 'foreign' antigens of a pathogen by producing specific antibodies. Antibodies are complex proteins that are released into the blood and carried to the site of infection. B cells do not fight pathogens directly.

An antibody, or **immunoglobulin**, is a Y-shaped protein molecule that is made by a B-lymphocyte in response to a particular antigen. Fig 34.14 shows the overall structure of an antibody molecule. Antibodies interact with the antigen and render it harmless.

Fig 34.14(a) **Antibodies are Y-shaped molecules consisting of four polypeptide chains, two heavy and two light. Much of the molecule is constant but the tips of the Y are variable and match precisely part of a particular antigen molecule**

Fig 34.14(b) **A computer generated 3-D image of an antibody molecule (green) bound to an antigen (red)**

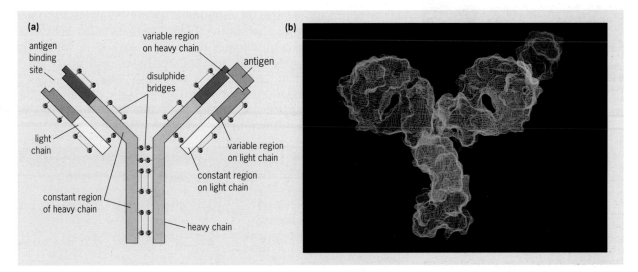

When a pathogen tries to invade the body for the first time, each of its antigens activates one B cell, which divides rapidly to produce a large population of cells. All the new cells are identical (we say they are **clones**) and they all secrete antibodies specific for the invading pathogen. When the infection is over, most of the newly made B cells die: their job is done. We describe this sequence of events as a **primary immune response.**

So that the body can respond more quickly next time, some of the activated B cells persist in the body for several years. These **memory cells** 'remember' what the pathogen is like and, if it tries to invade again, they all divide rapidly to produce an even greater number of active B cells, all capable of secreting specific antibody. This response is called a **secondary immune response** and is very much quicker and more effective than the primary response (Fig 34.15).

> A clone is a set of genetically identical individuals. In immunology, a clone refers to a population of B cells, all of which can produce the same antibody because they are genetically identical.

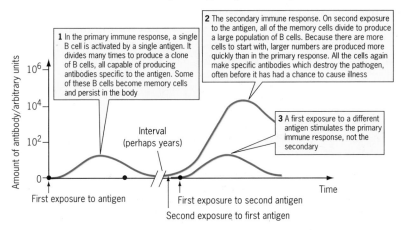

Fig 34.15 **The immune system can remember antigens**

1 In the primary immune response, a single B cell is activated by a single antigen. It divides many times to produce a clone of B cells, all capable of producing antibodies specific to the antigen. Some of these B cells become memory cells and persist in the body

2 The secondary immune response. On second exposure to the antigen, all of the memory cells divide to produce a large population of B cells. Because there are more cells to start with, larger numbers are produced more quickly than in the primary response. All the cells again make specific antibodies which destroy the pathogen, often before it has had a chance to cause illness

3 A first exposure to a different antigen stimulates the primary immune response, not the secondary

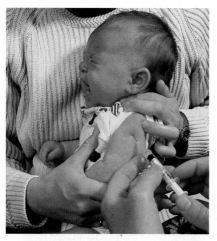

Fig 34.16 **Vaccinations are an effective way of stimulating the body's own defences so that we need not suffer the infectious diseases, such as measles, mumps and whooping cough, that used to be a common feature of childhood**

This ability of the immune system is central to **vaccination** (Fig 34.16). A vaccine stimulates the body to produce a primary immune response to a particular pathogen, without becoming infected by it. A subsequent booster produces a secondary response. Later, if the pathogen tries to invade, the body can mount a very fast response and we can avoid becoming ill.

Vaccinations

Several infectious diseases overwhelm the normal primary immune response and so can be fatal on first exposure. Thankfully we are able to speed up the specific immune response by giving vaccines against the pathogens that cause them. The basic idea behind a vaccine is that it contains some form of the pathogen, so that it stimulates memory cells to develop, ready to destroy the real pathogen should it be encountered. Obviously, the vaccine can't simply be the pathogen itself, or the toxins it makes. Somehow, the vaccine must be made less **virulent** – less able to produce disease. Examples of this are shown in Table 34.2.

Table 34.2 **Recommended vaccination schedule for children in the UK**

Vaccination	Type of vaccine	Age due (UK vaccination schedule)
Diphtheria	Killed organism	2, 3 and 4 months, 3 - 5 years
Tetanus	Modified toxin	2, 3 and 4 months, 3 - 5 years
Whooping cough	Killed organism	2, 3 and 4 months, 3 - 5 years
Polio	Live, non-virulent	2, 3 and 4 months, 3 - 5 years
Haemophilus influenzae type B (Hib) for meningitis	Purified bacterial capsule	2, 3 and 4 months, 3 - -5 years
Measles	Live, non-virulent	12–18 months, 3 - 5 years
Mumps	Live, non-virulent	12–18 months, 3 - 5 years
Rubella	Live non-virulent	12–18 months, 3 - 5 years
BCG (bacille Calmette-Guérin – for tuberculosis)	Live, non-virulent	10–14 years
Hepatitis B	Genetically engineered antigens	For people at risk, eg health professionals

See question 3. ■

Fig 34.17 **Babies get immunity from their mothers while their own immune system has a chance to develop. This baby is receiving antibodies in its mother's milk**

How mothers give immunity to their babies

When babies are born, they emerge from the protective environment of their mother's uterus. They are exposed to many potential pathogens. Healthy babies have a fully functional immune system and can mount primary immune responses to many different antigens straight away. However, babies also get a bit of help from their mother, who gives them some of her own immunity.

Antibodies pass across the placenta and, after birth, the supply continues through breast milk (Fig 34.17). Babies have very porous intestines that can absorb these large proteins directly into the bloodstream without digesting them. These large pores close by the age of one year. We call this kind of immunity – passed from one person to another – **passive immunity**. It does not last long, because the antibodies are broken down within a few days, but it can help a baby to fight off common pathogens.

Passive immunity is also used to treat some types of poisoning, such as snake bites. **Antiserum**, blood that contains antibodies specific to a particular snake venom, is produced in horses, purified and then given to people who have been bitten by a snake.

MONOCLONAL ANTIBODIES

FOR SEVERAL YEARS now, scientists have been trying to harness the power of the immune system to provide successful cancer treatments. One line of research has been the development of antibodies that can hunt for cancer cells and help to destroy them, leaving normal cells untouched.

In 1975, Georges Kohler and Cesar Milstein came up with a way of making large quantities of pure antibodies against particular antigens. They called these antibodies **monoclonal antibodies (MABS)**. Fig 34.18 shows how monoclonal antibodies are made.

Monoclonal antibodies are used as tools in different fields of medicine:

- **Medical diagnosis**. MABS can be used to detect cancer cells, pathogens such as viruses, and chemicals such as hormones. The use of MABS is pregnancy tests is described in Chapter 19.

- **Cancer treatment**. MABS can be attached to toxin molecules which kill cancer cells by binding to antigens on their surface. The antigens chosen are not present on normal cells and so these are unaffected by the treatment.

- **Transplant surgery**. MABS directed against the T cells that lead to the rejection of some transplants are a valuable alternative to powerful immuno-suppressive drugs (see page 519).

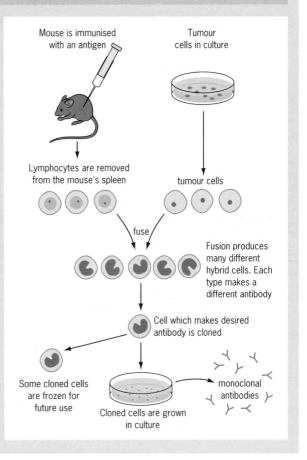

Fig 34.18 **The production of monoclonal antibodies (MABS)**

T cells and cell-mediated immunity

Like B cells, T cells respond to specific antigens. When a pathogen first infects the body, each individual antigen stimulates a single T cell. This divides to form a clone, in the same way that B cells do. Some of the activated T cells become memory cells and persist in the body, ready to mount a secondary response if the pathogen attacks again. The others, however, do not produce antibodies. They develop further to become one of three types of T cell:

- **Helper T cells** are so called because they help with, or rather control the rest of the specific immune response. They tell B cells to divide and then to produce antibodies, they activate the two other sorts of T cell (see below), and they activate macrophages, telling them to get ready to phagocytose pathogens and debris.

- **Killer T cells** attack infected body cells and the cells of some larger pathogens (eg parasites) directly. The two cells face each other, membrane-to-membrane, and the killer T cell punches holes in its opponent. The infected cell, or parasite, loses cytoplasm and dies.

- **Suppressor T cells** are a sort of safety cut-out mechanism. When the immune response becomes excessive, or when the infection has been dealt with successfully, these T cells damp down the immune response. Obviously, this is a good idea: if the body continued to make antibody and stimulate more and more T and B cells to divide, even when there was no need, this could damage the body and would be, at best, a waste of resources.

?

E What do you think it means when products such as make-up are labelled hypo-allergenic?

Fig 34.19 **A doctor performs skin tests to investigate the cause of this child's asthma. A small amount of an allergen such as pollen is placed under the skin with a sterile needle. If the patient responds to one allergen in particular, it is probably this substance that is causing the allergic reaction that produces the asthma symptoms. The child in this photograph is allergic to house dust, as this produces inflammation**

Allergies

An **allergy** or **hypersensitivity** is an overreaction to the presence of a normally harmless substance called an **allergen**. Some common allergens are pollen, house dust mites, animal fur and feathers, fungal spores, insect bites and penicillin (Fig 34.19).

The commonest symptoms of an allergy are sore eyes, runny nose, sneezing and asthma. Many of these symptoms result from an inflammation of the mucous membranes, caused by **mast cells,** which release chemicals such as histamine. Many anti-allergy treatments suppress mast cells or neutralise histamine – chemists sell many **antihistamines** in the pollen season. For more information about asthma see the Assignment in Chapter 8.

6 BLOOD TRANSFUSIONS AND ORGAN TRANSPLANTS

In the last 100 years, advances in modern medicine have led to the technology and knowledge that allows doctors to give one person's cells, tissues and organs to another person. One of the main objectives in developing these life-saving treatments has been to overcome the natural reaction of the immune system to destroy transplanted cells and tissue, which it 'sees' as non-self.

Blood grouping and transfusions

Towards the end of the last century, medical scientists realised that accidents were often fatal simply due to blood loss. Losing large volumes of blood sent victims into shock and they died, even though their injuries were otherwise not too severe. Many women, in particular, often died when they lost blood when giving birth. Different doctors tried transfusing blood from a healthy person into the injured person, to try to restore their blood volume.

Sometimes this worked and sometimes it didn't. When it failed, the results were disastrous. We now know that if the **blood groups** of the **donor** (the person giving the blood) and the **recipient** (the person receiving it) are different, the recipient's immune system reacts against the donated blood, producing a massive and deadly immune response.

In the early 1900s, an Austrian scientist, Karl Landsteiner, discovered that red blood cells from different people had different sets of antigens on their surface. The entire human population can be placed into one of four main blood groups: **A**, **B**, **AB** and **O**, according to the antigens on their red blood cells.

From Fig 34.20, you can see that people with blood group O have no antigens on the surface of their red blood cells. Blood from these people cannot cause an adverse immune response if it is transfused into people from the other three blood groups. We say that people with blood group O are **universal donors**. By the same reasoning, people with blood group AB cannot have any antibodies to either of the blood group antigens and so cannot mount an immune response if they are given blood by people in any of the other three groups. We say that people with AB blood are **universal recipients**.

Blood transfusions are dangerous when the *recipient has antibodies in their blood that react with antigens present on the surface of red cells in the donor blood* (Fig 34.21). So, for example, a person of blood group

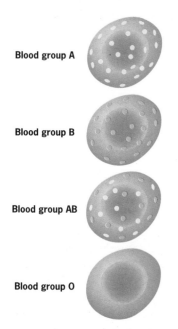

Blood group A

Blood group B

Blood group AB

Blood group O

Fig 34.20 **People can be placed in four groups according to which red blood cell antigens they have**

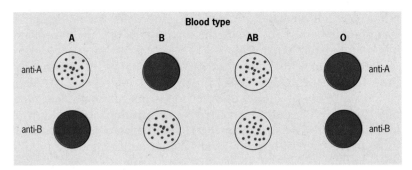

Fig 34.21 **We can find out what blood group someone has by mixing a sample of their blood with anti-A and anti-B antibodies. Blood from people with group A antigens agglutinates (forms clumps) with anti-A, blood from people with group B antigens agglutinates with anti-B. AB blood agglutinates with both, O with neither. In the diagram, the agglutination reaction produces a granular appearance as red cells clump together. In the smooth, dark samples, agglutination has not taken place**

A cannot donate blood to someone with group B. The recipient in this case has anti-A antibodies which react with the A antigens on the red blood cells of the donor blood. Table 34.3 gives a full list of possible transfusions.

F Why do you think some of the early transfusions worked, even though they were done before we knew about the importance of blood groups?

Blood group	Antigens present on red blood cells	Antibodies present in blood	Can donate blood to	Can receive blood from
A	A	anti-B	A, AB	A, O
B	B	anti-A	B, AB	B, O
AB	AB	none	AB	A, B, AB, O
O	none	anti-A, anti-B	A, B, AB, O	O

Table 34.3 **Which transfusions are possible?**

■ See question 4.

Transplants and grafts

Transplantation involves taking cells, tissues or organs from one individual and placing them into another individual. Since the first organ transplant operations in the 1950s, patients have received hearts, lungs, skin, corneas, kidneys and various other organs. The success rate of these operations has improved steadily.

G Look at Table 34.3. Why can someone with blood group AB donate blood only to someone else with the same blood group?

The biggest problem facing most transplant recipients is that of **rejection**. The cells of transplanted tissue are covered in antigens which stimulate the specific immune response of the recipient, notably the cell-mediated response brought about by T cells. Symptoms of rejection include the degeneration of blood vessels in the transplanted organ and the destruction of whole cells followed by their replacement with 'scar' tissue. A patient whose transplant is rejected becomes seriously ill: they lose the function of the organ concerned and the massive immune response puts an incredible amount of stress on their already weakened body.

H What is the difference between agglutination and blood clotting?

Rejection can be minimised by tissue typing, by using immunosuppressive drugs and by using monoclonal antibodies.

Tissue typing

Like red blood cells, other body cells also have surface antigens that are different in different people. Body cells, such as cells in the kidney or liver, have many antigens that differ, and so matching the tissue type of the donor and recipient is much more complex than matching blood groups. In practice, the cells from two people are never identical, so the aim is to match people who have as few differences as possible.

Immunosuppression

If the recipient of a transplant has a fully functional immune system, even a closely matched organ will be rejected to some extent. To counter this, transplant patients are given immunosuppressive drugs. The most effective is currently **cyclosporine**. This chemical, isolated from a fungus, inhibits the action of T cells and so has a profound

For examination purposes, a quick way to remember safe blood transfusions is as follows: O is the universal donor, so can be given to anyone. AB is the universal recipient, so can receive from anyone. Apart from that, only like-to-like transfusions (eg A to A) are safe. All others cause problems.

effect on cell-mediated immunity. Since the introduction of cyclosporine in the early 1980s, the success rates of some transplants has risen to over 90 per cent.

The problem with immunosuppressive drugs is, of course, that they reduce the body's ability to fight off infection. Transplant patients are more prone to infection from normally harmless microorganisms, particularly viruses. Although the dose of immunosuppressive drugs can be reduced after a few months, if there is little sign of rejection, transplantees must continue to take them for life.

See question 5. ■

Monoclonal antibody therapy

In the mid-1980s, a monoclonal antibody called OKT3, which attacks the T cells responsible for transplant rejection, was developed for use in kidney transplant patients. It is more effective than cyclosporine in reversing the acute rejection that occurs very soon after transplantation. Another monoclonal, Campath-IG, was developed in the mid-1990s. This also attacks T cells but is being used to treat bone marrow, after it has been taken from the donor, but before being put into the recipient. This technique seems to prevent rejection of the bone marrow, even when the patient and donor are not closely matched by tissue typing. It also allows them to avoid immunosuppressive drugs altogether.

7 ANTIBIOTICS AND THE TREATMENT OF INFECTION

Throughout human history, millions of people have died from bacterial infections. Today, we still suffer from such infections but we now have antibiotics, safe and effective drugs that can be used against bacteria that cause disease.

The first antibiotics developed in the 1930s and 1940s were chemicals that one microorganism produced to kill another. Penicillin, for example, the first antibiotic to be produced in large enough quantities to be used in combating bacterial infections, is produced a type of fungus. Since then, antibiotics have been extracted from a range of unusual sources including snowdrop bulbs and toad skin; yet others can be synthesised.

Antibiotics target processes such as protein synthesis that are subtly different in microorganisms compared to the cells of humans and other animals. Different antibiotics have different modes of action (Fig 34.23).

Fig 34.22 **In 1928 one of Sir Alexander Fleming's culture plates became contaminated with a species of *Penicillium* mould. This accident led to the discovery of the first antibiotic, penicillin. It was not, however until the early 1940s, however, that the compound could be produced in sufficient quantities to treat people with bacterial infections**

?

I Suggest why, although penicillin is effective against a wide range of bacterial diseases, it is no use for treating diseases caused by fungi.

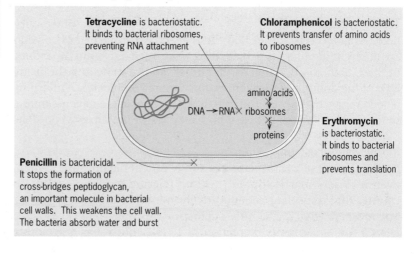

Tetracycline is bacteriostatic. It binds to bacterial ribosomes, preventing RNA attachment

Chloramphenicol is bacteriostatic. It prevents transfer of amino acids to ribosomes

amino acids

DNA → RNA × ribosomes

proteins

Erythromycin is bacteriostatic. It binds to bacterial ribosomes and prevents translation

Penicillin is bactericidal. It stops the formation of cross-bridges peptidoglycan, an important molecule in bacterial cell walls. This weakens the cell wall. The bacteria absorb water and burst

Fig 34.23 **Some antibiotics kill the microorganisms they target. These are said to be** bactericidal. Bacteriostatic **antibiotics do not kill microorganisms; they prevent them multiplying and give the immune system a better chance to overcome the infection**

ANTIBIOTIC RESISTANCE

IT HAS BEEN SAID that the most dangerous place for you if you are sick is in hospital! This, of course, is a great overstatement, but many bacteria that exist in hospitals are resistant to common antibiotics, and the problem is getting worse.

Antibiotics are undoubtedly one of the major medical advances of the twentieth century. They have saved the lives of millions of people. But nothing is perfect and there is not such thing as a wonder drug. Since the 1940s when antibiotics were first used widely, resistant strains of bacteria have arisen that can cause infections that are very difficult to treat. How did this happen?

Bacteria, like all living organisms, vary. In any population of bacteria, the vast majority can be killed by an antibiotic such as penicillin. However, there might be one that has the gene that allows it to produce the enzyme penicillinase. This enzyme breaks up penicillin molecules and bacteria that produce it are resistant to the action of penicillin. If such a penicillin-resistant bug can evade the body's defences long enough to reproduce, in a day or two – bang – a new infection that is even more difficult to treat.

Penicillin resistance has become common in the last few years and bacteria have also become resistant to several other antibiotics. We now recognise that over-

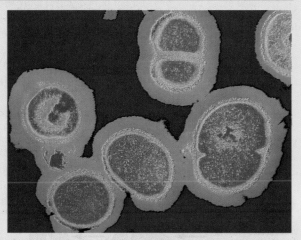

Fig 34.24 **Cells of this highly antibiotic-resistant strain of the bacterium *Staphylococcus aureus* are seen dividing. In hospitals, it infects the wounds of patients, causes internal abscesses and produces boils by entering the skin through hair follicles**

prescribing has played a role, and doctors are now much more reluctant to give antibiotics for every cough and cold. Medicine now faces the challenge of staying one step ahead to develop new antibiotics faster than bacteria evolve resistance.

SUMMARY

After studying this chapter, you should know and understand the following:

■ We are surrounded by potential **pathogens** (disease-causing organisms) such as bacteria and viruses which can cause disease if allowed to enter and multiply inside the body.

■ The body's defence system consists of barriers to keep microorganisms out, and mechanisms to detect and destroy those that do enter. Together, these mechanisms are the **immune response**.

■ The immune system is capable of non-specific responses and specific responses. Non-specific mechanisms (**inflammation** and **phagocytosis**) occur in response to any invading microorganism. Specific mechanisms allow the body to recognise and fight individual types of microorganism. There are two specific responses: **cell-mediated immunity** and **humoral immunity**.

■ **Lymphocytes** (white cells) are responsible for the specific immune response. **T cells** mature in the thymus gland and **B cells** come directly from bone marrow. Both are able to recognise foreign **antigens** that come from pathogens.

■ B cells are responsible for humoral immunity: they secrete specific proteins called **antibodies**.

■ T cells are responsible for cell-mediated immunity and help to control the overall specific response. **Killer T cells** attack pathogens or infected cells. **Helper T cells** activate B cells, telling them to secrete antibodies. **Suppressor T cells** damp down the immune response when the infection is over.

■ **Auto-immune diseases** occur when the body's immune system attacks its own tissues. Examples include rheumatoid arthritis, multiple sclerosis and myasthenia gravis. **Allergies** occur when the body 'over-reacts' to a normally harmless substance such as pollen.

■ When cells or tissues are taken from one individual and given to another, they are often recognised as 'foreign' and destroyed. This process is often called **graft rejection**. To minimise the chance of rejection in blood transfusions and organ transplants, it is important to match blood and tissue types.

■ Antibiotics are important drugs for treating infections caused by bacteria. In the time since antibiotics were first developed and used in the 1940s, bacteria have evolved to become resistant to them. Antibiotic resistant strains of many bacteria are becoming common and represent an on-going problem for doctors and researchers.

35 Lifestyle diseases

A cow with bovine spongiform encephalopathy becomes uncoordinated and is eventually unable to walk without stumbling. The disease affected many cows in the British beef herd, and many were destroyed to prevent infected meat entering the food chain. There is no evidence that milk from a cow infected by BSE is harmful

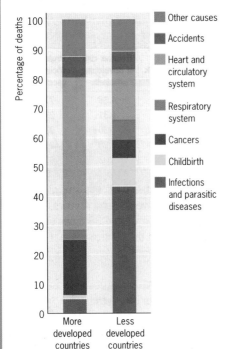

Fig 35.1 **This bar chart shows the percentage of deaths from a number of different causes in more developed and less developed countries**

IN 1996, A HANDFUL OF DEATHS were linked with a type of dementia called Creutzfeldt–Jakob disease (CJD). CJD has been known for a long time, but, with an average age of 27, these victims were much younger than the people who usually developed CJD. All the signs pointed to a new variant form.

When the UK government announced there was a chance that new variant CJD had been passed to people in the beef they ate, there was an incredible furore. Bovine spongiform encephalopathy (BSE), the bovine form of CJD, was known to be widespread in the British herd. Was everyone who ate beef regularly at risk of new variant CJD? The resulting debate demonstrates how difficult it is to pin-point a lifestyle risk factor for a disease that takes several years to develop.

BSE was first identified in cattle in 1986, and scientists concluded that this bovine form had developed from a similar disease in sheep, called scrapie. Why the scrapie agent jumped species isn't really understood. But the practice of recycling slaughtered sheep, and later, infected cows, to make protein supplements for cows, helped to establish a massive infection of BSE in British cattle. Many were culled to stem the epidemic and to lessen the possibility of infected meat entering the human food chain.

By the end of 1996, scientists had analysed the chemistry of the agent that causes new variant CJD and showed that it was actually closely related to BSE in cattle. Although we now know that the people who have developed new variant CJD got it from cattle, we cannot yet be sure how they got it. Was it from eating beef? Or was it by some other route? Were those few people just susceptible to infection? And are the majority of people resistant?

Answering these questions will probably take years of research but, by the end of 1998, the number of cases of new variant CJD was still very low and a widespread epidemic now seems unlikely.

1 WHAT ARE LIFESTYLE DISEASES?

In this chapter, we look at a group of diseases influenced by the things we eat and do. Such **non-transmissible** diseases or **lifestyle** diseases are the major cause of death in the more developed countries (Fig 35.1).

Lifestyle diseases include **cardiovascular disease**, which affects the heart and circulatory system, **cancers** and some diseases of the respiratory system.

In this chapter, our study of lifestyle diseases includes:

● The aspects of our lifestyle that are linked with disease: smoking, alcohol, poor diet and lack of exercise. These are all known to affect a person's chance of developing a particular condition, so they are often called **risk factors**.

● **Cardiovascular diseases**. In their most severe form, heart attacks and strokes can be fatal. Less severe forms of cardiovascular disease might not cause death but they can seriously reduce the quality of life of many middle-aged and older people in countries such as Britain.

● **Cancers**. These are tumours or swellings that can occur in any part of the body. A cancer is a mass of cells that multiply in an uncontrolled way, invading surrounding healthy tissue.

A Suggest why a higher percentage of people die from cancers in more developed countries.

Risk factors for lifestyle diseases

Some bacteria are responsible for specific forms of heart disease and some viruses are associated with particular types of cancer, but the main causes of non-transmissible disease are features of our lifestyle. A risk factor is an aspect of our lifestyle that increases our chances of developing a particular disease. Table 35.1 shows some of the most important ones.

Table 35.1 **Some examples of risk factors**

Factor	Level of risk
Smoking	There is good evidence that cigarette smoking is directly responsible for about 90 per cent of all cases of lung cancer. Smoking also makes us more likely to succumb to diseases such as bronchitis and emphysema and it increases the risk of developing cancers in other parts of the body.
Drinking alcohol	Regular drinking of small amounts of alcohol could have a beneficial effect on the circulatory system. However, there is plenty of evidence that shows that regular heavy drinking is linked to an increased risk of developing cancer of the mouth, throat, oesophagus and liver. Between 2 and 4 per cent of all cancers are thought to be due directly to alcohol consumption.
Exposure to ultraviolet light	Sunbathing is dangerous! Over the past 25 years, the incidence of skin cancers has doubled. Most of the increase is thought to be due to greater exposure to ultraviolet rays in sunlight (Fig 35.2). Pale-skinned people can burn easily and a handful of bad episodes of sunburn in childhood can double your chances of getting skin cancer later in life. Also, experts now believe that the protective ozone layer in the upper atmosphere, which acts as a barrier to the dangerous UV rays in sunlight, has been seriously eroded by pollution.
What we eat	The links between diet and heart disease and between diet and cancer are more complex. Some foods are associated with greater risk of developing some diseases. A high fat diet, for example, seems to increase risk of developing cardiovascular disease, but other environmental, lifestyle and genetic factors are also important, and teasing out which factors have the greatest effect is very difficult.

Fig 35.2 **In the last half of the twentieth century, the 'pale and interesting' look fell out of favour and hot, sunny beach holidays became more accessible**

B Suggest two reasons that could explain why skin cancer is more common today than it was in 1950.

Investigating risk factors

Proving that a risk factor causes a specific disease is quite tricky. It can appear that two things are linked or **correlated**, but this does not mean that one causes the other. Fig 35.3 shows what happens if we use data from a British seaside town in summer and plot the number of cases of sunburn against ice-cream sales.

The number of cases of sunburn seems to increase with the number of ice creams sold: there is a clear positive correlation between the two sets of data. However, we all know that eating ice-cream does *not* cause sunburn. A third factor, the hot weather, is involved. When it's hot, people are more likely to lie in the Sun and get sunburnt, and they are also more likely to buy ice-cream.

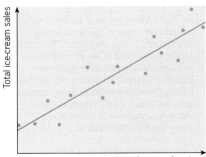

Fig 35.3 **Ice-cream sales plotted against number of cases of sunburn shows a correlation. But does it prove that ice-cream causes sunburn?**

Medical researchers try to identify factors that affect our risk of developing a particular disease. In Chapter 31, we saw how an increase in asthma in Barcelona was linked to unloading soya beans at the docks. In most cases, though, things aren't quite that simple. Pinning down risk factors to particular diseases is very difficult because we are continually altering our lifestyles. We might eat more healthily or take more exercise in our teens and then fall into bad habits in our twenties. Our exposure to environmental factors changes, often beyond our control. New preservatives could be added to food we eat, or there can be changes in the levels of pollution that contaminate the air we breathe. Any of these changes can alter our chances of developing a particular disease.

The Assignment at the end of this chapter looks at some of the methods that we can use to investigate risk factors.

2 CARDIOVASCULAR DISEASE

Cardiovascular disease is a major lifestyle disease in more developed countries. It claims about 175 000 lives per year in the UK and almost a million lives per year in the USA. As its name suggests it involves the cardiovascular system – the heart and blood vessels that transport blood throughout the body (see Chapter 9).

Cardiovascular disease is more common in people over 50, but it can happen in younger people. It shows itself in many ways but it has a common underlying physiological cause: the narrowing and clogging up of arteries as a result of **atherosclerosis**. We look at this process in detail later in the chapter, but first, Fig 35.4 presents an at-a-glance overview of the different forms of cardiovascular disease.

Fig 35.4 **Cardiovascular disease is a general term. The symptoms and effects of different forms depend on which blood vessel is affected. The severity of the condition depends on the extent to which the vessel is blocked. Generally, partial blockages in small arteries or veins are less life-threatening than complete blockage of a major vessel**

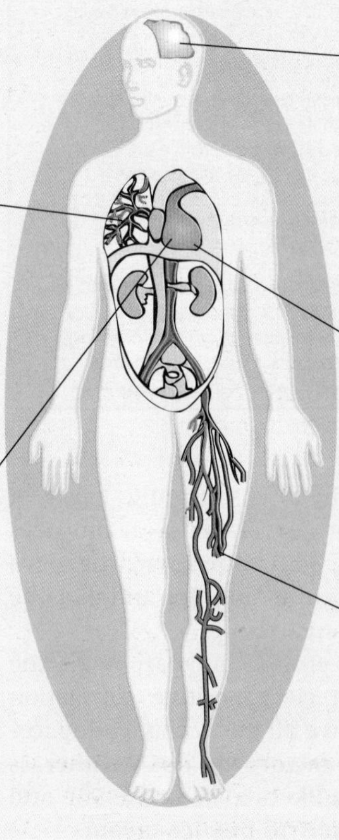

Fig 35.4(a) **An** embolism **occurs when a mobile blood clot (called an embolus, hence embolism) moves through the circulatory system and then lodges in a vein and blocks it. If a pulmonary vein (from lungs to heart) becomes blocked, this causes a** pulmonary embolism **that can be fatal**

Fig 35.4(b) **When the blood supply to the heart muscle is interrupted, the result can be a heart attack, or** myocardial infarct**. This leads rapidly to the death of the affected tissues and can produce ventricular fibrillation: instead of coordinated beating, the individual muscle fibres contract in an unsynchronised way. The affected part, usually the left ventricle, sometimes becomes unable to pump blood and the heart stops**

Fig 35.4(c) **A stroke results if the supply of blood to the brain is interrupted. The symptoms vary according to the size and location of the area affected. There might be no more than a tingling sensation or weakness down one side of the body, but at its most severe, a stroke can lead to permanent paralysis or death**

Fig 35.4(d) Angina pectoris **is a severe pain in the chest that occurs when the coronary arteries are unable to supply sufficient oxygen to the heart muscles**

Fig 35.4(e) **When a vein is damaged and develops a blood clot that blocks it, this condition is known as** thrombophlebitis. **Blood clots in the leg veins can occur in patients who are forced to remain in bed for long periods. If the clot breaks away, it can travel to the heart and then be pumped to the lungs, causing a pulmonary embolism. A** thrombosis **is the blockage of an artery caused by a blood clot. It can occur anywhere in the body**

HEART FAILURE

THE TERM HEART FAILURE suggests a condition in which the heart suddenly comes to a complete stop. This doesn't happen – heart failure usually develops slowly, often over years, as the heart gradually loses its pumping ability and works less efficiently. The underlying cause is usually coronary artery disease.

How serious the condition is depends on how much pumping capacity the heart has lost. Nearly everyone loses some pumping capacity as they get older but the loss is significantly more in heart failure.

There are two main types of heart failure:

● **Systolic heart failure** occurs when the heart's ability to contract decreases. The heart cannot pump with the force required to push enough blood out through the aorta. Blood coming into the heart from the lungs can back up and cause fluid to leak into the lungs, a condition known as **pulmonary congestion**.

● **Diastolic heart failure** occurs when the heart has a problem relaxing. The heart can't fill with blood because the muscles of the atria and ventricles have become stiff. People with this form of heart failure accumulate fluid, especially in the legs, ankles and feet. Their condition is called **oedema** (Fig 35.5).

Tiredness, breathlessness and excess fluid retention are common symptoms but, because heart failure usually develops slowly, these obvious signs don't often appear until the damage to the heart is serious. This is partly because the heart adjusts to cope with the effects of heart failure by getting bigger, developing thicker muscle and by contracting more often.

Heart failure is treated by a combination of lifestyle changes and drug treatment. Patients are advised to stop smoking, lose weight if necessary, avoid drinking alcohol, and change their diet to reduce the amount of salt and fat they eat. Regular, moderate exercise is also helpful, but this needs to be done only with a doctor's approval.

However, heart failure is always a serious condition because it poses an increased risk of sudden death, through cardiac arrest. Drugs are usually used to control the symptoms and to try to prevent them getting worse.

● Diuretics help reduce the amount of fluid in the body and are useful for patients with fluid retention and high blood pressure.

● The drug digitalis increases the force of the heart's contractions, helping to improve circulation.

● Recent studies have shown that drugs known as angiotensin-converting enzyme (ACE) inhibitors can be useful. ACE inhibitors improve survival among heart failure patients and can slow, or perhaps even prevent, the loss of heart pumping activity.

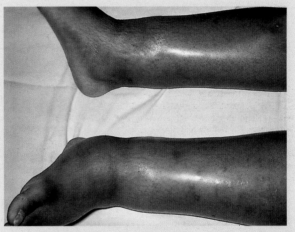

Fig 35.5 **This person has a bad case of oedema**

Atherosclerosis and cardiovascular disease

To understand how and why cardiovascular disease develops, we need to look at three things:

● How does an atheroma form?

● What are the risk factors associated with atherosclerosis and how do they increase the chance of atheroma formation?

● How does atherosclerosis lead to cardiovascular disease?

The formation of an atheroma

Atheromas, areas or **plaques** of fatty material, develop on the inside of arterial walls. Remind yourself of the structure of an artery wall by looking at Fig 35.6. As plaques build up, they can affect the circulatory system by:

● blocking the flow of blood through the artery.

● initiating the formation of blood clots. Blood clots that form inside coronary arteries are particularly dangerous. They can break free and lodge in the smaller coronary arteries, causing some of the

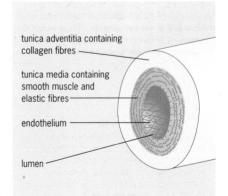

tunica adventitia containing collagen fibres

tunica media containing smooth muscle and elastic fibres

endothelium

lumen

Fig 35.6 **Cross-section through a large artery. The space through which the blood flows is called the** lumen. **Surrounding this is a layer of smooth flat cells called the** endothelium. **Outside this lining is the** middle layer **or** tunica media, **which contains a high proportion of smooth muscle and elastic fibres. Finally, there is an** outer layer **or** tunica adventitia **containing fibres of the structural protein collagen**

heart muscle to die. This is what initiates a heart attack. If blood clots interfere with the blood supply to the brain, the result is a stroke.

Arterial blood supplies all the body's organs with nutrients, including triglycerides and cholesterol. These are lipids and have several important functions in the body: they form cell membranes and they act as energy stores. Lipids combine with proteins to form **lipoproteins** before they can be transported in the blood.

Lipoproteins that contain cholesterol are engulfed by white cells called **phagocytes** that occur in the artery wall, just under the endothelium. If the blood contains large amounts of cholesterol-rich lipoproteins, large numbers of stuffed phagocytes accumulate. Their rich 'filling' gives them a foamy appearance and they are, in fact, often referred to as **foam cells**. When patches or streaks of foam cells start to appear on the inside of artery walls, this is recognised as the first sign of atherosclerosis. At this stage, there is little effect on the flow of blood through the artery and no discernible symptoms.

Gradually, usually over many years, muscle cells and fibres begin to grow over the affected patches, giving rise to a **fibrous plaque**. This bulges into the lumen of the artery and starts to obstruct the flow of blood (Fig 35.7). This can have one of several effects (Fig 35.8):

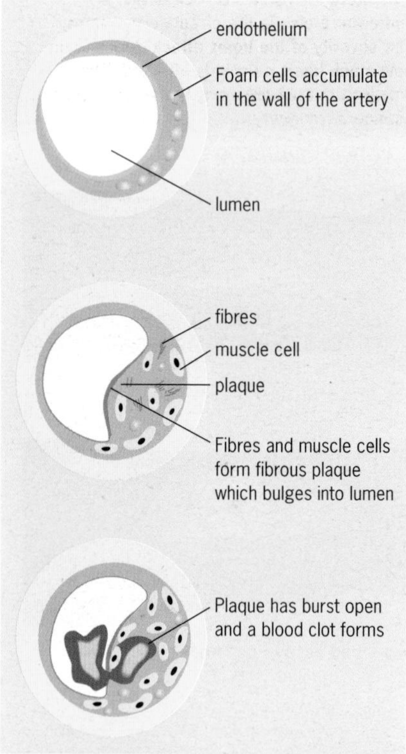

- The artery can become very narrow. Depending on the position of the narrowed artery, the oxygen supply to vital parts of the body can be reduced. If, for example, the arteries feeding the heart muscle are affected, the result can be **angina**.
- The blood vessel can become completely blocked. Blockage of an artery that takes oxygenated blood to a vital organ can lead to many different problems. If the coronary arteries become blocked, the result can be a heart attack. If arteries in the brain are affected, this can lead to a stroke.
- The blood vessel can burst. An burst artery, or **aneurysm**, inside the body is very serious. If one of the main arteries near the heart burst, this is usually fatal.
- The plaque can burst into the blood vessel, damaging the endothelium. When the cells underneath come into contact with blood, this triggers the blood-clotting mechanism. A blood clot or **thrombus** forms in the vessel. If this breaks free, it can travel around the body and block a small artery.

Fig 35.7 **Areas of fatty material called atheromas build up in the walls of arteries. This is one of the first signs of cardiovascular disease. Eventually it can almost block the artery**

endothelium

Foam cells accumulate in the wall of the artery

lumen

fibres

muscle cell

plaque

Fibres and muscle cells form fibrous plaque which bulges into lumen

Plaque has burst open and a blood clot forms

Risk factors for cardiovascular disease

As we have seen, the underlying process of atherosclerosis can have many different effects, depending on which part of the body is involved. The risk factors for **cardiovascular diseases** (CVDs) are also complex because many different factors affect the way this whole body system functions. We shall just look at the four main ones that have been identified.

Hypertension

Blood vessels are more than a series of pipes that take blood to and from different organs. They react to the changing needs of the body supplying, for example, more blood to the muscles during exercise. When we make our muscles work hard, they need more oxygen and more glucose and they also produce more carbon dioxide. The

C Explain why an exercising muscle needs more glucose.

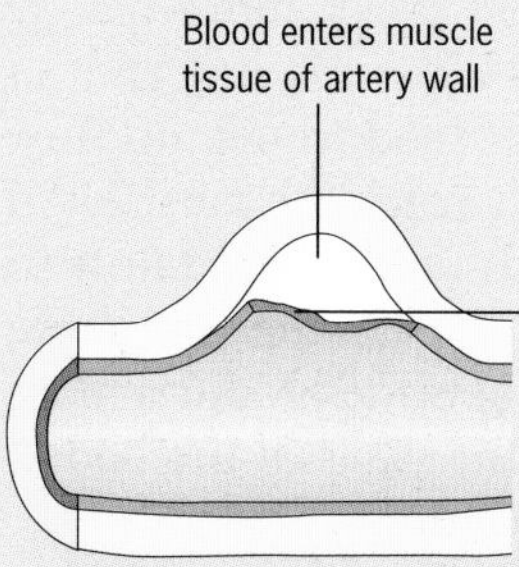

Blood enters muscle
tissue of artery wall

Tear in lining of
artery wall

An aneurysm **results from a tear in the wall of an artery. This allows blood to enter the muscle layer of the artery and produce a balloon-like swelling. If this swelling bursts, the consequences are usually fatal. An aneurysm can also occur in the wall of left ventricle after a heart attack**

A stroke **also has several possible causes. One of the arteries in the brain could have developed an atheroma that has blocked the vessel, cutting off the blood supply to part of the brain. Or a clot that developed elsewhere in the body could have travelled through the carotid artery and then lodged in a smaller artery, with the same result. More rarely, an atheroma can lead to an aneurysm that ruptures, causing blood loss and damage to brain tissue**

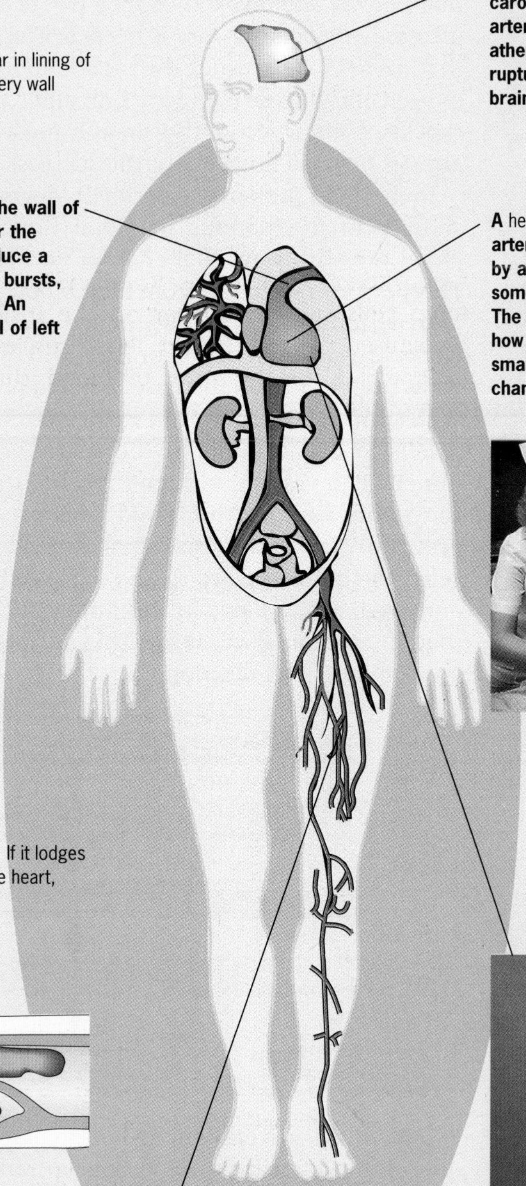

A heart attack **occurs when the coronary arteries become blocked by an atheroma or by a blood clot that has broken off somewhere else in the circulatory system. The severity of the heart attack depends on how much heart muscle is affected. The smaller the area affected, the greater the chance of recovery**

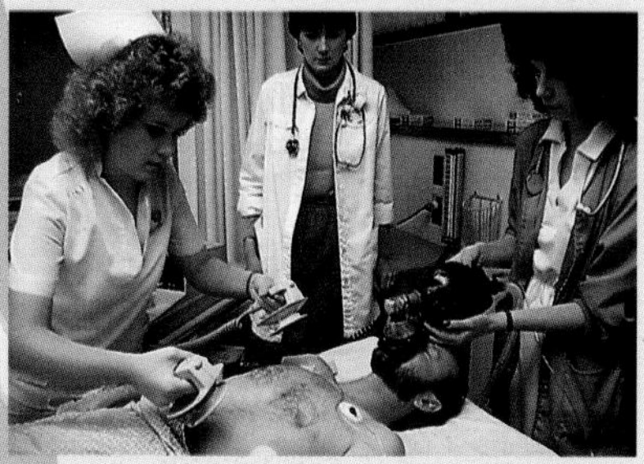

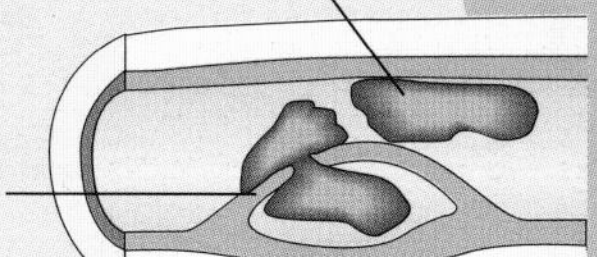

Thrombus may break free. If it lodges
in the arteries supplying the heart,
a heart attack will result

Plaque bursts
and thrombus
or blood clot
is formed

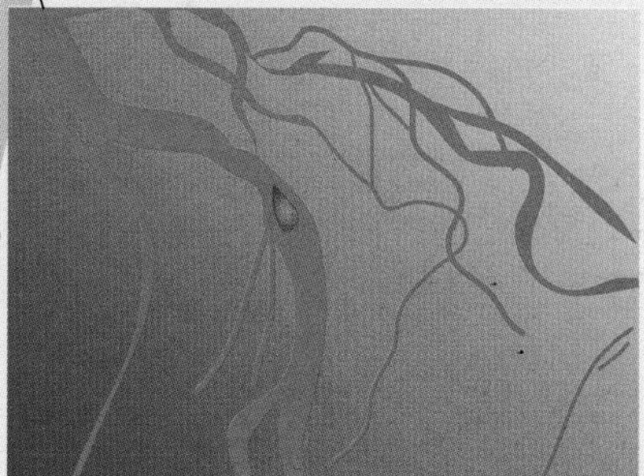

Angina **is often the result of an atheroma in the coronary arteries. As the arteries become narrower, the supply of oxygenated blood to the heart muscle becomes more erratic**

A blood clot or thrombus **can form when a fibrous atheroma plaque bursts into an artery. This caption points to the main femoral artery, but a thrombus can occur anywhere in the body. The cells underneath the endothelium are exposed to the blood and trigger the blood-clotting mechanism. A thrombus forms, and if this breaks free and lodges in one of the arteries supplying the heart, it can cause a heart attack. If it lodges in one of the arteries that leads to the brain, this can obstruct the blood flow and result in a stroke**

Fig 35.8 **The biological basis for the different forms of cardiovascular disease**

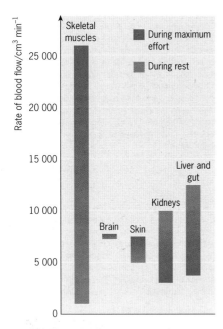

Fig 35.9 **The blood flow to different organs in the body changes as we exercise. The flow to muscles and the skin increases. The flow to the brain stays the same. The blood supply to organs like the kidneys and the gut decreases**

D Give the advantage to an exercising person of:

(a) an increased blood flow to the muscles,

(b) a decreased blood flow to the kidneys.

Fig 35.10 **This flow chart shows some of the ways in which hypertension can affect the body**

graph in Fig 35.9 shows how the supply of blood to different organs changes as we do different activities.

Think about an exercising muscle. How can it get more blood? Although blood vessels are flexible, they don't just expand and let more blood through. If this were to happen, the blood would just flow more slowly and no extra oxygen would actually reach the muscle. An increase in blood pressure is needed. This speeds up the blood flow so that more blood reaches the muscle in a given time, supplying more oxygen and glucose to muscle cells and removing waste products more rapidly. Contraction of the smooth muscle in the walls of the body's arteries helps to bring about this increase in pressure.

High blood pressure is perfectly normal as long as it results from short-term fluctuations in blood flow. Problems start when the blood pressure stays higher than normal while someone is at rest. This person is suffering from high blood pressure, or **hypertension**.

Hypertension sets up a sequence of events that becomes a viscious circle. This is summarised in Fig 35.10. Smooth muscle in the artery walls responds to an increase in blood pressure in the same way that skeletal muscle responds to extra exercise – it increases in size. This results in a thickening of the artery wall and a narrowing of the lumen. The volume of blood passing through the arteries remains the same, causing the blood pressure to rise further. At higher pressure, the blood also exerts a greater force on the endothelial cells which line the artery and these cells are more likely to become damaged. Small flaws in the lining of the artery increases the risk that an atheroma will form. This, in turn, makes it more likely that a thrombosis will develop.

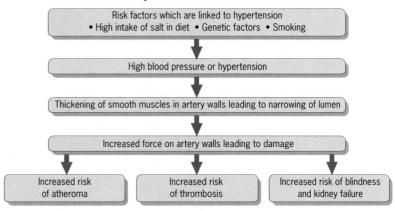

Two things increase the risk of hypertension. A high salt intake is one. If you increase the salt concentration of your blood, its water potential falls. A lower water potential means that blood volume increases as water is drawn from the cells of the body into the plasma. The higher the volume of the blood, the greater its pressure. The other is smoking.

The link between smoking and atherosclerosis

If you smoke, you are much more likely to develop atherosclerosis. However, there is no straightforward explanation of how smoking actually causes the formation of atheroma. Several mechanisms could be involved:

● Cigarette smoke contains nicotine. Nicotine is a **vasoconstrictor**: it narrows the blood vessels and leads to an increase in blood pressure. High blood pressure can damage the artery walls and so increase the likelihood of atheroma formation.

- Smoking increases the level of cholesterol-rich lipoproteins in the blood. This accelerates the formation of the fatty streaks which are the first sign of atherosclerosis.

- Smoking increases the amount of fibrinogen circulating in the blood. This increases the chance of a blood clot forming if an atheroma breaks up in the artery.

The link between diet and atherosclerosis

Although newspapers and magazines can give the impression that all cholesterol is bad, it is an essential component of cell surface membranes. Cholesterol is used by the body to synthesise hormones such as oestrogen and testosterone. It becomes a danger only when it is present in high concentrations in the plasma.

Cholesterol is a lipid and, like other lipids, it is insoluble in water. In order to be transported in the blood it is combined with a protein, forming a complex molecule called a **lipoprotein**. There are different sorts of lipoprotein that have varying amounts of cholesterol and other lipids. **Low-density lipoprotein** (LDL) is commonly called 'bad' cholesterol. This type of lipoprotein contains high amounts of cholesterol and can contribute to the formation of atheromas. High-density lipoprotein (HDL) is commonly called 'good' cholesterol because, not only does it not contribute to atherosclerosis, it actually helps to scavenge some of the cholesterol from LDL particles, making them less dangerous.

It is best to have a total blood cholesterol level of 200 milligrams per cubic centimetre of blood. Of this total, the amount of LDL should be low – between 100 and 130 mg cm^{-3}. HDL should make up the rest. If someone's total blood cholesterol is higher than 200 mg cm^{-3} or if LDL is over 130 mg cm^{-3}, or if HDL is lower than 35 mg cm^{-3}, they are at increased risk of developing atheromas.

?

E LDL receptors are proteins. Use this information to suggest why some people may be genetically more susceptible to heart attacks than others.

How does cholesterol cause atheromas?

The sequence of events is summarised in Fig 35.11. LDL particles that are excess to the requirements of the liver and other organs leave the blood and penetrate the walls of blood vessels, particularly in places where the wall has been damaged. Phagocytes then engulf them but are unable to break down the cholesterol that they contain. This accumulates in the cytoplasm of the phagocytes which then become foam cells (see page 528). The appearance of foam cells is the first recognisable sign of atherosclerosis.

Fig 35.11 **How LDL causes atheromas**

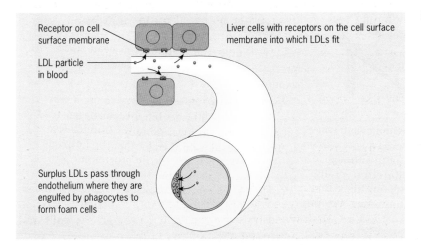

Receptor on cell surface membrane

Liver cells with receptors on the cell surface membrane into which LDLs fit

LDL particle in blood

Surplus LDLs pass through endothelium where they are engulfed by phagocytes to form foam cells

JIM'S STORY

25 March

31 March

Jim is 53. When he was younger, he used to play a lot of football. 'Actually, I haven't taken much exercise recently. I just don't have time. I've had a lot of stress at work in the last few years, trying to hang on to my job. OK, I smoke a bit too much, and I've put on a bit of weight, but that's normal in a man my age, isn't it? I can't remember when I last had to take a day off because I was ill, though.'

26 March

'I was mowing the lawn today when I suddenly got a really sharp pain in my chest. I felt quite funny for a bit and I had to sit down. I'll have to get some more indigestion tablets for when we have a good fried breakfast.' Later in the day, the pain came back every time Jim got up. By the evening he felt as if his chest was being crushed. His wife called the doctor and he was whisked into hospital with a suspected heart attack.

27 March, lunchtime

Jim is resting on the ward. The immediate danger has passed, but he has been having tests all morning and he still feels rough. Although a heart attack –a **myocardial infarction** – causes chest pain, it's not the condition thing that does. A diagnosis must be confirmed before treatment can begin.

Like Jim's, most heart attacks are not fatal. With careful treatment and good advice, the patient usually makes a full recovery and can lead a normal life. One of the most important aspects of treatment is advice on lifestyle changes that can reduce the risk of a further heart attack.

The most obvious first step for someone like Jim is to stop smoking. The link between this habit and heart disease has been established by a great weight of evidence. A middle-aged man who smokes is three times more likely to suffer a heart attack than a non-smoker.

'I've also got to change what I eat. Far less of the fried breakfasts from now on.'

Most blood cholesterol doesn't come directly from food. It's produced in the liver from saturated triglycerides that come from the animal fats in our diet. So it is sensible to reduce the intake of lipid and animal fats so that they make up no more than 25 to 30 per cent of the total energy intake.

1 September

Jim is now fully recovered from his heart attack, but there have been a lot of changes in the last six months.

'The thought of almost dying gave me a real scare. It's been difficult, but I've managed to stop smoking altogether and now eat much less fatty foods. I followed the diet that the hospital gave me and joined an exercise programme. I'm now two stones lighter and I actually feel better now than before the heart attack.'

Blood tests

Table 35.2 shows the results of some of Jim's blood tests. Jim has also had a chest X-ray (Fig 35.12) and an **electrocardiogram** (ECG). An ECG records the wave of electrical activity that travels through the heart during a cardiac cycle (see page 138 in Chapter 9).

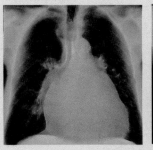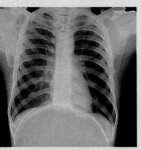

Fig 35.12 **A chest X-ray shows how Jim's heart and lungs, (a), differ in size and shape from those of a healthy person, (b)**

Life after a heart attack

There is strong evidence that strenuous exercise benefits the heart. Jim joined a study that looked at the effect of exercise on people who had recently had a heart attack. It showed that, after exercising for three months, he was 20 per cent less likely to die than someone who hadn't joined an exercise programme as part of their recovery.

Fig 35.13 **This graph shows how a programme of exercise affects fitness. The group that followed the programme suffered fewer deaths than the control group**

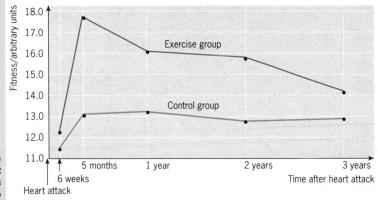

Table 35.2

Compound	Reason for test	Concentration in blood/arbitrary units	Normal range in healthy people/arbitrary units
Cholesterol	A high blood cholesterol concentration is often associated with an increased risk of heart disease.	8.2	3.6–6.7
Glucose	High blood glucose concentration is also a risk factor. This test showed that Jim had diabetes although he hadn't known it. People with diabetes are at a slightly higher risk of heart disease.	13.6	less than 7
Aspartate transaminase	Aspartate transaminase and lactate dehydrogenase are enzymes normally found inside heart muscle. During a heart attack the muscle is damaged	235	5–40
Lactate dehydrogenase	and some of the enzymes are released into the blood. Jim's test shows an elevated level of both enzymes, confirming his heart attack.	2266	300–650

Treating cardiovascular disease

Although emergency resuscitation is the first treatment needed to save the life of someone who has just suffered a major heart attack or stroke, there are many different treatments for underlying CVD.

Heart bypass surgery

A **coronary artery bypass graft operation**, sometimes called CABG or 'cabbage', is done to reroute blood around clogged coronary arteries to improve the supply of blood and oxygen to the heart. The surgeon sometimes takes pieces of vein from the leg of the patient, sews one end into the aorta and positions the other end to a coronary artery below the blocked area (Fig 35.14). In this way, the blockage is bypassed and normal blood supply is restored. It is also possible to detach one of the arteries that lead into the chest wall and then to graft it into the coronary artery. The number of grafts made is indicated by the way the surgeon describes the operation – a double heart bypass involves two grafts, triple and quadruple bypass operations use three and four.

F (a) Study Fig 35.13 and suggest how the control group was treated.

(b) Look at Table 35.2. Name two risk factors for heart disease.

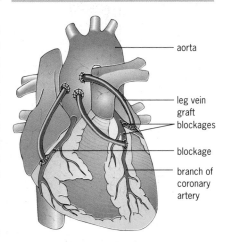

aorta

leg vein graft

blockages

blockage

branch of coronary artery

Fig 35.14 **The principle of a triple heart bypass**

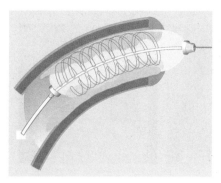

Fig 35.15 **Balloon angioplasty and stent**

Angioplasty

Angioplasty is used to treat angina or to reduce the risk of a heart attack. In **balloon angioplasty**, a thin tube with a deflated balloon at the end is introduced through the skin and fed into the coronary artery that has become narrowed by atherosclerosis (Fig 35.15). The surgeon checks its progress using a real-time high-speed X-ray image and, once the end of the tube is in the right place, the balloon is inflated gently. This pushes through the obstruction, making the lumen of the artery bigger.

Sometimes in balloon angioplasty, a short length of rigid tubing called a **stent** is left in the artery. This keeps the vessel open and prevents further narrowing at that point. Although angioplasty is most often done in coronary arteries, it can be carried out in any blood vessel in the body.

Common drug therapies

Hundreds of drugs are used in the treatment of heart disease. Some have already been discussed in more detail in this chapter, but Table 35.3 provides an overview of some of the main classes.

Table 35.3

Drug type	Drug	Used to treat	Effect on the body
Coronary vasodilator	Isosorbide dinitrate and other nitroglycerins	Angina	Relaxes veins in the body. The blood that returns to the heart is reduced and so that heart has to work less hard.
Beta blockers	Inderal	Angina; to prevent heart attacks in people with coronary artery disease; to reduce death rate in people who have suffered a heart attack.	Blocks receptors on the heart that respond to adrenaline and related hormones, preventing heart from working harder when these hormones are released.
Diuretics	Bemtanide	High blood pressure; congestive heart failure	Causes kidneys to retain less water, reducing excess fluid in the body.
Calcium channel blockers	Cardizem	Angina; coronary artery disease; high blood pressure	Relax the muscular walls of the arteries. In coronary arteries, this enables better blood supply to heart muscle. In arteries in the rest of the body, it reduces blood pressure.
Cholesterol-lowering drugs	Fluvastatin	High blood cholesterol and high LDL	Reduces total blood cholesterol, targeting LDL specifically by inhibiting an enzyme that is involved in LDL production. Used in combination with a low fat, low cholesterol diet.
Anti-arrhythmics	Digoxin	Heart rhythm problems; angina; high blood pressure	Different problems have different effects. Digoxin increases the strength of contraction of the heart muscle as well as helping to prevent abnormal heart rhythms.
Anti-coagulants	Warfarin, aspirin	Post-heart attack or stroke treatment to prevent further attacks	Makes blood less likely to clot, reducing the danger of thrombosis and thrombophlebitis.

SEND IN THE CLOT-BUSTERS

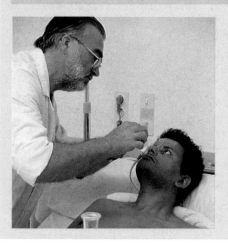

WHEN SOMEONE has had a heart attack or a stroke and is recovering, part of the treatment is designed to prevent further attacks. Giving up smoking, eating less fat and doing more exercise are priorities, but they take time to have an effect. While they are most at risk, heart attack or stroke victims, particularly those with quite severe underlying circulatory problems, are often prescribed clot-busting drugs such as warfarin or aspirin. These drugs 'thin' the blood so that it is less likely that a clot will form and break away from the blood vessel wall to cause an arterial blockage in the heart or brain. Several large clinical trials have shown that a small amount of warfarin or half an aspirin tablet taken every day, decrease the chance of having a second heart attack or stroke.

Fig 35.16 **Warfarin is widely used as a clot-busting treatment. It is also an ingredient of rat poison since large amounts of warfarin causes heavy internal bleeding – that's what kills the rats. Patients taking warfarin can be more prone to bleeding as a side-effect of their treatment, and so their condition and dose must be carefully monitored**

Preventing cardiovascular disease

■ See question 1.

Treatment for cardiovascular disease is improving constantly, but that old saying 'prevention is better than a cure' holds true. Once the arteries have become clogged with atheroma, the process is irreversible, and surgery and drug therapy can only treat the symptoms it causes.

Changing our lifestyle to prevent CVD in the long term is the answer, and education programmes and public health advice stress three main points:

- Never smoke. If you do, give up, or at least cut down.

- Eat a balanced diet that contains less than 25 per cent of its calories as fat. Saturated fat should be no more than 10 per cent.

- Take regular exercise. Exercise that is aerobic – that leaves you out of breath – should be done at least three times a week for at least 20 minutes.

3 CANCER

Cancer has been a major cause of death and disease throughout recorded history, but it is now the second biggest cause of death in the western world. There are two probable reasons for this apparent increase. First, many people are surviving longer, free from other diseases, and so have an increased chance of developing cancer. Second, we are now more likely to be exposed to carcinogens (cancer causing agents) such as cigarette smoke and sunlight.

Cancer is not a single disease: over 200 different types of cancer have been identified. This has led to some bewildering jargon, with tumours named according to the tissue they are in, their growth pattern, their effect on the patient, their response to treatment or the person who discovered them.

However, all cancers are basically similar: they all result from uncontrolled cell growth. The most common sites for cancer in the human body are shown in Fig 35.17. The number of people in the UK who die from each type each year is shown in Table 35.4.

Table 35.4 **Annual UK death rates in men and women for the ten most common types of cancer**

Men		Women	
lung	28312	breast	34604
skin (non-melanoma)	19934	skin (non-melanoma)	18265
prostate	15654	bowel	15651
bowel	15641	lung	13701
bladder	9232	ovary	5951
stomach	6920	stomach	4493
non-Hodgkin's lymphoma	3899	uterus	4191
oesophagus	3565	cervix	4173
pancreas	3373	bladder	3662
leukaemia	3238	pancreas	3564

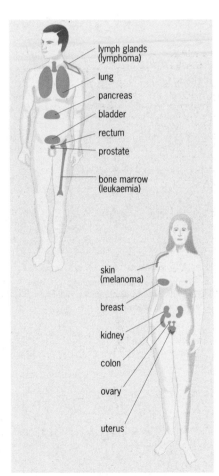

Fig 35.17 **Common cancer sites in males and females. Note that non-Hodgkin's lymphoma (Table 35.4) is a cancer of cells in the blood. The blood system and lymphatic system are both affected**

What is cancer?

Cancer occurs when there is a breakdown in the cellular control mechanism that puts the brakes on cell division. So cells that should be stable begin to divide, forming a tumour. A **tumour** is a swelling that can occur almost anywhere in the body. It is made up of a mass of abnormal cells that divide continuously.

Fig 35.18 **Benign tumours are not usually as big as this, but even those that are can be treated successfully by surgery**

Some tumours are **benign**. Although they can grow to the size of a grapefruit, they do not actually destroy the surrounding tissue or spread to other organs (Fig 35.18). Other tumours are **malignant**. They destroy the surrounding tissue and their cells often break away and spread through the blood or lymph system into other sites where they form **secondary tumours**. A malignant tumour is what we usually describe as cancer.

The rate of cell division varies greatly. Some tumours develop quickly, others can take ten years to reach a noticeable size. Cancer cells usually fail to differentiate: they cannot specialise for the particular function of the tissue they grow in.

Fig 35.19 shows how cancerous cells spread and look different from normal cells. In 35.19(b), notice that the cancerous cells are smaller, with clear, enlarged nuclei – the classic signs of active cell division.

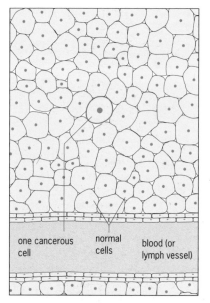

one cancerous cell normal cells blood (or lymph vessel)

Fig 35.19(a) **How cancers develop**
1 One cell starts to divide uncontrollably: it becomes cancerous

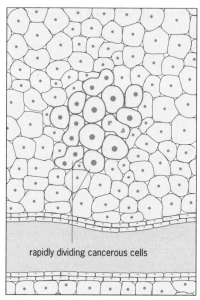

rapidly dividing cancerous cells

2 The cancerous cell divides rapidly, forming a mass of cells – the primary tumour – which squashes out the neighbouring normal cells

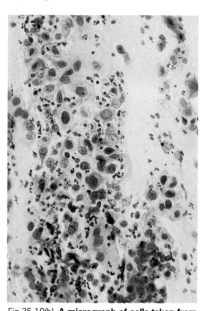

Fig 35.19(b) **A micrograph of cells taken from the cervix of a woman during a routine smear test. The pathologist then examines the tissue for signs of abnormal mitosis. Here, the cancerous cells are clearly seen because they stain red and have relatively large nuclei**

Why is cancer so dangerous?

Tumours interfere with the activity of the cells in the tissues of organs that surround them. Benign tumours can compress tissues, preventing normal blood flow or nerve function. Malignant tumours do even more damage, invading surrounding tissues and killing normal cells in the process. Cancer can also spread to other parts of the body by the process of **metastasis** (Fig 35.20).

Many tumours form from epithelial cells and, in order to break away, they must escape from the proteins that bind them to the underlying basement membrane that is characteristic of epithelial tissue (see Chapter 2). They do this by secreting protein-digesting enzymes. They also produce reduced amounts of the **adhesion molecules** that usually hold cells together.

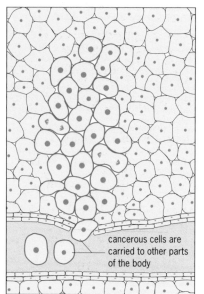

cancerous cells are carried to other parts of the body

Fig 35.20 **The process of metastasis. A clump of cells breaks free from the primary tumour and is then carried by the blood or lymphatic system to another part of the body. Fortunately, very few of these clumps of cells, about one in ten thousand, are able to establish themselves and form a secondary tumour, but that is enough. Nearly 60 per cent of people who are diagnosed with cancer are found to have well established secondary tumours**

The biological basis of cancer

Although there are many different types of cancer, we now know that all cancers are the result of DNA damage in cells. If this damage occurs in specific genes, control of cell division in that cell can be lost and it can begin to divide continuously (Fig 35.21).

Cell division is described in detail in Chapter 27. You might want to revise this before going on to look at how control of the process can break down.

Genes and the control of cell division

In Chapter 30 we saw that different types of cell in the body divide at different rates. Some, such as the neurons in the brain, do not divide at all after we reach the age of about 2. Others, such as liver cells, divide only rarely. Others, notably epithelial cells, divide as often as once every couple of days. Even so, all cell division is tightly controlled at the genetic level.

Some genes in the cell code for proteins that activate other genes to produce growth factors, chemicals that start a sequence of events that leads to cell division. Other genes code for proteins that inhibit cell division. Genes are turned on and off according to the cell type and its position in the body. Needless to say, the interplay of all the different genes involved is very complex. However, scientists have identified a few genes that, when they suffer a mutation, can cause the cell to lose control of its division pattern. These cells have been named **proto-oncogenes**, or just **oncogenes**.

The human genome contains many oncogenes. One, the *ras* oncogene is on chromosome 11. It codes for a chemical called G-protein which basically acts as an on-switch for cell division. Normal G-protein is usually inactivated very quickly by cellular enzymes. This means that the G-protein on-switch for cell division is only temporary and is switched off most of the time.

If the DNA in the *ras* oncogene is damaged, however, it codes for a G-protein that cannot be inactivated. In a cell with a defective *ras* gene, the switch for cell division always says 'on'. Cells that lose control of cell division can become cancerous, and *ras* mutations are associated with 20 to 30 per cent of all human cancers.

However, mutations in oncogenes are much more common than the cancer rate would suggest. This is because the cell has a second, back-up control. It also has genes called **tumour suppressor genes**. These prevent cells dividing too quickly, until the body's immune system can either kill the rogue cells, or until the damage in the DNA has been repaired by the cell's enzymes (Fig 35.22).

The tumour suppressor genes do an excellent job, but their DNA too can be damaged and they can suffer mutations. If these genes fail to work, this back-up system is lost and any cells that start dividing uncontrollably are more likely to go on to develop into cancers. A gene called *p53*, on chromosome 17, has been well studied, and about half of all human cancers have been linked with a mutation in this tumour suppressor gene.

Oncogenes and tumour suppressor genes can both mutate because of DNA damage caused during life. Occasionally, though, faulty forms are passed on to the next generation. People who are born with a gene defect in one of these genes are said to have a genetic predisposition to cancer. They are more likely to develop cancer and often do so at a relatively early age (see the Feature box on page 540).

G How might the ability of some cancers to spread very rapidly to other organs be related to the protein-digesting enzymes that the tumour cells contain?

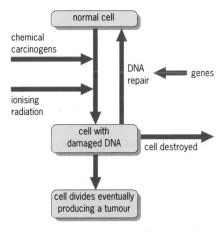

Fig 35.21 **Tumours arise when DNA in specific genes is damaged**

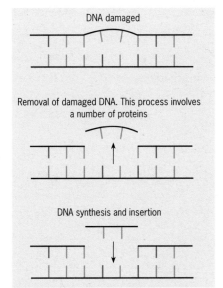

Fig 35.22 **Cells have many different ways of repairing DNA. This process is also controlled by genes. Mutations affect genes that repair other, damaged genes, as well as genes that control cell division**

Risk factors and cancer

So, two different classes of gene in the cell can fail and can result in cancer. But what does the damage to the DNA of these genes? Many things in our environment are carcinogenic. Some of the risk factors for cancer are connected with our lifestyle, others are present in our environment. In both cases, we can usually modify our cancer risk by changing our behaviour. This chapter looks at just a few of the major lifestyle and environmental factors that affect cancer risk.

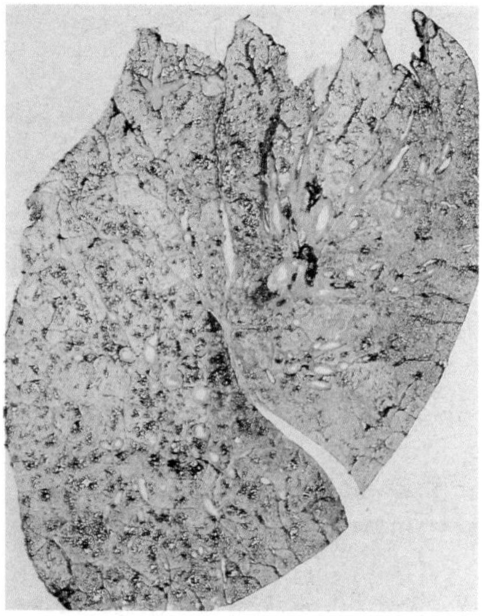

Smoking

Smoking is now one of the major causes of lung cancer in the developed world. Each year, lung cancer due to smoking leads to around 40 000 deaths from cancer in the UK (a large sports stadium holds about 40 000 people) and to around 175 000 deaths in the USA.

The link between smoking and lung cancer has been recognised for many years but it's only recently that scientists have identified a compound in tobacco smoke that binds to and damages the *p53* tumour suppressor gene. The longer a person smokes and the greater the number of cigarettes they have a day, the greater the chance that the *p53* gene will be damaged.

Fig 35.23 **If you smoke, your lungs probably look like this, clogged up with tar**

Diet

There is strong evidence that some of the things we eat and drink contain substances that are carcinogenic:

- Drinking alcohol is linked to 3 per cent of cancers in the western world, causing higher rates of mouth, throat and oesophageal cancer. Smokers who drink large amounts of alcohol are particularly at risk.

- Heavily salted and smoked foods also increase the risk of developing mouth, throat and oesophageal cancer. About 1 per cent of all cancers are due to eating salted and smoked foods regularly.

- Eating a diet very high in over-cooked meats is thought to increase the risk of stomach cancer, because charring red meat (such as blackening it on a barbecue) produces breakdown products that are carcinogenic.

- Food contaminated with aflatoxins, substances produced by the fungus *Aspergillus flavus*, have been linked to specific cases of liver cancer. In some parts of Africa, groundnuts (peanuts) contaminated with this fungus are a staple food and liver cancer is common. Aflatoxins are chemically modified by liver enzymes. This modified aflatoxin affects the base guanine, which is of the DNA molecule.

Apart from these specific examples, the cancer risk of eating specific foods is quite difficult to assess. People's diets are extremely varied and many studies involving hundreds of thousands of people are just beginning to uncover some definite patterns. However, the American Cancer Society states that, on the basis of current scientific evidence, one-third of the 564 800 cancer deaths that occurred in the US in 1996 were related to diet.

At the moment, the generally accepted view is that the diet to *avoid* is one high in red meat, particularly meat that is charred, high in animal fats, and low in fibre, fresh fruits and vegetables (see the next Feature box).

EAT THOSE GREENS

THERE IS A GREAT WEIGHT of evidence that eating lots of fresh fruit and vegetables decreases the risk of developing cancer – so much evidence, in fact, that governments and health organisations in Europe, the US and Australia all advise that we each eat 5 portions of fresh fruit or vegetables every day.

But why is fruit and veg that is so protective? One theory is that the high levels of vitamins they contain, particularly vitamins A and the B group, act as anti-oxidants. This type of chemical mops up free radicals that form in the body all the time, and that can damage DNA. The problem is that several large trials have shown that, while taking vitamin supplements does no harm, it doesn't actually lower cancer risk.

Other chemicals in fruit and vegetables are responsible. Three that have already been isolated are:

- Chemicals called dithiol-thiones that occur in vegetables such as broccoli, cauliflower and cabbage seem to have several effects on the body. They activate liver enzymes that deal with carcinogens and they inhibit tumour growth.

- Sulphorane, another chemical isolated from broccoli, also activates helpful liver enzymes.

- Genistein, a compound found in soya beans, prevents the formation of the blood vessels that tumours need to support their rapid growth.

Many studies are now planned, and scientists expect to develop new drugs based on natural plant compounds that can cut cancer deaths by prevention.

Fig 35.24 **These vegetables are known to help reduce the likelihood of cancer**

Exposure to radiation

We are all exposed to a common source of radiation – the Sun. Exposure to sunlight is linked with melanoma skin cancer which accounts for about 2 per cent of all cancer deaths. It occurs most often in fair-skinned people and is caused by the ultraviolet light in sunlight. Particularly dangerous are the higher-frequency UVB rays that can damage DNA: UVB radiation causes 90 per cent of all skin cancers.

The cancer risk posed by other forms of radiation seems to be low. Excessive radiation such as the fallout from the atomic bomb blasts at Hiroshima and Nagasaki led to cancers in the survivors, but the low levels emitted by nuclear power plants do not seem to increase cancer risk. There is also little evidence to support the theories that things like mobile phones, microwaves or overhead power lines increase the risk of cancer.

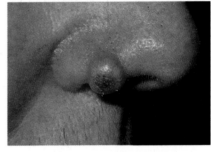

Fig 35.25 **Skin cancers are more likely to develop on parts of the body that are exposed to a lot of sunshine**

Exposure to chemical carcinogens

Substances like asbestos, benzene, methanal (formaldehyde) and diesel exhaust are all carcinogenic. We know this because of instances in the past where people have been exposed to very high levels – usually through their work. Today there is much greater control of work environments, but many cancers of this type are expected in the developing world as countries start industrialisation programmes.

Exposure to microorganisms

Specific microorganisms are linked to the development of some cancers, but this is quite rare.

- The Epstein–Barr virus, which causes glandular fever, has been linked to some cancers of the lymph gland in the western world and with nose and throat cancer in Asia.

- HIV leads indirectly to an increased cancer risk: people with AIDS have a much higher rate of a skin cancer called Kaposi's sarcoma because their immune system cannot weed out rogue cancer cells.

- Infection with genital wart viruses is associated with higher rates of cancer of the cervix. Women who have many sexual partners are more likely to be exposed to wart viruses and so run a higher risk of developing cervical cancer.

- Infection with the bacterium *Helicobacter pylori*, recently found to cause stomach ulcers and gastritis, also causes higher rates of stomach cancer.

CANCER AND HEREDITY

ABOUT 5 TO 10 PER CENT of cancers arise because of faulty genes that can pass from one generation to the next. In the last 10 years, several such genes have been identified.However, before the genes were recognised, it was clear that some cancers were hereditary:

- Some families have many members all with the same type of cancer.

- Cancer tends to develop early – when those affected are less than 50 years old.

Study of the DNA from members of such families have provided a lot of information about genes and cancer. Table 35.5 shows some of the genes that, when they are faulty, confer a very high cancer risk.

Table 35.5 **Genes and cancer risk**

Gene		Tumour type	Gene class
Breast cancer	BRCA1	breast, ovary	tumour suppressor
	BRCA2	breast (both sexes)	tumour suppressor
	p53	breast, sarcoma	tumour suppressor
Colon cancer	MSH2	colon, endometrium, other	mismatch repair
	MLH1	colon, endometrium, other	mismatch repair
	PMS1,2	colon, other	mismatch repair
	APC	colon	tumour suppressor
Melanoma	MTS1 (CDKN2)	skin, pancreas	tumour suppressor
	CDK4	skin	tumour suppressor
Neuroendocrine cancer	NF1	brain, other	tumour suppressor
	NF2	brain, other	tumour suppressor
	RET	thyroid, other	oncogene
Kidney cancer	WT1	Milms' tumour	tumour suppressor
	VHL	kidney, other	tumour suppressor
Retinoblastoma	RB	retinoblastoma, sarcoma, other	tumour suppressor

Cancer – the patterns of disease

In the nineteenth century, cancer caused the deaths of fewer people than it does today. We must be careful how we interpret this statement. A hundred and fifty years ago infectious diseases were much more important as causes of death than they are now. Because of this, the average age of death was lower than it is today, and fewer people lived to the ages where cancers become increasingly common. In addition, medical treatment was far more basic and methods of diagnosis meant that the disease could be recognised only when it had reached an advanced stage.

If you were to compare data that shows the deaths from different types of cancer in one particular year, with the corresponding figures from 10 years earlier, you would notice differences. Between 1996 and 1986, for example, stomach cancer, declined because of the discovery that various gastric problems, including cancer, were caused by infection by *Helicobacter pylori*. Treating people for this infection cut stomach cancer cases signficantly during this time.

You might also notice an increase in the incidence of some cancers. Breast cancer cases, for example, increased between 1986 and 1996 because of advances in methods of screening and diagnosis. Much of the apparent increase was probably due to doctors being able to detect the condition earlier than they would have.

However, often it is not as easy to account for differences. Most of the time, statistics on cancer death rates and cancer incidence are difficult to interpret.

It might be better to ask some questions: Are the differences actually statistically significant? An apparent increase in cases might be only a reflection of an underlying trend, or the differences could be just the result of chance. Has there been a change in exposure to particular risk factors? Could we explain any changes in the incidence of lung and bladder cancer in terms of changes in smoking habits? Is the picture incomplete, and does this confuse the issue? Why are some cancers more common in one sex than in the other?

Answers to questions like these are important if we are to understand the factors affecting the trends shown by such data. The Assignment at the end of this chapter provides some more information about how patterns in the incidence of a particular type of cancer can shed light on the factors that cause it.

The battle against cancer

Cancer prevention

Although drugs and food supplements to prevent cancer are a possibility, they are likely to take time to develop. In the meantime, changing your lifestyle could help you reduce the risk of cancer in later life. Four steps seem to be the most important:

1 Never smoke. If you smoke, stop.

2 Eat a well balanced diet in which animal fat makes up less than 25 per cent of the calories. Include at least 5 portions of fresh fruit of vegetables every day, and try to obtain most of your protein from plant sources (beans, peas, grains etc). Avoid very salty or smoked foods and charred meats, particularly red meats.

3 Take regular exercise and maintain a sensible weight.

4 Drink alcohol in moderation.

Screening for cancer

Successful cancer treatment often depends on getting an early diagnosis. The longer a malignant tumour is left, the larger it grows and the more likely it is to spread and affect other organs. Surgery can be used to remove a localised tumour but it is of little use if the condition has spread throughout the body.

Skin cancer is one of the commonest forms of cancer in younger people, but at least it is easily noticed. Surgical removal is very effective and results in 97 per cent of people surviving five years after the operation. Cancer of the colon is a very different matter. There are no visible signs and early symptoms are vague. Most people don't take them seriously and end up visiting their doctors for the first time when the cancer is already quite advanced. Not surprisingly, the five-year survival rate for this form of cancer is much lower, at only 37 per cent.

Regular screening can help to detect cancers early, but the advantages have to be weighed against its high financial cost.

?

H Look back at Section 1 in this chapter, and you will notice that the lifestyle recommendations that help to prevent cancer are much the same as those that help prevent heart disease. Why do you think large numbers of people in the UK and the US ignore them most of the time?

Fig 35.26(a) **A patient is undergoing routine mammography. This enables tumours to be identified early, and increases the chance of treatment being successful**

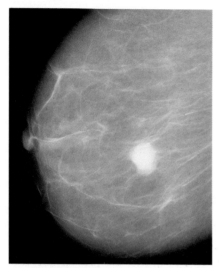

Fig 35.26(b) **This mammogram has been enhanced with the aid of a computer. The denser tissue of the tumour shows up as a white area**

?

I Explain why it is unlikely that there will be a national prostate cancer screening programme.

See questions 2 and 3. ■

The National Health Service has screening programmes for two of the most common female cancers, breast cancer and cervical cancer. Breast cancer screening is carried out every three years by mammography (Fig 35.26). It targets women over 50 years old since breast cancer occurs most often in women over this age and because breast tissue in younger women is too dense for mammography to be effective. The breast cancer programme has probably caused a 20 per cent fall in death rate, simply because cancers have been picked up and treated early enough.

The national cervical screening programme was launched in 1988 and has cut deaths from cervical cancer by about 30 per cent. The programme recommends that a smear is taken every three years. Cells are stained and examined under a microscope, and those likely to become cancerous can be identified (see Fig 35.19(b) on page 536). Early treatment is then possible.

Screening programmes for other cancers might not be not so effective. Prostate cancer is common, particularly in older men, and there methods of identifying it at an early stage. These include using ultrasound or testing the blood for high concentrations of prostate-specific antigen or PSA, an enzyme secreted by cells in the prostate. An enlarged prostate gland produces more PSA than a healthy one.

But is screening cost-effective? Can it be used to prevent cancer deaths? The answer to some extent depends on where you live and what kind of health-care system you have locally. In the US, where private medical insurance bears the cost, PSA screening is commonplace, even though there are some doubts about its reliability. Early treatment of prostate cancer is carried out routinely. In the UK, the cost of a general screening programme would be an enormous burden on the NHS and would save relatively few lives: most men with fully developed prostate cancer are very elderly and are usually at far higher risk of dying from some other condition.

Treating cancer

Surgery, chemotherapy and radiotherapy are important cancer treatments (Fig 35.27). They are all constantly being improved by research and have varying success rates, depending on the type of cancer and its site in the body. Drug treatment for leukaemia in children, for example, now cures 9 out of 10 children. Lung cancer can be much less easy to treat: the lungs cannot be completely removed by surgery, and the tissue is delicate and easy to damage.

New strategies are approaching cancer treatment from several different angles. Table 35.5 describes some of them.

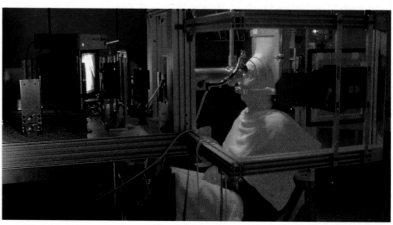

Fig 35.27 **This person has a tumour in the left eye and is undergoing proton radiation therapy.**

Table 35.6 **New strategies for treating cancer**

Treatment	What it does	Current status
Conformal radiation therapy	New technology allows radiation beams to be concentrated only on the tumour, avoiding damage to surrounding healthy tissue. The exact 3-D position of the tumour is found using CT scanning or MRI scanning, and the positioning of the treatment beams is determined by computer.	In relatively widespread use: new types of tumours are treated every year
Neutron and proton radiation therapy	X-rays and gamma rays are commonly used in radiotherapies; protons and neutrons can also be used.	Recent studies show that protons can treat small tumours of the spine that lie near vital organs, and that neutrons can treat salivary gland tumours.
Immunotherapy	New drugs to treat cancer are being developed all the time, but some of the most promising new drugs recognise substances found on the surface of tumour cells, and attack them while ignoring healthy cells.	Clinical trials are in progress. Some drugs are in use.
Differentiating agents	Drugs which do not kill tumour cells but force them to differentiate into cells with a specific function. When this happens, the cells stop dividing and tumour growth stops.	Several new drugs are in early clinical trials.
Angiogenesis inhibitors	Chemicals which stop the fast growth of new blood vessels into a tumour. Without a good blood supply, the tumour expands more slowly and can be easier to treat by radiotherapy or more traditional forms of chemotherapy.	Several angiogenesis inhibitors are undergoing clinical trials in patients, and other compounds that show potential are in an earlier stage of research.
Oncogene blockers	Scientists have been looking at how mutated oncogenes lead to cancer. They are trying to develop drugs to block the product of the damaged gene, so that the cell does not continue to divide uncontrolled.	Any cancer that arises because of a mutation in an oncogene.
Gene therapy	If a tumour suppressor gene becomes damaged, it no longer inhibits cancers from forming. In addition to treating the cancers that result, researchers are looking at the possibility of replacing the damaged gene in tumour cells.	Experimental studies using cells in culture have been quite successful. Clinical trials could begin in the early years of the twenty-first century.

SUMMARY

When you have studied this chapter, you should know and understand the following.

■ **Cardiovascular disease** (CVD) and **cancer** are the two main lifestyle diseases.

■ Lifestyle diseases are **non-transmissible**. They are associated with different risk factors such as smoking and diet.

■ The underlying cause of heart disease is the process of **atherosclerosis.** This is commonly called hardening of the arteries.

■ The major risk factors for CVD are hypertension (high blood pressure), smoking and diet.

■ Although many cases of CVD are preventable and new treatments are being developed all the time, this disease is a major killer in developed countries.

■ Cancer occurs when cells start to divide out of control. Many factors – genetic and environmental cause this to happen.

■ The major risk factors for cancer are smoking and diet. Many cases of cancer are preventable.

QUESTIONS

1 The table shows the results of some blood tests carried out on a patient admitted to hospital suffering from a suspected myocardial infarction (heart attack).

Substance	Concentration in patient's blood/ arbitrary units	Range of concentration in blood of healthy individual/ arbitrary units
Urea	5.7	2.5–6.7
Cholesterol	8.2	3.6–6.7
Lactate dehydrogenase enzyme	2263	300–600
Potassium	4.3	3.4–5.2

a) A myocardial infarction results in damage to the muscles of the heart.
    (i) Explain how a blood clot may cause damage to the muscle of the heart.
    (ii) Explain how the results of this blood test confirm that this patient had suffered a myocardial infarction.

b) Use the table to explain what is meant by 'a risk factor'.

[AEB Human Biology AS Specimen paper 2000, q.6]

2

a) The flow chart summarises one way in which tumours may arise.

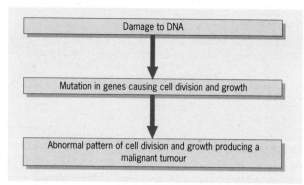

Fig 35.Q2(a)

(i) Use the information in this flow chart to help explain why there has been a recent increase in skin cancer in the United Kingdom.

(ii) Describe how a malignant tumour differs from a benign tumour.

b) Chemotherapy involves the use of drugs to kill tumour cells. The graph shows the effect of different doses of a drug used in chemotherapy on tumour cells and on healthy body cells.

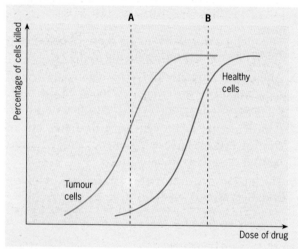

Fig 35.Q2(b)

(i) Which dose of the drug, **A** or **B**. would you use to treat a patient with a tumour? Give the reason for your answer.

(ii) Explain why it might be necessary to give the patient several sessions of treatment with the drug.

[AEB Human Biology Specimen paper 2000, q.7]

3 Read the following passage.

One aim of cancer therapy is to find a magic bullet that seeks out and kills tumour cells but leaves normal cells unharmed. For this to work, the bullet needs to be able to recognise a difference between the two types of cell. Some tumours grow so fast that they outgrow their blood supply and the oxygen concentration in their cells falls. Drugs are being developed that are effective only once they reach the low oxygen conditions inside a tumour cell. Here enzymes called 'reductase enzymes' activate the drug which then kills the cell.

Professor Stratford and his colleagues at Manchester are taking advantage of the fact that the P450 reductase gene is switched on only in an environment which is low in oxygen. His team have constructed the piece of DNA which is shown in Fig 35.Q3.

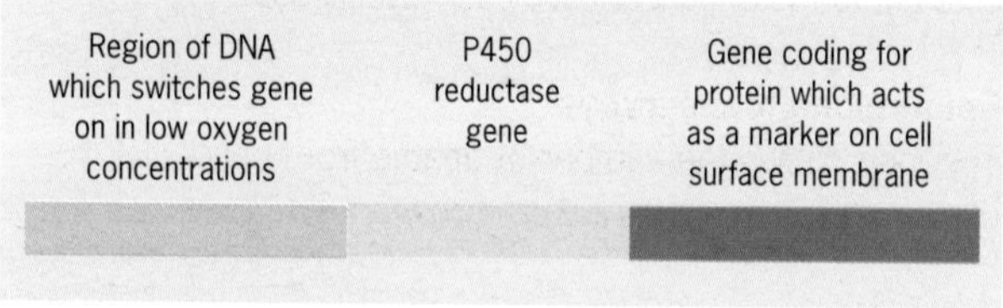

Region of DNA which switches gene on in low oxygen concentrations	P450 reductase gene	Gene coding for protein which acts as a marker on cell surface membrane

Fig 35.Q3

This piece of DNA was injected into breast cancer cells and the cells were grown in the laboratory. The marker protein was used to identify cells with the injected gene. When the oxygen concentration was reduced, the concentration of P450 reductase increased.

Use information from the passage and your own knowledge to answer the following questions.

a) Apart from the rates at which they grow, give **one** way in which tumour cells differ from normal cells.

b) Explain why the oxygen concentration in tumour cells may fall (sentence 1 of paragraph 2).

c) Explain why the drugs mentioned in this passage do not kill normal cells.

d) Name the type of enzyme used to:
   (i) remove the P450 reductase gene from the length of DNA;
   (ii) join the three pieces of DNA together.

e) (i) Explain why the investigators added the gene coding for the proteins which acts as a marker on the cell surface membrane to their specially constructed piece of DNA.

   (ii) Some antibodies fluoresce when illuminated with ultraviolet light. Suggest why these antibodies could be used to identify the cells which had the marker protein on their cells surface membranes.

[AEB Human Biology AS Specimen paper 2000, q.9]

Assignment

INVESTIGATING CANCER

Finding out what causes a particular cancer is not easy. Suppose we wanted to investigate whether hair dyes cause cancer. We could divide a randomly-chosen sample of people into experimental and control groups; expose the experimental group to the hair dye and then compare the number of cases of cancer in both groups.

Clearly, this approach is not acceptable ethically. In this Assignment we look at some ways in which the causes of different forms of cancer have been investigated.

Epidemiological surveys

The map shows the number of deaths from cancer of the stomach in England and Wales between 1921 and 1930.

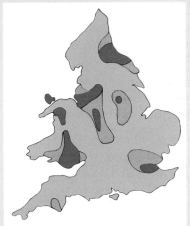

Fig 36.A1 **Deaths from stomach cancer in England and Wales between 1921 and 1930**

■ Extremely high
■ Moderately high

1

a) Explain why, in interpreting these figures, we should take into account the age and sex ratio of the population in different parts of the country.

b) Describe the general patterns in stomach cancer deaths shown in the map.

c) Suggest **three** possible explanations for this distribution.

The more information there is, the easier it becomes to link risk factors with particular cancers. Table 35.A1 summarises data showing stomach cancer deaths among people with different occupations.

Table 35.A1

Occupation	Number of people dying of lung cancer between the ages of	
	15 and 64	65 and 74
All occupations	100	100
Miners and quarry workers	141	127
Glass and ceramic makers	108	73
Labourers	113	108
Warehouse workers	107	108

2

a) Explain why it is necessary to use statistical tests when analysing data from Table 35.A1.

b) Suggest a hypothesis to account for the variation in the number of cases of stomach cancer in England and Wales between 1921 and 1930.

Case-controlled studies

Another approach is to use a **case-control study**. Two groups of people are used. One, the case group, consists of a large number of people who have a particular form of cancer. An equal number of people who do not have the cancer are chosen as the control group. Both groups should be roughly the same age and have the same proportion of males to females to make sure the study is valid. Both groups are then investigated to see if there is any differences between them.

A case-control study was carried out into the causes of childhood leukaemia. Mothers whose children had died from leukaemia formed the case group.

3 Describe the mothers who formed the control group.

Table 35.A2 shows how many mothers in the two groups had X-rays at various times before the birth of their children.

Table 35.A2

Time of X-ray examination	Case group	Control group	Ratio of case group to control group
Before marriage	44	26	1.69
Between marriage and conception	109	121	0.9
During pregnancy	178	93	1.91
At any time	296	215	1.38

4

a) Look at the last column in the table. What does a value of greater than 1.0 in this column mean?

b) What general conclusions can you draw from the results of this study?

Prospective studies

Once a risk factor has been identified, it can be investigated further using a **prospective study**. This identifies groups of people and monitors their health over months or years. If we wanted to investigate the effect of smoking on lung cancer, we would select a large group of people, with roughly equal proportions of smokers to non-smokers. All the people would be monitored carefully for several years and the number of cases of lung cancer would be recorded.

8 What would you expect the results of such as study to show. Give reasons for your answer.

36 Genetics, disease and the future

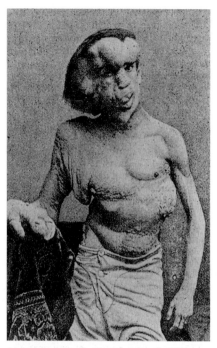

Joseph Merrick, the Elephant Man, in 1889

JOSEPH CARY MERRICK, better known as the Elephant Man, became well known when the story of his life was depicted in the film released in 1980. Merrick's physical problems were severe – his head was swollen and distorted and had many bony outgrowths. His skeleton was abnormal and he had difficulty walking because one of his legs was much shorter than the other. People at the time assumed he was also retarded and treated him as a curiosity and a freak, but he showed no signs of mental impairment.

Merrick died in 1890 and his remains have been studied by several generations of medical scientists. Many have assumed that Merrick suffered from neurofibromatosis, commonly called Elephant Man disease. People with mild forms of this genetic disorder, which occurs in about one in every 4000 births, sometimes have brownish spots on their faces or bodies. Merrick, however, had no skin markings, and over the years other inconsistencies have been noticed. Recently, studies were made of the skeletal remains of Merrick using the most up-to-date medical technology to see if the mystery could be solved.

The X-ray studies and CT scans showed that Merrick's skeleton was heavily disfigured and had large numbers of bony tumours. The pattern of bony growth suggests that Merrick actually suffered from the far more rare Proteus syndrome, of which only 120 cases have been reported worldwide. This disease, first identified 20 years ago, is triggered by dividing cells in bone that cause abnormal growths in the skull and other body tissues. No-one yet knows which genes are involved in Proteus syndrome and there is no effective treatment.

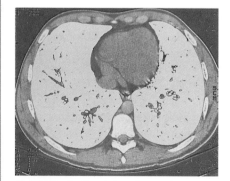

Fig 36.1 **Cystic fibrosis is a disease that results from a defect in a single gene. This CT scan shows the abnormal build-up of mucus in the bronchioles of a CF sufferer. Learn more about cystic fibrosis on page 478 and in the Assignment in Chapter 31**

1 GENETIC DISEASES

Some genetic diseases, such as cystic fibrosis (Fig 36.1), are caused by a single gene defect. Others are much more complicated. Not only are several genes involved; whether we develop that disease also depends on whether we are exposed to factors in our environment and in our lifestyle.

In this final chapter, we summarise the topics relevant to genetic diseases that we have covered elsewhere in the book (Table 36.1) and we look in detail at two other genetic conditions: **achondroplasia** (dwarfism), and **Tay-Sachs disease**. In the second section, the prospects that gene therapy offer for the treatment of genetic disease are explored, and in the final section in the book we highlight some of the ethical issues that are raised, not only by genetic disease, but also by the new treatments themselves.

Topic	Chapter	Page reference
Important concepts in genetics	26	386 to 393
Ethics of genetic engineering and gene cloning	26	388
	31	466
Genetic manipulation and cloning	31	471 to 472
Genes and cancer	35	537 and 540
The human genome project	31	467
Monohybrid inheritance in humans	29	428
Cystic fibrosis	29	429
	31	474
	32	478
Down's syndrome	27	405
	28	420
Sex-linked inheritance	29	434 to 435
Haemophilia	29	435
	34	511
Duchenne muscular dystrophy	29	435
Red-green colour blindness	29	435
Xeroderma pigmentosum	28	417
Sickle cell anaemia	28	423
Huntington's disease	29	424
Albinism	30	456
Phenylketonuria	32	483

Table 36.1 **Topics that you might find it useful to revise whilst reading this chapter**

Achondroplasia

Many conditions cause a lack of height. In some, the bones do not grow and develop normally. More than 100 specific **skeletal dysplasias**, as they are called, have been identified, and achondroplasia is the most common. It occurs about once in every 25 000 births in all races and with equal frequency in males and females.

Someone who has achondroplasia is disproportionately short. Their head is large and their arms and legs are short compared to their trunk. This shortness is particularly noticeable in the upper arms and thighs. Achondroplasiacs also tend to have a prominent forehead, a flat or even depressed area between their eyes, a protruding jaw, and crowded teeth. The typical signs of achondroplasia can be seen at birth, and the condition can be diagnosed early.

Achondroplasiacs, people with disproportionately short stature, often refer to themselves as dwarfs, little people, or short-statured persons. Intelligence and mental development is generally normal. Affected men average 131.6 cm in height, while women average 123.4 cm. There seems to be little or no relationship between the height of the parents and the adult height of their children with achondroplasia.

What causes achondroplasia?

This is a genetic condition that is due to a mutation of a single gene. The mutated gene is dominant. The condition can be passed from one generation to the next or it can result from a new mutation in a gene from average-sized parents. Nine out of ten children born with achondroplasia have average-sized parents, and no other family member is affected. The condition is all or nothing – a person with this mutation has achondroplasia, there are no mild forms.

In 1994, researchers discovered that a mutation of the fibroblast growth factor receptor-3 (FGFR3) gene on human chromosome 4 causes achondroplasia. Exactly how the mutations bring about the condition is not yet known.

The inheritance of achondroplasia

A couple in which one partner is achondroplasiac and the other is of average height has a one in two chance of having a child with achondroplasia. If the child does not inherit a mutated copy of the gene for achondroplasia, growth during childhood and adult height is usually within the normal range.

Average-sized children of parents with achondroplasia have no

GENETIC SCREENING AND DIAGNOSIS

THE OPENER IN CHAPTER 29 describes how the singer Woody Guthrie, along with other sufferers, only found that he had **Huntington's disease** when the symptoms appeared. By that time he was about 40, and had already had a son to whom he could have passed the disease. Now it is possible to screen newborn babies for an increasing list of genetic abnormalities.

All babies in the UK are tested routinely for phenyl-ketonuria. This condition and the heel test that detects it (Fig 36.2) is described on page 483. Cystic fibrosis, sickle cell anaemia (Fig 36.3) and **Tay-Sachs** disease (see page 549) can also be detected by screening techniques.

A simple blood test can tell Tay-Sachs carriers from non-carriers. Blood samples can be analysed by either enzyme assay or DNA studies. The enzyme assay is a biochemical test that measures the level of Hex-A in a person's blood. Carriers have less Hex-A in their body fluid and cells than non-carriers.

Fig 36.3 **Prospective parents can find out if they are likely to have children with sickle-cell anaemia. Electrophoresis is the technique used to screen their DNA for the sickle-cell gene**

DNA-based testing to test carriers of Tay-Sachs (people who show no symptoms but who are able to pass the recessive mutated gene on to their children) looks for specific mutations, or changes, in the gene that codes for Hex-A.

Since 1985, when the Hex-A gene was isolated, over 50 different mutations in this gene have been identified.

The limitation of DNA-based carrier testing for Tay-Sachs disease is that not all known mutations in the Hex-A gene are detected by the test, and others have yet to be identified. Since some carriers cannot be identified by DNA analysis alone, a blood test is always performed as well.

In fact, DNA testing can give more information when done in conjunction with the blood test, especially in cases where both members of a couple are carriers. The two tests complement each other and can reveal what type of Tay-Sachs disease is most likely to occur in the couple's child, if they decide to go ahead with a pregnancy after receiving genetic counselling. It also makes it easier to correctly diagnose Tay-Sachs in the developing fetus at an early stage of pregnancy.

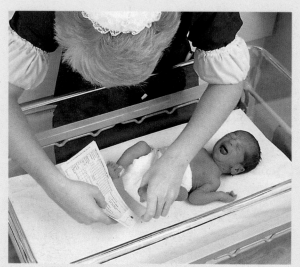

Fig 36.2 **Soon after birth, babies have the Guthrie test which screens for phenylketonuria. The brain damage this disease causes is avoided by having a special diet throughout life**

increased risk of having achrondroplasiac children themselves. If both parents have achondroplasia, there is a one in four chance that they will have baby who inherits two altered copies of the gene that leads to achondroplasia. These children are said to have homozygous achondroplasia and this condition is almost always lethal in the first year of life.

Treatment options

At present there is no specific treatment to promote growth in achondroplasia. Growth-hormone treatment seems to increase the rate of growth during the first year of treatment, but does not increase adult height. Surgery to lengthen the legs and arms of people with achondroplasia is being done experimentally in the US, but achondroplasiacs themselves have mixed views on whether this is a step in the right direction.

Tay-Sachs disease

Tay-Sachs disease (TSD) is a fatal genetic disorder in children that causes progressive destruction of the central nervous system. The disease was named after Warren Tay (1843–1927), a British doctor who identified the first patient with TSD in 1881, and after Bernard Sachs (1858–1944), an American neurologist who first described the cellular changes in Tay-Sachs disease.

Tay-Sachs disease is caused by the absence of a vital enzyme called **hexosaminidase A** (Hex-A). Without Hex-A, a fatty substance builds up to abnormally high levels in cells, especially in the neurones of the brain. This causes progressive damage to the cells. The destructive process begins in the fetus early in pregnancy, but the signs of the disease are not apparent until the child is several months old. By the time a child with Tay-Sachs disease is three or four years old, the nervous system is very badly damaged and all children with classical TSD die early in childhood, usually by the age of five.

There is no cure or effective treatment for Tay-Sachs disease. However, active research is being done in many laboratories around the world. The use of enzyme replacement therapy has been explored but its effectiveness has been very limited. Because the disease affects brain cells that are protected by the blood–brain barrier, the replacement enzyme cannot get to the cells that need it.

The genetics of Tay-Sachs disease

The genes that code for the enzyme Hex-A are on chromosome 15. If either or both Hex-A genes are active, the brain produces enough of the enzyme to prevent the abnormal build-up of the fatty deposits in its cells. Carriers of Tay-Sachs disease – people who have one mutated copy of gene as well as one copy of the active gene – are healthy. They do not have Tay-Sachs disease.

A carrier has a 50 per cent chance of passing the mutated gene on to his or her children; any child who inherits one mutated gene is a Tay-Sachs carrier like the parent. If both parents are carriers and their child inherits two copies of the mutated gene, that child will have Tay-Sachs disease since he or she cannot produce any functional Hex-A.

When both parents are carriers of the Tay-Sachs gene, there is a 1 in 4 chance with each pregnancy that their child will have Tay-Sachs disease, and a 3 in 4 chance (75 per cent) that their child will be healthy. Of their unaffected children, there is a 2 in 3 chance that each child will be a carrier, like the parents. This pattern of inheritance is called **monohybrid inheritance** (Fig 36.4 and see page 428).

Identifying carriers

Tay-Sachs often occurs in families with no history of the disease. The Tay-Sachs gene can be carried without being expressed through many generations. Before 1970, the only way to learn if you were a Tay-Sachs carrier was to be the parent of a baby with Tay-Sachs. Now, safe and reliable carrier testing is available to identify Tay-Sachs carriers. Most important, testing can identify carrier couples who are at risk for bearing a child with Tay-Sachs – before a tragedy occurs (see the Feature box on the opposite page).

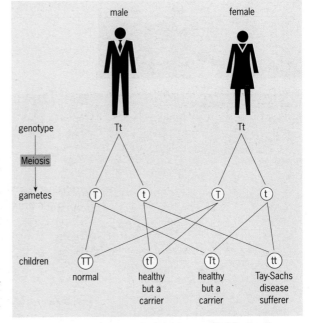

Fig 36.4 **The inheritance of Tay-Sachs disease. If both parents are carriers, there is a 25 per cent chance that any child will be a sufferer. There is also an equal chance that a child will inherit no faulty alleles and that his/her descendants will be completely free from the disease. There is a 50 per cent chance of the child being a carrier**

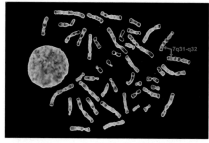

Fig 36.5 Cause and treatment of cystic fibrosis

Top: **Chromosome pair 7 is highlighted in pink, with the location of the cystic fibrosis gene mutation arrowed**

Centre: **Electrophoresis of the gene sequence shows that base sequence TTC is present in an unaffected person (left), and absent in someone with cystic fibrosis.**

Bottom: **The healthy gene is isolated before being inserted into a virus. Cystic fibrosis patients use an inhaler with the virus that delivers the healthy gene to lung cells**

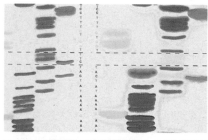

2 GENE THERAPY

The new technology of gene therapy promises to revolutionise medicine in the twenty-first century (Fig 36.5). The idea that it can be used to treat genetic diseases such as cystic fibrosis and Tay-Sachs disease has been around for several years.

Recently, though, scientists have begun to look beyond just the diseases caused by a defect in a single gene and are considering gene therapy as a potential treatment for all sorts of problems, from cancer and heart disease to AIDS (see the Feature box on the page opposite).

Gene therapy for genetic diseases

The first person to be given gene therapy was Ashanti DeSilva. She became the first patient with severe combined immunodeficiency (SCID) (Fig 36.6) to be treated in September 1990, when she was four years old. A team in the USA led by French Anderson gave Ashanti four infusions of cells containing the working gene that she lacked. Over the four months of the treatment, her condition improved.

With the help of follow-up treatments she has now been transformed from a small girl who was constantly ill and could not leave the house, to a normal, healthy and lively teenager. She continues to do well and, apart from needing extra treatment now and again, she lives a normal life.

Many children with various forms of SCID have since been treated using a similar method of gene therapy. French Anderson is now trying to find a way to perform gene therapy in the uterus for some types of SCID that kill an affected baby before it is born (see the Feature box on page 553).

36.6(a) **A SCID baby lives in a sterile compartment and is handled by medical staff wearing full body gowns to avoid passing on any infections**

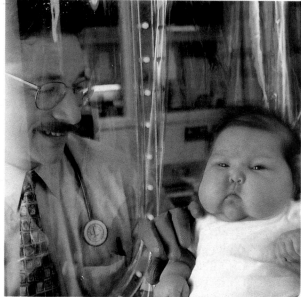

36.6(b) **This Apache baby is having gene therapy: some of his stem bone marrow cells are removed, the gene for the enzyme he lacks is inserted into them and they are returned to his body. Then he can produce normal immune cells**

GENE THERAPY FOR HIV INFECTION

IN 1998, SCIENTISTS at the National Institute of Allergy and Infectious Diseases in the USA reported their first findings from a study that set out to determine whether virus-fighting immune cells can be genetically altered using gene therapy, to boost the immune system's response to HIV infection.

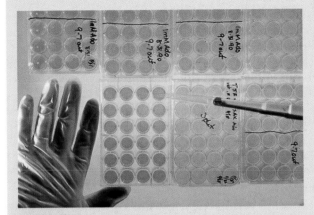

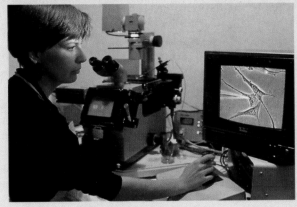

The NIAID scientists studied 30 sets of identical twins in which one twin was infected with HIV and the other was uninfected. HIV-fighting T helper cells were taken from the healthy twin and genetically altered to produce an extra receptor that helps the cells recognize HIV-infected cells. The engineered cells were then infused into the HIV-infected twin, to try to restore their ability to fight the virus.

Transfer of genetically altered T helper cells into 30 sets of HIV-infected twins seems to be safe and has produced no adverse reactions. By trying different combinations of T helper cells, the researchers learned that a mix of genetically modified cells proved to be the most long-lived. Researchers followed the cells by giving them a gene which produced a detectable marker on their surface. Results from these 'tracking' experiments showed that the T helper cells not only persisted at high levels in the bloodstream for at least 100 days after being introduced into the body, but also divided. This was quite an unexpected finding because it had been assumed that cells transferred into the body were quite short-lived and could not multiply.

NIAID scientists are now conducting the next stage of clinical trials of the experimental treatment. The new therapy provides hope that it will one day be possible to treat and cure AIDS using T helper cells removed from the patient. By genetically modifying these cells to allow them to deal with the virus, and then giving them back to the same person, problems of rejection would be minimised.

Fig 36.7(a) Upper: **Cells from a patient with a faulty immune system are genetically altered, cultured and returned in large numbers to the patient**

Fig 36.7(b) Lower: **When someone's T cells cannot produce the enzyme adenosine deaminase they need in order to mature, the correct gene can be inserted into these cells using micro techniques**

How gene therapy works

One of the main problems that make gene therapy difficult is finding out how to transfer genetic material containing corrected, healthy genes into cells that contain the faulty gene. Scientists have now developed several methods.

Modified viruses are often used as carriers, or **vectors**. Viruses are useful because they are able to penetrate cells and to introduce their genetic material into the host cell. Before researchers can use viruses in therapy, however, they must remove the genes that code for proteins that viruses use to reproduce themselves. When those genes are replaced with corrective genes, the vector virus becomes a delivery unit that is identical to the original virus on the outside, and can transport useful genes into cells, but that cannot cause illness.

Three main approaches to gene therapy are currently possible.

Ex vivo gene therapy

The most commonly used technique in gene therapy is known as **ex vivo** because the gene transfer is done outside the body. Cells with defective genes are removed from the patient and given normal copies of the affected DNA before being returned to the body. This therapy has generally targeted blood cells because many genetic defects alter the function of one or other of the blood cells. However, the problem with this approach is that blood cells have limited life spans. Corrected cells can be introduced into the body and they can work well, but after a few weeks they die and the patient's original problems return, making it necessary for many repeat treatments to keep the person healthy.

A better approach is to target the stem cells of the bone marrow. Stem cells survive as long as the patient does, so if these immature cells that give rise to the full array of blood cells in the circulation can be given the correct gene, the treatment should last indefinitely.

Although scientists have been able to obtain stem cells from human bone marrow, it has proved difficult to get genes into these cells. But advances have been made recently. In 1993 a group in Los Angeles treated three newborn babies with SCID (page 550) by inserting genes into their stem cells. Their blood cells began producing the critical enzyme that they lacked and they grew into healthy children.

In situ gene therapy

In **in situ** treatment, vectors carrying the corrective genes are introduced directly into the tissue where the genes are needed. This approach makes most sense for problems that are localised, but it cannot correct disorders that affect several systems of the body.

In situ gene therapy is being explored for several diseases. For cystic fibrosis, workers have introduced gene vectors containing healthy copies of the cystic fibrosis gene into the lining of the bronchial tubes. As a first step towards treating muscular dystrophy, other researchers have injected a gene directly into muscle tissue in animals to investigate the possibility of re-engineering the body to make normal muscle proteins. Several teams have also inserted 'suicide' vectors into tumours. These carriers contain a gene that is intended to make cancer cells self-destruct when treated with certain chemotherapy drugs.

In vivo gene therapy

This third approach is still in the experimental stages but it is likely to be the gene therapy of the future. Gene carriers will simply be injected into the body, in much the same way that drugs are given now. Once in the body, the carriers will find their target cells, ignore other cell types, and transfer their genetic information efficiently and safely.

3 PROBLEMS AND ETHICS

The people who have been trying to develop gene therapy techniques have had to overcome many technical problems. In situ therapy, for example, is still hampered by a lack of safe ways to implant corrected genes into various organs. Ex vivo therapies do not have this problem, but the genes that are transferred do not always yield good quantities of the encoded proteins. And, in both forms of treatment, once the genes enter the cells, they can insert themselves randomly into the DNA of the chromosomes. Sometimes this is harmless, sometimes it is not. If the gene sent in as therapy manages to disrupt a tumour suppressor gene (see page 537) that normally protect the body against cancer, a tumour could result.

Living things are comp
which have 'skeleton
with itself repeatedly,

The carbon at

Atoms have electrons ir
eight electrons fill the s
Atoms with a less than
others in order to fill th

Carbon has an atomi
protons and 6 neutr
shell, the other 4 in
full, and therefore st
more. So, carbon for
shared between the

Fig A2.1 **Dot and cross di
(a) Methane**

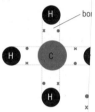

**Notice how the carbon a
an electron with each o
hydrogen atoms to achi
outer shell of eight elec**

Carbon can form
form stable chains
compounds which
saturated. C=C
possible. Compoui
unsaturated. Th
hydrogen (more
multiple bonds). T
fatty acids – see p

Carbon atoms (
nitrogen, oxygen
make up the vast

GENE THERAPY BEFORE BIRTH

FRENCH ANDERSON is currently working towards carrying out gene therapy on a fetus that is affected by a genetic disease that is usually fatal before birth.

Because of the enormity of the ethical issues involved, he has raised the possibility of in utero gene therapy for public discussion, even before the techniques that would enable it to go ahead have been developed. He argues that if society decides that this sort of treatment should not be allowed, there is little point in spending the time and the money to work on it.

Within the next 5 years, Anderson hopes that it will be possible to take some of the bone marrow stem cells from a fetus with untreatable SCID or who has α-thalassaemia, transform the cells using a corrective gene, and then reintroduce the cells into the fetus at about 15 weeks. Fig 36.8 shows a 15-week-old fetus.

If the therapy works, the child would be born healthy and would never suffer from the symptoms of the disease it has been treated against. However, the problem is that it might not work – if the baby dies before birth, you might say that nothing has been lost. However, the possibility exists that some sort of partial cure could result and the baby could be born, only to suffer a short life of severe illness and disability.

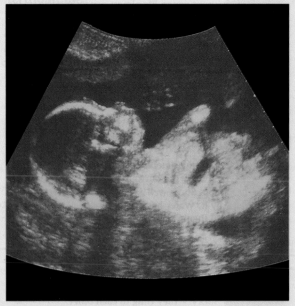

Fig 36.8 **The CT scan of a 15-week-old fetus**

Development of in vivo gene therapy has been slow because researchers have found it difficult to create delivery units that will allow the genes to find their way into the correct cell and that can avoid the immune system of the body.

Cost is also a potential hurdle to gene therapy. Because we are still at an early stage of gene therapy research, it is still quite expensive to do and is carried out only in the major medical centres of the developed world. However, French Anderson, the pioneer of the first gene therapy procedure (page 550) thinks that, ultimately, gene therapy will become relatively inexpensive and, in perhaps only 20 years, will be used regularly and routinely to cure many different medical conditions.

However, the financial and technical challenges of gene therapy are nothing compared to the ethical dilemmas that crop up. So many different questions need to be raised and considered. We end this book with just a few for you to mull over:

- Should gene therapy be used to prevent disease?

- All the gene therapy carried out so far has affected only the somatic cells, not the germline cells – the eggs or the sperm. Do you think germline gene therapy could ever be justified?

- Where should the line be drawn for genetic therapy? How will it be possible to avoid people trying to use gene therapy to practise eugenics (altering racial characteristics)?

- In 1998, scientists managed to grow human stem cells from embryos in culture for the first time. This raises the possibility of engineering an entire human being. Do you agree that this type of research should continue?

There is no life v
water. Most bioo
organisms which
their own inter
survive dehydrat
allow a state of
available once a
percentage of v
human brain, is

The propert

All cells are ti
exchange mater
take place in solι
also surrounded
properties, many

The water mol

The water molec
atom of oxygen.
Many compounc
its molecules are

Electrons are
attract and like
oxygen atom ef
oxygen bonds. T
shows. The regio
charge, while th
charge. Molecul
charge are said t

2 pairs
of electrons

δ⁺ H

Electroι
to tɦ
with

Fig A1.1 **Water is a ρ
negative charge**

Attractive forces
on the hydrogei
charge on the ox
forces are called
water molecule
each other, wate

H

O

H

Fig A1.2 **Weak attract
molecule and the neg
of water are due to tɦ**

ACKNOWLEDGEMENTS

The publisher thanks examination boards for their permission to reproduce examination questions. Questions are acknowledged as follows:

AQA: AEB — The Associated Examining Board
NEAB — Northern Examinations and Assessment Board
OCR: UCLES — University of Cambridge Local Examinations Syndicate
Edexcel: ULEAC — University of London Examinations and Assessment Council

Every effort has been made to trace the holders of copyright. If any have been overlooked, the publisher will be pleased to make the necessary arrangements at the earliest opportunity.

Photographs

The publishers thank the following for permission to reproduce photographs:

A–Z Botanical Collection: M Nimmo 30.6 (right); Allsport: A Bello 9.A1, T Duffy 3.18, B Heer 3.3(a), J Herbert 13.1, M Hewitt 13.4, G Mortimore 13.A1, M Powell p 192 (top), 22.28, B Radford 13.7, 13.A2, 32.1; Animal Photography: S A Thompson 17.2; Heather Angel 21.12 (centre), 22.34, 24.22; Aquarius Library 31.1; Ardea: S Gooders 21.2 (left), D & K Urry 24.6; BBC Natural History Unit: L M Stone 16.9, D Wechsler 24.23; Biophoto Associates: 1.3, 1.8, 1.13, 1.14, 1.15, 1.17, 1.19, 1.26, 1.29, 1.32, 1.33, 1.A2, 2.6, 5.15(b), (c), 5.20, 5.A1, 6.4, 6.15, 9.13, 9.17, 9.21(a), 10.5(d), 10.7, 16.8, 18.3(b), 18.10, 18.Q1, 20.16(a), 22.3 (inset), 22.6, 24.15 (bottom), 24.17 (bottom), 27.1(a), 27.6, 27.9, 27.10, 27.11, 27.13, 27.16(b), 32.4, 33.11, 33.17 (centre left), 34.1(b), 34.3(a, b, d), 34.7, 35.18; Biofotos: B Rogers 21.12 (top), I Tait 21.14; John Birdsall Photography 8.6, 12.13(a), 12.A2, 14.A3, p 296, 20.1 (right), 24.12, p 408, 29.1, 34.3(c); from The Discovery of Insulin by Michael Bliss, published by Paul Harris Publishing, Edinburgh, 1983 p 154; Adam and Eve by Titian, Prado, Madrid/Index/Bridgeman Art Library, London, p 442; The Three Graces by Reubens, Prado, Madrid/Bridgeman Art Library, London, 3.13(a); Chris Bonington Picture Library: D Scott 8.8; Bubbles Photo Library 16.A2; Cephas: F B Higham 22.36 (top right), M Rock 22.36 (top left), 22.38, Wine Magazine 22.36 (centre right); Cleveland Museum of Natural History, Ohio 26.2 (top), 30.23; Bruce Coleman Ltd: A Baccella 30.8, J & D Bartlett p 372, J Burton 24.18, P Davey 30.22, D Davies 22.7 (left), J Grayson 28.8 (centre left), S J Krasemann 30.17 (left), F Labhardt 28.8 (centre), W Layer 20.1 (left), G McCarthy p 524, A J Purcell 28.8 (top right), H Reinhard 30.14(a), Dr F Sauer 22.7 (far left), J Shaw 22.3 (bottom), K Taylor 24.10, 24.13, 28.8 (top left), 28.13(b), 30.4, 30.14(b), 30.19, R Williams 24.14 (bottom right); Collections: G Howard 20.11(a), S Lousada 16.1, A Sieveking p 278; Corbis: Hulton-Deutsch Collection p 26 (centre); Corbis/Bettmann: UPI 33.2; Brian Culcheth 21.3; from Damasio H, Grabowski T, Frank R, Galaburda AM, Damasio AR, 'The return of Phineas Gage: Clues about the brain from a famous patient', Science, 264:1102–1105, 1994, Dept of Neurology and Image Analysis Facility, Univ Iowa p 226 (top); Environmental Images: R Brook 25.6 (top right & bottom left), 25.12 (left), G Burns 25.13, P Glendell 25.12 (right); Mary Evans Picture Library 12.5, 29.14; photo courtesy of Eyam Parish Church/Mike Williams 33.1; Joyce M Filer p 476; from the F G Banting Papers, Thomas Fisher Rare Book Library, Univ Toronto 10.3; FLPA: E & D Hosking 30.16, P Perry 24.9, M Ranjit 14.A1, M J Thomas 29.7, T Whittaker 30.17 (right); The Fotomas Index 30.3; Geophotos: T Waltham 30.6 (left); The Ronald Grant Archive 28.13(a); Green Moon Ltd 12.3; Sally & Richard Greenhill 3.8, 3.16, 6.1, 20.2, 20.3; Holt Studios International: R Anthony 25.5 (left), N Cattlin 22.A1, 23.8, 25.4, 25.5 (centre & right), G Roberts 25.1; Hulton Getty Picture Collection p 26 (top), 20.14, 32.12; Bill Indge 34.A1 & A2; Andrew Lambert 3.3(b, c, d), 3.5, 3.9, 3.12, p 58, 4.A1, 6.6, 6.7, 6.10, 6.12, 6.13, 11.2, 13.8, 22.25, 35.24; Life Sciences Images 14.8; Lomond Mountain Rescue Team: G Baird 12.A1; Muscular Dystrophy Group 29.15; NASA p 266; Nature & Science AG: F-L Vaduz 11.7(a); NHM Photo Library 15.9, 30.1; NHPA: A Bannister 12.1, B Wood 12.6; Novo Nordisk A/S 10.A1, 10.A2; OSF: G Bernard 22.1, D Bromhall 9.24, G Gardner p 205, M Hill p 310, D Macdonald 21.13, S Osolinski 20.7 (right), 21.6(b), R Packwood 20.7 (left), R Redfern 21.9, T de Roy 21.6(a), D H Thompson 21.11, 30.5, W Shattil & B Rozinski 24.11; Panos Pictures: H J Davies 33.19, J Hammond 33.23, P Harrison 32.2, T Page 6.11, 22.23, C Stowers p 252 (top); Dr C Paterson, Ninewells Hospital, Dundee 3.32; Planet Earth Pictures: A Bartschi 24.14 (centre right), M Conlin 22.17, N Garbutt 24.14 (left), F J Jackson 30.14(d), A Mounter 17.13, Purdy & Matthews 22.3 (top), J Scott 13.A3, M Snyderman 30.14(c), M Welby 24.5; Popperfoto p 240; PPL Laboratories, Edinburgh p 466; Professional Sport International p 153; Professional Sport International: T Hindley 16.A1; Punch Ltd 32.3; Redferns: Michael Och's Archive p 424; Carol Redhead et al 14.6; Reuter: Popperfoto 6.A1; Rex Features Ltd 3.13(b), 6.9, 11.1; Roslin Institute, Edinburgh 27.1(b & c), 31.6; Royal Holloway College, Univ London, EM Unit 1.22, 1.30; Science Museum 1.1; Science Photo Library 2.3, 2.A3, p 102, 7.5, 7.8, 9.5(d), 27.5, 29.2, 33.17 (top right), p 508, 35.12; Science Photo Library: P Andrews 22.2, P Arnold 33.4, J Ashton 35.19(b), A Bartel 26.2 (centre right), St Bartholomew's Hospital 36.3, W Baumeister 4.8, Biophoto Associates 9.5(c), C Bjornberg p 475, Dr V Bradbury 34.8, Breast Screening Unit, Kings College Hospital, London 35.26(b), Dr A Brody 34.9, BSIP/LECA 2.A4, 35.16, BSIP/VEM 14.18, S Camazine 32.17, D Campione p iv (centre right), R U Chandran/TDR/WHO 9.16, M Clarke10.4, 19.16, Dr R Clark & M R Goff 12.2, CNRI 1.4 (left), 1.5, 7.9 (bottom), p 394 (top), 26.1, 27.7, 20.16(b), 29.12, 32.11, 32.19, T Craddock 25.7, 30.10, A Crump 33.A4, V Deemen 34.1(d), M Devlin 2.A1 (centre), Deep Light Productions p 180, 32.5, A Dex, Publiphoto Diffusion 32.13, M Dohrn 33.6, M Dorhn/RC Surgeons p 87, 8.1(b), B Dowsett p 488, J Durham 8.A2, Eye of Science 33.7, 33.17 (centre), D Fawcett 1.12, 1.24, 34.11, S Ford 14.A2, 33.16, 33.21, S Fraser 24.7, 25.9, 25.6 (top left), GCa/CNRI 10.5, Grapes/Michaud 35.8, E Grave 24.15 (top), 33.17 (top left), 33.17 (centre right), K Guldbrandsen 36.6(a), P Jude 17.5, Y Hamel 25.6 (bottom right), A Hart Davis 29.16, J Holmes 26.2 (bottom left), 31.5, J Howard 22.3 (centre), G Jerrican 34.17, 35.26(a), A Kage/CNRI 7.14, M Kage 7.9 (centre left), 14.2, 17.3, 22.37, J King-Holmes 34.19, J King-Holmes/D Mercer p 88, M Kulyuk 36.8, Dr A Lesk 3.33, Dr K Lounatmaa 34.24, D Lovegrove 9.9(a), A McClenaghan 17.15, Dr P Marazzi 12.16, 20.16(c), 25.11), p 394 (centre), 35.25, 30.20, P Menzel 18.12, 26.2 (bottom right), 36.6(b), Prof O Miller 28.3(b), Montreal Neuro Institute/McGill Univ/CNRI p 226 (centre & bottom), Moredun Animal Health Ltd 1.6, 24.17 (top), H Morgan 19.A1, H Morgan p 166, Prof P Motta, Dept of Anatomy, Univ la Sapienza, Rome 8.9, 11.7(b), 17.8, 18.2(c), 19.3(b), Prof P M Motta & Prof J van Blerkom 27.2, Prof P Motta, G Macchiavelli, S Nottola 19.6(b), 34.4, Prof P Motta, Prof K Porter, Dr G Murti p 1, 28.A1 (above left), National Library of Medicine 30.2, B Nelson 2.5, B Nelson/Custom Medical Stock p 72, NIBSC 34.6, Dr Y Nikas 19.A2, NOAA 25.10, R T Nowitz 36.7(a), Omikron 28.A1 (below left), 34.1(c), D Parker 12.9, 24.21, Perquis/G Watson 15.1, Petit Format p 277, Petit Format/Nestle 26.2 (centre left), D Phillips 1.10, H Pincis 33.3, P Plailly 18.8, 35.27, 36.7(b), A Pol/CNRI 9.4, Quest 1.A3, 7.16(a), 8.7, John Radcliffe Hospital 2.A1 (bottom), R G Rawlings/Custom Medical Stock p 2, 2.1, Dr M Read p 309, J C Revy cover photograph, 4.2, 13.2, 34.14(b), 36.1, 36.5, H C Robinson 16.3, P Saada/Eurelios 32.18, T van Sant, Geosphere Project, Santa Monica p 318 (top), Saturn Stills 4.16, 34.16, 32.16, D Scharf 8.A1, 14.3, 27.3, K F R Schiller 7.9 (top & centre right), Prof K Seddon & Dr T Evans p 385, J Selby 14.20, M Sklar 27.4, S Stammers 2.A1 (top), 33.22, 33.A1, J Stevenson 35.23, R Sutherland 36.2, A Syred 18.2(b), S Terry 19.8, G Tompkinson 17.7, A Tsiaras p 120, 19.A3, P Tweedie p iv (bottom right), B P Wolff 34.1(a), P Wolff p iv (centre left), M Wurtz/Biozentrum/Univ Basel 1.4 (right), H Young 11.9, 27.16(a), 32.6; Science & Society Picture Library 8.11; Science & Society Picture Library: NMPFT 34.22; South American Pictures: T Morrison 22.36 (centre); Still Pictures: F Bavendam 21.6(c), M & C Denis-Huot p 34, N Dickinson 21.4, M Edwards 11.3, p 206, R Giling 6.2, G Moti 25.2, M Nicolotti p 132, K Schafer 21.6(d), H Schwarzbach 24.A1; Tony Stone Images 13.5, 13.10, 16.2, 17.1, p 356, 26.4, 35.2; C&S Thompson 11.4; US National Marine Fisheries Service, Southwest Fisheries Science Center: Robin Westlake p 348; John Walmsley 21.2 (right); The Wellcome Trust 5.6, 6.A2, 12.13(b), 28.11, 32.7, 34.10, 35.5, p 546; Woodfall Wild Images: B Gibbons 24.24, David Woodfall 24.3; Zeneca Plant Science 31.7.

ANSWERS TO SELF-TEST QUESTIONS

Chapter 1

A Mice, attracted by the smell, came to eat the wheat.

B (a) The full stop is about 0.3 mm = 300 μm. (b) 1 million.

C Dispersal of oil pollution, breakdown of waste plastic.

D 1000.

E They are all secretory and would have a lot of ER.

F The ER is a network of thin membranes that would not show up under a light microscope.

G All its tissues would be destroyed.

H We could destroy tumours.

I It needs a lot of energy for its long swim to the egg.

J (a) 500 (b) 200 (c) 5000.

K The root system, because it receives no light.

Chapter 2

A Ciliated epithelium.

B Smooth muscle because its contraction is slowest.

C It fatigues too quickly.

D The reproductive system.

Chapter 3

A Sucrose consists of glucose and fructose, both reducing sugars. In the sucrose molecule, the reactive groups of the two component sugars point inwards and so are 'hidden' in the three dimensional structure. They only become available to react with Benedict's solution when the sucrose has been split apart.

B The glycosidic link, which is between carbon 1 and carbon 4.

C Links: sucrose 1,2; lactose 1,4.

D Drinking artificial milk in which lactose had been replaced by glucose.

E It would cause an osmotic imbalance – the cytoplasm would become too concentrated.

F Because few organisms have the enzymes that can degrade cellulose.

G Oleic and linoleic acid.

H Oleic and linoleic. They are both unsaturated.

I 46 x 5 cm = 2.3 metres.

J Size, bases, lifetime, sugar.

K TAGCAATGG.

Chapter 4

A Protein synthesis.

B Protein, lipid, nucleic acids.

C Because they all contain the same food type: starch.

D Tears, sweat, in the lysosomes of some white cells.

E It stimulates glycogen production.

F Peptide bonds.

G Proteins are denatured. This is irreversible.

H Because that is the optimum temperature for enzymes.

I It would be denatured by the acid.

J Competitive inhibitors compete with the normal substrate for the active site; non-competitive inhibitors bind away from the active site. Competitive inhibitors do not affect the ability of the enzyme to bind to the normal substrate; non-competitive inhibitors modify the enzyme so that it cannot recognise the substrate.

K There would be no activity at all: the plot would go along the x-axis.

Chapter 5

A 1 mm.

B Water-loving, water-hating. In a phospholipid, the phosphorus end is attracted to water, the lipid end repels water.

C Because the atoms/molecules in solids aren't free to move.

D To maintain the diffusion gradient.

E The jam is hypertonic to the inside of the bacterial cell, bacteria lose water by osmosis and die.

F You must have a solution to compare it to: Hypertonic to what?

G Cyanide blocks aerobic respiration, so virtually no ATP is available.

Chapter 6

A Yes. They might get enough energy but lack essentials such as vitamins or minerals.

B Small organisms have a large surface area:volume ratio and so lose heat faster than larger organisms. They compensate by using most of the energy produced by respiration to maintain their body temperature.

C (a) Women store more fat since they have evolved to bear and feed children. (b) Carbohydrate is used quickly or converted to something else.

D (a) Rice. (b) Inhabitants in the cold Arctic would have mainly a fish and meat diet (very poor in carbohydrates).

E The symptoms are those of vitamin A poisoning.

Chapter 7

A (a) Entrance blocked by soft palate. (b) Entrance blocked by epiglottis. (c) Way back out blocked by the curled tongue.

B Upper jaw: 2 incisors, 1 canine, 2 premolars, 3 molars. Lower jaw: 2 incisors, 1 canine, 2 premolars, 3 molars. The notation is for just one side of the mouth.

C The would have to eat smaller, more frequent meals as they would have no stomach to store large meals.

D (a) You would expect rennin production to decrease with age. It does. Young mammals on milk diets need it but older animals do not. (b) No, since only mammals produce milk.

E Almost 11 hours (6.35 m at a rate of 1 cm min⁻¹ = 635/60 = 10.58 h).

F So they do not self-digest the organs that produced them.

G An endopeptidase acts in the middle of the polypeptide, breaking it into smaller fragments; an exopeptidase works at the ends of a polypeptide, stripping off amino acids.

H The small intestine is a very absorptive region and can take in many of the nutrients produced by bacteria; only a small amount of absorption takes place in the colon, so nutrients produced here will just be lost in the faeces.

Chapter 8

A 7.5 litres min⁻¹.

B One five thousandth.

C Dead space would be enormous and diver would breathe the same air in and out. Also, the chest muscles are not strong enough to keep expanding the lungs against the water pressure at that depth.

Chapter 9

A (a) 6 and (b) 15 litres per minute.

B (a) Increase. (b) Decrease.

C 120 mmHg = systolic pressure, 70 mmHg = diastolic pressure.

D (a) Increase. (b) Decrease.

E Hb 'steals' oxygen from an oxygen-rich environment and delivers it to tissues that are oxygen-poor.

Chapter 10

A Because they detect a change in internal conditions and then adjust it back to the normal level.

B The urine of diabetic dogs has high concentrations of sugar. This is very attractive to flies.

C They are metabolically active and are involved in a rapid exchange of large amounts of materials.

D Any excess of the daily need would be deaminated and excreted.

Chapter 11

A 7.5 litres.

B 14.4 times per hour.

C Blood pressure would drop, filtration would become ineffective, kidneys would suffer damage and fail.

D The loop of Henle is shaped like a hairpin, and it behaves in the same way as a countercurrent multiplier.

E Dark yellow urine indicates dehydration. Brown = trouble!

F Much more water is lost in faeces and dehydration can follow.

G Proximal and distal tubules, ascending limb of loop of Henle.

H It makes the blood more concentrated, either by adding salt or removing water or both.

Chapter 12

A Humidity changes our ability to lose heat by sweating. Low humidity means a higher upper lethal temperature because we can sweat and lose heat by evaporation more effectively.

B They must be below, otherwise they would disappear as old skin was shed and replaced by new skin.

C Piloerection because we have so few hairs.

D To replace the sodium and chloride ions that they lose constantly by sweating.

Chapter 13

A We are using the lactic anaerobic system. This means that respiration can't be completed and so the intermediate compound (lactate) builds up.

Chapter 14

A (a) Active transport mechanism needs specific membrane protein and ATP. (b) It moves particles against a diffusion gradient rather than down a diffusion gradient.

B (a) To ensure that a neurone responds to a specific minimum level of stimulation. (b) Continuous depolarisation and fatigue of neurone.

C Myelinated nerves conduct impulses more quickly.

D Narrow to make diffusion of the neurotransmitter as fast as possible and high resistance to ensure that the action potential can't jump across.

E Because the transmitter is only made on one side.

F No action potential.

Chapter 16

A The digestive juices undiluted by food would attack the stomach lining. This could cause an ulcer.

Chapter 17

A (a) Photoreceptor; exteroceptor. (b) Chemoreceptor; exteroceptor. (c) Chemoreceptor; exteroceptor.

B Exteroceptors since they respond to stimuli from outside the body.

C The light sensitive film.

D (a) The ciliary muscles relax. (b) The lens becomes thinner due to tension on the suspensory ligaments. (c) The pupil dilates.

E Night sight (ie vision in dim light) needs rods that work properly. Vitamin A is necessary for the formation of retinal in rhodopsin.

F (a) Blue, green and red cones. (b) Blue cones only.

Chapter 18

A They have to support more weight.

B The spinal cord.

C Ligaments connect bones together at joints; ligaments need to be flexible to allow the joined bones to move.

D A tendon joins a muscle to a bone. As the muscle contracts it exerts a force on the bone, moving it. If the tendon were elastic, the bone would move less than the change in muscle length.

E (a) smooth (b) smooth (c) skeletal (d) cardiac (e) skeletal.

Chapter 19

A It would secrete more GnRF, thus stimulating more gonadotrophin release and therefore more oestrogen release.

G Day 21.

Chapter 20

A The growth spurt that occurs at puberty.

B No. Growth occurs evenly in the crocodile's body, so the ratio stays the same.

C (a) The brain is small at birth to allow the baby's head to pass through its mother's pelvis. It then develops rapidly to enable the child to become independent and have a better chance of survival. (b) The reproductive organs do not mature until we are large enough and emotionally mature enough to have children. In our society, cultural pressures delay childbearing until well past the age when females are physiologically ready to have babies.

D Between the ages of 2 and 3 years.

E Human tissue.

F **(i)** The person might have a condition that prevents them from making it. **(ii)** The hormone is produced in pulses and the sample was taken between pulses. **(iii)** The hormone is quickly converted into another form or is broken down as soon as it is released.

G The number of younger people in the population is growing at an equivalent rate.

Chapter 21

A Temperate deciduous forest.

B Abiotic: **(a)**, **(c)**, **(d)**. Biotic: **(b)**, **(e)**, **(f)**.

C Habitat is where an organism lives, the term ecological niche also describes what it does and how it relates to its physical environment and to other organisms.

Chapter 22

A We still need to maintain body temperature, breathe, keep our hearts beating and keep cellular processes going. All this requires energy.

B False. Some heterotrophs eat other heterotrophs (eg lions eat gazelles).

C The transfer of energy from one organism to another.

E Glycerate-3-phosphate + 9ADP + 8P$_i$ + 6NADP.

F Glycerate 3-phosphate.

G X light, Y carbon dioxide, Z temperature.

H These more metabolically active cells need more ATP to power their processes and therefore need more mitochondria (the sites of ATP synthesis in the cell).

I It would take too long to break glucose down. ATP releases energy when one phosphate group is removed.

J These more metabolically active cells need more ATP to power their processes and therefore need more mitochondria (the sites of ATP synthesis in the cell).

K **(a)** Each one of the 20 amino acids has a different number of carbon atoms. In respiration, the reactions which break each one down produce different volumes of CO_2 and use different volumes of O_2. **(b)** CO_2 is produced but no O_2 is used: any value divided by 0 is infinity.

L The end-products are different (though the processes are similar).

Chapter 24

A Intra: within a species. Inter: between different species.

B The oak wood.

Chapter 25

A Sustainable: used but continually replaced, eg solar power, hydroelectric power. Unsustainable: when used it is gone forever, eg coal deposits, oil, gas.

Chapter 26

B Tall.

Chapter 27

A DNA contains the genetic code, RNA is built up on it, then RNA is used as the instructions for making proteins.

B $n = 23$.

C G_1, S and G_2.

D The amount of DNA would be halved at each division.

E **(a)** In ovaries and testes respectively. **(b)** Because it reduces the chromosome number (by half).

F In a human cell there are 23 bivalents.

Chapter 28

A DNA is double stranded, contains the sugar deoxyribose and the bases C, A, T, & G. RNA is single stranded, contains ribose and the bases C, A, U & G.

B **(a)** UAUGCGAUA. **(b)** 3. **(c)** tyrosine, alanine, isoleucine.

C They eat and digest protein, absorb amino acids, transport them in the circulation from the gut to cells.

D Anabolic.

E 61 different RNAs carry only 20 different amino acids.

F Six: all ^{14}N strands, two hybrids.

G With no spindle, the chromosomes double but cannot separate, so a cell with twice the normal number of chromosomes results.

Chapter 29

A 50:50 (half Yy and half yy).

B If you use a plant that is TT, all the offspring of a cross with a Tt or a TT plant will be tall and you wouldn't gain any information.

C 700 000/2500 = 280.

D In previous centuries, villagers married people from the next village. Because we now have more transport, people move around the world more easily and so people meet and mate with people to whom they are unlikely to be related.

E X^hX^h. Female haemophiliacs must be a double recessive whose mother is a female carrier and whose father is a male sufferer. The chance of two such people meeting and marrying is unlikely and male haemophiliacs often don't live long enough to reproduce.

F One in two chance.

G No, If the mother is O and the father is A, the child could be only A or O.

H 9 agouti, 3 black, 4 albino.

Chapter 30

A Acquired characteristics don't affect an organism's DNA and so cannot be passed on to the next generation.

B Insects have short life-spans and high reproductive capacity; whales and rhinos have long life-spans and only produce a handful of offspring during their life time.

C Peppered moth, *Biston betularia*: selection of melanic form following industrial revolution. Hawks: selection for great visual acuity. Cheetah: selection for high speed running.

D **(a)** Analogous, **(b)** Homologous.

Chapter 31

A It performs transcription in reverse: it makes DNA from RNA.

B Because the smaller recognition site is more common in any DNA sequence.

Chapter 32

A We now have vaccines and antibiotics, a better standard of living, less poverty and sanitation.

C Knowing what to avoid would help you avoid it.

D Larger population, more crowded cities.

E **(a)** Most protein digestion occurs in the small intestine, before the pancreatic duct.
(b) Pancreas does not produce as much trypsin when it starts to fail because of pancreatitis.

F Reduce the amount of phenylalanine and increase the amount of tyrosine.

G Less messy, more convenient to carry around.

H **(a)** Endoscopy. **(b)** X-ray.

Chapter 33

A Infection with *Salmonella typhimurium* does not always cause the disease – some people are carriers.

B **(a)** Gloves, masks, sterile instruments.
(b) Gloves, mask.

C There are now effective vaccines for measles, mumps and diphtheria, but none for the common cold.

D **(a)** In winter people crowd together in centrally heated buildings.
(b) In summer it is hotter and people are more likely to eat food that has been lying around in the heat.

E If you don't have the gene to produced the cell receptor for a particular virus, it cannot get into your cells.

F Antibiotic to kill the bacteria, antitoxin to neutralise the toxin they release when they die.

G Bacteria do not divide as quickly at lower temperatures so curve will be lower and flatter.

H Salmonellosis develops when the bacteria that cause it multiply inside the cells that line the intestine. Other types of food poisoning are caused by bacteria that do not multiply in the body. They simply release toxins, and so cause symptoms more quickly.

I Attack from antibodies and cells of the immune system.

J More people are in the water to get infected in the day.

K No problems from osmosis.

Chapter 34

A Warm, moist, lots of food.

B Mucous secretions containing dust and dead cells, that should be swept by cilia to the exterior, remain trapped in the airways to cause congestion and infection.

C Anticoagulants stop blood clotting so that leech can feed freely. Uses: microsurgery, plastic surgery, dispersal of blood clots, treatment of chronic skin ulcers.

D Two basophils, a lymphocyte and a monocyte.

E Not likely to cause an allergic reaction.

F Some of the donors would have been blood group O, some of the recipients would have been blood group AB, and so no adverse reactions would have occurred.

G AB blood has A and B antigens on the surface of the red blood cells. These react with antibodies present in the blood of people from the other three groups.

Chapter 35

A They live longer.

B Ozone hole, people holiday in more exotic locations.

C It uses more energy than a muscle at rest.

D **(a)** More oxygen and fuel to muscles.
(b) Don't have to go to the loo as often.

E The interplay between control genes and genes that code for the LDL receptor might result in a lower risk of developing high blood cholesterol.

F **(a)** They were told to resume their normal lifestyle.
(b) Hypertension and smoking.

G Enzymes digest adhesion proteins that hold normal cells together, making it easier for tumour cells to grow.

H Doing things that are bad for you is fun. Eating lettuce, drinking mineral water and doing lots of strenuous exercise, generally, isn't fun.

I Cost:benefit ratio is poor.

INDEX

As most of this book concerns humans and their activities, the index entry **humans** contains only basic information on humans as a species. For more information on human activities and biology, use more specific headings, eg blood, growth, nervous system.